Hydropower Station

Yongguang Cheng · Kai Su · Jinping Li ·
Juntao Chen · Weijia Yang
Editors

Hydropower Station

Concepts, Theories, and Design Principles

中国水利水电出版社
China Water & Power Press

Springer

Editors
Yongguang Cheng
School of Water Resources
and Hydropower Engineering
Wuhan University
Wuhan, Hubei, China

Kai Su
School of Water Resources
and Hydropower Engineering
Wuhan University
Wuhan, Hubei, China

Jinping Li
School of Water Resources
and Hydropower Engineering
Wuhan University
Wuhan, Hubei, China

Juntao Chen
School of Water Resources
and Hydropower Engineering
Wuhan University
Wuhan, Hubei, China

Weijia Yang
School of Water Resources
and Hydropower Engineering
Wuhan University
Wuhan, Hubei, China

ISBN 978-981-96-2382-2 ISBN 978-981-96-2383-9 (eBook)
https://doi.org/10.1007/978-981-96-2383-9

Jointly published with China Water & Power Press
The print edition is not for sale in China (Mainland). Customers from China (Mainland) please order the print book from: China Water & Power Press.
ISBN of the Co-Publisher's edition: 978-7-5226-3669-6

This Springer imprint is published by the registered company Springer Nature Singapore Pte Ltd.
The registered company address is: 152 Beach Road, #21-01/04 Gateway East, Singapore 189721, Singapore

Preface

Hydropower is the most mature and economic renewable energy in the world. During the past 20 years, a series of the world's largest hydropower stations have been built in China, and hydropower technology has been in great progress. In the future, many world-class hydropower stations will be developed in Africa, South America, and East South Asia. Although the research and practice are developing fast, there are only a few English textbooks published on hydropower technology. This textbook Hydropower Station is edited based on the former Chinese editions, which have been published for many versions since 1987 by China Water & Power Press and were used in teaching the specialty course Hydropower Station for thousands of senior undergraduates majoring in Hydropower Engineering at Wuhan University and other universities.

It covers most of the basic concepts, basic analysis methods, basic design principles, main industry standards, and many new examples in hydropower station design, building, and management. The contents include four units, hydro-turbine, water delivery system, hydraulic transients, and power houses, which are composed of 13 chapters. All the ideas, methods, and facts are from the engineering practice and research frontiers.

This textbook is edited by the faculty members of Wuhan University, who teach and research in hydropower engineering for many years. Chapters 1 to 5 and 9 to 10 are edited by Prof. Dr. Yongguang Cheng and Associate Prof. Dr. Jinping Li; Chapters 6 to 9 and 11 to 13 are edited by Prof. Dr. Kai Su and Associate Prof. Dr. Juntao Chen, and these 13 chapters were improved by Prof. Dr. Weijia Yang.

This textbook is for senior undergraduates majoring in hydropower engineering and also has referential value for undergraduates and postgraduates in the related specialties and engineers and researchers in hydropower engineering.

The editors would like to thank Profs. Jiandong Yang, Ming Xiao, Hegao Wu, and Xu Lai for discussing the book structure, contents, and expressions and providing new materials. Thanks also go to the graduate students guided by the editors, for reading and checking the draft version of this textbook.

Wuhan, China

Yongguang Cheng
Kai Su
Jinping Li
Juntao Chen
Prof. Weijia Yang

Contents

Chapter 1
Introduction to Hydropower Engineering

Yongguang Cheng

Abstract In this chapter, we first introduce the basic principles of hydropower generation and the types of hydropower stations and then summarize the development histories, achievements, and prospects of hydropower in the world, especially in China. The important roles of hydropower in the power grid and the world sustainable developments are emphasized. Students are required to understand the general development schemes of hydropower, recognize the irreplaceable place of hydropower in the modern energy system and economic development, and build up the confidence of joining the journey of developing this green renewable energy for the world.

Keywords Hydropower principles · Hydropower resources · Achievements and prospects · Sustainable development

1.1 Hydropower Generation

1.1.1 Basic Principles of Hydropower Generation

Water energy is one of the clean and renewable energy resources, and hydropower plays an irreplaceable role in the modern power systems. Peak load regulation, frequency modulation, phase modulation, emergency standby, and black-starting are the basic functions of modern hydropower in power grids. With the rapid development of solar and wind powers, hydropower is not only a power supply to increase electricity quantity but also an important regulator for safe, stable, and efficient operation of power grids. The basic principle of hydropower generation is to convert the potential energy and kinetic energy of water flow to the rotating mechanical energy through hydraulic turbine and then convert the rotating mechanical energy to the electrical energy through the generator.

Y. Cheng (✉)
Wuhan University, Wuhan, China
e-mail: ygcheng@whu.edu.cn

Y. Cheng et al. (eds.), *Hydropower Station*,
https://doi.org/10.1007/978-981-96-2383-9_1

In order to harvest water energy in a river, it is often necessary to block the river with a dam (water-retaining buildings) and build water conveyance structures for concentrating water head and transporting water flow. When the transported water flows through the turbine-generator units, its mechanical energy is converted to electrical energy, which is then supplied to power grid through transformers and switches. Therefore, a hydropower station (or hydroelectric power station) is a complex of equipment and buildings, which are composed of a water conveyance system, mechanical system, electrical system, structural system, etc. In addition to the equipment and buildings needed for power generation, some hydropower stations often have other buildings for comprehensive utilization purposes, such as flood control, irrigation, shipping, wood crossing, and fish culture. The complex of all the above buildings is called a hydropower project or a water conservancy project.

The water conveyance system of a hydropower station usually includes intake, water conduit, pressure regulating facility, tail water outlet, and so on. The composition and layout of the water conveyance system vary with the development mode, installed capacity, and number of units. The main function of the system is to ensure the smooth and efficient flow (with small head loss) of water and the safe and stable operation of hydropower station.

The mechanical system (or the main machine and auxiliary equipment system) mainly includes inlet valve (if any); turbine; generator; exciter; speed governor; oil, gas, and water auxiliary systems; lifting equipment. Its main functions are to convert water energy into electric energy, adjust output power, and provide high-quality electricity according to power grid dispatching.

The electrical system consists of two parts: primary circuit transmission and transformation equipment and secondary circuit control equipment. The primary electrical system includes current and voltage transformers; disconnecting switches; circuit breakers; main transformers. Its main functions are to raise the voltage and transfer the electric energy to power grid. The secondary electrical system is connected to the central control room and normally includes control, protection, detection, monitoring, communication, dispatching, and DC system equipment. Its main functions are to monitor the hydropower generating process and equipment operating status and to provide protection for possible operation restrictions and accidents.

The structural system (hydropower station buildings) includes the main powerhouse and auxiliary houses for arranging various mechanical and electrical equipment and the buildings involved in the water conveyance system. Its main functions are to meet the needs of installation, operation, and maintenance of various equipment, so as to provide good working conditions for operation employees.

1.1.2 Types of Hydropower Stations

According to the different development modes of water head formation, hydropower stations can be classified into three types: dam-type, conduit-type (or diversion-type), and mixed-type. However, from the point of view of hydropower station layout, water conveyance system arrangement, and powerhouse location, the conduit-type and the mixed-type are the same. Therefore, the types of hydropower stations can be classified into two categories: dam-type hydropower stations and conduit-type hydropower stations.

1. Dam-Type Hydropower Station

The dam-type hydropower station is built by blocking the river with a dam and forming the head by storing water in the reservoir. The reservoir makes the potential energy consumed by water head losses in the original river reach preserved, and the water surface in front of the dam is higher than that in back of the dam. The higher the dam, the deeper the water, the smaller the velocity, the smaller head losses, and the higher the water head for the hydropower station.

In the upper reaches of most rivers, the hydraulic gradient is large, the river bed is narrow, and the riverside mountains are high and steep; therefore, high dams should be built. Because of the narrow river bed, the layout of water-retaining buildings, flood discharge buildings, and water conveyance buildings (usually no navigable buildings) is difficult. In this situation, the water conveyance system is often arranged by passing the dam through diversion tunnels, and the powerhouse is placed on the riverbank downstream of the dam. This is called the riverbank hydropower station with ground powerhouses, e.g., Liyuan Hydropower Station (Fig. 1.1). The powerhouse can also be placed within the riverside mountains. This is called riverbank hydropower station with underground powerhouses, e.g., Xiluodu Hydropower Station (Fig. 1.2). No matter with ground or underground powerhouse, the dam-type hydropower stations in the upper reach of rivers usually have high water heads and install mixed-flow turbines (Francis turbines).

In the lower reach of rivers, the hydraulic gradient is small, and the water head that can be concentrated is very limited, usually smaller than 40 m. Concentrating higher water heads is not economical because of large inundation loss and mass migrations, which are related to larger reservoir areas. For such low-head hydropower stations, the powerhouse itself can withstand the upstream water pressure and may be designed as an integral part of the water-retaining buildings. The powerhouse building can be arranged in alignment with water-retaining, flood discharge, and navigation buildings on the broad river bed. This hydropower station is called water-retaining-type, e.g., Gezhouba Hydropower Station (Fig. 1.3). Water-retaining-type hydropower stations are usually equipped with through-flow turbines (e.g., Feilaixia Hydropower Station shown in Fig. 1.4) or axial flow turbines (Kaplan turbines) because of lower water head and large flow rate.

In the middle reach of rivers, the values of hydraulic gradient and river width are between those in the upper reach and lower reach of rivers. It is possible to make a complete layout of water-retaining, flood discharge, power generation, and navigation buildings on the wider river bed. However, because the water head at the dam site is high and the size of powerhouse is relatively small, the powerhouse cannot be designed as a part of the retaining building. The powerhouse can be placed behind the dam; therefore, it is called the hydropower station at the dam-toe, e.g., the

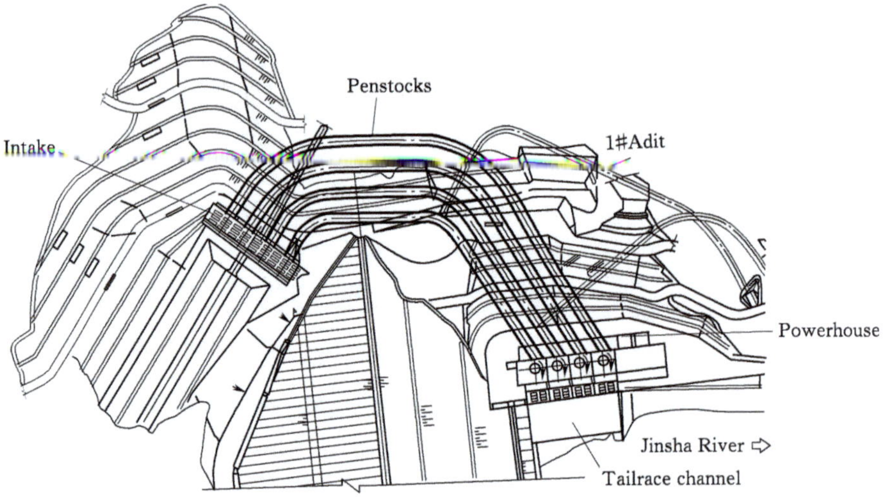

Fig. 1.1 Liyuan Hydropower Station, China. Liyuan Hydropower Station, is the third step in the hydropower development plan of "one-reservoir-eight-levels" in the middle reach of Jinsha River, China. The project is mainly composed of water retaining, flood discharge and sand discharge, water conveyance, and powerhouse buildings. In the riverbank ground powerhouse, four 612 MW units are installed with a single-penstock-single-turbine scheme. The total installed capacity is 2448 MW

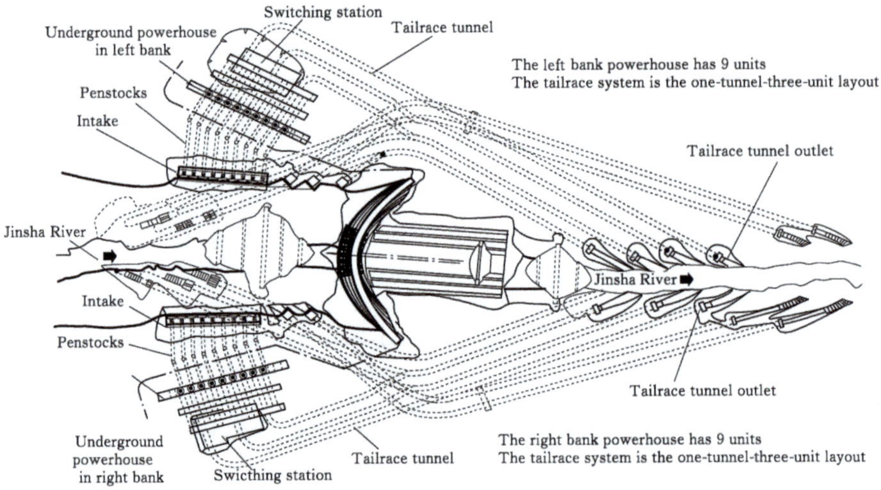

Fig. 1.2 Xiluodu Hydropower Station, China. Xiluodu Hydropower Station, is the third step of the four cascade hydropower steps in the lower reach of Jinsha River, China. In each of the two riverbank underground powerhouses, nine 770 MW units are installed. The total installed capacity is 13860 MW. The tailrace water system adopts a one-tunnel three-turbine scheme

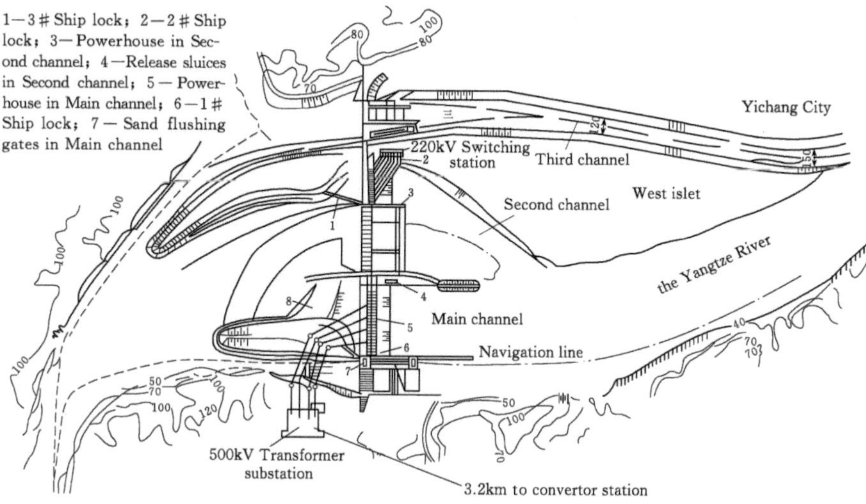

Fig. 1.3 Gezhouba Hydropower Station, China. Gezhouba Hydropower Station, is a water-retaining-type hydropower station. The water conservancy project consists of the ship lock, powerhouse, flood sluice, sand sluice, and water-retaining structures. Twenty-one axial flow turbine-generator units are installed in the two powerhouses, forming a total installed capacity of 2715 MW

Fig. 1.4 Feilaixia Hydropower Station, China. Feilaixia water conservancy project, consists of the earth dam, concrete overflow dam, powerhouse, and ship lock. In the powerhouse four 35 MW through-flow turbine-generator units are installed, forming a total installed capacity of 140 MW

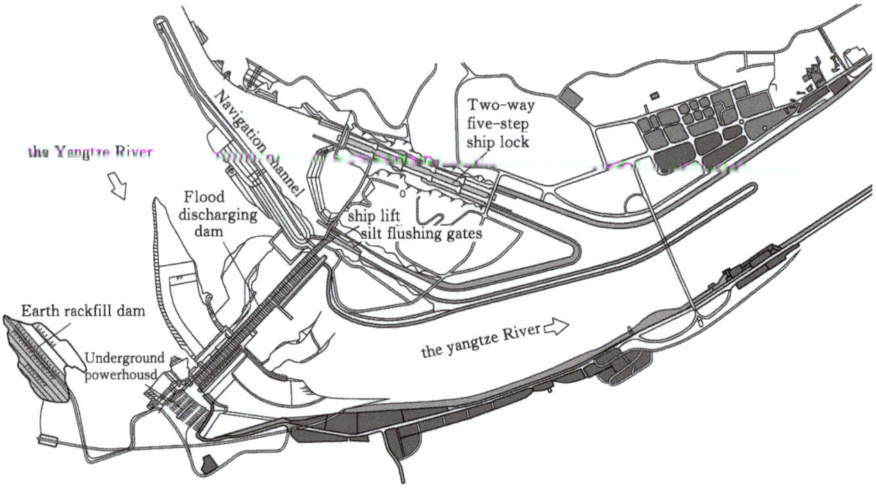

Fig. 1.5 The Three Gorges Project, China. The overall layout scheme of the Three Gorges Project is that the flood discharge dam section is located in the middle of the river bed, the power station dam section and the non-overflow dam section are arranged on the two sides, the five-stage lock and the ship lift are arranged on the left bank, and an underground power station is located on right bank. There are thirty-two 700 MW mixed-flow turbine-generator units (including 14 in the left bank powerhouse, 12 in the right bank powerhouse, and 6 in the underground powerhouse) and two additional 50 MW power supply units, forming a total installed capacity of 22500 MW

Three Gorges Project (Fig. 1.5). The hydropower stations at dam-toe usually install mixed-flow hydraulic turbines (Francis turbines).

If it is impossible to completely arrange water-retaining, flood discharge, power generation, and navigation buildings owing to the narrow river bed, some special arrangement schemes can be selected. Before the 1980s, when the underground cavern excavation and support technology was underdeveloped, engineers created some space-saving layout patterns to solve the problem, by combining powerhouse with other buildings. The typical examples include the powerhouse within the dam (Fengtan Hydropower Station shown in Fig. 1.6), the powerhouse with overflow on the roof (Xinanjiang Hydropower Station shown in Fig. 1.7), and the powerhouse with overflow over the roof (Wujiangdu Hydropower Station shown in Fig. 1.8). However, due to the poor operating conditions of these types of powerhouses, they have been rarely used in the past 30 years. Instead, a new type, called the underground powerhouse with dam intake, was used in some projects. For example, in Longtan Hydropower Station, the intake is arranged on the dam axis and the powerhouse is placed on the downstream bank, as shown in Fig. 1.9. This layout type should still be classified as a riverbank hydropower station.

2. Conduit-Type Hydropower Station

This type of hydropower station collects water from the sluice gate or low dam built in the upstream of river and transports the water downstream through a long

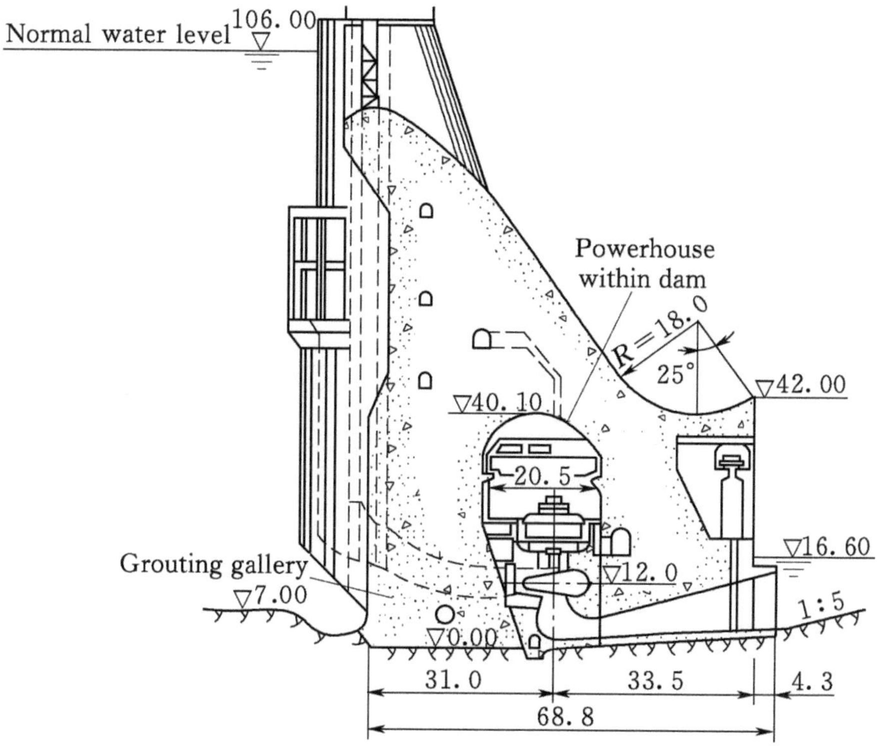

Fig. 1.6 Fengtan Hydropower Station, China. Fengtan hydropower station, is composed of a concrete hollow gravity dam, powerhouse within the dam, spillway, flood discharge bottom hole, irrigation culvert, and crossing boat and crossing raft passages. The powerhouse is arranged in the empty stomach of the 9–12 dam section, equipped with four 100 MW mixed-flow turbine-generator sets

conduit of thousands of meters to the powerhouse. The water head is concentrated mostly by the conduit; however, when the dam is high, the water head is concentrated partly by the conduit and partly by the dam. The principle of forming the water head by the diversion conduit is that the head loss in the artificial diversion conduit is smaller than that in the original river. The longer the diversion conduit, the higher the water head.

Compared to the dam-type hydropower station, the conduit-type hydropower station has the following three characteristics:

(1) The conduit-type hydropower station is suitable for hydropower development in the river reach with a large slope, in the place where conduit can be a shortcut across river bend, and in the situation of cross-basin water transfer. In these situations, the obtained water head is higher. Tianwan River Cascade Hydropower Stations (Fig. 1.10), Jinping II Hydropower Station (Fig. 1.11), and Sima Hydropower Station (Fig. 1.12) are typical examples. This type

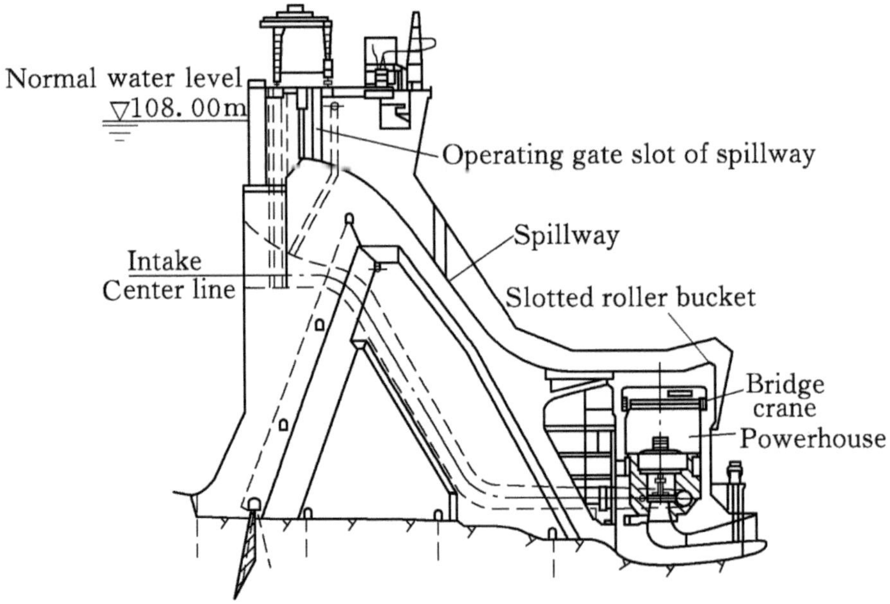

Fig. 1.7 Xin'anjiang Hydropower Station, China. Xin'anjiang Hydropower Station, was built in April 1957. It is the first large-scale hydropower station in China after the founding of the People's Republic of China. The roof of the powerhouse is overflow bottom plate. Nine mixed-flow turbine-generator units are installed and the total installed capacity is 855 MW

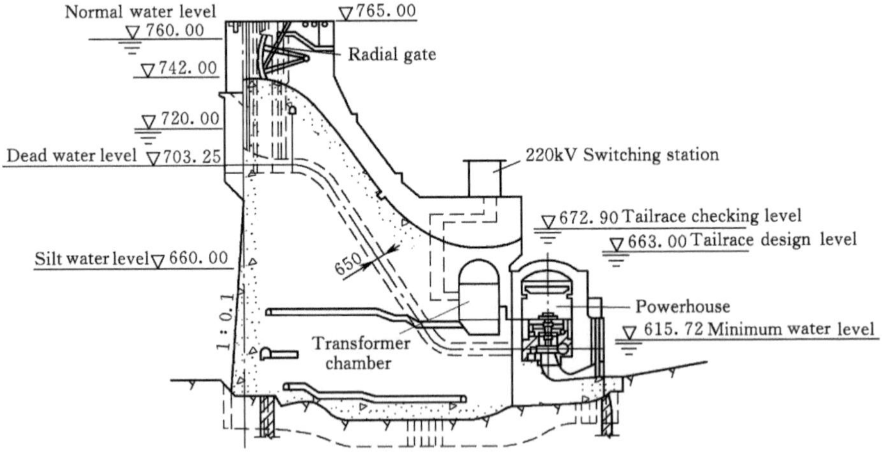

Fig. 1.8 Wujiangdu Hydropower Station, China. Wujiangdu Hydropower Station, is the first large hydropower station built in the typical karst development area in China. The powerhouse is under the overflow nappe. The original three 210 MW units were installed, and recently, the units were updated to 250 MW and two more units were placed

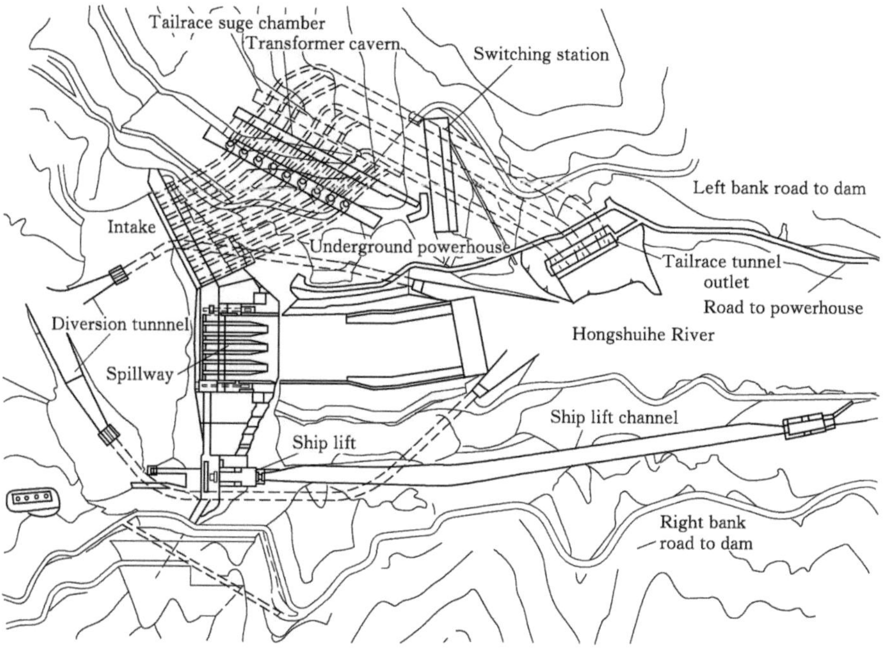

Fig. 1.9 Longtan Hydropower Station, China. Longtan Hydropower Station, is composed of a concrete gravity dam, an underground powerhouse in the left bank, and a vertical ship lift in the right bank. The powerhouse is located in the mountain behind the left bank dam, and the tailrace system uses the one-tunnel three-turbine pattern. Nine 600 MW mixed-flow turbine-generator units are installed

of hydropower station usually has high head and small discharge, which are suitable for installing impulse turbines or mixed-flow turbines.

(2) The conduit-type hydropower station simplifies the layout of the dam complex, which is beneficial to the layout of water-retaining, flood discharge, and navigable buildings. However, on the other hand, the diversion open channel or pressure tunnel is thousands of meters long, and it is necessary to build large pressure regulating facilities, such as a forebay and daily regulating pond for open-channel diversion, or surge chamber for pressure diversion to meet the requirements of regulation guarantee and regulation quality.

(3) Because of the long diversion conduit and huge investment, it is impossible to adopt the single-tunnel-single-turbine arrangement. Therefore, this type of hydropower station usually adopts a one-tunnel-multi-turbine pattern of the water conveyance system, which is much more complicated than the single-tunnel-single-turbine arrangement for dam-toe-type and riverbank-type hydropower stations. There are hydraulic interference problems between the multiple turbine-generator units.

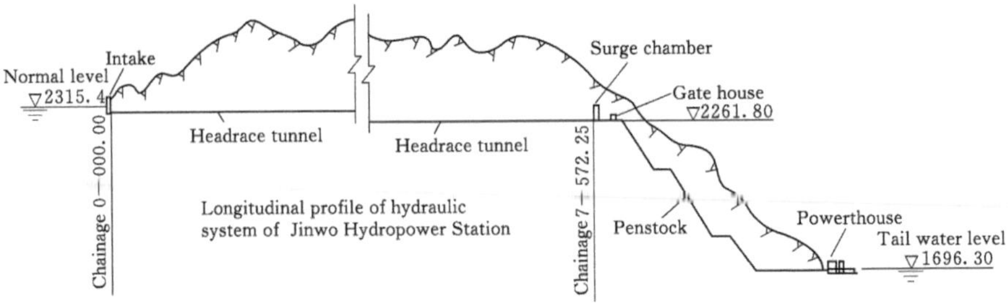

Fig. 1.10 Jingwo Hydropower Station, a part of Tianwan River Cascade Hydropower Stations, China. Tianwan River Cascade Hydropower Stations, consists of Renzonghai, Jinwo, and Dafa hydropower stations. Each hydropower station has the same arrangement of water conveyance system, that is, the intake, diversion tunnel, upstream surge chamber, penstock, bifurcation, branch pipes, and two 120 MV vertical shaft impulse turbine-generator units installed in the underground powerhouse. The diversion tunnels are 7.5, 7.6, and 9.4 km, respectively

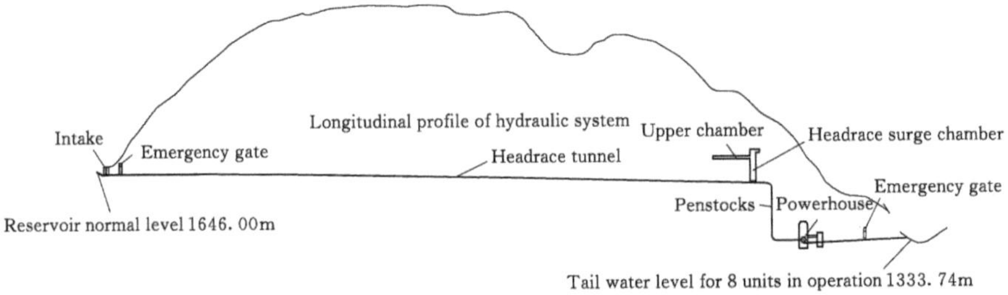

Fig. 1.11 Jinping II Hydropower Station, China. Jinping II Hydropower Station, takes advantage of the natural river drop in the 150 km bend on Yalong River to build four 16.7 km long straight headrace tunnels for power generation. Four differential surge chambers are set between the tunnels and penstocks. Eight 600 MW mixed-flow turbine-generator units are installed in the underground powerhouse, and the total capacity is 4800 MW

3. Composition of Water Conveyance System of Hydropower Station

For conduit-type hydropower stations, the water conveyance system usually consists of buildings such as an intake, diversion tunnel, upstream surge chamber, penstock, bifurcation, and tailrace channel. In the case of the underground powerhouse, it should also include tailrace water system buildings, such as tailrace branch tunnels, tailrace gate chambers, downstream surge chambers, and tailrace tunnels.

For dam-type hydropower stations, the water conveyance systems of the dam-toe-type and water-retaining-type are relatively simple. The former has only the intake and the penstock, while the latter has only the intake. The water conveyance system of the riverbank-type is complex and related to the development modes (the front section development, the middle section development, and the rear section development), the number of units, and even the pressure regulation scheme of the hydropower station. For example, Pengshui Hydropower Station (Fig. 1.13) uses a

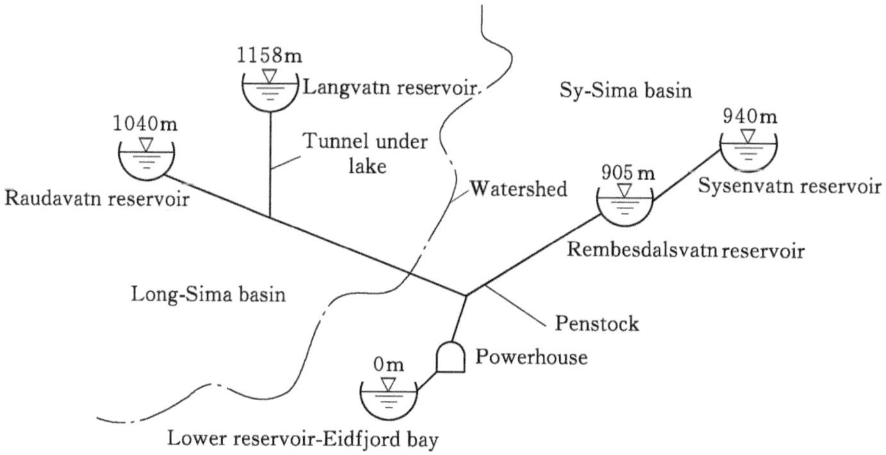

Fig. 1.12 Sima Hydropower Station, Norway. Sima Hydropower Station, is located on the Hordaland Plateau in southwest Norway and is an integral part of the Eidfjord Hydropower Development Project. The total length of the diversion tunnel is 44 km, the total installed capacity is 1120 MW, and the 5-nozzle impulse turbine-generator units are adopted

sloping ceiling tailrace tunnel, and Wudongde Hydropower Station (Fig. 1.14) uses a pressurized and free-surface combined tailrace tunnel.

The types of hydropower stations are summarized in Table 1.1. The layout characteristics of different types of hydropower stations are further explained. For the dam-toe-type and water-retaining-type hydropower stations, the system usually adopts the "single-penstock-single-turbine" type because of the short water conveyance system, and the layout is always arranged in "1"-shape, e.g., the Three Gorges

Fig. 1.13 Pengshui Hydropower Station, China. Pengshui Hydropower Station, is the 10th step in the "12-step cascade development of Wujiang Hydropower Base", China. The project is composed of a dam, underground powerhouse, navigable building, and so on. Five 350 MW mixed-flow turbine-generator units are installed, forming a total installed capacity of 1750 MW

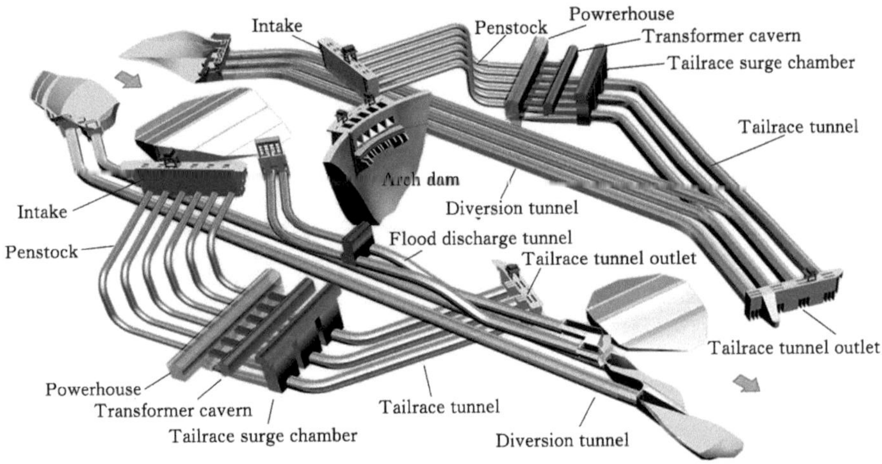

Fig. 1.14 Wudongde Hydropower Station, China. Wudongde Hydropower Station, is the first step of four world-class giant hydropower stations in the lower reach of Jinsha River, China. The project is composed of an arch dam, flood discharge tunnel, underground powerhouse, and so on. Two powerhouses are arranged on the left and right sides of the dam, and each installed six 850MW mixed-flow turbine-generator units, forming a total installed capacity of 10200 MW

Project (Fig. 1.5) and Gezhouba Hydropower Station (Fig. 1.3). For the riverbank hydropower stations, diversion tunnels or buried penstocks are needed because the water conveyance system has to bypass the dam abutment. To avoid the interference of flood discharge operation to tailrace tunnel outlet operation, the length of the water conveyance system is about 1000 m and usually arranged in the "7"-shaped layout, e.g., Liyuan Hydropower Station (Fig. 1.1) and Xiluodu Hydropower Station (Fig. 1.2). A few riverbank hydropower stations are also in "1"-shaped layout, e.g., Geheyan Hydropower Station (Fig. 1.15) and Nuozhadu Hydropower Station (Fig. 1.16). For the conduit-type hydropower stations, because the water conveyance system is longer than thousands or even tens of thousands of meters, the "single-tunnel-multiple-turbines" layout should be chosen, and the headrace surge chamber or tailrace surge chamber or pressure pool (for the open-channel diversion-type) are normally set.

Table 1.1 Types of hydropower stations

Head concentrating-type	Dam-type			Conduit-type	Mixed-type
Hydropower station type	Dam-type			Conduit-type	
	Riverbank-type	Dam-tore type	Water-retaining-type	Conduit-type	
Powerhouse type	Underground/ground	Ground	Ground	Underground/ground	
Turbine type	Mixed-flow	Mixed-flow/axial-flow	Through-flow/axial-flow	Mixed-flow/impulse	
Feature of water conveyance system	About 1 km long; "7"-shaped layout; can use free-surface pressurized tunnel	Short flow passage; "1"-shaped layout	Shorter flow passage; "1"-shaped layout	Very long flow passage; should set headrace surge chamber, or headrace and tailrace double surge chambers	

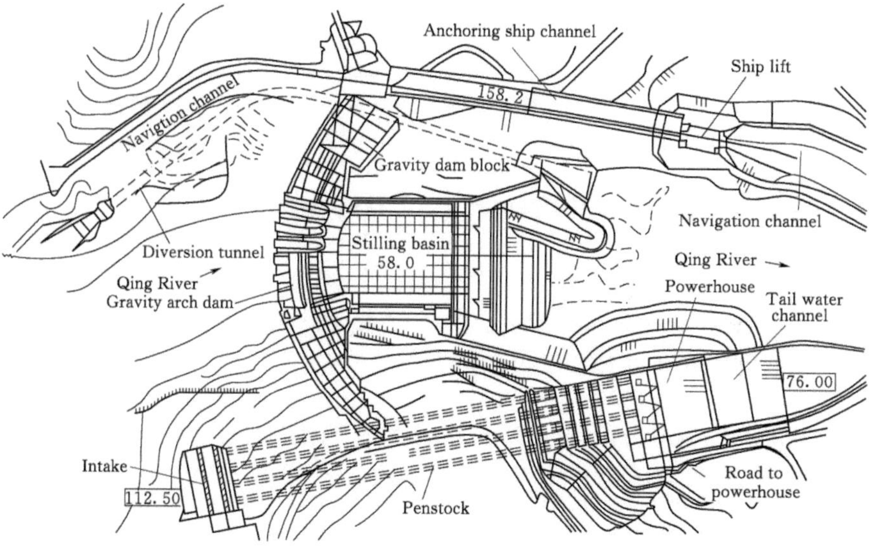

Fig. 1.15 Geheyan Hydropower Station, China. Geheyan Hydropower Station, is the backbone of the cascade development of Qingjiang River, China. The project is composed of a concrete dam, power plant, and ship lift. In the riverbank ground powerhouse, four 300 MW units are installed, forming a total installed capacity of 1200 MW

Fig. 1.16 Nuozhadu Hydropower Station, China. Nuozhadu Hydropower Station, is the key hydropower project in the lower reach of the Lancang River, China. The project is composed of a concrete core rockfill dam, left bank spillway, left bank conduit-type power station, and so on. Nine 650 MW units are installed in the underground powerhouse, forming a total installed capacity of 5850 MW

1.2 Achievements and Prospects of Hydropower Development

1.2.1 Achievements and Prospects of Hydropower Development in the World

The first hydropower station in the world, with an installed capacity of 25 kW, was built in 1878 in France. The first hydropower station in China, Shilongba Hydropower Station with an installed capacity of 480 kW, was built in 1912 in Kunming, Yunnan Province. Since the 1930s, because of the development of long-distance electricity transmission technology, the hydropower electricity generated in remote areas could be supplied to large cities, and the development of hydropower became rapid and the capacity became larger. Building large dams is necessary for most hydropower stations. With the increase of hydropower installation, the reservoirs are becoming larger and larger, forming many artificial lakes. After World War II, dams were widely used in the world to develop hydropower. In the 1970s, the western developed countries entered the peak of hydropower development, and almost every year two to three new dams were constructed. In that period, hydropower in China was still in the initial stage, and many small dams were generally used to develop small-sized and medium-sized hydropower stations.

Hydropower is a renewable energy, and hydroelectricity is low in cost and clean for environment and ecology. According to the statistics of the International Energy Agency in 2018, fossil fuels generate 63.9% of total electricity, hydropower accounts for 16.2%, nuclear power accounts for 10.1%, biomass and waste power account for 2.4%, and others such as geothermal, solar, and wind powers account for 7.4%. That is, nearly a fifth of the world's electricity generation comes from hydropower. In fact, twenty-four countries around the world rely on hydropower to provide more than 90% of their energy, such as Brazil and Norway. Fifty-five countries rely on hydropower to provide more than 50% of their energy, such as Canada, Switzerland, and Sweden. In some countries, hydropower is the only source of electricity, such as Paraguay. Table 1.2 is a list of countries with a large percentage of hydropower in total electricity supply.

1. Development Potential and Geographical Distribution of Hydropower in the World

Theoretically, about 41914 TWh/yr hydropower resources are available worldwide; however, the distribution is uneven and the development degree varies from region to region. As shown in Table 1.3, up to 2019, the global level of hydropower development had reached 27.3% of the technical potential and 44.7% of the economic potential. Asia has the largest share of the world's total technical potential, at around 50.7%, and the followed is South America, at around 18.1%. The developed percentages of technical potentials were 54.7% in Europe, 38.7% in North and Central America, 26.2% in Asia, 24.0% in South America, 22.3% in Oceania, and 8.3% in Africa. Therefore, by 2019, the untapped percentages of technical potential were 91.7% in Africa, 77.7% in Oceania, 76.0% in South America, 73.8% in Asia, 61.3% in North and Central America, and 45.3% in Europe. Even in the most industrialized parts of the world, the untapped potential remains high.

For nearly a decade, the global installed hydropower capacity has been growing, with average an increase of 24.2 GW per year. By 2011, 1067 GW (including pumped

Table 1.2 Countries with a large percent of hydropower in total electricity supply

Country	Percent of hydropower/%	Country	Percent of hydropower/%
Paraguay	100	Sri Lanka	94
Norway	99	Brazil	87
Zambia	99	Colombia	78
Ghana	99	Portuguesa	77
Albania	97	New Zealand	75
Mozambique	96	Latvia	70
Zaire	95	Iceland	70
Switzerland	74	Venezuela	73
Nepal	74	Austria	67
Canada	67		

Table 1.3 Development potential and status of hydropower in the world

Region	Theoretical potential*	Technical potential*	Economic potential*	Share of world total technical potential*	Developed until 2019**	Developed percent of technical potential	Developed percent of economic potential
	TWh/yr	TWh/yr	TWh/yr	%	TWh/yr	%	%
Global	41914	15778	9624	100	4306	27.3	44.7
Asia	18248	8000	4786	50.7	2096.51	26.2	43.8
South America	7848	2859	1728	18.1	685.72	24.0	39.7
Africa	4423	1647	1124	10.4	137.66	8.3	12.2
North and central America	7601	1891	1045	12.0	731.74	38.7	70.0
Europe	3136	1195	852	7.6	653.1	54.7	76.6
Oceania	658	186	89	1.2	41.27	22.3	46.4

Note *Data from IJHD World Atlas and Industry Guide 2017; **Data from IHA 2020 Hydropower Status Report

storage) was in operation, and by 2019, the value reached 1308 GW (Fig. 1.17). Within the top 20 hydropower countries, China has the largest installed capacity (including pumped storage), and the followed are Brazil, USA, Canada, India, Japan, Russia, Norway, etc. (Fig. 1.18).

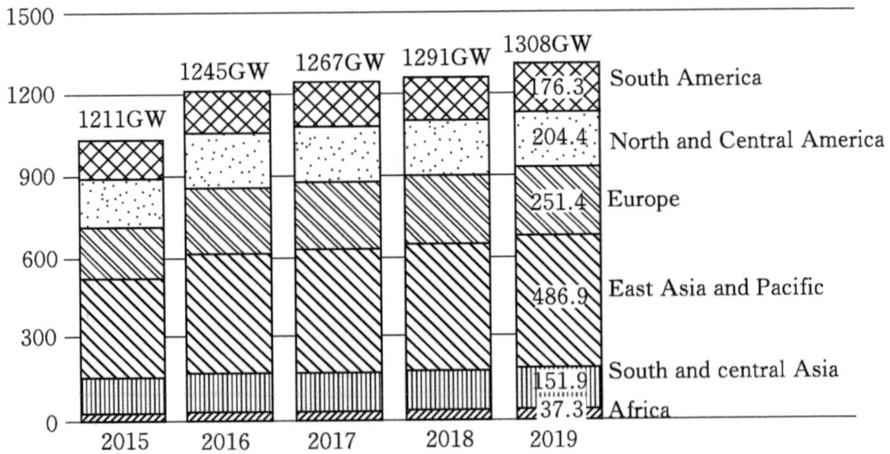

Fig. 1.17 Hydropower installed capacity from 2015 to 2019. Data from IHA 2020 Hydropower Status Report

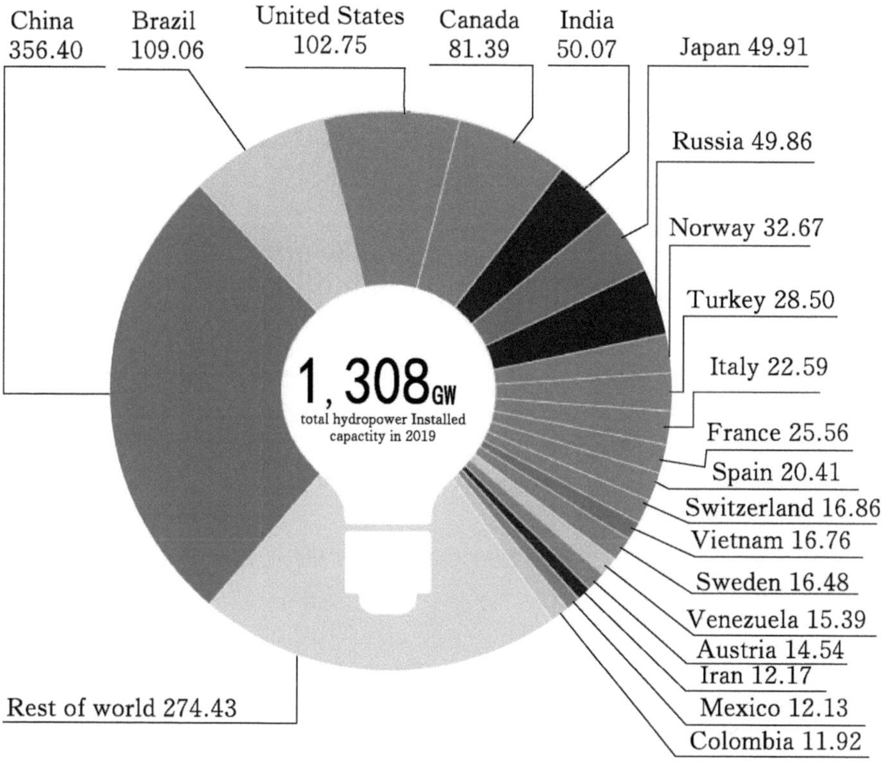

Fig. 1.18 Hydropower installed capacity by 2019, including pumped storage. Data from IHA 2020 Hydropower Status Report

Based on *IEA 2012 energy technology outlook*, to achieve the objective of reducing carbon dioxide emissions, the global hydropower capacity was projected to reach 1947 GW in 2050, and the electricity generation was projected to reach 7100 TWh/yr (Fig. 1.19), both of which are twice as large as in 2012. According to *IHA 2020 Hydropower Status Report*, in order to limit the global temperature rise to below 2 °C above pre-industrial levels, International Renewable Energy Agency in its "Transforming Energy Scenario" suggests that global hydropower capacity would need to increase by 25% by 2030 and by 60% by 2050, reaching 2147 GW. This equates to around 850 GW in additional installed capacity over the next 30 years. To reach the 2050 target, the yearly average growth in hydropower capacity would need to reach an estimated 2.0% a year on average.

Geographical distributions of hydropower potential and development are explained as follows (the data is combined from different sources).

Africa's mainland has the greatest untapped hydropower potential, with only 8.3% developed by 2019. Much of the potential for development lies in many parts of Africa and transboundary basins, including the Congo River, the Nile River, the Niger River,

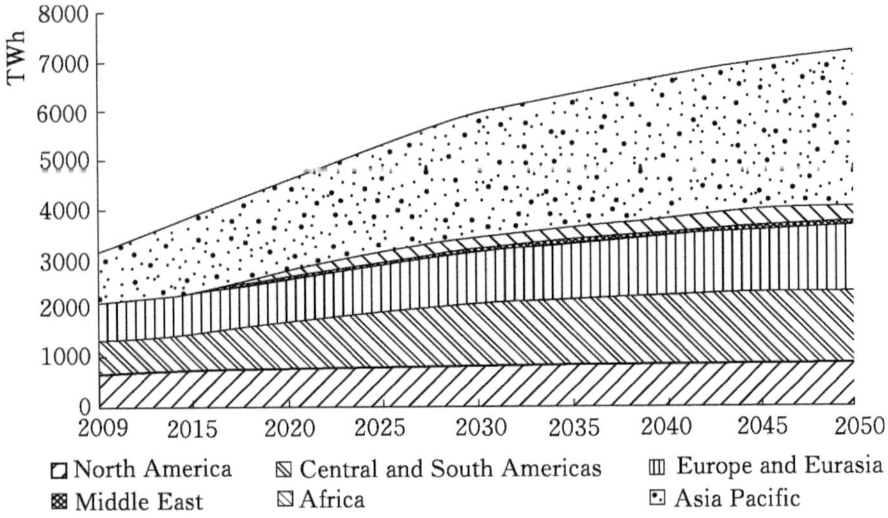

Fig. 1.19 Development prospects until 2050 (Generation, TWh/yr). Data from *IEA 2012 energy technology outlook*

and the Zambezi River. The total hydropower capacity in Africa is projected to be 88 GW and the hydropower generation to 500 TWh/yr by 2050.

The hydropower development in South America has been remarkable, especially since 1970. The total installed hydropower capacity has reached 176.3 GW by 2019. About half of the electricity in this area comes from hydropower. The total hydropower capacity in South America is expected to reach 240 GW by 2050, of which 130 GW will be in Brazil. The hydropower generation will reach 1350 TWh/yr, with more than half in Brazil.

The hydropower development in North America will depend on the growth of small hydropower in the USA and the untapped potential in Canada. The total hydropower capacity in North America is projected to reach 215 GW and hydropower generation to 1220 TWh/yr by 2050.

The hydropower development in Asia will depend mainly on China and India. The hydropower generation in China rose sharply from less than 400 TWh/yr in 2005 to 735TWh/yr in 2011, 1100 TWh/yr in 2017, and 1302 TWh/yr in 2019. It could be over 1500 TWh/yr by 2035. India has mapped a hydropower development plan for 399 hydropower stations (total capacity 148.7GW) and 56 pumped storage projects (94 GW). And in May 2003, India also launched 162 projects under the "*50 GW hydropower program*". The estimated total hydropower capacity in Asia by 2050 is 852 GW. China accounts for half of that, and India accounts for a quarter. In that time, Asia's hydropower generation will reach 3800 TWh/yr.

The common goal of EU member states for renewable energy usage is to reach 20% by 2020. Under this background, reservoir hydropower stations and pumped storage power stations will promote the utilization of renewable energy and play an important role in the development of wind power and photovoltaic power. The European hydropower capacity is projected to reach 310 GW and hydropower generation to 930 TWh/yr in 2050.

Russia's energy strategy by 2020 is that the share of hydropower in the mixed energy system will remain at about 20%. Tajikistan and Kyrgyzstan have great potential for hydropower development. Russia and the Eurasian mainland are expected to generate hydropower generation of 510 TWh/yr with an installed capacity of 145 GW in 2050, of which about 75% will be in Russia.

2. Large Hydropower Stations in Big Rivers

The comprehensive development of big rivers requires constructing dams, which form reservoirs. Up to now, more than 50,000 dams have been built around the world. The number of dams higher than 100 m is 712, of which 293 are in Asia, 203 in Europe, 117 in North America, 53 in South America, 29 in Africa, and 17 in Oceania. The number of hydropower stations with installed capacities of more than 300 MW is 418, with which 143 in Asia, 100 in Europe, 78 in North America, 71 in South America, 17 in Africa, and 9 in Oceania. The number of reservoirs with a capacity larger than 3 billion cubic meters is 289, with which 85 in Asia, 71 in North America, 56 in South America, 50 in Europe, 22 in Africa, and 5 in Oceania. For many large hydropower stations, power generation is their dominant function, and most of them also have flood control, irrigation, water supply, entertainment, shipping, aquaculture, and other functions. The top ten hydropower stations in the world are shown in Table 1.4.

3. Small-sized and Medium-Sized Hydropower Development

Small hydropower stations (power station capacity up to 10 MW) are generally built on small rivers, which are normally located in remote mountainous areas. Because

Table 1.4 Top ten hydropower stations in the world (data in May 2015)

Order	Name	Finishing time	Installed capacity/ MW	Power generation/ (GWh/yr)	Country
1	Three Gorges	2012	22500	88200	China
2	Itaipu	1991	14000	90000	Brazil/ Paraguay
3	Xiluodu	2014	13860	57120	China
s	Guri	1986	10305	51000	Venezuela
5	Tucurui	1984	8370		Brazil
6	Xiangjiaba	2014	7750	30747	China
7	Grand Coulee	1980	6494	20200	The USA
8	Sayano-Shushenskaya	1987	6400	23500	Russia
9	Longtan	2009	6300	18700	China
10	Krasnoyarsk	1972	6000	20400	Russia

they normally belong to the conduit diversion-type, ecological or environmental negative effects, such as flow cut-off or reduction and soil erosion always, cause concerns of society. Also, the economy of small hydropower is not comparable to that of large hydropower. Therefore, the development of small hydropower stations has always been ignored.

With the increasing energy crisis, renewable energy development becomes urgent, and small hydropower encounters new development opportunities. As a well-developed small-scale renewable energy technology that contributes to improved access to electricity in rural areas, small hydropower is one of the programs of the United Nations Industrial Development Organization (UNIDO) to address inclusive and sustainable industrial development. The most important challenge in hydropower development is the financing problem; however, small hydropower has solved this problem well, because it needs relatively little investment. Small hydropower is a locally accessible renewable energy source that is clean, efficient, and safe for electrification of the countryside.

At present, 152 countries and regions have small hydropower stations. The global small hydropower potential is about 229 GW, of which more than half is in Asia and about one-third in Europe and the Americas. There are possibilities to obtain more information on small hydropower potential in Africa and the Americas in future. By the end of 2019, the total installed capacity of small hydropower worldwide was about 78 GW, and the development rate was about 34%. The distribution of small hydropower (power station capacity up to 10 MW) in the world is shown in Tables 1.5.

By the end of 2019, 47498 small hydropower stations (station capacity up to 50 MW) had been built in China. The total installed capacity is more than 79.30 GW, three times of the installed capacity of the Three Gorges Hydropower Station. China's small hydropower technical potential is 128 GW; therefore, the current development ratio is about 61.9%.

4. Pumped Storage Hydropower Stations

Pumped storage power is the largest and best regulator in power grid for its irreplaceable capacities of peak fitting, valley filling, frequency regulating, phase modulating,

Table 1.5 World distribution of small hydropower (power station capacity up to 10 MW)

Continent	Asia	Europe	Africa	Oceania	South America	North and Central America	Total in the world
Potential capacity/MW	138276	37554	10240	1208	28483	13378	229139
Installed capacity/MW	51069	19699	595	441	806	5375	77985
Development percent/%	36.9	52.5	5.8	36.5	2.8	40.2	34.0

Note Data from the United Nations Industrial Development Organization, 2019

spinning reserving, and black-starting. With the development of intermittent renewable energies, the complementary functions of pumped storage become more and more important. Different from normal hydropower stations, most of which are in remote areas, pumped storage power stations can be located in or near power load centers and energy bases, which are normally closed to large cities.

The first pumped storage power plant was built in Switzerland in 1882, with a history of more than 130 years. In the early years, the main function of pumped storage power stations was pumping water during the flood season to adjust the seasonal imbalance of conventional hydropower stations. Since the 1950s, the development of pumped storage hydropower has been speeded up; during the 1960–1980s, the construction of pumped storage power stations came into the golden period, with the average annual capacity increase of 1259 MW in the 1960s, 3015 MW in the 1970s, and 4036 MW in 1980s. By the end of 2019, the global pumped storage capacity had grown up to 157.994 GW. The pumped storage hydropower capacities of the top 10 countries are shown in Fig. 1.20, and the top 10 pumped storage power stations in the world before 2020 are shown in Table 1.6.

China has been also vigorously developing pumped storage power stations (Fig. 1.21). As of the end of 2019, there were 34 stations in operation and 32 under construction, corresponding to the installed capacities of 30.29 and 43.21 GW, respectively. According to the *13th Five-Year Plan for Hydropower Development* issued by the State Energy Administration, the accumulative installed capacity of pumped storage power in China should reach 40 GW by the end of 2020 and 90 GW by 2025.

5. Prospects for World Hydropower Development

Hydropower can bring important synergies to an energy system of the future. For example, hydropower will play an important role in helping to integrate higher shares of variable renewable energy (VRE). In the Transforming Energy Scenario, hydropower capacity would increase by two-thirds to 2147 GW by 2050 compared to current levels. In other words, around 850 GW of new installed capacity is required in the next 30 years. Of that, pumped storage would need to more than double to 325 GW, providing storage and energy flexibility. In addition to the new capacity, the current aged hydropower estimated at more than 600 GW will require refurbishment, which offers a timely opportunity to modernize hydropower to provide greater flexibility to support the variable sectors.

Hydropower also can play the additional roles of regulating river flows and reducing flooding. Increasing hydropower capacity does not entail only building new dams; options also exist to upgrade turbines and systems in existing stations and to electrify non-power dams. For the new hydropower stations, planners need to consider local environmental impacts and to negotiate with communities in the impacted areas. Hydropower stations will need operational changes that reflect the changing power system needs, including faster and more frequent ramping, and planning practices that include evaluating the impacts of climate change on water supply and reservoir storage requirements. Due to longer planning cycles for hydropower projects, policymakers and planners would need to start thinking now

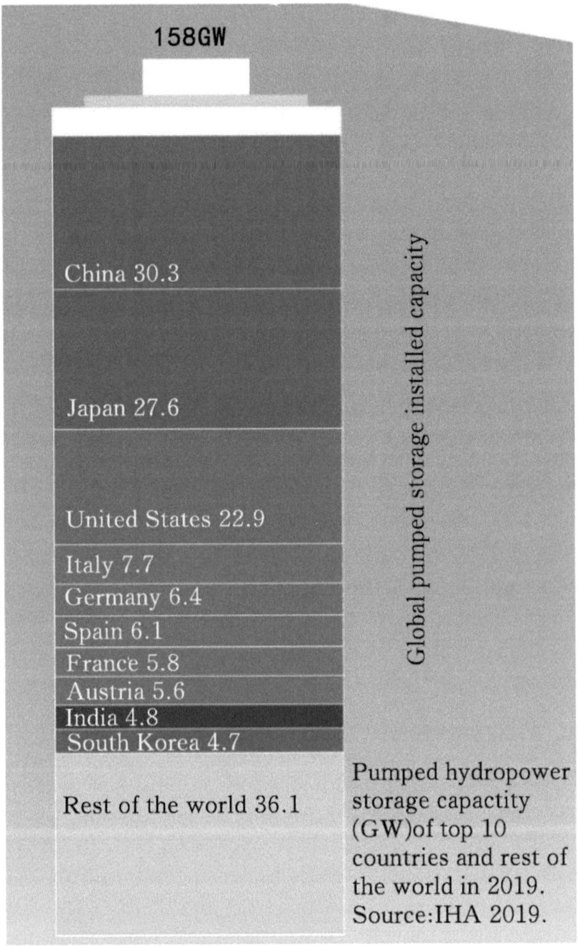

Fig. 1.20 Pumped storage hydropower capacities (GW) of the top 10 countries and the rest of the world in 2019. Data from *IHA 2020 Hydropower Status Report*

Table 1.6 Top 10 pumped storage power stations in the world (data at the end of 2020)

Order	Name	Finishing time	Installed capacity/MW	Country
1	Fengning	2022	3600	China
1	Bath County	1985	3003	The USA
2	Kannagawa	2020	2820	Japan
3	Dniester	2017	2947	Ukraine
4	Huizhou	2010	2448	China
5	Guangzhou	2000	2400	China
6	Okutataragi	1998	1932	Japan
7	Ludingtron	1973	1872	The USA
8	Tianhuangping	2004	1836	China
9	Jixi	2020	1800	China

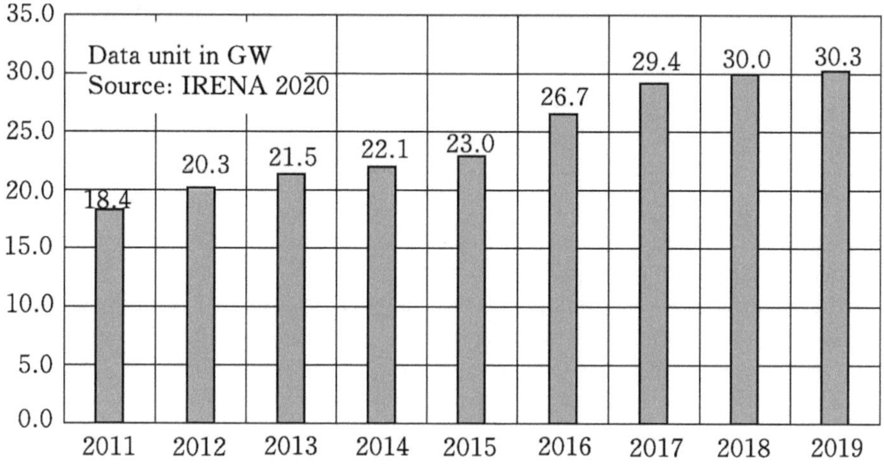

Fig. 1.21 Installed pumped storage hydropower in China from 2011 to 2019

about new projects. With enabling policies, market conditions, common frameworks to scale investments, and well-prepared projects, hydropower will be one of the key technologies in coming decades.

Looking forward to the future, the world hydropower has the following development trends:

(1) The utilization of renewable hydropower energy has many irreplaceable advantages, and the priority development of hydropower has brought great benefits to social and economic development. Therefore, active development of hydropower with priority has become one of the basic national policies of many countries. The active participation of China's hydropower construction team and the export of hydropower technology have brought about the vigorous development of hydropower construction in Southeast Asia, Africa, and South America.
(2) Under the background of "Halving Global Greenhouse Gas Emissions by 2050", countries all over the world attach great importance to large-scale development of renewable energies, such as wind and solar. Wind power and solar power are both random and intermittent in nature, and their large-scale participation in power grids requires the regulation and joint operation of hydropower, especially pumped storage power. Therefore, the technology and construction of pumped storage power stations have been paid more attention to and will have rapid development. Large capacity (single unit capacity 500 MW), high head (900–1000 m) single-stage pump-turbine, and variable speed reversible unit are expected to break through in recent years and be used in engineering.
(3) The development of environmentally affordable hydropower and the comprehensive utilization of hydropower projects are the important development directions. Hydropower projects not only provide low-carbon and renewable energy to the world but also bring great economic benefits. They also play a great

role in preventing flood and drought disasters, irrigation, water supply, shipping, cross-basin water transfer, land and water resources development, environmental pollution control, and so on. But hydropower development does face the problem of environmental carrying capacity. In addition to water pollution, there are also problems such as earthquakes induced by reservoir water storage, inundation of land, migration, swamping, and soil fertility decline. Dam construction also influences river hydrological conditions, sediment transport, the living environment of flora and fauna, forest vegetation conditions, and the natural landscape. Therefore, it is necessary to deepen the research on the environmental impacts and comprehensive benefits of hydropower development, to minimize the negative impacts and formulate more stringent and objective standards.

(4) The renovation and full use of existing hydropower projects are also important. Early hydropower stations in the world have been operating for more than 50 years, and some are close to their service life. The renewal and transformation of old hydropower stations can not only prolong life and reduce cost but also adopt new technology to improve efficiency and increase unit output. Comprehensive development can also be carried out, such as adding water supply pipeline systems to meet the needs of residents. In addition, the demolition and refurbishment of dams and hydropower stations are included in the planning and design vision of many countries and the prospect is broad.

(5) Hydropower development depends on scientific and technological progress, which can not only reduce the cost, improve operation efficiency, and prolong service life, but also promote the development of hydropower stations with poor conditions and increase the technical potential capacity. In addition, the progress of science and technology can better realize the superior role of hydropower units in power grid regulation, namely peak shaving, frequency modulation, spinning reserve, and black start. It is expected that with the development of ultra-high head and super capacity hydropower units, variable speed constant frequency units, intelligent control of hydropower systems, and new building materials and technology, the hydropower industry will usher in a more brilliant tomorrow.

1.2.2 Achievements and Prospects of Hydropower in China

The first hydropower station in China mainland is Shilongba Hydropower Station built on Tanglang Chuan stream in Yunnan Province in 1912, with an original capacity of 480 kW. Later, it was rebuilt and expanded to 6000 kW. Before the founding of the People's Republic of China in 1949, only 42 hydropower stations were built or partially built throughout the country, with a total capacity 360 MW and annual electricity generation of 1200 GWh (excluding Taiwan). Since 1950, the construction of hydropower has been greatly developed. Hydropower stations in China were classified as large (with installed capacity larger than 250 MW), medium (25–250 MW), and small (below 25 MW) sized, and all were promoted. A number of large backbone hydropower stations have been built. One of the largest is the Three Gorges Project on the Yangtze River. A large number of medium-sized hydropower stations have been built, and many of them are also cascaded. A large number of small-sized hydropower stations have been built on some small and medium rivers and ditches. On August 25, 2010, the Unit 4 (single capacity 700 MW) in Xiaowan Hydropower Station, Yunnan Province, was officially put into production. It is the iconic unit that claimed China's hydropower installation of 200 GW ranks the first in the world. By the end of 2022, China's installed capacity of hydropower reached 413.50 GW.

1. Hydropower Potential and Watershed Distribution

With a vast territory of 9.6 million km^2, China is rich in hydropower resources. The latest survey of hydropower resources in 2004 provided the following facts: The total theoretical potential of hydropower in China mainland is 694.4 GW on 3886 rivers that have 10 MW potential or more, corresponding to a total theoretical power generation potential of 6082.9 TWh/yr. The total technical potential is 541.64 GW for the 13286+28/2 hydropower stations (28/2 means that the 28 stations are on international boundary rivers) with capacities larger than 0.5 MW, corresponding to a total technical power generation potential of 2474.0 TWh/yr. The economic potential is 401.795 GW for the 11653+27/2 hydropower stations (27/2 means that the 27 stations are on international boundary rivers) with capacities larger than 0.5 MW, corresponding to a total economic power generation potential of 1753.4 TWh/yr.

In addition, the survey in 1995 provided the facts for Taiwan Province that the theoretical potential, technical potential, and economic potential are 11.652, 5.048, and 3.835 GW, respectively. The corresponding power generation potentials are 102.07, 20.15, and 13.83 TWh/yr.

The detailed data are listed in Tables 1.7 and 1.8.

Table 1.7 Results of the national survey of hydropower resources in China mainland (2004)

Order	Item		Unit	Data
1	Theoretical potential	Annual generation	TWh/yr	6082.9
		Average capacity	GW	694.4
2	Technical potential	Number of stations	Site	13286+28/2
		Installed capacity	GW	541.64
		Annual generation	TWh/yr	2474.0
3	Economic potential	Number of stations	Site	11653+27/2
		Installed capacity	GW	401.795
		Annual generation	TWh/yr	1753.4
4	Constructed	Number of stations	Site	6053+4/2
		Installed capacity	GW	130.980
		Annual generation	TWh/yr	525.9

Note The statistical range in the table includes the rivers with theoretical potential larger than 10 MW and hydropower stations with capacity larger than 0.5 MW, excluding Hong Kong, Macao, and Taiwan.

Table 1.8 Statistics of hydropower stations classified by capacity (2004)

Order	Item			Unit	Data
1	Large-sized hydropower stations (300 MW or larger)	Technical potential	Number	Site	263+10/2
			Installed capacity	GW	388.7
			Annual generation	TWh/yr	1792.0
		Economic potential	Number	Site	175+10/2
			Installed capacity	GW	276.082
			Annual generation	TWh/yr	1205.0
		Constructed	Number	Site	65+3/2
			Installed capacity	GW	90.102
			Annual generation	TWh/yr	357.1
2	Middle-sized hydropower stations (50 MW to 300 MW)	Technical potential	Number	Site	785+10/2
			Installed capacity	GW	87.73
			Annual generation	TWh/yr	392.7
		Economic potential	Number	Site	631+9/2
			Installed capacity	GW	68.252
			Annual generation	TWh/yr	293.3
		Constructed	Number	Site	162+1/2
			Installed capacity	GW	17.693
			Annual generation	TWh/yr	70.5
3	Small-sized hydropower stations (0.5 MW to 50 MW)	Technical potential	Number	Site	12238+8/2
			Installed capacity	GW	65.21
			Annual generation	TWh/yr	289.3
		Economic potential	Number	Site	10847+8/2
			Installed capacity	GW	57.461
			Annual generation	TWh/yr	255.1
		Constructed	Number	Site	5826
			Installed capacity	GW	23.185
			Annual generation	TWh/yr	98.3

Note The statistical range in the table includes the rivers with theoretical potential larger than 10 MW and hydropower stations with capacity larger than 0.5 MW, excluding Hong Kong, Macao, and Taiwan

According to the flow direction, China's rivers can be divided into two categories: outflow rivers flowing to the ocean and inland rivers that do not communicate with the ocean. The outflow river basins account for about 2/3 of the country's land area. Most of the outflow rivers flow from west to east or southeast to the Pacific Ocean, mainly including the Heilongjiang River, Liaohe River, Haihe River, Yellow River, Huaihe River, Yangtze River, Pearl River, Lancang River, Qiantang River, Minjiang River, etc. Nujiang River, Yaluzangbu River, and other rivers flow south into the Indian Ocean; Ertix River flows into the Arctic Ocean through Kazakhstan and Russia.

Ranked by length, the top three rivers are the Yangtze River 6300 km, the Yellow River 5464 km, and the Heilongjiang River 4341 km (including foreign parts). Ranked by basin area, the top three are the Heilongjiang River 1.855 million km^2 (including foreign parts), the Yangtze River 1.8085 million km^2, and the Yellow River 0.7947 million km^2. Ranked by natural annual average runoff, the top three are the Yangtze River 961.3 billion m^3, the Heilongjiang River 355 billion m^3 (including foreign parts), and the Pearl River 328 billion m^3. Ranked by the technical potential of hydropower development, the top three are the Yangtze River Basin with 256.2 GW, the Yaluzangbu River Basin with 67.8496 GW (only the domestic part), and the Yellow River Basin with 37.3425 GW.

According to hydropower potential and geographical location, the rivers in China are divided into ten parts in the national survey of hydropower resources in 2004: the Yangtze River Basin, the Yellow River Basin, the Pearl River Basin, the Haihe River Basin, the Huaihe River Basin, the Rivers in Northeast China, the Southeast Coastal Rivers, the Southwest International Rivers, the Yaluzangbu River and other rivers in Xizang, and the Rivers in the northern inland and Xinjiang. Details are given in Tables 1.9 and 1.10.

2. Characteristics of Hydropower Resources and Planning for Hydropower Development in China

China's hydropower resources have the following characteristics.

(1) The regional distribution of hydropower resources is uneven, and "West-East Electricity Transmission" is necessary. Due to the vast territory of China and the large gaps between topography and rainfall, the distribution of hydropower resources is unbalanced, with more in the west and less in the east. The economy is relatively developed in the east and relatively underdeveloped in the west. Therefore, the development of hydropower resources in the west should consider both western and eastern power markets, and the "West-East Electricity Transmission" should be implemented.
(2) The time distribution of hydropower resources is uneven, and it is necessary to build reservoirs for regulation. In most rivers, the annual and interannual runoff distributions are uneven, with runoff varied greatly in rainy and dry seasons. Therefore, it is necessary to build reservoirs to regulate runoff, improve hydropower quality, and meet power market needs.
(3) The hydropower resources are concentrated in the mainstream of the large rivers; therefore, establishing hydropower bases to strategically develop the large hydropower stations is possible. The hydropower resources are enriched in the 13 hydropower bases, namely Jinsha River; Yalong River; Dadu River; Lancang River; Wujiang River; Upper Yangtze River; Nanpan River and Honghe River; Upper Yellow River; rivers in west Hunan Province; rivers in Fujian, Zhejiang, and Jiangxi provinces; rivers in Northeast China; the north main stream of the Yellow River; Nujiang River. The total technical capacity of these hydropower bases accounts for about 50.9% of the national technical capacity.

Table 1.9 Summary of hydropower resources in China mainland (classified according to basin in 2004)

Order	Basin	Theoretical potential		Technical potential			Economic potential			Constructed		
		Annual generation	Average capacity	Number of station	Installed capacity	Annual generation	Number of station	Installed capacity	Annual generation	Number of station	Installed capacity	Annual generation
		TWh	GW	Site	GW	TWh/yr	Site	GW	TWh/yr	Site	GW	TWh/yr
1	The Yangtze River Basin	2433.598	277.8080	5748	256.2729	1187.899	4968	228.3187	1049.834	2441	69.7271	292.496
2	The Yellow River Basin	379.413	43.3121	535	37.3425	136.096	482	31.6478	111.139	238	12.0304	46.479
3	The Pearl River Basin	282.394	32.2367	1757	31.2880	135.375	1538	30.0210	129.768	957	18.1007	78.578
4	The Haihe River Basin	24.794	2.8303	295	2.0295	4.763	210	1.5100	3.501	123	0.8034	1.950
5	The Huaihe River Basin	9.800	1.1185	185	0.6560	1.864	135	0.5565	1.592	75	0.3103	0.958
6	The Rivers in Northeast China	145.480	16.6074	644+26/2	16.8208	46.523	510+26/2	15.7291	43.382	269+4/2	6.3968	15.174
7	The Southeast Coastal Rivers	177.611	20.2753	2558+1/2	19.0749	59.339	2535+1/2	18.6483	58.135	1388	11.6537	36.308
8	The Southwest International Rivers	863.007	98.5168	609+1/2	75.0148	373.182	532	55.5944	268.436	313	9.3227	44.277

(continued)

Table 1.9 (continued)

Order	Basin	Theoretical potential		Technical potential			Economic potential			Constructed		
		Annual generation	Average capacity	Number of station	Installed capacity	Annual generation	Number of station	Installed capacity	Annual generation	Number of station	Installed capacity	Annual generation
		TWh	GW	Site	GW	TWh/yr	Site	GW	TWh/yr	Site	GW	TWh/yr
9	The Yaluzangbu River and other rivers in Xizang	1403.482	160.2148	243	84.6636	448.611	130	2.5955	11.969	52	0.3466	1.155
10	The Rivers in the northern inland and Xinjiang	363.357	41.4791	712	18.4716	80.586	616	17.1740	75.639	270	2.2902	8.510
Total		6082.9	694.400	13286+28/2	541.640	2474.0	11653+27/2	401.795	1753.4	6053+4/2	130.980	525.9

Note The statistical range in the table includes the rivers with theoretical potential larger than 10 MW and hydropower stations with a capacity larger than 0.5 MW, excluding Hong Kong, Macao, and Taiwan

Table 1.10 Summary of hydropower resources in China mainland (classified according to basin and power station size in 2004)

Order	Basin	Large-sized power station (larger than 300MW)					
		Technical potential		Economic potential		Constructed	
		Capacity	Generation	Capacity	Generation	Capacity	Generation
		GW	TWh/yr	GW	TWh/yr	MW	TWh/yr
1	The Yangtze River Basin	187.3070	861.972	168.6450	769.827	51.965	211.085
2	The Yellow River Basin	30.6645	106.472	25.8945	85.709	10.1225	37.780
3	The Pearl River Basin	17.4675	73.911	17.4675	73.911	12.0335	51.757
4	The Haihe River Basin						
5	The Huaihe River Basin						
6	The Rivers in Northeast China	8.2640	22.624	8.2640	22.624	3.7640	8.406
7	The Southeast Coastal Rivers	4.1625	11.161	4.1625	11.161	3.8625	10.801
8	The Southwest International Rivers	62.4860	308.145	47.3460	225.798	7.5750	35.684
9	The Yaluzangbu River and other rivers in Xizang	74.0770	392.784	0.3320	1.736		

(continued)

Table 1.10 (continued)

Order	Basin	Large-sized power station (larger than 300MW)					
		Technical potential		Economic potential		Constructed	
		Capacity	Generation	Capacity	Generation	Capacity	Generation
		GW	TWh/yr	GW	TWh/yr	MW	TWh/yr
10	The Rivers in the northern inland and Xinjiang	4.2700	14.954	3.9700	14.254	0.7800	1.618
Total		388.700	1792.0	276.082	1205.0	90.102	357.1

Middle-sized power station (50MW to 300MW)						Small-sized power station (0.5–50 MW)					
Technical potential		Economic potential		Constructed		Technical potential		Economic potential		Constructed	
Capacity	Generation	Capacity	Generation	Capacity	Generation	Capacity	Generation	Capacity	Generation	Capacity	Generation
GW	TWh/yr	GW	TWh/yr	GW	TWh/yr	GW	TWh/yr	GW	TWh/yr	GW	TWh/yr
38.8073	181.741	33.2469	152.821	7.8786	35.819	30.1589	144.188	26.4277	127.185	9.8834	45.595
4.0821	17.141	3.4395	14.264	1.1435	4.973	2.5959	12.485	2.3138	11.169	0.7644	3.722
6.0084	27.178	5.5594	25.239	2.5770	11.883	7.8121	34.284	6.9940	30.618	3.4904	14.938
0.7490	1.147	0.6170	1.021	0.2970	0.589	1.2805	3.616	0.8930	2.480	0.5064	1.361
0.0505	0.110	0.0500	0.110	0.0	0.0	0.6060	1.754	0.5065	1.482	0.3103	0.958
5.1170	12.993	4.6510	11.735	1.8655	4.301	3.4399	10.908	2.8142	9.024	0.7674	2.467
4.9312	13.835	4.6812	13.141	2.5962	7.535	9.9812	34.343	9.8046	33.833	5.1949	17.973
8.9290	45.833	5.3685	27.418	0.6265	2.918	3.5998	19.204	2.8799	15.220	1.1212	5.675
9.7805	51.566	1.7365	7.681	0.2125	0.499	0.8061	3.962	0.5271	2.552	0.1341	0.656
9.2715	41.112	8.9015	39.843	0.4970	1.963	4.9301	24.521	4.3025	21.542	1.0132	4.930
87.730	392.7	68.252	293.3	17.693	70.5	65.210	289.3	57.461	255.1	23.185	98.3

Note The statistical range in the table includes the rivers with theoretical potential larger than 10 MW and hydropower stations with a capacity larger than 0.5 MW, excluding Hong Kong, Macao, and Taiwan

(4) Large-sized hydropower stations account for a large proportion of the total installed capacity, and small and middle-sized hydropower stations spread everywhere. The proportions of the installed capacity and annual power generation of the large-sized hydropower stations (installed capacity larger than 300 MW) are 71.76 and 72.43%, respectively, among which the proportions of the super large hydropower stations (1000 MW or above) are both more than 50%. The number of small-sized hydropower stations accounts for 92.1% of the total number in the country. Although the installed capacity and annual power generation are not large, small-sized hydropower stations are valuable resources to solve local electricity shortage.

The present situation and planning of hydropower development in China are as follows.

In 2013, with the commissioning of the large turbine-generator units in Xiluodu and Xiangjiaba hydropower stations in Jinsha River, the total installed hydropower capacity in China reached 280.02 GW, with which 258.49 GW were from conventional hydropower stations. By 2015, the conventional hydropower capacity reached 271 GW, which was 28.6% of the total installed power capacity of China. By 2022, the hydropower capacity reached 413.50 GW, accounting for 16.2% of the total installed power capacity.

Based on a new version planning, by 2025, the total installed hydropower will be 459 GW, including conventional hydropower 392 GW and pumped storage hydropower 68 GW. By 2035, the total installed hydropower will be 700 GW, including large- and middle-sized hydropower 360 GW, small-sized hydropower 100 GW and pumped storage hydropower 80–100 GW, and the annual generation will be 2100 TWh/yr, corresponding to development level 70%. By 2050, the total installed hydropower will be 800 GW and the annual generation will be 2500 TWh/yr, corresponding to a development level of 83%.

Starting in 2020, the main battlefield of hydropower development will gradually shift to the upper reaches of the Jinsha River, Lancang River, and Nujiang River, and the strategic "Xizang Electricity Delivery" project will start. There are many rivers in the Xizang Autonomous Region, providing abundant water resources. The theoretical potential of hydropower resources in the whole region accounts for 29% of the country's potential, sitting the first place in the country. The technical potential accounts for 20.3% of the country's potential, ranked second in the country. The development of the Yaluzangbu River in southern Tibet is important and relatively difficult. The mainstream from Qusong to Milin reach (about 5000 MW), the mainstream big bend (about 48000 MW, with water head up to 2000 m or more), and the tributary Palongzangbu River (about 7000 MW) are the key sites of the development. Relying on technological innovations in hydropower and transmission engineering, and the expansion of electricity transmission corridors and electricity markets through international cooperation, we should plan scientifically to exploit these hydropower rich mines as soon as possible for the better world.

3. Achievements in Hydropower Research and Construction

From the 1980s to the beginning of the twenty-first century, China's hydropower construction has made rapid progress, and a number of world-class giant hydropower stations have been built, such as the Three Gorges, Xiangjiaba, Xiluodu, Longtan, Xiaowan, and Nuozhadu. The construction of these hydropower stations is of great significance for China to implement the western development strategy, alleviate the increasingly prominent power shortage due to rapid economic development, effectively improve the power supply structure of China, and promote the complementary and coordinated development of the eastern, central, and western regions. Moreover, the construction has also promoted China's scientific and technological level in hydropower by remarkable achievements. The main indexes to measure the achievements of hydropower station construction and the progress of science and technology can be summarized as the single unit capacity, rated head, runner diameter, the scale of a powerhouse (span, height, and total length), the scale of pressure regulating buildings (such as the cross-sectional area and height of surge chamber and the dimensions of slope ceiling tailrace tunnel), the length and cross-sectional area of tunnels and penstocks, and the complexity of the hydraulic system with multiple units. According to the statistics of the main technical parameters of some large underground hydropower stations in China (Tables 1.11, 1.12, 1.13, 1.14), we can make the following summaries:

(1) The single unit capacity of Baihetan Hydropower Station is 1000 MW, the largest in the world. The second is 850 MW in Wudongde Hydropower Station, and the third is 800 MW in Xiangjiaba Hydropower Station. The water head of Baihetan Hydropower Station is 202 m, followed by Xiluodu Hydropower Station and Goupitan Hydropower Station. The turbine runner diameter of the Three Gorges Hydropower Station has exceeded 10 m, to 10.25 m, and is also the world's largest.
(2) The dimensions of underground powerhouse are very large (Table 1.13 and Fig. 1.22). The height of the Wudongde powerhouse is 89.8 m; the spans of the Xiangjiaba and Baihetan are 33.4 m/31.4 m and 34.0 m/31 m, respectively; the lengths of the Xiluodu and Longtan are 443.34 and 398.2 m, respectively, with nine units which are arranged for each.
(3) The dimensions of the surge chambers are large (Table 1.12 and Fig. 1.23). The highest surge chamber is in Jinping II hydropower station, with a height of 136.8 m. The followed are those in the Wudongde and Goupitan hydropower stations. The largest cross-sectional area is in Xiluodu hydropower station, with 2500 m^2 for each of the three corridor-shaped chambers, which are 317 m long and 26.5 m wide. As for structure type or shape, the one in Baihetan is cylinder-shaped, with a diameter of 49.0 m and a height of 105.0 m. The diameter and height of Jinping II are 28.0 and 136.8 m, respectively. The shape of the Wudongde surge chambers is half-cylinder, which has the largest diameter 53 m and a height just shorter than that of Jinping II.
(4) Tailrace tunnel with sloping ceiling. The sloping ceiling tailrace tunnel is a new type of pressure suppress structure that first be used in Hepin Hydropower

Table 1.11 Main technical parameters of large underground hydropower stations in China (incomplete statistics)

Station	Single unit capacity/MW	Rated head/m	Runner diameter/m	Main powerhouse/m				Plane dimensions of spiral casing/m	
				Total length	Length of machine section	Span	Height	Transversal direction	Stream direction
The Three Gorges	700	85	9.6 to 10.25	311.3	38.3	32.6/31	87.3	34.105 to 34.429	29.448 to 30.449
Xiangjiaba	800	100	9.68 to 9.801	255.0	40	33.4/31.4	85.5	31.483 to 31.825	27.162 to 27.819
Xiluodu	770	197	7.15 to 7.65	443.34	34	31.9/28.4	75.6	21.62 to 23.93	19.474 to 21.65
Baihetan	1000	202	8.6	294.8	38	34.0/31	88.8	27.2	24.9
Wudongde	850	137	8.9	333.0	37	32.5/30.5	89.8	31.79	28.06
Ertan	550	165	6.366	280.3	31	30.7/25.5	65.38	20.11	17.876
Xiaolangdi	300	110	6.3	251.5	26.5	26.2/25	61.44	19.637	17.09
Longtan	700	140	7.9	398.2	32.5	30.7/28.9	76.4	25.402	22.378
Pengshui	350	67	7.68	252.0	35	30.0/28.5	78.5	27.13	23.709
Goupitan	600	175.5	7.0	230.45	31	27.0/25.3	75.32	22.16	19.75
Xiaowan	700	216	6.7	326.0	35	30.6/28.3	65.6	20.80	17.875
Laxiwa	700	205	6.9	311.75	39	30.0/27.8	74.84	21.686	19.56
Pubugou	600	156.7	7.18	294.1	34	30.7/26.8	70.1	21.673	19.410

Note The rated head of turbines in the left bank powerhouse of the Three Gorges Project is 80.6 m, and in the right bank and underground powerhouses is 85 m

Station in Vietnam. It was developed in the Three Gorges Project, Penshui, and Xiangjiaba hydropower stations in China (Table 1.13). This structure is a good option for ensuring the lowest draft tube pressure in the allowable range for large underground hydropower stations. When the tail water level is lower than the ceiling, the water flow in the sloping ceiling tunnel is divided into an open-channel flow section and a pressurized flow section (Fig. 1.24).

The purpose of using the sloping ceiling in the tailrace tunnel is to solve the problem that the lowest pressure at the inlet section of the draft tube exceeds the requirement of specifications. Evaluating from the actual engineering point of view, the prerequisite for adopting is that the pressurized tailrace system reaches the condition of needing a tailrace surge chamber, but there is no place to arrange it. The applicable condition is that the tailrace tunnel length is between

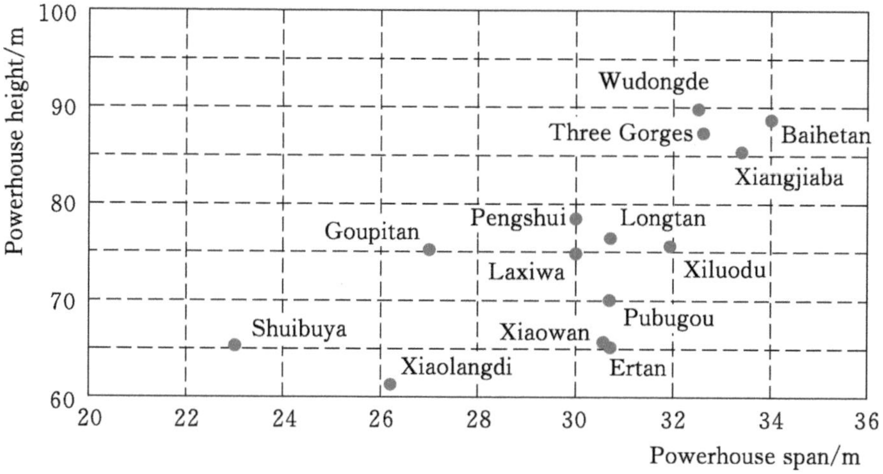

Fig. 1.22 Spans and heights of underground powerhouses

Table 1.12 Scale of surge chambers of some underground power stations in China (incomplete statistics)

Station	Unit capacity/MW	Surge chamber structure	Surge chamber dimension/m			Rock column thickness/m	Distance of adjacent chambers/m	Geological conditions
			Length	Width	Height			
Wudongde	850	Half cylinder	252.0	42.0/39.3 ((D53.0)	113.5	31.0	45.0/40.0	Thick limestone, marble and e thick dolomite
Xiluodu	770	Corridor	317.0	26.5	95.0	—	49.85	Basalt
Baihetan	1000	Cylinder	D42.0 to D49.0		105.0	78.8 to 83.3	76.75	Basalt
Ertan	550	Corridor	101.5	19.8	59.8	17.2	—	Ortholite and altered basalt
Longtan	700	Corridor	236.6	21.58	89.7	—	—	Sandstone and mudstone
Laxiwa	700	Cylinder	D32.0		71.58	—	25.1	Granite
Goupitan	600	Corridor	158.0	18.35	113.0	16.0	24.75	Limestone
Pubugou	600	Corridor	80.0	17.4	53.35	19.13	32.7	Granite
Jinping I	600	Cylinder	D41.0		80.5	59.1.0	49.7	Marble
Jinping II	600	Cylinder	D28.0		136.8	—	—	Marble
Guandi	600	Corridor	223.0	21.5	76.5	17.0	48.8	Basalt

Table 1.13 Main technical parameters of underground power station with sloping ceiling tailrace tunnel

Station		Unit	Hepin	Pengshui	Three Gorges underground	Xiangjiaba underground
Number of units			8	5	6	4
Single unit capacity		MW	240	350	700	800
Total capacity		MW	1920	1750	4200	3000
Runner diameter		m	5.67	7.68	10.25	9.30
Highest head		m	97.50	81.60	113.00	114.20
Lowest head		m	67.50	53.60	71.00	86.10
Rated head		m	88.00	67.00	85.00	100.00
Rated discharge		m^3/s	300.00	589.14	991.80	890.00
Type and size of headrace system	Layout type of headrace system		One tunnel one turbine, circular	One tunnel one turbine, circular	One tunnel one turbine, circular	One tunnel one turbine, circular
	Length of headrace system	m	About 270	460.01	298.64	328.58
	Diameter of circular tunnel	m	8.00	14.00	13.50	14.40
Type and size of headrace system	Layout type of tailrace system		One tunnel two turbines, archway	One tunnel one turbine, archway	One tunnel one turbine, archway	One tunnel two turbines, archway
	Length of tailrace system	m	432.10	480.55	276.62	385.83
	Section dimensions of the sloping ceiling tunnel (width× height)	m× m	12.0×10.0 to 12.0× 14.8	12.6× 23.46 to 12.6×27.5	15.0×24.5	20.0×32.0 to 20.0×34.0
	Ceiling slope	%	2.0	6.37	4.7	4.0
	Bottom slope	%	1.9	5.19	4.7	3.0
Finishing year			1993	October 2009	August 2012	May 2013

Note The lengths of headrace and tailrace systems are from the intake to turbine center and from turbine center to the outlet, respectively

150 and 600 m and the variation of tail water level is relatively large. Compared with the surge chamber scheme, the sloping ceiling tailrace tunnel can reduce construction costs and ensure the safe and stable operation of the power station.

(5) The dimensions of headrace tunnels and penstocks (Table 1.14). The headrace tunnel of Futang Hydropower Station is the longest, 19.3 km; the diameter of penstocks of Xiangjiaba underground power station is the largest, 14.40 m;

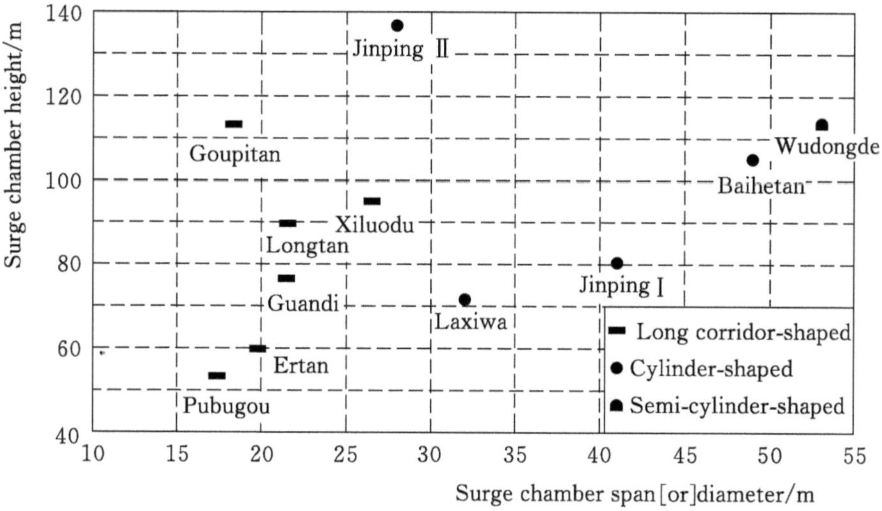

Fig. 1.23 Heights and spans or diameters of surge chambers

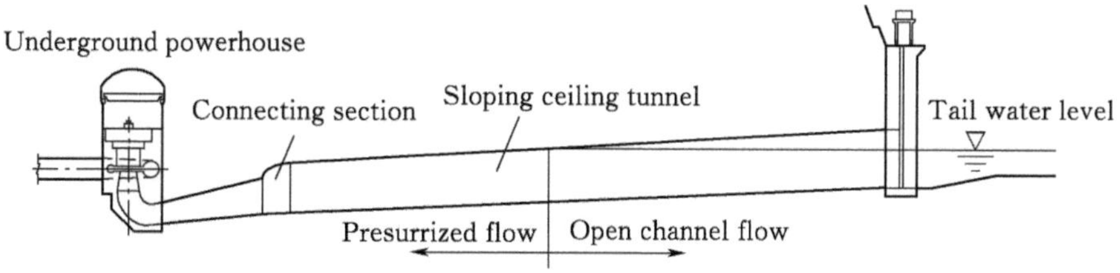

Fig. 1.24 Schematic of the tailrace tunnel with sloping ceiling

the HD value of the turbine spiral casing of Wudongde Hydropower Station is the largest, up to 2823 m^2. The headrace tunnel of Jinping II power station has the characteristics of deep burying, long tunnel line, large tunnel diameter, high-level rock stress, complex karst hydrogeological conditions, and difficult construction arrangement. In hydropower engineering, it is the largest and most difficult tunnel project in the world.

4. Role of Hydropower in Energy and Sustainable Development

As for the total amount of energy resources, China is one of the most abundant countries in the world, with hydropower resources ranked first, proven coal resources ranked third, proven oil resources ranked tenth, and proven natural gas resources ranked eighteenth in the world. The total proven resources of China's conventional energy resources (including coal, oil, natural gas, and hydropower resources, in which hydropower is calculated according to the use of 100 years because it is a renewable) are about 845 billion tons of standard coal (technical exploitable) and 159 billion

Table 1.14 Parameters of headrace tunnel and penstock of underground power stations (incomplete statistics)

Station	Installed capacity/MW	Layout type	Headrace length/m	Diameter of headrace tunnel/m	Diameter of penstock/m	Maximum pressure at spiral casing/m	HD value at spiral casing inlet/(m·m)
Three Gorges underground	6×700	One tunnel one turbine	298.64	13.50	13.50 to 12.40	156.00	1934
Xiangjiaba underground	4×800	One tunnel one turbine	342.68	14.40	14.40 to 11.40	158.00	1801
Xiluodu	18×700	One tunnel one turbine	444.29	10.00	10.00 to 7.74	287.00	2221
Baihetan	16×1000	One tunnel one turbine	570.42	11.00	10.20 to 8.60	320.00	2752
Wudongde	12×850	One tunnel one turbine	572.25	13.50	13.50 to 11.50	245.50	2823
Pengshui	5×350	One tunnel one turbine	460.00	14.00	14.00 to 10.50	140.00	1470
Jinping I	6×600	One tunnel one turbine	597.68	9.00	9.00 to 6.40	310.00	1984
Jinping II	8×600	One tunnel two turbines	17.19×10^3	11.8 to 11.2	6.50	410.00	2665
Futang	4×90	One tunnel two turbines	19.30×10^3	10.40 to 9.00	5.20	258.00	1341

(continued)

Table 1.14 (continued)

Station	Installed capacity/MW	Layout type	Headrace length/m	Diameter of headrace tunnel/m	Diameter of penstock/m	Maximum pressure at spiral casing/m	HD value at spiral casing inlet/(m·m)
Maoergai	3×140	One tunnel three turbines	16.15×10^3	8.60	7.00	329.00	2303
Jinwo	2×120	One tunnel two turbines	7.94×10^3	5.81 to 5.25	3.60 to 2.00	734.00	1468
Guandi	4×600	One tunnel one turbine	478.09	11.80	11.80 to 9.60	148.00	1421
Liahekou	6×500	One tunnel one turbine	458.69	7.50	7.50 to 5.90	311.00	1835

Note The lengths of headrace system are from the intake to turbine center

tons of remaining standard coal (economic exploitable), accounting for about 2.6 and 11.5% of the world's total, respectively.

The energy shares in the total proven reserves are as follows: raw coal 85.1%, hydropower 11.9%, crude oil 2.7%, and natural gas 0.3%. The compositions of the remaining total energy reserves are as follows: raw coal 51.4%, hydropower 44.6%, crude oil 2.9%, and natural gas 1.1%.

However, from the point of view of per capita energy resources, China is one of the countries with relatively poor energy resources in the world, and the per capita value is only half of the world average. Per capita oil recoverable reserves in 2000, for example, were only 2.6 tons, per capita natural gas recoverable reserves 1074 m^3, and per capita coal recoverable reserves 90 t, respectively, being the 11.1, 4.3, and 55.4% of the world average.

(1) Giving priority to the development of hydropower is the inevitable choice for optimizing and adjusting the energy structure of China. Hydropower resources are the only high-quality renewable energy resources that can be developed on a large scale in China. The per capita generation is also close to the world average (calculated based on 91% of the economic potential). The earlier the hydropower energy is developed, the better. All over the world, especially in developed countries, utilizing hydropower resources is the priority. At the end of the twentieth century, the developed percentages of hydropower were 82%

in the USA, 84% in Japan, 73% in Germany, 65% in Canada, 80% in France, and 69% in Austria.

Although the final development of hydropower is limited, the renewable nature of hydropower resources determines that the hydropower industry will not be exhausted. Hydropower will continue to operate with the maximum capacity of development, and the aging of mechanical and electrical equipment and civil engineering can be upgraded. On the contrary, thermal power, which depends on coal, oil, and natural gas, cannot be produced once the fuel is exhausted.

(2) Vigorously developing hydropower is the inevitable choice for the development of China's economy and society. In order to realize the Chinese dream of "the Great Rejuvenation of Chinese Nation" and the two-one-hundred-year plans, China is in the rapid development period of modernization. With the continuous improvement of people's living standards, energy demand and electricity consumption will maintain a high growth rate. Now, the per capita installed capacities are 3.5 kW in the USA, 2.5 kW in Japan, around 1.5–2.0 kW in European countries, and 0.9 kW in China. According to experts, China needs 2.0 kW per capita by 2050, when the population will reach 1.5 billion, which will require a total installed power capacity of 3000 GW. Of the total installed 3000 GW, the total share of hydropower is only 500 GW (excluding pumped storage), and hydropower is still the most matured, the most economical, and the most flexible renewable green energy. As a result, it can be expected that the next 20–30 years will be the golden period for China's hydropower construction to achieve leapfrog development. At present, the degree of hydropower resource development in the western region of China is still low, and 75% of the large hydropower stations with good economic benefits are concentrated in the west. The development of hydropower resources in the west can transform the advantages of western resources into economic advantages, driving regional economic development, and speeding up the development of the western region.

(3) Developing clean and renewable green hydropower is the inevitable choice of environmental protection in China. In conventional coal-fired power generation, coal emits a large number of harmful substances (sulfur dioxide, nitrogen oxides, carbon dioxide, soot, etc.) during combustion, which causes serious pollution of the atmospheric environment and leads to many environmental problems such as acid rain and "greenhouse effect" and so on. At present, 80% of China's electricity generation depends on coal, and carbon dioxide emissions are the first in the world, accounting for about 1/5 of the world's total emissions. In order to achieve the goal of "Halving the Total Global Greenhouse Gas Emissions by 2050", the development of electric power in China must take structural adjustment as the premise, take effective environmental protection measures on the one hand, and vigorously develop clean and renewable green hydropower on the other hand.

(4) Hydropower development is the main way of managing rivers and the important content of comprehensively utilizing water resources in China. Uneven rainfall distributions in time and space and frequent water disasters in China are the important reasons for the low degree of comprehensive management of rivers.

Hydropower development is an important part of the comprehensive utilization of water resources. The construction of key water conservancy projects can not only develop hydropower resources but also effectively control and regulate floods to ensure the safety of the middle and lower reaches. Hydropower projects can transfer water to water shortage areas and optimize allocation of water resources, to mitigate the uneven distribution of rcgional water resources. It can also bring many other benefits, such as irrigation, shipping, soil and water resources development, and environmental pollution control.

(5) Developing pumped storage hydropower is the inevitable choice for adjusting the energy structure and guaranteeing the safe operation of the power grid in China. The energy structure of China presents a trend of clean and low carbonization. The share of non-fossil energy in energy consumption will rise from 15% in 2020 to about 38% in 2050. In order to meet the need for energy structure adjustment and to cooperate with the large-scale development of wind power and nuclear power, it is necessary to build a number of pumped storage power stations with better-regulating performance. In addition, the western region of China is rich in water energy, wind energy, and coal resources; therefore, developing the "West to East Electricity Transmission" project is inevitable. In order to ensure the safe operation of the large power grid, it is also necessary to build a certain scale pumped storage power station at the receiving end and the sending end.

The development planning for pumped storage power is as follows: 30 GW by the end of 2015, 70 GW by 2020, 110 GW by 2030, and 160 GW by 2050. The pumped storage power shares in 2050 within the total installed power capacity is only 5.3%, still lower than 8–10% in the developed countries. Therefore, there is a long way to go.

(6) Guaranteeing the sustainable development of small hydropower needs scientific ideas and effective policies. China's development rate of small hydropower is less than 50%, even if the installed capacity and power generation are the largest in the world. Compared to the average 70–80% development rates in developed countries, the small hydropower potential in China is still great. Under the background of promoting energy saving and emission reduction, developing low-carbon economy, and improving people's livelihood, the sustainable development of small hydropower should be guaranteed with scientific ideas and effective policies and measures in future. Many changes should be made, such as from emphasizing the full utilization of hydropower to requiring restrainedly, orderly, and compensatively development of hydropower, from emphasizing the function of power generation to paying more attention to the ecological function and environmental effect, from isolated development to joint development with wind, solar, and biomass energies.

In short, hydropower resources are irreplaceable high-quality energy for human beings to make full use of natural resources and maintain the coordinated development of human society, economy, and environment. Hydropower is a basic industry and public welfare undertaking with great social benefits. It plays an important role and has broad prospects in the sustainable development of the economy and society.

Questions

(1) Please make brief descriptions of the basic principles of hydropower generation and the basic functions of different components in hydropower systems.
(2) What is hydropower energy? How about the development types of hydropower?
(3) What is a dam-type hydropower station? Please explain the applicable conditions of dam-type hydropower stations and their sub-types.
(4) What is a conduit- (or diversion-) type hydropower station? What are the characteristics and applicable conditions of the conduit-type hydropower station, compared with the dam hydropower station?
(5) Please briefly describe the composition and characteristics of the water conveyance system of a hydropower station, dam-type, or conduit-type.
(6) Please make a comprehensive analysis of hydropower station type, powerhouse type, turbine type, and water conveyance system characteristics under different hydropower development modes.
(7) Please extensively collect and summarize information on recent world hydropower achievements.
(8) Please make a brief description of the characteristics of the regional distribution of water energy resources in the world.
(9) What are the comprehensive development methods of the world's hydropower resources on great rivers? What are the characteristics?
(10) Try to analyze the achievements and characteristics of large hydropower stations, small and medium-sized hydropower stations, and pumped storage hydropower stations in the world.
(11) Please describe the trends and prospects of world hydropower development.
(12) Please analyze the hydropower potentials and development situations in China.
(13) How about the characteristics of hydropower resources in China? Please describe the development mode and prospect plan for hydropower development in China.
(14) How about the major achievements in hydropower development in China? What are the main scientific and technological progress?
(15) Please analyze the position of hydropower in the energy industry and explain the necessity and urgency of vigorously developing hydropower from the perspective of sustainable development.

Bibliography

1. I.H.A. 2020. IHA 2020 Hydropower Status Report. International hydropower association.
2. IRENA. 2020. Global Renewables Outlook: Energy transformation 2050. International renewable energy agency.
3. REN21. 2020. Renewables 2020 global status report 2020, REN21.
4. UNIDO. 2019. The World Small Hydropower Development Report (WSHPDR) 2019, United Nations Industrial Development Organization and International Center on Small Hydro Power.
5. I. E. A. 2020. Electricity Market Report. International Energy Agency.
6. Jinsheng, Jia. 2016. A technical review of hydro-project development in China. *Engineering* 2: 302–312.
7. Dong, Han, Hongwei Fang, Bingzhong Yan, Xingya Xu. 2014. China's hydropower status in 2013. *Journal of Hydroelectric Engineering* 33 (5): 1–5.
8. Xiaolin, Chang, Xinghong Liy, and Wei Zou. 2010. Hydropower in China at present and its further development. *Energy* 35: 4400–4406.
9. China Society of Hydroelectric Engineering. World hydropower development status. *China Three Gorges* 2011 (1): 54–58.
10. Chao, Zhang. 2005. *Hydropower development and unitization*, 1. Beijing: Chemical Industry Press.
11. Cunhong, Zhao, Zhenhong Zhu, Duanzhuang Zou. 2000. *Rivers and dams in the world*. China Water and Power Press.
12. Wei, Li and Jianfeng Li. 2008. Analysis for the development of pumped storage power station. *Jilin Water* 2008 (7): 52–54, 57.
13. Binru, Qiu. 2006. *New development of pumped storage power in the world*, 1. Beijing: China Electric Power Press.
14. Kaygusuz, Kamil. 2004. Hydropower and the world's energy future. *Energy Sources* 26: 215–224.
15. Bartle, Alison. 2002. Hydropower potential and development activities. *Energy Policy* 30: 1231–1239.
16. Yüksel. 2009. Dams and Hydropower for Sustainable Development. *Energy Sources, Part B: Economics, Planning, and Policy* 4: 100–110.
17. Jianping, Zhou, and Gangliang Qian. 2011. Planning and development status of China's 13 hydropower bases. *Water Conservancy and Hydropower Construction* 1: 1–7.
18. Xinliang, Dong, Chuanqing Xia, et al. 2013. Hydropower in the sustainable energy development strategy. *China Three Gorges* 3: 38–42.
19. Shouren, Zhen. 2007. Development and unitization of hydropower resources in China: opportunity and Challenges. *Journal of Hydraulic Engineering (Shui Li Xue Bao)* supplement issue(10): 1–5.
20. Shucheng, Wang. 2013. Development trends of electrical power in China. *Journal of Hydroelectric Engineering* 32 (3): 1–3.
21. Zhi, Zheng. 2007. China's records in hydropower. *Guizhou Water Power* 21 (5): 1–5.
22. Jiandong, Yang, Ming Xiao, Hegao Wu, and Yongguang Cheng. 2017. Hydropower station, 3rd ed. Beijing: China Water and Power Press.

Chapter 2
Type and Structure of Hydro Turbine

Jinping Li and Yongguang Cheng

Abstract In this chapter, we will discuss operating parameters, main types, basic structures, and main components of modern hydro turbines, along with their application scopes and model designation rules. Students should be acquainted with these aspects by reading this textbook and observing the teaching models of various hydro turbines.

Keywords Operating parameters · Main types · Basic structure and components · Application scopes · Designation rules

2.1 Operating Parameters of Hydro Turbine

A hydro turbine or hydraulic turbine is a machine that can convert the kinetic energy and potential energy of water flow to the rotating mechanical power of the machine. The turbine drives a generator through a transmission device, such as a shaft, and the generator converts the rotating mechanical power into electrical power. The turbine runner and generator rotor are coupled and rotated together, and the synthetic of the two machines is referred to turbine-generator unit or hydropower unit, which works for converting water power to electrical power. The hydropower units are the main power equipment of a hydropower station.

The operating state of hydro turbine (referred to as the operating condition or working condition) and its operating performance under this condition can be described by the operating parameters and their relations of the turbine. The operating parameters include water head, discharge, opening, output, efficiency, speed, and torque. The meanings of these parameters are explained as follows.

J. Li (✉) · Y. Cheng
Wuhan University, Wuhan, China
e-mail: lukeping@whu.edu.cn

Y. Cheng et al. (eds.), *Hydropower Station*,
https://doi.org/10.1007/978-981-96-2383-9_2

2.1.1 Water Head

The water head of hydro turbine, also known as the operating head or net head, refers to the difference of total energy per unit volume of water flow between the inlet section and outlet section of the turbine. It is usually symbolled by H, with unit in m. According to specification *Model Acceptance Tests of Hydraulic Turbines, Storage Pumps, and Pump-Turbines* [1], and as shown in Fig. 2.1, the definition of turbine operating head is

$$H = E_1 - E_2, \tag{2.1}$$

while

$$E_1 = Z_1 + \frac{P_1}{\rho g} + \frac{V_1^2}{2g}, E_2 = Z_2 + \frac{P_2}{\rho g} + \frac{V_2^2}{2g}, \tag{2.2}$$

where E_1 is the total energy in per unit volume of water, m; Z is the position height of the section center from the reference plane, which is usually referred to as the sea level, m; P is the relative pressure at the section center, N/m^2 or Pa; V is the section area average velocity of water flow, m/s; ρ is the water density, related to the local temperature, 998.203 kg/m^3 at 20 °C; g is the gravity acceleration, related to the local latitude and altitude elevation, usually 9.81 m/s^2. The subscripts 0, 1, 2, and 3 indicate the upstream inlet section, the volute inlet section, the draft tube outlet section, and the downstream river section, respectively.

As can be seen from Fig. 2.1,

$$Z_0 = \Delta h_{0-1} + E_1, Z_3 = E_2 - \Delta h_{2-3} \tag{2.3}$$

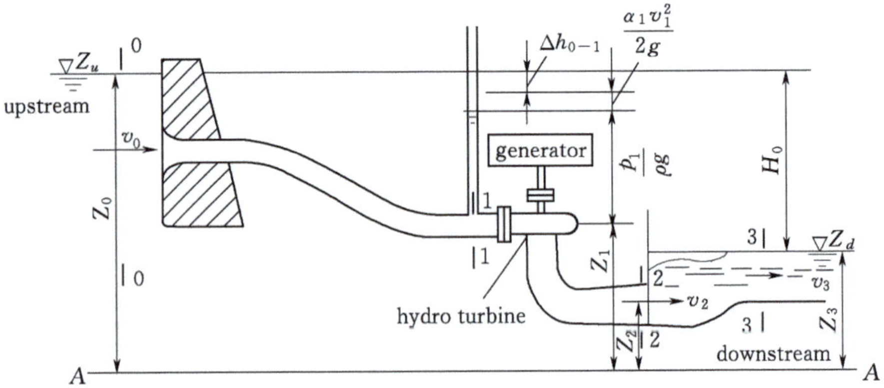

Fig. 2.1 Definition of the operating head of a hydro turbine

where, Z_0 is the water surface height of the upstream reservoir from the reference plane (A-A section), m. Usually the average flow velocity v_0 of the upstream reservoir section is negligible. Z_3 is the water surface height of the downstream river section 3-3 from the reference surface (A-A section), m. If the average flow velocity of the section is not negligible, Z_3 should be replaced by $Z_3 + V_3^2/2g$.

Δh_{0-1} is the total head loss between section 0–0 and section 1–1, including minor head loss and frictional head loss, m. It is the head loss on the upstream side of the unit, including the head losses at the intake and in the penstock. Δh_{2-3} is the total head loss between Sects. 2.2 and 3.3, including minor head loss and frictional head loss, m. It is the head loss on the downstream side of the unit, including the head losses in the tailrace conduit and at the tailrace conduit outlet.

Substituting Eqs. (2.3) into (2.1), we have

$$H = Z_0 - \Delta h_{0-1} - Z_3 - \Delta h_{2-3} = H_0 - (\Delta h_{0-1} + \Delta h_{2-3}), \tag{2.4}$$

where $H_0 = Z_0 - Z_3$ is called the gross head of the hydropower station. Therefore, the operating head, or the net head, of the turbine is equal to the gross head minus the total head loss in the water delivery systems except the head loss in the turbine. Moreover, the net head varies with the change of the upstream and downstream water levels of the hydropower station. There are four kinds of water heads to characterize the operation conditions of hydro turbines. They are the important parameters for the choice of turbine type.

(1) Maximum Head: the maximum net head that allows the turbine to operate. It is one factor for determining the strength of a hydro turbine.
(2) Minimum Head: the smallest net head that guarantees safe and stable operation of a hydro turbine.
(3) Weighted Average Head: the average head taking into account the duration time of various heads. It can be calculated by

$$H_a = \frac{\sum H_i T_i}{\sum T_i}, \text{ based on operating time}$$

$$\text{or } H_a = \frac{\sum H_i T_i N}{\sum T_i N_i}, \text{ based on output power,}$$

where T_i and N_i represent the corresponding duration time and power output of operating head, respectively. Note that a hydro turbine should operate as far as possible in the high-efficiency zone to ensure larger output power. Namely, we should select an appropriate H_{a} to let the turbine operate for most time at or near the highest efficiency point.
(4) Rated Head: the minimum net head while the turbine can generate the rated output power. Below this head, the turbine can only generate smaller output power.

Strictly speaking, the operating head of a turbine is not equal to the effective head that a turbine can be used to perform mechanical work. The difference is the head loss in the turbine (from the volute inlet Sect. 1.1 to the draft outlet Sect. 1.2), Δh_{1-2} in m. This loss includes minor local loss, frictional loss, striking loss at the runner inlet, and whirling vortex loss at the runner outlet, which vary with turbine operating conditions. The effective head can be expressed as

$$H_e = H - \Delta h_{1-2}. \tag{2.5}$$

Defining the hydraulic efficiency of turbine as $\eta_h = \frac{H_e}{H} = 1 - \frac{\Delta h_{1-2}}{H}$, we have

$$H_e = H\eta_h \tag{2.6}$$

2.1.2 *Discharge and Guide Vane Opening*

The discharge of a hydro turbine refers to the volume of water passing through a turbine per second, usually expressed by Q, with unit. Obviously, when the opening of the water guiding mechanism (guide vane opening for short) is constant, the higher the operating head of the turbine, the larger the discharge of the turbine. It is usually defined that the minimum head required for a hydropower unit to generate the rated output is the rated head, and the corresponding volume flow rate is the rated discharge, also known as the design discharge. In other words, the rated discharge is the maximum volume flow rate required for the turbine to produce a rated output, and the corresponding guide vane opening is the maximum opening of the turbine operation.

In addition, the discharge through the turbine runner is not equal to the discharge into the turbine runner chamber. The reason is that there are gaps between the moving parts and the fixed parts, and there is leakage through the gaps. Therefore, the effective discharge of doing work can be expressed as $Q_e = Q - \Delta q$.

We can define the volumetric efficiency $\eta_{vol} = \frac{Q_e}{Q} = 1 - \frac{\Delta q}{Q}$, which usually has a value of 98–99% because the leakage is small. Thus, the effective discharge is

$$Q_e = Q\eta_{vol}. \tag{2.7}$$

It must be noted that the discharge is controlled by the guide vane opening and influenced by the head. Under the condition with the rated head and rated output, the discharge should reach the maximum.

2.1.3 Output and Efficiency

The water flowing through a turbine gives the turbine an input power N_w, which can be expressed in unit kW as

$$N_w = \gamma QH = 9.81QH, \tag{2.8}$$

where $\gamma = \rho g$ is the bulk density of water, usually with a value of 9.81 kN/m^3. From this formula, we know that the head and discharge are the two major elements of hydropower.

The effective power that the turbine receives from water is

$$N_e = \gamma Q_e H_e = 9.81QH\eta_h\eta_{vol}. \tag{2.9}$$

However, N_e is not equal to the output power of the turbine, N, which is the power transmitted to from the turbine to the generator by the shaft. Because there is mechanical loss $\Delta N_{mec} = N_e - N$ due to rotation, the mechanical efficiency is usually defined as $\eta_{mec}=N/N_e=1-\Delta N_{mec}/N_e$. Because the mechanical loss is small, η_{mec} normally has a value of 98–99%. Thus

$$N = N_e\eta_{mec} = \gamma QH\eta_{vol}\eta_h\eta_{mec} = \gamma QH\eta \tag{2.10}$$

with

$$\eta = \eta_{vol}\eta_h\eta_{mec}. \tag{2.11}$$

That is to say, the efficiency of hydro turbine is equal to the product of hydraulic efficiency, volumetric efficiency, and mechanical efficiency and is expressed by percentage. The maximum efficiency of modern large turbines can reach 94–96%.

2.1.4 Torque and Speed

The output power of hydro turbine can also be expressed in the form of rotating mechanical energy:

$$N = \gamma QH\eta = M\omega = M\frac{2\pi n}{60}, \tag{2.12}$$

where M is the rotating torque of shaft, in N·m; it is transferred to the generator by the main shaft to overcome the generator retarding torque. ω is the rotational angular speed of the turbine, in rad/s. n is the rotational speed of the hydropower unit, in r/min.

Table 2.1 Relationship of synchronous speed (n) and magnetic pole pair (p)

p	3	4	5	6	7	8	9
$n/(r/\min)$	1000	750	600	500	428.6	375	333.3
$n/(r/\min)$	10	12	14	16	18	20	22
$n/(r/\min)$	300	250	214.3	187.5	166.7	150	136.4
$n/(r/\min)$	24	26	28	30	32	34	36
$n/(r/\min)$	125	115.4	107.1	100	93.8	88.2	83.3
$n/(r/\min)$	38	40	42	44	46	48	50
$n/(r/\min)$	79	75	71.4	68.2	65.2	62.5	60

Obviously, the torque M acting on the generator rotor is smaller than the effective torque M_e that the turbine has and smaller than the torque M_w acting on the turbine by the water flow. Given constant rotational speed n, the relations among M , M_e and M_w are consistent with the relations among N, N_e, and N_w, namely

$$\eta_{mec} = \frac{M}{M_e}, \eta_{vol}\eta_h = \frac{M_e}{M_w}. \tag{2.13}$$

For large and medium-sized hydropower units, the turbine's main shaft is directly connected to the generator's main shaft; therefore, turbine speed must be equal to the synchronous speed of the generator. That is, the following relationship must be satisfied:

$$f = \frac{np}{60} \tag{2.14}$$

in which f is the frequency specified by the electricity grid, in Hz. In China, it is 50 Hz. p is the pole pairs of generator, which must be an integer.

Thus, the speed of unit must be satisfied.

$$n = 3000/p. \tag{2.15}$$

For generators with different pole pairs, the synchronous speeds are shown in Table 2.1.

2.2 Main Types of Hydro Turbine

Due to a wide variety of hydropower resources, the heads and discharges of hydropower stations vary widely. Each hydropower development site is unique, and each hydropower station is unique. In order to fully and effectively utilize

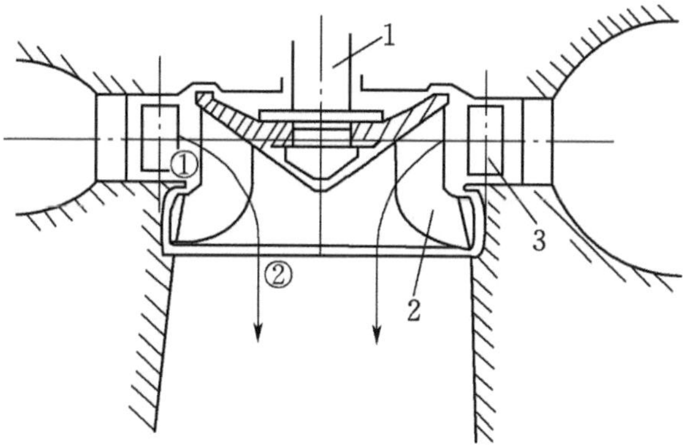

Fig. 2.2 Mixed-flow turbine 1—main shaft; 2—blade; 3—guide vane

hydropower, many types of hydro turbines such as the Francis, Kaplan, and Pelton were invented for different discharge Q and net head H conditions.

According to the characteristics of energy conversion from water flow, hydro turbines can be divided into two categories: reaction turbines and impulse turbines. Each type can be also divided into various forms according to the characteristics of the water flow in the runner. The main component of the turbine to achieve energy conversion is the runner; therefore, the type of turbine can be distinguished according to the working principle of water flow in the runner.

As shown in Fig. 2.2, we can select the runner inlet point 1 and the runner exit point 2 to explore the characteristics of turbine energy conversion. The energy per unit volume of water used by the runner can be expressed as the difference between the total energy values at point 1 and point 2:

$$H = \left(Z_1 + \frac{p_1}{r} + \frac{\alpha_1 V_1^2}{2g}\right) - \left(Z_2 + \frac{p_2}{r} + \frac{\alpha_2 V_2^2}{2g}\right). \tag{2.16}$$

Rewriting Eq. (2.16), we can get α_1.

$$\frac{\left(Z_1 + \frac{p_1}{r}\right) - \left(Z_2 + \frac{p_2}{r}\right)}{H} + \frac{\alpha_1 V_1^2 - \alpha_2 V_2^2}{2gH} = 1,$$

where α_1 and α_2 are the velocity distribution coefficients at the runner inlet and the runner outlet, respectively.

Let $\frac{(Z_1+\frac{p_1}{r})-(Z_2+\frac{p_2}{r})}{H} = E_P$ and $\frac{\alpha_1 V_1^2-\alpha_2 V_2^2}{2gH} = E_C$, the their relation satisfies

$$E_P + E_C = 1. \tag{2.17}$$

Equation (2.17) shows that the energy utilized by the turbine is the sum of the potential energy and the kinetic energy. E_p is the ratio of the potential energy of water flow, and E_c is the ratio of the kinetic energy of water flow.

If $E_P = 0$ and $E_C = 1$, the turbine only utilizes the kinetic energy of water flow to do work and is called an impulse turbine.

If $0 < E_P < 1$ and $E_P + E_C = 1$, the turbine utilizes not only the kinetic energy but also the potential energy of water flow to do work and is called a reaction turbine.

According to the water flow direction in the turbine runner, the turbine can be divided into two categories: conventional turbine and reversible turbine. With the development of pumped-storage power stations and tidal power stations, the reaction turbine that can operate both normally and reversely is called a reversible turbine, while the impulse turbine is not reversible. The reversible turbine for pumped-storage power stations is characterized by the turbine flow direction with forward rotation when converting water power into electricity and the pump flow direction with reverse rotating when converting electrical energy to water potential energy for energy storage. The characteristics of the reversible turbine for tidal power stations are that both forward and reverse rotations can convert water power into electricity.

2.2.1 Reaction Turbines

The runner of a reaction turbine, which is completely submerged in the pressurized water flow, contains a number of rigid blades with spatially torsional surfaces. When water flows through the runner, the blades restrain the water flow by continuously changing its direction and velocity, which means that the runner gives a force to the water flow. Conversely, the water flow uses its pressure and kinetic energy to give the runner a reaction force, forming a moment and making the runner rotate. In the reaction turbines, two effects make the energy converted from the water flow into the mechanical rotating energy on the main shaft. Firstly, there is a pressure drop from the inlet to the outlet of the runner, which is denoted as the reaction part of the energy conversion. Secondly, there is a velocity direction change through the blade passages, which is denoted as the impulse part of the energy conversion. The pressure drop from the inlet to the outlet of the runner is obtained because the runners are completely filled with water.

According to characteristics of the water flow direction in the runner or runner chamber relative to the main shaft, the reaction turbines can be divided into four types: mixed flow, axial flow, diagonal flow, and cross flow.

1. Mixed Flow Turbine

As shown in Fig. 2.2, the water flow enters the runner in the radial direction and then flows out of the runner approximately in the axial direction. The direction of the water flow in the runner is changed from the radial direction to the axial direction; therefore, it is called a radial-axial turbine or mixed-flow turbine. The runner of the mixed turbine is simple in structure, high in integrated strength, stable in operation, high in efficiency, and wide in application (head range about 30–700 m). It is the most widely used turbine in modern hydropower stations.

The mixed-flow turbine was invented by American engineer Francis in 1849; thus, it was also known as the Francis turbine. The maximum efficiency of this type of turbine can reach 95–96%. The highest head Francis turbine is installed in the Rosshack Hydroelectric Power Plant in Austria. Put into operation in 1967, it has a water head of 672 m and a single unit capacity of 58.4 MW. The Francis turbine with the largest runner size is installed in the Three Gorges Hydropower Station in China. Put into operation in 2003, its runner diameter is about 10.40 m, unit capacity is 700 MW, rated head is 80.6 m, and rotational speed is 75 r/min. The largest output Francis turbine is installed in Xiangjiaba Hydropower Station in China. Put into operation by the end of 2012, its maximum capacity is 800 MW, runner diameter is 9.30 m, rated head is 100.0 m, and rotational speed is 71.4 r/min. At present, giant Francis turbines of unit capacity 1000 MW will be designed for the under construction Baihetan Hydropower Station, China.

2. Axial-Flow Turbine

As shown in Fig. 2.3, the water flows radially through the stay vanes and guide vanes into the runner chamber and then flows in parallel to the main shaft through the runner. Because the flow direction from the inlet to the outlet of the runner is in the axial direction, this type of turbine is called the axial-flow turbine. The blades are cantilever structure, and the integrated strength is low; therefore, the axial-flow turbine is suitable for medium or low head and large discharge hydropower stations. The application head range is about 3–80 m.

The axial-flow turbine can be divided into fixed-blade and adjustable blade subtypes according to whether the blades of the runner can rotate during operation. The blades of the axial-flow fixed-blade turbine are fixed on the wheel hub, and the blade angle cannot be changed during operation; therefore, the structure is simple and the cost is low, but the efficiency declines drastically when the operating condition deviates from the design condition. Therefore, the turbine is generally

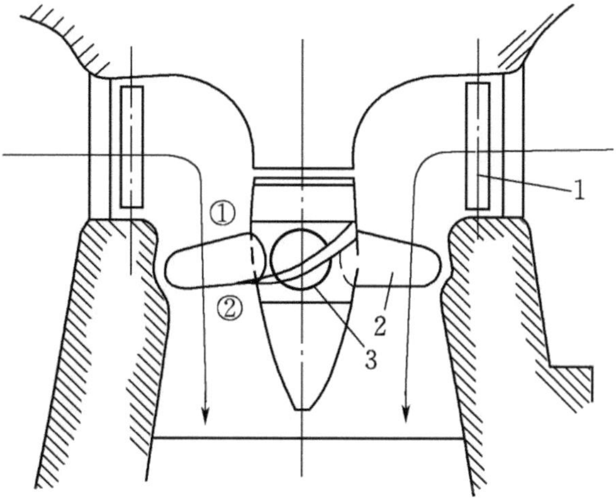

Fig. 2.3 Axial-flow turbine 1—guide vane; 2—blade; 3—wheel hub

used in a hydropower station with lower head, smaller output, and smaller head variation. The blades of the adjustable axial-flow turbine can rotate, thereby expanding the high-efficiency zones and improving operation stability. However, the turbine needs to add a mechanism for adjusting the blade angle, which makes the structure more complicated and the cost high. It is generally used for large and medium-sized hydropower stations with large variations in head and output.

The blade-adjustable axial-flow turbine was invented by Austrian engineer Kaplan in 1920; therefore, it was also called as the Kaplan turbine. The runner blades are generally operated by a hydraulic servomotor mounted in the wheel hub and can be adjusted according to the head and load variation to maintain the optimal coordination of the guide vane angle and blade angle, thereby improving average efficiency. The maximum efficiency of the Kaplan turbine has exceeded 94%. In the 1980s, the largest Kaplan turbine in size was manufactured by Dongfang Electric Machinery Factory in China and installed in Gezhouba Hydropower Station. With unit a capacity of 170 MW, rated head of 18.6m, synchronous speed of 54.6 r/min, and runner diameter of 11.3 m, it was put into operation in 1981. The Kaplan turbine with the largest head was installed at the Nambia Hydropower Station in Italy. With a rated head of 88.4 m, unit capacity of 13.5 MW, and synchronous speed of 375 r/min, it was put into operation in 1959.

3. Diagonal flow turbine

As shown in Fig. 2.4, the water flows through the runner in the direction at an angle to the main shaft; therefore, it is called the diagonal flow turbine. Compared with the axial-flow turbine, the runner is also cantilevered structure and adjustable, but the approximately spherical wheel hub can hold more blades; therefore, the diagonal flow turbine can be applied to hydropower stations with higher heads. The application head range is about 40–200 m.

The diagonal flow turbine was invented in 1956 by Swiss engineer Deriaz; therefore, it was also known as the Deriaz Turbine. As the head and load change, the hydraulic servomotors in the wheel body operate the blades mounted obliquely on the wheel hub to rotate about their axes. The highest efficiency of this type is slightly

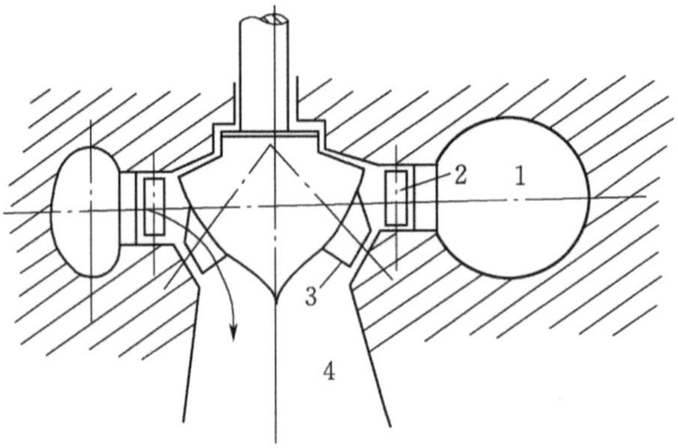

Fig. 2.4 Diagonal flow turbine 1—spiral case; 2—guide vane; 3—runner blade; 4—draft tube

lower than that of the Francis type, but the average efficiency is much higher than that of the Francis type.

Due to the complicated structure and high cost, the diagonal flow turbine is generally only used when the mixed-flow or axial-flow turbine is not suitable or ideal. Therefore, only a limited number of hydropower stations have adopted this type of turbinc. The largest diagonal flow turbine in unit capacity, with a unit capacity of 215 MW and rated head of 78.5 m, was installed in Jieya Hydropower Station in the former Soviet Union. Diagonal flow turbine is abandoned gradually by engineering practice.

4. Tubular flow turbine

The tubular flow (or through flow) turbine is a horizontal-axis turbine that does not have a spiral case and the entire flow passage is like a straight pipe. The blades are similar to those of the axial-flow turbine and can be fixed or adjustable. According to the different forms of generators, they are divided into two types: full-through flow and semi-through flow turbine.

The generator rotor of a full-through flow turbine is mounted directly on the outer edge of the runner blades, as shown in Fig. 2.5. Its advantages are small hydraulic loss, large discharge, high efficiency, and compact structure. However, since the circumferential velocity of the outer edge of the runner blade is large, the rotating sealing is very difficult. Therefore, the through flow turbine is seldom applied.

At present, the bulb tubular turbine, one of the through flow turbines, is widely used, as shown in Fig. 2.6. This type has compact structure, good stability, and high efficiency. The generator of the bulb tubular unit is arranged in a steel bulb surrounded by water flow, and the turbine and generator can be directly coupled or connected by a speed-increasing device.

Since the 1960s, bulb turbines have been widely adopted by large and medium-sized hydropower stations with water head below 25 m around the world, and the unit capacity has been growing. In 1989, the unit capacity of tubular turbine in Zhijian Hydropower Station, Japan, reached 65 MW, and the runner diameter was 6.7 m. In 1985, the runner diameter of the tubular turbines in Sidney Murray hydropower station in the USA reached 8.2 m and the unit capacity was 24 MW. In China, the largest unit capacity of the tubular turbines is 57 MW in Qiaogong Hydropower Station, Guangxi Province. The largest diameter of the tubular turbines is 7.5 m in Changzhou Hydropower Station, Guangxi Province, and the highest applied head of the tubular turbine is 27.3 m in Hongjiang Hydropower Station, Hunan Province.

2.2.2 *Impulse Turbine*

The impulse turbine is mainly composed of nozzles and a runner. The high-pressure water from the steel pipe flows through the injector at the end of the nozzle into the free jet flow. Before the water flows into the runner, the potential energy is converted into kinetic energy, which forms a high-speed free jet flow. The jet impinges on some

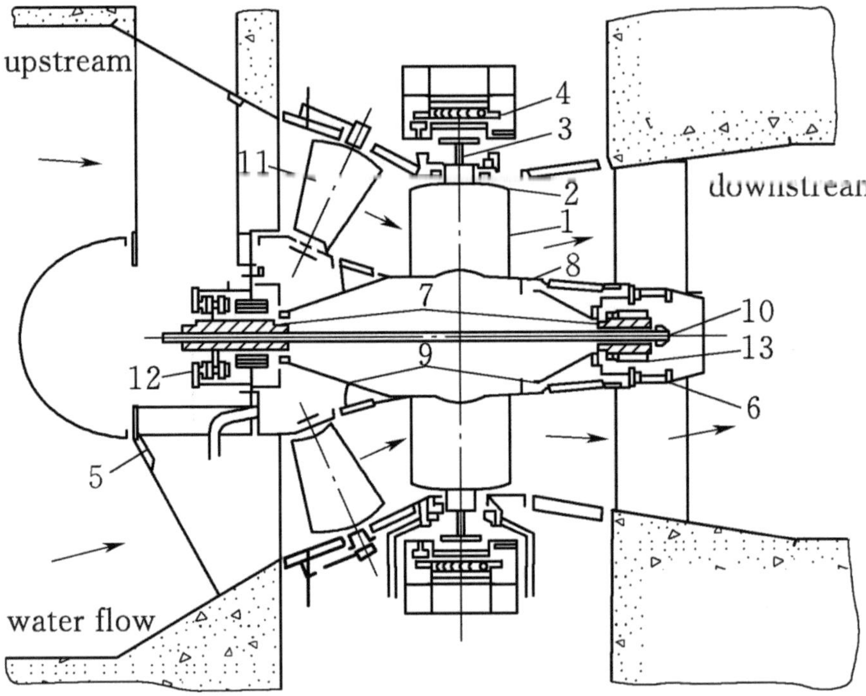

Fig. 2.5 Full-through flow turbine 1—runner blade; 2—runner rim; 3—generator rotor rim; 4—generator stator; 5 and 6—pillar; 7—journal; 8—wheel hub; 9—tapered insert; 10—tighten rod; 11—guide vane; 12—thrust bearing; 13—guide bearing

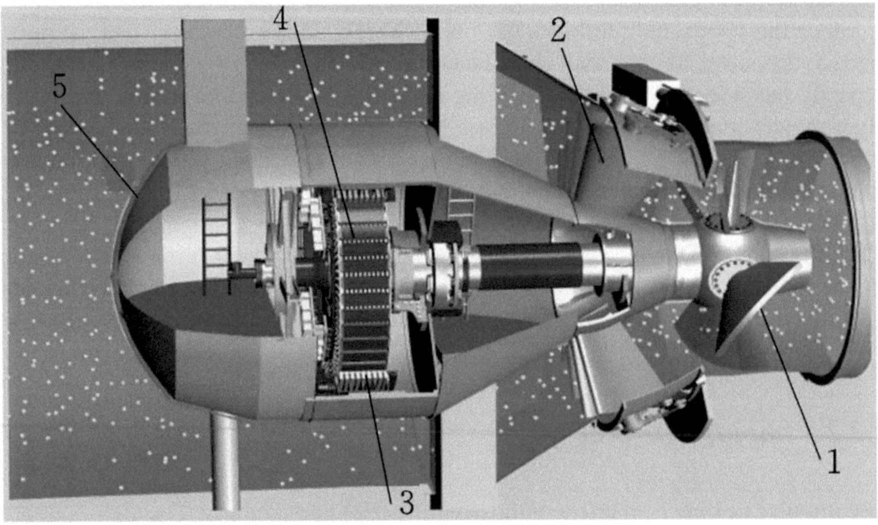

Fig. 2.6 Bulb tubular turbine 1—runner blade; 2—guide vane; 3—generator stator; 4—generator rotor; 5—bulb body

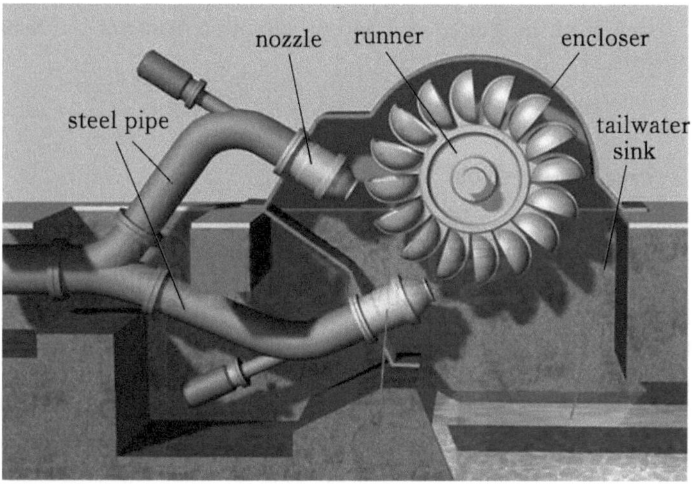

Fig. 2.7 Impulse turbine

buckets attached to a wheel and drives the wheel to rotate. The velocity and direction of the jet flow have to be changed in the buckets, converting most of the kinetic energy to the rotating energy of the wheel (Fig. 2.7).

Basically, the water energy inputted to the impulse turbines is completely converted to kinetic energy before the jet strikes the runner. This means that the water jets, coming into and out of the runner buckets, have no difference in pressure. Therefore, it is the impulse forces that only can be transferred to the buckets through the changes in velocity direction and magnitude.

Impulse turbines can be divided into the Pelton turbine, Turgo turbine, and Banki turbine, according to the manner the jets strike the runner. The Turgo turbine and Banki turbine are less efficient and are therefore largely eliminated by engineering practice.

The head range of large Pelton turbines is about 300–2000 m and that of small Pelton turbines is about 40–250 m. The Pelton turbine is particularly suitable for high-head, low-discharge hydropower stations. It will play an increasingly important role in the future application of high-head hydropower development.

The Pelton turbine was invented by American engineer Pelton in 1889. In the early 1980s, the largest unit capacity of the Pelton turbine was installed in Sima Hydropower Station, Norway. Its unit capacity is 315 MW, rated head is 885 m, and synchronous speed is 300 r/min. It was put into operation in 1980. The highest head of the Pelton turbine was installed in Lessek Hydropower Station, Austria. Its unit capacity is 22.8 MW, rated speed is 750 r/min, and rated head is 1767 m. It was put into operation in 1959. In China, Tianhu Hydropower Station installed Pelton turbines of the rated head 1022.4 m. Yele Hydropower Station installed two Pelton turbines of rated unit capacity 120 MW. Those Pelton turbines have six nozzles and a vertical shaft. The maximum head is 644.8 m, the rated speed is 375 r/min, the maximum diameter of the runner is 3.346 m, the pitch diameter is 2.6 m, and the bucket number is 21.

2.2.3 Reversible Turbine

With the development of pumped-storage power stations and tidal power stations, reversible turbines are getting more and more applications. The common types of reversible turbines for pumped-storage power stations are the mixed-flow and diagonal flow turbines, and those for tidal power stations are normally the bulb turbines.

The reversible turbines in pumped-storage power stations are also called pump-turbines. The basic functions of pumped-storage power stations include shaving the peak and filling the valley of the grid load. That is, when the grid load is higher than the basic load, the pump-turbines work as turbines to generate electricity for balancing the peak load; when the grid load is lower than the basic load, the pump-turbines work as pumps to lift water from the lower reservoir to the upper reservoir for storing potential energy. In addition, pumped-storage power stations play the roles of fast responses for grid dispatching, such as frequency regulation, phase modulation, and spinning reserve, which are essential for the operation safety, flexibility, and reliability of the power grid.

For the diagonal flow-type pump-turbines, because the blades of runner are adjustable, their performance is good despite the changes in head and load. Due to the limitation of hydraulic characteristics and material strength, the highest head for this type of pump-turbines was only 136.2 m in the early 1980s (Takagen First Power Station in Japan). For higher head, the mixed-flow type of pump-turbines is required.

The largest mixed-flow pump-turbine was installed in Bremen Pumped-Storage Power Station, Germany. Its turbine head was 237.5 m, generator output was 660 MW, pump head was 247.3 m, motor power was 700 MW, and speed was 125 r/min. In 1977, Bayina Basta Power Station in Yugoslavia, with the highest head at that time, was put into operation. Its turbine head was 600.3 m, unit capacity was 315 MW, pump head was 623.1 m, and speed was 428.6 r/min.

In China, the construction of pumped-storage power stations started later, but the starting point was higher. Several large-scale pumped-storage power stations that have been built and under construction are at the advanced level in the world. For example, the total installed capacity of Guangzhou and Huizhou pumped-storage power stations has reached 2,400 MW, respectively. A batch of pumped- storage power stations, such as Guangzhou Pumped-Storage Power Station (Fig. 2.8), have unit capacity larger than 300 MW and rated head higher than 500 m, reaching at the world advanced level of single-stage reversible pump-turbines. In Jixi, Dunhuang, and Yangjiang pumped-storage power stations, the maximum heads are higher than 700 m and the unit capacities are 350 and 400 MW. The maximum head of single-stage reversible pump-turbines in Changlongshan Pumped-Storage Power

Station is 724 m, higher than those in Kawakawa and Shinikawa Pumped-Storage Power Stations in Japan. This is marking the rapid development and comprehensive improvement of pumped-storage technology in China.

The bulb tubular pump-turbine is suitable for the tidal power stations with pumping and storage functions. In addition to the one-way power generation and pumping functions, two-way power generation, two-way pumping, and two-way draining functions are sometimes required. On August 4, 1980, the first single-reservoir two-way tidal power station in China, Jiangxia Tidal Power Test Station (Fig. 2.9), was officially put into operation. With an installed capacity of 3,000 kW and an average annual power generation of 10.7 million kWh, it is second only to the Rance Tidal Power Station in France.

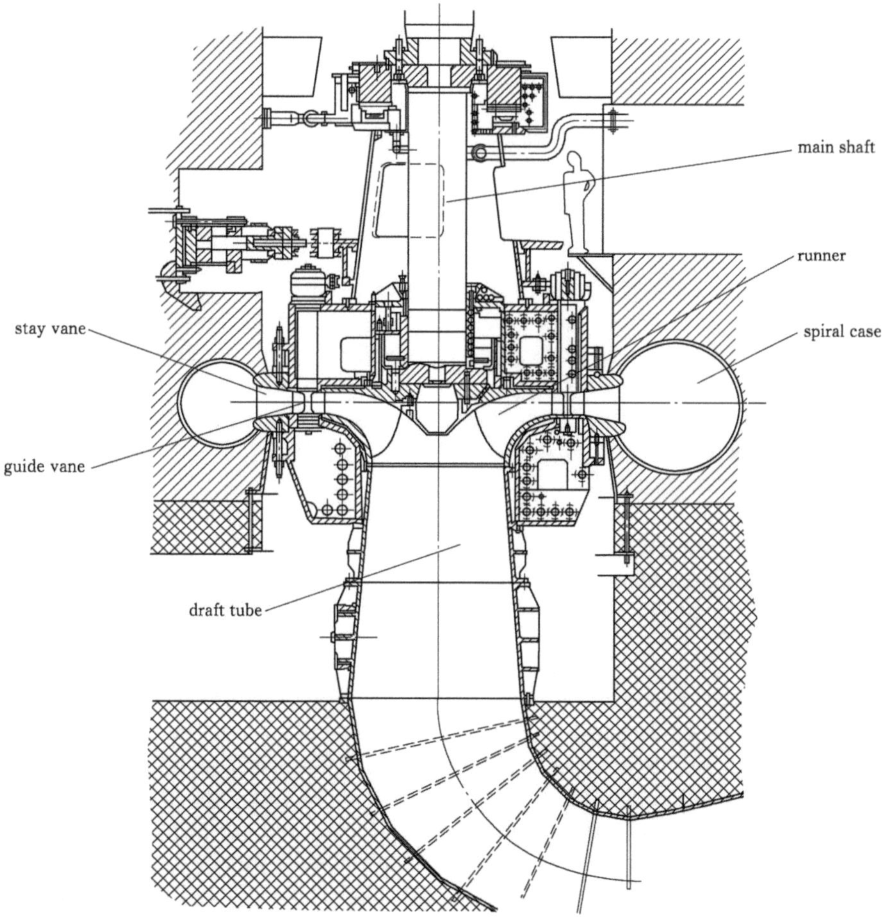

Fig. 2.8 Francis pump-turbine (Guangzhou Pumped-Storage Power Station)

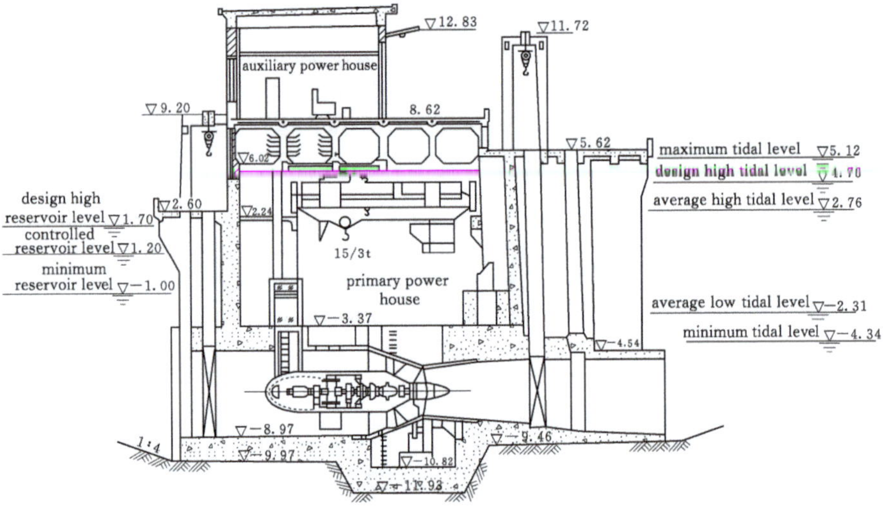

Fig. 2.9 Bulb tubular pump-turbine (Jiangxia Tidal Power Test Station)

2.3 Basic Structure of Hydro Turbine

The reaction turbine is generally composed of a water intake part, a water guiding part, a working part (runner), and a water discharging part. Different types of hydro turbines have different types of components and different features. For different reaction turbine types, the working parts are quite different not only in shape but also in characteristics. The Runner, which converts water power to rotating mechanical power, is the heart of a turbine. This working part determines the performance, structure, and dimensions of the hydro turbine.

The structure of the impulse turbine is relatively simple and is mainly composed of the water distribution ring, nozzle, needle valve, runner, and deflector. The runner is still the core component.

The four components of the reaction turbine are as follows.

1. Spiral Case

The spiral case of the reaction turbine is designed in such a way that the cross-sectional area goes on reducing uniformly around the circumference. The cross-sectional area is the maximum at the entrance and the minimum at the tip. Due to its spiral shape, it is called the spiral case (casing) or scroll case (casing). The function of the spiral case is mainly to make the water uniformly and symmetrically flow into the water guiding mechanism and the runner, with small hydraulic losses. The spiral case surrounds the stay vane ring and is connected with the outer rim of the ring. The water from the penstock moves along the spiral case and flows into the guide vane in the radial direction. The metal or reinforced concrete spiral case is selected according to the design head of the hydropower station. Because the operating heads of large and medium mixed-flow turbines are usually above 40 m, metal spiral cases

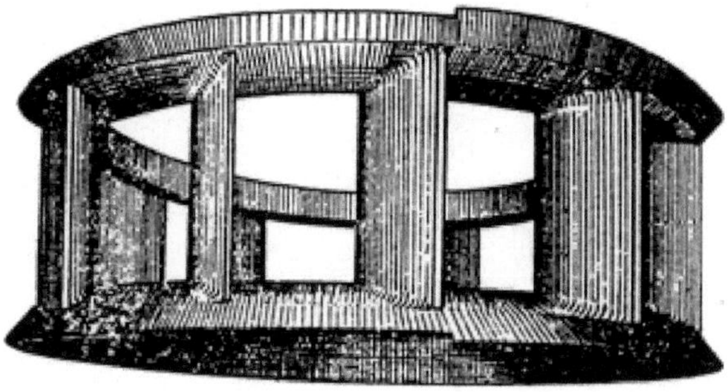

Fig. 2.10 Structure of stay vane ring

are used. The metal spiral case is subjected to a large internal water pressure, and the wrap angle is generally 345°.

2. Stay Vane Ring

The stay vane ring consists of upper and lower steel rings and a number of fixed and evenly distributed vanes, as shown in Fig. 2.10. The outer edges of the upper and lower rings are fixedly connected with the outlet of the spiral case, the top of the upper ring supports the concrete pier of the generator, the inner edge of the upper ring is fixed with the top cover of the turbine, and the bottom of the lower ring is the foundation concrete. Therefore, the functions of the stay vane ring are to support the weight of the turbine and generator, the axial water thrust of the turbine, the weight of the top cover, and the weight of the concrete upper the spiral case and to transmit this force to the foundation of the powerhouse through the stay vanes. Thus, the stay vane ring must have sufficient strength and stiffness. Since the stay vanes stand in the water flow passage, in order to reduce hydraulic loss, the sectional shape of the stay vanes is formed into a wing shape and placed according to the streamlines. The number of stay vanes is usually half of the number of adjustable guide vanes, which are controlled by the guide mechanism.

3. Guiding Mechanism

The guiding mechanism is composed of the guide vanes (index 3) and the operating mechanisms (7–13), as shown in Fig. 2.11. The guide vanes are evenly distributed circumferentially in the annular space between the stay vane ring and the runner, and the upper and lower axle journals are respectively supported by the sleeves in the top cover (6) and the bottom ring (21) and are rotatable about their pivots. In order to reduce hydraulic loss, the sectional shape of the vanes is designed as a wing shape. The guide vanes are also called active guide vanes to distinguish them from the fixed guide vanes.

The guide apparatus has two functions. First, it guides water flowing into the runner in a certain direction to form a certain velocity moment to the runner shaft. Secondly, the position of guide vanes is made adjustable so that they can open or

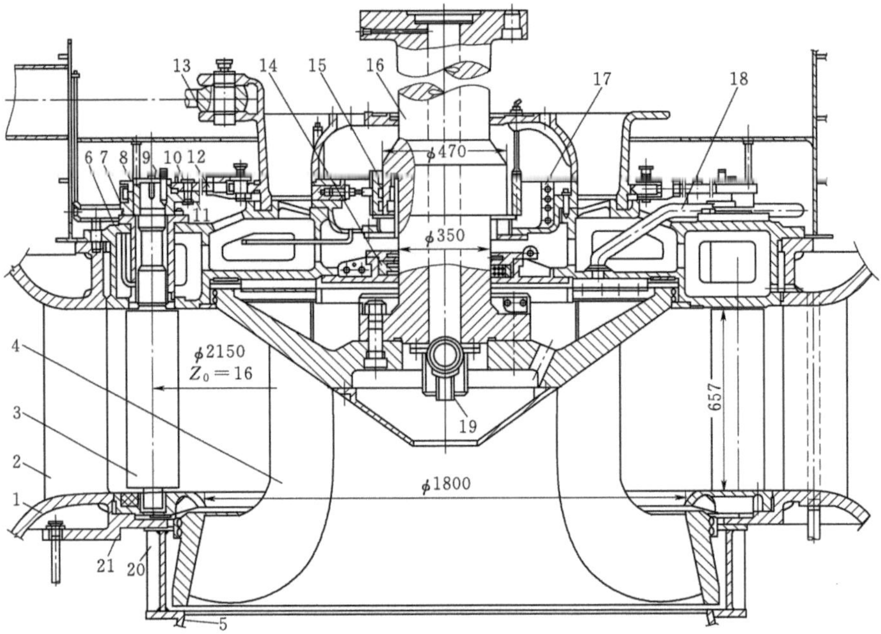

Fig. 2.11 Structure diagram of Francis turbine 1—spiral case; 2—stay vane ring; 3—guide vane; 4—runner; 5—draft tube; 6—turbine cover; 7—upper bush; 8—connecting plate; 9—half split key; 10—shear pin; 11—crank arm; 12—connecting rod; 13—control ring; 14—sealing device; 15—guide bearing; 16—shaft; 17—oil cooler; 18—cover drain pipe; 19—air supply device; 20—foundation ring; 21—bottom ring

close the water flow passage and thereby govern the flowrate of the turbine according to a given load condition. In the closed position, the water flow can be cut off to stop the turbine.

The characteristic parameter characterizing the position of guide vanes during the flow regulation process is called the vane opening a_0, commonly in unit mm. The vane opening a_0 is the shortest distance between any two adjacent vanes. The maximum mechanical opening a_{0max} of guide vanes corresponds to the position in which the vanes are radially placed, as shown by the dashed line in Fig. 2.12. However, in this maximum opening, the guide vanes may penetrate the runner; therefore, the maximum operating opening a_{max} allowed for the turbine to operate is less than a_{0max}. The maximum operating opening depends not only on turbine mechanical design, but also on turbine factors such as head, efficiency, output, and cavitation. In general, the higher the operating head of the turbine, the smaller the ratio of the maximum operating opening to the diameter D_1 of the runner. The turbine can be shut down when the opening is zero, in which the guide vanes are connected end by end, the flow passage is closed, and the discharge is zero.

The guide vanes are controlled by the guiding mechanism, as shown in Fig. 2.11. The upper axle journal of each vane penetrates the turbine cover (6) and is fixed to

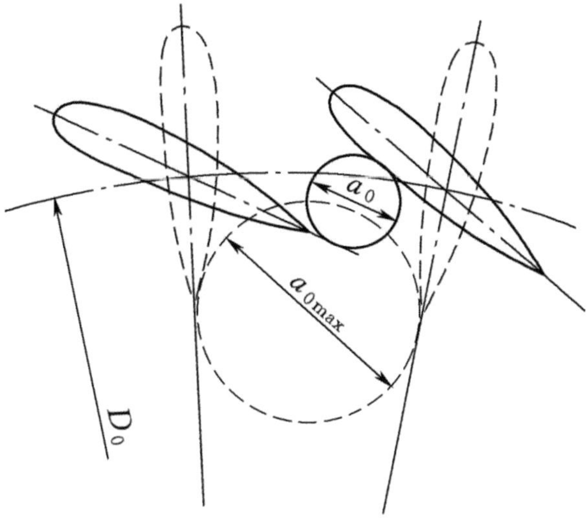

Fig. 2.12 Guide vane opening

the crank arm (11) by the key (9). The crank arm is connected to the control ring 13 via a connecting plate (8) and a connecting rod (12). The transmission principle of the guide vane operating mechanism is shown in Fig. 2.13. When the piston of the servomotor moves, the push–pull rod drives the control ring to rotate, so that the guide vanes rotate and adjust the opening a_0.

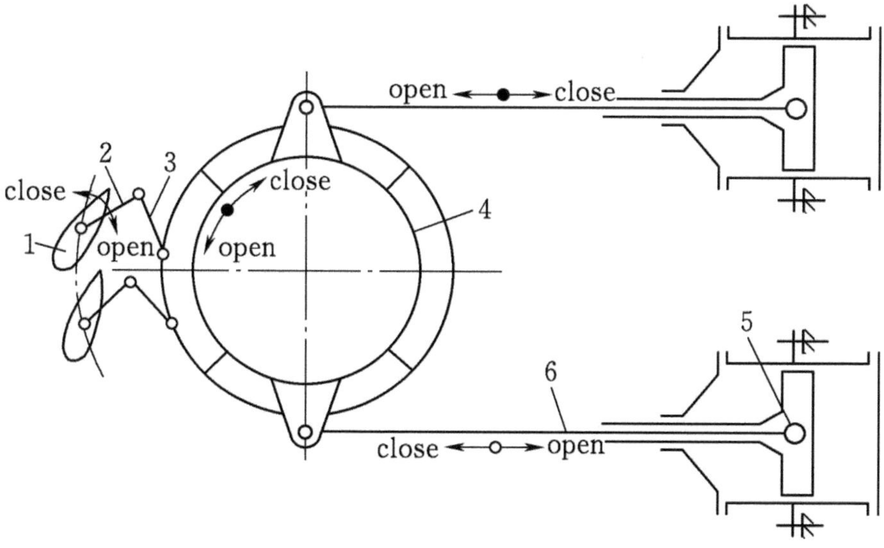

Fig. 2.13 Transmission principle diagram of guide vane operating mechanism 1—guide vane; 2—crank arm; 3—link rod; 4—control ring; 5—servomotor piston; 6—push–pull rod

When the vane is stuck by some debris and cannot be closed, it will seriously affect the operation of the turbine. For this reason, the shear pin (10) is connected between the crank arm (11) and the connecting plate (8). When some individual vanes are stuck, the shear pin on the vane is cut off, so that the guide vane is separated from the operating mechanism. And the remaining guide vanes can still be closed normally. The main geometric parameters of the guide vanes are as follows:

(1) The number of guide vanes Z_0 is generally related to the diameter of the runner. When the diameter of the runner $D_1 = 1.0$ to 2.25 m, Z_0 is normally 16; when $D_1 = 2.5$ to 8.5 m, Z_0 is normally 24.
(2) The relative height of guide vanes b_0/D_1 is mainly related to the turbine type. The smaller the turbine with the higher head, the smaller the relative height b_0/D_1. Generally for Francis turbines, $b_0/D_1 = 0.1$ to 0.39; for axial-flow turbines, $b_0/D_1 = 0.35$ to 0.45.
(3) The diameter of the guide vane distribution circle D_0 should satisfy that the guide vanes cannot extend into the runner when it is at the maximum operating opening. Generally, $D_0 = (1.13$ to $1.16)D_1$.

4. Draft tube

The draft tube has three shapes: straight cone shaped, curved cone shaped, and curved elbow shaped. The first two shapes are suitable for small turbines, and the elbow shape is suitable for large and medium-sized turbines. In addition to guiding the water flowing out of the runner to the tailrace channel, the main function of the draft tube is to recover the potential energy and a part of the kinetic energy, which has a great influence on the efficiency of the turbine, especially for the low-head hydropower stations.

2.3.1 *Francis Turbine*

Figure 2.11 is a structural view of a Francis turbine. The water from the penstock flows through the water-passing components of the hydro turbine, namely, the spiral case (1), stay vane ring (2), guide vane (3), runner (4), and draft tube (5), and flows downstream. In order to seal the water and support the guiding mechanism, a top cover (6) is set up on the upper portion of the runner and fixed on the stay vane ring. The lower end of the main shaft (16) of the turbine is connected to the runner by a flange and bolts, and the upper end is coupled to the generator shaft.

The runner of Francis turbine consists of the upper crown (1), blade (2), lower ring 3, anti-leakage rings (4) and (5), and discharge cone (6), as shown in Fig. 2.14. The upper crown is in the form of a curved round truncated cone, the upper end of which is connected to the main shaft by a flange, and the lower end is connected with a discharge cone by screw or welding. There are several pressure-relief holes around

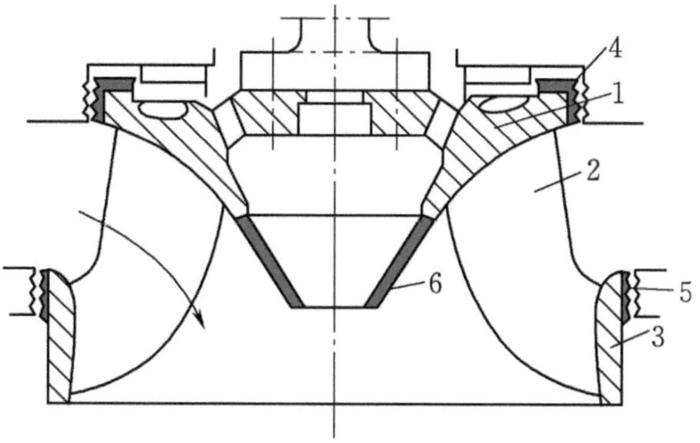

Fig. 2.14 Runner of Francis turbine 1—upper crown; 2—blade; 3—lower ring; 4 、5—anti-leakage ring; 6—discharge cone

the flange to discharge the accumulated water that leaks into the upper space of the upper crown through the outer edge of the upper crown, into the draft tube, so as to reduce the axial water pressure acting between the runner and top cover. For large turbines, air supply device is usually set at the main shaft end, which is connected to the upper crown. By supplying air to the low-pressure region of the lower side of the discharge cone, cavitation vortex instability can be mitigated. The function of the discharge cone is to smoothly direct the radial flow of water to the axial flow, so as to eliminate the water impact and vortex flow of the radial flow.

The runner blades, with wing shape, are twisted surfaces fixed uniformly between the upper crown and the lower ring along the circumference. The three components are welded as a whole, the water inlet side has a small degree of bending, and the water outlet side has a large degree of bending. The number of blades is usually between 12 and 21 pieces.

The anti-leakage ring, also called the labyrinth ring, consists of a fixed portion and a rotating portion mounted on the outer edge of the crown and the lower ring of the runner. A series of small grooves or labyrinth-shaped right-angle turns are formed between the rotating portion and the corresponding fixed one to increase the seepage path and increase the resistance, thereby reducing the leakage loss.

Due to the structural characteristics of Francis turbine runners, such as the integrated structure connection and the large number of blades, Francis turbines can be adopted by hydropower stations with higher heads. The higher the water head, the longer the flow path of the runner, and the shorter the runner's height; on the contrary, as the water head decreases, the discharge increases, the runner's height increases, the flow path length becomes shorter and shorter, and the blade number gradually reduces, as show in Fig. 2.15.

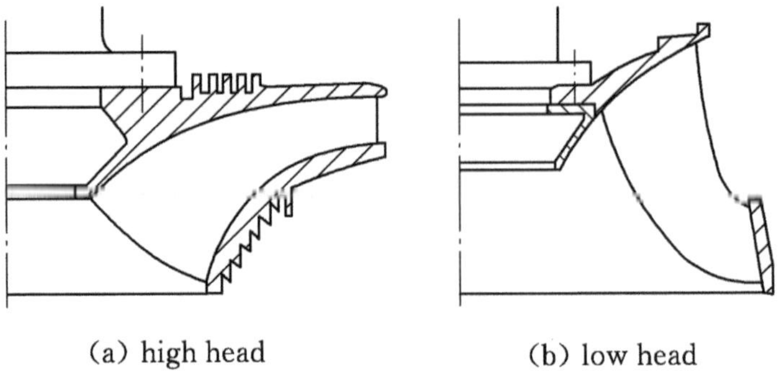

Fig. 2.15 Francis turbine runner for high head and low head

2.3.2 *Axial-Flow Turbine*

Figure 2.16 is a structural view of a Kaplan turbine. The structures of the most components of this turbine are basically the same as those of a Francis turbine. The main differences are shown in the runner (including the flow passage of the runner, blades, and blade-adjust mechanism) and the runner chamber.

The runner of the axial-flow turbine is composed of the blades (12), wheel hub (13), and discharge cone (16). The blades are evenly distributed around the wheel hub in cantilever form, with wing-shaped twisted surfaces. The shape of the blade is similar to the propeller. In the root, the arc is short and the curvature and thickness are large; in the tip, the arc is long and the edge is thin and flat. The number of runner blades is generally 4–8, depending on the operating head. For lower operating head turbines, the number of blades is less; for higher head turbines, the number of blades is more.

For the fixed-blade axial-flow turbine (also called propeller turbine), the blades are fixed to the wheel hub. The angle φ of blades is always fixed at the optimum position in the design condition, which is usually defined as $\varphi = 0°$.

For the adjustable blade axial-flow turbine, the blades are connected to the wheel hub by a spherical flange. The blade angle can be adjusted according to flow and output conditions, so as to maintain a relatively optimal setting angle. The blade angle $\varphi > 0°$means that the blade placement slope is greater than the optimal placement slope of the design condition, indicating that the blade rotates in the opening direction. When $\varphi < 0°$ the blade rotates in the opposite direction. Blade angle φ is generally between –15° and +20°, as shown in Fig. 2.17.

The operating mechanism of blade rotation is installed in the wheel hub. The transmission principle is shown in Fig. 2.18. When the oil pressure in operating oil pipe in the main shaft changes, the piston of the runner servomotor (8) moves up and down, thus driving the link rod (6) and the crank arm (5) to rotate the blade (1). The rotation of the blade and the rotation of the guide vane are coordinated under the control of the governor to achieve optimal operating condition.

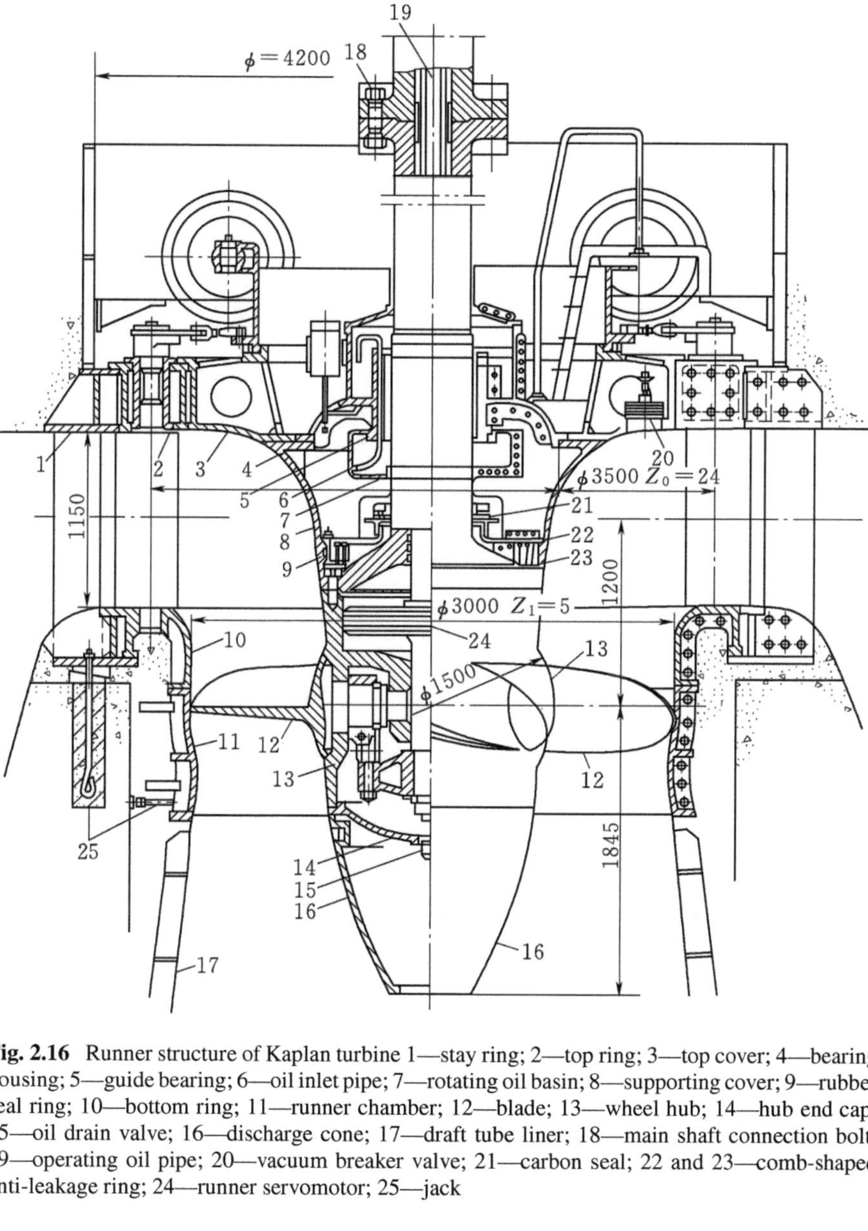

Fig. 2.16 Runner structure of Kaplan turbine 1—stay ring; 2—top ring; 3—top cover; 4—bearing housing; 5—guide bearing; 6—oil inlet pipe; 7—rotating oil basin; 8—supporting cover; 9—rubber seal ring; 10—bottom ring; 11—runner chamber; 12—blade; 13—wheel hub; 14—hub end cap; 15—oil drain valve; 16—discharge cone; 17—draft tube liner; 18—main shaft connection bolt; 19—operating oil pipe; 20—vacuum breaker valve; 21—carbon seal; 22 and 23—comb-shaped anti-leakage ring; 24—runner servomotor; 25—jack

Since increasing the diameter of the wheel hub will decrease the size of the water flow passage and deteriorate water flow patterns, the ratio of the wheel hub diameter d_g to the runner diameter D_1 (referred to as the hub ratio) is generally limited to the range of 0.33–0.55. Because the operating mechanism of the blade rotation is

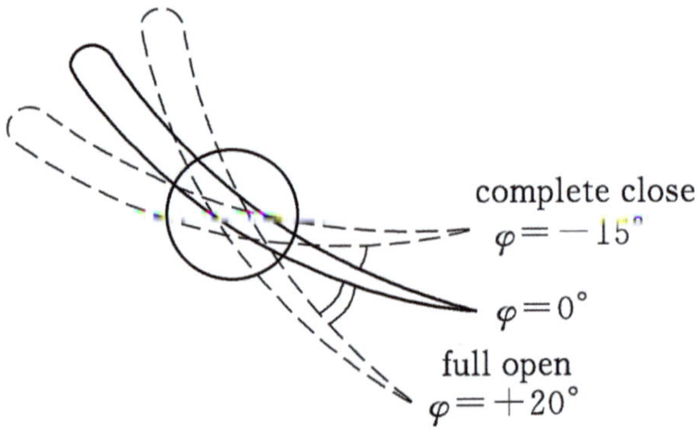

Fig. 2.17 Blade angle for Kaplan turbine

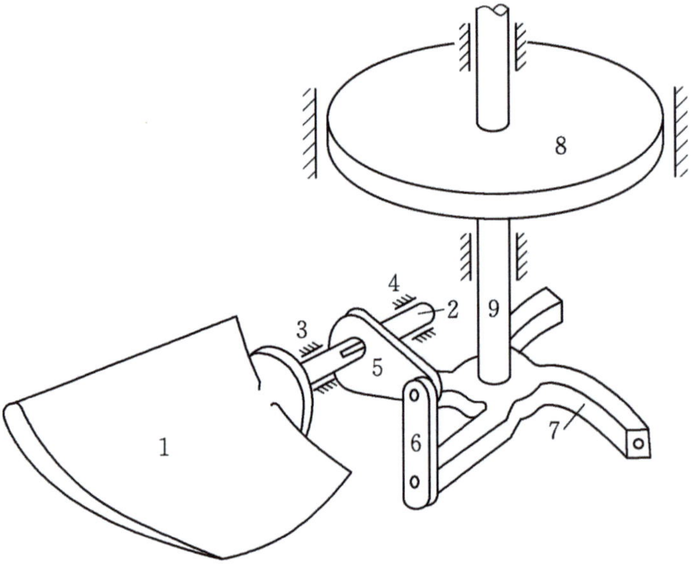

Fig. 2.18 Operating mechanism of blade rotation 1—blade; 2—blade pivot; 3 and 4—bearing; 5—crank arm; 6—link rod; 7—operating rack; 8—piston of runner servomotor; 9—piston lever

complicated and the installation is difficult, Kaplan turbine is generally only used for large and medium-sized hydropower stations.

The inner wall of the runner chamber (11) (Fig. 2.16) of the axial-flow turbine is often subjected to large pulsating water pressures. Therefore, the reinforcing steel bars are often arranged on the outer side and fixed to the peripheral concrete by tensioners or jacks (25). The inner surface of the runner chamber above the axis of blade rotation is generally cylindrical to facilitate the installation and disassembly of the runner. The inner surface of the runner chamber below the axis of blade rotation is usually formed into a spherically curved surface to ensure the clearance of the outer edge of the rotating blade, thereby reducing the leakage of water flow.

2.3.3 *Diagonal Flow Turbine*

Figure 2.19 is a structure view of a diagonal flow turbine (also called the Deriaz turbine). The embedded components: spiral case (1), stay ring (2), guiding mechanism (4), draft tube (29), main shaft (27), and guide bearing (21), are basically similar to those for Francis turbine and high-head Kaplan turbine. The runner is composed of the blades, wheel hub, and rotating mechanism therein. The differences with Kaplan turbine are that the blade rotation axis of the Deriaz turbine is 45–60° with the main shaft, the number of blades is about 8 to14 (between those for Francis turbine and Kaplan turbine), and the outer surface of the wheel hub (26) and the inner surface of the runner chamber (5) are substantially spherical curved surfaces.

Since the Deriaz turbine is equipped with more rotatable blades, it can adapt to higher head and larger head changes than Kaplan turbine. It is more adaptable to load changes than Francis turbine and can maintains a wide, high-efficiency zone.

The blade adjusting mechanism of the diagonal flow turbine is commonly used in two forms. One is an operating mechanism of a piston servomotor similar to that of Kaplan turbine; the structure is complicated and less applied. The other is to use the scraper servomotor (28) or the annular servomotor to drive the operation plate (7) to rotate, and then drive the blade (6) to rotate through the slider (10) and the rotating

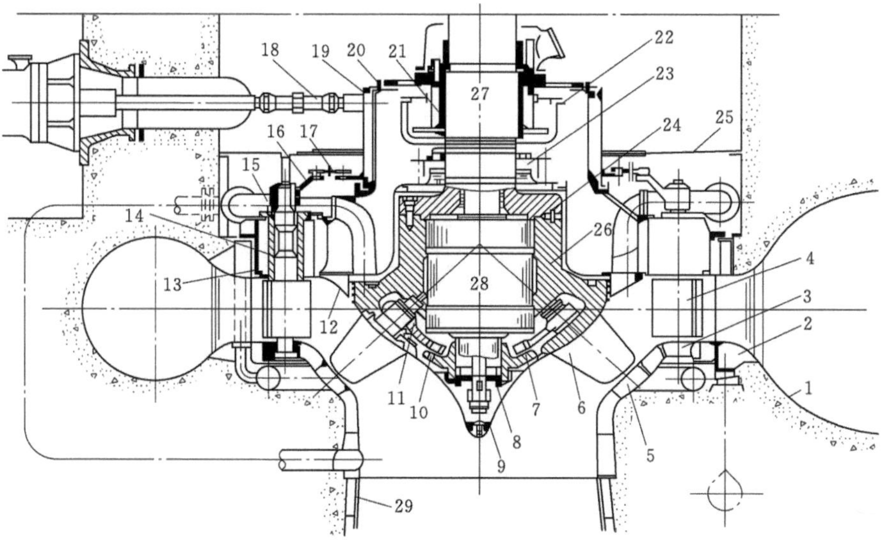

Fig. 2.19 Structure diagram of diagonal flow turbine 1—spiral case; 2—stay ring; 3—bottom ring; 4—guide vane; 5—runner chamber; 6—blade; 7—operation plate; 8—lower end plate; 9—discharge cone; 10—slider; 11—rotating arm; 12—top cover; 13—top ring; 14—bush; 15—hydraulic balancing pipe; 16—crank arm; 17—link rod; 18—push-pull lever; 19—control ring; 20—support rack; 21—guide bearing; 22—oil basin; 23—main shaft seal; 24—key; 25—cover; 26—wheel hub; 27—main shaft; 28—scraper servomotor; 29—draft tube

arm (11). The structure of the scraper servomotor is relatively simple and currently has many applications, but the sealing requirements of the oil circuit are high.

The inner wall of the runner chamber of the diagonal flow turbine is also made spherical and lined with steel plates to ensure a minimum clearance with the outer edges of blades to reduce water leakage loss. However, it should be noticed to prevent the runner from colliding with the runner chamber due to the axial displacement caused by axial water thrust and temperature changes. The measure taken for this is to install a signal relay protection device for the axial displacement to automatically stop the turbine in an emergency when the allowable displacement value is exceeded.

2.3.4 Bulb Tubular Turbine

Figure 2.20 is the structure diagram of a typical bulb tubular turbine. This type of machine is essentially an axial-flow turbine without volute, and with an elbow-shaped draft tube and horizontal shaft. The generator is mounted in the bulb-shaped housing (15), so that the main shaft is short and the structure is very compact. The housing (15) is fixed to the outer casing by the front pillar (16) and rear pillar (4). The front pillar is hollow at the top and the manhole (13) is placed inside. The guide vane (2) is arranged in an oblique conical shape. The blades (1) are fixed or adjustable. The shape of the blades and their adjusting mechanism are similar to those of Kaplan turbine, and the number of blades is usually four. The rotating part of the unit is supported by the radial guide bearings (6, 7), and the axial displacement is limited by the thrust bearing (8). The inlet pipe (17) is approximately a tapered circular straight pipe, and the draft tube (20) is approximately a gradually expanding straight pipe.

The bulb housing can be placed on the upstream or downstream side of the runner. When the head is low, the unit with the bulb body placed on the upstream side is more efficient; when the head is high, the unit with the bulb body on the downstream side is better in terms of structure strength and running stability.

When the head is low and the unit capacity is large, if the turbine is directly connected to the main shaft of the generator, the generator will have a larger diameter due to the lower speed, which will cause the bulb body size too large and the flow hydraulic loss to increase. For this reason, a gear speed-increasing transmission mechanism is often arranged between the turbine and the generator so that the speed of the generator is 3–10 times larger than the turbine speed, thereby reducing the size of the generator and the size of the bulb body, and improving the flow patterns in the turbine passage. However, this speed-increasing device is complicated, the requirement of processing technology is high, and the transmission efficiency is generally low, therefore, it is currently only applied to small-sized tubular units.

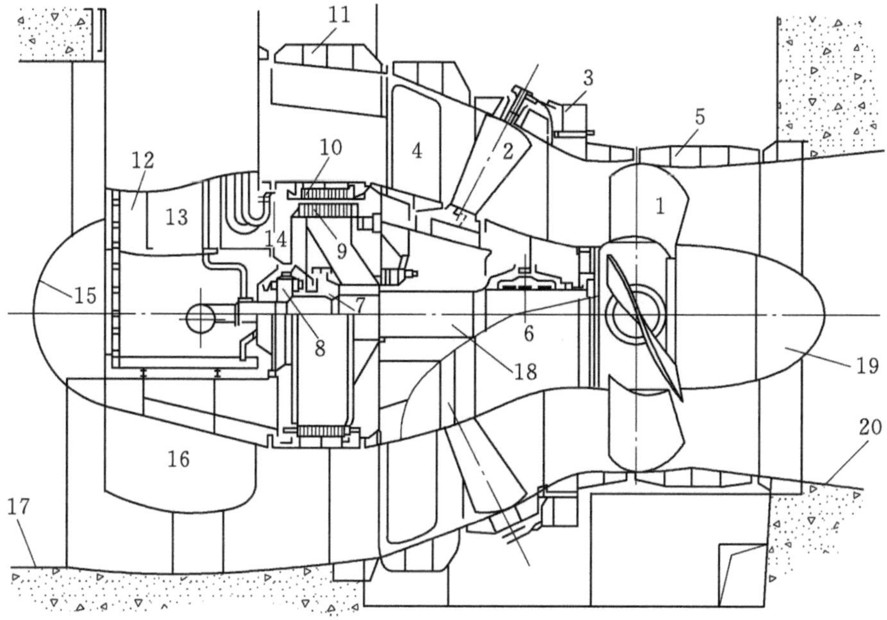

Fig. 2.20 Structure diagram of bulb tubular turbine 1—blade; 2—guide vane; 3—control ring; 4—rear pillar (stay vane); 5—runner chamber; 6—turbine guide bearing; 7—generator guide bearing; 8—generator thrust bearing; 9—generator rotor; 10—generator stator; 11—maintenance manhole; 12—auxiliary pipeline passage; 13—manhole in front pillar; 14—cable outlet channel; 15—bulb-shaped housing; 16—front pillar; 17—inlet pipe; 18—main shaft; 19—discharge cone; 20—draft tube

2.3.5 Pelton Turbine

The structure of the Pelton turbine is relatively simple, mainly composed of the water distribution ring, nozzle, needle valve, runner, and deflector. The function of the nozzle is to guide the pressure water to flow uniformly and convert it into a high kinetic energy-free jet by nozzle contraction. The function of the needle valve is to control the discharge to meet the needs of output.

Figure 2.21 shows the structure of a two-nozzle Pelton turbine. The high-pressure flow from the penstock (1) is converted by the nozzles (3) to high-speed jets. The jets impinge on the runner (6) to make it rotate. After that, the water is discharged downstream through the tail sink (9).

The discharge adjustment of the Pelton turbine is realized by the needle valve, which is composed of needle (4) and nozzle (3). When the needle moves, the annular cross-sectional area of the nozzle outlet changes. And when the needle moves to the front, it can cut off the water flow. The movement of the needle is controlled by the servomotor (10) and its transmission mechanism. In front of the nozzle, a deflector (8) is installed. When the load is suddenly rejected by the unit, in order to avoid the runner from running away, the deflector acts immediately, and the jet is completely

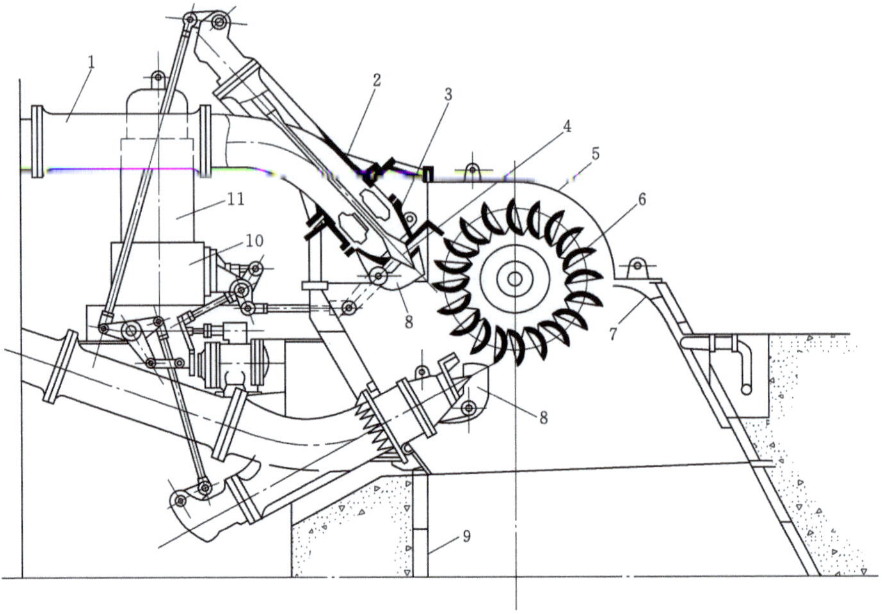

Fig. 2.21 Structure diagram of the Pelton turbine 1—penstock; 2—nozzle tube; 3—nozzle; 4—needle; 5—casing; 6—runner; 7—guide baffle; 8—deflector; 9—tail sink; 10—servomotor; 11—governor

deviated from the runner within 1–2s. Then, the needle can be slowly closed in 5–10s or longer to avoid excessive water hammer pressure in the penstock.

The nozzle and runner are positioned within the casing (5) to prevent water from splashing into the plant. In order to prevent the water from splashing to the top of the runner or the back of the buckets, causing additional losses, a guide baffle (7) is provided on the lower right side of the casing.

The runner (6) is composed of a wheel and many buckets, which are uniformly distributed along the circumference. The front surface of the bucket (Fig. 2.22) consists of two half-spoon-shaped inner surfaces (1), which have slightly oblique water outlet edges (5). In the centerline, a water dividing edge (6) is placed to divide the jet into symmetrical two parts. In order to prevent the front bucket from affecting the jet impulse the behind working bucket during rotation, a notch (2) is placed at the front end of the bucket and a slot at the back. In order to enhance the strength of the bucket, transverse rib (7) and longitudinal ribs (8) are added to the back of the bucket. The buckets and wheels of large and medium-sized units are often integrally cast or welded.

In order to increase the discharge and rotating speed, two or more nozzles are often installed on one runner. Sometimes two or more runners are mounted on one shaft to increase the unit capacity. Large and medium-sized Pelton turbines are mostly

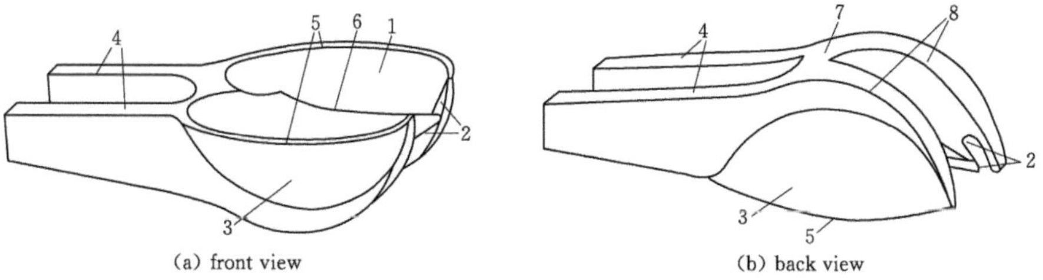

Fig. 2.22 Bucket structure of the Pelton turbine runner 1—inner surface; 2—notch; 3—back surface; 4—bucket handle; 5—outlet edge; 6—water dividing edge; 7—transverse rib; 8—longitudinal ribs

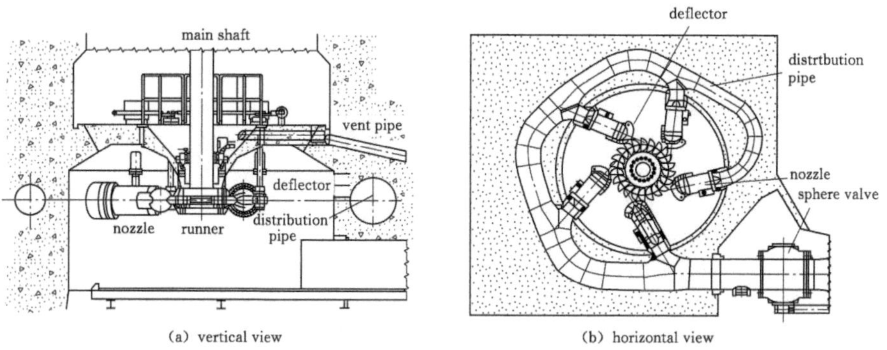

Fig. 2.23 Multi-nozzle Pelton turbine with vertical axis arrangement 4. Application scope and designation of hydro turbine

arranged in a vertical arrangement, which not only reduces the space of the powerhouse but also facilitates the installation of multiple nozzles, as shown in Fig. 2.23. Small and medium-sized Pelton turbines are usually arranged in a horizontal arrangement, which can simplify the structure, reduce the cost, and is easy to install and maintain.

2.4 Application Scope and Designation of Hydro Turbine

2.4.1 Application Scope of Turbine

The main types and basic structures of hydraulic turbines are introduced in the previous sections, which provide basics to understand the application scope of turbines. One of the purposes of knowing turbine application scope is to make full use of the diversity of natural water resources, to meet the rational, economic, and safe development principles, and to select appropriate turbines.

Figure 2.24 shows the application scope for various types of turbines drawn by GE Company. The figure is expressed in logarithmic coordinates, with the horizontal ordinate the discharge of the turbine and the longitudinal ordinate the working head of the turbine. The oblique line represents the contour of the output in MW, and the region is divided into four zones according to the Pelton turbine, Francis turbine, Kaplan turbine, and bulb tubular turbine.

The following can be seen from this figure.

(1) The Pelton turbine can be applied in the highest head. For medium and large-sized turbines with a unit capacity of 50 MW or more, the operating head is more than 500 m, and the maximum operating head can reach more than 1000 m. When the unit capacity reaches 800 MW, not only the working head is close to 1000 m, but also the discharge is around 90 m^3/s, and multiple nozzles need to be installed.
(2) The Francis turbine can be applied in the widest working head. For medium and large turbines with a unit capacity of 50 MW or more, the working head is more than 60 m, and the maximum working head can be up to 700 m. Francis turbine has the widest capacity range, up to 1000 MW. When the working head is 200 m, the largest discharge is over 500 m^3/s.
(3) The Kaplan turbine can be applied in a narrow working head. For medium and large-sized turbines with a unit capacity of 50 MW or more, the working head is between 15 and 70 m. When the unit capacity reaches 200 MW, not only the operating head is close to 70 m, but also the discharge is 300 m^3/s.
(4) The tubular turbine can be applied in the lowest working head, usually within 17 m. For medium and large-sized turbines with a unit capacity of 50 MW or more, not only the working head is close to 17m, but also the discharge is about 300 m^3/s. Therefore, it is very difficult for a tubular turbine to have a unit capacity of more than 150 MW.

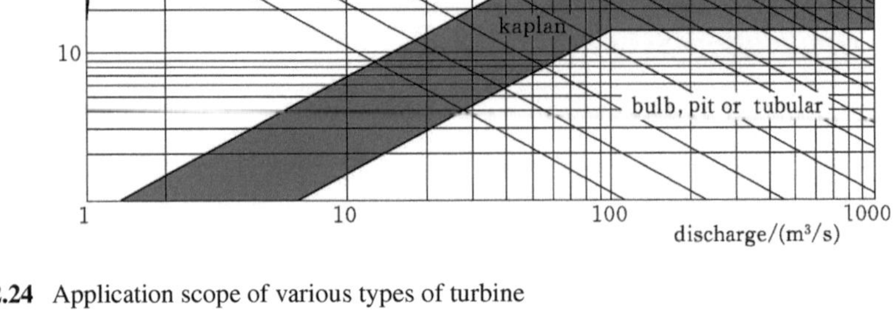

Fig. 2.24 Application scope of various types of turbine

It can be seen that in the development of medium and large hydropower stations, the Francis turbine still plays a leading and irreplaceable role. In addition, it should be pointed out here that under certain Q and H conditions, turbine design and selection should meet the following requirements:

a. High efficiency and anti-cavitation performance.
b. Sufficient mechanical strength.
c. Large discharge capacity, expecting larger discharge in a certain runner diameter condition. In other words, under a certain output, the larger discharge capacity means smaller diameter. This can reduce the size of the turbine and save investment in electromechanical equipment and civil engineering.

2.4.2 *Designation of Turbine*

The designation of turbine type, also known as the label name of the turbine, has specialized in unified standards for improving the qualities of products and making them easy to choose and use. According to the Chinese Code *Model compilation method of hydro turbine, storage pump and pump-turbine* GB/T 28,528—2012, the product model of the hydro turbine, storage pump, and pump-turbine is composed of three or four sections. The fourth section is only used for storage pump and pump-turbine. The different sections are separated by a "—", as shown in Fig. 2.25.

The first section consists of the type code and the runner/impeller code of the hydro turbine, storage pump, and pump-turbine models. The type code of the turbine is expressed in the Chinese phonetic alphabet, such as the mixed-flow type HL, the diagonal flow type XL, the axial-flow adjustable blade type ZZ, the tubular flow adjustable

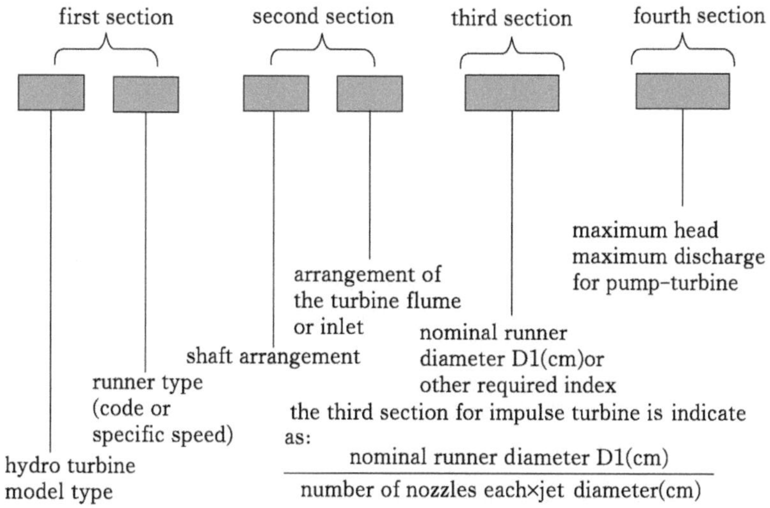

Fig. 2.25 Specification of product model for hydro turbine, storage pump, and pump-turbine

blade type GZ, and the impulse/bucket type CJ. The type code of the reversible pump-turbine is indicated by the letter "N" before the above Chinese phonetic alphabet, such as the mixed-flow pump-turbine NHL, the axial-flow adjustable blade pump-turbine NZZ, and the tubular flow adjustable blade pump-turbine NGZ. For turbines and pump-turbines, the runner code is represented by the serial number of model runner and/or the specific speed of the prototype turbine under the rated operating condition. The serial number of the model runner is separated from the specific speed by the "/" symbol. The specific speed is expressed in Arabic numerals and labeled by units m·kW.

The second section consists of the main shaft arrangement type and structural features of turbine, storage pump, and pump-turbine. The type code of the main shaft arrangement is represented by the Chinese phonetic alphabet, such as vertical axis L, and horizontal axis W. The code name of the structural characteristics of the turbine is also expressed in Chinese Pinyin letters, such as metal volute J, concrete volute H, full tubular flow Q, and bulb type P.

The third section consists of turbine runner diameter D_1 (cm) or runner diameter and other parameters, expressed in Arabic numerals. The schematic diagram of turbine runner diameter D_1 is shown in Fig. 2.26. Or pump impeller diameter D_1 (cm) for the storage pump is used; or both the turbine diameter D_1 and pump impeller diameter D_1 (cm) for the combined pump-turbine are used. For the impulse turbine, the third section of is shown as runner diameter/number of nozzles × jet diameter.

The fourth section is just for the pump-turbine, and the maximum head (m) and maximum discharge (m^3/s) in pump mode for the actual power station are expressed.

The examples of the nameplate of turbine models are as follows.

Example 1

$$\text{HL A153}/\times\times-\text{LJ}-300$$

It indicates that the turbine type is the Francis turbine, the serial number of the model runner is A153, the specific speed under rated operating condition is C, the main shaft is vertical, the volute is metal spiral case, and the runner diameter is 300 cm.

Example 2

$$\text{ZZ 560A}/\times\times-\text{LH}-400$$

It indicates that the turbine type is the Kaplan turbine, the serial number of the model runner is 560 A, the specific speed is $\times\times$ m · kW for the prototype turbine under the rated operating condition, the main shaft is vertical, the volute is concrete spiral case, the runner blade is adjustable, and the runner diameter is 400 cm.

Example 3

$$\text{GZ}\times\times/\times\times-\text{WP}-450$$

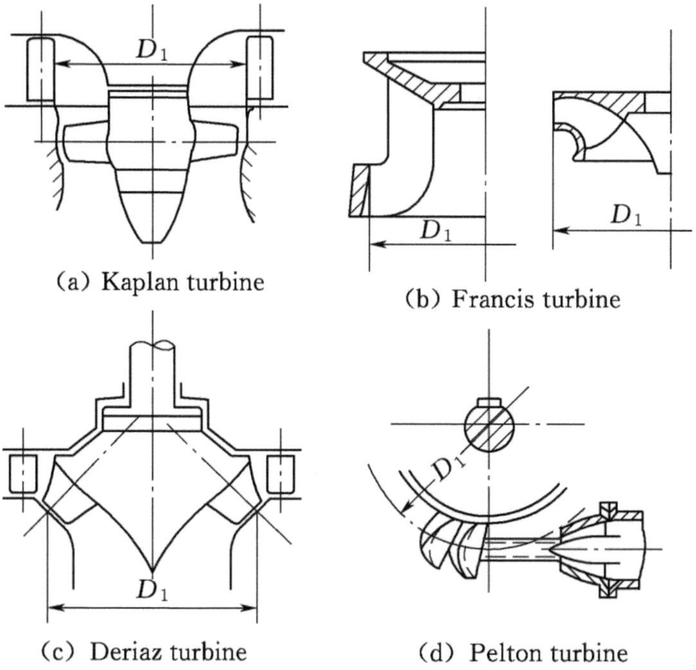

Fig. 2.26 Schematic diagram of various types of turbine runner diameter D_1

It indicates that the turbine type is the bulb tubular flow turbine, the serial number of the model runner is × × and/or the specific speed under the rated operating condition is × × m · kW, the main shaft is horizontal, the runner blade is adjustable, and the runner diameter is 450 cm.

Example 4

$$CJA475/\times\times - W - 120/2 \times 10$$

It indicates that the turbine type is the Pelton turbine, the serial number of the model runner is A475 and/or the specific speed under rated operating condition is × ×m·kW, the main shaft is horizontal, the runner diameter is 120cm with two nozzles, and the jet diameter is 10 cm.

Example 5

$$\mathrm{GZ006}/\times\times - \mathrm{WZ} - 275$$

It indicates that the turbine type is shaft extension tubular flow turbine, the serial number of the model runner is 006 and/or the specific speed under rated operating condition is $\times \times$ m · kW, the main shaft is horizontal, the runner blade is adjustable, and the runner diameter is 275 cm.

The examples of the nameplate for pump-turbine models are as follows.

Example 1

$$\text{NHL} \times \times / \times \times - \text{LJ} - 408 - 615/57.7$$

It indicates that the turbine type is the reversible Francis turbine, the model runner number is $\times\times$ and/or the specific speed of turbine prototype under the rated operating condition is $\times \times$ m · kW, the main shaft is vertical, the volute is metal spiral case, the runner diameter is 480 cm, the maximum head in pump mode in the actual operation is 615m, and the maximum discharge is 57.7 m^3/s.

Example 2

$$\text{NXL} \times \times / \times \times - \text{LJ} - 250 - 59/38.5$$

It indicates that the turbine type is the reversible Deriaz turbine, the model runner number is $\times\times$ and/or the nominal specific speed is $\times \times$ m · kW, the main shaft is vertical, the vertical shaft, the volute is metal spiral case, the runner diameter is 250 cm, the maximum head in pump mode in the actual operation is 59 m, and the maximum discharge is 38.5 m^3/s.

Example 3

$$\text{NZZ} \times \times / \times \times \text{J} - 450 - 30/64$$

It indicates that the turbine type is the reversible Kaplan turbine, the model runner number is $\times\times$ and/or the nominal specific speed is $\times \times$ m · kW, the main shaft is vertical, the volute is metal spiral case, the blade is adjustable, the runner diameter is 450 cm, the maximum head in pump mode in the actual operation is 30 m, and the maximum discharge is 64 m^3/s.

Example 4

$$\text{NGZ} \times \times / \times \times - \text{WP} - 610 - 10.2/330$$

It indicates that the turbine type is the reversible bulb tubular turbine, the model runner number is $\times\times$ and/or the nominal specific speed is $\times \times$ m · kW, the main shaft is horizontal, the runner blade is adjustable, the runner diameter is 610 cm, the maximum lift of the pump in the actual operation is 10.2 m, and the maximum discharge is 330 m^3/s.

Questions

1. What are the main working parameters of hydraulic/hydro turbines? What is the meaning of each parameter?
2. What are the basic types of hydro turbines? Please give a brief introduction to thc classification mcthod and application scope of hydro turbines.
3. Please briefly describe the basic structure of hydro turbines. What are the differences between different types of hydro turbines? Please describe the characteristics of main components of the Francis turbine, Kaplan turbine, and Pelton turbine with comparison in detail.
4. Please explain the indication method of turbine models and the significance of codes for each part of the designation.

Bibliography

1. Zhongyuan, Jin. 1992. *Hydraulic machinery*, 2nd ed. Beijing: Water Resources and Electric Power Press.
2. Qizhao, Liu, H. Ming. 2010. *Hydropower station*, 4th ed. Beijing: China Water and Power Press
3. Ruyun, Luo. 1990. *Design manual for power equipment of hydropower station*. Beijing: Water Resources and Electric Power Press.
4. *Mechanical and electrical design manual of hydropower station, hydraulic machinery*. 1983. Beijing: Water Resources and Electric Power Press.
5. GB/T 28528–2012. 2012. *Compilation method for designation of hydraulic turbine, storage pump and pump*. Beijing: China Standards Press.
6. Yuan, Zheng. 2011. *Hydraulic turbine*. Beijing: China Water and Power Press.
7. Dakai, Liu. 1996. *Hydraulic turbine*, 3rd ed. Beijing: China Water and Power Press.
8. Xilin, Sha. 1999. *Tubular hydropower station*. Beijing: China Water and Power Press.
9. Wenwu, Song. 2015. *Theory and application of high head tubular turbine (in Chinese)*. Beijing: Science Press.
10. Zanpei, You. 2009. *Bulb tubular hydropower station*. Beijing: China Water and Power Press.
11. Guoxuan, Liu. 2006. *Operation and maintenance of bulb tubular turbine generator*. Beijing: China Water and Power Press.
12. Bo, Yu. 2008. *Principle and operation of hydraulic turbine*. Beijing: China Electric Power Press.
13. Enqi, Mao. 2000. *Type and structure of turbine runner*. Beijing: China Electric Power Press.

Chapter 3
Working Principle of Hydro Turbine

Jinping Li and Yongguang Cheng

Abstract In this chapter, we will discuss the following key points: the movement of water in the reaction turbine; the basic equation and efficiency of the hydro turbine; the working principle of the draft tube; the concept of cavitation, the suction head and installation elevation of the hydro turbine; the pressure pulsation characteristics of a hydro turbine. Students should be acquainted with the basic equation of turbine and the working principle of the draft tube; understand the optimum work conditions and the influencing factors of efficiency; master the concept of cavitation and the method to determine the installation elevation of the hydro turbine; be familiar with the flow patterns and pressure pulsation characteristics of a hydro turbine.

Keywords Working principle · Velocity triangle · Basic equation · Best working condition · Cavitation · Suction head · Installation elevation · Pressure pulsation

3.1 Motion of Water in Reaction Turbine Runner

When a turbine is running, regardless of whether the working condition is steady or not, the working parameters, such as head H, discharge Q, output N, and speed n, depend on the movement of the water flowing in the turbine runner. Therefore, it is necessary to explore the water flow velocity in the turbine runner, and then establish the equations between the water flow velocity and the turbine operating parameters. This needs to understand the working principle of the turbine.

3.1.1 Synthesis and Decomposition of Water Flow Motion

The motion of water in a turbine runner is a complex three-dimensional flow state. When water flows through the runner passage, on the one hand, the water flows

J. Li (✉) · Y. Cheng
Wuhan University, Wuhan, China
e-mail: lukeping@whu.edu.cn

Y. Cheng et al. (eds.), *Hydropower Station*,
https://doi.org/10.1007/978-981-96-2383-9_3

relative to the blades, that is the relative motion; on the other hand, the water rotates together with the runner, that is, the circumferential motion or convected motion. Therefore, the velocity of the water flow at any point in the turbine runner $\vec{v}$, also known as the absolute velocity, can be described by the combination of the relative velocity $\vec{w}$ and the convected velocity $\vec{u}$ at that point. This can form a velocity triangle as shown in Fig. 3.1 and Eq. (3.1). In Fig. 3.1, the direction angles α and β are the angles of the absolute velocity $\vec{v}$ and the relative velocity $\vec{w}$ intersecting with the convected velocity $\vec{u}$, respectively.

$$\vec{v} = \vec{w} + \vec{u} \tag{3.1}$$

where the unit of velocity is m/s.

The circumferential velocity can be expressed as

$$u = \frac{\pi Dn}{60} \tag{3.2}$$

where D is the circumference diameter of the inspection point, m; n is the rotational speed of runner, r/min.

In practical applications, for convenience of analysis, the absolute velocity $\vec{v}$ is orthogonally decomposed in the direction of circumferential velocity and the direction perpendicular to the circumferential velocity, thus the following two projection velocities are obtained: one is the circumferential projection velocity $\vec{v}_u$ of the absolute velocity $\vec{v}$, called the absolute circumferential projection velocity; the other is the axial projection velocity $\vec{v}_m$ of the absolute velocity $\vec{v}$, called the absolute axial projection velocity because it is on the axial plane. Then we have the velocity vector relation

$$\vec{v} = \vec{v}_u + \vec{v}_m \tag{3.3}$$

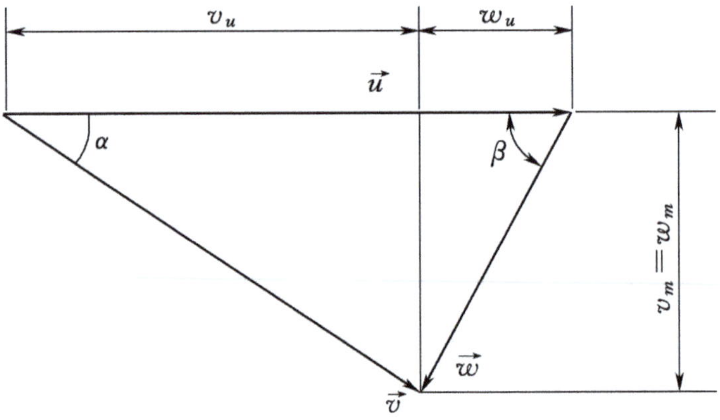

Fig. 3.1 Velocity triangle

Further decomposition of the axial velocity $\overrightarrow{v}_m$ yields

$$\vec{v}_m = \vec{v}_r + \vec{v}_z \tag{3.4}$$

Substituting Eqs. (3.4) into (3.3), we get

$$\overrightarrow{v} = \overrightarrow{v}_u + \overrightarrow{v}_m = \overrightarrow{v}_u + \overrightarrow{v}_r + \overrightarrow{v}_z \tag{3.5}$$

where $\overrightarrow{v}$ is the absolute velocity of water at a certain point in the runner, m/s; $\overrightarrow{v}_r$ is the radial projection component of the absolute velocity $\overrightarrow{v}$ at this point, m/s; $\overrightarrow{v}_z$ is the axial projection component of the absolute velocity $\overrightarrow{v}$ at this point, m/s; $\overrightarrow{v}_z$ is the circumferential projection component of the absolute velocity $\overrightarrow{v}$ at this point, m/s; $\overrightarrow{v}_m$ is the axial plane projection component of the absolute velocity $\overrightarrow{v}$ at the point, m/s.

Similarly, the relative velocity can also be decomposed as follows

$$\vec{w} = \vec{w}_u + \vec{w}_m = \vec{w}_u + \vec{w}_r + \vec{w}_z \tag{3.6}$$

The velocity vectors in the velocity triangle, with their axial direction projection and coordinate direction projections, are shown in Fig. 3.2. This figure clearly expresses the synthesis and decomposition of velocity triangle.

3.1.2 Velocity Triangle at the Inlet and Outlet of the Runner of Francis Turbine

In Francis turbine, the change of flow direction from the radial direction to the axial direction is achieved in the runner. For convenience of research, the cylindrical coordinate system is usually used. We can define three directions: the axial direction of the turbine, Z; the radial direction, r; the circumferential direction, ϕ (or the tangential direction), as shown in Fig. 3.3. This tangential direction is perpendicular to the meridional plane, which is formed by the radial direction and the axial direction. Since the plane passes through the axis of the turbine, it is also called the axial plane. The shaded part in Fig. 3.3 is an axial plane. And the following premises are assumed.

(1) The runner is composed of an infinite number of infinitely thin blades. This assumption means that the water flow in the runner is totally uniform and axisymmetric. The relative motion trajectory (the motion trail of a certain fluid particle) coincides with the centerline of a blade hydrofoil (also known as the bone line).
(2) The viscosity of water is ignored. This assumption means that the stream surfaces do not interfere with each other. In Fig. 3.3, the curve a-0-1-2 is a streamline on the stream surface. The streamline is the direction line of a certain instantaneous flow of water. That is, at any instant, the velocity direction of the

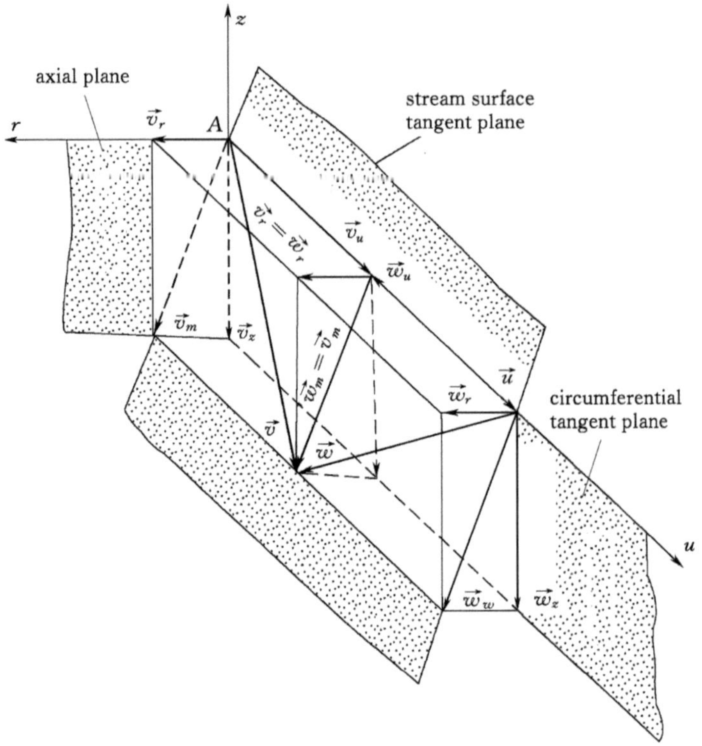

Fig. 3.2 Velocity and its projection velocities in the velocity triangle

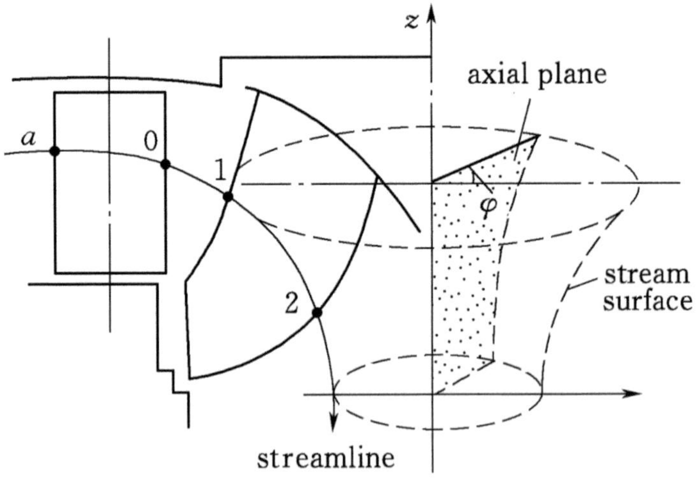

Fig. 3.3 Stream surface and axial surface in the runner of Francis turbine

fluid particle at each point on the line coincides with the tangential direction of the line at the same point.

(3) The working parameters of the turbine do not change with time. This means the head H, discharge Q, output N, and rotational speed n remain unchanged, and the water flow in each passage of the turbine is steady. In this condition, the streamlines of the relative motion and the absolutc motion of the water flow in the runner coincide with the trajectory lines of fluid particles. It should be pointed out that the streamlines or trajectories of the relative motion are different from those of the absolute motion.

According to the above assumptions, for Francis turbine, it can be considered that the motion of any fluid particle in the runner is a helical curve along a stream surface, a trumpet-shaped spatially curved surface. The axisymmetric helical curve (streamline or trajectory line) constitutes the stream surface as Fig. 3.3, and there are numerous neatly arranged and non-interfering stream surfaces in the whole runner.

Take a certain stream surface as example to analyze one of the streamlines, such as the one along curve a-0-1-2 in Fig. 3.3. Point a and Point 0 on the streamline are at the inlet and outlet of the guide vane, respectively, and Point 1 and Point 2 on the same streamline are at the inlet and outlet of the runner, respectively. Revolving the streamline along the axle forms the stream surface, which intersects with the guide vane and runner. Expanding the stream surface flat, we have the diagram shown in Fig. 3.4. In this figure, the angle between the bone line of the foil section of a blade and the circumferential tangent of the runner is called the blade inlet place angle, denoted by β_{e1} at the entrance of the runner; similarly, the blade outlet place angle denoted by β_{e2} at the outlet of the runner. Since the analysis is based on steady flow conditions, only the velocity triangles of the inlet and outlet of Francis turbine runner need to be discussed.

1 Inlet velocity triangle

(1) Circumferential velocity $\vec{u}_{1i}$.

$$u_{1i} = \frac{\pi D_i n}{60} \tag{3.7}$$

where u_{1i} is the circumferential/peripheral velocity at the inspection point i on the inlet side, m/s; D_i is the diameter of the circumference at the inspection point i, m; n is the turbine rotational speed, r/min.

(2) Axial plane velocity $\vec{v}_{m1i}$.

$$\vec{v}_{m1i} = \frac{Q\eta_{vol}}{F_{1i}} \tag{3.8}$$

where Q is the discharge of hydro turbine, m^3/s; η_{vol} is the volumetric efficiency of hydro turbine; F_{1i} is the cross-sectional area of the water flow at the inspection point i, m^2.

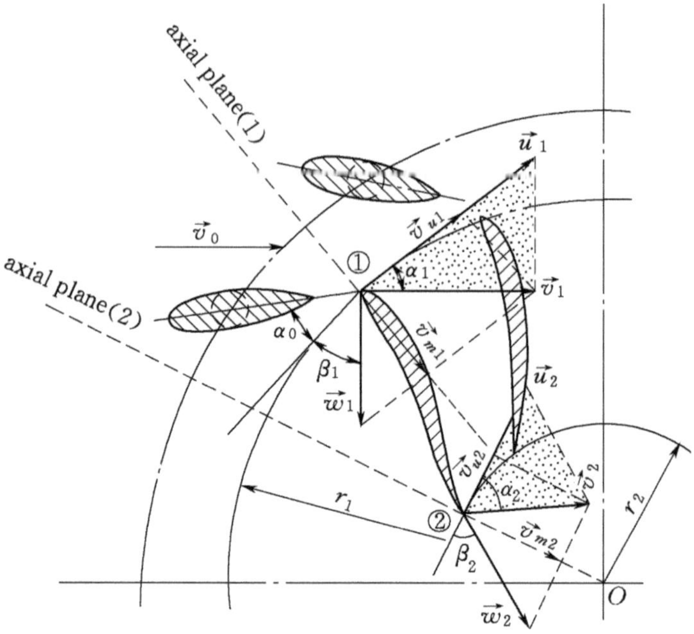

Fig. 3.4 Expansion of a stream surface

In order to obtain the cross-sectional area F_{1i} of the water flow passing through the inspection point (point i in Fig. 3.5), a line segment ae that is perpendicular to the streamlines such as A, B, C, D, and E at the axial plane through the inspection point can be drawn. Revolving ae around the axis, we obtain a circular surface. The area of this surface is F_{1i}. By Guldinus theorem, we know

$$F_{1i} = 2\pi R_g L_{ae} \tag{3.9}$$

where R_g is the radius of the circle that is formed by revolving the gravity center of the line segment ae, m; L_{ae} is the length of the line segment ae, m.

If blade thickness and its occupation are considered, the actual cross-sectional area of all axial plane water flow is:

$$F'_{1i} = F_{1i} - \delta_1 Z_1 L_{ae} = 2\pi R_g L_{ae}\left(1 - \frac{\delta Z_1}{2\pi R_g}\right) = 2\pi R_g L_{ae}\varphi_1 \tag{3.10}$$

where δ_1 is the thickness of the runner blade at the inlet side, m; Z_1 is the number of runner blades; φ_1 the blade occupation coefficient of the runner inlet,. The more the blade occupies the real water flow passage, the smaller the coefficient is. If blade occupation is not considered, or assuming infinitely thin blades, $\varphi_1 = 1$.

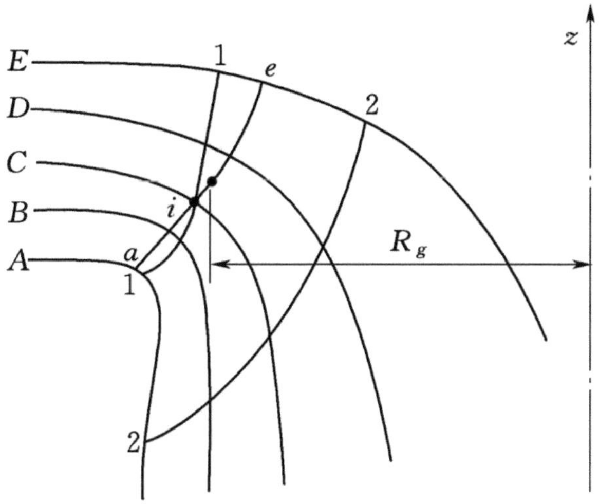

Fig. 3.5 Streamline on the axial plane

(3) The circumferential component of absolute velocity $\vec{v}_{u1i}$.

The direction angle of the absolute velocity is angle α_1, which is between absolute velocity $\vec{v}_1$ and circumferential velocity $\vec{u}_1$. For a low specific speed Francis turbine with a higher head, the inlet side of the runner blade approaches the outlet side of the guide vane, therefore the absolute velocity angle at the runner inlet can be considered to be approximately equal to the absolute velocity angle at the outlet of the guide vane. For the Francis turbine and the Kaplan turbine with medium and high specific velocity and lower head, the velocity moment remains the same from the outlet of the guide vane to the inlet side of the runner, according to the momentum moment theorem, then:

$$v_{u0}\frac{D}{2} = v_{u1i}\frac{D_i}{2} \tag{3.11}$$

where v_{u0} is the circumferential component of the absolute velocity v_0 at the outlet of the guide vane, m/s; D_0 is the diameter of the circle at the outlet of the guide vane, m; v_{u1i} is the circumferential component of the absolute velocity v_{1i} at point i at the inlet edge of the runner blade, m/s; D_i is the diameter of the circle where the point i is located at the inlet edge of the runner blade, m.

Besides, using the velocity triangle at the outlet of the guide vane, as shown in Fig. 3.6, the velocity of v_{u0} can be obtained.

$$v_{u0} = v_r \cot\alpha_0 = \frac{Q}{\pi D_0 b_0}\cot\alpha_0 \tag{3.12}$$

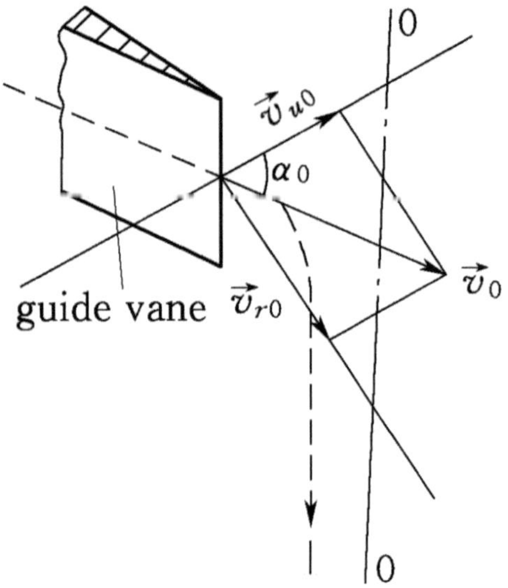

Fig. 3.6 Velocity triangle at the outlet of the guide vane

Substituting formula (3.12) into formula (3.11), it can be get

$$v_{u1i} = \frac{v_{u0}D_0}{D_i} = \frac{Q}{\pi D_i b_0}\cot\alpha_0 \tag{3.13}$$

where b_0 is the height of guide vane, m; α_0 is the place angle at the outlet of the guide vane, angle degree.

With the above three velocities, the velocity triangle at a certain point on the inlet side of the runner can be expressed. And then the absolute velocity $\vec{v}_{1i}$ and relative velocity $\vec{w}_{1i}$ as well as other velocity components can be attained.

2 **Outlet velocity triangle**

(1) Circumferential velocity $\vec{u}_{2i}$.

$$u_{2i} = \frac{\pi D_i\prime n}{60} \tag{3.14}$$

where $D_i\prime$ is the diameter of the circle at which a certain inspection point is located at the outlet side blade of the runner, m.

(2) Axial plane velocity.$\vec{v}_{m2i}$

$$\vec{v}_{m2i} = \frac{Q\eta_{vol}}{F_{2i}\varphi_2} \tag{3.15}$$

where φ_2 is the blade occupation coefficient at the outlet of the runner; F_{2i} is the cross-sectional area of the inspection point i at the outlet of the runner blade, m^2, and its calculation method is the same as F_{1i}.

(3) Place angle of relative velocity $\vec{w}_2$.

Denoted by β_2, this angle is between the relative velocity $\vec{w}_2$ and circumferential velocity $\vec{u}_2$. According to the assumption that the blade is infinitely thin, the angle is equal to the blade place angle β_{e2}, which is the angle between the bone line at the outlet side of the runner blade and the circumferential tangent of rotating circle of the runner. In fact, the blade number is finite. In the flow passage between the blades, due to its own inertia, only the water closely attached to the blade meets the above assumption, and the rest of the water flow does not change the direction of motion according to the direction of the blade twist. Therefore, the relative velocity direction at the outlet of the runner is not completely consistent with the direction of the blade. Thus, the relative velocity angle β_2 and the blade place angle at the outlet side of the runner are always somewhat different.

With the above two velocities and one velocity angle, the velocity triangle at the runner outlet can be obtained, with $\vec{v}_{2i}$, $\vec{v}_{u2i}$, $\vec{w}_{2i}$, and α_{2i} known.

3.1.3 *Velocity Triangle at the Inlet and Outlet of Kaplan Runner*

For the axial flow turbine shown in Fig. 3.7, the water flows into the runner along the axial direction and flows out of the runner also in the axial direction. Therefore, the cylindrical coordinate system can be used for flow analysis. It is assumed that the water flow in the runner is on a cylindrical surface with the main shaft as the central axis. When the viscosity of the water flow is neglected, each layer flow along a cylindrical surface does not interfere with each other. The radial velocity of the water flow $\vec{v}_r$ is zero, only the axial velocity $\vec{v}_z$ is not zero on the axis plane. Therefore, the triangle velocity expression for any point on each cylindrical surface can be simplified as

$$\vec{v} = \vec{v}_u + \vec{v}_z,\ \vec{w} = \vec{w}_u + \vec{w}_z \tag{3.16}$$

Axial plane velocity

$$\vec{v}_m = \vec{v}_z = \vec{w}_m = \vec{w}_z \tag{3.17}$$

Then, the velocity at any point in the runner can be determined by the axial and circumferential velocity components. The stream surface along the moving cylindrical surface is cut out by the blades. The expanded planar cascade flow diagram is as Fig. 3.8. At the inlet and outlet of the runner, the velocity triangle can also be drawn on the plane cascade, where

$$\vec{u}_1 = \vec{u}_2 = \vec{u}_0 \tag{3.18}$$

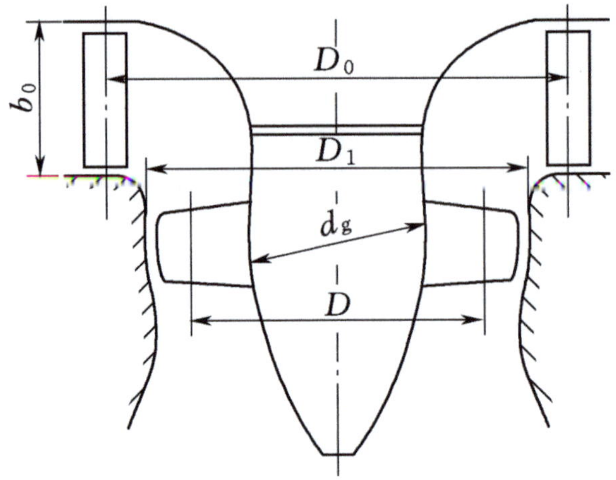

Fig. 3.7 Cylindrical surface of axial flow runner

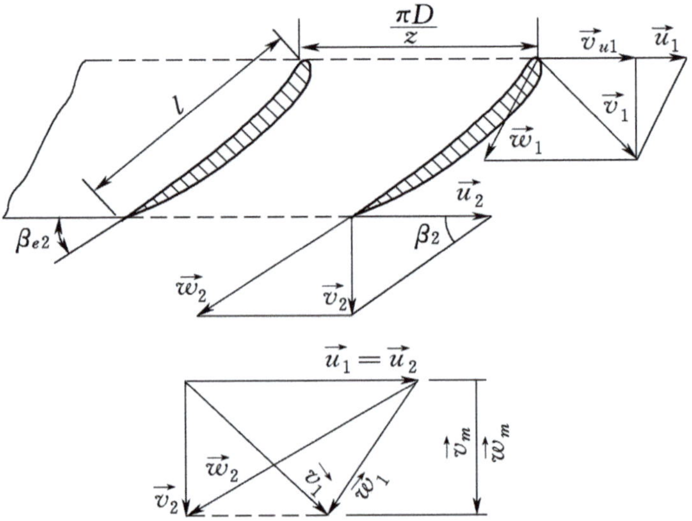

Fig. 3.8 Flow diagram of plane cascade and velocity triangle at inlet and outlet

The velocity triangle at the inlet and outlet of the runner for the axial flow turbine runner can be constructed according to Eqs. (3.17) and (3.18), and the following conditions:

(1) The velocity on an axial plane is evenly distributed, and the velocity on the cylindrical surfaces with different radius are equal, then

$$v_m = v_{m1} = v_{m2} = \frac{Q}{\left(D_1^2 - d_g^2\right)\pi/4} \tag{3.19}$$

where D_1 is the runner diameter (Fig. 3.7), m; d_h is the wheel hub diameter, m.

(2) The peripheral velocity at thc inlct and outlet of the runner is determined by

$$u_1 = u_2 = \frac{\pi D_i n}{60} \tag{3.20}$$

where D_i is the diameter of the inlet and outlet of the runner, which are located on the same cylindrical stream surface, m.

(3) The quantity of $\vec{v}_{u1}$ and the direction of $\vec{w}_2$(expressed by β_2) can be determined by $v_{u1} = \frac{Q}{\pi D_i b_0} \cot \alpha_0$ and $\beta_2 = \beta_{e2}$, according to the same analytical method as that for Francis runners.

3.1.4 Correlation Between the Inlet and Outlet Velocity Triangle and the Operating Parameters of Hydro Turbine

The shape and size of a triangle can be uniquely determined by its two sides and the angle between them. For the inlet velocity triangle, u_1 is equal to $r_1\omega$, the direction is tangent to the circumference, so u_1 depends on the rotational speed n and the runner diameter D_1. The relative velocity w_1 depends on the operating head H, the guide vane opening α, the rotational speed n, and the runner diameter D_1; and its direction angle β_1 is also determined by the guide vane opening α. Thus the relative velocity $\vec{w}_1$ is a function of the turbine operating parameters H,α, n and D_1, expressed by $\vec{w}_1 = f_{w1}(H, \alpha, n, D_1)$. Because $\vec{v}_1 = \vec{u}_1 + \vec{w}_1$, it can be stated that the shape and size of the inlet velocity triangle also depend on H, α, n, and D_1.

For a certain runner, a certain outlet velocity triangle is always corresponding to a certain inlet velocity triangle. In practice, the velocity triangle at any point in the runner is determined by the four operating parameters of hydro turbine, expressed by $\vec{v} = f_v(H, \alpha, n, D_1)$. This means the discharge Q, the efficiency η, and the output N of the turbine are also the function of the four operating parameters, H, α, n, and D_1. They can be expressed by $Q = f_Q(H, \alpha, n, D_1)$, $\eta = f_\eta(H, \alpha, n, D_1)$, and $N = f_N(H, \alpha, n, D_1)$.

Generally, the shape and magnitude of velocity triangle at the inlet and outlet can be determined, as long as D_1, n, H, and α are known. That means these four parameters can define the working condition of the turbine. Other parameters, such as N, Q, η, etc., are also the functions of D_1, n, H, and α.

However, because the flow regime in the flow passage of the runner is very complex, the previous expression cannot obtained by theoretical method. And there is neither a precise equation nor an accurate solution for the turbine flow. In practice, experimental methods must be utilized to draw various characteristic curves to express the functions of all the turbine parameters.

3.2 Basic Equation and Efficiency of Hydro Turbine

3.2.1 *Basic Equation of Hydro Turbine*

The basic equation of hydro turbine can be derived by the theorem of the moment of momentum. It is the basics of the design and operation of a turbine runner. The theorem states that the change rate of the moment of momentum of the entire flow in the runner about the main shaft is equal to the torque produced by all external forces on the same shaft.

It is assumed that the motion of the water flow in the turbine runner is axisymmetric, and the absolute velocity is equal to the sum of the three velocity components, expressed by $\vec{v} = \vec{v}_u + \vec{v}_z + \vec{v}_r$. Because the radial velocity $\vec{v}_r$ passes through the axis, and the axial velocity $\vec{v}_z$ is parallel to the main shaft, these two velocity components will not produce a velocity moment about the main shaft. Therefore, the expression is

$$\frac{\mathrm{d}(mv_u r)}{\mathrm{d}t} = \sum M_\omega \tag{3.21}$$

where $\sum M_\omega$ is the total torque of all external forces about the main shaft acting on the mass of the water body, N m; m is the total mass of water passing through the turbine runner in dt time period, kg m can be calculated by

$$m = \rho Q_e \mathrm{d}t = \frac{\gamma Q_e}{g}\mathrm{d}t \tag{3.22}$$

where Q_e is the effective discharge of the hydro turbine, namely the flow rate passing through the runner, m^3/s.

According to the continuity theorem, the difference between the momentum moment $\frac{\gamma Q_e}{g} v_{u1} r_1$ flowing into the inlet edge of the runner per unit time and the momentum moment $\frac{\gamma Q_e}{g} V_{u2} r_2$ flowing away from the outlet edge of the runner per unit time is equal to the increment of momentum moment $\frac{\mathrm{d}(mv_u r)}{dt}$ of the total water mass m passing through the runner per unit time. This may be written in the form of

$$\frac{\mathrm{d}(mv_u r)}{\mathrm{d}t} = \frac{\gamma Q_e}{g}(v_{u2} r_2 - v_{u1} r_1) \tag{3.23}$$

The external torque includes the gravity torque M_g, the pressure torque M_p, the friction torque M_f, and the interaction torque M_0 from the runner acting on the water, namely

$$\sum M_\omega = M_g + M_p + M_f + M_0 \tag{3.24}$$

About the main shaft, it is obvious that $M_g = M_p = 0$, and the friction torque is negligible, $M_f \approx 0$, therefore

$$\sum M_\omega = M_0 = -M_e = -\frac{N_e}{\omega} = -\frac{\gamma Q_e H_e}{\omega} = -\frac{\gamma Q_e H \eta_h}{\omega} \tag{3.25}$$

where M_e is the reaction torque acting on the water from the blades of the runner, N m; ω is the angular speed of the runner rotation, rad/s.

Substituting Eqs. (3.25) and (3.23) into (3.21) yields

$$-\frac{\gamma Q_e H \eta_h}{\omega} = \frac{\gamma Q_e}{g}(v_{u2} r_2 - v_{u1} r_1) \tag{3.26}$$

According to Eq. (3.26), there are four kinds of expressions for the basic equation of a hydro turbine, namely

$$\begin{aligned} H\eta_h &= \frac{\omega}{g}(v_{u1} r_1 - v_{u2} r_2) \\ H\eta_h &= \frac{1}{g}(v_{u1} u_1 - v_{u2} u_2) \\ H\eta_h &= \frac{1}{g}(v_1 u_1 \cos\alpha_1 - v_2 u_2 \cos\alpha_2) \\ H\eta_h &= \frac{\omega}{2\pi g}(\Gamma_1 - \Gamma_2) \end{aligned} \tag{3.27}$$

where $\Gamma_1 = 2\pi v_{u1} r_1$, denoting the velocity circulation of the water flow at the inlet of the runner; and $\Gamma_2 = 2\pi v_{u2} r_2$, denoting the velocity circulation of the water flow at the outlet of the runner.

The basic equations for turbine can be interpreted as follows:

(1) The energy conversion is achieved by the change of velocity moment or velocity circulation. If $v_{u1} r_1 - v_{u2} r_2 = 0$, it is obvious there is no energy to be converted. In order to make the conversion efficiency η_h higher, Γ_1 is required to be larger as far as possible, and $\Gamma_2 = 0$. As for $\Gamma_1 = 2\pi v_1 \cos\alpha_1 r_1$, it requires $\cos\alpha_1 = 1$, that is $\alpha_1 = 0°$. But it is impossible for $\alpha_1 = 0°$, because on this condition, the water cannot enter into the runner according the velocity triangle at the inlet. Therefore, in practice it can be assured there is no water impact at the inlet, and the relative velocity angle equals the blade place angle, namely $\beta_1 = \beta_{e1}$. On the other hand, $\Gamma_2 = 0$ and $\cos\alpha_2 = 0$ are applicable in practice, which needs

$\alpha_2 = 90°$. This is often called the normal outlet, because the absolute velocity is normal to the outlet of the runner.

For large and medium-sized reaction turbines, the velocity moment at the runner inlet is formed by the spiral case and guide vanes. For turbines without spiral cases and guide vanes, the velocity circulation at the runner inlet is formed only by the runner itself. When the water flows through the runner blades, the velocity moment is gradually reduced, finally to $v_{u2}r_2$ at the runner outlet. $v_{u2}r_2$ is small or even zero. Therefore, the reasonable shape of the runner blade is the key factor to ensure the smooth reduction of the velocity moment.

(2) When the application head of the turbine is high, to seek high efficiency it is necessary to try to make $u_1 > u_2$, which results in the inlet diameter D_1 larger than the outlet diameter D_2.
(3) When deviating from the optimal inlet and outlet conditions, the efficiency of a hydro turbine η_h may be very low, even if the turbine is just rotating with no output.
(4) According to the velocity triangles at the inlet and outlet, the following relationship can be obtained

$$w_1^2 = v_1^2 + u_1^2 - 2u_1 v_1 \cos\alpha_1$$

$$w_2^2 = v_2^2 + u_2^2 - 2u_2 v_2 \cos\alpha_2$$

The basic equation of the turbine can be rewritten as

$$H\eta_h = \frac{v_1^2 - v_2^2}{2g} + \frac{u_1^2 - u_2^2}{2g} - \frac{w_1^2 - w_2^2}{2g} \tag{3.28}$$

The first two terms on the right side of Eq. (3.28) are called the kinetic heads, and the third one is called the potential head. The potential head is mainly used to overcome the centrifugal force generated by the rotation of the water flow and to accelerate the relative motion of the water flow in the runner.

For axial flow turbines, the efficiency of η_h depends on the absolute velocity and relative velocity due to $u_1 = u_2$. However, the absolute velocity difference and the relative velocity difference between the runner inlet and outlet cannot be excessively increased, otherwise, the hydraulic loss will be increased, which limits the head scope of application for the axial flow turbine.

Finally, it should be noted that in the basic equation of a hydro turbine, the relationship between the flow velocity and the head H is not obvious, so it is needed to do some transformation according to the sine theorem of the triangle.

From $\frac{v}{\sin\beta} = \frac{w}{\sin\alpha} = \frac{u}{\sin(180^\circ - \alpha - \beta)}$, the following formulas can be obtained

$$v_1 = u_1 \frac{\sin\beta_1}{\sin(\alpha_1 + \beta_1)} \quad v_2 = u_2 \frac{\sin\beta_2}{\sin(\alpha_2 + \beta_2)}$$

Combining with $\omega = \frac{u_1}{r_1} = \frac{u_2}{r_2}$ and eliminating v_1, v_2 and u_2 from the basic equation, we get

$$u_1 = k_{u1}\sqrt{2gH\eta_h} \tag{3.29}$$

In which the velocity coefficient can be written as

$$k_{u1} = \sqrt{\frac{1}{2\left[\frac{\sin\beta_1 \cos\alpha_1}{\sin(\beta_1+\alpha_1)} - \left(\frac{r_2}{r_1}\right)^2 \frac{\sin\beta_2 \cos\alpha_2}{\sin(\beta_2+\alpha_2)}\right]}}$$

In the similar way, the similar relations of velocities u_2, v_1, v_2, w_1, and w_2 with head H can be obtained.

Obviously, we have

$$k_{u1} = f_{u1}\left(\alpha, \beta, \frac{r_2}{r_1}\right) \tag{3.30}$$

Analysis of the velocity coefficient as follows.

(1) Equation (3.29) is consistent with the expression of orifice flow, then the symbol k_{u1} is called velocity coefficient. But the expressions of its coefficients is much more complicated for (3.30). They not only related to the velocity triangle, but also related to the type of runner and the ratio of the inlet and outlet diameters.
(2) For a certain turbine, as the operating condition varies, the velocity direction angles α and β change, leading to the change of k_{u1}.
(3) For two turbines with geometry similarity, if the velocity triangles are similar, the velocity coefficients are equal.
(4) For a certain turbine, the velocity coefficient depends on the velocity triangle, which in turn depends on H, n and a, therefore $k_{u1} = f_{u1}(H, n, \alpha, D_1)$.
(5) Using two velocity coefficients, such as k_{u1} and k_{v1}, can determine the shape of the velocity triangle, namely the operating conditions of hydro turbine.

3.2.2 Efficiency and Optimal Operating Condition of Hydro Turbine

In the process of hydro turbine converting water energy into rotational mechanical energy of the turbine, there are hydraulic losses, volume losses, and mechanical losses. The corresponding efficiency is called hydraulic efficiency, volumetric efficiency, and mechanical efficiency, respectively.

1. **Hydraulic efficiency**

The specific water energy during water flowing through the hydro turbine cannot be fully converted into the rotational mechanical energy, which is output by the main shaft. Some of the energy is consumed by the head loss Σh_t in overcoming various hydraulic resistances, such as frictional resistance along the path and local resistance due to vortex and separation. The hydraulic efficiency η_h is the ratio of the effective head $H_e = H - \Sigma h_t$ and the working head H of the hydro turbine, namely

$$\eta_h = \frac{H_e}{H} = 1 - \frac{\Sigma h_t}{H} \tag{3.31}$$

2. **Volumetric efficiency**

The discharge Q flowing into the hydro turbine does not fully enter the runner blades, and a small portion of the flow Σq leaks from the gap between the rotating and stationary parts of the hydro turbine. The gap includes the leakage-stop ring gap of the Francis turbine and the gap between the runner blades and runner chamber of the Kaplan turbine. The volumetric efficiency η_{vol} is equal to the ratio of the effective discharge $Q_e = Q - \Sigma q$ entering the runner blades and the total discharge Q entering the turbine, namely

$$\eta_{vol} = \frac{Q_e}{Q} = 1 - \frac{\sum q}{Q} \tag{3.32}$$

3. **Mechanical efficiency**

A portion of the effective output N_e of water to runner is consumed by various mechanical losses ΔN_{mec}, which includes the friction at the bearing and shaft seals, friction at the outer surface of the runner from the surrounding water (wheel loss), etc. The mechanical efficiency η_{mec} is equal to the ratio of the main shaft output N and the effective output N_e of the runner, namely

$$\eta_{mec} = \frac{N}{N_e} = 1 - \frac{\Delta N_{mec}}{N_e} \tag{3.33}$$

Turbine efficiency is equal to the product of hydraulic efficiency, volumetric efficiency, and mechanical efficiency, namely

$$\eta = \eta_h \eta_{vol} \eta_{mec} \tag{3.34}$$

Figure 3.9 shows the relation between turbine efficiency and output when the reaction turbine changes its discharge under the condition with speed n, runner diameter D_1, and working head H given. The figure also shows the variation of various losses with the output. It can be seen that the hydraulic efficiency changes greatly with the output, while the volumetric efficiency and mechanical efficiency change little. There is a maximum output point of total efficiency. The turbine efficiency is a comprehensive index to measure the performance of turbine energy conversion. It is related to turbine type, structure dimension, manufacturing technology, and operating conditions. The highest efficiency of modern hydro turbines can reach 0.90–0.95, which is usually obtained by theoretical calculation based on turbine model test results.

4. **Optimal operating conditions of hydro turbine**

During the operation of hydro turbine, with the variation of external conditions (head, load), the parameters such as the working head, discharge, and output of the turbine change. The flow patterns in the runner passage also change. Therefore, the turbine can be operated under different conditions, and the working condition corresponding to the highest efficiency is called the optimal working condition. It is known that the main factor affecting efficiency is hydraulic loss. The impact loss and vortex loss account for the main part of hydraulic loss, and their quantity depends on the direction of the inlet flow and outlet flow of the runner.

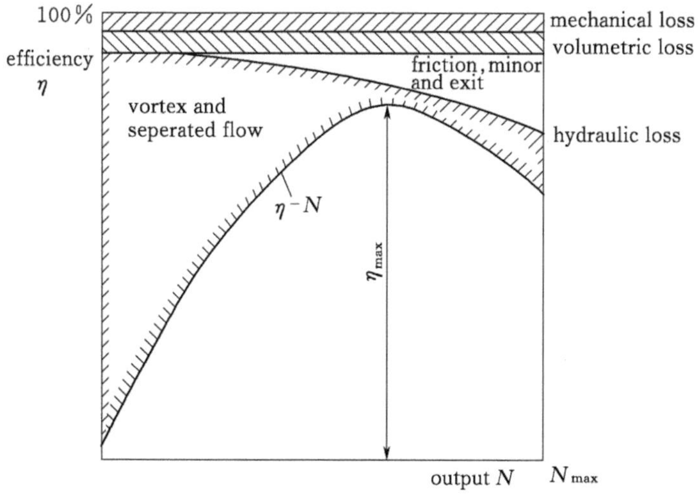

Fig. 3.9 Schematic of turbine efficiency versus output

In the analysis of the basic equations, it was pointed out that the optimal condition for the inlet flow is no impact, i.e., $\beta_1 = \beta_{e1}$, and for the outlet flow is a normal outlet, i.e., $\alpha_2 = 90°$. Under this condition, there is no the impact loss and the vortex loss. Therefore, this operating condition is the optimal operating condition of the hydro turbine, and the efficiency is the maximum.

For Kaplan turbines, the blade angle can be adjusted as the working condition changes. Therefore, this type of turbine is easy to reach or close to the optimal working condition and its high-efficiency area is relatively wide.

The inlet and outlet conditions of the optimal operating point of hydro turbine are obtained under the ideal assumptions mentioned above. But in practice, there are some differences, and most of the operating conditions are non-optimal conditions. How to widen the high-efficiency zone of the turbine is one of the key issues in the hydraulic design of hydro turbines.

(1) Optimal inlet. When the relative velocity $\vec{w}_1$ of the water flow at the inlet of the runner is inconsistent with the tangential direction of the blade bone line at the same point, an attack angle α of water flow is formed with positive or negative value, as shown in Figs. 3.10 and 3.11.

$$\alpha = \beta_1 - \beta_{e1} \tag{3.35}$$

In the case of an impact inlet, separation flow and vortex flow appear in the inlet region of the blade, resulting in impact loss. The impact component is $\vec{w}_c$ and the impact loss can be estimated by $w_c^2/2g$. Generally, when the attack angle α is small, the impact loss is small. Therefore, in the hydraulic design of a

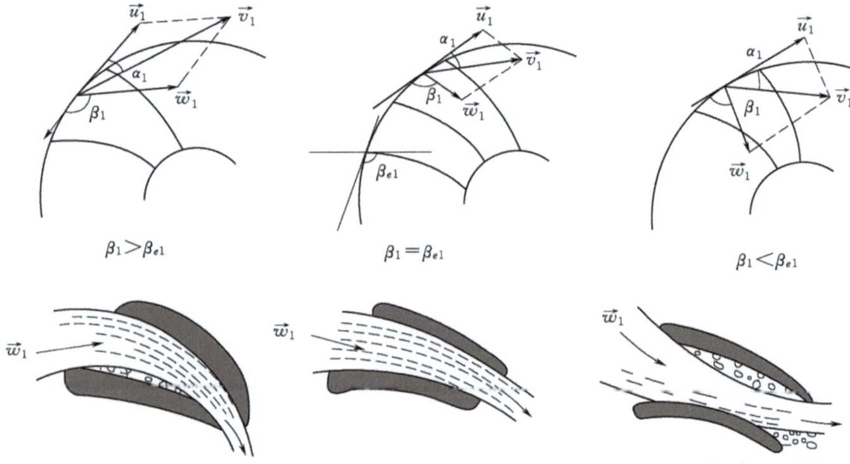

Fig. 3.10 Flow patterns at the inlet of the runner blade

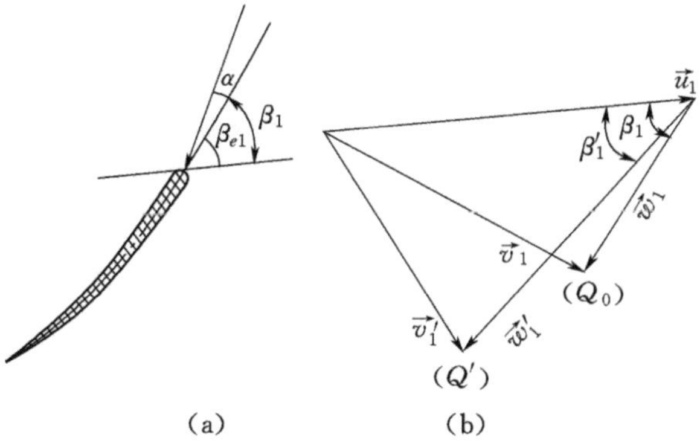

Fig. 3.11 Attack angle at the round edge of the blade inlet

hydro turbine, the inlet blade angle is generally slightly smaller than the relative velocity angle, as shown in Fig. 3.11a. Usually $\alpha = 3°$ to $10°$ is taken.

A certain positive attack angle is considered to improve the hydraulic performance of a hydro turbine under large discharge condition, because the turbine is operated longer in this area. The turbine blades are generally of hydrofoil type, and the leading edge of the blade is a round head. When the inlet water flow has little impact on the front of the blade, the hydraulic impact loss is small. However, on the back of the blade, even if the inlet flow impact is not large, the flow of water will be separated on the front side of the blade, which will deteriorate the flow patterns in the flow channel and greatly increase the hydraulic loss. If the hydro turbine is designed strictly according to no-impact condition $\beta_{e1} = \beta_1$, when running with a large discharge deviating from the design condition, the discharge is larger than the design discharge $Q\prime > Q_0$, and the inlet relative velocity angle is smaller than the design relative velocity angle $\beta_1\prime < \beta_1$, as shown in Fig. 3.11b, which will lead to a sharp drop in efficiency because of the impact on the back of the blade, or negative impact.

Therefore, a small positive attack angle $\alpha > 0°$ is usually used for the design condition. Although the hydraulic efficiency of the turbine under the design condition is slightly decreased, the hydraulic performance under large discharge condition is significantly improved.

In addition, a small positive attack angle can reduce the bending of the entire blade and help to improve the hydraulic performance when the blade inlet angle β_{e1} is slightly smaller than the inlet relative velocity angle β_1.

(2) Optimal outlet. It is theoretically known that when the absolute velocity angle is equal to 90°, or the whirl velocity component is zero, the hydraulic efficiency η_h is the maximum. The following is a further analysis from the perspective of the outlet velocity circulation.

a. Influence on the energy utilization of draft tube. The greater the circulation of the water flow at the outlet of the runner, the more violent the water whirling flow in the draft tube. The kinetic energy at the draft tube inlet is composed of two parts: rotational kinetic energy and axial plane kinetic energy, namely

$$\frac{v_2^2}{2g} = \frac{v_{u2}^2}{2g} + \frac{v_{m2}^2}{2g} \tag{3.36}$$

It is obvious that the axial plane velocity v_{m2} of the water flow is inversely proportional to the square of the radius of the draft tube, and the circumferential velocity v_{u2} of the water flow is only inversely proportional to the radius. When the water flows along the draft tube, the decay of axial plane velocity v_{m2} is much faster than circumferential velocity v_{u2}, indicating that the draft tube can more effectively utilize the kinetic energy of the motion on the axial plane. Therefore, when the outlet velocity circulation or v_{u2} is larger, the effect of the energy recovery from the draft tube is poor, and the utilization of the kinetic energy of the draft tube is low.

b. Influence on the inlet flow of the runner. It can be known from formula $H\eta_h = \frac{\omega}{2\pi g}(\Gamma_1 - \Gamma_2)$ that if the outlet circulation Γ_2 of the runner increases, the inlet circulation Γ_1 of the runner is needed to increase accordingly. Due to the increase of v_{u2}, the water can flow close to the wall of the diffusion-shaped draft tube, driven by the centrifugal force. It is not easy to cause flow separation or stagnant region, which can decrease the energy loss in the draft tube. At the same time, due to the increase of the inlet and outlet velocity circulations of the runner, the relative velocities w_1 and w_2 will decrease accordingly, thereby reducing the hydraulic loss in the runner. Of course, this can result in an increase in hydraulic loss in the spiral case of a hydro turbine. However, for medium and low-head Francis turbines and Kaplan turbines, the hydraulic loss in the spiral case accounts for a small proportion of the total energy loss, and the main losses occur in the runner and the draft tube. Therefore, it is beneficial for the runner outlet with a slightly positive velocity circulation v_{u2}, namely α_2 is slightly less than 90°, as shown in Fig. 3.12b. For high head Francis turbine, a slightly positive velocity circulation v_{u2} can also improve the flow patterns in the draft tube, but the correspondingly v_{u1} and v_1 will increase, leading to the increase of hydraulic loss in the spiral case and guide vanes. For this type of turbine, the hydraulic loss in the spiral case and guide vanes accounts for a large proportion of the total hydraulic loss of the turbine. Therefore, the normal outflow condition is advantageous and can minimize the total hydraulic loss in the turbine.

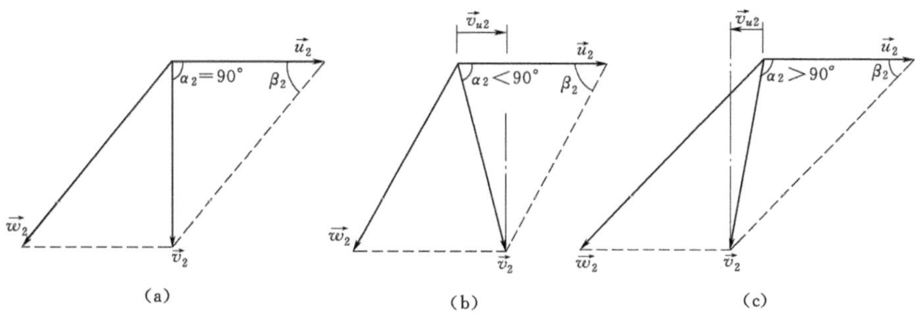

Fig. 3.12 Comparison of velocity triangle at the outlet of runner blade

3.2.3 *Working Principle of Impulse Turbine*

1. **Operating characteristics of impulse turbine**

The impulse turbine only uses the kinetic energy of water flow to do work. The nozzles at the end of the penstock convert the pressure energy of water flow into the kinetic energy, forming high-speed free jets impacting the runners of the turbine, thereby converting the water energy into the mechanical energy output of the runner shaft. Compared with the reaction turbine, the impulse turbine has the following characteristics.

(1) For the impulse turbine, the nozzle is like the water guiding mechanism of the reaction turbine. Its function is to guide the flow of water, regulate the discharge, and convert the water energy into the kinetic energy of the free jet. While for the reaction turbine, in addition to guiding the water flow and adjusting the discharge, the water guiding mechanism can also form a certain velocity circulation before entering the runner to meet the requirements of turbines with different specific speeds.
(2) In an impulse turbine, the entire flow of water from the nozzle to the outlet of the runner is always exposed in the air. Therefore, the pressure of the water flow in each part remains unchanged, and it is the atmospheric pressure. While in the reaction turbine, from the inlet of the spiral case to the outlet of the draft tube, the water flow in each part is always pressurized and the pressure is variable along streamlines.
(3) For an impulse turbine, the velocity of the water leaving the runner is very small, and it is under the atmospheric pressure, therefore, there is no need to install a draft tube. While the reaction turbine needs to be equipped with a draft tube to restore the pressure, to reduce the kinetic energy loss at the outlet of the runner, and to further utilize the water energy between the runner and the downstream water surface. The impulse turbine cannot utilize this portion of the water energy between the runner and the downstream surface, but the reaction turbine can.
(4) The runner of the impulse turbine is only partially operated by the water bucket, therefore the discharge is small, and the output is small under a certain head and

runner diameter. And the rotating speed is relatively low, due to the large absolute velocity and small peripheral velocity at the runner inlet. The combined effect of the two aspects makes the impulse turbine suitable for hydropower stations with high head and small discharge.

2. **Velocity triangle at the inlet and outlet of the runner**

The flow patterns of the free jet impinging into the runner bucket are extremely complicated. Therefore, to establish the velocity triangle at the inlet and outlet of the runner bucket, the following assumptions are made.

(1) The velocity triangle is established by taking the water flow particle located at the nominal diameter D_1 of the runner as the representative point.
(2) The movement of the bucket entering the jet action zone is considered to be a straight path motion parallel to the jet axis, and the bucket is considered to receive all of the flow from the nozzle.
(3) The jet flow in the bucket is considered to be an in-plane motion parallel to the runner axle and the jet axis.

Based on the above assumptions, the velocity triangle at the inlet and outlet of the runner bucket can be obtained, as shown in Fig. 3.13.

The inlet velocity triangle can be determined by the following conditions.

(1) The convected velocity or circumferential velocity on the inlet edge of the bucket at the nominal diameter D_1 of the runner can be written in the form of $u_1 = \pi D_1 n/60$, and the direction corresponds to the rotational speed ω.
(2) The absolute velocity of this point, subject to the above assumption, will be given similarly as $v_1 = 4Q_j/\pi d_0^2$, in which Q_j is the discharge through one nozzle and d_0 is the jet diameter. The direction of v_1 is the same with u_1.

Thus the velocity triangle at the runner inlet can be determined by absolute velocity $\vec{v}_1$ and convected velocity $\vec{u}_1$.

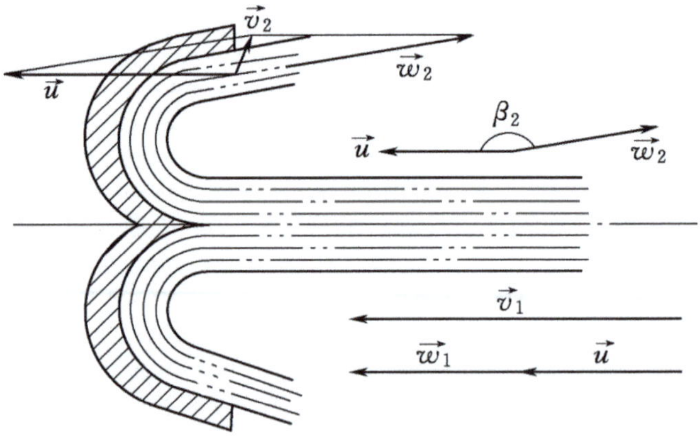

Fig. 3.13 Velocity triangle at the inlet and outlet of the runner bucket

The outlet velocity triangle can be determined by the following conditions.

(1) The circumferential velocity on the outlet edge of the bucket at the nominal diameter D_1 of the runner can be written as $u_2 = u_1 = \pi D_1 n/60$, and the direction is the same as $\vec{u}_1$.
(2) Given the friction loss of the water flow on the bucket is neglected, then there exists $w_2 \approx w_1$, and the water flow leaves the bucket closely against the outlet edge of the bucket. Therefore, the relative velocity angle β_2 at the outlet is determined by the geometric angle of the bucket working surface at the bucket outlet. Therefore, for a bucket with a certain shape and size, the relative velocity angle β_2 is known.

Thus the velocity triangle at the runner outlet can be determined by displacement velocity $\vec{u}_2$ and relative velocity $\vec{w}_2$.

3. **The basic equation for Pelton turbine**

It can be seen from Fig. 3.13 that the velocity quantity and direction of the jet flow are changed from entering to leaving the bucket. Known by Newton's theorem, the water body must be affected by external forces. Obviously, this external force is the force of the bucket on the jet flow.

Let the force of the jet flow acting on the bucket is $\vec{F}_R$, then the force of the bucket acting on the jet flow is $-\vec{F}_R$. Assuming that the runner rotates at a constant speed within a relatively short time dt, which can be considered as the time between the jet flow entering and leaving the bucket. According to the momentum theorem, there is $-F_R = \rho Q(w_2 \cos\beta_2 - w_1)$, then

$$F_R = \rho Q(w_1 - w_2 \cos\beta_2) \tag{3.37}$$

where w_1 is the relative velocity of the water flow when the jet is near the inlet of the bucket but has not yet reached the bucket, m/s; w_2 is the relative velocity of the water flow as it exits the bucket but is still close to the outlet edge, m/s; ρ is the water density, kg/m^3; Q is the discharge of a single nozzle, m^3/s.

Given the hydraulic friction loss of the jet flow on the working face of the bucket is ignored, then

$$w_2 = w_1\ w_1 = v_1 - u_1;\ u = u_1 = u_2 \tag{3.38}$$

Substituting Eqs. (3.38) into (3.37), we can get

$$F_R = \rho Q(v_1 - u)(1 - \cos\beta_2) \tag{3.39}$$

Equation (3.39) shows that, in the case of a certain discharge Q, jet velocity v_1 and the bucket outlet edge angle β_2, the force F_R of the jet acting on the bucket, also called the circumferential force, is linear with the circumferential velocity u of the runner. When the bucket meets with the jet flow but has not rotated, the runner receives the maximum circumferential force, namely $u = 0$, and $F_R = F_{R,\max}$. When

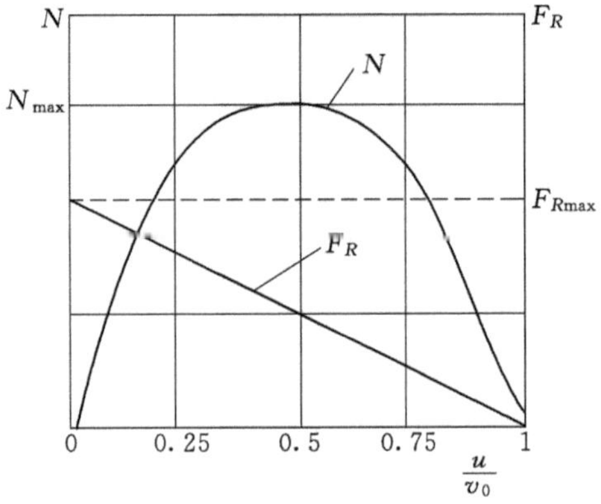

Fig. 3.14 Relation between circumferential force, shaft output and speed ratio

the circumferential velocity of the runner is increased to the same velocity as the jet, the force of the jet flow on the bucket is zero, namely $u = v_1$, $F_R = 0$, as shown in Fig. 3.14.

When the runner rotates at a circumferential velocity u under the action of a circumferential force F_R, the shaft output is:

$$N = F_R u = \rho Q(v_1 - u)(1 - \cos\beta_2)u \tag{3.40}$$

The above equation shows the relationship between the output of the runner and the velocity and bucket angle at the inlet and out edge, namely, it is the basic equation for the impulse turbine. The left side of the equation represents the shaft output of the runner, and the right side of the equation represents the variation of energy before and after the jet flowing around the bucket.

Substituting $N = \rho g Q H\eta$ into Eq. (3.40) yields

$$H\eta = \frac{1}{g}(v_1 - u)(1 - \cos\beta_2)u \tag{3.41}$$

Comparing the basic Eq. (3.41) of impulse turbine with the basic Eq. (3.42) of reaction turbine

$$H\eta_h = \frac{1}{g}(v_1 u_1 \cos\alpha_1 - v_2 u_2 \cos\alpha_2) \tag{3.42}$$

We know that when $u = u_1 = u_2$, $\alpha_1 = 0$,$u_2 \cos\alpha_2 = u - w_2 \cos\beta_2\prime$,$\beta_2\prime = \pi - \beta_2$,$w_2 = w_1 = v_1 - u$, it can be get that, $v_1 u_1 \cos\alpha_1 = v_1 u$, $v_2 u_2 \cos\alpha_2 = [u + (v_1 - u)\cos\beta_2]u$. Substituting them to Eq. (3.42), we obtain

$$H\eta_h = \frac{1}{g}[u(v_1 - u)(1 - \cos\beta_2)] \tag{3.43}$$

Equation (3.43) is consistent with Eq. (3.41), indicating that the reaction turbine and the impulse turbine not only have the same basic principle but also have the same basic equation.

4. **Optimal velocity ratio**

For a certain impulse turbine, the discharge Q, jet velocity v_1 and bucket outlet edge angle β_2 are determined. It can be seen from Eq. (3.40) that the runner output N converting from the jet flow is a quadratic function of the circumferential velocity u. Obviously, let $\mathrm{d}N/\mathrm{d}u = 0$, the u value when the shaft output can reach the maximum can be found. According to the equation $\frac{\mathrm{d}N}{\mathrm{d}u}\left(uv_1 - u^2\right) = v_1 - 2u = 0$, we can get

$$u = v_1/2 \tag{3.44}$$

Equation (3.44) shows that when the circumferential velocity of the runner is half of the jet velocity, the energy obtained by the runner from the free jet flow reaches the maximum.

The relationship between the shaft output N and the velocity ratio in the form of $\psi = u/v_1$ is also plotted in Fig. 3.14. Base on the curve we have the following conclusions: when the circumferential velocity $u = 0$, and $u = v_1$, the runner output $N = 0$; when peripheral velocity $u = v_1/2$, the output is maximum.

In the previous discussion, the friction loss of the jet flow on the bucket working surface was neglected. In fact, the friction loss exists due to the viscosity of the water, therefore $w_1 \neq w_2$. In addition, in order to ensure good drainage condition, the bucket outlet edge angle β_2 is less than 180°, generally $\beta_2 = 175^{\circ}$ to 176°, which will reduce the efficiency. Therefore, the optimal velocity ratio of the Pelton turbine is not equal to 0.5. According to the test data, the optimal speed ratio ψ is between 0.46 and 0.49. For Pelton turbines with high specific speed n_s, ψ get smaller values; and for Pelton turbines with low specific speed n_s, ψ get larger values.

It must be noticed that, according to recent relevant reports, German researchers have found that the optimal velocity ratio ψ is between 0.54 and 0.58 through experimental research on hydraulic loss in the bucket, breaking the traditional theoretical constraints and increasing the velocity ratio, which is of great value for reducing the weight of the unit and reducing the cost of the product.

Pelton turbine efficiency theoretically may be expressed by the following formula

$$\eta = 2\psi\varphi^2(1 - \psi)(1 - \zeta\cos\beta_2) \tag{3.45}$$

where ψ is the velocity ratio, φ^2 is the efficiency of the nozzle, ζ is the energy loss coefficient of the jet on the surface of the bucket, and β_2 is the bucket outlet edge angle.

It can be seen from Eq. (3.45) that the efficiency of Pelton turbines can be related to the speed ratio, nozzle efficiency, energy loss coefficient of the jet on the surface of the bucket, and the bucket outlet edge angle. Under normal conditions, the total efficiency of Pelton turbines is 85% the 90%, and the efficiency varies gently in the operating range.

3.3 Working Principle of Draft Tube

3.3.1 Function of Draft Tube

The draft tube is an important part of the reaction turbine, and its performance directly affects the efficiency and stability of the turbine. The shape of the draft tube is related to the development mode of hydropower, ground or underground, and the water supply system arrangement.

To illustrate the working principle of the draft tube, Fig. 3.15 shows three different turbine installations: (1) without draft tube; (2) with a cylindrical draft tube; (3) with a diffusion draft tube.

In the above three cases, the water energy that the runner can utilize can be expressed as

$$\Delta E = E_1 - E_2 = \left(H_d + \frac{p_a}{\rho g}\right) - \left(\frac{p_2}{\rho g} + \frac{\alpha_2 v_2^2}{2g}\right) \quad (3.46)$$

where ΔE is the specific energy difference before and after the runner, m; H_d is the static head at the entrance of the runner, m; p_a is the atmospheric pressure, N/m^2; p_2 is the pressure at the outlet of the runner, N/m^2; v_2 is the velocity at the outlet of the runner, m/s.

However, the pressure p_2 and velocity v_2 at the outlet of the runner are different in the three cases, therefore, the energy used by the runner is also different.

1. **No draft tube**

This situation is shown in Fig. 3.15a. The pressure of the runner outlet is the atmospheric pressure, $p_2 = p_a$. Substitute this into Eq. (3.46) to give the runner flow energy utilization of $\Delta E'$ as

$$\Delta E' = H_d - \frac{\alpha_2 v_2^2}{2g} \quad (3.47)$$

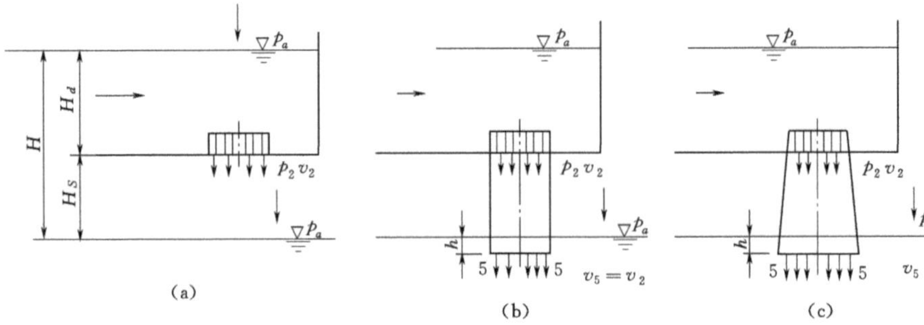

Fig. 3.15 Schematic for analyzing the function of draft tube

Equation (3.47) shows that the runner only utilizes the H_d part of the total head of the hydropower station without a draft tube, the elevation difference H_s between the runner outlet and the downstream water surface is not utilized, and the kinetic energy of $\frac{\alpha_2 v_2^2}{2g}$ at the outlet of the runner is all lost.

2. **A cylindrical draft tube**

This situation is shown in Fig. 3.15b. In order to find the pressure p_2 at the outlet of the runner, the Bernoulli equation between the runner outlet Sect. 3.2 and the draft tube outlet Sect. 3.5 is established as

$$\frac{p_2}{\rho g} + H_s + h + \frac{\alpha_2 v_2^2}{2g} = \frac{p_a}{\rho g} + h + \frac{\alpha_5 v_5^2}{2g} + h_w \tag{3.48}$$

where h is the elevation difference between the downstream water surface and the exit Sect. 3.5 of the draft tube, m; h_w is the hydraulic loss in the draft tube, m; v_5 is the average velocity of the draft tube outlet Sect. 3.5.

For a cylindrical draft tube, there exists $v_2 = v_5$. Substitute this into Eq. (3.48), then get

$$\frac{p_a - p_2}{\rho g} = H_s - h_w \tag{3.49}$$

This means the static vacuum $\frac{p_a - p_2}{\rho g}$ is formed by H_s under the action of the cylindrical draft tube.

Substituting Eqs. (3.49) into (3.46), we obtain the energy $\Delta E''$ utilized by the runner when the cylindrical draft tube is set.

$$\Delta E'' = (H_d + H_s) - \left(\frac{\alpha_2 v_2^2}{2g} + h_w\right) \tag{3.50}$$

Compared Eq. (3.50) with Eq. (3.47), the cylindrical draft tube utilizes the elevation difference H_s, also known as the suction head, between the exit of the runner and the downstream water surface. However, the kinetic energy $\frac{\alpha_2 v_2^2}{2g}$ at the exit of the runner is still lost, and there is a hydraulic loss h_w in the draft tube, but h_w usually much smaller than H_s. Therefore, a cylindrical draft tube can make more energy of $H_s - h_w$ by forming the static vacuum.

3. **A diffusion draft tube**

In this case, as shown in Fig. 3.15c, in order to obtain the pressure p_2 at the outlet of the runner, it is still necessary to use the Bernoulli equation between the runner outlet Sect. 3.2 and the draft tube outlet Sect. 3.5, as shown in Eq. (3.48).

For the diffusion draft tube $v_2 > v_5$, the static vacuum Eq. (3.48) can be rewritten as

$$\frac{p_a - p_2}{\rho g} = H_s + \left(\frac{\alpha_2 v_2^2 - \alpha_5 v_5^2}{2g} - h_w\right) \tag{3.51}$$

Comparing Eqs. (3.51) and (3.49), it is known that, in addition to using $H_s - h_w$ to form a static vacuum at the exit of the runner, a kinetic vacuum of value $\frac{\alpha_2 v_2^2 - \alpha_5 v_5^2}{2g}$ is added, which is generated by the diffusion draft tube, the velocity is reduced from the inlet v_2 to the outlet v_5.

Substituting Eqs. (3.51) into (3.46) to obtain the energy $\Delta E''$ utilized by the runner when using the diffusion draft tube, namely

$$\Delta E''' = (H_d + H_s) - \left(\frac{\alpha_5 v_5^2}{2g} + h_w\right) \tag{3.52}$$

Comparing Eqs. (3.52) and (3.50), the exit kinetic energy loss is reduced from $\frac{\alpha_2 v_2^2}{2g}$ to $\frac{\alpha_5 v_5^2}{2g}$ when using a diffusion draft tube to replace a cylindrical draft tube. The value of $\frac{\alpha_2 v_2^2 - \alpha_5 v_5^2}{2g}$ is known as the additionally utilized kinetic vacuum at Sect. 3.2. Note that the hydraulic loss in this type of draft tube also exists, therefore the actually recovered kinetic energy in Sect. 3.2 is $\frac{\alpha_2 v_2^2 - \alpha_5 v_5^2}{2g} - h_w$.

In summary, the draft tube of a reaction turbine can reduce the energy loss in the way of recovering the potential energy and part of the kinetic energy by reducing the pressure energy at the outlet of the runner. The functions of the draft tube are summarized as the following three points: ① directing the water flow at the outlet of the runner to the downstream; ② using the elevation difference between the downstream water level and the outlet section of the runner to form a static vacuum at the outlet of the runner; ③ using the kinetic energy at the outlet of the runner and converting it to a kinetic vacuum at the outlet of the runner.

3.3.2 *Kinetic Recovery Coefficient of Draft Tube*

It should be pointed out that adding H_s is meaningless for recovering more lost energy. As can be seen from Fig. 3.15, the turbine working head $H = H_d + H_s$, increasing H_s results in a decrease in H_d. The role of the draft tube is only to ensure the recovery of the potential energy H_s. The suction head H_s depends on the turbine installation elevation but has no direct relationship with the performance of the draft tube. Therefore, the main index for measuring the performance of the draft tube is the kinetic energy recovery coefficient of the draft tube.

In order to evaluate the effectiveness of the kinetic energy recovery of the diffusion draft tube, one may make assumptions that there is no hydraulic loss in the draft tube $h_w = 0$; the outlet section is infinite, namely, there is no kinetic energy loss $\frac{\alpha_5 v_5^2}{2g} \approx 0$. Thus the ideal kinetic vacuum at Sect. 3.2 is equal to the whole kinetic energy $\frac{\alpha_2 v_2^2}{2g}$ at the runner outlet, which is called the ideal kinetic energy recovery.

The ratio of the actual kinetic energy recovery of the draft tube to the ideal kinetic energy recovery is called the kinetic energy recovery coefficient η_w of the draft tube, denoted as

$$\eta_w = \frac{\frac{\alpha_2 v_2^2}{2g} - \left(\frac{\alpha_5 v_5^2}{2g} + h_w\right)}{\frac{\alpha_2 v_2^2}{2g}} \tag{3.53}$$

Equation (3.53) shows that the smaller the head loss and the outlet kinetic energy in the draft tube, the higher the recovery coefficient of the draft tube. Therefore, the recovery coefficient characterizes the hydraulic performance of the draft tube, reflects its ability to recover energy, and is sometimes referred to as the efficiency of the draft tube.

The total head loss of the draft tube is the sum of the internal hydraulic loss h_w and the outlet kinetic energy loss $\frac{\alpha_5 v_5^2}{2g}$. Usually, there is another index for evaluating the performance of the draft tube, which is the relative hydraulic loss coefficient ζ_w of the draft tube. The relative hydraulic loss ζ_w of the draft tube is equal to the ratio of its total head loss to the turbine working head H, and can be defined as

$$\zeta_w = \frac{\frac{\alpha_5 v_5^2}{2g} + h_w}{H} = (1 - \eta_w)\frac{\alpha_2 v_2^2}{2gH} \tag{3.54}$$

It can be seen from Eq. (3.54) that the kinetic energy recovery coefficient η_w of the draft tube is different from the relative hydraulic loss coefficient of the draft tube. The former only reflects the effect of kinetic energy recovery. For example, two turbines with different specific speeds have the same draft tube recovery coefficient, but their relative head loss coefficients of the draft tube are different because their runner outlet kinetic energy $\frac{\alpha_2 v_2^2}{2g}$ accounts for different proportions of the total heads. The kinetic energy of the runner outlet of the high specific speed turbine accounts for about 40% of the total head, while that of the low specific speed turbine is less than 1%. The estimated recovery coefficients of the two draft tubes are equal to 75%, the relative head loss coefficient of the high specific speed turbine draft tube reaches $\zeta_w = 10\%$, while the relative head loss coefficient of the low specific speed one is only about $\zeta_w = 0.25\%$. It can be concluded that the draft tube plays an important role in high-specific-speed turbines. Therefore, the effect of draft tube on the Kaplan turbines is more important than on the Francis turbines.

3.4 Cavitation Performance, Suction Height, and Installation Elevation of Hydro Turbine

3.4.1 Generation and Type of Cavitation and Cavitation Erosion

1. **Generation of cavitation and cavitation erosion**

The phenomena of cavitation and cavitation erosion of hydraulic turbines are special phenomena happened during the energy conversion process of water flow. Nearly at the beginning of the twentieth century, it was discovered that high-speed metal propellers of ships could be destroyed in a short time. Later, the runner blades of hydro turbines could be damaged during a certain time. Then cavitation phenomenon began to be discovered and taken seriously.

Cavitation is the formation of cavities in a liquid to destroy the continuity of the liquid phase fluid flow. This phenomenon occurs in a flow region where the pressure drops to a certain critical value. The cavities are filled with liquid vapor and gas released from the solution. When these cavities stay in the low-pressure region for a longer time, they begin to grow into larger bubbles. If the bubbles are carried by the fluid into the region where the pressure is higher than the critical value, they will collapse precipitately. This process of cavity generation and development is called cavitation. Cavitation can occur inside the liquid or on a fixed boundary. Cavitation erosion refers to the material damage on the overflow surface due to the collapse of the bubbles. In the process of bubble collapsing, mechanical, electrochemical, thermal, chemical and other processes are accompanied. Cavitation erosion is a direct consequence of cavitation, and erosion occurs only on solid boundaries.

The formation of cavitation is closely related to the vaporization of water. At a given temperature, the critical pressure at which water begins to vaporize is called the vaporization pressure of water. The vaporization pressure values of water at various temperatures are shown in Table 3.1. For convenience of application, the vaporization pressure is expressed in terms of its derived unit mH_2O (1 mH_2O = 9806.65 Pa).

It can be seen from the above that for a certain temperature of the water, when the pressure drops to a certain vaporization pressure, the water begins to vaporize. When the water flows through the turbine, if the velocity increases in some places, it

Table 3.1 Vaporization pressure of water

Water temperature /°C	0	5	10	20	30	40	50	60	70	80	90	100
Vaporization pressure/mH_2O	0.06	0.09	0.12	0.24	0.43	0.72	1.26	2.03	3.18	4.83	7.15	10.33

is known from the energy equation of hydraulics that the local pressure at that place will be reduced. If the velocity increases to a condition that the pressure is reduced to the vaporization pressure at the water temperature, the water in the low pressure zone begins to vaporize, and cavitation occurs, which may further cause cavitation erosion.

At present, it is believed that cavitation erosion on the surface of metal materials is caused by three kinds of effects: mechanical, chemical and electrochemical, with the mechanical one prominent.

(1) Mechanical effect. The local pressure drop may occur as the water flows in the turbine runner. When the local pressure is getting lower to the vaporization pressure, the water begins to vaporize, and the tiny air bubbles (about 10^{-5} to 10^{-4} mm in diameter) originally dissolved in the water begin to accumulate and escape. Thus, a large amount of bubbles with diameters smaller than 0.1 to 2.0 mm appear in the water. These bubbles are a mixture of air and water vapor. When the bubbles flow with the water into the region where the pressure is higher than the vaporization pressure, on one hand, the external water pressure outside the bubbles increases, and on the other hand, the water vapor in the bubbles rapidly condenses to make the inside pressure very low. The dynamic pressure difference between the inside and outside of the bubbles is much larger than the surface tension that maintains the bubbles in spherical shapes, resulting in the instantaneous collapse of the bubbles with a cracking time of only a few hundredths or a few thousandths of a second. At the sharp moment of bubble collapsing, the surrounding water will produce a great velocity to the bubble center under the action of extremely high pressure difference, forming a huge impact pressure, which can reach tens or even hundreds of the atmosphere pressure. Under the huge impinging pressure, the air in the original bubbles is completely dissolved into the water, and the surrounding water contracts together to form a high-energy water core. The water core then rapidly expands and impacts the surrounding water body, and the impact is transmitted to the surface of the overflow component, causing the overflow surface to be hit by a small high-speed jet. This impact phenomenon is caused by the continuous generation and collapse of bubbles in the moving water flow. It has the characteristics of high-frequency pulses, which cause material erosion to the overflow surface of the turbine. This destructive effect is called the mechanical effect of cavitation erosion.
(2) Chemical effect. When cavitation and cavitation erosion occur, the high temperature on the surface of the metal material caused by the air bubbles is the main reason for the chemical action. This locally occurred high temperature may be the heat released by the bubbles when it is compressed in the high-pressure zone, or the heat released by the high-velocity jet impinging on the surface of the overflow component. According to the test, when the bubble is condensed, the local instantaneous high temperature can exceed 300°C, and under the action of the high temperature and high pressure, the oxidative corrosion effect of the bubble on the surface of the metal material is promoted.

(3) Electrochemical effect. In the event of cavitation and cavitation erosion, a local temperature difference occurs between the locally heated material and the surrounding low temperature material. This difference can form a thermocouple, and an electric current flowing through the material can cause a thermoelectric effect and bring galvanic corrosion, which can destroy the surface of the metal material by making it dark and rough, accelerating mechanical erosion.

According to the observations on cavitation and erosion, it is believed that cavitation and cavitation erosion are mainly mechanical damage, and chemical and electrochemical effects are secondary. When mechanical erosion happens, chemical and electrochemical corrosion can accelerate the mechanical erosion process. At the beginning of the damage, cavitation and cavitation erosion are generally characterized by luster loss and darkening of the metal surface, followed by roughening and developing into pitting, which is generally pinhole-like with a depth of 1–2 mm. It can further make the metal surface very loose and spongy, also known as honeycomb, with a depth of 3 mm to several tens of millimeters. When the cavitation erosion is severe, perforation damage to the turbine blades may occur. The existence of cavitation and cavitation erosion is extremely unfavorable for turbine operation, and its effects are mainly reflected in the following aspects.

a. Damage the overflow components of the turbine, such as vanes, runners, runner chambers, upper and lower leakage rings, and draft tubes.
b. Reduce the output and efficiency of the turbine, because the cavitation and cavitation erosion will destroy the normal operation and energy conversion of the water flow, and increase the leakage and hydraulic losses of the water flow.
c. When cavitation and cavitation erosion are severe, the turbine may generate strong vibration, noise, and load fluctuations, resulting in the turbine generator unit not being able to operate safely and stably.
d. The maintenance cycle of the unit has to be shortened, while the complexity of the unit maintenance is increased. The overhaul for cavitation and cavitation erosion not only consumes a lot of steel but also extends the construction period and affects electricity production.

2. **Type of cavitation and cavitation erosion**

Since the water flow in the hydraulic machine is relatively complicated, the cavitation phenomenon can occur in different parts and under different conditions. From observations and laboratory tests of various types of hydro machine cavitation zones, cavitation often occurs in the low-pressure zone around the overflow surface or the sharply changing region of flow velocity direction, while the zone with the maximum cavitation erosion is located near the downstream end of the average cavitation length, but the entire cavitation erosion zone is a relatively wide range, extending upstream and downstream from the maximum cavitation erosion point. Therefore, the cavitation erosion position of the overflow surface is not the low-pressure point that causes the cavitation phenomenon. The low-pressure point is upstream of the cavitation erosion region, namely, at the upstream end of the cavitation zone.

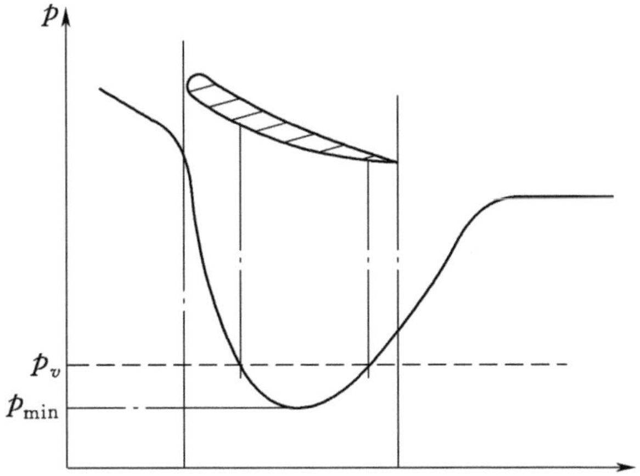

Fig. 3.16 Pressure distribution along the back of the blade

According to the conditions and locations of cavitation and cavitation erosion, they can be generally divided into the following four types.

(1) Hydrofoil cavitation and cavitation erosion. The hydrofoil cavitation and cavitation erosion are caused by a pressure drop when the water flows around the blades. The pressure on the back of the blade is often negative, and the pressure distribution is shown in Fig. 3.16. Closely related to the geometry of the blade foil section, cavitation and cavitation erosion occur when the pressure in the low-pressure zone on the back side drops below the ambient vaporization pressure. Thus this type is called hydrofoil cavitation and cavitation erosion, which is the main cavitation and cavitation erosion forms of reaction turbines. And it is related to operating conditions. When the turbine is operating in non-optimal conditions, hydrofoil cavitation and cavitation erosion are induced or exacerbated.

    From the investigation of many hydro turbines in China, the hydrofoil cavitation and cavitation erosion of Francis turbines may occur mainly in the four areas, denoted by *A* to *D* as shown in Fig. 3.17b. *A* is the outlet side of the lower part of the blade back; B is the vicinity of the blade back and the lower ring; *C* is the inner side of the lower ring facade; *D* is the junction of the blade back and the upper crown. The hydrofoil cavitation and cavitation erosion of the axial flow turbine mainly occur near the outlet of the blade back, and the junction of the blade and the hub, as shown in Fig. 3.17a.

(2) Gap cavitation and cavitation erosion. Gap cavitation and cavitation erosion are a form that occurs when water passes through a narrow channel or gap, causing a local velocity rise and pressure decrease to a certain vaporization pressure, as shown in Fig. 3.18. Gap cavitation and cavitation erosion mainly occur at the gap of the upper and lower labyrinth rings of the Francis turbine runner, the gap between the outer edge of the Kaplan turbine blade and the runner chamber, the gap between the blade root and the hub, and the gap of the guide vane end faces.

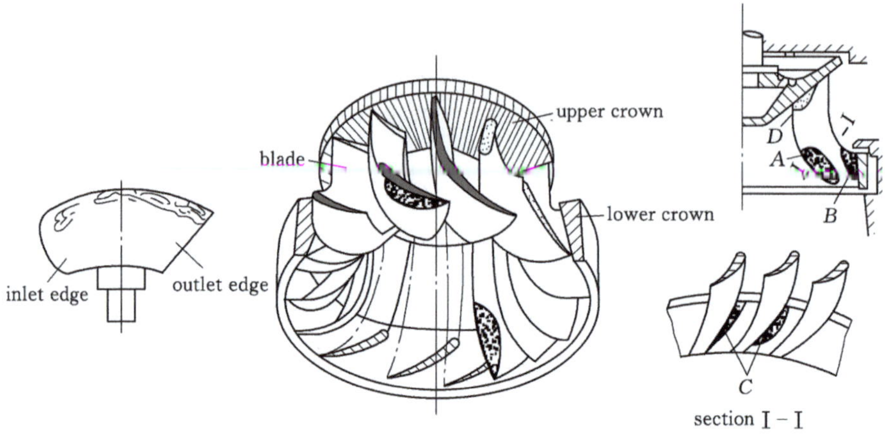

(a) Axial flow runner (b) Francis turbine runner

Fig. 3.17 Main locations of turbine hydrofoil cavitation

Fig. 3.18 Gap cavitation and cavitation erosion

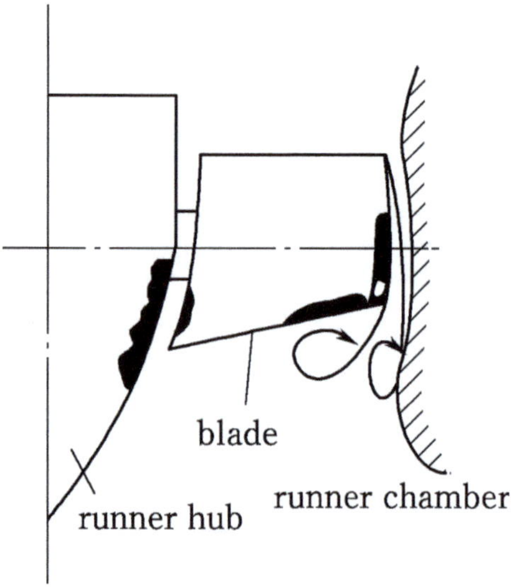

(3) Local cavitation and cavitation erosion. Local cavitation and cavitation erosion are mainly caused by sudden changes in local flow patterns due to defects in the casting and machining to form surface irregularities, blisters, pores, etc. For example, local cavitation and cavitation erosion of a Kaplan turbine generally occur at a non-smooth step connected to the runner chamber or behind a local pit; and it may also occur at the blade fixing screws and sealing screws due to the indentation or protrusion of the screws. The cavitation and cavitation erosion

after the crown drain hole on the Francis turbine runner is also a kind of local cavitation and cavitation erosion.

(4) Cavity cavitation and erosion. Cavity cavitation and erosion are unique kinds of vortex cavitation to reaction turbines, especially to Francis turbines. When the turbine is operated under normal working conditions, the runner outlet always has a certain circumferential velocity, consequently, the rotating water in the draft tube forms a vacuum vortex. When the negative pressure in the center of the vortex is less than the vaporization pressure, this kind of cavitation will occur, and the rotating vortex rope will periodically collide with the draft tube wall, causing cavitation and cavitation erosion in the draft tube wall.

The occurrence of cavity cavitation and cavitation erosion is generally related to operating conditions. In the case of a large load, the shape of the vortex rope almost coaxial with the centerline of the draft tube is column-shaped, as shown in Fig. 3.19b, and its diameter is small and stable, especially in optimal conditions, the vortex rope can even disappear. However, in low load conditions, the cavity vortex is thick, spiral, and also rotating itself. This eccentric spiral vortex is extremely unstable in space, and strong cavity cavitation and cavitation erosion will occur, as shown in Fig. 3.19a, c.

In general, the cavitation and cavitation erosion of Francis turbines are mainly hydrofoil cavitation and cavitation erosion, while the gap and local cavitation are only secondary. The cavitation and cavitation erosion of the Kaplan turbine is mainly caused by gap cavitation. The cavitation and cavitation erosion of the impulse turbine mainly occur at the nozzle and the needle, and cavitation often occurs at the water-dividing edge of the bucket due to the high-speed jet flow. In the above four kinds of cavitation and cavitation erosion, gap cavitation and local cavitation generally only occur in a small local range, and hydrofoil cavitation is the most common and serious one. Besides, cavity cavitation may be serious for some hydropower stations, which may affect the stable operation.

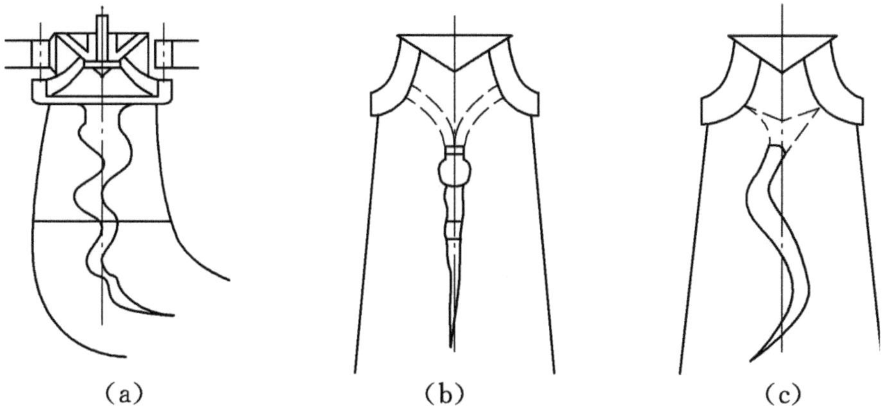

Fig. 3.19 Shape of the vortex cavity

Table 3.2 Cavitation erosion ratings list

Cavitation erosion ratings	Cavitation index K_h		Cavitation degree
	10^{-4} mm/h	mm/a	
I	<0.0577	<0.05	Slight
II	0.0577 to 0.115	0.05 to 0.1	Moderate
III	0.115 to 0.577	0.1 to 0.5	Slightly serious
IV	0.577 to 1.15	0.5 to1.0	Serious
V	≥1.15	≥1.0	Extremely serious

Regarding the criteria for assessing the cavitation and cavitation erosion of turbines, in addition to the maximum and average values of the erosion area and erosion depth of the cavitation erosion site. In China, the cavitation erosion index is currently used to reflect the degree of cavitation erosion. It refers to the average depth of cavitation erosion per unit area of the back of the blade per unit time, denoted by the symbol K_h

$$K_h = \frac{V}{FT} \tag{3.55}$$

where V is the cavitation erosion volume, m^2; T is the effective operating time, excluding phase modulation time, h; F is the total area of the blade back, m^2; K_h is the cavitation index of a hydraulic turbine, 10^{-4} mm/h.

In order to distinguish the cavitation and cavitation erosion of various turbines, cavitation erosion can be divided into five levels according to the value of K_h, as shown in Table 3.2.

3.4.2 *Cavitation Coefficient of Hydro Turbine*

There are two important parameters for measuring the performance of a turbine. One parameter is efficiency, which is energy performance, and the other is cavitation coefficient, which indicates cavitation and cavitation erosion performance. Therefore, a good turbine runner must have both good energy performance and cavitation performance, that is, it must have high efficiency, can fully utilize water energy, and have a small cavitation coefficient, less prone to cavitation erosion during operation. This section will analyze and derive the conditions with no cavitation for the runner blade and the cavitation coefficients that characterize the cavitation performance of the turbine.

Figure 3.20 is a schematic diagram of the flow path of a hydro turbine. The lowest pressure point is point k, and the pressure is p_k. Point 2 with pressure p_2 is on the outlet edge of the blade. Point a with pressure p_a is on the downstream water surface. And if the downstream is open channel flow, p_a is the atmospheric pressure. The

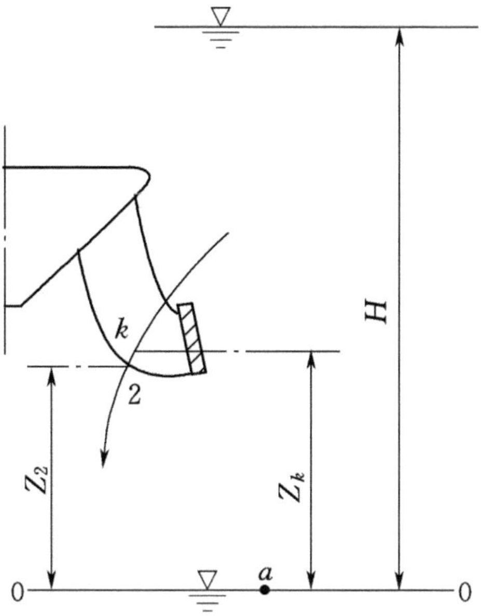

Fig. 3.20 Analysis of hydrofoil cavitation conditions

Bernoulli equation for the relative motion of point k and point 2 is listed as

$$Z_k + \frac{p_k}{\rho g} + \frac{w_k^2}{2g} - \frac{u_k^2}{2g} = Z_2 + \frac{p_2}{\rho g} + \frac{w_2^2}{2g} - \frac{u_2^2}{2g} + h_{k-2} \tag{3.56}$$

where h_{k-2} is the hydraulic loss from point k to point 2.

Since point k and point 2 are very close, it can be approximated that $u_k = u_2$, then the above formula can be rewritten as

$$\frac{p_k}{\rho g} = \frac{p_2}{\rho g} + \frac{w_2^2 - w_k^2}{2g} + (Z_2 - Z_k) + h_{k-2} \tag{3.57}$$

In order to find the pressure at point 2, we can take the Bernoulli equation of the absolute motion of water between point 2 of the blade exit and the point a of the downstream section.

$$Z_2 + \frac{p_2}{\rho g} + \frac{v_2^2}{2g} = Z_a + \frac{p_a}{\rho g} + \frac{v_a^2}{2g} + h_{2-a} \tag{3.58}$$

where h_{2-a} is the head loss from point 2 to point a. Since the exit velocity of the draft tube is small, it can be considered that $v_a \approx 0$, then the above formula can be written as

$$\frac{p_2}{\rho g} + Z_2 = \frac{p_a}{\rho g} + Z_a + h_{2-a} - \frac{v_2^2}{2g} \tag{3.59}$$

Substituting Eqs. (3.59) into (3.57) yields

$$\frac{p_k}{\rho g} = \frac{p_a}{\rho g} - (Z_k - Z_a) - \left(\frac{v_2^2}{2g} + \frac{w_k^2 - w_2^2}{2g} - h_{k-a}\right) \tag{3.60}$$

where, $h_{k-a} = h_{k-2} + h_{2-a}$, then the vacuum value of pressure at point k is

$$\frac{p_v}{\rho g} = \frac{p_a}{\rho g} - \frac{p_k}{\rho g} = (Z_k - Z_a) + \left(\frac{v_2^2}{2g} + \frac{w_k^2 - w_2^2}{2g} - h_{k-a}\right) \tag{3.61}$$

From Eq. (3.61), the vacuum of pressure at point k consists of two parts: (1) The dynamic/kinetic vacuum $h_v = \frac{v_2^2}{2g} + \frac{w_k^2 - w_2^2}{2g} - h_{k-a}$, which is formed by the runner and the draft tube of the turbine. It is related to the velocities at the runner and draft tube, the geometry of the runner blades and the draft tube, namely, related to the structure parameters and operating conditions of the turbine. (2) The static vacuum $H_s = Z_k - Z_a$, also known as suction head/height, which is related to the turbine installation elevation. It depends on the height of the device relative to the downstream water surface, regardless of the turbine type.

From Eq. (3.60), by subtracting $p_v/\rho g$ from both sides and dividing each side by the head H, it can be obtained that

$$\frac{p_k - p_v}{\rho g H} = \left(\frac{p_a}{\rho g} - \frac{p_v}{\rho g} - H_s\right)/H - \left(\frac{w_k^2 - w_2^2}{2gH} + \eta_w \frac{v_2^2}{2gH}\right) \tag{3.62}$$

where $\eta_w \frac{v_2^2}{2g} = \frac{v_2^2}{2g} - h_{k-a}$ is the recovery factor of the draft tube.

Let

$$\sigma = \frac{w_k^2 - w_2^2}{2gH} + \eta_w \frac{v_2^2}{2gH} \tag{3.63}$$

$$\sigma_p = \left(\frac{p_a}{\rho g} - \frac{p_v}{\rho g} - H_s\right)/H \tag{3.64}$$

where σ is the cavitation coefficient of hydro turbine; σ_p is the cavitation coefficient of hydropower station installation. Then Eq. (3.62) can be rewritten as

$$\frac{p_k - p_v}{2gH} = \sigma_p - \sigma \tag{3.65}$$

It can be seen from Eq. (3.63) that σ is the relative value of the dynamic vacuum and is a dimensionless quantity, which is related to the geometry of the turbine runner hydrofoil, operating conditions, and draft tube performance. For a certain given geometry of a turbine, the cavitation coefficient σ is fixed in a given operating condition. For turbines with similar geometry, including draft tube similarity, according to the similarity theory, the velocity triangle is similar at the similar operating point, $n_{11} = \frac{nD_1}{\sqrt{H}}$, and the relative values of the velocities are equal. In the similar operating condition, the recovery coefficient of draft tube is also equal, thus σ is identical. It can be seen that σ is a similar criterion reflecting the cavitation performance of the turbine.

It can be seen from Eq. (3.64) that when the pressure on the downstream water surface is the atmospheric pressure, the hydropower station cavitation coefficient σ_p, only representing the characteristic value leaving the cavitation starting point, depends only on the relative height of the runner to the downstream water surface.

It can be known from Eq. (3.65) that when the pressure p_k at point k falls to the vaporization pressure p_v of the corresponding temperature, the cavitation of the turbine is in a critical state, and $\sigma_p = \sigma$. When $\sigma_p > \sigma$, then the pressure at the lowest pressure point in the runner must satisfy the inequality of $p_k > p_v$, there will be no cavitation in the runner. When $\sigma_p < \sigma$, on the contrary, $p_k < p_v$, the cavitation will occur in the runner. From the above analysis, it is known that the runner can be assured to operate without cavitation by selecting an appropriate H_s.

3.4.3 Suction Head and Installation Elevation of Hydro Turbine

1. **Suction head of hydro turbine**

Under a certain working condition, the dynamic vacuum at the lowest pressure point k of the turbine is constant, but the static vacuum is related to the installation elevation of the hydro turbine. Therefore, the total vacuum at point k can be controlled by selecting a suitable suction height H_s to achieve the purpose of avoiding cavitation and cavitation erosion. In order to avoid cavitation on the runner blades, the pressure at point k must be greater than the saturated vaporization pressure of the water flow, namely

$$\frac{p_k}{\rho g} = \frac{p_{\min}}{\rho g} \geq \frac{p_v}{\rho g} \tag{3.66}$$

Substituting the above formula into Eq. (3.64) yields

$$\sigma_p = \left(\frac{p_a}{\rho g} - \frac{p_v}{\rho g} - H_s\right)/H \geq \sigma \tag{3.67}$$

Rewriting Eq. (3.67) yields

$$H_s \leq \frac{p_a}{\rho g} - \frac{p_v}{\rho g} - \sigma H \tag{3.68}$$

where $p_a/\rho g$ is the atmospheric pressure head at the turbine installation location, mH_2O. Considering that the average atmospheric pressure of the standard sea level is 10.33 mH_2O, within the altitude of 3000 m, the atmospheric pressure will decrease by 1 mH_2O while the altitude increase every 900 m. Therefore, when the altitude of the turbine installation position is ∇ m, the atmospheric pressure will be $\frac{p_a}{\rho g} = 10.33 - \frac{\nabla}{900}$ (mH_2O). $P_v/\rho g$ is the vaporization pressure head corresponding to the average water temperature, mH_2O. Considering that the water temperature in the penstock of the hydropower station is generally 5–20 °C, for clean water with less impurities, it can be taken as $\frac{p_v}{\rho g} = 0.09$–$0.24$ (mH_2O). σ is the cavitation coefficient of the turbine at the practical operation condition, and it is usually obtained from the model test. However, with the error of the cavitation test of the turbine model and the influence of the size difference between the model and the prototype considered, the model cavitation coefficient σ_m needs to be corrected by $\sigma = \sigma_m + \Delta\sigma$ or $\sigma = K_\sigma \sigma_m$.

In practical application, Eq. (3.68) is usually simplified as

$$H_S \leq 10.0 - \frac{\nabla}{900} - (\sigma_m + \Delta\sigma)H \tag{3.69}$$

or

$$H_S \leq 10.0 - \frac{\nabla}{900} - K_\sigma \sigma_m H \tag{3.70}$$

where ∇ is the altitude above sea level of hydro turbine installation location, m, and it may be taken as the altitude of the downstream average water level in the initial calculation. σ_m is the model cavitation coefficient, and its values at various working conditions can be found from the model comprehensive characteristic curve for the hydro turbine. $\Delta\sigma$ is the correction value of the cavitation coefficient, and it can be found in Fig. 3.21 according to the design head H_r. H is the head of the turbine, m, and it is generally taken as the design head H_r. For the axial flow turbines, the minimum head $H_{\min}$ and the corresponding value of σ_m should also be used in the calculation of H_s for comparison. And for the Francis turbines, the maximum head $H_{\max}$ and the corresponding value of σ_m should also used for comparison. K_σ is the cavitation safety factor of the turbine, and its selection of value is based on the technical specifications. For Kaplan turbines, $K_\sigma = 1.1$ can be taken; for Francis turbines, the data in Table 3.3 can be used.

Of course, the final determination of the suction height H_s must also consider the infrastructure conditions, investment size, and operating conditions, through technical and economic comparison. If the amount of sand in the water is large, in order to avoid the interaction and joint action of cavitation and sediment wear, the suction height H_s should be taken for safety.

2. **Specification of suction height**

The exact definition of the suction height H_s of the hydro turbine is the vertical height from point k (the lowest pressure point at the blade back) to the downstream water level. However, the position of point k is difficult to determine in actual calculations,

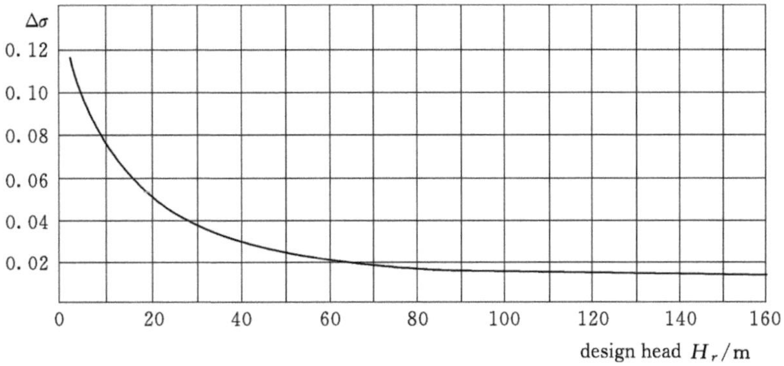

Fig. 3.21 Correlation curve between cavitation coefficient $\Delta\sigma$ and head H_r

Table 3.3 Relationship between head and cavitation safety factor K_σ

Head H, m	30 to 100	100 to 250	More than 250
Cavitation safety factor K_σ	1.15	1.20	1

and its position is also changed at different working conditions. Therefore, in order to facilitate uniformity in engineering, the suction heights of hydro turbines with different types or different arrangements (Fig. 3.22) is specified as follows.

(1) The suction height H_s of the Kaplan turbine is the vertical distance from the downstream level to the rotation centerline of the runner blades.
(2) The suction height H_s of the Francis turbine is the vertical distance from the downstream level to the lower ring plane of the guide vanes.
(3) The suction height H_s of the Deriaz turbine is the distance from the downstream level to the intersection of the blade rotation axle and the interior surface of the runner chamber.
(4) The suction height H_s of the horizontal reaction turbine is the distance from the downstream level to the highest point of the runner blades.

If the suction height H_s is a positive value, then the runner is above the downstream water level; if it is negative, then the runner is below the downstream water level, and H_s absolute value is often referred to as the submerged depth.

3. **Installation elevation of hydro turbine**

For the vertical reaction turbine, installation elevation is the center elevation of guide vanes; for the horizontal turbine, it is the center elevation of the main shaft. The calculation method of installation elevation of the hydro turbine with different arrangements (see Fig. 3.23) is as follows.

(1) Vertical Francis turbine

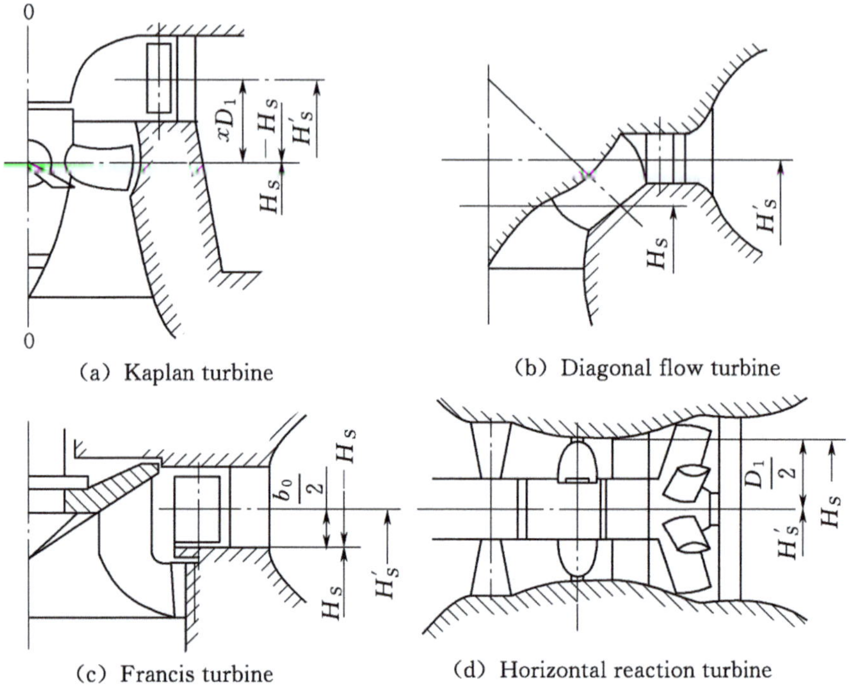

Fig. 3.22 Suction height of turbine with different types and arrangements

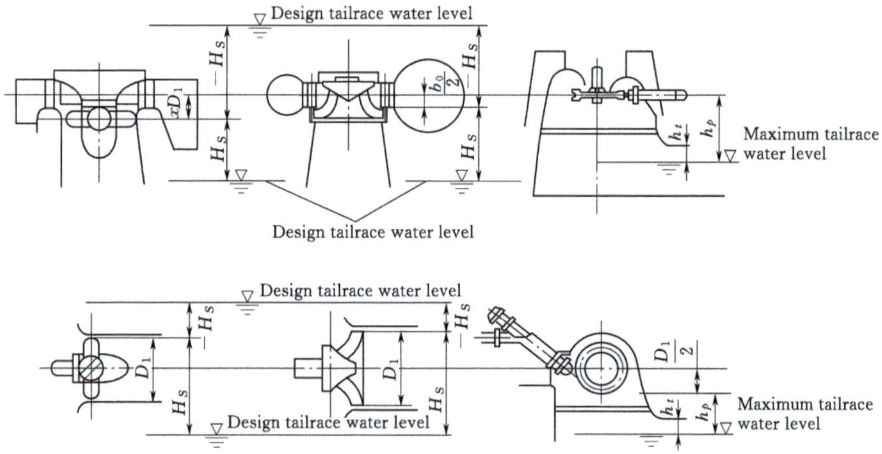

Fig. 3.23 Turbine installation elevation diagram

$$\nabla_{inst} = \nabla_w + H_s + \frac{b_0}{2} \tag{3.71}$$

where ∇_w is the tailrace water level, m; b_0 is the height of guide vanes, m.

(2) Vertical Kaplan turbine

$$\nabla_{inst} = \nabla_w + H_s + xD_1 \tag{3.72}$$

where D_1 is the diameter of runner, m; x is the structure height coefficient of Kaplan turbine, and it can be obtained by Table 3.4.

(3) Horizontal reaction turbine

$$\nabla_{inst} = \nabla_w + H_s - \frac{D_1}{2} \tag{3.73}$$

(4) Penton turbine

Vertical shaft

$$\nabla_{inst} = \nabla_{wm} + h_p \tag{3.74}$$

Horizontal shaft

$$\nabla_{inst} = \nabla_{wm} + h_p + \frac{D_1}{2} \tag{3.75}$$

where ∇_{wm} is the maximum tailrace water level, m; h_p is the discharge height, m, as shown in Fig. 3.23. It is the necessary height to ensure the safe and stable operation of the turbine, avoid the surge under variable load conditions, ensure ventilation, and prevent runner energy loss caused by water splashing and eddy current in the tailrace channel. According to experience statistics, generally $h_p = (0.1 \sim 0.15)D_1$, and it is taken as the larger value for the vertical shaft unit and the smaller value for the horizontal shaft unit. When determining the discharge height, it must be attention to ensure the necessary ventilation height h_t (see Fig. 3.23), so as to avoid excessive surge and eddy current in the tailrace channel. Generally, the ventilation height should not be less than 0.4 m.

The design tailrace water level is usually used to determine the installation elevation of the hydro turbine. The design tailrace water level can be obtained from the downstream water level and discharge curve according to the discharge of the turbine.

Table 3.4 Structure height coefficient of Kaplan turbine

Turbine designation	ZZ360	ZZ440	ZZ460	ZZ560	ZZ600
x	0.3835	0.3960	0.4360	0.4085	0.4830

Table 3.5 Turbine discharge used for determining the design tailrace water level

Number of units	Turbine discharge
1 or 2	50% rated discharge of one turbine
3 or 4	Rated discharge of one turbine
Over 5	Rated discharge of 1.5 to 2 turbines

In general, the discharge of the turbine can be selected according to the number of installed units in the hydropower station, as shown in Table 3.5.

The installation elevation of the hydro turbine directly affects the excavation volume of civil engineering and the cavitation performance of turbine operation. Therefore, the installation elevation of turbines in large and medium-sized hydropower stations should be determined after economic and technical comparison according to the operation conditions of the unit.

3.4.4 Measures for Anti-cavitation of Hydraulic Turbine

When cavitation develops to a certain stage in the turbine, the flow around the blade will deteriorate, thereby reducing the water torque, and causing the turbine output and the efficiency to decrease. With the occurrence of cavitation, cavitation erosion is inevitably formed on the overflow components of the hydro turbine. When it is slight, there are only a few erosion etch spots. When it is serious, the metal materials in the erosion area are ablated or denuded, resulting in a honeycomb like surface, and even the phenomenon of blade perforation or edge falling. Besides, the noise and pressure pulsations occur, and the pulsating vortex in the draft tube is especially harmful. Once the pulsating frequency coincides with the natural vibration frequency of the relevant component, it will inevitably cause resonance, resulting in the vibration of the unit and the swing of the output, which seriously threatens the safe operation of the unit.

Therefore, improving the cavitation performance of hydro turbines has become an important task for hydraulic machinery design and operation. How to prevent and avoid the occurrence of cavitation and cavitation erosion? After more than 40 years of practice in turbine operation in China, more mature experiences and measures to prevent and reduce cavitation and cavitation erosion have been achieved. However, the problem has not been solved fundamentally. Therefore, comprehensive research on turbine hydraulics, structure and materials to prevent the cavitation of hydro turbines is still one of the important topics nowadays.

1. **Improve the hydraulic design of the turbine**

Hydrofoil cavitation and cavitation erosion are the main types of cavitation and cavitation erosion of hydro turbines. They are related to many factors, such as the parameters of the hydrofoil shape, the parameters of blade cascade composition, and the operating conditions of the turbine.

In terms of hydrofoil design, it is necessary to design and test a runner with good cavitation performance. The main way is to make the lowest pressure on the blade back located at the trailing edge, so that the bubble collapse occurs in the area outside the blade, avoiding damage from the cavitation and cavitation erosion. When the cavitation and cavitation erosion occur on the back of the runner blade, a large amount of bubbles will be formed in the lowest negative pressure zone, as shown in Fig. 3.24a. The length l_c of the bubble region is smaller than the length l of the blade. Both the cavitation and cavitation erosion of the blade surface by the bubble suddenly collapsing, and the recovery of the continuity of the water flow occur near the tail point A of the bubble region, therefore the hydrofoil type cavitation and cavitation erosion are mostly found in the middle and rear part of the blade back. If the blade design of the runner can be changed, as shown in Fig. 3.24b, the bubble collapse and the water continuity recovery will occur after the blade tail, with $l_c > l$, thus avoiding serious damage to the blade. Engineering practice has proved that when the blade shape is designed reasonably, cavitation and cavitation erosion can be avoided or mitigated.

In the design of the turbine, it is necessary to reasonably determine the suction height H_s of the turbine, the specific speed n_s, and the cavitation coefficient σ. The higher the specific speed, the larger the cavitation coefficient and the deeper the runner is under the water. The experience shows that these three parameters should be optimally matched. For turbines operating in a heavy sediment flow, selecting a lower specific speed runner and a larger turbine diameter, and lowering the suction height will help in mitigating the combined effects of cavitation erosion and sediment abrasion.

2. **Improve manufacturing level and use anti-corrosion material**

The level of processing and manufacturing technology directly affects the turbine performance of cavitation and cavitation erosion. When the quality of processing and manufacturing is poor, some problems that may lead to cavitation and erosion will appear, such as poor blade leading profile (usually square head), large difference in blade opening, uneven thickness of trailing edge, local bulge, and large profile deviation. The profile of the runner blade after casting and processing should be as consistent as possible with the design model diagram to ensure that the prototype is similar to the model turbine.

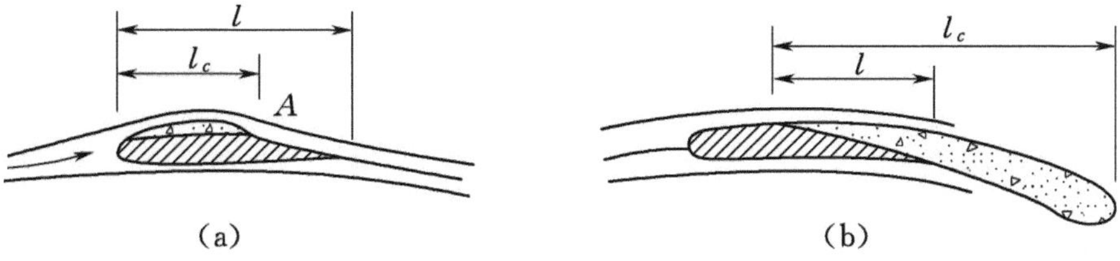

Fig. 3.24 Cavitation flow around the hydrofoil

Another effective measure to improve the corrosion resistance of the runner is to use good anti-erosion materials to manufacture turbine components, or to increase the anti-erosion performance of the overflow surface and to apply a protective layer to it. Generally, stainless steel has superior anti-cavitation performance than carbon steel, such as chrome molybdenum stainless steel (Cr8CuMo), low nickel stainless steel (0Gr13Ni), and 13 chrome (Cr13). However, the cost of using anti-erosion materials is higher.

3. **Improve operating conditions and adopt appropriate operational measures**

The cavitation and cavitation erosion are closely related to the operating conditions of the turbine. Near the design operating conditions, there will be no severe cavitation. However, when the turbine operates under off-design conditions for a long time, the flow conditions of the hydrofoil and the outlet flow conditions of the runner will change greatly, and then the hydrofoil cavitation and cavity cavitation will be exacerbated. Therefore, in order to avoid cavitation and cavitation erosion, it is necessary to make reasonable operation mode of the hydropower plant and keep the units operating in the optimal condition. And try to avoid running in the serious cavitation area to ensure the stable operation of the turbine.

When the turbine operates under off-design conditions, the method of supplementing air into the lower part of the runner can be adopted to destroy the cavity and reduce cavitation, cavitation erosion, and turbine vibration. At present, small and medium-sized units often adopt two methods: natural air supplement and forced air supplement.

The form and location of the natural air supply device are as follows.

(1) Air admission to main shaft center. As shown in Fig. 3.25, the structure of air admission to the main shaft center hole is simple. When the vacuum in the draft tube reaches a certain value, the air supply valve automatically opens, through which the air enters the lower part of the runner from the main shaft center hole to improve the vacuum at this place, thereby reducing cavity cavitation. However, because this method is difficult to replenish air to the cavitation part of hydrofoils and the lower ring, it is not good for improving the hydrofoil cavitation and cavitation erosion. And the air supply volume is small, which is often insufficient to eliminate the pressure pulsation caused by the draft tube vortex. Also the air admission may cause loud noise.
(2) Draft tube air supply. It is well known that in some operating conditions, at the center of straight cone section of the draft tube, the cavitation water flow forms a vortex rope which will cause pressure pulsations. The pressure pulsations can be restrained to a certain extent by adding air into the draft tube, and the suppression effect depends on three factors: the amount of air supply, the position of the air supply, and the structure shape of the air supply device. Details are as follows.

The amount of air supply directly affects the air supply effect. Some tests show that, as shown in Fig. 3.26, when there is enough air supply, the pressure pulsations in the draft tube can be effectively reduced. But too much air supply is not beneficial to reducing the pressure pulsation. On the contrary, because the average pressure

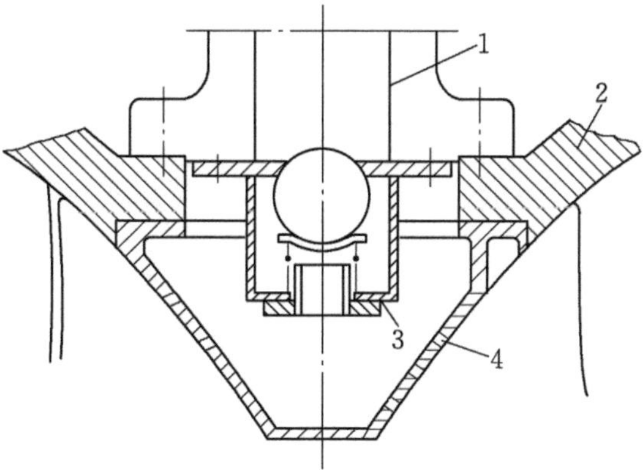

Fig. 3.25 Air admission to the center hole of main shaft 1-main shaft; 2-runner; 3-air supply valve; 4-discharge cone with air admission hole

in the draft tube rises, the unit efficiency may decrease. Generally, the amount of air supply that most effectively eliminates the pressure pulsations of the draft tube is called the optimal air supply, and it varies with the changes in turbine operating condition. According to many test data, the optimal air supply (free air volume) is approximately 2% of the turbine design discharge.

There are two common structure shapes of air supply device: the cross pipe air supply and the short pipe air supply, as shown in Fig. 2.27. For the cross pipe air supply device (Fig. 2.27a), when negative pressure is generated on the back of the runner blade, air enters the air equalization tank 4 from the intake pipe 5, and then enters the center body 2 through the cross pipe 1, thereby breaks the vacuum at the lower portion of the runner. For small and medium-sized units, an air supply pipe is installed in the upper part of the draft tube at the time of manufacture. Generally, the distance of the cross pipe from the lower ring of the runner is f_b = (1/3 to 1/4)D_1, the angle between the cross pipe and the horizontal plane is about α = 8° to 11°, and there are 3 to 4 cross pipes with the diameter of d_1 = 100 to 150 mm. The small air vent on the cross pipe should be opened on the back side to prevent water from entering the cross pipe.

Figure 3.27b shows the short pipe air supply device. The notch and air vent of the short pipe should be arranged on the back side of the water flow, and its optimal

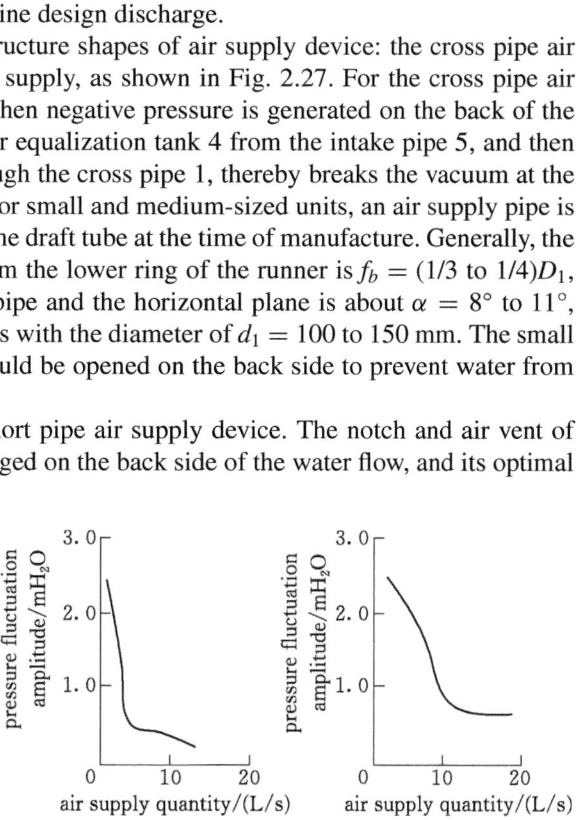

Fig. 3.26 Influence of air supply quantity on pressure pulsations in the draft tube

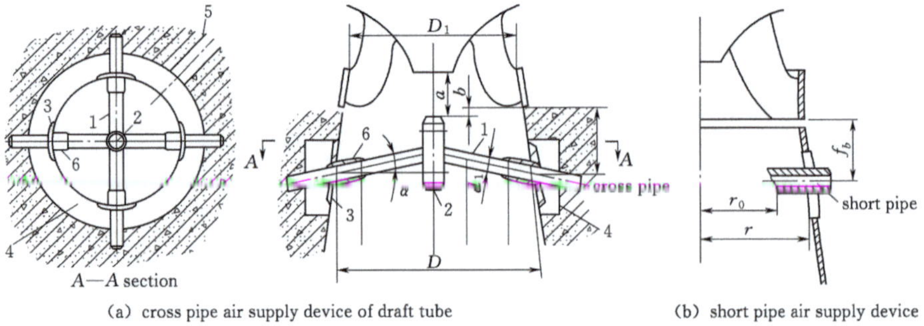

Fig. 3.27 Draft tube air supply device 1- cross pipe; 2-central body; 3-lining plate; 4-air equalizing tank; 5-intake pipe; 6- stainless steel bushing

radius is about $r_0 = 0.85r$, where r is the diameter of the draft tube. The short pipe should be close to the lower part of the runner and also the distance can be $f_b =$ (1/3 to 1/4)D_1. The forced air supply device is used when the suction height H_s is small and the natural air supply is difficult. The draft tube jet pump and the top cover compressed air are used to force air supply.

Figure 3.28 shows the device of the jet pump air supply at the draft tube. The working principle is as follows. The upstream pressure water flow is injected into the air pipe at a high velocity from the jet nozzle, which is installed at the inlet of the air pipe. The jet causes a negative pressure at the air pipe inlet, which draws the air into the draft tube. This method can save compressed air equipment and is generally applicable to hydropower stations m with $H_s = -4$ to -1 m.

The research on the effects of air supply on unit efficiency is currently insufficient. On one hand, the air supply weakens the pressure pulsations of the draft tube vortex rope and stabilizes the unit operation, and can improve the unit efficiency. On the other hand, the air supply reduces the vacuum of the draft tube and the air supply structure increases the resistance of the water flow, which reduces the efficiency of the unit. The comprehensive result shows that the efficiency of the unit is improved under the optimal air supply quantity and reasonable air supply structure, which are confirmed by the operation experience of many power stations.

3.5 Pressure Pulsation Performance of the Hydro Turbine

The increasing demands for peak shaving, frequency regulation, and spinning reserve by the power grids have caused hydro turbines to frequently start, stop, and change operating conditions, and work in off-design conditions. In the conditions deviating from the design condition, hydro turbines have to experience higher pressure pulsations. In recent years, pressure pulsation characteristics have begun to become an important indicator to measure the performance of large hydropower units, and

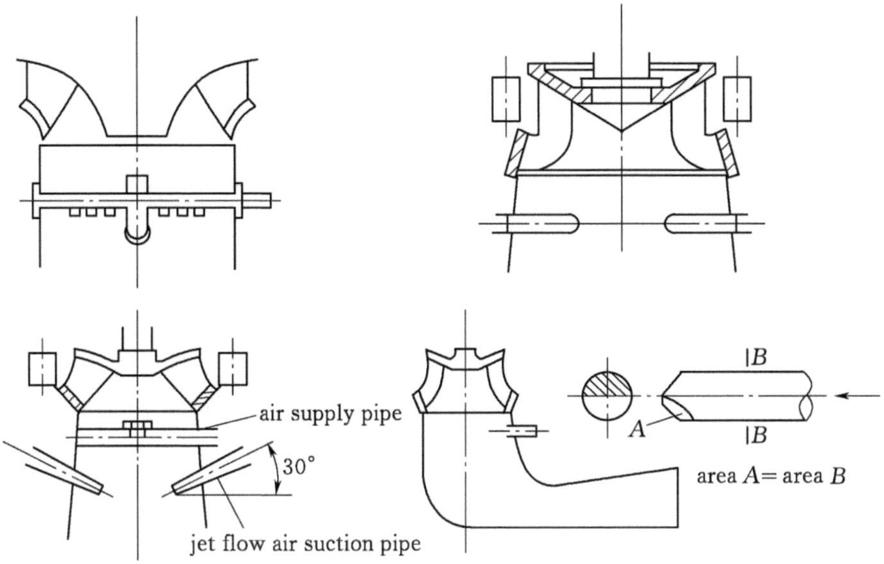

Fig. 3.28 Several air supply structures

reducing pressure pulsation level has become one of the goals of hydraulic design of hydro turbines.

Usually, a measured pressure signal p can be divided into two parts, $\overline{p}$ and $\tilde{p}$, according to Eq. (3.76).

$$p = \overline{p} + \tilde{p} \tag{3.76}$$

where $\overline{p}$ is the time mean value of pressure, and $\tilde{p}$ is the pulsating value of pressure.

When the turbine is running steadily, $\overline{p}$ is a constant value; during the transient process, $\overline{p}$ is a function of time, characterizing water hammer pressure. The pressure pulsating value $\tilde{p}$ includes a periodic component and a non-periodic component. The periodic component has a clearly corresponding frequency, and the non-periodic portion is randomness, which is usually caused by factors such as turbulence and eddies. In this section, the discussion focus is on the periodic part of pressure pulsations, because it is the main component of pressure pulsations.

3.5.1 Causes of Pressure Pulsations

There are many unstable phenomena caused by hydraulic factors during the operation of the turbine. The manifestations include the pressure pulsations in the spiral case, the draft tube, and the vaneless space. Pressure pulsations can cause variations of dynamic loads of the runner, which may lead to resonance and fatigue failure of the unit and affect the operating life of the unit. For Francis turbines, the causes of pressure pulsations mainly include the rotor–stator interaction, the draft tube vortex rope, and the rotating stalls in the pump-turbine when operating far from the design condition.

1. **Rotor-stator interaction**

The rotor–stator interaction refers to the interference between the runner blades and the guide vanes. As shown in Fig. 3.29, the pressure distribution in the vaneless space changes with time and space. It is the most important component of pressure pulsations during steady-state operation of the hydro turbine, and its frequency is high, which is related to the combination of the number of runner blades and the number of guide vanes. Equations (3.77) to (3.78) respectively lists the runner rotation frequency f, the blade rotation frequency f_b, and the relative rotation frequency f_g of the guide vanes (Fig. 3.29)

$$f = n/60 \tag{3.77}$$

$$f_b = z_b f \tag{3.78}$$

$$f_g = z_g f \tag{3.79}$$

where n is the rotational speed of turbine, rpm; z_b is the number of runner blades; and z_g is the number of guide vanes. When the pressure sensor is arranged in the vaneless space, the measured dynamic and static interference frequency follows Eq. (3.78); when the pressure sensor is arranged on the runner blade, rotating with the blade together, the measured dynamic and static interference frequency follows Eq. (3.79).

2. **Draft tube vortex rope**

When a Francis turbine is operated at partial load, the draft tube will have a vortex rope due to the revolving cavitation, as shown in Fig. 3.30. The draft tube vortex rope frequency f_{vortex} is an important component of the low frequency in the pressure pulsation spectrum. When the turbine is running away from the optimal operating condition, its amplitude will increase significantly. The frequency is close to the natural frequencies of the hydraulic system of hydropower stations, and there is a

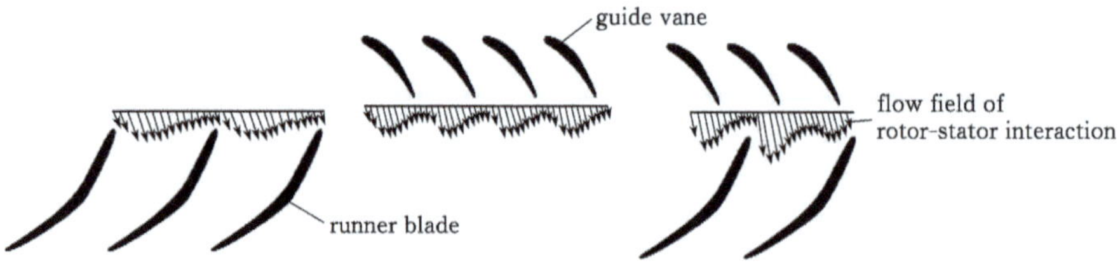

Fig. 3.29 Schematic of the interference between blade flow patterns and guide vane flow patterns

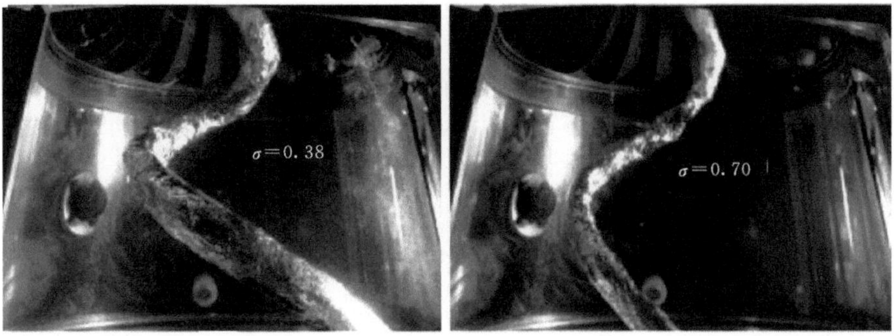

Fig. 3.30 Draft tube vortex rope of Francis turbine

risk of resonance, which is a hidden danger affecting the stability of the hydropower stations.

$$f_{vortex} = (0.2 \text{ to } 0.4)f \tag{3.80}$$

3. **Rotating stall**

A rotating stall is a unique source of pressure pulsations for pump-turbines. It is expressed in the form of uneven distribution of runner discharge and blockage of local flow channels by vortex, when the unit is running away from the design condition, as shown in Fig. 3.31. The periodic local high pressure occurs, and the discharge of each channel is unbalanced along the circumference. It may appear both in turbine operating condition and pump operating condition. The corresponding frequency is (Eq. 3.81).

$$f_{rs} = (0.3 \text{ to } 0.7)f \tag{3.81}$$

Fig. 3.31 Flow separation caused by rotating stall in a low discharge condition

3.5.2 Harmfulness of Pressure Pulsations

Periodic signals have strong physical effects from both the time and frequency domains, and the pressure pulsations in turbines are periodic. The components of the pressure pulsations discussed previously can be regarded as a kind of excitation source with the known frequency and position, so that the harm to the safe operation of hydropower station can be discussed from the perspective of time domain and frequency domain, according to its location and frequency of action.

1. **Harm of rotor-stator interaction**

The origin of rotor–stator interaction is located in the vaneless space between the runner blades and the guide vanes. The amplitude is large, the frequency is high, and the time domain features are obvious. The frequency is not in the same order of magnitude as the natural frequency of the power station, but its harmonics are close to the natural frequency of the runner.

Rotor–stator interaction is more harmful to the guide vanes and the runner. It not only propagates along the path of the vaneless space → guide vanes → stay vanes → spiral case, causing the strong mechanical vibration of the guide vanes, but also propagates into the flow path of the runner blades, so that the runner is subjected to high-frequency dynamic load variation and eventually breaks as shown in Fig. 3.32. Researchers have measured the vibration signal of the same frequency as the rotor–stator interaction in the turbine guide bearing through the vibration sensor, which indicates that the dynamic and static interference propagates along the main shaft of the turbine, causing the structure of the guide bearing to vibrate. In addition, the guide vane vibration also affects the pulsating amplitude of the rotor–stator interaction in the vaneless space.

2. **Harm of draft tube vortex rope**

The vortex rope is a low-frequency excitation source caused by cavity cavitation, and its harm is mainly manifested in the frequency domain characteristics. Its frequency is in the range of the natural frequencies of the hydropower plant, making resonance more likely to occur. When the vortex rope frequency is consistent with the system frequency, the resonance will cause the unit discharge and output to oscillate, as shown in Fig. 3.33, which will do great harm to the operation stability of the power grid. CFD studies have shown that the source of the vortex rope excitation is located at the elbow of the draft tube.

3. **Harm of rotating stall**

The unique source of pressure pulsation for such pump-turbine may occur in both turbine and pump operating conditions. In the turbine operating condition, it occurs in the turbine braking zone after the runaway point.

Due to the S-shaped characteristics of pump-turbine, the prototype unit cannot be stabilized in the region where the rotating stall occurs. In normal operation, there

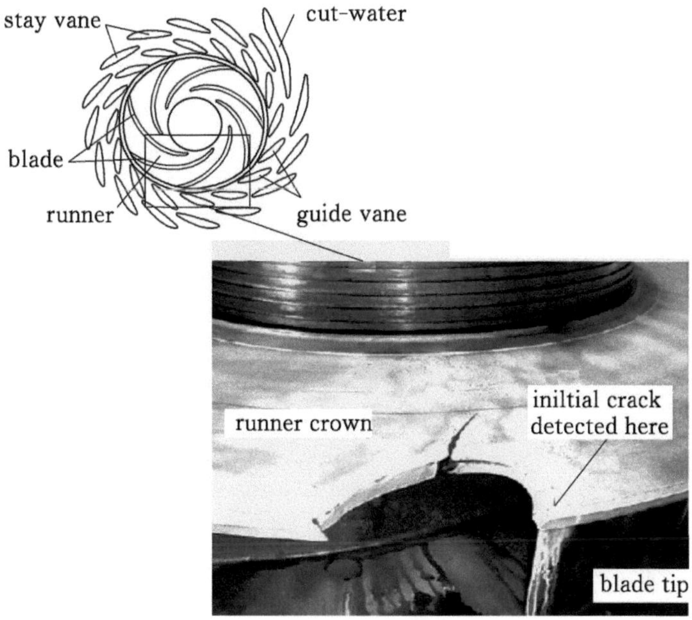

Fig. 3.32 Damage of rotor–stator interaction on the runner

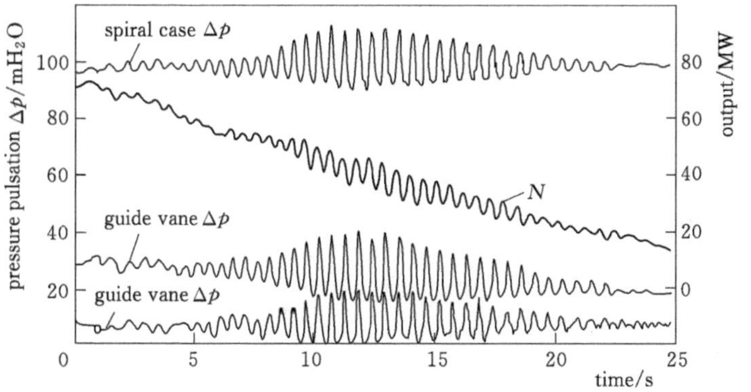

Fig. 3.33 Draft tube vortex band that causes swing output

should have no rotating stall, and only during transient processes passing through the runaway point to the turbine braking zone the rotating stall can appear in a short time. Model experiments show that the rotating stall will produce local high pressure in the runner channels, as shown in Fig. 3.34, which means that the rotating stall is the reason for the low frequency and high amplitude of pressure pulsations at the spiral casing and the vaneless space.

Fig. 3.34 Rotating stall that causes local high-pressure in the runner

Questions

1. How to describe the flow patterns of the hydro turbine with a velocity triangle? What parameters determine its shape and size?
2. How to get the basic equation of a hydro turbine from the velocity triangle? Is there any difference between the basic equation of the reaction turbine and the impulse turbine?
3. What are the influencing factors of turbine energy loss? How to define the optimal operation condition? What is the difference between the optimal conditions of an impulse turbine and a reaction turbine?
4. Try to analyze the working principle of the draft tube. How to evaluate its performance?
5. What is the cause of cavitation? What are the types of cavitation and cavitation erosion of hydro turbines? How to avoid cavitation erosion?
6. What is the cause of pressure pulsations of the hydro turbine? What are the types? What are the characteristics of each type?

Bibliography

1. Zhongyuan, Jin. 1992. *Hydraulic machinery*, 2nd ed. Beijing: Water Resources and Electric Power Press.
2. Qizhao, Liu and H. Ming. 2010. *Hydropower station*, 4th ed. Beijing: Water Resources and Electric Power Press.
3. Ruyun, Luo. 1990. *Design manual for power equipment of hydropower station*. Beijing: Water Resources and Electric Power Press.
4. *Mechanical and electrical design manual of hydropower station, hydraulic machinery*. 1983. Beijing: Water Resources and Electric Power Press.

5. Yuan, Zheng. 2011. *Hydraulic turbine.* Beijing: China Water and Power Press.
6. Dakai, Liu. 1996. *Hydraulic turbine*, 3rd ed. Beijing: China Water and Power Press.
7. Changzhi, Zhang. 1988. *Hydraulic turbine: principle and mathematical model.* Wuhan: Huazhong Institute of Technology Press.
8. Rangqin, Gong. 2006. *Basic technical conditions of hydraulic turbine.* Beijing: China Standards Press.
9. Yuan, Zheng, and Zhang Qiang. 2003. *Power equipment of hydropower station.* Beijing: China Water and Power Press.
10. Quanlong, Wang. 2009. *Basic technical conditions of Francis pump turbine.* Beijing: China Standard Press.
11. Xilin, Sha. 1999. *Tubular hydropower station.* Beijing: China Water and Power Press.
12. Wenwu, Song. 2015. *Theory and application of high head tubular turbine.* Beijing: Science Press.
13. Zanpei, You. 2009. *Bulb tubular hydropower station.* Beijing: China Water and Power Press.
14. Guoxuan, Liu. 2006. *Operation and maintenance of bulb tubular turbine generator.* Beijing: China Water Resources and Hydropower Press.
15. Bo, Yu. 2008. *Principle and operation of hydraulic turbine.* Beijing: China Electric Power Press.
16. Wentong, Zhou. 2007. *Basic theory and design of Pelton turbine.* Beijing: China Water and Power Press.
17. Xinfang, Chen. 1994. *Structural analysis of hydraulic turbine.* Beijing: China Water and Power Press.
18. Wentao, Su. 2016. *Hydraulic stability of hydraulic turbine.* Harbin: Harbin Institute of Technology Press.
19. Jinshi, Chang. 2016. *Cavitation and cavitation erosion of turbine and pump.* Beijing: Science Press.
20. Yulin, Wu. 2007. *Cavitation and solid-liquid two-phase hydrodynamics of hydraulic machinery.* Beijing: China Water and Power Press.
21. Shuhong, Liu. 2007. *Fundamentals of hydromechanics of hydraulic machinery.* Beijing: China Water and Power Press.
22. Hongquan, Xu and W. Wanpeng. 2010. Cavitation coefficient of hydraulic turbine and its influence on hydraulic performance. *Large Electric Machine Technology* 5: 44–47.
23. Evaluation of cavitation erosion of reaction turbine: GB/T 15469–1995. 1995. Beijing: China Standards Press.
24. Rongsheng, Nie. 1985. *Cavitation and cavitation erosion in hydraulic turbine.* Beijing: Water Resources and Electric Power Press.
25. Shuguang, Dai. 2003. Cavitation safety factor and specific speed of hydraulic turbine. *Water Conservancy and Hydropower Technology* 2: 52–53.
26. Earls Brennan, Christopher. 1995. *Cavitation and bubble dynamics.* Oxford: Oxford University Press.

Chapter 4
Turbine Characteristics and Selection

Jinping Li and Yongguang Cheng

Abstract In this chapter, we will discuss the following key points: mechanical similarity, unit parameters, and specific speed of hydro turbine; efficiency conversion and unit parameter correction of hydro turbine; model test and comprehensive characteristic curves of turbine; performance and operation characteristic curves of turbine; selection of turbine types and parameters; and dimension determination of spiral case and draft tube. Students should be acquainted with the similarity law and the unit parameters of hydro turbine; master the efficiency conversion and unit parameter correction method; be familiar with the model test and characteristics of hydro turbine; master the selection method of turbine; be familiar with determining the dimensions and drawing profiles of spiral case and draft tube.

Keywords Similarity · Unit parameter · Efficiency correction · Model test · Comprehensive characteristics · Turbine selection · Dimension determination

4.1 Similarity Law, Unit Parameters, and Specific Speed of Hydro Turbine

In Chap. 3, the working principle of turbine has been introduced and the basic equation of turbine has been obtained. The premise of the equations assumes that the water viscosity is ignored and the water flow is uniformly axisymmetric. In fact, the flow of water in the turbine runner is viscous turbulence, especially in the non-optimal conditions. Because the hydraulic phenomenon is more complicated, it is impossible to theoretically derive the actual operating characteristics of the turbine, namely, the relations between various operating parameters (such as head, discharge, speed, output, efficiency, and cavitation coefficient) under different working conditions. Therefore, the combination of model tests, and theoretical analysis and calculation are widely used to study and acquire the characteristics of the turbine.

J. Li (✉) · Y. Cheng
Wuhan University, Wuhan, China
e-mail: lukeping@whu.edu.cn

Y. Cheng et al. (eds.), *Hydropower Station*,
https://doi.org/10.1007/978-981-96-2383-9_4

Turbine model tests use the model runners with smaller diameters and lower heads. It has the advantages of low cost, accurate measurement, and arbitrary change of operating parameters. However, how to apply the test results of the model turbine to the prototype turbine is the problem studied by the similarity theory of the turbine. Specifically, it is to find the similarity constants under certain conditions and establish the similarity formulas.

4.1.1 Similarity Condition of Hydro turbine

To make the model turbine similar to the prototype turbine, it is necessary to make the water flow motion of the two meet the similarity conditions of fluid mechanics, namely, the three conditions: geometric similarity, kinematic similarity, and dynamic similarity must be satisfied.

1. **Geometric similarity**

Geometric similarity means that all corresponding angles of the geometry of the two turbine flow channels are equal, and all corresponding dimensions are proportional, as shown in Fig. 4.1. Geometrically similar turbines are called the same series of turbines or homologous turbines.

$$\beta_{e1} = \beta_{e1M};\ \ \beta_{e2} = \beta_{e2M};\ \ \phi = \phi_M \tag{4.1}$$

$$\frac{D_1}{D_{1M}} = \frac{b_0}{b_{0M}} = \frac{a_0}{a_{0M}} = \cdots = \text{const.} \tag{4.2}$$

where β_{e1}, β_{e2}, ϕ are the blade leading edge angle, the blade trailing edge angle, and the rotation angle of rotatable runner blades, respectively; D_1, b_0, a_0 are the nominal diameter of the runner, the height of guide vanes, and the opening of guide vanes, respectively. The subscript M means the parameter is for the model turbine, otherwise the parameter is for the prototype turbine, and the following follows this denotation.

2. **Kinematic similarity**

Kinematic similarity means that for the geometrically similar/homologous turbines, the directions of flow velocities in the corresponding points in the flow passage are the same, and the magnitudes of velocities are proportional. That is, the velocity triangles of the corresponding points are similar and in an isogonic state, as shown in Fig. 4.1.

$$\alpha_1 = \alpha_{1M}\ \ \beta_1 = \beta_{1M};\ \ \alpha_2 = \alpha_{2M} \tag{4.3}$$

$$\frac{v_1}{v_{1M}} = \frac{u_1}{u_{1M}} = \frac{w_1}{w_{1M}} = \frac{v_2}{v_{2M}} = \cdots = \text{constant} \tag{4.4}$$

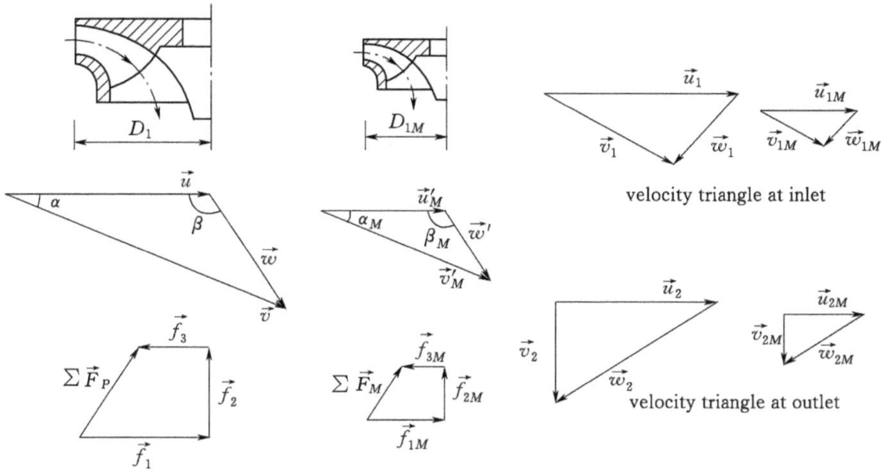

Fig. 4.1 Similar velocity triangles

3. **Dynamic similarity**

Dynamic similarity means that for the geometrically similar/homologous turbines, the flow force directions at the corresponding points in the flow passages, such as pressure, inertial force, gravity, viscous force, and friction force, are the same, and the force magnitudes are proportional.

It should be pointed out that it is very difficult to fully satisfy the above three similarity conditions when testing the turbine model and converting the model parameters into the prototype turbine parameters. For example, the similarity of the relative roughness, written as $\frac{\Delta}{D_1} = \frac{\Delta_M}{D_{1M}}$, is very difficult to achieve. Therefore, the influence of secondary factors such as relative roughness, water viscous force, and gravity has to be neglected to grasp the main factors and obtain preliminary similarity formulas. After that, the deficiencies can be corrected, making it easier and more reasonable to convert the model test results to the prototype turbine parameters.

4.1.2 Similarity Law of Hydro turbine

The fixed relations between the parameters of turbines that are under the homologous and isogonic conditions are called the similarity law of the hydro turbine, which is a concrete manifestation of the similarity principle.

The nominal diameters of the two turbines with similar geometry are D_1 and D_{1M}, respectively, and the guide vane opening angles and the runner blade angles are correspondingly equal. In order to make the operating conditions of the two turbines similar, the fixed relations between the discharge Q and Q_M, the rotational speeds n and n_M, and the output N and N_M should be found.

Under any homologous and isogonic conditions, the velocity triangles at each point in the flow passages have the same proportional relation, which can be transformed by the triangular sinusoidal relation

$$\frac{v}{\sin\beta} = \frac{w}{\sin\alpha} = \frac{u}{\sin(180^\circ - \beta - \alpha)}. \tag{4.5}$$

From Eq. (4.5), the following relation can be obtained

$$v_1 = u_1 \frac{\sin\beta_1}{\sin(\beta_1 + \alpha_1)} \tag{4.6}$$

$$v_2 = u_2 \frac{\sin\beta_2}{\sin(\beta_2 + \alpha_2)} \tag{4.7}$$

and the angular velocity of the runner rotation has the following relation with the linear velocity

$$\omega = \frac{u_1}{r_1} = \frac{u_2}{r_2}. \tag{4.8}$$

Substituting Eqs. (4.6)–(4.8) into the basic equation of hydro turbine and eliminating v_1, v_2, and u_2 yield

$$u_1 = k_{u1}\sqrt{2gH\eta_h} \tag{4.9}$$

where the velocity coefficient k_{u1} can be formed as

$$k_{u1} = \sqrt{\frac{1/2}{\frac{\sin\beta_1\cos\alpha_1}{\sin(\beta_1+\alpha_1)} - \left(\frac{r_2}{r_1}\right)^2 \frac{\sin\beta_2\cos\alpha_2}{\sin(\beta_2+\alpha_2)}}}.$$

Using the same method, the relations among u_2, v_1, v_2, w_1, w_2 and H are formed. Obviously

$$k_{u1} = f_{u1}\left(\alpha, \beta, \frac{r_2}{r_1}\right). \tag{4.10}$$

Because Eq. (4.9) is consistent with the orifice outflow formula in hydraulics, the coefficient k_{u1} is called the velocity coefficient. But the expression of the coefficient is much more complicated, because it is not only related to the velocity triangles, but also to the ratio of the runner inlet and outlet diameters, along with the type of runner.

For a known turbine, when the operating condition varies, velocity angles α and β vary, thereby the coefficient k_{u1} changes. Two coefficients, such as k_{u1} and k_{v1}, can determine the shape of the velocity triangle, which represents an operating condition of a hydro turbine. Since the velocity coefficient depends on the velocity triangle, and the velocity triangle depends on the parameters of H, n and α, thus $k_{u1} = f_{u1}(H, n, \alpha, D)$.

For two turbines with similar geometry, if the velocity triangles are similar, namely meet the homologous and isogonic condition, the velocity coefficients of the two turbines are equal.

1. **Speed similarity law**

Rewriting Eq. (4.9), we obtain the following form

$$k_{u1} = \frac{u_1}{\sqrt{2gH\eta_h}} = \frac{D_1}{2}\frac{2\pi n}{60} \Big/ \sqrt{2gH\eta_h}$$

that is

$$\frac{nD_1}{\sqrt{H\eta_h}} = 84.6k_{u1}. \tag{4.11}$$

Under homologous and isogonic condition, the velocity coefficients k_{u1} are the same, therefore

$$\frac{nD_1}{\sqrt{H\eta_h}} = \left(\frac{nD_1}{\sqrt{H\eta_h}}\right)_M = \text{Const.} \tag{4.12}$$

Equation (4.12) is called the similarity law of the turbine speed, which means that under an isogonic state, the speed of the homologous turbines is inversely proportional to the runner diameter, additionally proportional to the square root of the working head.

2. **Discharge similarity law**

The effective discharge through the turbine runner can be calculated as follows

$$Q\eta_{vol} = v_{m1}F_1 = v_1 \sin\alpha_1 \cdot \pi D_1 b_0 f = v_1 \sin\alpha_1 \cdot \alpha D_1^2 \tag{4.13}$$

where b_0 is the height of guide vanes, m; f is the blade extrusion coefficient of the runner inlet; α is the comprehensive coefficient, $\alpha = \pi f b_0/D_1$,

and

$$k_{v1} = \frac{v_1}{\sqrt{2gH\eta_h}}. \tag{4.14}$$

Substituting Eqs. (4.13) into (4.14) yields

$$\frac{Q\eta_{vol}}{D_1^2\sqrt{H\eta_h}} = 4.44\alpha \sin\alpha_1 k_{v1}. \tag{4.15}$$

Under homologous and isogonic condition, the velocity coefficients k_{v1} are the same and $\alpha \sin \alpha_1 = (\alpha \sin \alpha_1)_M$, thus

$$\frac{Q\eta_{vol}}{D_1^2\sqrt{H\eta_h}} = \left(\frac{Q\eta_{vol}}{D_1^2\sqrt{H\eta_h}}\right)_M = \text{Const.} \tag{4.16}$$

Equation (4.16) is called the discharge similarity law of hydro turbine. It means that when the homologous turbines operate under the isogonic state, the effective discharge is proportional to the square of the runner diameter, and additionally proportional to the square root of the working head.

3. **Output similarity law**

The turbine output expression is expressed as

$$N = \gamma Q H \eta \tag{4.17}$$

Substituting Eqs. (4.15) into (4.17) yields

$$\frac{N}{D_1^2(H\eta_h)^{3/2}\eta_{mec}} = 9.81 \times 4.44\alpha \sin \alpha_1 k_{v1} \tag{4.18}$$

In the same way, it is concluded that

$$\frac{N}{D_1^2(H\eta_h)^{3/2}\eta_{mec}} = \left(\frac{N}{D_1^2(H\eta_h)^{3/2}\eta_{mec}}\right)_M = \text{Const.} \tag{4.19}$$

Equation (4.19) is called the output similarity law of hydro turbine. It means that when the homologous turbines operate under the isogonic state, the output of the hydro turbine is proportional to the square of the runner diameter, and additionally proportional to the 3/2 power of the working head.

4.1.3 Unit Parameters

The turbines that satisfy the homologous condition can be made in different sizes and can operate in different heads, speeds, and outputs. Therefore, to characterize the characteristics of hydro turbines with similar geometries, there must be common measurement standards, which are the unit. parameters.

If the similarity laws of hydro turbine are directly quoted as a measurement standard, there are two problems as follows.

(1) The similarity laws of hydro turbine contain the hydraulic efficiency η_h, volumetric efficiency η_{vol} and mechanical efficiency η_{mec} of turbines, which are difficult to separate from the overall efficiency η of the hydro turbine, and the

overall efficiency of the prototype turbine is unknown. Therefore, the direct use of the similarity laws cannot obtain accurate results, and it is impossible to use it as common measurement standards of hydro turbine.

(2) Due to the different test equipment conditions and test requirements of research institutions, the runner diameter D_{1M} and test head H_M of the model turbine are different, thus the model parameters obtained from model tests are not uniform. This problem hinders not only turbine development and applications, but also the comparison of different model test results for the same geometric similar turbines, and even makes it impossible to compare the performance of different homologous turbines. Therefore, under the premise assumption that the prototype and model turbines are equal in efficiency, the parameters obtained by any model tests are converted into the parameters under the standard conditions of $D_{1M} = 1\text{m}$ and $H_M = 1\text{m}$ according to the similarity laws.

The parameters obtained by the conversion under the standard conditions are called unit parameters, namely

$$n_1' = \frac{nD_1}{\sqrt{H}} \quad (\text{r/min}) \tag{4.20}$$

$$Q_1' = \frac{Q}{D_1^2\sqrt{H}} \quad (\text{m}^3/\text{s}) \tag{4.21}$$

$$N_1' = \frac{N}{D_1^2 H^{3/2}} \quad (\text{kW}) \tag{4.22}$$

where n_1', Q_1', and N_1' are the unit speed, unit discharge, and unit output, respectively.

Obviously, the values of the unit parameters are the same for homologous turbines under similarity conditions. However, for the turbines without geometric similarity, even under characteristic conditions (such as the optimal condition, and the limited condition), they are different. Therefore, comparing the unit parameters under characteristic conditions can be used to evaluate the performance of different turbines with different geometric features for the design and selection of hydro turbines.

4.1.4 Specific Speed

Since the unit parameters of unit speed $n_1' = \frac{nD_1}{\sqrt{H}}$, unit discharge $Q_1' = \frac{Q}{D_1^2\sqrt{H}}$, and unit output $N_1' = \frac{N}{D_1^2 H^{3/2}}$ contain the diameter D_1 of the runner, it is still inconvenient to characterize the feature of a kind of geometrically similar hydro turbines. Therefore, it is needed to remove D_1 and introduce the concept of specific speed n_s, with

$$n_s = n_1'\sqrt{N_1'} = \frac{nD_1}{\sqrt{H}}\sqrt{\frac{N}{D_1^2 H^{3/2}}} = \frac{n\sqrt{H}}{H^{5/4}} \quad (\text{m} \cdot \text{kW}) \tag{4.23}$$

This definition states that the specific speed of a hydro turbine is the rotational speed of the hydro turbines that satisfy geometric similarity and operate in the specific condition of head $H = 1$ m and output $N = 1$ kW.

Substituting $N = \gamma Q H \eta$, $n = n'_1 \sqrt{H}/D_1$ and $Q = Q'_1 D_1^2 \sqrt{H}$ into Equation (4.23), we have other two expressions about the specific speed

$$n_s = 3.13 \frac{n\sqrt{Q\eta}}{H^{3/4}} \quad (\text{m} \cdot \text{kW}) \tag{4.24}$$

$$n_s = 3.13 n'_1 \sqrt{Q'_1 \eta} \quad (\text{m} \cdot \text{kW}) \tag{4.25}$$

1. **Constant property of specific speed**

For the geometrically similar turbines under similar operating conditions, the specific speed is a constant, independent of the runner diameters of the turbines. It reflects the relation between the turbine speed n, working head H, and output N, and it is an important comprehensive parameter representing the characteristics of hydro turbines.

Although the hydro turbines are geometrically similar, namely, the homologous turbines, the specific speed n_s is different if the operating conditions are different, because the velocity triangles are different, namely not in an isogonic state. Therefore, in order to use n_s to represent the turbine series, the representative working conditions should be selected. The design condition is usually selected, namely, the specific speed under the operating condition with the design or rated head, rated speed, and rated output are taken as the representative characteristic parameters of the geometrically similar turbines. Some manufacturers also use the specific speed under the optimal operating conditions as a representative.

2. **Unit property of specific speed**

It can be seen from Eq. (4.25) that increasing the specific speed n_s of the hydro turbine under a certain working head means that the values of $Q'_1 \eta$ and/or n'_1 increase. When $Q'_1 \eta$ increases, the output of the turbine increases because $N = 9.81\, D_1^2 H^{3/2} Q'_1 \eta$. Increasing n'_1 means the runner diameter of the turbine can be reduced, when the head and output remain unchanged. Then the unit size, the unit weight and the investment of the hydropower station can also be reduced. Therefore, increasing specific speed is beneficial to the dynamic economic and technical indicators of the unit and hydropower station, especially for large units and large hydropower stations. In a word, higher specific speed hydro turbines should be selected as far as possible.

3. **Relationship between specific speed and cavitation coefficient**

For a hydro turbine with a certain discharge, increasing the specific speed n_s will reduce the runner diameter D_1 and increase the runner outlet velocity v_2. According to the expression of cavitation coefficient $\sigma = \left(\frac{\alpha_2 v_2^2}{2g} - \frac{\alpha_5 v_5^2}{2g} - h_{2-5} \right) \Big/ H$, the larger the velocity v_2, the larger the cavitation coefficient σ. That is, the larger the specific speed n_s, the larger the cavitation coefficient σ. To increase the cavitation coefficient, the turbine runner must be buried deeper below the downstream water level. For the ground hydro power station, it will increase the construction difficulty and civil construction investment. Therefore, the cavitation conditions limit the application of high specific speed turbines in higher head hydropower stations.

4. **Relationship between specific speed and characteristics of runner shape**

It can be seen from Eq. (4.25) that when the unit speed n_1' is fixed, the unit discharge Q_1' is larger if the specific speed n_s is larger. Therefore, the relative height b_0/D_1 of the guide vanes is also higher, while the number of runner blades is reduced. And because the cavitation coefficient increases with the increase of specific speed, for high specific speed turbines, in order to reduce the cavitation coefficient, it is necessary to reduce the runner outlet velocity v_2, and accordingly increase the diameter ratio D_2/D_1 of the outlet to the inlet of Francis turbines, or reduce the hub ratio d_g/D_1 of axial flow turbines.

In short, the specific speed is the most important representative parameter for the comprehensive performance of hydro turbines, and it has important guiding functions in the design, manufacture, selection, and operation of hydro turbines. At present, the specific speed is used as the basis for the classification of hydro turbines, namely, Pelton turbine $n_s = 10$ to 70, Francis turbine $n_s = 60$ to 350, Diagonal flow turbine $n_s = 200$ to 450, Axial flow turbine $n_s = 400$ to 900, and tubular turbine $n_s = 600$ to 1100. With the continuous development and application of new technologies, new processes, and new materials, the specific speed values of various types of turbines are still further increased.

4.2 Efficiency Conversion and Correction of Unit Parameters of Hydro Turbine

4.2.1 Efficiency Conversion of Hydro turbine

The unit parameter expressions (4.20)–(4.22) are obtained under the assumption that the efficiency values of the model turbine and the prototype turbine are equal under the similarity conditions, but the efficiency values of the two are different in practice. Therefore, it is necessary to convert the efficiency of the model turbine obtained from the model test, and consider the influence factors to obtain the efficiency of the prototype turbine.

The main reason for the efficiency difference between the prototype and model turbines is that the runner diameters and the working heads are quite different. Therefore, the relative roughness and relative viscous force of the prototype turbine are smaller, thus the relative hydraulic loss is smaller. That is the efficiency of the prototype turbine is always higher than that of the model turbine, usually with more than 2%. However, in the past ten years, with the improvement of the technology level of the model turbine, the model turbine efficiency has been significantly improved. For example, the maximum efficiency of the model turbine for the Goupitan Hydropower Station is 95.17%; the maximum efficiency of the model turbine for the Three Gorges Left Bank Power Station is 95.26%; the maximum efficiency of the model turbine for the Pengshui Hydropower Station is 94.92%; and the maximum efficiency of the model turbine for the Longtan Hydropower Station is 94.76%.

In the total efficiency of hydro turbine, the hydraulic loss is dominant, and the volume loss and mechanical loss are small. In order to simplify the efficiency conversion, ignoring the volume loss and the mechanical loss, the hydraulic efficiency η_h can approximately replace the total efficiency η of hydro turbine.

Assuming that the hydraulic viscous friction loss in the turbine is similar to the pipeline flow, the hydraulic loss under any condition can be expressed as

$$\Delta h = \lambda \frac{l}{d} \frac{V^2}{2g} + \zeta \frac{\omega^2}{2g} \tag{4.26}$$

where λ is the resistance coefficient along the path, $\lambda = f\left(Re, \frac{\Delta}{D_1}\right)$. The larger Reynolds number Re and the smaller the relative roughness $\frac{\Delta}{D_1}$, the smaller the λ; l and d are the linear length and equivalent diameter of the runner flow, respectively, m, which can be expressed as $k = l/d$; V is the flow velocity, m/s; ζ is the vortex loss coefficient; and ω is the vortex water flow velocity, m/s.

Under the optimal condition, the turbine flow passage has no vortex, and its surface can be regarded as hydraulically smooth. At this time, λ is only related to Re, which can be expressed as

$$\lambda = 0.316/(Re)^m \tag{4.27}$$

where m is the index.

Therefore, the hydraulic frictional friction loss of the hydro turbine under the optimal working condition is

$$\Delta h = \frac{0.316k}{R_e^m} \frac{V^2}{2g} \tag{4.28}$$

For a model turbine satisfying geometrical similarity, the same expression exists under the optimal condition

$$\Delta h_M = \frac{0.316k_M}{R_{eM}^m} \frac{V_M^2}{2g} \tag{4.29}$$

Comparing the above two equations and considering that V^2 is proportional to the head H, we have

$$\frac{\Delta h}{\Delta h_M} = \left(\frac{\mathrm{Re}_M}{\mathrm{Re}}\right)^m \frac{H}{H_M} \tag{4.30}$$

Because of $\Delta h/H = 1 - \eta_{h\max}$, rewrite Eq. (4.30), and get

$$\frac{1 - \eta_{h\max}}{1 - \eta_{h\max M}} = \left(\frac{\mathrm{Re}_M}{\mathrm{Re}}\right)^m \tag{4.31}$$

Based on Eq. (4.31) and assuming $\eta = \eta_h$, the formula for the turbine efficiency conversion recommended by the International Electrotechnical Commission (IEC) in the 1980s is as follows:

$$\eta_{\max} = 1 - (1 - \eta_{\max M})\left[0.3 + 0.7\left(\frac{\mathrm{Re}_M}{\mathrm{Re}}\right)^{0.16}\right] \tag{4.32}$$

Substituting Reynolds number $\mathrm{Re} = \frac{VD_1}{\nu} = \frac{D_1\sqrt{2gH}}{\nu}$, where ν is the viscosity coefficient of water, into Eq. (4.31) and making an empirical correction, we can derive the following formulas for turbine efficiency conversion, which were recommended by IEC in *Code for Acceptance of Hydraulic Turbine Model Test* in 1963.

For Francis turbines

$$\eta_{\max} = 1 - (1 - \eta_{\max M})\sqrt[5]{\frac{D_{1M}}{D_1}} \tag{4.33}$$

For Kaplan turbines

$$\eta_{\max} = 1 - (1 - \eta_{\max M})\left(0.3 + 0.7\sqrt[5]{\frac{D_{1M}}{D_1}}\sqrt[10]{\frac{H_M}{H}}\right) \tag{4.34}$$

Thus, the efficiency correction value for the maximum efficiency point of the prototype turbine is:

$$\Delta\eta = \eta_{\max} - \eta_{\max M} \tag{4.35}$$

The efficiency correction values for non-optimal working conditions are assumed to be the same with $\Delta\eta$ under the optimal working condition, therefore, $\eta = \eta_M + \Delta\eta$ for all working conditions.

For Kaplan turbines, different blade angles have different correction values, thus $\eta_\varphi = \eta_{\varphi M} + \Delta\eta_\varphi$.

For Pelton turbines, a reasonable diameter ratio is $D_1/d_0 = 10$ to 20, and in this range the efficiency does not change significantly with different turbine sizes, therefore no correction is needed, namely $\eta = \eta_M$.

Equations (4.32)–(4.34) are suitable for the optimal conditions. Actually, the conversion of turbine efficiency for other conditions was already studied, such as Miyagi formula

$$\eta = \eta_M + \left(1 - \eta_{\max M}\frac{Q_M}{Q_{\max M}}\right)^{\alpha}\left[1 - \left(\frac{D_{1M}}{D_1}\right)^{1/4}\left(\frac{H_M}{H}\right)^{1/8}\right] \tag{4.36}$$

where $\alpha = 1$, 3/4 and 1/3, corresponding to Francis turbines, fixed-blade propeller turbines, and Kaplan turbines, respectively.

4.2.2 Correction of Unit Parameters

Since the efficiency of the prototype turbine differs from the efficiency of the model turbine, the unit parameters n_1' and Q_1' should be corrected. Still assuming $\eta = \eta_h$, we can obtain

$$\frac{Q_{1M}'}{\sqrt{\eta_M}} = \frac{Q_1'}{\sqrt{\eta}} \tag{4.37}$$

$$\frac{n_{1M}'}{\sqrt{\eta_M}} = \frac{n_1'}{\sqrt{\eta}} \tag{4.38}$$

Under the optimal condition

$$Q_{10}' = Q_{10M}'\sqrt{\frac{\eta_{\max}}{\eta_{\max M}}} \tag{4.39}$$

where subscript 0 indicates the optimal condition. Then

$$\Delta Q_1' = Q_{10}' - Q_{10M}' = Q_{10M}' = \left(\sqrt{\frac{\eta_{\max}}{\eta_{\max M}}} - 1\right) \tag{4.40}$$

Similarly, it can be obtained that

$$\Delta n_1 = n_{10}' - n_{10M}' = n_{10M}'\left(\sqrt{\frac{\eta_{\max}}{\eta_{\max M}}} - 1\right) \tag{4.41}$$

The unit parameter corrections $\Delta Q_1'$ and $\Delta n_1'$ under non-optimal operating conditions can be equal to the corrections under the optimal condition, namely $Q_1' = Q_{1M}' + \Delta Q_1'$ and $n_1' = n_{1M}' + \Delta n_1'$.

In general, relative to the unit discharge Q_1', the correction $\Delta Q_1'$ is small, thus the correction can be canceled in practical applications. For the unit speed n_1', when the correction value $\Delta n_1' < 0.03 n_{10M}'$, the correction can also be canceled.

4.3 Model Test and Comprehensive Characteristic Curve of Hydro turbine

The model test of the hydro turbine is the work of reducing the prototype turbine to a model turbine at a certain geometric scale, and measuring the parameters of the model under a lower head and smaller discharge condition. The model test can provide the data for various working conditions. From the data, we can obtain the comprehensive parameters of the turbine, such as unit speed n_1', unit discharge Q_1', efficiency η, and cavitation coefficient σ by using the similarity formulas, along with the conversion and correction methods. Based on these parameters, we can draw the comprehensive characteristic curves, runaway characteristic curves, etc., for the model turbine.

It should be pointed out that if the normal operating conditions of the hydropower station are completely modeled in the model test, namely, the rotational speed is constant and the operating conditions are defined by different working heads and outputs, the test device should be very complicated or even difficult to implement. Therefore, the test process for different working conditions is generally achieved by changing the rotating speed and the guide vane opening (discharge) but keeping the working head fixed. The above two different modes of operation still constitute similar working conditions according to the similarity theory. Thus the results of the model test can be converted to parameters of the prototype turbine operation according to the similarity laws.

The task of the turbine model test is to measure the energy characteristics, cavitation characteristics, pressure pulsation characteristics, runaway characteristics, axial thrust, and mechanical characteristics of flow components under various working conditions.

4.3.1 Model Test Device for Hydro turbine

Before the 1990s, the energy performance and cavitation performance tests of the turbine were carried out by energy test rigs and cavitation test rigs, respectively. Nowadays, the high-head hydraulic machinery model test rig is usually used to test all the performance data of the hydro turbine.

Figure 4.2 is a high-head hydraulic machinery model test platform built in China in 2000. It is a high-parameter, high-precision hydraulic machinery universal test equipment. Its main function includes: carrying out hydraulic performance test of the reaction hydraulic machine; providing efficiency, cavitation, runaway, force characteristics, runner blade stress, and pressure pulsation; and observing flow patterns in various parts. There are two test benches provided, and the model tests can be carried out about the tubular flow, axial flow, and mixed flow hydraulic machines, respectively. The test benches can be adapted to the vertical shaft units or horizontal shaft units.

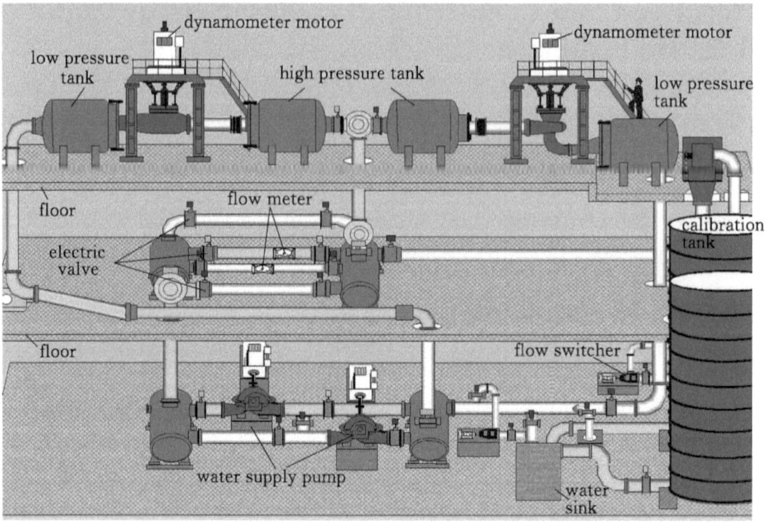

Fig. 4.2 High-head hydraulic machinery model test rig

The structure of the test platform is characterized by a set of power equipment and a computer data acquisition system for the two test benches A and B. When bench A is in the testing process, bench B can install another model unit for testing. When the testing in bench A ends, the system can switch to bench B for testing. The whole pipeline forms a closed circulation system. The names, parameters, and functions of the main components in the system are as follows.

(1) Pressure water tanks: The high pressure water tank is a horizontally installed cylindrical structure with inner diameter φ 2500 mm. The low-pressure tail water tank is also a horizontal installed cylindrical tank with diameter φ 3500 mm and length 4200 mm.
(2) Oil supply device: There are 8 sets of JG80/10 static pressure oil supply device, two of which are for backup. The oil supply pressure is 25 kg/cm^2, and the oil supply discharge is 7 L/min.
(3) Vacuum tank and vacuum pumps: They are the device for generating vacuum. There are 3 sets of H-70 valve type vacuum pumps.
(4) Water supply pumps: There are 2 sets of 24SA-10B double suction centrifugal pumps. The two pumps can be operated in series, parallel or single according to test requirements.
(5) Electric valves: The butterfly valve with diameter φ 500 mm and φ 600 mm are used to switch on/off the pipes of the system to meet the requirements of various operating modes of the test bench.
(6) Calibration tanks: Two calibration tanks with an effective volume 100 m^3 for each are installed. The channel between the two calibration tanks is connected by a pipe, on which a switching valve is installed. The water level is measured by the water level gauges or differential pressure sensors.

(7) Flow meters: There are electromagnetic flow meters with diameters φ 300 mm and φ 500 mm, for calibrating small and large discharges, respectively. By pipeline and valve switching, the flow meter can guarantee one-way flow operation under any working conditions. The flow meters are manufactured by ABB, Germany, and the model number is PROMAG33, which can measure two-way flows in a discharge range 0–2 m^3/s with an accuracy of $\pm 0.15\%$.
(8) Water sink (Reservoir): The system circulating water is supplied by the water sink with a storage capacity of $750m^3$.
(9) Flow switcher: It is used to switch the water flow when calibrating the discharge. The action time of one stroke is 0.02 s, and the servomotor is driven by compressed air.
(10) Dynamometer motors: The model number of the DC dynamometers is ZC56/32-4. With a capacity of 600 kW, they can be operated according to motor or generator mode with the maximum speed of 2500 r/min.

The highest head of the test rig is 160 m, the model runner diameter is 300–500 mm, and the overall efficiency error is less than 0.20%.

4.3.2 *Measurement of Test Parameters*

The parameters required to measure in turbine model tests include head, discharge, speed, output, guide vane opening, pressure pulsations at some related measuring points (such as the inlet of spiral case, the vaneless space, the upstream and downstream sides of the draft tube straight cone, and the downstream of draft tube elbow) and internal flow patterns (such as separation and cavitation on the blade back, blade passage vortex, and draft tube flow and vortex rope).

(1) Head H (mH_2O): It is defined as the specific energy difference between the spiral case inlet and the draft tube outlet

$$H = \frac{1000 p_{1-2}}{\rho g} + \frac{Q^2}{2g}\left(\frac{1}{A_1^2} - \frac{1}{A_2^2}\right) \tag{4.42}$$

where p_{1-2} is the pressure difference between the inlet of the spiral case and the outlet of the draft tube, kPa; p_{1-2} can be measured by a high-precision differential pressure sensor; ρ is the average density of water at the inlet of the spiral case and at the outlet of the draft tube, kg/m^3; g is the gravitational acceleration, m/s^2; Q is the discharge of hydro turbine, m^3/s; A_1 and A_2 are the cross-sectional areas of the spiral inlet and the draft tube outlet, respectively, m^2.
(2) Discharge Q (m^3/s): It is determined by an electromagnetic flow meter.
(3) Speed n(r/min): It is usually measured by a photoelectric encoder or a photoelectric speedometer.

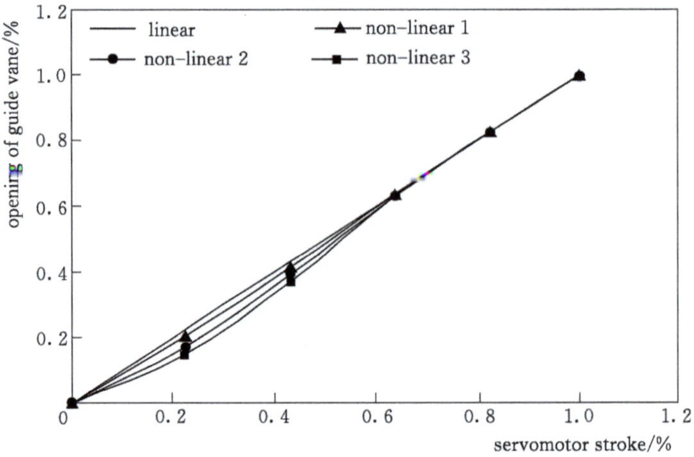

Fig. 4.3 Relationship between the servomotor stroke and the guide vane opening

(4) Input of hydro turbine N_w (kW): It is calculated by

$$N_w = 0.001\rho_1 QH \tag{4.43}$$

where ρ_1 is the water density at the spiral case inlet, kg/m^3.

(5) Output of hydro turbine N (kW): It is calculated by

$$N = 0.001\pi nT/30 \tag{4.44}$$

where T is the torque of shaft, N·m. T can be measured by the dynamometer motor.

(6) Opening of guide vane a in mm or (°): The guide vane opening and the servomotor stroke are linear or nonlinear as shown in Fig. 4.3. The high-precision stroke sensor can be used to measure the servomotor stroke to calculate the corresponding vane opening.

(7) Pressure pulsation p (kPa): It is usually measured by a high-precision pressure pulsation sensor, and the amplitude and frequency of the pressure pulsation are statistically analyzed.

(8) Observation of flow patterns: It is captured and recorded by industrial endoscopes and stroboscopes.

4.3.3 Test Method and Data Recording

The experimental data recording and sorting of turbine model tests are automatically completed by the computer data acquisition system. The manual intervention is mainly to determine the test range and the spacing between the various operating conditions.

1. **Efficiency test**

The efficiency test is carried out under conditions in which the head is constant and there is no cavitation. The purpose is to draw the comprehensive characteristic curves of the turbine model, so the test range should include and moderately exceed the full operating range of the prototype turbine. The test operating points are placed into grids according to the fixed guide vane opening and the fixed unit speed, namely, the guide vane opening changes from small to large, for example, 7–40 mm with the opening difference 3 mm, and the unit speed also changes from small to large, for example, 55–90 r/min with the unit speed difference 5 r/min. The acquisition time of each working point is 60 s. The calculation methods of efficiency, unit speed, and unit discharge are as follows

$$\eta_M = \frac{N_M}{N_{wM}} = \frac{0.001\pi nT/30}{0.001\rho_1 QH} = 0.10472\frac{nT}{\rho_1 QH} \tag{4.45}$$

$$n'_{1M} = \left(\frac{nD_1}{\sqrt{H}}\right)_M \quad (\text{r/min}) \tag{4.46}$$

$$Q'_{1M} = \left(\frac{Q}{D_1^2\sqrt{H}}\right)_M \quad (\text{m}^3/\text{s}) \tag{4.47}$$

For Kaplan turbines, in every fixed-blade angle, the above process is repeated one time to obtain the efficiency curve for this blade angle.

2. **Cavitation experiment**

The cavitation test is carried out by changing the pressure of the tail water tank to change the cavitation coefficient under the condition of constant guide vane opening and fixed unit speed. That is, the vacuum of the tail water tank is gradually changed, making cavitation to happen on the turbine runner, until reaching the limit of the vacuum equipment. Every change of the vacuum in the tail water tank corresponds a change of the turbine working head. In order to maintain the unit speed, the speed of the dynamometer motor needs to be adjusted, and the pressure difference p_{1-2} (kPa) between the spiral case inlet and the draft tube outlet, discharge Q (m^3/s), rotational speed n (r/min), shaft torque T (N m), and tail water absolute pressure p_2 (kPa) need to be measured accordingly. The parameters $H, N_w, N, \eta_M, n'_{1M}$, and Q'_{1M} are calculated according to Eqs. (4.42)–(4.47), respectively. And the cavitation coefficient σ_p of the hydro power station and unit output N'_{1M} can be calculated by Eqs. (4.48) and (4.49), respectively.

$$\sigma_p = \frac{p_2/\rho_2 g - p_v/\rho_2 g - H_s}{H} \tag{4.48}$$

where ρ_2 is the water density at the draft tube outlet, kg/m^3; p_v is the vaporization pressure associated with water temperature, kPa; H_s is the vertical distance from the tail water level to the model turbine installation elevation, m, namely the geometric suction height.

$$N'_{1M} = \left(\frac{N}{D_1^2 H^{3/2}}\right)_M \quad (4.49)$$

Then, for each operating point of a fixed guide vane opening and a fixed unit speed, draw the test curves of $\eta_M = f(\sigma_p)$, $Q'_{1M} = f(\sigma_p)$, and $N'_{1M} = f(\sigma_p)$ as shown in Fig. 4.4. The critical cavitation coefficient σ_c for each operating point is derived according to the code requirements of IEC.

With σ_c,n'_{1M} and Q'_{1M} of each operating point known, the contour of cavitation coefficients can be drawn on the comprehensive characteristic plane of the model turbine. Thus the isoclines of cavitation coefficients are the contour of the critical cavitation coefficients.

3. **Runaway experiment**

The speed of the turbine generator unit is constant during normal operation. When an accident occurs, the unit suddenly rejects all the load, and the load becomes zero, but the output of the turbine still exists. If the guide vanes refuse to close in this situation, then the water flow cannot be cut off, thus the unit speed will rise rapidly and reach the maximum value until the turbine output becomes zero, when the energy of water inflow is equal to the energy dissipated in the turbine. This speed is called the runaway speed n_f of the hydro turbine. Obviously, under different heads and different guide vane openings, the runaway speed is different. Therefore, it is necessary to obtain the unit runaway speed $n_{1fM}{}'$ under different vane openings by model tests.

$$n'_{1fM} = \left(\frac{n_f D_1}{\sqrt{H}}\right)_M \quad (4.50)$$

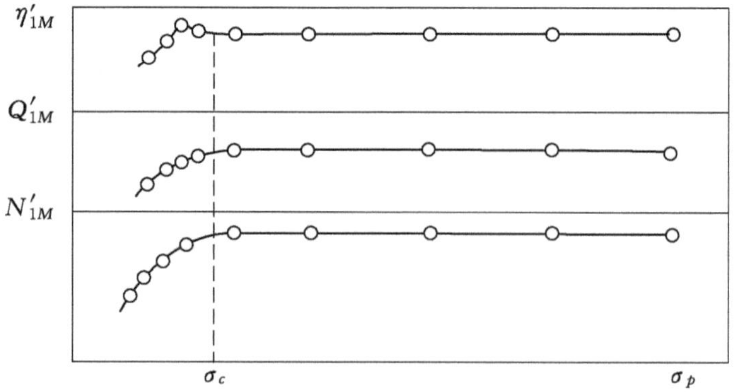

Fig. 4.4 Determination of critical cavitation coefficient σ_c

The process of testing the runaway speed characteristics is that with a given guide vane opening and dynamometer motor speed, adjust the test head to make the output torque approximately zero, and collect experimental data to calculate the unit runaway speed value. Then, the relations between the unit runaway speed and the unit runaway discharge corresponding to different vane openings can be plotted as a relationship curve, namely, the turbine runaway characteristic curve.

Due to the space limitations, the pressure pulsation tests, axial water thrust tests, etc., are not introduced in this textbook.

4.3.4 *Comprehensive Characteristic Curves and Runaway Characteristic Curve of Hydro turbine Model*

1. **Comprehensive characteristic curves of hydro turbine model**

Figure 4.5 shows the comprehensive characteristic curves of a model Francis turbine based on model test results. The characteristic curves are drawn in the plane with the unit speed and unit discharge as the vertical and horizontal coordinates, respectively. In the plane, there are the contours of guide vane opening, describing the unit discharge values at different openings; the contours of efficiency, describing the energy performance; the contour of critical cavitation coefficient, describing the cavitation performance; the lines of blade passage vortex development, blade passage vortex nascent, cavitation inception on the blade inlet front, and cavitation inception on the blade inlet back, describing the internal flow characteristics of the turbine; the boundaries of the normal working region, limiting the variation ranges of water head and output variation.

It can be seen from Fig. 4.5 that from left to right, the guide vane opening increases from small to large, and the unit discharge also increases accordingly. The shape and slope of the guide vane opening contours depend on the specific speed of the hydro turbine, namely the flow passage shape of the runner. As shown in Fig. 4.6, the discharge characteristics of a low specific speed turbine show the trend that the unit Q_1' discharge decreases as the unit speed n_1' increases, because the guide vane opening contours incline to the left. The trend for high specific speed turbine is opposite, the unit discharge Q_1' increases as n_1' increases, and the guide vane opening contours incline to the right. The guide vane opening contours of a medium specific speed turbine are substantially perpendicular to the Q_1' axis of the abscissa. The shape and slope of the guide vane opening contours will affect the water hammer pressure during transient processes and the selection of guide vane closing/opening law. For example, when a low specific speed turbine rejects the load and the guide vane begins closing, because of the left inclining guide vane opening contours, as the speed increases, n_1' increases, but Q_1' decreases, that is, the positive water hammer pressure will be larger. This is because the increase of speed will further accelerate the reduction of turbine discharge. Therefore, in order to reduce the positive water hammer pressure, it is usually appropriate to select the first-slow-then-fast closing law of guide vanes.

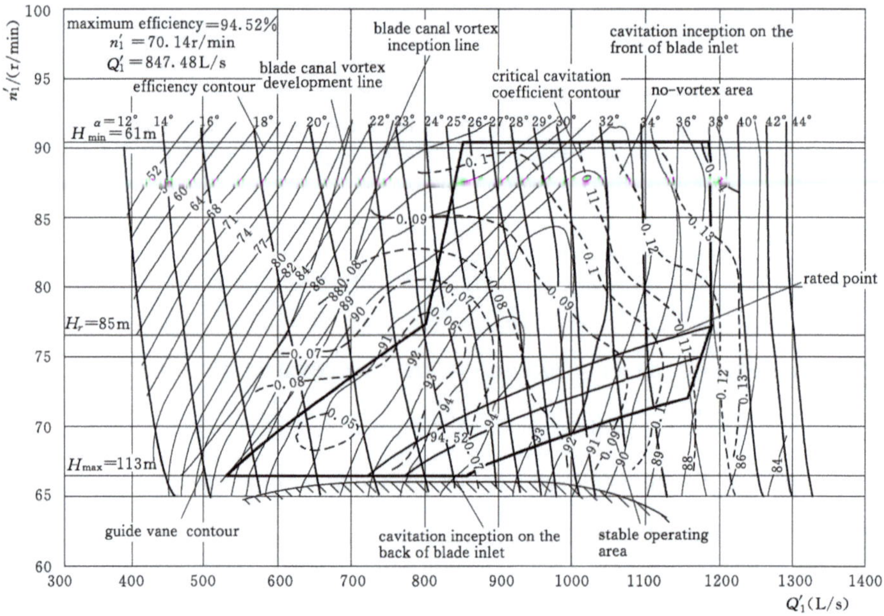

Fig. 4.5 Comprehensive characteristic curves of a certain model of Francis turbine

Fig. 4.6 Discharge characteristic curves of turbines with different specific speed, *a*-Pelton turbine; *b*-low specific speed Francis turbine; *c*-medium–high specific speed Francis turbine; *d*-fixed-blade propeller turbine; *e*-Kaplan turbine

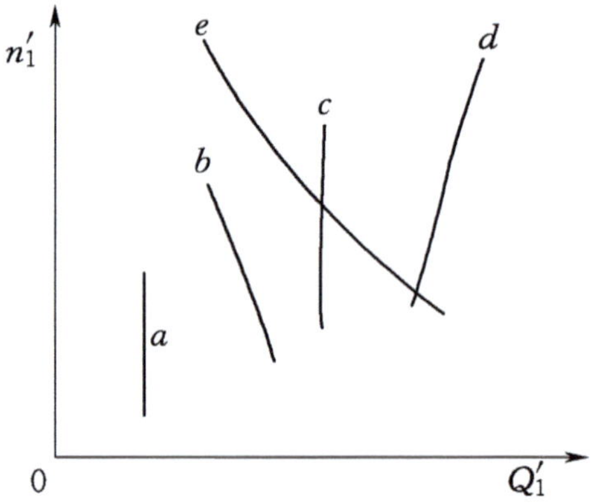

The efficiency contours are closed loops. The maximum efficiency of the model turbine is 94.52%, the operating point is $n_1' = 70.14$ r/min and $Q_1' = 847.48$ L/s. Under the premise of the same rated output, the higher the efficiency, the smaller the discharge of the turbine, and the more economical the operation of the hydropower station. The shape and gradient of the efficiency contours also depend on the specific speed of the turbine. As shown in Fig. 4.7, the shape of efficiency contour of a medium–high specific speed mixed flow turbine is close to an ellipse, the difference between the major axis and minor axis is small, and the major axis is at a small angle to the horizontal coordinate Q_1'. The sensitivity of efficiency to the variations of Q_1' (or output N) and n_1'(or head H) is moderate among all types of turbines. The 95% output limit line is close to the optimal efficiency zone. This kind of turbine is suitable for hydropower stations with medium and low head, small discharge, large head variation, and small load variation. The shape of efficiency contours of low specific speed Francis turbine is similar to a flat ellipse. The major axis and the minor axis have a great difference, and the long axis and the Q_1' coordinate axis are nearly parallel. The efficiency of the turbine is not sensitive to changes in Q_1' or output N, but is sensitive to variations in n_1' or head H. This type of turbine is suitable for hydropower stations with high head, small flow, small head variation, and large load variation. The shape of efficiency contours of Pelton turbine is flatter. The major axis and the minor axis have greater differences, and the long axis is parallel to the Q_1' coordinate axis. The efficiency of turbine is sensitive to variation of n_1' or head H, but is not sensitive to change of Q_1' or output N. Namely the output of this type of turbine can vary over a wide range, with efficiency η changed a little. This type of turbine is also suitable for hydropower stations with high head, small discharge, small head variation, and large load variation. The shape of the efficiency contours of Kaplan turbine is similar to an ellipse. The major axis and the minor axis have little difference, and the long axis is closely parallel to the Q_1' coordinate axis. When Q_1' (or output N) changes over a large range, efficiency η does not change much. This type of turbine is suitable for hydropower stations with low head, large discharge, large load variation, and large head variation. The shape of the efficiency contours of the fixed-blade axial flow turbine is a narrow and long ellipse. The major axis and the minor axis have great differences, and the major axis and the Q_1' coordinate axis form a considerable inclination angle. This reflects that the efficiency is insensitive to variation of n_1' (or head H), but when Q_1' (or output) varies slightly the efficiency decreases dramatically. This type of turbine is suitable for hydropower stations with low head, large discharge, large head variation, and small load variation.

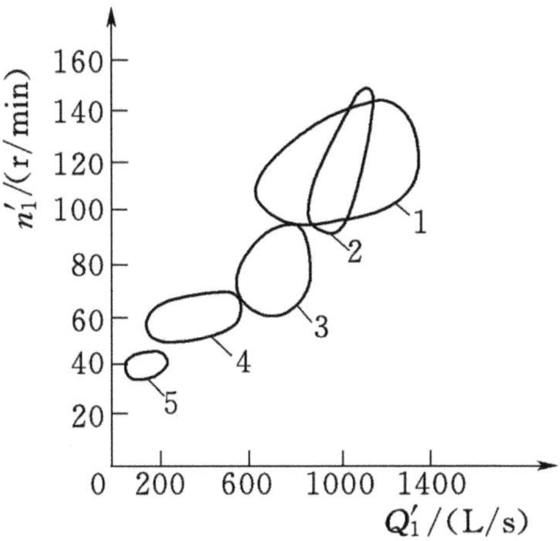

Fig. 4.7 Characteristics of efficiency curves for turbines with different specific speed, 5-Pelton turbine; 4-low specific speed Francis turbine; 3-medium–high specific speed Francis turbine; 2-fixed-blade propeller turbine; 1-Kaplan turbine

The critical cavitation coefficient contours are also the closed loops, as shown in Fig. 4.5. The minimum critical cavitation coefficient contour of the model turbine is 0.05, which is located on the lower left. Along the direction to the upper right, the critical cavitation coefficient gradually increases to 0.14. The distribution of the critical cavitation coefficient contours also depends on the specific rotational speed of the hydro turbine.

The turbine torque characteristic curves can be converted from the comprehensive characteristic curves, with formula

$$M_1' = \frac{30 \times 1000}{\pi} g\eta \frac{Q_1'}{n_1'} = 9549.30 \frac{gQ_1'\eta}{n_1'} \tag{4.51}$$

The torque characteristic curves mainly describe the change of M_1' with turbine unit speed n_1'. When $M_t = 0$, or $\eta = 0$, the turbine speed reaches the maximum value. Therefore, the runaway condition curve and the torque characteristic curves of small guide vane opening have a great influence on the process of turbine speed change.

2. **Runaway characteristic curve**

Figure 4.8 is the runaway characteristic curve of a Francis turbine according to the model test results. The characteristic curve takes the unit speed and the unit discharge as the vertical and horizontal coordinates, and uses the guide vane opening degree as a parameter to draw the $\eta = 0$ efficiency contour, which is the runaway curve.

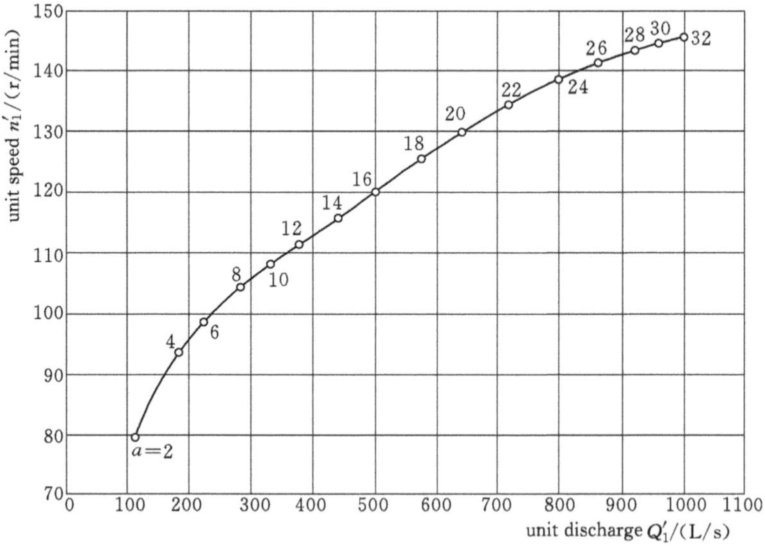

Fig. 4.8 Runaway characteristic curve of a model Francis turbine

The maximum unit speed on the runaway characteristic curve is called the maximum unit runaway speed $n'_{1f\,\max}$, which corresponds to a certain guide vane opening. The maximum runaway speed of the prototype turbine

$$n_{f\,\max} = n'_{1f\,\max}\frac{\sqrt{H_{\max}}}{D_1}\quad (\text{r/min}) \tag{4.52}$$

There are three key points on the runaway characteristic curve.

(1) Maximum runaway unit speed $n'_{1f\,\max}$

The ratio of the maximum runaway unit speed to the rated unit speed is called the runaway factor, it is $k_f = n'_{1f\,\max}/n'_{10}$.

For the reaction turbine, $k_f = 1.6$ to3.0, with 1.6 corresponding to low specific speed turbine and 3.0 corresponding to high specific speed turbine; for Pelton turbine, $k_f = 1.7$ to 2.0.

(2) No-load opening a_x

The guide vane opening corresponding to the intersection of the runaway characteristic curve and rated unit speed n'_{10} is called the no-load opening. The no-load opening can be obtained by the following formula

$$\begin{aligned} n_{s0} &= 3.65n'_{10}\sqrt{Q'_{10}\eta_{\max}} \\ a_x &= 0.0195n_{s0} - 0.5635 \end{aligned} \tag{4.53}$$

Usually for low specific speed turbines, $a_x = 10\%\ a_{\max}$, and for high specific speed turbines, $a_x = 15\%\ a_{\max}$.

(3) Runaway characteristic curve begins from the origin point

With the above three points, the runaway characteristic curve can be drawn in the shape of the zero efficiency contour. Usually, the manufacturer gives the runaway characteristic curve with large guide vane openings, we need to find the no-load guide vane opening, and extend the curve according to its trend.

The runaway condition is an accident condition. When the unit operates in a normal condition, its speed remains at the rated speed. The running away phenomenon can only occur when the unit rejects the load and the guide vanes cannot close. Under this condition, the turbine output is zero and the input water energy still exists. Then the unit speed increases rapidly until the mechanical energy loss is balanced with the input water flow energy, and the speed of the turbine generator unit reaches a certain stable maximum value, that is, the runaway speed.

Long time runaway will cause damage to the rotating parts of the unit and strong vibration of the power house. Therefore, the rapidly closing valve is often set in hydropower stations, and it can cut off the flow within 2 min, so that the water flow energy can be reduced to zero, which further causes the unit speed to drop rapidly and finally stop rotating.

4.4 Performance and Operation Characteristic Curves of Hydro turbine

The relations between the parameters of hydro turbine are called the characteristics of hydro turbine, which can be drawn in the form of characteristic curves. However, the characteristic curves of turbine have their unique characteristics, mainly because its parameters are too many. The common parameters include the following.

(1) Structural parameters: Guide vane height b_0, runner diameter D_1, guide vane opening a, blade angle ϕ, etc.
(2) Operating parameters: Working head H, turbine discharge Q, turbine speed n, suction height H_s, etc.
(3) Comprehensive parameters: Unit speed n_1', unit discharge Q_1', turbine output N, turbine efficiency η, turbine cavitation coefficient σ, etc.

When discussing the velocity triangle in the first section of Chap. 3, it has been pointed out that the relations between the above parameters can be expressed by functional relations as follows

$$\begin{aligned} Q &= f_Q(H, a, n, D_1, \phi) \\ \eta &= f_\eta(H, a, n, D_1, \phi) \\ N &= f_N(H, a, n, D_1, \phi) \\ M &= f_M(H, a, n, D_1, \phi) \end{aligned} \tag{4.54}$$

However, at present, these functional relations are difficult to express by exact mathematical formulas, and the relationship curves are usually drawn according to turbine model tests or CFD numerical simulations. But, there are too many turbine parameters, so that when drawing the characteristic curves according to the test results, some parameters must be fixed. Under this condition, the relations between specific parameters are reflected.

For a group of geometrically similar hydro turbines, their model comprehensive characteristic curves and the runaway characteristic curve are unique. They are based on the results of model tests and can express the energy characteristics, cavitation characteristics, pressure pulsation characteristics, etc. And from the comprehensive characteristic curves of turbine model, we can obtain the performance characteristic curves, the speed characteristic curve, the operating characteristic curve, and the operating comprehensive characteristic curves of a particular prototype turbine.

4.4.1 Performance Characteristic Curve of Hydro turbine

When parameters H, n, D_1, and ϕ are constant, changing guide vane opening a is equivalent to changing Q, which causes output N and efficiency η to change. Thus we get relations $N = f_N(a)$ and $\eta = f_\eta(a)$. Therefore, the relations between output N

and efficiency η can be obtained for different types of turbines. This curve is called the performance characteristic curve, as shown in Fig. 4.9.

It can be seen from Fig. 4.9 that the performance characteristic curve for fixed-blade axial flow turbine has a steep slope and a narrow high-efficiency zone, and when the working point deviates slightly from the optimal working condition, the efficiency drops dramatically. The performance characteristic curve for Kaplan turbine has a relatively wide high-efficiency zone, and the efficiency η varies gently. The curve for Francis turbine has higher efficiency, but the high-efficiency zone is not as wide as that for Kaplan turbine and that for Pelton turbine. The curve for Pelton turbine has relatively lower efficiency, but the efficiency changes gently and has wide high-efficiency zone. Therefore, the Kaplan and Pelton turbine generator units are suitable for carrying load that changes frequently and greatly.

In addition, it can be seen from Fig. 4.9 that the performance characteristic curves of the mixed flow and fixed-blade axial flow turbines do not pass through the origin of coordinates. The reason is that the turbine needs to consume a certain power ΔN_x when it runs at the no-load condition, at which the unit rotates at the rated speed. The corresponding operating point a is called the speed no-load point, and the corresponding guide vane opening $a \neq 0$ is called the idling guide vane opening, as shown in Fig. 4.10. Point b in Fig. 4.10 is called the maximum efficiency point; point c is the maximum output point; and point d is the mechanical maximum guide vane opening point.

It can be seen from Fig. 4.10 that the feature in the section of c to d on the performance characteristic curve is that as the guide vane opening a increases, the turbine output N does not increase, but decreases, resulting in the operation becomes unstable. The regulation mechanism of hydro turbine is mechanical rather than intelligent. The rule is that when the guide vane opening increases, the turbine output increases; when the guide vane opening decreases, the turbine output decreases. However, if the turbine is working near the maximum output N_{limit}, the external requirement is to increase the output. According to the rules of the regulation mechanism, the opening of the guide vanes will increase, but if it slips over N_{limit}, the

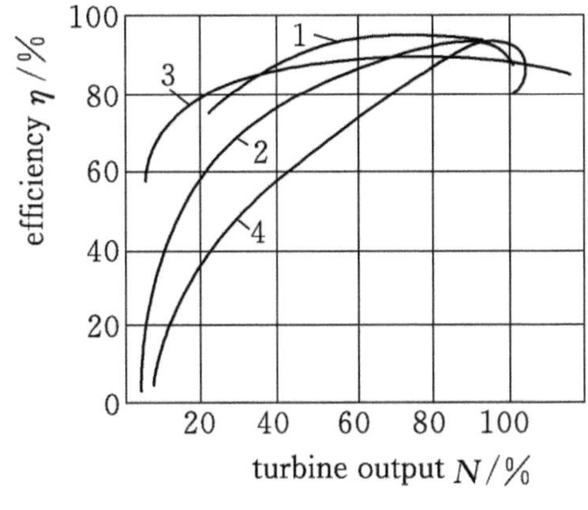

Fig. 4.9 Performance characteristic curve for different specific speed turbines, 1-Kaplan turbine, $n_s = 625$; 2-Francis turbine, $n_s = 300$; 3-Pelton turbine, $n_s = 20$; 4-Fixed-blade propeller turbine, $n_s = 570$

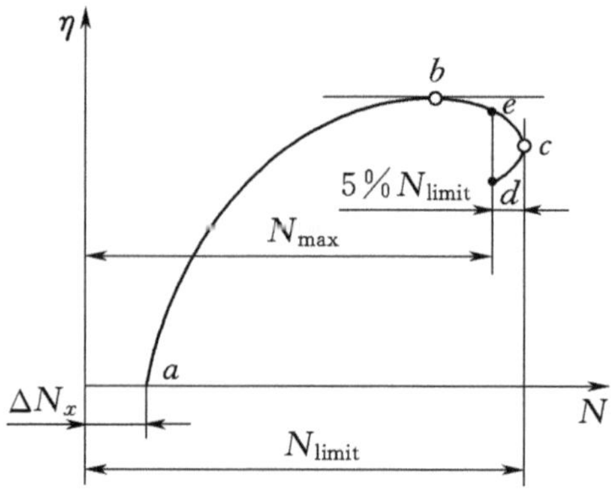

Fig. 4.10 Schematic diagram of the 95% output limit

output will decrease. As a result, the guide vane opening is further increased, while the output is further reduced. This not only leads to unstable working condition, but also make the turbine unable to operate. Therefore, this working condition is very dangerous and not allowed in the practical hydropower stations. To prevent this problem, the practical maximum output N_{max} is limited within 95%N_{limit}, and point *e* is called as the 95% output limit point.

It should be pointed out that modern Francis turbine has a maximum output limit N_{limit} that is much larger than the practical maximum output N_{max} required for turbine design and operation. And the 95% output limit line is not marked on the comprehensive characteristic curve of the turbine model. The old-fashioned Francis turbine is marked with a 95% output limit line for indicating the allowable turbine operation region.

4.4.2 *Speed Characteristic Curves of Hydro turbine*

When parameters H, a, D_1, and ϕ are constant, changing speed n can get the speed characteristic curves or the main characteristic curves, which are $Q = f_1(n)$, $\eta = f_2(n)$, and $N = f_3(n)$, as shown in Fig. 4.11.

1. **Characteristics of** $\eta = f_2(n)$ **and** $N = f_3(n)$

It can be seen from Fig. 4.11 that the turbines with different specific speeds have the expressions $\eta = f_2(n)$ and $N = f_3(n)$ as concave parabolic curves. The following three feature points on the parabolic curve are worthy of attention.

(1) When $n = 0$, there is $N = 0$ and $\eta = 0$.
(2) When $n = n_0$, there is $\eta = \eta_{max}$, and the speed n_0 is called the optimal speed. In order to let the hydro turbine operate at high-efficiency point, its rated speed should be as close as possible to the optimum speed.
(3) When $n = n_f$, the water energy is consumed by the head loss and mechanical loss, thus the output $N = 0$ and efficiency $\eta = 0$. The speed n_f is called the

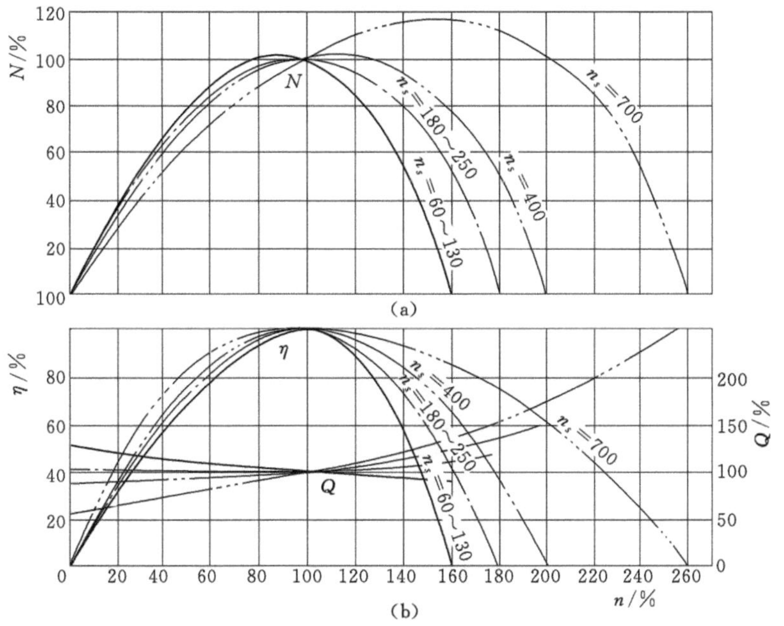

Fig. 4.11 Comparison of speed characteristic curves of turbine with different specific speeds

runaway speed for a certain guide vane opening, and is generally 1.8 to 2.4 times the optimal speed. In the event of a runaway, the turbine generator unit generates strong vibration, therefore, measures should be taken to prevent the unit from the runaway accident.

2. **Characteristics of** $Q = f_1(n)$

It can be seen from Fig. 4.11 that for the reaction turbine, when speed n changes, the velocity triangle of turbine runner changes accordingly, and the discharge Q changes obviously. For the high specific speed turbine, the discharge increases with the increase of the speed. For the low specific speed turbine the discharge decreases with the increase of the speed. For Pelton turbine, the discharge is almost constant with the increase of the turbine speed, but it obviously changes when the needle valve opening changes.

4.4.3 Operating Characteristic Curve of Hydro turbine

The performance characteristic curve is drawn for a certain head. If head H is used as a parameter, a series of $\eta = f(N)$ curves are drawn on the same diagram. This is called the operating characteristic curve, as shown in Fig. 4.12.

The following can be known from Fig. 4.12.

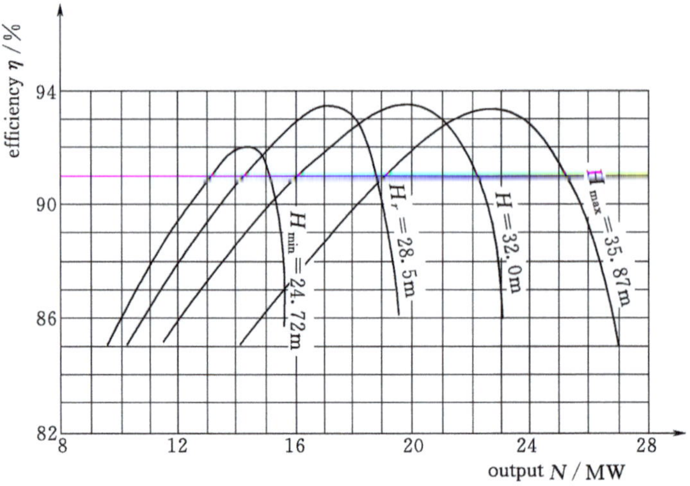

Fig. 4.12 Operating characteristic curve of hydro turbine

(1) The higher the head, the greater the change range of output.
(2) For a certain turbine model, the parameters and their corresponding head range and output range are determined. If the actual operation exceeds the above ranges, the operating efficiency will be greatly reduced.
(3) For different heads, the maximum efficiency points on the operating characteristic curve correspond to different outputs. The lower the head, the smaller the corresponding output at the maximum efficiency point. In practical operation, the management and dispatch of hydropower stations should fully consider its characteristics.

In engineering practice, the hydro turbine is operating with constant speed n and the runner diameter D_1 is fixed. But the head H is variable in a year. Due to the discontinuity of H in the operating characteristic curve, it is inconvenient to use. Therefore, it is necessary to convert the operating characteristic curve into the operating comprehensive characteristic curves, as shown in Fig. 4.13.

On the operating comprehensive characteristic curve plane, the rated head is the minimum head that the generator generates the rated output. Obviously the corresponding turbine discharge is the maximum.

In addition, there are two important limit lines on the operating comprehensive characteristic curve plane: the turbine output limit line and the generator output limit line. These lines give the boundaries of the normal operation. The head corresponding to the intersection of the two limit lines is the rated head. Sometimes the contours of the suction height are also drawn on this diagram.

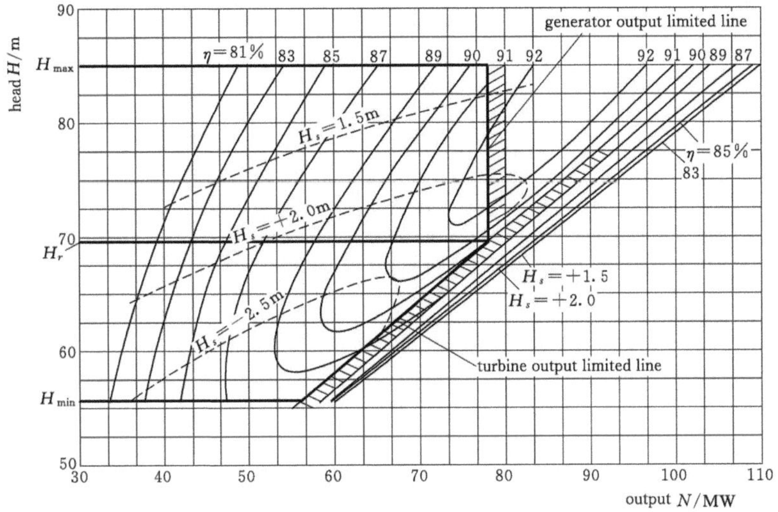

Fig. 4.13 Operating comprehensive characteristic curves of hydro turbine

4.4.4 Operating Comprehensive Characteristic Curve of Hydropower Station

For a hydropower station with multiple turbines, in addition to mastering the characteristics of a single turbine, understanding the turbine energy index of the entire hydropower station is very important for selecting the installed number of turbines and guiding the economic operation of the hydropower station. Therefore, it is necessary to draw the operation comprehensive characteristic curves of hydropower stations.

The operation comprehensive characteristic curve of a hydropower station is formed by superimposing the comprehensive characteristic curves of several single turbines according to the optimal principle. When the characteristics of each unit in the hydropower station are the same, the principle of optimal load distribution is the distribution with equal loads. Therefore, when drawing the comprehensive characteristic curves of a hydropower station, we can first draw the operating comprehensive characteristic curve of a single turbine, and then multiply the horizontal coordinate of each contour and the output limit line by 2, 3,…., n times. The vertical coordinate (head H) is unchanged. Through these treatments, the comprehensive characteristic curves of the hydropower station that has $Z = 2, 3, \ldots, n$ turbines can be obtained, as shown in Fig. 4.14. The logarithmic coordinates are often used to reduce the overlap of partial curves, making the operation comprehensive characteristic curves of hydropower stations clearer, as shown in Fig. 4.15.

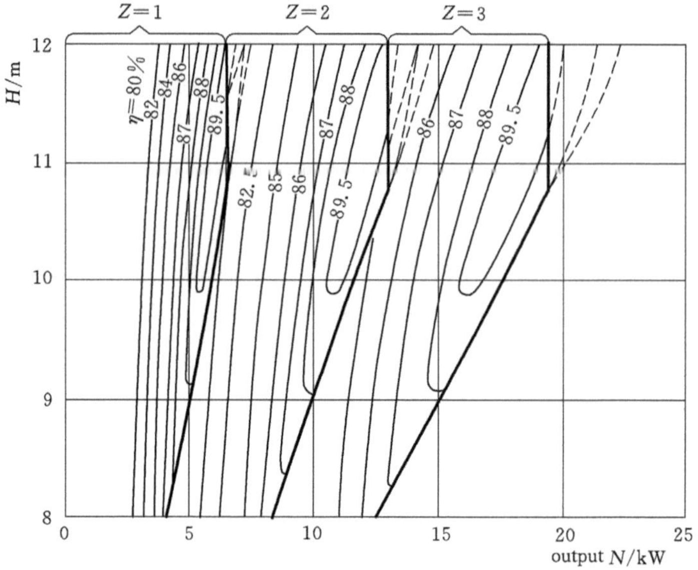

Fig. 4.14 Operation comprehensive characteristic curve of hydropower station

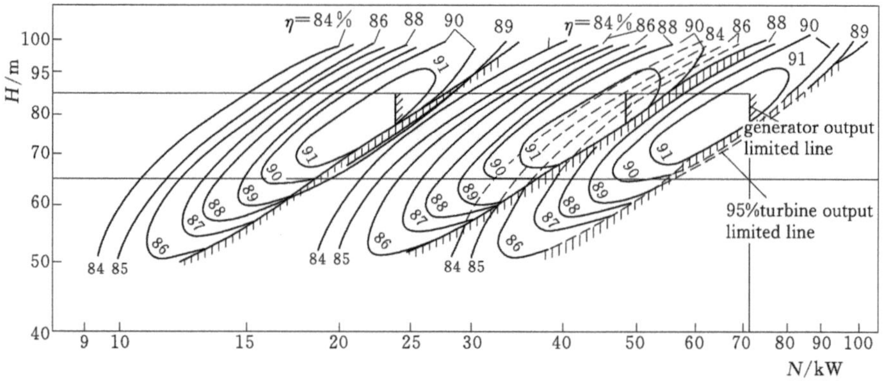

Fig. 4.15 Operation comprehensive characteristic curves of hydropower station with a logarithmic coordinate

4.5 Type Selection of Hydro turbine

Turbine type selection is an important task in the design of hydropower stations. The reasonable selection of type and parameters of the turbine has important influences on the energy and economy indicators, operation stability, and reliability of hydropower station.

The type selection of turbine is generally based on the development mode of the hydropower station, the dynamic energy parameters, the layout of the hydraulic structure, etc. The parameters of existing turbines and the production levels of manufacturing factories should be also considered. After comprehensive comparisons of technology and economy, the optimal type of the corresponding parameters of the turbine can be finally determined.

4.5.1 Contents, Requirements, Data, and Basic Methods for Turbine Selection

1. **Main contents of turbine selection**

   (1) Determine the number of turbines and unit capacity.
   (2) Select the type and installation mode of turbine.
   (3) Calculate the runner diameter and speed of turbine. For the impulse turbine, the jet diameter and nozzle number are also included.
   (4) Determine the suction height and installation elevation of turbine.
   (5) Draw the operation comprehensive characteristic curves of hydro turbine.
   (6) Determine the type and main dimensions of the spiral case and draft tube.

2. **Basic requirements for turbine selection**

   (1) Ensure that the turbine's rated output can occur under the rated head, and the unit's blocked capacity is as small as possible when the head is lower than the rated one.
   (2) Select the suitable turbine type and parameters according to the head variation and operation mode of the hydropower station. The average efficiency of operation should be as high as possible.
   (3) Ensure the performance and structure of the turbine should meet the requirements of flexible and stable operation of the hydropower station, and the anti-cavitation performance should be good. In hydropower stations on heavy sediment-laden rivers, the turbines should also have good performance of anti-abrasion and anti-cavitation erosion.
   (4) Ensure the structure of turbine should be advanced and reasonable, and the consumable parts should be interchangeable and easy to replace, which provides easy for operation, installation, and maintenance.
   (5) Consider the transport mode and transport feasibility for the largest and heaviest components of the turbine.

3. **Basic data required for turbine selection**

   (1) Basic technical data of hydropower stations.

a. Development plan of water conservancy project, reservoir water levels, regulation performance, development mode, and layout of hydropower station, topography and geology of hydro project, water quality, sediment of the river, etc.

b. The installed capacity of the hydropower station and its role in the power system (such as peak shaving, base load, phase modulation, accident standby, and parallel operation mode with other hydropower stations), power system load composition, design level annual load diagram, typical daily load diagram, prospective load, etc.
c. The characteristic heads of the hydropower station, such as the maximum head H_{max}, the minimum head H_{min}, the weighted average head H_{av}, and the rated head H_r. The weighted average head is the head that occurs most frequently during the operation of the hydropower station and has the longest operating time; the rated head is the minimum head required for the turbine to generate the rated output, and its exact value can only be determined after the turbine is selected. The initial evaluation can use the values as follows.

   In run-of-river hydropower stations, $H_r = 0.9H_{av}$; in dam hydropower stations, $H_r = 0.95H_{av}$; in diversion hydropower stations, $H_r = H_{av}$
d. Characteristic discharges of the hydropower station, such as the maximum discharge Q_{max}, the minimum discharge Q_{min}, and the average discharge Q_{av}. Head-discharge hydrograph for typical year (such as design level year, wet year, and dry year) and tail water level-discharge curve, in which the discharge is not only the hydro turbine discharge, but also includes the discharge from other water passages.
e. Transportation and installation conditions: The water and land traffic conditions to the hydropower station, such as the carrying capacity, tonnage and size of the road, waterway and port, and the on-site assembly conditions of the equipment, should be known.

(2) Technical data of hydro turbine product.

a. Type spectrum of hydro turbine series, mainly including the application range of the turbine head, the optimal unit speed, the maximum unit discharge, and the model cavitation coefficient.
b. Standard diameter D_1: Large and medium-sized turbines are customized according to actual sizes, but for small-sized turbines the diameter normally should be selected within standard diameters.
c. Synchronous speed n: The synchronous speed of unit is related to the number of magnetic pole pairs of generator. The synchronous speed is calculated according to Eq. (2.15).
d. Model comprehensive characteristic curves of hydro turbines in the plane of unit speed n_1' and unit discharge Q_1'.The runaway characteristic curve should also be included.
e. Application range of a certain series of homologous hydro turbines, as shown in Fig. 4.16. By this figure, D_1,n, and H_s can be found based on the rated head H_r and unit capacity N_r. However, the results are not precise enough, and only suitable for the planning stage or the selection of small turbines.

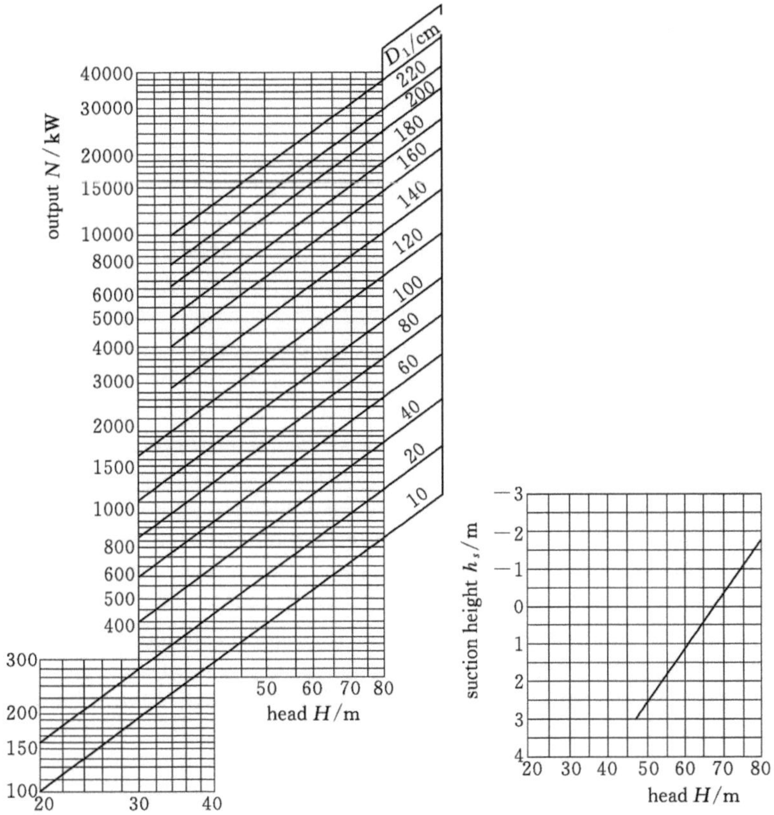

Fig. 4.16 Application range diagram of HL220 turbine series

(3) Similar project data.

Collect relevant data on turbines and hydropower stations of the similar type that are being designed, constructed, and in operation for use. Including turbine parameters and operating experience, existing problems, etc.

4. **Basic methods of turbine selection**

There are three main methods for turbine selection at present.

(1) Apply statistics data to select turbine: The method is based on the collection and statistics of the basic parameters of the hydro turbines of the hydropower stations that have been built, and the data can be classified according to the turbine type, the application head, and the unit capacity. Based on the collected data, mathematical statistics methods are used to formulate the relations between the specific speed, the unit parameters, and the applied water head, such as $n_s = f(H)$,$n_1' = f(H)$ and $Q_1' = f(H)$, the relations between the installation cavitation coefficient of hydropower station and the specific speed $\sigma_p = f(n_s)$,

and so on. Or use the least squares method to derive semi-theoretical semi-empirical expressions of these parameters. After the basic parameters such as the head and installed capacity of the proposed hydropower station are determined, the type and basic parameters of the hydro turbine can be determined according to the above statistical curves or empirical formulas, and the turbine selection process is completed. This method is widely used at present.

(2) Select turbine according to turbine series spectrum: In some countries, turbine equipment has been serialized, generalized, and standardized. The model spectrum of hydro turbines has been established, and for each head range, one or more turbine runners have been developed. And through model tests, the basic parameters and model comprehensive characteristic curves of various series turbines are obtained. Therefore, the turbine selection of the turbine model and parameters of the proposed hydropower station can be based on the turbine model spectrum and the model's comprehensive characteristic curves. The turbine spectrum provides convenience for the turbine selection. However, the turbine spectrum needs to be continuously updated. Otherwise, it is difficult to keep up with the technological development of the turbine industry and it is difficult to meet the needs of the turbine selection of new hydropower stations.
(3) Select turbine according to similar projects. This method directly applies the turbine models and parameters of the built hydropower stations which are similar to the basic parameters (head and capacity) of the proposed hydropower station. Although the direct application makes the selection design of the turbine simpler, the difference of design ages often makes the turbine parameters lower. Therefore, in practice, the parameters of the proposed hydropower station and the built hydropower station should be compared and analyzed in detail. If necessary, the turbine parameters of the built hydropower station should be properly corrected before being applied.

In the past, the method used in China was normally to select the turbine according to the turbine spectrum. However, with the progress of the times, the original turbine spectrum can no longer meet the needs of the current hydropower station design. Therefore, in the selection of turbines, different selection methods should be used to verify each other to keep the scientific and rationality of the turbine selection.

4.5.2 Selection of the Number of Units

For a hydropower station with a certain installed capacity, the number of units and the consistency of single unit capacity not only are related to turbine manufacturing level, transportation and installation conditions, but also affect directly the project investment, energy economic index, and operation flexibility and reliability of the hydropower station. Therefore, when selecting the number of units, it is necessary to conduct a comprehensive technical and economic evaluation from the following aspects.

1. Influence of the number of units on equipment manufacturing, transportation, and project investment

When the number of units increases, the single unit capacities of the turbine and generator are reduced, and the corresponding sizes are reduced, which is convenient for the manufacture, transportation, and installation of the unit, but the number of civil works is inevitably increased. For huge hydropower stations, the total installed capacity is very large. Only by maximizing the capacity of single unit, the limitations of topographic and geological conditions on the layout of the water transmission and power generation systems can be overcome, and the project investment can be reduced. The typical examples include Baihetan Hydropower Station and Wudongde Hydropower Station in China. The future development of hydropower in Xizang is also facing an urgent need to increase the capacity of single unit, especially for the high-head and large-capacity impulse turbine generator sets. The maximum unit capacity depends not only on the hydraulic and structural design of the hydro turbine, but also on the allowable value of the heat dissipation of the generator and the maximum linear speed of the generator rotor.

2. Influence of the number of units on the operation efficiency and flexibility of hydropower station

Units with large single unit capacity are usually more efficient. Therefore, when the hydropower station undertakes the base load in the power system and the units are often fully loaded, selecting fewer units can maintain a high average efficiency of the hydropower station. However, for a hydropower station undertaking variable load, if the number of units is too small, it is often operating in a partial load state, which not only reduces the average efficiency of the hydropower station, but also reduces the flexibility of the hydropower station operation. Therefore, it is also necessary to increase the number of units and even units with different capacities to improve the flexibility and average efficiency of hydropower station. However, when the number of units increases to a certain amount, the average efficiency of the hydropower station will not increase significantly.

In addition, due to the different types of turbines, the number of units has different effects on the operating efficiency of hydropower stations. For fixed-blade turbines, especially fixed-blade axial flow turbines, the efficiency curve is steep, and when the output varies, the efficiency changes drastically. If the number of units is large or the capacity of single unit is inconsistent, the low-load operation can be avoided by adjusting the number of the operating units, thereby improving the average efficiency of the operation of the hydropower station. However, for a Kaplan turbine or a multi-nozzle Pelton turbine, the hydro turbine can maintain high efficiency by changing the blade angle or the number of nozzles. Therefore, under this condition, the number of units has little effect on the average efficiency of hydropower station operation.

3. Influence of the number of units on the operation and maintenance of hydropower station and power system

When the number of units is large, the advantages are that the operation mode is flexible, the impact on the hydropower station itself and the power system is small when an accident occurs, and the maintenance is easy to arrange. The disadvantages are that the times of start-stop operations increase as the unit number increases, and the probability of accidents increases. The number of operators also needs to increase, and the operating costs also increase. Therefore, it is not appropriate to use too many units.

The above various factors related to the number of units depend on each other and affect each other, and it is impossible to satisfy them all. Therefore, the selection of the possible number of units should be further analyzed and compared according to the specific situation. When the technical and economic conditions are similar, as far as possible, the number of units with fewer units is used, but generally not less than two. At the same time, for the convenience of manufacturing, installation, operation, and maintenance, and spare parts supply, the same type of units (with one to two different capacities) should be selected as far as possible. In addition, for large and medium-sized hydropower units the enlarged unit connection type (except for huge units) is normally used, and the hydraulic units are symmetrically arranged, thus the number of units is usually even.

4.5.3 Turbine Type Selection

With the continuous improvement of the design and manufacture level of hydro turbines, the applicable head ranges of various types of turbines are also expanding, and there is a certain intersection between the applicable head ranges of different types. Therefore, it is necessary to analyze and compare the alternative turbines according to the specific conditions of the proposed hydropower station to finally determine the turbine type.

1. **Scope of application of various types of turbines**

Figure 4.17 shows the application ranges of head and discharge of various types of turbines. In fact, the application ranges are also related to turbine capacities. Even for turbines with the same type and specific speed, the low head one has a small capacity and the high head one has a large capacity. Therefore, Fig. 4.17 shows the turbine types and application scopes of the large and medium-sized turbines with the head and output as the vertical and horizontal coordinates, respectively. The corresponding parameters are shown in Table 4.1.

Different types of turbines have different application scopes and characteristics, which are summarized as follows.

(1) The impulse turbine has a lower specific speed and is suitable for heads above 300 m. The current maximum head of Pelton turbine can reach 1700 m. The efficiency of this type of turbine changes gently, the adaptability to load change is strong, and the number of nozzles can be adjusted to operate with high efficiency. The deflector can alleviate the contradiction between the increase of water hammer pressure and the increase of speed, which is beneficial to

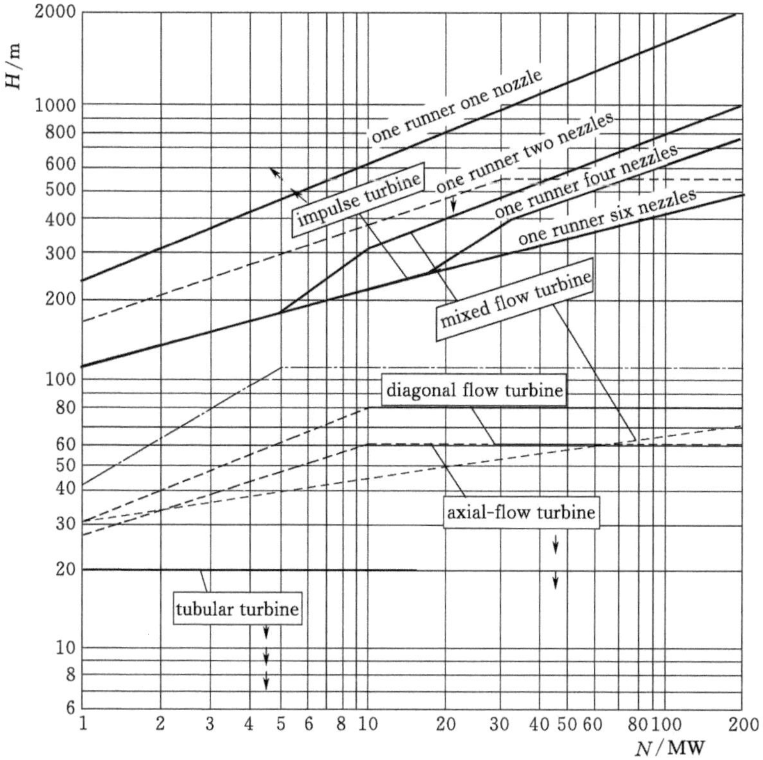

Fig. 4.17 Application scopes of various types of turbines

Table 4.1 Types and application scopes of hydro turbine

Turbine type			Application head range/m	Specific speed range n_s/m · kW
Energy conversion method	Water flow mode	Structure type		
Reaction turbine	Tubular flow	Bulb shaft extension	<20	600–1000
	Axial flow	Propeller Kaplan	3–80	200–850
	Diagonal flow	Deriaz	40–180	150–350
	Mixed flow	Francis	30–700	50–300
Impulse turbine	Jet flow	Pelton	300–1700	10–35 single nozzle

the design of guaranteed regulation of hydropower station. However, the flow passage components are prone to wear and the single unit capacity is small, therefore, the recent goal is to further increase the capacity and application head of this type of turbine to meet the needs of hydropower development.

(2) The mixed flow turbine has a wide range of specific speed, and its application head range is 30–700 m. The structure of Francis turbine is simple and the draft tube can reduce the energy loss of the runner outlet. The disadvantage is that there are no other means for regulation other than the vanes. It is necessary to further improve the operation stability and reduce the pressure pulsations and unit vibration of this type of turbine.

(3) The axial flow turbine has a higher specific speed and a larger discharge, with a head range of 3–80 m. The Kaplan turbine operates in a coordinated manner between the guide vanes and the runner blades, which ensures high-efficiency operation when the head and load are variable. The fixed-blade propeller turbine has a simple structure and high reliability and is suitable for a hydropower station with low head and base load. However, in recent decades, the axial flow turbines for low head have been largely replaced by the bulb turbines, and the axial flow turbines for higher head have been replaced by Francis turbines.

(4) The specific speed of the diagonal flow turbine is between Kaplan turbine and Francis turbine. The blades can be adjusted to achieve high-efficiency operation when the head and load are varied. Compared with Kaplan turbine, the number of blades is larger, and the application head is higher. Currently, it is commonly used in pumped-storage hydropower stations with medium and low heads.

(5) The tubular turbine has a higher specific speed and a lower head, with the allocation head range of less than 20 m. Increasing the unit capacity of a tubular turbine is also an inevitable requirement for hydropower development.

2. **Selection of turbine in the overlapping range of head**

(1) Comparison between the tubular turbine and axial flow turbine: It can be seen from Table 4.1 that the tubular turbine and the axial flow type share a large overlap of the application head ranges. The flow path of tubular turbine is smooth, there is no spiral case and elbow draft tube, the discharge is large, the head loss is small, and the operation efficiency is high. Compared with the axial flow turbine, the installation of the tubular unit not only reduces the depth of underwater excavation, but also reduces the height of the entire plant, resulting in a reduction in the amount of civil works. In addition, the tubular turbine generator unit is also suitable for tidal power stations, which can meet the needs of forward and reverse electric generation, forward and reverse pumping, and forward and reverse water discharge. However, the commonly used bulb tubular turbine generator sets are all under water, requiring strictly closed structure, and good ventilation and moisture protection measures. It is also difficult to maintain and repair the tubular turbine.

(2) Comparison between the axial flow turbine and Francis turbine: The Kaplan turbine is suitable for hydropower stations with large head and load variations. It can operate stably and efficiently in a wide working area, and its average efficiency is higher than that of Francis turbine. Under the same head, the specific speed of an axial flow turbine is higher than that of a mixed flow turbine, which is beneficial to reduce the size of the unit. However, the cavitation coefficient of the axial flow turbine is large, it is about twice of that of Francis turbine within the same head range. In order to prevent cavitation and erosion, it is necessary to increase the amount of underwater excavation of the power plant. When the draft tube is long, the axial flow turbine is more prone to occur the lifting accidents due to the opposite water hammer than Francis turbine. The axial water thrust coefficient of the axial flow turbine is about 2–4 times of the mixed flow turbine, and the thrust bearing has a large load and is easily damaged.

(3) Comparison between Francis turbine and the diagonal flow turbine: Under the same head and same output condition, the specific speed of the diagonal flow turbine is higher than the mixed flow type, therefore the rotation speed of the diagonal flow turbine can be higher than that of Francis turbine, which can reduce the size of the generator. However, the runaway speed of the Francis turbine is about 15% higher than that of the diagonal turbine, and Francis turbine is required to have higher strength. Although the maximum efficiency of Francis turbine is 0.5–1% higher than that of the diagonal flow turbine, the average efficiency is much lower than that of the diagonal flow type. On the other hand, within the same working parameters, the cavitation coefficient of the diagonal flow turbine is greater than that of the mixed flow turbine, thus in order to prevent cavitation and erosion, the diagonal flow turbine requires a lower installation elevation and deeper underwater excavation. The structure of the diagonal flow turbine is complicated, and the operation reliability is not as good as that of the mixed flow turbine.

(4) Comparison between the diagonal flow turbine and axial flow turbine: Under the same head and same output condition, the specific speed of the axial flow turbine is higher than that of the diagonal flow turbine, therefore, the speed of the axial flow turbine can be higher than that of the diagonal flow turbine, which leads to smaller unit size of the former. However, when the head is higher to a certain extent, the weight of the axial flow turbine generator unit will exceed that of the diagonal flow turbine due to the strength and structural requirements. The efficiency values of the two kinds of turbines are similar, and the efficiency of the diagonal flow is slightly higher only under the partial load operation condition.

(5) Comparison between Francis turbine and the impulse turbine: Under the same head and same output condition, the specific speed of Francis turbine is 10–20% higher than that of the impulse turbine, which can reduce the size of the unit. The maximum efficiency of Francis turbine is higher than that of the impulse turbine, but the adaptability to load variation is not as good as that of the impulse turbine. In contrast, the cavitation and erosion

phenomenon of the impulse turbine is lighter, and only occurs in the needle valve, nozzle, and water bucket, which are easier to be repaired and replaced. The impulse turbine has no axial water thrust and the bearing structure is simple. In the case of heavy sediment-laden water flow, the abrasion and wear phenomenon of the impulse turbine is severe, resulting in a big decrease in efficiency.

In short, for choosing a turbine in the overlapping range of head, attention should be paid to the pros and cons. In order to make a reasonable choice, the comparisons should include many aspects: technical, economic, operation, maintenance, etc.

4.5.4 Selection of Specific Speed of Hydro turbine

The specific speed of hydro turbine includes three basic parameters: speed, output, and head, which comprehensively reflect the basic characteristics of the hydro turbine. A right selection of the specific speed plays an important role in ensuring good energy, cavitation, pressure pulsation, and dynamic characteristics of the hydro turbine.

1. **Relationship between efficiency and specific speed of hydro turbine**

Turbines with different specific speeds have different performance characteristic curves or efficiency curves. It can be seen from Fig. 4.18, as for Francis turbines, the efficiency curves are more gradual when the specific speeds are lower, and the efficiency curves are steeper when the specific speeds are higher. Especially for the Francis turbine with a super-high specific speed, the high-efficiency area is very narrow, and its efficiency drops sharply as long as it deviates from the design condition.

Different types of turbines have their own optimal specific speeds. It can be seen from Fig. 4.19, the highest unit efficiency of Pelton turbine is corresponding to the specific speed n_s in the range of 13–18 m kW, the highest unit efficiency of Francis

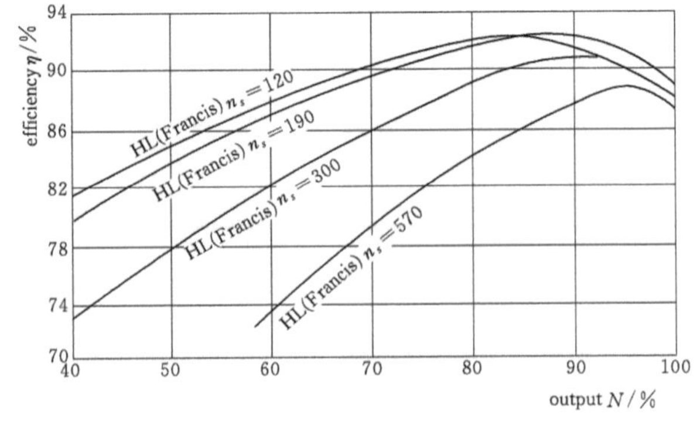

Fig. 4.18 Efficiency curves of Francis turbines with different specific speeds

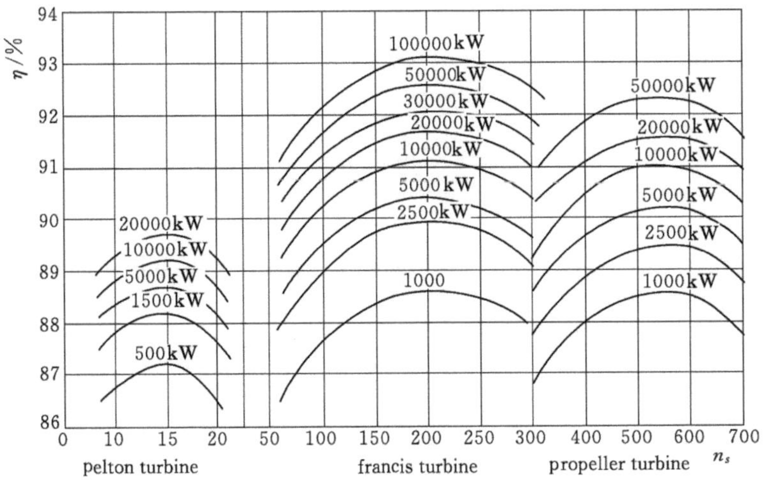

Fig. 4.19 Relations between the optimal turbine efficiency and specific speed

turbine is corresponding to the specific speed in the range of 150–250 m·kW, and the highest unit efficiency of the propeller turbine is corresponding to the specific speed in the range of 500–600 m kW. Therefore, when selecting a turbine, the specific speed should be controlled within the optimum range of the corresponding type of turbine.

2. **Relationship between cavitation coefficient and specific speed of hydro turbine**

The cavitation coefficient of hydro turbine has been expressed as Eq. (3.63). Theoretically, the relations between the cavitation coefficient and the specific speed of hydro turbine can be derived as

$$\sigma = K_0 n_s^{4/3} \tag{4.55}$$

where K_0 is a constant.

This formula shows that the turbine cavitation coefficient is proportional to the specific speed.

According to the statistic results of the model test data of some excellent Francis turbines, the relations between the model turbine cavitation coefficient and the specific speed can be obtained as follows

$$\sigma = 3.5 \times 10^{-5} n_s^{1.5}. \tag{4.56}$$

In addition, various forms of the relations of hydropower station cavitation coefficient and the turbine specific speed have been proposed. Examples are as follows: US Bureau of Reclamation $\sigma_p = 2.56 \times 10^{-5} n_s^{1.64}$, Japan Electrical Society

$\sigma_p = 3.46 \times 10^{-6} n_s^2$, Italy $\sigma_p = 7.54 \times 10^{-5} n_s^{1.41}$, and Swedish Electric Company $\sigma_p = 8 \times 10^{-5} n_s^{1.4}$. According to the statistics of 32 large hydropower stations in China, $\sigma_p = 4.56 \times 10^{-5} n_s^{1.54}$. Although these relations are different due to technical levels of different ages, it is consistently shown that the hydropower station cavitation coefficient is a function of the specific speed of turbine and is proportional to the specific speed. Therefore, the influence of cavitation coefficient should be considered when selecting the specific speed of hydro turbine.

3. **Relationship between pressure pulsation distribution and specific speed of hydro turbine**

Figure 4.20 shows the contour distribution of the relative amplitudes of pulsation pressure in the spiral case and the draft tube, which are obtained by the model tests of a pump turbine. The pressure pulsation characteristics of the hydro turbine not only depend on the type of turbine, but also have an inherent connection with the specific speed. Therefore, the following study should be conducted: collect the pressure pulsation curves of turbines with different specific speeds; select the relative amplitudes of pressure pulsation at representative operating points as the basis for numerical fitting and analysis; and establish the relation curves or the mathematical expressions between the contour distribution of pressure pulsation and the specific speed.

4. **Relationship between runaway characteristic curve and specific speed of hydro turbine**

The ratio of the highest runaway unit speed $n'_{1f\,\max}$ to the rated unit speed n'_{10} is called the runaway coefficient, namely $k_f = n'_{1f\,\max}/n'_{10}$. For a reaction turbine, $k_f = 1.6$ to 3.0, with 1.6 corresponding to a low specific speed and 3.0 corresponding to a high specific speed. For a Pelton turbine, $k_f = 1.7$ to 2.0.

The guide vane opening corresponding to the intersection point of the runaway curve and n'_{10} is called the no-load opening, and it can be calculated by $a_x = 0.0195 n_{s0} - 0.5635$, in which $n_{s0} = 3.65 n'_{10} \sqrt{Q'_{10}} \eta_{\max}$. Usually for low specific speed turbine $a_x = 10\% a_{\max}$, and for high specific speed turbine $a_x = 15\% a_{\max}$.

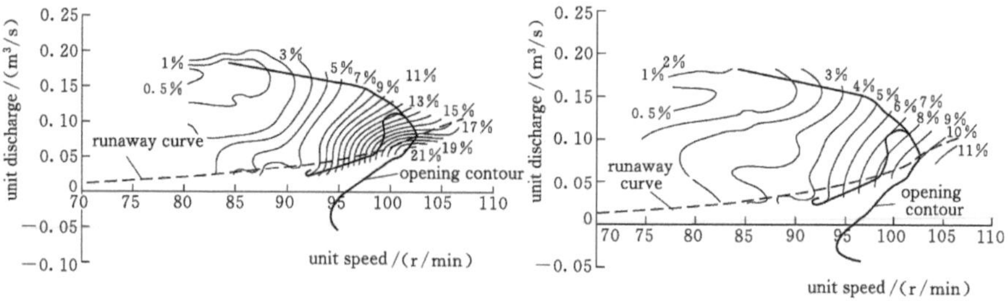

(a) In the spiral case (b) In the draft tube

Fig. 4.20 Contours of the relative amplitudes of pulsating pressure of the model test of a pump turbine

There are some problems needed to be studied: collect the runaway characteristic curves of different specific speed turbines, and numerically fitting each runaway curve; determine the maximum runaway speed $n'_{1f\,\max}$ and no-load opening a_x of each runaway curve; establish the relations between the fitting coefficient of the runaway characteristic curve and the specific speed, and give the mathematical expression of the runaway curve of any specific speed; establish the mathematical expression of runaway coefficient k_f of any specific speed; establish the mathematical expression of the no-load opening a_x of any specific speed; and investigate whether the runaway curve passes the origin.

5. **Empirical formulas between specific speed and working head of various hydro turbines**

The major countries of turbine manufacturing have divided the application ranges of specific speed of various hydro turbines according to their design level and production capacity. Empirical formulas describing the relations of specific speed and working heads of various turbines have be obtained by mathematical statistics. These useful formulas are expressed as follows.

(1) Relations between specific speed and working head for the axial flow turbine

$$\text{China}: \quad n_s = \frac{2300}{\sqrt{H}} \ (\text{m} \cdot \text{kW}) \tag{4.57}$$

$$\text{Japan}: \ n_s = \frac{20000}{H + 20} + 50 \ (\text{m} \cdot \text{kW}) \tag{4.58}$$

$$\text{The Soviet Union}: n_s = \frac{2500}{\sqrt{H}} \ (\text{m} \cdot \text{kW}) \tag{4.59}$$

(2) Relations between specific speed and working head for Francis turbines

$$\text{China}: \quad n_s = \frac{2000}{\sqrt{H}} - 20 \ (\text{m} \cdot \text{kW}) \tag{4.60}$$

$$\text{Japan}: \ n_s = \frac{20000}{H + 20} + 30 \ (\text{m} \cdot \text{kW}) \tag{4.61}$$

$$\text{United States}: \ n_s = \frac{2105}{\sqrt{H}} \ (\text{m} \cdot \text{kW}) \tag{4.62}$$

(3) Relations between specific speed and working head for the diagonal flow turbine

$$\text{Japan}: \ n_s = \frac{20000}{H} + 40 \ (\text{m} \cdot \text{kW}) \tag{4.63}$$

$$\text{The Former Soviet Union}: \ n_s = \frac{1400 \text{ to } 1500}{H_{\max}^{0.4}} \ (\text{m} \cdot \text{kW}) \tag{4.64}$$

(4) Relations between specific speed and working head for the tubular flow turbine are shown in Fig. 4.21.
(5) Relations between specific speed and working head for the impulse turbine are shown in Figs. 4.22 and 4.23.

When the impulse turbine has multi-runner or multi-nozzle, its specific speed should be calculated by

$$n_s = n_{s1}\sqrt{K_r Z_0} \tag{4.65}$$

where n_{s1} is specific speed of turbine with single runner and single nozzle; K_r is the number of runners; and Z_0 is the number of nozzles per runner.

For the multi-runner or multi-nozzle impulse turbine, n_{s1} should be selected within the optimal range, generally $n_{s1} = 10$ to 20 m kW, and the overall specific speed value of the turbine should not exceed 60 m kW.

6. **Selection of specific speed of hydro turbine in heavy sediment-laden river**

The above specific speed selection formulas of hydro turbines are based on the premise that the water quality in the hydropower station is good. For the hydropower

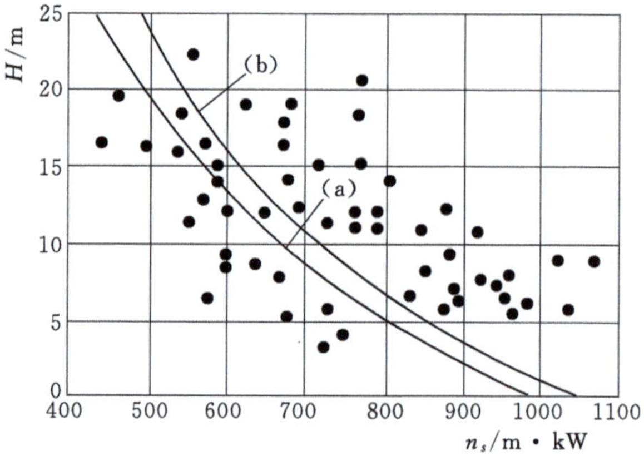

Fig. 4.21 Relations between specific speed of and working head for the tubular flow turbine, **a** $n_s = \frac{20000}{H+20}$; **b** $n_s = \frac{20000}{H+20} + 50$

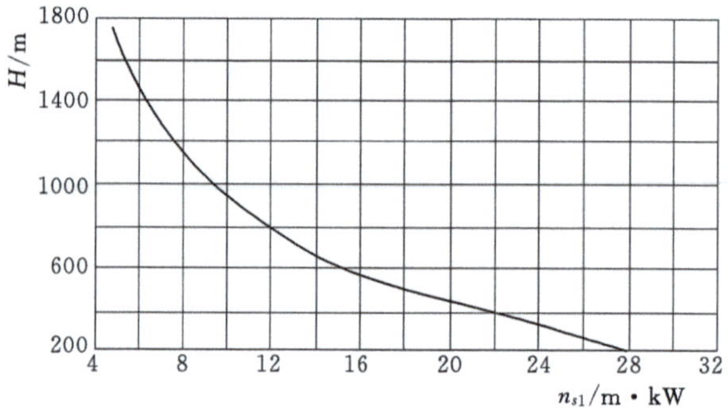

Fig. 4.22 Relations between specific speed and working head of the single-nozzle impulse turbine

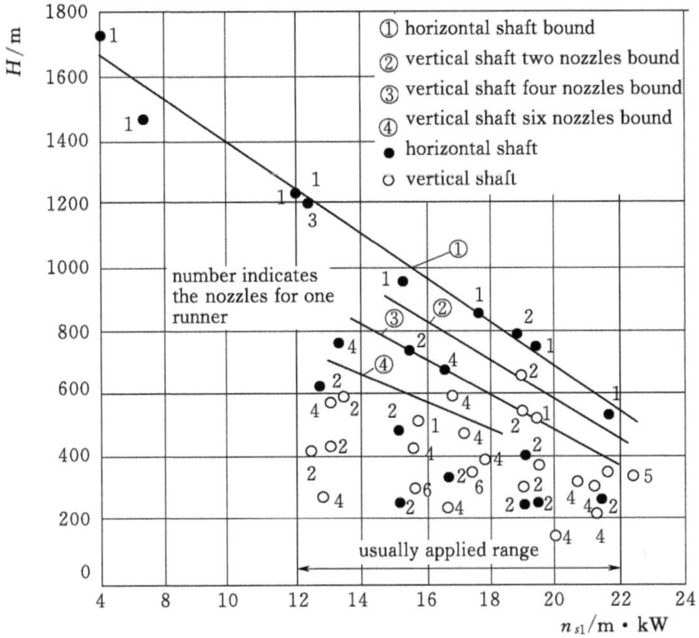

Fig. 4.23 Relations between specific speed and working head of the multi-nozzle impulse turbine

station on the heavy sediment-laden river, in order to reduce the combined effect of sediment abrasion and cavitation erosion, the specific speed should be reduced. Its purpose is to reduce the relative velocity in the runner passage for reducing abrasion and erosion. According to some engineering data, for the hydropower stations on sand enriched rivers, when the specific speed is lower, such as in the range of $n_s = \frac{1320 \text{ to } 1360}{\sqrt{H_r}}$, the wear and damage are relatively light.

4.5.5 *Calculation of Basic Parameters of Reaction Turbine*

1. **Calculate basic parameters of hydro turbine based on turbine spectrum and model comprehensive characteristic curves**

(1) Calculation of runner diameter D_1

According to the turbine output formula $N = 9.81QH\eta = 9.81Q_1' D_1^2 H^{3/2}\eta$, the nominal diameter can calculated by

$$D_1 = \sqrt{\frac{N}{9.81Q_1' D_1^2 H^{3/2}\eta}} \tag{4.66}$$

The parameters in Eq. (4.66) are explained as follows.

a. N is the rated output of the hydro turbine N_r, with $N_r = N_f/\eta_f$, in which N_f is the rated output of the generator or unit capacity, and η_f is the efficiency of generator.
b. H is the design head or rated head H_r.
c. Q_1' is the maximum unit flow $Q'_{1\,\max}$ recommended in the turbine spectrum sheet, or Q_1' corresponding to the intersection of the optimal unit speed n'_{10} and the 95% output limit line on the model comprehensive characteristic curve chart. When the excavation depth of the hydropower station is limited, it shall be determined according to the allowable suction height.
d. efficiency $\eta = \eta_M + \Delta\eta$. On the model comprehensive characteristic curve chart, the model efficiency η_M is corresponding to the intersection of $Q'_{1\,\max}$ and n'_{10}. The efficiency correction $\Delta\eta$ is usually estimated as 1% to 3% in the initial parameter calculation.
e. Calculate D_1 from Eq. (4.66), and slightly enlarge or apply a larger standard diameter, so that the hydro turbine has a certain margin capacity.
f. Substitute the preliminarily selected parameters D_1 and $\eta_{M,\max}$ into the empirical formula for efficiency correction, and see whether the resulting $\Delta\eta$ is close to the initial estimation. If they are inconsistent, repeat the above three steps of (4), (5), and (6).

(2) Calculation of rotational speed n.

According to the unit speed formula,$n_1' = \frac{nD_1}{\sqrt{H}}$, the speed can be calculated by

$$n = \frac{n_1'\sqrt{H}}{D_1} \tag{4.67}$$

where D_1 is the already known diameter of the runner; H is the weighted average head H_{av}; and n_1' is the optimal unit speed of the model plus the correction value, namely $n'_{10} = n'_{10M} + \Delta n_1'$ and $\Delta n_1' = n'_{10M}\left(\sqrt{\frac{\eta_{\max}}{\eta_{M,\max}}} - 1\right)$.

Base on the calculated value by Eq. (4.67), the rotational speed n is usually selected to be equal to the standard synchronous speed that close but larger than the calculated value.

(3) Check the working range on the model comprehensive characteristic curve chart

$$\left.\begin{aligned} Q'_{1\,\max} &= \frac{N_r}{9.81 D_1^2 H_r^{3/2}\eta} \\ n'_{1\,\min} &= \frac{nD_1}{\sqrt{H_{\max}}};\quad n'_{1M,\,\min} = n'_{1\,\min} - \Delta n_1' \\ n'_{1\,\max} &= \frac{nD_1}{\sqrt{H_{\min}}};\quad n'_{1M,\,\max} = n'_{1\,\max} - \Delta n_1' \end{aligned}\right\} \tag{4.68}$$

On the model comprehensive characteristic curve chart, three straight lines are drawn according to Eq. (4.68) to determine the working range of the hydro turbine. If the range is to the left of the 95% output limit line and includes the high-efficiency zone of the model comprehensive characteristic curves, then the selected D_1 and n are considered reasonable. Otherwise, the value of D_1 or n should be adjusted appropriately to move the working range to the high-efficiency zone.

(4) Calculate the suction height and determine the installation elevation

The maximum allowable suction height of the hydro turbine can be calculated by

$$H_s = 10 - \frac{\nabla}{900} - k\sigma H \tag{4.69}$$

where the downstream water level ∇ can be determined according to the relations between the downstream water level and the discharge; the safety factors are normally $k = 1.1$ to 1.2, but k can be larger than 1.5 for huge hydro turbines.

Several heads such as H_r, $H_{\max}$ and $H_{\min}$ may be selected to calculate H_s and their corresponding cavitation coefficients, which are values at the actual output limit points corresponding to these heads. In the calculation, it is necessary to calculate n_1' and Q_1' at the output limit point first, and then find the corresponding σ on the model comprehensive characteristic curve chart.

The installation elevation ∇_{inst} of the hydro turbine depends not only on the suction height, but also on the turbine type and the arrangement of the main shaft, as shown in Fig. 4.24.

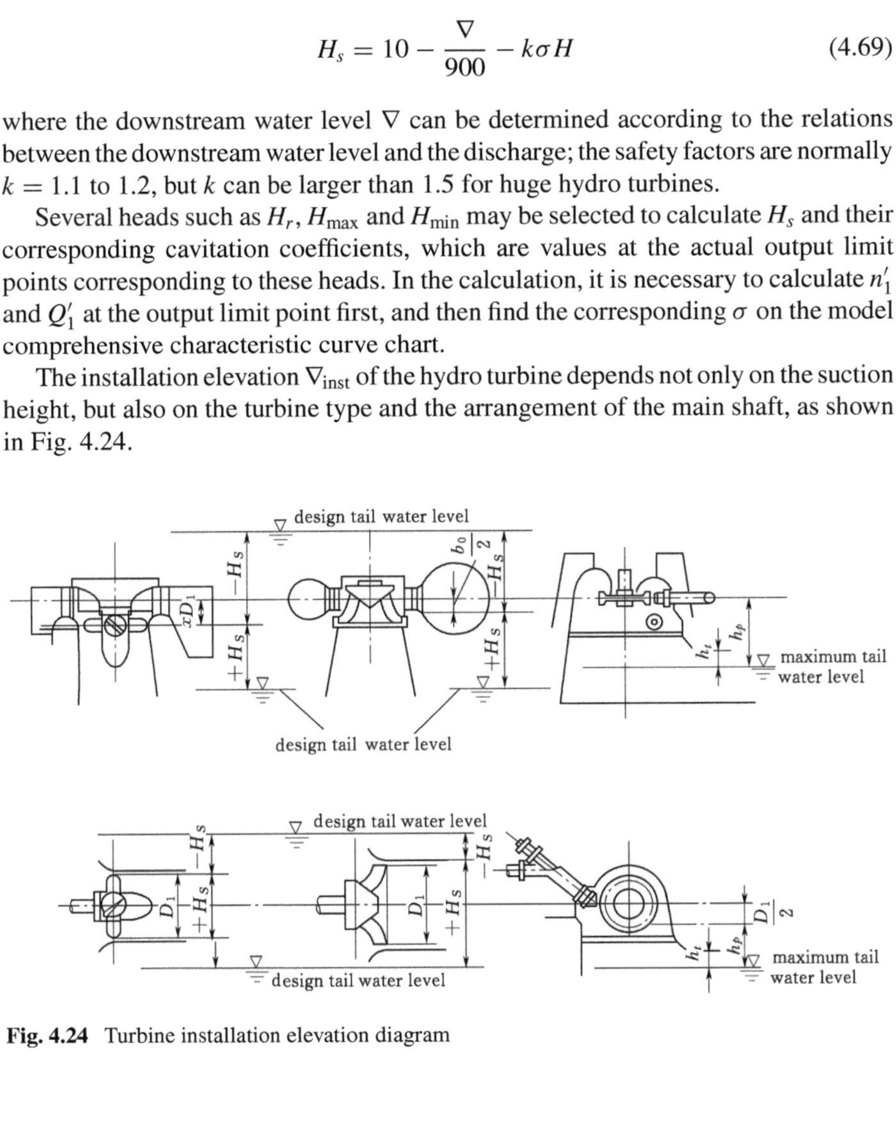

Fig. 4.24 Turbine installation elevation diagram

a. For vertical shaft Francis turbines

$$\nabla_{\text{inst}} = H_s + \nabla_w + b_0/2 \tag{4.70}$$

where ∇_w is the design tailrace water level, m and b_0 is the guide vane height, m.

b. For vertical shaft axial flow turbines

$$\nabla_{\text{inst}} = H_s + \nabla_w + xD_1 \tag{4.71}$$

where x is the axial flow turbine height coefficient.

c. For horizontal shaft reaction turbines

$$\nabla_{\text{inst}} = H_s + \nabla_w - D_1/2 \tag{4.72}$$

It should be noted here that for hydropower stations with long tailrace water channels, the turbine installation elevation depends not only on the suction height determined by the cavitation characteristics, but also on the minimum pressure at the draft tube inlet caused by load rejection transient processes. If necessary, the installation elevation can be moderately reduced to meet the requirements of guaranteed regulation.

2. **Estimate basic parameters of hydro turbine based on specific speed and statistics data**

This method is based on statistical curves or empirical formulas. First, determine the n_s value of the hydro turbine according to the design head using the $n_s = f(H)$ curves (Fig. 4.25) or the empirical formulas (Eqs. (4.57)–(4.64)). Secondly, based on $n_1' = f(H)$, $Q_1' = f(H)$ and $\sigma_p = f(n_s)$ (Figs. 4.26, 4.27, 4.28, 4.29, and 4.30), the unit speed, unit discharge, and hydropower station cavitation coefficient are obtained respectively. Finally, based on these parameters, the basic operating parameters, D_1,n, and H_r of the hydro turbine can be calculated.

(1) Calculation of runner diameter D_1

$$D_1 = \sqrt{\frac{N_r}{9.81 Q_1' H_r^{3/2} \eta}} \ \text{(m)} \tag{4.73}$$

(2) Calculation of speed n

$$n = n_s H_r^{5/4} / \sqrt{N_r} \ \ \text{(r/min)} \tag{4.74}$$

(3) Calculation of the maximum allowable suction height

$$H_s = 10 - \frac{\nabla}{900} - \sigma_p H \ \ \text{(m)} \tag{4.75}$$

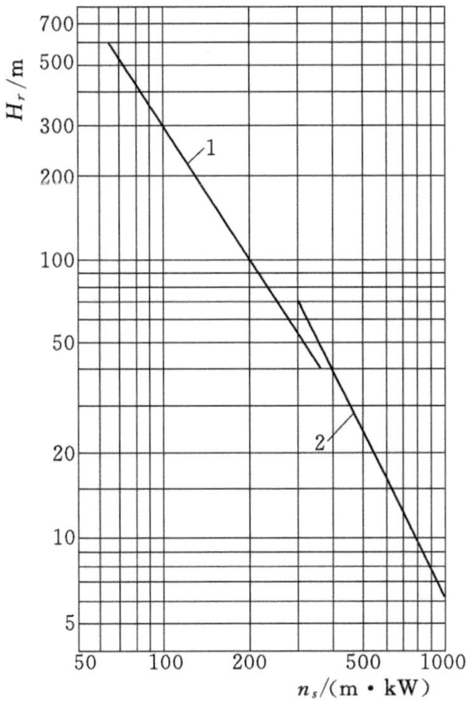

Fig. 4.25 $n_s = f(H)$ curve for Francis and Kaplan turbines 1-Francis turbine, $n_s = 3470H_r^{-0.625}$; 2-Kaplan turbine, $n_s = 2419H_r^{-0.489}$

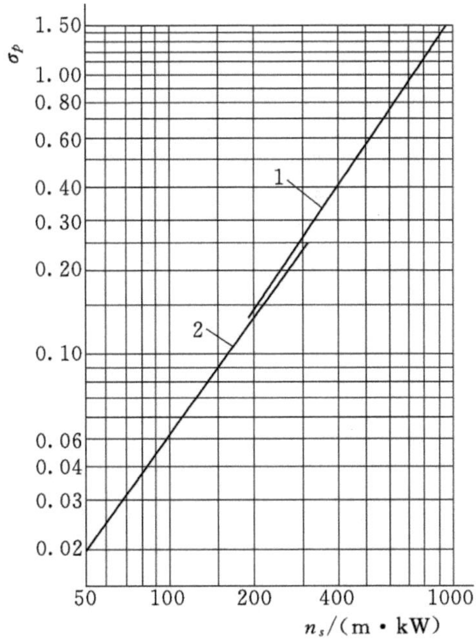

Fig. 4.26 Curve for Francis and Kaplan turbines Kaplan turbine, $\sigma_p = 6.4 \times 10^{-5} n_s^{1.46}$; 2-Francis turbine, $\sigma_p = 7.54 \times 10^{-5} n_s^{1.41}$

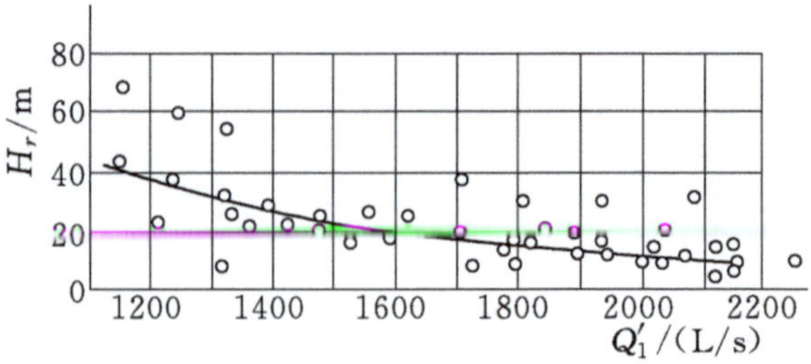

Fig. 4.27 $Q'_1 = f(H)$ curve for Francis turbine

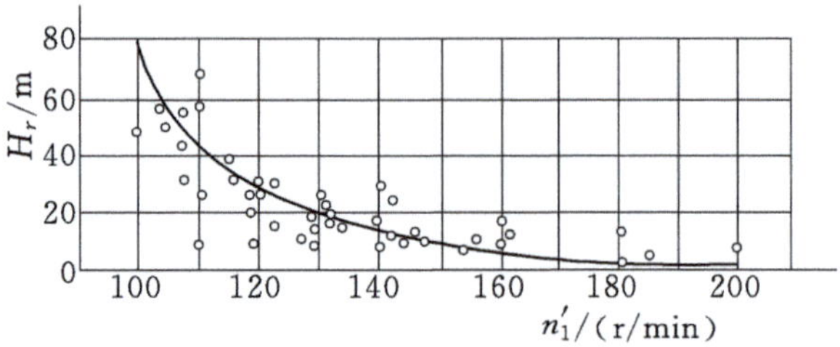

Fig. 4.28 $n'_1 = f(H)$ curve for Francis turbine

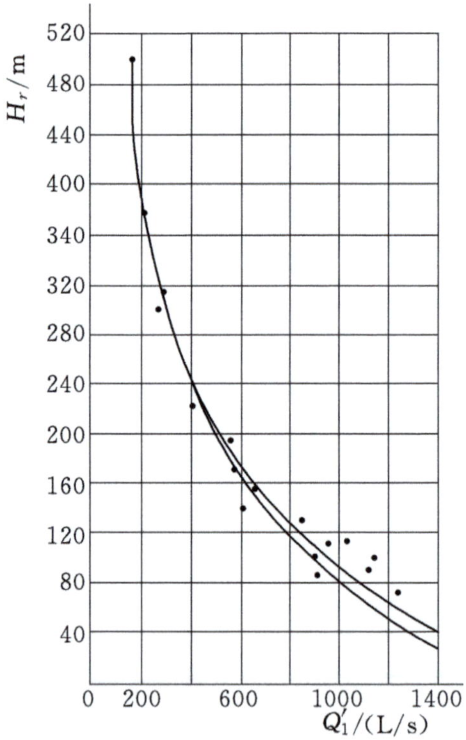

Fig. 4.29 $Q'_1 = f(H)$ curve for Kaplan turbine

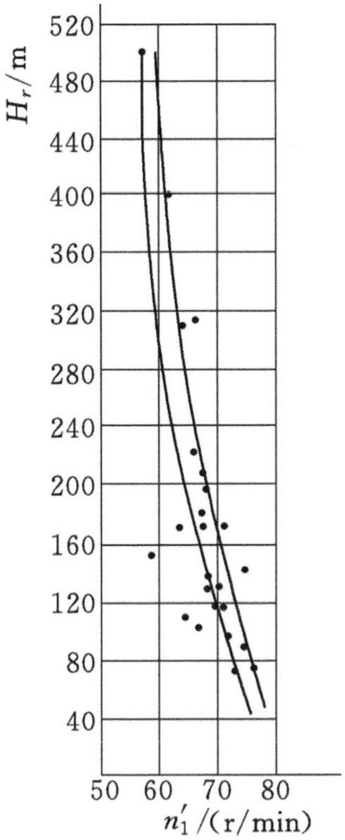

Fig. 4.30 $n_1{}' = f(H)$ for Kaplan turbine curve

4.5.6 Calculation of Basic Parameters of Pelton Turbine

Equation (4.65) shows that the specific speed of Pelton turbine is related to the number of runners and the number of nozzles. If the runner number, nozzle number, and d_0/D_1 of the prototype turbine to be calculated are the same as those of the model turbine, the calculation methods of the basic parameters are basically the same as those for the reaction turbine. That is, applying the model comprehensive characteristic curve chart to select the relevant parameters under the design working condition, such as n'_{1r}, Q'_{1r}, and η_r, then calculating the basic working parameters such as D_1 and n according to the calculation methods for the reaction turbine. The nozzle diameter d_0/D_1 can be converted from the d_0/D_1 value of the model.

If the runner number, nozzle number, and d_0/D_1 of the prototype turbine to be calculated are different from those of the model turbine, different d_0/D_1,Z_0, and K_r can be used to obtain the specific speed required by the prototype turbine. The calculation methods are as follows.

(1) Calculation of speed n

$$n = n_{s1} H_r^{5/4} / \sqrt{N_{r1}} \quad (\mathrm{r/min}) \tag{4.76}$$

where n_{s1} is the specific speed corresponding to the turbine with single rotor single nozzle, m kW and N_{r1} is the output corresponding to the turbine with single runner single nozzle, kW, $N_{r1} = N_r/(Z_0 K_r)$.

(2) Calculation of jet diameter d_0

$$d_0 = \sqrt{\frac{4Q}{\pi Z_0 K_r \varphi \sqrt{2gH_r}}} \approx 0.545\sqrt{\frac{Q}{Z_0 K_r \sqrt{H}}} \quad \text{(m)} \tag{4.77}$$

where Q is the turbine discharge, m^3/s and ϕ is the nozzle jet flow velocity coefficient, $\phi = 0.97$ to 0.98.

(3) Calculation of runner diameter D_1

$$D_1 = \frac{(30 \sim 40)\sqrt{H_{av}}}{n} \quad \text{(m)} \tag{4.78}$$

where H_{av} is the weighted average head, m.

(4) Inspection of D_1/d_0 and estimation of turbine efficiency

From the calculated D_1 and d_0, D_1/d_0 can be obtained. In order to ensure high efficiency of the hydro turbine, D_1/d_0 is generally in the range of 10–20. When D_1/d_0 of the prototype turbine is the same as the model turbine or the D_1/d_0 is in the range of 10–20, the efficiency correction may not be performed. Otherwise, please refer to Table 4.2 to estimate the efficiency of the prototype turbine.

(5) Estimation of bucket number Z_1

The number of buckets can generally be estimated by

$$Z_1 = 6.67\sqrt{\frac{D_1}{d_0}} \tag{4.79}$$

Table 4.2 Expected efficiency of Pelton turbine%

D_1/d_0	Load percentage				D_1/d_0	Load percentage			
	100%	75%	50%	25%		100%	75%	50%	25%
50	85.1	86.2	83.3	72.6	12	86.5	88.0	87.0	81.5
40	85.5	86.6	83.8	73.8	11	86.5	88.0	87.0	82.0
30	85.9	86.9	84.4	74.8	10	86.3	87.8	86.8	82.0
20	86.3	87.8	85.7	77.0	9	85.8	87.4	86.3	81.9
18	86.5	88.0	86.2	78.2	8	84.6	86.5	85.0	81.0
16	86.5	88.0	86.8	79.3	7	82.7	84.7	83.0	79.3
14	86.5	88.0	87.0	80.4	6	80.0	81.5	80.0	76.5
13	86.5	88.0	87.0	80.9	5.5	77.7	78.5	77.7	74.5

For multi-nozzle Pelton turbines, the angle of jet flows should be avoided to be the integer times of the angle between adjacent buckets.

(6) Determination of installation elevation of Pelton turbine

The runner of Pelton turbine is installed above the downstream water level (Fig. 4.24), but the installation elevation must be appropriate. If it is too high, some energy or the working head will be wasted; if it is too low, the tail water column or the tail water splashing will cause additional head loss. Therefore, the drainage height h_p can be reasonably determined according to the empirical Eq. (4.80). The relationship between the drainage height h_p and the runner diameter D_1 is as follows.

$$h_p \geq (1.0 \text{ to } 1.5)D_1 \tag{4.80}$$

In the selection of h_p, the vertical shaft installation needs a larger value and the horizontal shaft installation needs a smaller value.

4.5.7 An Example of Turbine Selection Design

The parameters of the concerned hydropower station are as follows: the maximum head $H_{\max} = 73.6$ m, the minimum head $H_{\min} = 46.8$ m, the rated head $H_r = 59.8$ m, the weighted average head $H_{av} = 65.76$ m, the installed capacity 450 MW, the unit capacity 112.5 MW, and the installation elevation $\nabla = 96$ m.

1 Calculation of basic parameters of hydro turbine by turbine spectrum and model comprehensive characteristic curve chart

(1) Selection of turbine type: The head of this hydropower station varies from 46.8 to 73.6 m. Thus the appropriate models in the turbine spectrum are HL220 and HL230, because the applicable head range of HL220 is 50–85 m, and the applicable head range of HL230 is 35–65 m. Model HL220 has a small cavitation coefficient. Considering the above factors, we choose HL220 runner.
(2) Calculation of basic parameters:

a. Calculation of the runner diameter D_1: Taking the intersection point of the optimal unit speed $n'_{10} = 70$ r/min and the 95% output limit line as the operating point of the design/rated condition, we find the unit discharge $Q'_1 = 1.15$ m^3/s and the corresponding model efficiency, $\eta_M = 0.889$. With the estimated efficiency correction, $\Delta\eta = 3\%$, the estimated efficiency for the prototype turbine is $\eta = \eta_M + \Delta\eta = 0.889 + 0.03 = 0.919$ under the rated condition.

Turbine rated output

$$N_r = \frac{N_f}{\eta_f} = \frac{112500}{0.96} = 117187.5 \text{ (kW)}$$

Runner diameter

$$D_1 = \sqrt{\frac{N_r}{9.81\eta Q_1' H_r^{3/2}}} = \sqrt{\frac{117187.5}{9.81 \times 0.919 \times 1.15 \times 59.8^{3/2}}} = 4.94(\text{m})$$

According to the diameter series of runners specified in China, the calculated value is between 450 and 500 cm, thus it is appropriate to take the standard value of 5 m.

b. Calculation of the efficiency: The maximum efficiency of prototype turbine

$$\eta_{\max} = 1 - (1 - \eta_{M\max}) \times \sqrt[5]{\frac{D_{1M}}{D_1}} = 1 - (1 - 0.915) \times \sqrt[5]{\frac{0.46}{5}}$$
$$= 0.9473$$

$$\Delta\eta = \eta_{\max} - \eta_{M\max} = 0.9473 - 0.915 = 0.0323.$$

The efficiency of the prototype turbine under the rated condition is $\eta = \eta_M + \Delta\eta = 0.889 + 0.0323 = 0.9213$. Because this value is very close to the estimated one, there is no need to iterate the above process to adjust the efficiency correction value.

c. Calculation of the rotational speed:

$$\Delta n_1' = n_{10}'\left(\sqrt{\frac{\eta_{\max}}{\eta_{M\max}}} - 1\right) = 70 \times \left(\sqrt{\frac{0.9213}{0.889}} - 1\right) = 1.26(\text{r/min})$$

$$n_1' = n_{10M}' + \Delta n_1' = 70 + 1.26 = 71.26(\text{r/min})$$

$$n = n_1' \frac{\sqrt{H_{av}}}{D_1} = 71.26 \times \frac{\sqrt{65.76}}{5} = 115.5(\text{r/min})$$

Checking the generator standard synchronous speed series, we know that the most closed synchronous speed of the generator is 115.4 r/min, which can be selected as the rotational speed of this turbine.

d. Calculation of the maximum allowable suction height H_s: Under the rated working condition, the cavitation coefficient of the model turbine is $\sigma_M = 0.133$. The maximum allowable suction height of the prototype turbine is

$$H_s = 10 - \frac{\nabla}{900} - K\sigma_M H_r = 10 - \frac{96.0}{900} - 1.15 \times 0.133 \times 59.8 = 0.75\ (\text{m}).$$

e. Calculation of the axial water thrust: According to the technical data of the HL220 model runner and the relations between the axial force coefficient of the prototype turbine and the model turbine, the range of the axial water thrust coefficient K_t of the runner is 0.28–0.34.

When the runner diameter is small and the gap of the anti-leakage ring is large, the axial water thrust coefficient is corresponding to the higher value. In this hydropower station, the runner diameter is large, but there is a certain sediment-laden in the water, and the gap of anti-leakage ring is appropriately larger, thus we can take $K_t = 0.29$. Then the axial water thrust of the hydro turbine runner is

$$F_t = 9810 K_t \frac{\pi}{4} D_1^2 H_{\max}$$

$$F_t = 9810 \times 0.29 \times \frac{\pi}{4} \times 5^2 \times 73.6 = 4.11 \times 10^6 \text{ (N)}.$$

(3) Check of the working range of the prototype turbine.

a. The maximum discharge $Q_{\max}$

$$Q'_{1\max} = \frac{N_r}{9.81 H_r^{1.5} D_1^2 \eta} = \frac{117187.5}{9.81 \times 59.8^{1.5} \times 5^2 \times 0.9213} = 1.122 \left(\text{m}^3/\text{s}\right)$$

$$Q_{\max} = Q'_{1\max} D_1^2 \sqrt{H_r} = 1.122 \times 5^2 \times \sqrt{59.8} = 216.8 \ \left(\text{m}^3/\text{s}\right).$$

b. The unit speed corresponding to the maximum and minimum head

$$n_{1\min}' = \frac{nD_1}{\sqrt{H_{\max}}} = \frac{115.4 \times 5}{\sqrt{73.6}} = 67.3 \quad (\text{r/min})$$

$$n'_{1\max} = \frac{nD_1}{\sqrt{H_{\min}}} = \frac{115.4 \times 5}{\sqrt{46.8}} = 84.3 \quad (\text{r/min}).$$

c. The unit speed corresponding to the rated head

$$n'_{1r} = \frac{nD_1}{\sqrt{H_r}} = \frac{115.4 \times 5}{\sqrt{59.8}} = 74.6 \quad (\text{r/min}).$$

The lines of $Q'_{1\max} = 1122\,\text{L/s}$, $n'_{1\max} = 84.3\,\text{r/min}$, and $n'_{1\min} = 67.3\,\text{r/min}$ are drawn on the comprehensive characteristic curve chart of HL220 turbine model. It can be seen that the working range of the hydro turbine enclosed by the three straight lines basically includes the high-efficiency zone of the characteristic curves, therefore it is reasonable and feasible to choose HL220 turbine (Fig. 4.31).

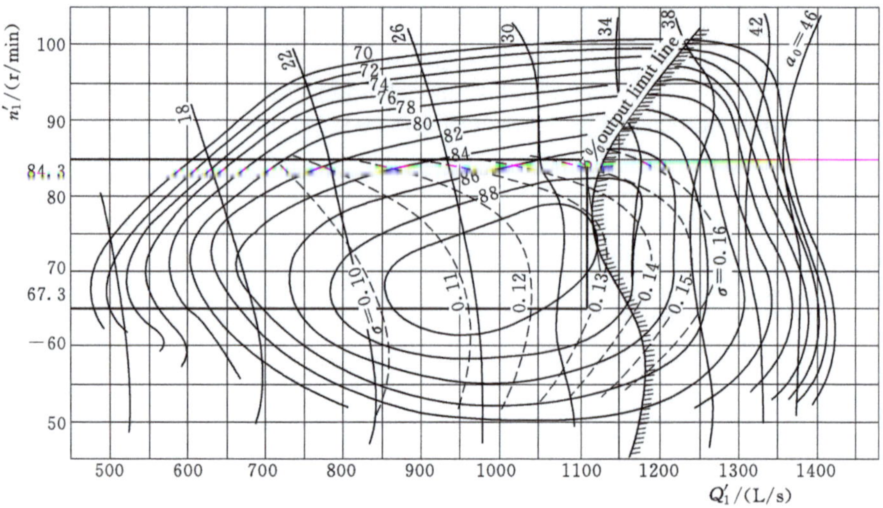

Fig. 4.31 Check of working range for HL220 turbine

2. **Estimating the basic parameters of a turbine according to the specific speed and statistics**

(1) Estimation of the specific speed: According to the relations between the rated head and the specific speed recommended by the turbine spectrum in China, the specific speed n_s of the turbine is

$$n_s = \frac{2000}{\sqrt{H_r}} - 20 = \frac{2000}{\sqrt{59.8}} - 20 = 238.6(\mathrm{m \cdot kW}).$$

(2) Estimation of unit parameters: Referring to the statistical empirical formula of the rated unit speed, rated unit discharge about the specific speed in China, we can estimate the unit speed and unit discharge of the prototype turbine as follows:

$$\begin{aligned} n_1' &= \left[146.7 \times 10^4/(482.6 - n_s)\right]^{0.5} \\ &= \left[146.7 \times 10^4/(482.6 - 238.6)\right]^{0.5} \\ &= 77.54(\mathrm{r/min}) \end{aligned}$$

$$Q_1' 0.1134\left(\frac{n_s}{n_1'}\right)^2 = 0.1134\left(\frac{238.6}{77.54}\right)^2 = 1.074\left(\mathrm{m^3/s}\right).$$

(3) Estimation of efficiency: Referring to the development of model turbines and prototype turbines that have been put into operation in recent years, the rated

efficiency values of prototype Francis turbines are not less than 91%, and the highest efficiency values are not less than 94%.

(4) Estimation of cavitation coefficient:

Cavitation coefficient of the model turbine

$$\sigma_m = 0.036\left(\frac{n_s}{100}\right)^{1.5} = 0.036\left(\frac{236.8}{100}\right)^{1.5} = 0.131.$$

Cavitation coefficient of the hydropower station

$$\sigma_p = 8 \times 10^{-6} n_s^{1.8} + 0.01 = 8 \times 10^{-6} 236.8^{1.8} + 0.01 = 0.160.$$

(5) Estimation of turbine parameters:

The rated output

$$N_r = \frac{N_f}{\eta_f} = \frac{112500}{0.96} = 117187.5(\text{kW}).$$

The rated speed

$$n = \frac{n_s H_r^{5/4}}{\sqrt{N_r}} = \frac{238.6 \times 59.8^{5/4}}{\sqrt{117187.5}} = 115.9(\text{r/min}).$$

Looking up the table for the synchronous speeds of generator, we can select the rotational speed 115.4 r/min.

The runner diameter

$$D_1 = \sqrt{\frac{N_r}{9.81\eta Q_1' H_r^{3/2}}} = \sqrt{\frac{117187.5}{9.81 \times 0.91 \times 1.2 \times 59.8^{3/2}}} = 4.86(\text{m}).$$

The runner diameter is selected from the standard runner series and takes a diameter of 5 m.

The specific speed

$$n_s = \frac{n\sqrt{N_r}}{H^{5/4}} = \frac{115.4\sqrt{117187.5}}{59.8^{5/4}} = 237.6(\text{m} \cdot \text{kW}).$$

The unit speed

$$n_1' = \frac{nD_1}{\sqrt{H_r}} = \frac{115.4 \times 5}{\sqrt{59.8}} = 74.6 \quad (\text{r/min}).$$

The unit discharge

$$Q'_{1\max} = \frac{N_r}{9.81 H_r^{1.5} D_1^2 \eta} = \frac{117187.5}{9.81 \times 59.8^{1.5} \times 5^2 \times 0.91} = 1.14 \quad (\text{m}^3/\text{s}).$$

The maximum discharge of the prototype turbine

$$Q_{\max} = Q'_{1\max} D_1^2 \sqrt{H_r} = 1.14 \times 5^2 \times \sqrt{59.8} = 219.5 \quad (\text{m}^3/\text{s}).$$

According to the selected specific speed and unit parameters, the rated speed of the selected unit is 115.4 r/min, and the diameter of the runner is 5 m. At the corresponding rated operating point, the specific speed is 237.6 m·kW, the unit speed is 74.6 r/min, and the maximum unit discharge is 1.14 m^3/s.

The suction height can be calculated from the rated head according to the selected cavitation coefficient and safety factor. The maximum allowable suction height of the turbine is

$$H_s = 10 - \frac{\nabla}{900} - \sigma_p H_r = 10 - \frac{96.0}{900} - 0.16 \times 59.8 = 0.33\,(\text{m}).$$

The results obtained by the two selection methods are similar.

4.6 Type and Dimension Determination of Spiral Case and Draft Tube

4.6.1 Spiral Case

For the reaction turbine, a spiral case or volute is arranged between the end of the penstock and the turbine stay vane ring, as shown in Fig. 4.32. The function is to make the water flow from the penstock evenly and axial symmetrically flows into the guide vanes with a small head loss.

According to the turbine type, the working head, and the discharge, the spiral cases can be generally classified into the following two types.

1. **Concrete spiral case**

Usually, when the maximum working head of a hydro turbine is below 40 m, in order to save steel materials, the reinforced concrete is used to make the spiral case, which is referred to as the concrete spiral case. For the convenience of construction, the cross-sectional shape of this spiral case is generally trapezoidal. Since the trapezoidal section can extend upward or downward in the axial direction, the radial dimension is smaller than the circular shape on the premise that the sectional area is equal, which is advantageous for reducing the plane size of the power house. Therefore, the

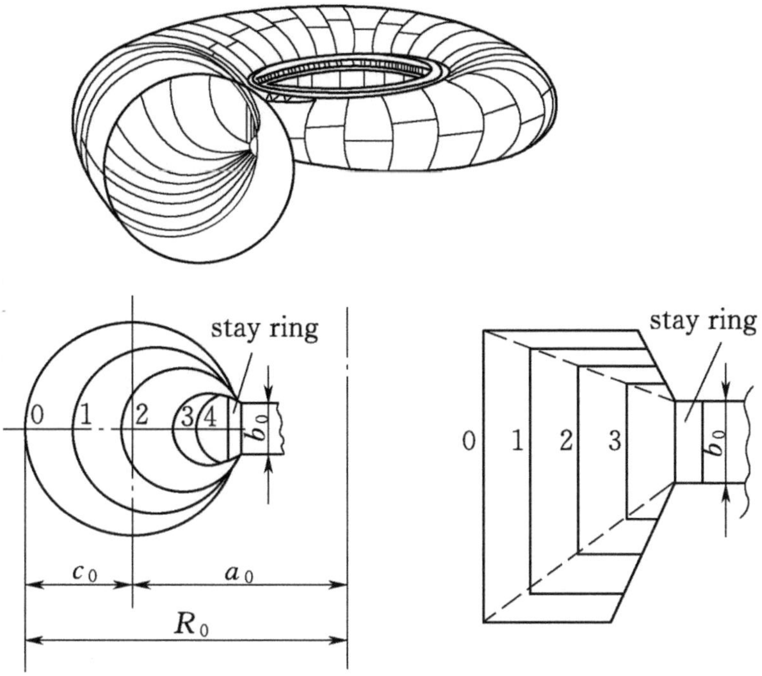

Fig. 4.32 Scheme diagram of spiral case

concrete spiral case is especially suitable for axial flow turbines with low head and large discharge.

The trapezoidal section of the concrete spiral case may have four forms as shown in Fig. 4.33. In order to facilitate the placement of the servomotor on the top of the spiral case, two forms of $m = n$ and $m > n$ are generally used. When the hydropower station has an upstream reservoir with a high dead water level and the rock foundation, in order to reduce the amount of excavation at the inlet section, form of $m < n$ is often used. The flat-top spiral case of $n = 0$ can reduce the amount of concrete in the lower part of the hydropower plant, but excessive extension of the section will form a water flow dead zone, affecting the flow velocity and flow patterns in the spiral case. Therefore, when selecting the trapezoidal section form and sizes, the following conditions should be met as far as possible, along with satisfying the requirements of the power house design.

When $n = 0$ or $m = 0$, there are $\frac{b}{a} = 1.5$ to 1.7 or even to 2.0, $\delta = 30°$, and $\gamma = 10°$ to 15°.

When $n > 0$, there are $\frac{b-n}{a} = 1.5$ to 1.7 or even to 1.85, $\delta = 20°$ to 30°, and $\gamma = 10°$ to 20°.

When $m \leq n$, there are $\frac{b-m}{a} = 1.2$ to 1.7 or even to 1.85, $\delta = 20°$ to 30°, and $\gamma = 10°$ to 35°.

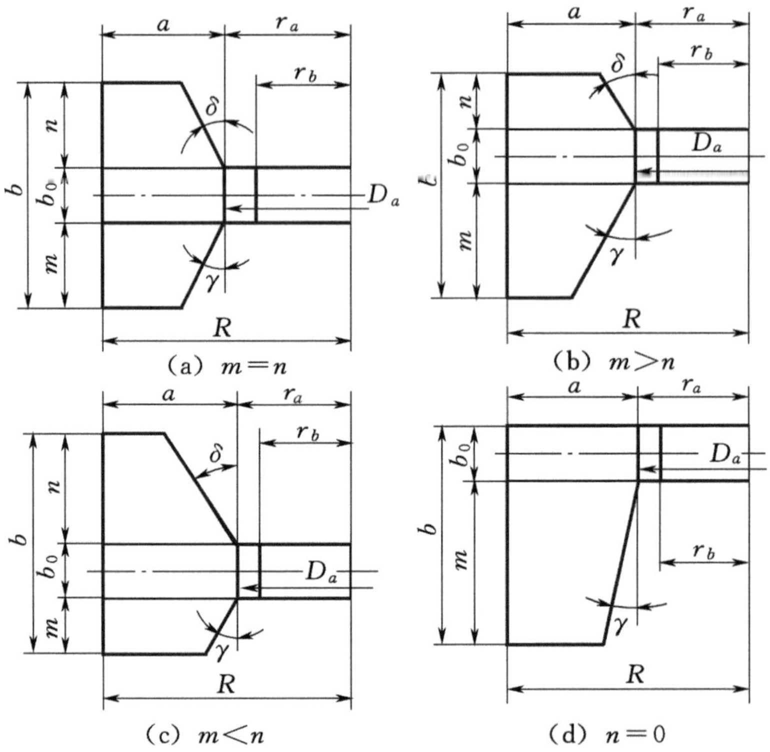

Fig. 4.33 Four forms of trapezoidal section of the concrete spiral case

Due to the large discharge of the axial flow turbine, in order to reduce the size of the spiral case, the wrap angle of the trapezoidal section concrete spiral case generally adopts $\phi_0 = 180°$ to 270°, with $\phi_0 = 180°$ more common, as shown in Fig. 4.34. The wrap angle of the spiral case is defined as the angle from the nose tip to the inlet section. The spiral case with wrap angle $\phi_0 = 180°$ has a part of water flows directly from the penstock into the guide vanes, forming an asymmetric inflow, which is not conducive to the energy conversion performance of the runner. Therefore, dense stay vanes are placed in the upper 1/4 circumference from the nose tip and their waterfoil shape is curved suitable for flow circulation.

The concrete spiral case can also be used in hydropower station with head of more than 40 m (up to 80 m has been used). In this case, the inner wall of the spiral case needs to be the steel plate lining to prevent seepage and reduce friction loss.

2. **Metal spiral case**

When the maximum head of hydro turbine is above 40 m, the metal spiral case is usually used, that is, the spiral case is welded by steel plate or steel casted. In order to improve the stress conditions of the spiral case structure, the cross-sectional form is usually circular or three-centered circular, as shown in Figs. 4.35 and 4.36. The

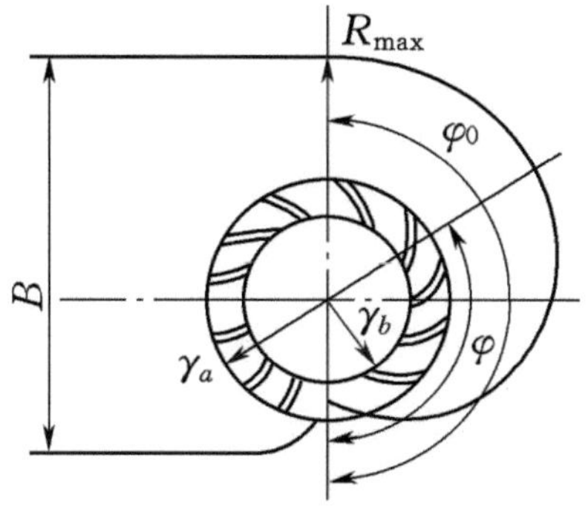

Fig. 4.34 Arrangement of stay vanes for the concrete spiral case of $\phi_0 = 180°$

metal spiral case is usually used for Francis turbines with medium and high heads. There are two ways to connect the metal spiral case to the stay vane ring:

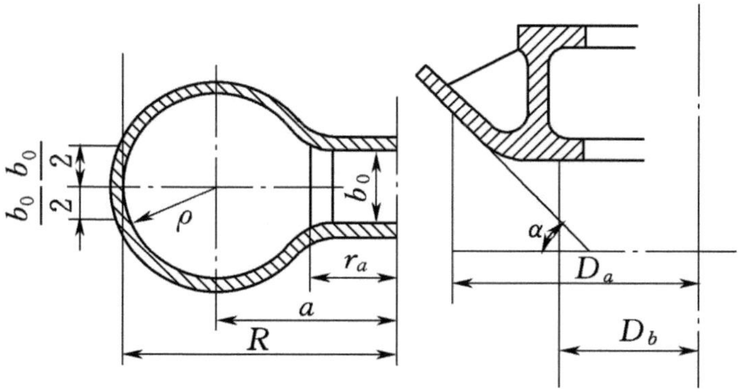

Fig. 4.35 Diagram of the circular metal spiral case

Fig. 4.36 Schematic diagram of the three-centered circular section of metal spiral case

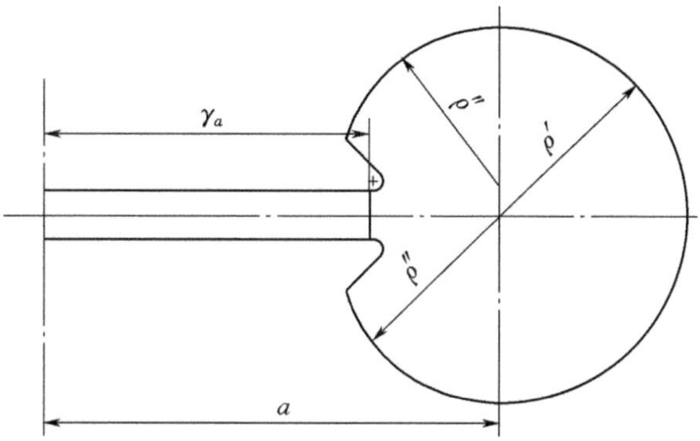

(1) dish edge connection, the angle α between the tangential line of the dish edge of stay vane ring and the horizontal line is generally 55°
(2) direct welding, this connection method has good stress condition, which is currently used in the medium and large Francis turbines. In order to obtain better hydraulic performance, the metal spiral case usually has a wrap angle of 345°.

4.6.2 *Estimation of Dimensions of Spiral Case*

1. **Selection of the inlet section average velocity v_c of spiral case**

The larger the average velocity v_c at the spiral case inlet section, the smaller the spiral case size, but the greater the head loss. Therefore, during the design phase, it is necessary to coordinate the needs of many aspects and reasonably select the average velocity at the inlet section of the spiral case. Based on the statistics of some operating turbines, the curve in Fig. 4.37 is recommended. The average velocity at the inlet section of the spiral case is obtained according to the design head of the hydro turbine. Generally, the intermediate value in the figure can be used. For steel lined concrete spiral cases and metal spiral cases, the upper limit may be taken; if the spiral case size is not the control factor of the layout of the hydropower plant, the lower limit may be taken.

2. **Motion law of water flow in the spiral case**

The motion law of the water flow in the spiral case is the base to determine the size of any section of the spiral case. After the water flows into the spiral case, there exists a spiral motion of water flow. Before entering the stay vanes, the water flow has a certain amount of circulation, so that the water can flow through the stay vanes and the guide vanes with small impacts. The velocity of the water flows in the spiral case can be decomposed into two components: the radial velocity v_r and the circumferential

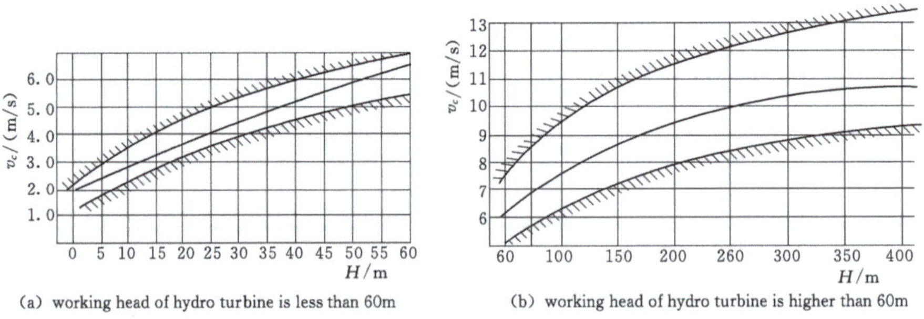

Fig. 4.37 Relations between turbine the design head H_r and average velocity at the inlet section of spiral case

velocity v_u. According to the requirement of uniform axisymmetric inflow into the stay vane ring, the radial velocity v_r should be constant, namely

$$v_r = \frac{Q_{\max}}{\pi D_a b_0} \tag{4.81}$$

where D_a is the outer diameter of stay vane ring, m and b_0 is the height of guide vanes, m.

For the variation law of the circumferential velocity v_u, there are different assumptions in the application.

(1) Velocity moment is constant, $v_u r = \text{const}$. This assumption indicates that the circumferential velocities v_u of the water flow are the same at each point in the circular line of the same radius r from the turbine axis, and v_u decreases as the radius r increases. Because the radius of the spiral case centerline about the turbine rotating axis is gradually reduced from the inlet section to the nose tip, the circumferential velocity at the centerline of the spiral case inlet section is the smallest and at the nose tip section is the largest.
(2) Circumferential velocity is constant, $v_u = \text{const}$. This assumption indicates that the circumferential velocity v_u on any section of the spiral is constant and equal to the average flow velocity of the spiral case inlet section.

Under this assumption, the velocity is smaller and the cross section is larger than those of the former assumption that the velocity moment is constant. It is beneficial to reduce the head loss and the complexity of manufacturing. However, when the spiral case size is the control factor of the power house, it will increase the plant size and increase the investment in civil construction.

3. **Hydraulic calculation and size determination of the spiral line**

(1) Hydraulic calculation of the metal spiral case: According to the assumption that the circumferential velocity is constant, the following calculations can be performed:

At the inlet section

$$\left.\begin{aligned} F_c &= \frac{Q_c}{v_c} = \frac{Q_{\max}\varphi_0}{360 v_c} \\ \rho_{\max} &= \sqrt{\frac{F_c}{\pi}} = \sqrt{\frac{Q_{\max}\varphi_0}{360\pi v_c}} \\ R_{\max} &= r_a + 2\rho_{\max} \end{aligned}\right\} \tag{4.82}$$

At any section

$$\left.\begin{aligned} Q_i &= \frac{Q_{\max}\varphi_i}{360} \\ \rho_i &= \sqrt{\frac{Q_{\max}\varphi_i}{360\pi v_c}} \\ R_i &= r_a + 2\rho_i \end{aligned}\right\} \quad (4.83)$$

where φ_i is the wrap angle at any section.

According to the above calculation results, the plane profile and the cross-section profile of metal spiral case can be drawn, as shown in Fig. 4.38.

(2) Hydraulic calculation of the concrete spiral case: The hydraulic calculation of the concrete spiral case is usually carried out by a semi-graphic method. The steps are as follows.

a. Calculate the area of spiral case inlet section according to Eq. (4.84).

$$F_c = \frac{Q_c}{v_c} = \frac{Q_{\max}\varphi_0}{360v_c} \quad (4.84)$$

b. According to the practical condition of the hydropower station, the trapezoidal section type and the proportion of each part of the section are selected to meet the area requirements of F_c. Then the inlet section is drawn on the upper right of Fig. 4.39.
c. Select the variation rule of the top corner point and the bottom corner point, such as a straight line or a parabola (a straight line change rule is used in Fig. 4.39), which is indicated by a dotted line. And draw a number of intermediate sections (such as 1, 2, 3 and other sections in Fig. 4.39).
d. Calculate the area of each section according to the graph, and correspondingly plot the relationship curve $F = f(R)$ on the below of Fig. 4.39.

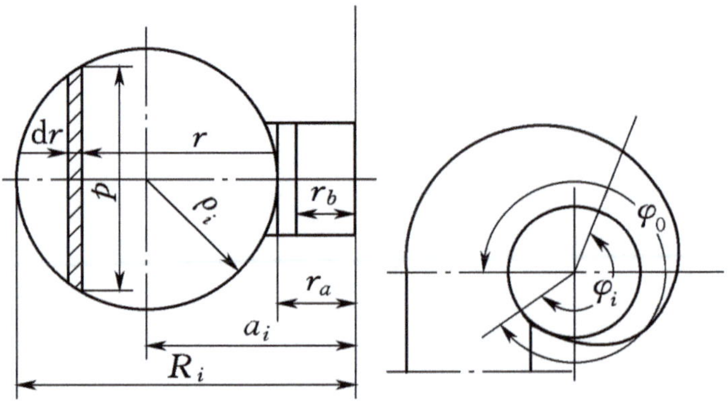

Fig. 4.38 Profile of the metal spiral case by hydraulic calculation

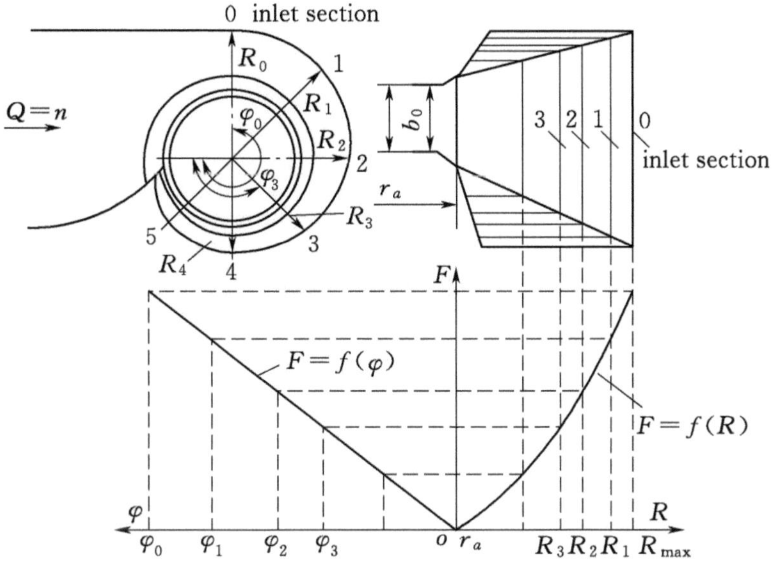

Fig. 4.39 Semi-graphic calculation of the concrete spiral case

e. Calculate according to Eq. (4.85), and draw the relationship curve of $F = f(\phi)$ on the lower left of Fig. 4.39.

$$F_i = \frac{Q_{\max}\varphi_i}{360v_c} \tag{4.85}$$

f. According to the required wrap angle ϕ_i, the corresponding cross-sectional area F_i, the outer radius R_i, and cross-sectional dimension can be obtained in Fig. 4.39. This can be used to draw the plane profile and the cross-sectional profile of the spiral case.

4.6.3 Draft Tube Type

The type and size of the draft tube have great influences on the recovery of the kinetic energy at the turbine runner outlet, the size of the foundation excavation, and the amount of the lower block concrete. Especially in the underground power house, the type and size of the draft tube need to be connected to the tail water tunnel with smooth arrangement and proper size. Therefore, determining the dimensions of the draft tube should be based on the practical requirements of the hydropower station layout, and its hydraulic calculation and design should be carried out to keep higher efficiency of the hydro turbine.

Small hydro turbines often use straight conical draft tube or curved conical draft tube. Large-sized and medium-sized turbines usually use elbow-shaped draft tube. The form and main dimensions of the draft tube are as follows.

1. **Straight conical draft tube**

The axis of the draft tube is a straight line, called a straight conical draft tube, as shown in Fig. 4.40a. This kind of draft tube adopts the simplest diffusion shape, and the water flow is uniform, the resistance is small, that is, the head loss is small, and the recovery coefficient η_w is relatively high, generally up to 83%. The outline of the straight cone draft tube is mostly straight line, and sometimes is curved to make the draft tube horn-shaped.

2. **Curved conical draft tube**

For the convenience of layout, curved conical draft tubes are often used in small-scale horizontal shaft Francis turbines, as shown in Fig. 4.40b. The draft tube consists of an equal diameter 90° elbow and a straight conical tube. Due to the large flow velocity of the elbow, the velocity distribution is not uniform, and the head loss is large, so that the recovery coefficient η_w is low, generally 40–60%.

3. **Elbow draft tube**

For large and medium-sized turbines, in order to reduce the excavation depth of the powerhouse, or to connect with the tail water tunnel, the elbow-shaped draft tube is always used. This kind of draft tube consists of three parts as shown in Fig. 4.40c, namely the inlet straight cone section, the elbow section, and the outlet diffusion section. However, for the ground power houses and the underground powerhouse, and for the conventional turbines and the reversible pump turbine, the three parts are quite different in shape and size, and the design should focus on the key problems

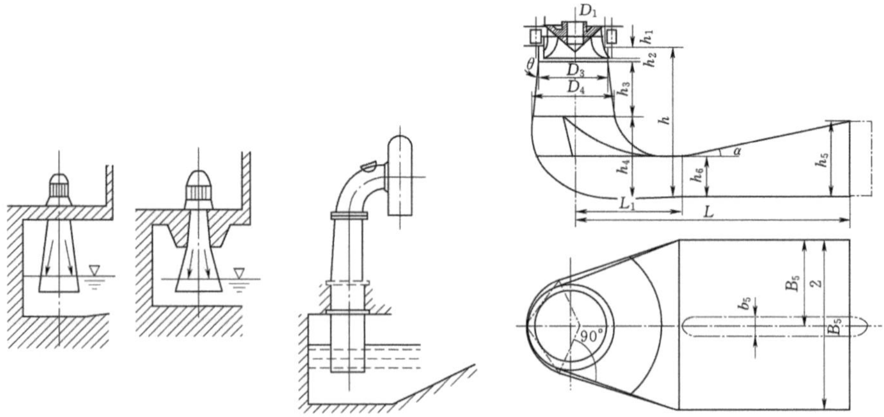

(a) straight conical draft tube (b) curved conical draft tube (c) elbow shape draft tube

Fig. 4.40 Diagram of different draft tubes

of the respective conditions. The energy recovery coefficient η_w of the draft tube is usually 60–75%.

4.6.4 Dimension Determination of Draft Tube

1. **Straight conical draft tube**

The straight cone draft tube shown in Fig. 4.40a has a simple structure. The method of determining the external dimensions is as follows:

(1) Draft tube inlet diameter D_3

$$D_3 = D_1 + (0.5\,\text{to}\,1.0) \tag{4.86}$$

where D_1 is the nominal diameter of the turbine runner, cm.

(2) Draft tube outlet diameter D_5

$$D_5 = \sqrt{\frac{4}{\pi}\frac{Q}{v_5}} = 1.13\sqrt{\frac{Q}{v_5}} \tag{4.87}$$

where v_5 is the outlet velocity of the draft tube, m/s, and it is calculated according to the empirical formula $v_5 = 0.008H + 1.2$.

(3) Cone angle θ and tube length L

According to the principle of minimum head loss in the diffuser, the cone angle $\theta = 12°$ to $16°$ is generally selected, thus the tube length can be calculated by

$$L = \frac{D_5 - D_3}{2}\cot\frac{\theta}{2} \tag{4.88}$$

2. **Elbow-shaped draft tube**

(1) Inlet straight cone section: The inlet straight cone section is a vertical conical diffuser with D_3 the inlet diameter. For Francis turbines, the straight cone section is connected to the base ring, $D_3 = D_2$(runner exit diameter) and $\theta = 14°$ to $18°$. For axial flow turbines, the straight cone section is connected to the lining of the runner chamber, $D_3 = 0.973D_1$ and $\theta = 16°$ to $20°$ can be taken. The height of the inlet straight cone section is h_3. Increasing h_3 can reduce the inlet velocity of the elbow section, and further reduce the head loss. However, if h_3 is too high, it will increase the depth of excavation.

(2) Elbow section: The elbow section is a 90° variable section elbow with a circular cross section at the inlet. For a ground powerhouse, the exit section is rectangular, as shown in Fig. 4.40c. For an underground powerhouse, the exit section can be rectangular with two semi-circular or circular, as shown

in Fig. 4.41a, b. The water flow changes the direction in the elbow section, and the centrifugal force makes the pressure and velocity distribution quite uneven. After the direction turn, the water flows to the outlet diffusion section placed horizontally or upward, so that a large head loss is formed in the elbow section. The main factors affecting the head loss value are the curvature radius of the elbow section and the area variation from the inlet section to the outlet section. It is generally recommended to use a radius of curvature $R = (0.6 \text{ to } 1.0)D_4$, in which D_4 is the diameter of the elbow section inlet. And let the curvature radius R_6 of the outer wall is the upper limit, the curvature radius R_7 of the inner wall is the lower limit, and the ratio of the area of the inlet and exit of the elbow section is about 1.3.

(3) Outlet diffusion section: For ground powerhouses, the outlet diffusion section is a rectangular diffuser with a cross section whose outlet width is consistent with the inlet width. When the width is large, the middle pier should be added according to the structural design requirements (Fig. 4.42). The top plate is upward, the elevation angle is $\alpha = 10^\circ$ to 13°, the length $L_2 = (2 \text{ to } 3)D_1$, and the bottom plate is usually horizontal, in order to reduce the excavation of the rock foundation of the powerhouse. The bottom plate can also be upward with an elevation angle of $\alpha_d = 6^\circ$ to 10°. For underground powerhouses, the outlet diffusion section usually consists of an inlet cross section that is combined by a rectangular and two semi-circulars, and an arch gate-shaped exit section. The inlet shape is gradually varied to the exit shape. Other parameters are basically similar to the outlet diffusion section for ground powerhouses. There are also outlet diffusion sections with a circular cross section.

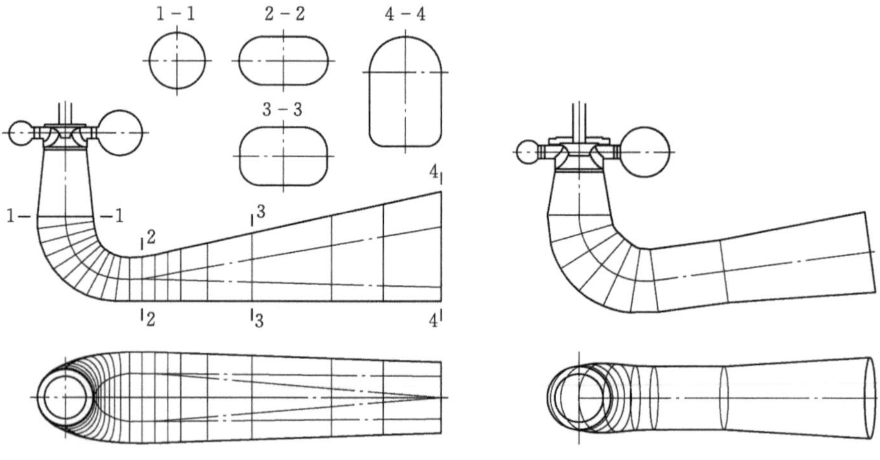

(a) Narrow-high draft tube of Goupitan Hydropower Station

(b) Round draft tube of Pengshui Hydropower Station

Fig. 4.41 Schematic diagram of the draft tubes in underground power stations

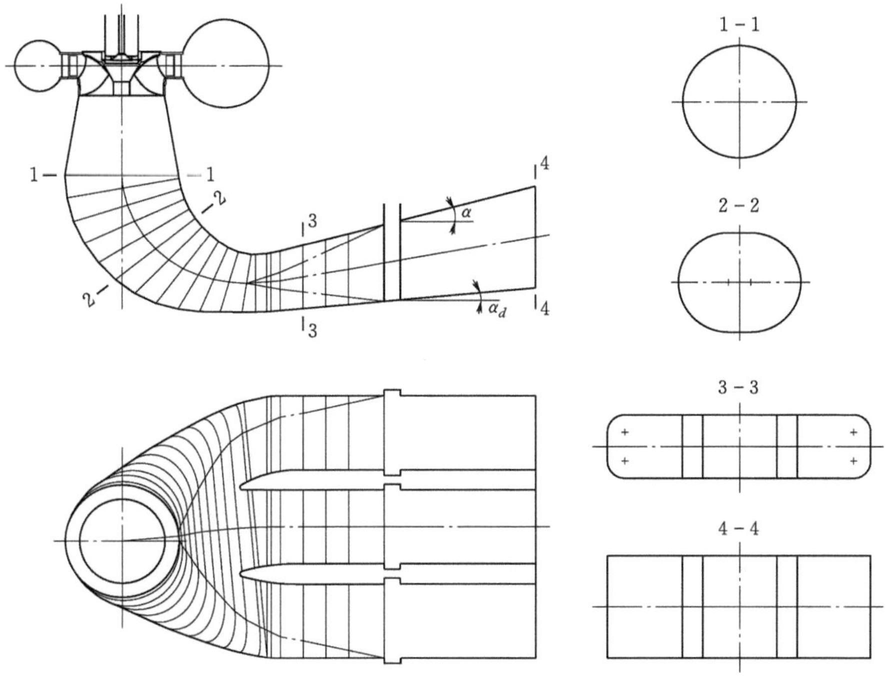

Fig. 4.42 Outlet diffusion section of draft tube with middle pier (draft tube of Three Gorges Power Station)

In summary, for ground powerhouses, the shape and dimension of the draft tube can be selected by referring to the draft tube size recommended in Table 4.3 (Fig. 4.40c) and the standard elbow sizes given in Table 4.4 (Fig. 4.43). For underground powerhouses, the dimension selection of the draft tube should consider the requirements of the arrangement and connection to the tail water tunnel, and detailed analysis and design should be conducted.

The dimensions in Table 4.3 are normalized by the turbine runner diameter D_1. The sizes of the considered turbine can be obtained by multiplying the data in Table 4.3 by D_1. The dimensions of Table 4.4 are the main dimensions of the various geometric curved profiles of standard concrete elbow, as shown in Fig. 4.43. The data listed in

Table 4.3 Recommended draft tube size

h	L	B	D	h	h	L_1	h	Elbow type	Scope of application
2.2	4.5	1.808	1.00	1.10	0.574	0.94	1.30	Metal-lined	Francis $D_1 > D_2$
2.3	4.5	2.420	1.20	1.20	0.600	1.62	1.27	Standard concrete	Axial flow
2.6	4.5	2.720	1.36	1.35	0.675	1.82	1.22	Standard concrete	Francis $D_1 < D_2$

Table 4.4 Standard concrete elbow size

Z	y_1	x_1	y_2	x_2	y_3	x_3	R_1	R_2	F
50	–11.90	605.200			94.36	552.89		579.61	79.61
100	41.70	569.450							
150	124.56	542.450							
200	190.69	512.720			94.36	552.89		579.61	79.61
250	245.60	479.770			94.36	552.89		579.61	79.61
300	292.12	444.700			94.36	552.89		579.61	79.61
350	331.94	408.130			94.36	552.89		579.61	79.61
400	366.17	370.440			94.36	552.89		579.61	79.61
450	295.57	331.910			94.36	552.89		579.61	79.61
500	420.65	292.720	–732.66	813.12	94.36	552.89	1094.52	579.61	79.61
550	441.86	261.180	–457.96	720.84	99.93	545.79	854.01	571.85	71.65
600	459.48	209.850	–344.72	679.36	105.50	537.70	761.82	563.69	63.69
650	473.74	168.800	–258.78	646.48	111.07	530.10	696.36	555.73	55.73
700	484.81	128.090	–187.07	618.07	116.65	522.51	645.77	547.77	47.77
750	492.81	87.764	–124.36	592.50	122.22	514.92	605.41	539.80	39.80
800	497.84	47.859	–67.85	568.80	127.79	507.32	572.92	531.84	31.84
850	499.94	7.996	–15.75	546.65	133.36	499.73	546.87	523.88	23.88
900	500.00	0	33.40	525.33	138.93	492.13	525.40	515.92	15.92
950	500.00	0	81.50	504.36	144.60	484.54	510.00	507.96	7.96
1000	500.00	0	150.07	476.95	150.07	476.95	500.00	500.00	0

Table 4.4 and Fig. 4.43 are the normalized values with $h_4 = D_4$. Multiplying h_4 or D_4 with the data can obtain the sizes for the considered turbine.

Questions

1. What is the similarity of hydro turbine? What are the characterization parameters of the similarity law? What are the characteristics?
2. What is the physical meaning of specific speed of hydro turbine? What are the effects on runner structure, applicable head, performance, and cavitation characteristics?
3. Why should turbine efficiency be corrected? What are the basic principles and methods of amendment? Does the unit parameter needs to be corrected?
4. What parameters need to be measured in hydro turbine model test? What kind of model tests are needed?
5. What is the comprehensive characteristic curve of hydro turbine model and what characteristics can it reflect? What is the law of these characteristics changing with the specific speed of the hydro turbine?
6. What are runaway characteristic curve, performance characteristic curve, and speed characteristic curve of hydro turbine? What are the special points?

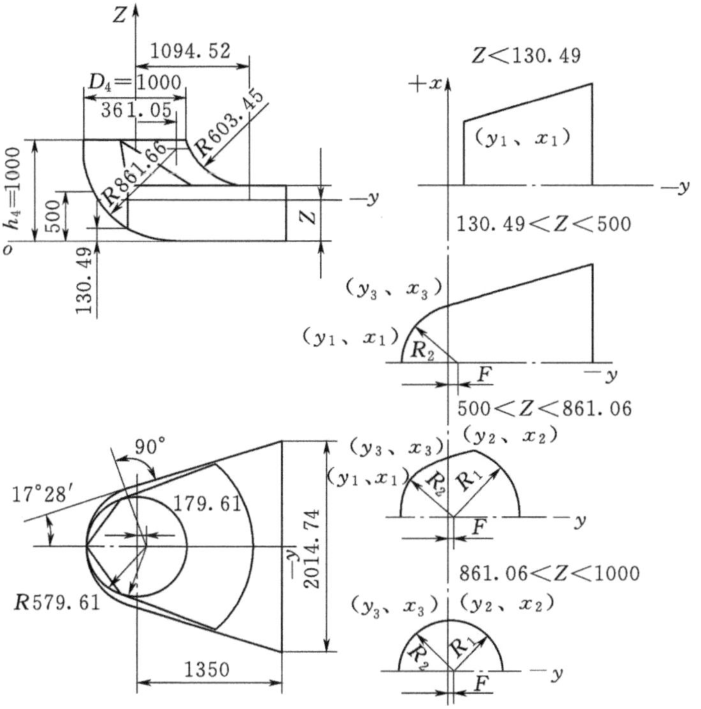

Fig. 4.43 Standard concrete elbow

7. What is the operation characteristic curve of hydro power station? How to determine the output limit line of turbine and generator? Why is it said that the design head is the minimum head for the rated output, and the flow of the turbine is the maximum at this time?
8. In a hydro power station, the total capacity is 60 MW, and the number of units is 4. The maximum head $H_{\max} = 56$ m, the minimum head $H_{\min} = 38$ m, the average head $H_{av} = 50.84$ m, and the design head $H_r = 48.30$ m. The relation of downstream water level-discharge is illustrated in the following table.

Water level /m	68.50	69.20	69.50	69.75	69.90	70.20
Discharge/(m^3/s)	0.0	40.0	80.0	120.0	160.0	200.0
Water level/m	70.95	71.60	72.20	74.50	76.30	78.50
Discharge/(m^3/s)	400.0	600.0	800.0	1000.0	1200.0	1400.0

The tasks are as follows.

(1) To determine the turbine type and arrangement/setting of shaft.
(2) To determine the turbine runner diameter D_1 and speed n, to check the operating range of the turbine and the rated output of turbine.

(3) To calculate the suction head H_s under the condition of design head with rated output, to determine the installation elevation of hydro turbine. One question: is this condition the most dangerous condition of cavitation? Why?
(4) The metal spiral casing with circular cross section is utilized, and the volute angle $\varphi_{\max} = 345°$, and the relative height of guide vane $b_0 = 0.224$. Please calculate the size of the spiral case and draft tube contour and draw the diagram of the spiral case, draft tube by CAD.

Bibliography

1. Zhongyuan, Jin. 1992. *Hydraulic machinery (in Chinese)*, 2nd ed. Beijing: Water Resources and Electric Power Press.
2. Qizhao, Liu and H. Ming. 2010. *Hydropower station (in Chinese)*, 4th ed. Beijing: China Water and Power Press.
3. Ruyun, Luo. 1990. *Design manual for power equipment of hydropower station (in Chinese).* Beijing: Water Resources and Electric Power Press.
4. *Mechanical and electrical design manual of hydropower station, hydraulic machinery (in Chinese).* Beijing: Water Resources and Electric Power Press, 1983.
5. Yuan, Zheng. 2011. *Hydraulic turbine (in Chinese).* Beijing: China Water and Power Press.
6. Dakai, Liu. 1996. *Hydraulic turbine (in Chinese)*, 3rd ed. Beijing: China Water and Power Press.
7. Changzhi, Zhang. 1988. *Hydraulic turbine: principle and mathematical model (in Chinese).* Wuhan: Huazhong Institute of Technology Press.
8. Rangqin, Gong. 2006. *Basic technical conditions of hydraulic turbine (in Chinese).* Beijing: China Standards Press.
9. Yuan, Zheng, and Z. Qiang. 2003. *Power equipment of hydropower station (in Chinese).* Beijing: China Water and Power Press.
10. Qizhang, Li, Z. Qiang and Y. Jixing et al. 2014. *Study on hydraulic stability of Francis turbine (in Chinese).* Beijing: China Water and Power Press.
11. Quanlong, Wang. 2009. *Basic technical conditions of Francis pump turbine (in Chinese).* Beijing: China Standard Press.
12. Xilin, Sha. 1999. *Tubular hydropower station (in Chinese).* Beijing: China Water and Power Press.
13. Wenwu, Song. 2015. *Theory and application of high head tubular turbine (in Chinese).* Beijing: Science Press.
14. Zanpei, You. 2009. *Bulb tubular hydropower station (in Chinese).* Beijing: China Water and Power Press.
15. GuoXuan, Liu. 2006. *Operation and maintenance of bulb tubular turbine generator (in Chinese).* Beijing: China Water and Power Press.
16. Shutang, Tian. 2000. *Tubular turbine generator set and its selection method (in Chinese).* Beijing: China Electric Power Press.
17. Yuanchu, Fu. 2002. *Start-up test code for reversible pumped storage units (in Chinese).* Beijing: China Standard Press.
18. Bo, Yu. 2008. *Principle and operation of hydraulic turbine (in Chinese).* Beijing: China Electric Power Press.
19. Wentong, Zhou. 2007. *Basic theory and design of Pelton turbine (in Chinese).* Beijing: China Water and Power Press.
20. Wentao, Su. 2016. *Hydraulic stability of hydraulic turbine (in Chinese).* Harbin: Harbin Institute of Technology Press.

21. Shuhong, Liu. 2007. *Fundamentals of hydromechanics of hydraulic machinery (in Chinese).* Beijing: China Water and Power Press.
22. Jinshi, Chang. 2005. *Transition process of hydraulic machinery (in Chinese).* Beijing: Higher Education Press.
23. Bryan R., Kendra V. Sharp. 2013. Impulse (Turgo and Pelton) turbine performance characteristics and their impaction pico-hydro installations. *Renewable Energy* 50: 959–964.
24. Helmut, K., and S. Mirjam. 2008. Thirty years of numerical flow simulation in hydraulic turbo machines. *Acta Mechanica* 201: 211–229.
25. Hongquan, Xu, W. Wanpeng, and L. Tieyou. 2011. On the relationship between the selection of specific speed of hydro turbine and the stability of hydropower station (in Chinese). *Journal of Hydropower* 10: 220–223.

Chapter 5
Regulation of Hydro Turbine

Jinping Li and Yongguang Cheng

Abstract In this chapter, we will discuss the following key points: basic concepts of turbine governor; governing principles and type of turbine governor; pressure oil supply system; selection of governing equipment. Students should be acquainted with the basic concepts of turbine governor; be familiar with the governing principles and type of turbine governor; master the selection method of turbine governor; and be familiar with the pressure oil supply system of the hydro turbine.

Keywords Turbine regulation · Turbine governor · Governing principle · Pressure oil supply system · Governor equipment

5.1 Basic Concepts of Turbine Governing

5.1.1 Turbine Regulating Task

When the hydropower station supplies electricity to the electric power system, in addition to meeting the safety and reliability requirements of the users, the frequency and voltage of the electric energy are required to remain within the allowable ranges near the rated values. If the frequency or voltage deviates too much from the rated value, it will adversely affect or even harm the operation of the electrical equipment. In China, the power system has a definite stipulation: the grid frequency is rated at 50 Hz, and the deviation shall not exceed ±0.5 Hz, but for large power grids, it must be within ±0.2 Hz. In addition, the error between the electric clock indication and the standard time should not be greater than 60 s at any time and should not be greater than 30 s for large power grids.

The frequency stability of a power system depends primarily on the balance of active power within the system. Since the changing load of the power system cannot be completely predicted at any time, the variation period is several seconds or tens

J. Li (✉) · Y. Cheng
Wuhan University, Wuhan, China
e-mail: lukeping@whu.edu.cn

Y. Cheng et al. (eds.), *Hydropower Station*,
https://doi.org/10.1007/978-981-96-2383-9_5

of minutes, and the amplitude can reach 2% to 3% of the total capacity of power system (load variation of a small system or an isolated system may be greater than this value). At the same time, the electric energy cannot be stored, so that the voltage U, current I, and frequency f of the system also change. Therefore, the task of turbine regulation is to quickly change the output of the unit to adapt to changes in system load, so that the turbine rotational speed, corresponding to the frequency of the power system, can be restored and maintained within the specified permit range.

The regulation of turbine speed is done by the governor of hydro turbine. The way of regulation can be explained as follows.

The motion equation of the turbine-generator unit is

$$J\frac{d\omega}{dt} = M_r - M_g, \tag{5.1}$$

where $J = \frac{GD^2}{4g}$ is the moment of inertia of the turbine-generator unit, N m s^2; $\omega = \frac{2\pi n}{60} = \frac{\pi n}{30}$ is the angular speed of the unit, rad/s; n is the rotational speed, r/min; M_t is the driving torque from the turbine, N · m; M_g is the retarding torque of the generator (load torque), N m.

Only when $M_t = M_g$, the turbine output is balanced with the generator load and $\frac{d\omega}{dt} = 0$; then $\omega = \text{const} \to n = \text{const} \to f = \text{const}$.

When $M_t > M_g$, the hydro turbine has excessive energy and causes the generator to accelerate, $\frac{d\omega}{dt} > 0$; then the speed increases, $n \uparrow$.

When $M_t < M_g$, the hydro turbine has not enough energy and causes the generator to decelerate, $\frac{d\omega}{dt} < 0$; then the speed decreases, $n \downarrow$.

Therefore, M_t must be regulated to balance with M_g, while

$$M_t = \gamma Q H \eta / \omega. \tag{5.2}$$

It can be known from Eq. (5.2) that changing the turbine efficiency η, working head H, and unit angular velocity ω is almost unrealistic and uneconomical. The best and most effective way to change M_t is to change the turbine discharge Q, which requires adjusting the opening a of the guide vanes. The device for making this regulation is called the turbine governor.

The turbine governor is based on the deviation of the rotational speed to achieve the regulation of the guide vane opening. The characteristics are as follows.

(1) According to the turbine output equation $N = 9.81QH\eta$, in order to generate a larger active power, it is necessary to regulate the opening of guide vanes, for getting a larger discharge through the hydro turbine. While adjusting the guide vane opening to regulate the hydro turbine, it is necessary to give the guiding mechanism a powerful operating work. This requires the turbine governor to be equipped with multi-stage hydraulic amplification components and powerful external energy sources. The nonlinearity and time delay of the hydraulic amplifying component may deteriorate the regulation quality of the turbine governing system.

(2) Hydro power station usually requires a long pressure pipe to convey water into the turbine. The longer the pipe, the greater the water inertia. When the opening of the guide vane changes, the inertia of the water flow changes accordingly, and the water hammer pressure will be generated in the pressure pipeline. The water hammer pressure is opposite to the turbine regulating function, which will seriously affect the regulating quality of thc turbinc rcgulating system. In addition, in order to limit the maximum water hammer pressure in the pressure pipe, the regulating speed of the guiding mechanism must be limited, thereby weakening the effect of the turbine regulation.
(3) The turbine-generator unit has the characteristics of quick startup and rapid adaptation to load changes. Therefore, in the power system it is often undertaken tasks of peak shaving and accident backup. With the expansion of power grid and the improvement of automation degree, the turbine governor is required to have more and more automatic operation and automatic control functions. This makes the turbine governor a very important equipment in the operation of hydropower stations. Its functions and reliability not only affect the safety, stability, and economy of the unit operation, but also directly affect the hydraulic design and structural design of the water delivery system of the hydropower station.
(4) In order to improve the operating efficiency of the turbine or to enhance the safety of the operation of the unit, some turbines have a dual regulation mechanism. For example, for the blade-adjustable turbine, the guide vanes and the blades can be regulated coordinately, and for the Pelton turbine, the needle valve opening and the deflector position can be adjusted. But at the same time, a set of regulator and actuator have been added, and the complexity of the turbine governor is increased.

5.1.2 Characteristics of Turbine Regulating System

When the retarding torque (or load) of the generator changes, the first result is the deviation of the unit speed from the rated one. Sensing this, the governor begins changing the servomotor stroke and the guide vane opening, making that the driving torque of the turbine and the retarding torque of the generator reach a new balance in a short period of time and the speed returns to the rated value or the allowable range. The above process is referred to as the transient process of the turbine governing system, a part of the transient processes of the hydropower station. In this process, not only the guide vane opening, the turbine output, and the rotational speed vary with time, but also the discharge and pressure at any section in the water delivery piping system changes with time. Therefore, the working state of the turbine governing system can be summarized into two types: the steady state before and after the regulation, and the transient state from the beginning to the end of the regulation. For a turbine governing system, the former state can be described and measured using static characteristics related indicators, while the latter state can be described and

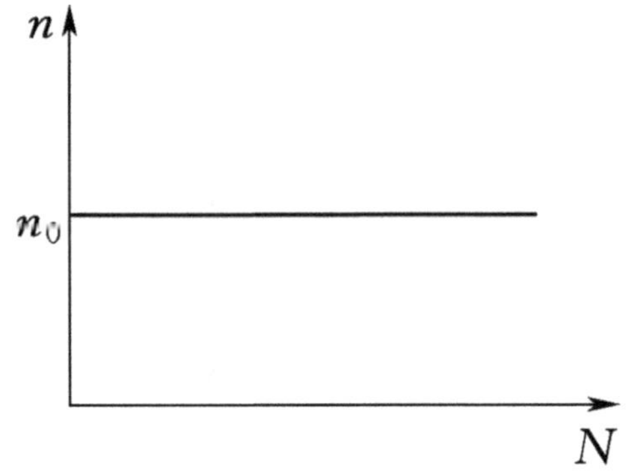

Fig. 5.1 Static characteristics with no-difference regulation

measured using dynamic characteristics related indicators. Similarly, the steady state of the water delivery system of hydropower station can be described by a steady flow, and the transient state can be described by an unsteady flow.

1. **Static Characteristics of the Regulation System**

The static characteristic of the turbine governing system refers to the relationship between speed and output under the condition that the governor does not act and the guide vane opening is constant.

If the speed n is independent of the magnitude of the output N, that is, the speed is kept constant at the rated speed, which is called no-difference regulation (Fig. 5.1). This type of regulation is feasible in a single machine operation, but when multiple units are operated in parallel, the problem of unclear load distribution will be occurred. The reason is that the sensitivity of each governor cannot be exactly the same. Therefore, when the load changes, the response sequence and action speed of each unit are different. Some units have larger loads, but some have smaller loads. In this situation, not only the loads cannot be distributed according to requirements, but also the loads may automatically shift between the units, leading to unstable operation of units.

In order to solve the above problem, a difference regulation is employed. The static characteristics with difference regulation are a linear relationship between the speed n and the output N (Fig. 5.2). Each output N corresponds to a speed n, and the difference regulation rate is defined as

$$e_p = \frac{n_{\max} - n_{\min}}{n_0} \times 100\%, \tag{5.3}$$

where $n_{\max}$ is the maximum rated speed of the unit corresponding to the no-load condition, r/min; $n_{\min}$ is the minimum rated speed of the unit corresponding to the maximum output condition, r/min; n_0 is the rated speed of the unit corresponding to frequency of power system, r/min.

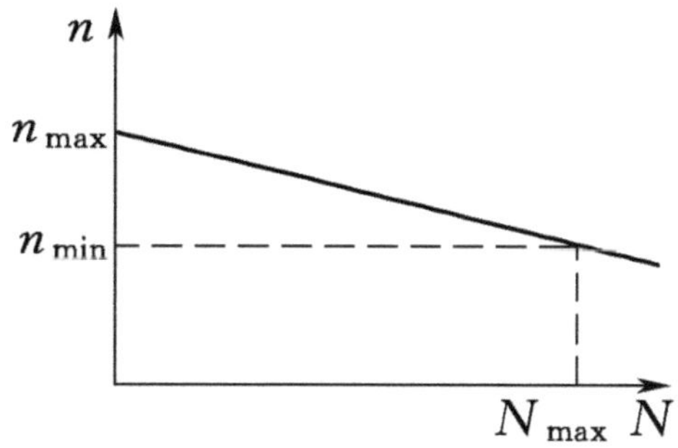

Fig. 5.2 Static characteristics of difference regulation

In practical operation, $e_p = 0$ to 8% is generally used. When $e_p = 0$, it is no-difference regulation. The differential regulation mechanism in the governor is used to adjust or tune e_p.

It should be pointed out that when multiple units are operating in parallel, if the principle of equal load distribution is adopted, the difference regulation rate e_p of each governor is required to be the same, and otherwise the conditions for synchronous parallel operation at the same speed cannot be satisfied. Under this equal e_p condition, the hydropower station can operate at the highest efficiency. If the units are required to work in different load distribution modes, the unit with larger e_p will have smaller load change. As shown in Fig. 5.3, the different e_p values of the two unit are corresponding to different load variations.

2. **Dynamic Characteristics of the Turbine Regulating System**

The dynamic characteristics of the turbine governing system mainly refer to the speed change with time in the regulation process, which is usually expressed by the dynamic characteristic curve of $n = f(t)$. In practical operations, not only the dynamic characteristic curve is required to decay and converge rapidly within the allowable range of speed variation, but also the dynamic quality indicators can meet the requirements of the relevant standards. The common indicators are as follows.

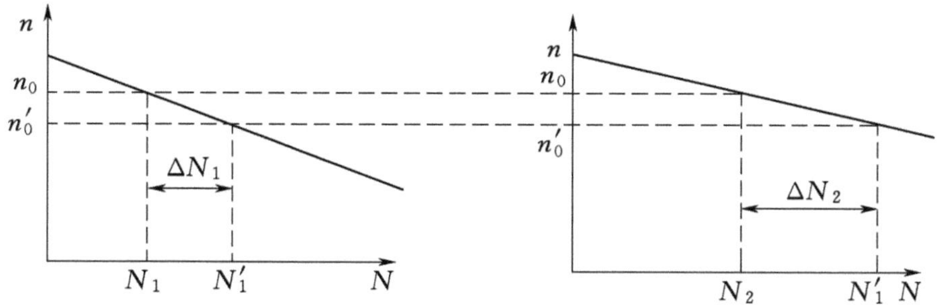

Fig. 5.3 Load distribution between two units with different difference regulation rates in parallel operation

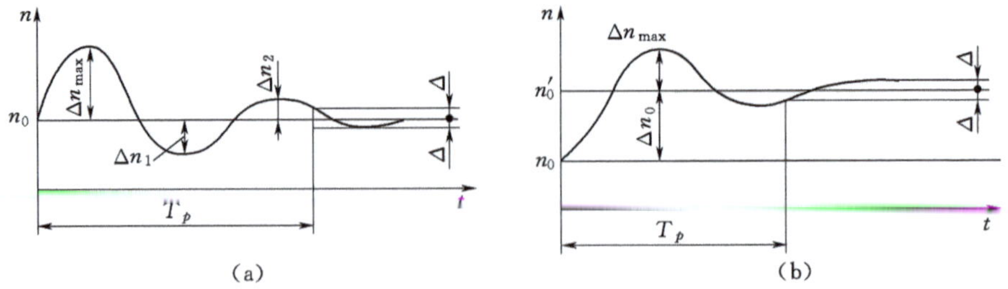

Fig. 5.4 Relevant indicators for dynamic quality of turbine regulating system

(1) Regulating Time T_p. It is the time elapsed from the start of the speed change to the restoration of speed into the allowable range of the speed deviation. That is, the rotational speed n after T_p is in the range of $n_0' \pm \Delta$. The Δ value is 0.2% n_0' for the large speed control system and 0.4% n_0' for the small speed control system, as shown in Fig. 5.4.
(2) Maximum Speed Deviation $\Delta n_{\max}$. It is the deviation of the first peak of the speed history curve during the transient process (Fig. 5.4). That is $\Delta n_{\max} = n_{\max} - n_0$ also can be expressed by a relative value $\frac{\Delta n_{\max}}{n_0}$.
(3) Overshoot of Speed Regulating δ. For the no-difference regulation, it is the ratio of the first trough deviation value Δn_1 (Fig. 5.4) and the maximum speed deviation $\Delta n_{\max}$ during the transient process, namely

$$\delta = \frac{\Delta n_1}{\Delta n_{\max}} \times 100\%. \tag{5.4}$$

For the differential regulation, it is usually expressed as a percentage of the maximum speed deviation $\Delta n_{\max}$ in the given variation amplitude Δn_0 (Fig. 5.4), namely

$$\delta = \frac{\Delta n_{\max}}{\Delta n_0} \times 100\%. \tag{5.5}$$

(4) Number of Oscillation X. It is usually expressed as a half of the number of oscillation peaks and troughs of the speed history appearing during the regulating time.
(5) Attenuation ψ. It is usually expressed as the relative value of the deviation differences between the second peak and the first peak amplitudes, namely:

$$\psi = \frac{\Delta n_{\max} - \Delta n_2}{\Delta n_{\max}}. \tag{5.6}$$

Among the above indicators, the regulating time T_{p}, the maximum speed deviation $\Delta n_{\max}$ and the oscillation number X, are more commonly used. In order to ensure the turbine regulating system has good dynamic quality, it is generally through setting the relevant parameters of the turbine governor to achieve the goal that when a step-shaled load disturbance happens from the no-load condition, the regulating time T_p is short, the overshoot δ is less than 30%, and the number of oscillation X in not more than 2 times.

After a load rejection, the dynamic quality of the unit should meet the following requirements.

(1) After 25% rated load rejection, the delay time of servomotor shall not exceed 0.2 s for large-scale electro-hydraulic regulating system; it shall not exceed 0.3 s for medium-sized electro-hydraulic regulating system. The delay time of servomotor is defined as the time from the stator current disappearance of the generator to the obvious action of the servomotor, when the generator rejects 25% load.
(2) After 100% rated load rejection, the peaks that exceed 3% of the rated speed in the oscillating speed history curve shall not exceed 2 times.
(3) After 100% rated load rejection, the regulating time, from the time at which the servomotor first moves toward the opening direction to the time at which the relative deviation within 0.5% of the rated speed, shall be not more than 40 s.
(4) After 100% rated load rejection, the volute pressure rise rate, the turbine speed increase rate, and the draft tube inlet vacuum should meet the requirements of the calculation of guaranteed regulation.

At the same time, in the speed-no-load automatic operation condition, the deviation of speed (or frequency) of the unit should also meet the following requirements: for large-scale electro-hydraulic regulation system, not exceed ±0.15%; for medium-scale electro-hydraulic regulation system, not exceed ±0.25%; for small-scale electro-hydraulic regulation system, not exceed ±0.35%.

5.2 Governing Principles and Types of Turbine Governor

5.2.1 Working Principle of Turbine Governor

1. Compositions of Turbine Regulating System

The turbine governor and the regulated object (unit, water delivery system, and load) are often referred to as a turbine regulating system. A simplified block diagram of a turbine regulating system is shown in Fig. 5.5. A turbine governor with automatic regulation function is generally composed of the following segments or components.

(1) Frequency Measurement Segment (Sensitive Component). Its function is to convert the frequency signal of the unit into the corresponding displacement signal (of mechanical regulator) or voltage signal (of electrical regulator) or digital signal (of microcomputer regulator).

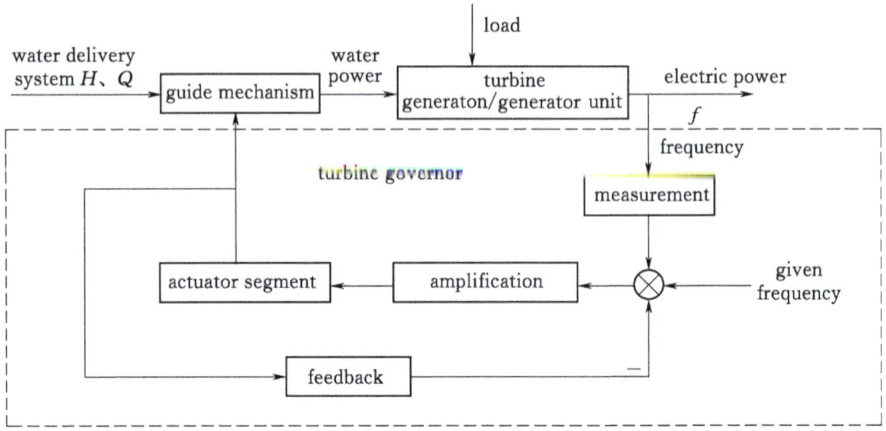

Fig. 5.5 Simplified block diagram of turbine regulating system

(2) Integrated Segment (Comparison Component). Its function is to compare the frequency signal measured by the frequency measurement segment with the given frequency signal and the feedback signal to obtain the frequency deviation value and sign and issue the regulation commands in a proper sequence. For microcomputer regulator and the full electrical regulator, the regulation commands are issued by a certain regulating law according to the frequency deviation.
(3) Amplification Segment (Amplifying Component). Its function is to amplify the power of regulation commands to reach the extent that the actuator can push the servomotor at the required speed.
(4) Actuator Segment (Executive Component). Its function is to push the guide vane control mechanism (guiding mechanism, blade adjusting mechanism for blade-adjustable turbine), to change the vane opening degree, and realize the adjustment of turbine discharge.
(5) Feedback Segment (Feedback Component). Its function is to return the guide vane opening information to the integrated segment (negative feedback signal).

2. Working Principle of the Turbine Regulating System

The output regulation, operating condition transition, and startup and shutdown operations of the hydraulic turbine-generator unit are realized under the control of the turbine governor with corresponding functions. For example, when the load changes, the turbine governor controls the guiding mechanism to change the discharge entering the turbine runner according to a certain regulation law, so as to restore the torque balance and speed stability of the unit, and to form no-difference or differential regulation characteristics. In order to explain the working principle of the speed control system, an intuitive and easy-to-understand single-regulation mechanical hydraulic governor is taken as an example. Figure 5.6 shows its schematic diagram. This type of turbine governor is mainly composed of centrifugal pendulum, pilot valve, auxiliary servomotor, main distributing valve, main servomotor, buffer, differential regulation mechanism, manual control hand wheel, and lever system.

The centrifugal pendulum is used to measure the speed of the unit and convert the speed signal into the displacement signal of the pilot valve. Since the load capacity of the centrifugal pendulum is very small, to push the bulky turbine guiding mechanism, an amplifying device must be added. Therefore, the pilot valve and the auxiliary servomotor constitute the first-stage hydraulic amplifying device, and the main distributing valve and the main servomotor constitute the second-stage hydraulic amplifying device. The output signal of the auxiliary servomotor is sent back to the input terminal of the pilot valve as the internal feedback of the first-stage hydraulic amplifying device; the output signal of the main servomotor goes through the buffer

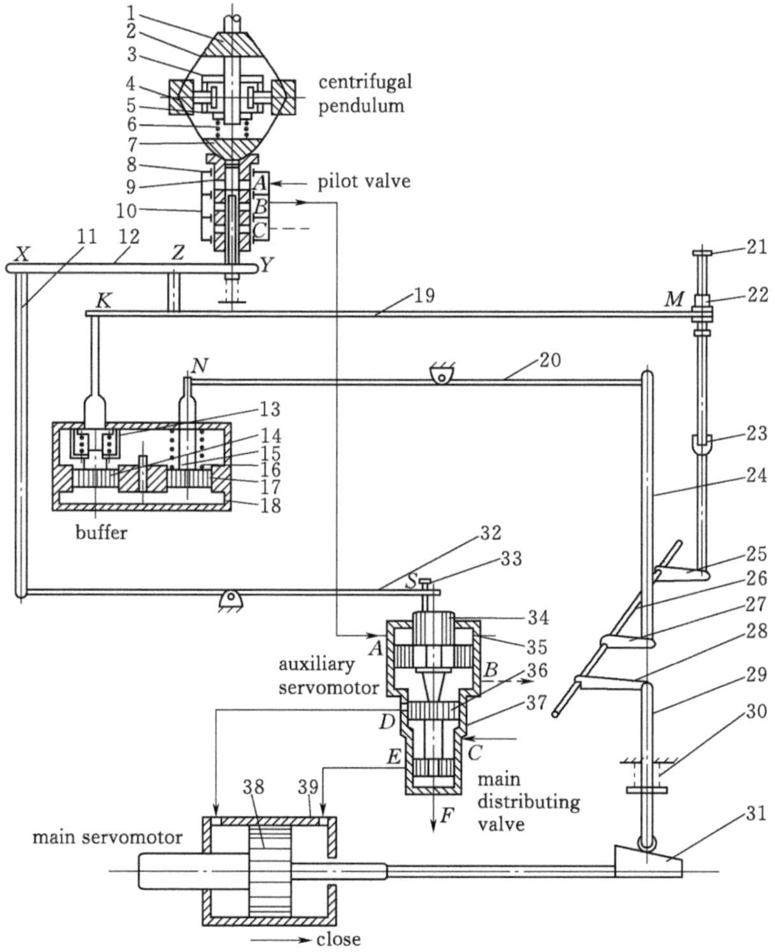

Fig. 5.6 Schematic diagram of single-regulation mechanical hydraulic turbine governor

and the differential regulating mechanism to the input terminal of the pilot valve, in the form of the main feedback signal.

(1) Action Principle of the Centrifugal Pendulum

As shown in Fig. 5.6, the centrifugal pendulum is composed of steel strip, heavy blocks (or weights), lower support block, displacement restricted brackets, adjustable nut, compression spring, rotating shaft, and upper support block.

The steel strip wraps and fixes to the outer circle of the lower support block. The two heavy blocks are hung on both sides of the steel strip, and the upper support block and the rotating shaft are connected by a fixing pin, and the upper part of steel strip is fixed on the upper support block by screws. The compressed spring is located between the adjustable nut and the lower support block, and changing the position of the adjustable nut is equivalent to changing the given speed. The centrifugal pendulum is mounted and rotated with the motor by the rotating shaft, and the centrifugal force generated by the heavy blocks causes the steel strip to expand outward, and drives the lower support block to move upward and compress the spring. When the centrifugal pendulum is at a certain position, the various forces acting on the centrifugal pendulum are well balanced, and the centrifugal pendulum is operated at a certain equilibrium state. If the rotating speed increases at this time, the centrifugal force of the weights also increases, so that the support block moves up against the spring resistance; if the rotating speed decreases, the centrifugal force decreases, and the spring force causes the lower support block to move downward. The displacement of the lower support block is the output signal of the centrifugal pendulum. Under the premise of ignoring inertial force and hydraulic friction, the output of the centrifugal pendulum is proportional to the input; thus, the centrifugal pendulum is a proportional element.

There are two kinds of power supply to drive the centrifugal pendulum: ① one is from the permanent magnet generator which is coaxial with turbine-generator; ② another is from the voltage transformer at the generator terminal. The frequency of the power supply system reflects the speed of the unit, so the speed change of the centrifugal pendulum also reflects the speed change of the unit. In the past, the permanent magnet generator was generally used as the power supplier of the centrifugal pendulum in large and medium-sized water turbine-generator units. This is because the permanent magnet generator can still ensure the normal operation of the centrifugal pendulum and the governor, even if the excitation of the main unit is cut off. Thus, the reliability of the centrifugal pendulum can be improved and the external interference can be reduced. However, the use of permanent magnet generation will increase the investment of the main generator structure. Therefore, the power supply mode of terminal voltage transformer is generally used in small turbine-generator units.

(2) Hydraulic Amplifying Device

The first-stage hydraulic amplifying device is composed of the pilot valve and auxiliary servomotor, but the pilot valve is integrated with the centrifugal pendulum, and the auxiliary servomotor is integrated with the main distributing valve. Between them, the oil flow is used to transmit the regulation signal, and the lever is used to transmit the feedback signal.

As shown in Fig. 5.6, the pilot valve is composed of valve housing, rotating sleeve, needle valve and so on. In the mechanical hydraulic governor, the rotating sleeve is connected with the lower support block of the centrifugal pendulum and rotates with the centrifugal pendulum; thus, its position can reflect the level of the unit rotation speed. There are three rows of oil holes on the rotating sleeve, the upper oil hole is connected with the pressure oil, the lower oil hole is connected with the oil discharge, and the middle oil hole is connected with the auxiliary servomotor piston working chamber by the oil pipe. The needle valve has upper and lower valve discs, and when it is in equilibrium, two valve discs just cover the upper and lower oil holes of the rotating sleeve. At this time, the oil in the intermediate B chamber has a certain oil pressure, and the oil pressure is also transmitted to the auxiliary servomotor working chamber. When the rotating sleeve moves up as the speed of centrifugal pendulum increases, the upper oil hole is closed, the lower oil hole is opened, and the intermediate B chamber is connected with the oil discharge, and oil in the auxiliary servomotor working chamber is drained and the oil pressure decreased. Conversely, when the rotating sleeve moves downward as the rotation speed of the centrifugal pendulum decreases, the lower oil hole is closed, and the upper oil hole is opened, so that the intermediate B chamber is connected with the upper oil hole with pressure oil, and the oil pressure in auxiliary servomotor working chamber increased. Therefore, the function of the pilot valve is to convert the displacement of the rotating sleeve (displacement of the support block under the centrifugal pendulum) into a change of the oil pressure and transmit it to the auxiliary servomotor to control the movement of the piston.

The movement of the auxiliary servomotor piston is driven by a pressure difference, because the lower chamber of the piston is connected to the oil discharge, where the pressure is zero, and the upper chamber of the piston (working chamber) has a certain oil pressure p_i from the intermediate B chamber of the pilot valve. Thus, there is a downward force with value

$$P_A = p_i F_A, \tag{5.7}$$

where p_i is the oil pressure output from the pilot valve middle chamber, N/m^2; F_A is the annular area of the upper surface of the auxiliary servomotor piston, m^2; P_A is the oil force/thrust acting on the auxiliary servomotor piston, N.

The second-stage hydraulic amplifying device is composed of the main distributing valve and the main servomotor. Since the main distributing valve body and the auxiliary servomotor piston are integrated, the movement of the auxiliary servomotor piston drives the main distributing valve body to move. The upper and lower valve discs of the main distributing valve body are not equal in diameter, and the upper valve disc is large and the lower valve disc is small. There is pressure oil between the two valve discs (there is also three oil holes on the main distributing valve housing, the middle hole C is connected with the pressure oil, the upper hole D is connected with the left chamber of the piston of the main servomotor piston through the oil pipe, and the lower hole E is connected with the right chamber of the main servomotor piston). Therefore, the thrust is generated by the pressure oil acting on both the lower surface of the upper valve disc and the upper surface of the lower valve disc. However, due to the oil receiving area of the upper valve disc which is larger, the combined force is an upward force, which is

$$P_M = p(F_1 - F_2), \tag{5.8}$$

where P_M is the resultant force of the pressure oil acting upward on the valve disc of the main distributing valve, N; p is the pressure of pressure oil, N/m^2; F_1, F_2 are the oil receiving areas of the upper and lower valve discs, respectively, m^2.

The state of the auxiliary servomotor and the main distributing valve depends on the comparison of the two forces between P_M and P_A (ignoring the piston and valve body weights). When $P_M = P_A$, the auxiliary servomotor piston and the main distributing valve body are not moved (in an equilibrium state); when $P_M > P_A$, the auxiliary servomotor piston and the main distributing valve body move upward; when $P_M < P_A$, the auxiliary servomotor piston and the main distributing valve body move downward.

The movement of the auxiliary servomotor piston is transmitted back to the needle of the pilot valve by the adjustable screw, pull rod, and lever. This action constitutes an internal feedback correction of the first-stage hydraulic amplifying device. Because it uses a lever system to transmit signals, it is a proportional element.

When the rotation speed increases to a certain value, the pilot valve rotation sleeve is also moved up to a certain position. The lower hole on rotation sleeve is open, and the oil in the upper chamber of the auxiliary servomotor is connected with the oil discharge through the rotation sleeve lower hole. And the auxiliary servomotor piston and the main distributing valve body are moved up, and the needle of pilot valve is also moved up by the lever system. When the needle is moved up to the intermediate position of the rotating sleeve, the lower hole on the rotating sleeve is closed again by the lower valve disc of the needle, and the oil pressure output to the auxiliary servomotor from the pilot valve is also restored to the original value. Then the force on the auxiliary servomotor piston is balanced with the force on the main distributing valve body, and the piston stops moving. The displacement m_A of the auxiliary servomotor piston is proportional to the displacement S of the pilot valve rotation sleeve (or needle), and they have the same direction. The proportional coefficient is the transmission ratio K of the lever system and $m_A = KS$. K can be adjusted between 2 and 10, so as to enlarge the displacement of the support block under the centrifugal pendulum. At the same time, due to the large area of the auxiliary servomotor piston, it can overcome the dry friction of several tens or even several hundreds of Newton without causing excessive insensitivity.

However, such a large force is still not enough to drive the bulky turbine guiding mechanism; therefore, a second-stage hydraulic amplifying device is required. Since the main distributing valve body of the second-stage hydraulic amplifying device is integrated with the output component auxiliary servomotor piston of the first-stage hydraulic amplifying device, the main distributing valve moves together with the auxiliary servomotor piston. The main distributing valve controls the main servomotor, and the intermediate oil hole C of the main distributing valve bushing is connected with the pressure oil, the top and bottom of the bushing is connected with the oil discharge. The upper and lower oil holes of the bushing are connected with the oil cylinders on both sides of the main servomotor piston. When the main distributing valve body is in the middle position, the upper and lower valve discs just cover the D and E holes. At this time, the oil pressures on both sides of the main servomotor piston are substantially equal, and the piston does not move. When the main distributing valve body is moved upward, the D hole is connected with the pressure oil, the E hole is connected with the oil discharge, and the main servomotor piston is moved to the right under the action of the oil pressure to close the turbine guide vanes. When the main distributing valve body moves downward, the D hole is connected with the oil discharge, the E hole is connected with the pressure oil, and the piston of the main servomotor moves to the left under the action of the oil pressure to open the turbine guide vanes.

(3) Buffer

The buffer is a feedback correction device whose performance directly affects the stability of the turbine regulating system. As shown in Fig. 5.6, the housing of the buffer is like a connector filled with oil. The active piston is coupled to the main servomotor piston by lever, pull rod, crank arm, recovery shaft, crank arm, pull rod, and ramp block. The driven piston is connected to the pilot valve needle by levers. The driven piston is positioned by the spring and its bracket, normally in an intermediate position. When the active piston is forced to move downward due to the right movement of the main servomotor piston, since the oil is incompressible, and the oil in the lower part of the active piston is too late to flow from the orifice to the upper part, the oil pressure in the lower part of the piston rises. The oil pressure acts on the lower surface of the driven piston, pushing the driven piston up, thereby transmitting a signal to the pilot valve. The driven piston moves up to compress the spring. Subsequently, the oil in the lower part of the piston slowly flows through the orifice to the upper part of the piston. When the active piston stops moving downward, the driven piston slowly returns to the intermediate position under the action of the spring reaction force. When the active piston moves upward, a negative pressure is generated in the lower part of the piston. Under the action of atmospheric pressure, the driven piston moves downward, thereby transmitting a signal to the pilot valve. The spring on the slave piston is also compressed at this time. Subsequently, as the oil in the upper part of the piston slowly flows through the orifice to the lower portion, when the active piston stops moving upward, the driven piston slowly returns upward to the intermediate position under the action of the spring reaction force.

The buffer operating characteristics can be adjusted by the throttle pin plug. If the throttle pin plug is fully opened, the upper and lower chambers of the piston are connected, and the oil can flow quickly. No matter how the active piston moves up or down, the slave piston is always in the middle position, that is, the signal output is zero, and it is equivalent to the removal of the buffer. If the unit is running at no-load condition or with an isolated load, the turbine regulating system will be unstable.

This is because the turbine regulating system contains a pressure diversion system. During the regulation process, water hammer is generated due to the inertia of the water flow, which causes the anti-regulation power (anti-regulation effect), resulting the regulation system to overshoot. If there is no calibration device like buffer, the regulation system will be unstable, which has been confirmed in engineering practice. When the unit is operating independently (before the unit is juxtaposed with the electric grid), if the buffer is cut off, the oscillation of the regulation system can be observed. Therefore, when the unit is running alone, it is not allowed to cut off the buffer. However, if the throttle pin plug is completely closed and the upper and lower chambers are completely isolated, the slave piston will not automatically return to the intermediate position after the active piston deviates from the intermediate position, but keep the output signal unchanged. That is to say, the feedback signal generated by the buffer not only exists during the regulating process, but also remains after the regulating process ends. At this point, the buffer becomes proportional feedback, namely hard feedback. In general, the buffer must have a feedback of 20–30% in order to stabilize the operation after the unit is isolated from the system. When the buffer becomes a hard feedback, that is, there will be a static error of 20–60% after the regulation, which is obviously too large for stable operation. Therefore, in engineering practice, the feedback signal of the buffer is required to exist only during the regulating process, and is always zero after the regulating process ends, that is, the buffer needs to function as a transient feedback. For this reason, when the throttle pin plug is placed at a certain position in the middle, the buffer will work as described above, that is, after the active piston of the buffer moves, the slave piston first deviates from the intermediate position, thereby outputting a feedback signal. Then slowly returns to the middle position, the output signal becomes zero again.

(4) Operating Principle of Turbine Speed Control System

The operating principle of the entire speed control system is described in Fig. 5.6. When the unit is operated alone with load, if the load suddenly decreases, the driving torque of the turbine is greater than the retarding torque of the load. The unit speed starts to increase, and the rotational speed of the centrifugal pendulum motor driven by the permanent magnet generator (or the voltage transformer) also increases. The support block moves upward with the centrifugal pendulum, and the rotating sleeve of the pilot valve also moves upwards. The upper oil hole on the rotating sleeve of the pilot valve is closed, and the lower oil hole is opened, so that the middle oil hole is connected with the lower oil hole. The upper chamber oil pressure of the auxiliary servomotor piston decreases, and the auxiliary servomotor piston moves up together with the main distributing valve body, driven by the upward thrust of the main distributing valve. And through the first-stage internal feedback lever system, namely lever, pull rod and lever, the pilot valve needle moves up, and the relative position of the rotating sleeve is restored, so that the auxiliary servomotor output oil pressure is restored to the original value, and the auxiliary servomotor piston and the main distributing valve body stop moving up. While the main distributing valve body moves up, the upper oil hole D is connected to the pressure oil, and the lower oil hole E is connected to the oil discharge. Therefore, the pressure oil enters the left chamber of the main servomotor piston, and the right chamber is connected to oil discharge. The piston of the main servomotor is then moving to the right to close the turbine guiding mechanism. The discharge of the turbine decreases, the driving torque reduces, and the unit gradually stops acceleration, and then the unit decelerates, so that the speed is restored to the rated value. In addition, during the movement of the main servomotor piston to the right, the ramp block is also moved to the right, and the recovery shaft is rotated counterclockwise by pull rod and crank arm, which drives the active piston of buffer down. The oil pressure in the lower part of the buffer piston rises, forcing the buffer slave piston to move up, and the pilot valve needle is also moved up by levers, so that the rotating sleeve occupies a lower position relative to the needle. The upper oil hole is opened, and the lower oil hole is closed, so that the middle oil hole and the upper oil hole or pressure oil are connected. The output oil pressure of the pilot valve increases, so that the auxiliary servomotor piston and the main distributing valve body move downward, and gradually return to the intermediate position. The main servomotor piston also stops moving to the right. As the guide vane is closing, the turbine speed gradually decreases and the rotating sleeve moves down. At the same time, the buffer driven piston returns to the intermediate position under the action of the spring reaction force, and through the lever the pilot valve needle moves down to restore the relative position of the rotating sleeve, so that the speed control system enters a new equilibrium state.

5.2.2 Classification of Turbine Governors

The mechanical hydraulic governor came into being in the late nineteenth century. It has been perfect in the 1930s and is still used in countries all over the world. The electrical hydraulic governor began in the 1940s, developed in China in the 1960s, and applied in the 1970s. In the 1980s, with the rapid development of electronic technology and computer technology, the microcomputer governor was born. The turbine governor equipment consists of three parts: the speed control cabinet, the main servomotor, and the oil pressure device. It has a wide variety and various types, which can be defined according to different classification methods.

1. **Classification by Working Capacity**

The working capacity of the governor refers to the operating capacity of the actuator, namely the servomotor torque (N m) to drive the guiding mechanism of hydro turbine. Since the outputs of the controlled turbines are generally from several hundred kW to several hundred MW, the governors are divided into different scales, as follows.

Micro speed governor: working capacity is less than 3 kN m.

Small speed governor: working capacity is about 3–15 kN m.

Medium speed governor: working capacity is about 15–50 kN m.

Large and giant governor: working capacity is more than 50 kN m.

The working capacity for large and giant governors is calculated according to the diameter of the main distributing valve of the amplifying actuator (80, 100, 150, 200, and 250mm).

2. **Classification by Oil Supply**

Since the pressure oil supply mode of the governor has two kinds: direct and indirect, the governor can be divided into through-flow type and pressure oil tank type.

(1) Through-Flow Type of Governor. The oil pump runs continuously and directly supplies oil to the governor during the regulating process. When the regulating process is stopped, all the oil output from the oil pump is discharged back to the collecting tank by the pressure limiting relief valve. During the working process, the oil flow circulates repeatedly and the oil quality is easy to deteriorate. However, the equipment is simple and the cost is low. It is mainly used for small and micro governors.
(2) Pressure Oil Tank Governor. There is a special oil pressure equipment, for which the oil pump runs intermittently to maintain the pressure and oil level of the pressure oil tank. The pressure oil tank provides oil to the regulating process of turbine governor at any time. The equipment is complicated and the cost is high. It is mainly used for medium and large governors.

Pressure oil tank governors are further divided into the combined governor type and separated type. For the combined governor type, the whole governor (including the servomotor) and the pressure oil equipment are combined together. It is mainly used for medium and small speed control equipment. For the separated governor type, the main servomotor and pressure oil equipment of the governor are separately set up. It is mainly used for large and giant speed control equipment.

3. **Classification by Number of Regulation Mechanisms**

According to the number of regulation mechanisms, governors can be divided into two types: single-regulating governor and double-regulating governor. Double regulation means that there are two flow regulating mechanisms, such as the blade-adjustable turbine governor, the tubular-flow turbine governor, and the impulse turbine governor.

4. **Classification by Component Structure**

According to the different component structure, the governor can be divided into two categories: mechanical hydraulic governor and electric hydraulic governor. The electric hydraulic governor is further divided into analog electric hydraulic governor and microcomputer electric hydraulic governor.

(1) Mechanical Hydraulic Governor

The mechanical hydraulic governor is referred to as the mechanical regulation system, and its main functional components are mechanical and hydraulic components. In the mechanical hydraulic governor, the frequency measurement segment adopts the centrifugal pendulum. The integrated segments of the signal use the pilot valve and lever. The amplification segment is the hydraulic amplifier and usually has two stages of hydraulic amplification: the first stage consists of a pilot valve and an auxiliary servomotor; the second stage consists of a main distributing valve and a main servomotor. The feedback segment uses a buffer and a hard feedback differential mechanism, which are also mechanical hydraulic components.

(2) Analog Electric Hydraulic Governor

The analog electric hydraulic governor is referred to as the electric regulation system, and it generally adopts the full electric regulator and the electro-hydraulic servomotor system. For the relevant components of the governor, one part is an electrical component, and the other part is a mechanical hydraulic component. In regard to the integrated component, some adopt integrated amplifiers, and some use electric integral circuits. The soft feedback component uses an actual differential circuit. The frequency measurement component uses a frequency measuring circuit, and some also have a measuring acceleration circuit. In addition, the hard feedback differential component and the power output given are also electrical circuits. However, the actuator still uses a second-stage hydraulic amplifier. In addition, an electrical integrated amplifier and electro-hydraulic conversion component are added to connect the electrical part and the hydraulic part. And the displacement sensor converts the mechanical displacement of the main servomotor into an electrical signal.

(3) Microcomputer Electric Hydraulic Governor

The microcomputer electric hydraulic governor is referred to as the microcomputer regulation system. This kind of governor only changes the electrical components of the analog electric hydraulic governor to a microcomputer or a single-chip microcomputer, and the mechanical hydraulic components are basically without any difference. Of course, in addition to the electro-hydraulic servomotor system, there are many other types of servo system. The control mode of the microcomputer regulator usually adopts parallel PID or series–parallel PID type, and the implementation of these regulation rules is implemented by software (mathematical calculation).

In addition, the classification of the turbine governor can also be divided into PI type and PID type according to the different regulation rules. The PI regulation law means that the regulation signal output of the governor is the superposition of the proportional amplification and the integral calculation of the frequency difference signal. The PID regulation law adopts the superposition signal of the differential calculation, the integral calculation, and the proportional amplification of the frequency difference signal to control the guide vanes.

Moreover, according to the different positions that differential feedback signal taken from, governors can also be divided into three types: intermediate servomotor, auxiliary servomotor, and electrical regulator.

5.3 Pressure Oil Supply System

5.3.1 Composition of Pressure Oil Device

The pressure oil device of the medium sized and large-sized turbine governing systems usually consists of a pressure oil tank (or pressure oil drum), an oil sump (or return oil tank), an oil pump and its driving device, a protection device, a control device, a compressed air supply device, and so on, as shown in Fig. 5.7.

1. **Pressure Oil Tank**

The pressure oil tank is cylindrical, the lower part is used to store the pressure oil, the pressure oil is about 1/3 of the tank volume, and the 2/3 volume of the upper part of the tank is filled with compressed air. The pressure change caused by the volume change of the compressed air is much smaller than the pressure change caused by the volume change of the compressed liquid. During the action period of governor and the intermittent period of oil supply, even if the oil level in the pressure oil tank drops a certain height, the oil pressure can still be kept within the normal operating range, thus reducing the times of oil pump startup and shutdown.

2. **Oil Sump**

The oil sump is a storage device for the pressure oil in the turbine governing system. The oil sump is usually a box structure for collecting the returned oil and leakage oil of the turbine governor. The oil sump is divided into a returned oil zone and a

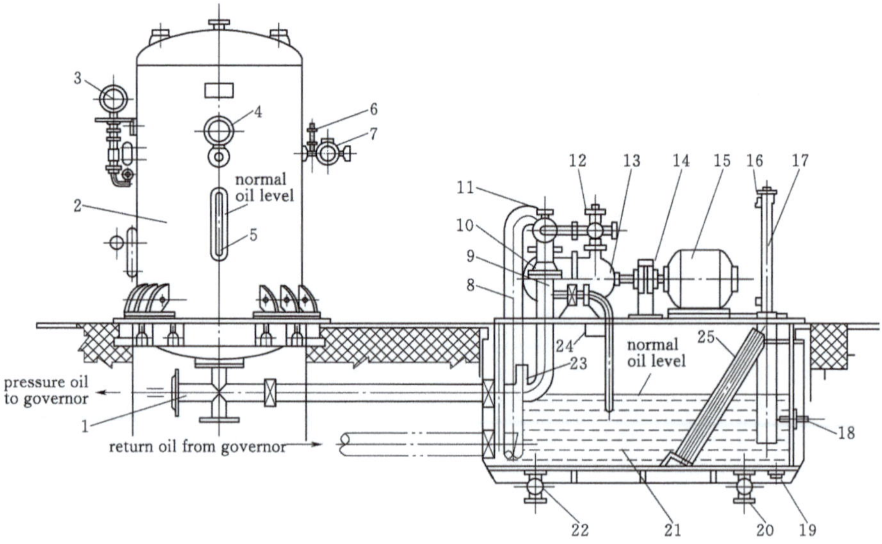

Fig. 5.7 Schematic diagram of pressure oil device

clean oil zone by a filter net. Generally, the returned oil zone is smaller than the clean oil zone, and the oil pump is placed in the clean oil zone. The oil sump not only can accommodate all the returned oil of the governor, but also has a certain margin. During normal operation, the storage capacity of the oil sump should be larger than the volume of 5 to10 minutes of pumping operation of one oil pump. The oil sump should be equipped with an oil level gauge to observe the oil level. If the oil level exceeds its threshold, the corresponding treatments should be done accordingly.

3. **Oil Pump and its Driving Device**

The oil pump used in the pressure oil system of turbine regulating system is usually a screw pump or a gear pump, and the driving device is an asynchronous motor. The oil pump and its driving device are generally arranged in a vertical shaft and located at the top of the oil sump. The small speed governor usually only has one set of oil pump and its driving device, while the large and medium speed governors can be equipped with two sets of oil pump and its driving device, one for operation and the other for standby.

4. **Protection and Control Device**

The oil pressure device is equipped with a check valve, a safety valve, a relief valve, etc., for protecting the oil circuit. An electric contact pressure gauge is installed to monitor the change of the oil pressure in the pressure oil tank, thereby controlling the start and stop of the oil pump. An electric contact pressure signal meter is equipped to monitor the pressure reduction when accident occurs to send a fault signal. If the fault cannot be eliminated in time, and the oil pressure continues to decrease to the lowest accident oil pressure, the electric contact pressure signal device will start the accident relay and operate the electromagnetic valve for emergency stop.

5. **Compressed Air Supply Device**

Although the pressure oil tank takes a variety of sealing measures, the leakage and take-away of the compressed air cannot be completely avoided. Therefore, it is necessary to automatically supply compressed air to the pressure oil tank. When the compressed air in the pressure oil tank is reduced due to consumption, the oil level in the pressure oil tank rises, and after being detected by the automatic detecting device, the electromagnetic air valve and its compressed air circuit are turned on, so that the air compressor can replenish the pressure oil tank with compressed air until the oil level drops to the specified height. At this point, the detecting device cuts off the electromagnetic air valve control circuit and stops the air supply.

5.3.2 Classification of Pressure Oil Device

The pressure oil devices can be divided into the separated and combined types according to their arrangement. The separated type means that the pressure oil tank and the oil sump are arranged separately, and the combined type means that the pressure oil tank and the oil sump are assembled. The pressure oil devices for small and medium-sized governors are usually the combined type.

In China, the pressure oil devices have been standardized, as shown in Table. 5.1.

The nameplate of the pressure oil device consists of four parts.

(1) The first part is the classification code. YZ is the separated type, and HYZ is the combined type.
(2) The second part is the fraction of Arabic numerals. The numerator indicates the total volume of the pressure oil tank, m^3, and the denominator indicates the number of pressure oil tanks (no denominator means one-pressure oil tank).
(3) The third part is also Arabic numerals indicating the rated oil pressure of the pressure oil device, MPa. The rated operating oil pressure of small and medium-sized mechanical hydraulic governors is usually 2.5 MPa. At present, many microcomputer hydraulic governors use an oil pressure of 4.0 MPa, 6.3 MPa or even higher.
(4) The fourth part is the code of the manufacturer or the code that indicates the characteristics or series of the product, which are specified by the manufacturer. If there is no manufacturer code, it means that the product is produced according to the unified design drawing.

For example, HYZ-4-2.5 is a combined pressure oil device, the total capacity of the pressure oil tank is 4 m^3, and one tank with the rated oil pressure is 2.5 MPa. YZ-20/2-4.0-02 is a separated pressure oil device, the total volume of the pressure oil tank is 20 m^3, two tanks of 10 m^3 with rated oil pressure 4.0 MPa, and it is the 02 series products of the unified design.

5.4 Selection of Turbine Speed Governor

The turbine speed governor usually consists of a speed regulation cabinet, a main servomotor, and a pressure oil device. The small and medium-sized governors combine these three parts together as a whole equipment, as shown in Fig. 5.8. Based on the operating capacity of the main servomotor (or speed regulation work), a series of products are formed. Therefore, it is only necessary to calculate the speed regulation work of the turbine, and select the required governor in the governor series spectrum. For large turbine governors, the speed regulation cabinet (including the main distributing valve), the main servomotor, and the pressure oil device are separated. Because there is no fixed match between the three parts, they need to be chosen separately. The large governor is characterized by the diameter of the main distributing valve. Therefore, the volume of the main servomotor needs to be obtained first, then the diameter of the main distributing valve is calculated, and the corresponding governor is selected.

Table. 5.1 Standard specifications of pressure oil device

Separated type	YZ-1	YZ-1.6	YZ-2.5	YZ-4	YZ-6	YZ-8	YZ-10	YZ-12.5	YZ-16/2	YZ-20/2
Combined type	HYZ-0.3	HYZ-0.6	HYZ-1.0	HYZ-1.6	HYZ-2.5	HYZ-4.0				

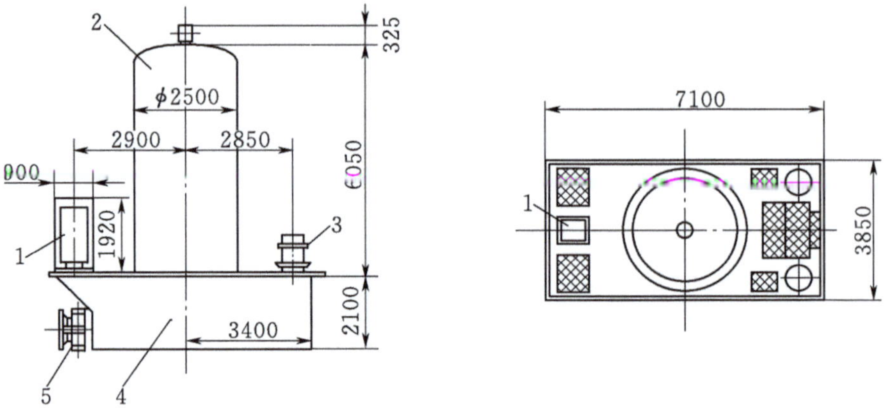

Fig. 5.8 YT-type governor diagram

5.4.1 Medium and Small Turbine Governor Selection

The speed regulation work of the small and medium governors refers to the product of the oil force working on the piston of the servomotor and its stroke. In *The series spectrum of governor and pressure oil device of hydro turbine* (JB/T 7072-2004), the small and medium-sized governors (with pressure oil tank and servomotor) is characterized by the working capacity (the speed regulation work) of the servomotor under the minimum required operating oil pressure. Therefore, only need to calculate the operating capacity of the servomotor to find the corresponding governor in the series spectrum.

There are many empirical formulas for calculating speed regulation work. The commonly used calculation formulas are as follows.

For reaction turbines

$$A = K_i \frac{N_{\max}}{H_r}, \tag{5.9}$$

where A is the speed regulation work, N m; $N_{\max}$ is the maximum output of turbine, kW; H_r is the rated head, m; K_1 is the coefficient, 20 for Francis turbine and 25 for Kaplan turbine.

Another formula is

$$A = K_2 Q \sqrt{H_{\max} D_1}, \tag{5.10}$$

where Q is the corresponding discharge under the maximum head and rated output, m^3/s; $H_{\max}$ is the maximum head, m; K_2 is the coefficient, 200–250.

For impulse turbines

$$A = 9.81 Z_0 \left(d_0 + \frac{d_0^3 H_{\max}}{6000} \right), \tag{5.11}$$

where Z_0 is the number of nozzles; d_0 is the jet diameter at the rated discharge, cm.

Considering the factors such as the manufacturing quality, installation quality, and working conditions, it is generally advisable to add an additional 30%.

5.4.2 *Selection of Large Turbine Governor*

Large turbine governors (without pressure oil tank and servomotor) are characterized by the diameter of the main distributing valve. The main distributing valve has a series of diameters, such as 80/100/150/200/250 mm. The following principles should be followed when selecting the operating capacity.

(1) The design oil velocity in the external pipe matched with the governor is generally not more than 5 m/s.
(2) When calculate the oil pressure of the governor capacity, the lower limit of the normal operating oil pressure should be considered.
(3) The maximum pressure drop through the main distributing valve and the connecting pipes does not exceed 20–30% of the rated oil pressure.
(4) The minimum closing time of the servomotor should meet the requirements of the calculation guaranteed regulation of the hydro power station.

The specific selection steps are as follows.

1. **Diameter Selection of the Main Distributing Valve**

In preliminary selection, the diameter of the main distributing valve can be calculated by

$$d = \sqrt{\frac{4V}{\pi v_m T_s}}, \tag{5.12}$$

where V is the total volume of the servomotor of guide vane or the deflector, m^3; v_m is the oil velocity in the pipeline, m/s; T_s is the time of servomotor closing, s.

According to the calculation result, the diameter of the main distributing valve can be selected in the series diameter list.

2. **Calculation of the Capacity of the Servomotor**

(1) Calculation of Guide Vane Servomotor

Large governors usually use two servomotors to operate the guiding mechanism. When the rated oil pressure of the pressure oil device is 2.5 MPa, the diameter d_s of each servomotor can be calculated according to the following empirical formula:

$$d_s = \lambda D_1 \sqrt{\frac{H_{\max} b_0}{D_1}} (m), \tag{5.13}$$

where λ is the calculation coefficient, as shown in Table 5.2; b_0 is the guide vane height, m.

When the rated oil pressure of the hydraulic device is 4.0 MPa, the diameter d_s' of each servomotor can be calculated according to the following empirical formula:

$$d_s' = d_s \sqrt{1.05 \frac{2.5}{4.0}} = 0.81 d_s (m). \tag{5.14}$$

From the above calculations, we obtain the values of d_s or d_s', based on which we can select the standard diameter in the standard series of servomotor given in Table. 5.3.

The volume of a single servomotor V_s is

$$V_s = \frac{\pi}{4} d_s^2 S_{\max} (m^3), \tag{5.15}$$

where $S_{\max}$ is the maximum stroke of the servomotor, m. It can be calculated according to the following empirical formula:

$$S_{\max} = (1.4 \text{ to } 1.8) a_{0\max} (m), \tag{5.16}$$

where $a_{0\max}$ is the maximum opening of the guide vanes, m. When the diameter of the runner is less than 5 m, a smaller factor is used.

The total volume of two servomotors is

$$V = \frac{\pi}{2} d_s^2 S_{\max} (m^3). \tag{5.17}$$

Table. 5.2 Coefficient λ for servomotor capacity calculation

Number of guide vanes z_0	16	24	32
Standard positive curvature guide vane	0.031 to 0.034	0.029 to 0.032	
Standard symmetrical guide vane	0.029 to 0.032	0.027 to 0.030	0.027 to 0.030

Note 1. If b_0/D_1 value is the same, for different runner, take the large value when Q_1' is large
2. For the same Kaplan runner, take a large value when the volute wrap angle is large and the standard symmetrical guide vane is used, take a small value when the wrap angle is large, but the standard positive curvature vane is used

Table. 5.3 Standard series of servomotor

Servo diameter/mm	200	225	250	275	300	325	350	375	400	450
	500	550	600	650	700	750	800	850	900	

(2) Calculation of the Kaplan Turbine Servomotor

For the double-regulating governor of the Kaplan turbine, in addition to selecting the guide vane servomotor, it is also necessary to select the servomotor for rotating the blade, which is installed in the hub, and its diameter d_c, maximum stroke $S_{c\max}$, and volume V_c can be estimated according to the following empirical formula:

$$d_c = (0.3 \text{ to } 0.45)D_1\sqrt{\frac{25}{p_0}}(m), \tag{5.18}$$

where p_0 is the rated oil pressure of the governor pressure oil device, MPa.

$$S_{c\max} = (0.036 \text{ to } 0.072)D_1 \quad (m), \tag{5.19}$$

$$V_c = \frac{\pi d_c^2 S_{c\max}}{4} \quad (m^3). \tag{5.20}$$

When $D_1 > 5$ m, smaller coefficients are used in Eqs. (5.18) and (5.19).

Since the movement of the runner blade servomotor is much slower than that of the guide vane servomotor, the main distributing valve that can meet the requirements for the movement of the guide vane servomotor can also meet the requirements for the movement of the runner blade servomotor.

3. **Calculation of Oil Pressure Device**

The operating capacity of the oil pressure device is characterized by the total volume of the pressure oil tank. The volume V_K can be estimated according to the following empirical formula.

For Francis turbines

$$V_K = (18 \text{ to } 20)V(m^3). \tag{5.21}$$

For Kaplan turbines

$$V_K = (18 \text{ to } 20)V + (4 \text{ to } 5)V_c(m^3). \tag{5.22}$$

If the oil pressure device needs to supply oil to other servomotor such as the inlet valve and empty valve at the same time, the volume required for the inlet valve and the empty valve servomotors can be appropriately added into the above total volume.

When the rated oil pressure is 2.5 MPa, the total volume of the pressure oil tank can be estimated according to the above formulas. Then, the oil pressure device with closer and larger capacity can be selected, according to Table. 5.1.

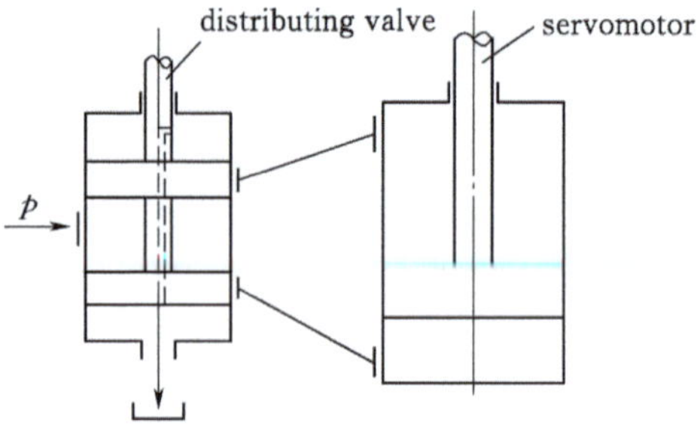

Fig. 5.9 Schematic diagram of the hydraulic amplification system of the governor

5.4.3 *Nomograph Method for Governor Selection*

The diameter of the main distributing valve does not fully reflect the operating capacity of the governor, because the main distributing valves with the same diameter but different structures have different maximum strokes, window sizes and on/off time setting modes, along with different friction losses and possible maximum outputs. In order to obtain the nomograph for governor selection, the following three assumptions are made when calculating the oil discharge through the main distributing valve due to the movement of servomotor piston.

(1) The main distributing valve is a three-position four-way spool valve structure, and the servomotor is a double-sided differential pressure structure (Fig. 5.9).
(2) The oil flow is turbulent flow.
(3) The window of the main distributing valve for controlling the oil flow is rectangular.

Therefore, according to the orifice outflow formula, the discharge is

$$Q = \mu\omega\sqrt{2g \times 0.5 \times \Delta p_2/\gamma} \times 10^3 \quad (L/s), \tag{5.23}$$

where μ is the velocity coefficient; ω is the window area of the main distributing valve for controlling the oil flow, m^2; g is the acceleration of gravity, m/s^2; γ is the specific gravity of oil, kg/m^3; Δp_2 is the oil flow pressure loss through the main distributing valve, MPa; 0.5 is the pressure loss considering the secondary cutting of the oil flow.

Letting $\mu = 0.6$ to 0.7, $\omega = 0.1885d^2$, $\gamma = 900$ kg/m^3, $\Delta p_2 \approx 0.23p_0$, with d is the piston diameter and p_0 is the operating oil pressure, and substituting them into Eq. (5.23), we have

$$Q = (0.0057 \text{ to } 0.0065) \times d^2\sqrt{p_0} \times 10^3 \quad (L/s). \tag{5.24}$$

It can be seen from the above formula that the oil discharge through the main distributing valve is proportional to the square of the diameter of the main distributing valve and the square root of the operating oil pressure. Considering $p_0 = 2.5$ MPa, 4.0 MPa, 6.3 MPa and taking their lower limit of 2.3 MPa, 3.6 MPa, 5.7MPa, we can

Table. 5.4 Allowable oil discharge of the main distributing valve of the governor

Main distributing valve diameter d /mm	12	35	50	80	100	150	200	250
Series spectrum specifies oil discharge Q /(L/s)	0.5	≤5	5–10	10–25	25–50	10–100	100–150	–
Recommend the allowable oil discharge Q /(L/s)	≤0.6	≤5.5	≤11	≤28	≤46	≤105	≤180	≤280

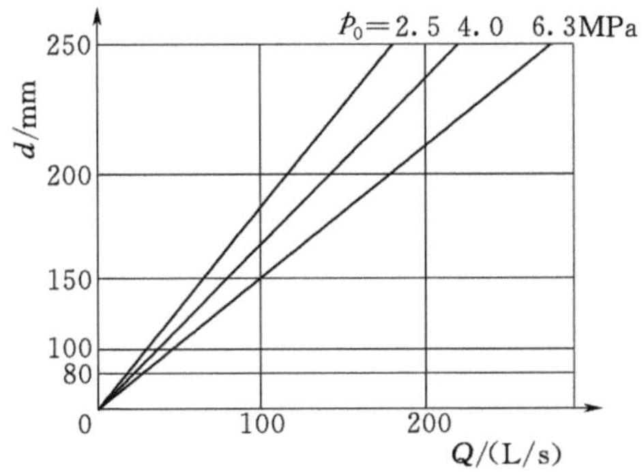

Fig. 5.10 Nomograph for governor capacity selection of large turbines

calculate the oil discharge Q for the different diameters d of the main distributing valve by Eq. (5.24). Then compare the formula calculations with the actual test results of the power station. Table 5.4 shows the allowable oil delivery selection table for the main distributing valve of the turbine governor. The nomographs for large and medium-sized governor capacity selection are shown in Figs. 5.10 and 5.11, respectively. The governor can be selected according to the operating oil pressure p_0 and the required oil discharge Q (the total working volume of the servomotor divided by the guide vane closing time), which is convenient for engineering and practical applications.

Questions

1. What is the basic task of hydro turbine regulation? How about the way and equipment to realize the regulation?
2. What are the components of the hydro turbine governing system? What is the basic working principle of hydro turbine governor?
3. What are the static and dynamic characteristics of turbine governors? How to describe them?
4. What are the types of turbine governors?
5. What are the functions of the oil pressure device? What are the main components and layout/setting modes?

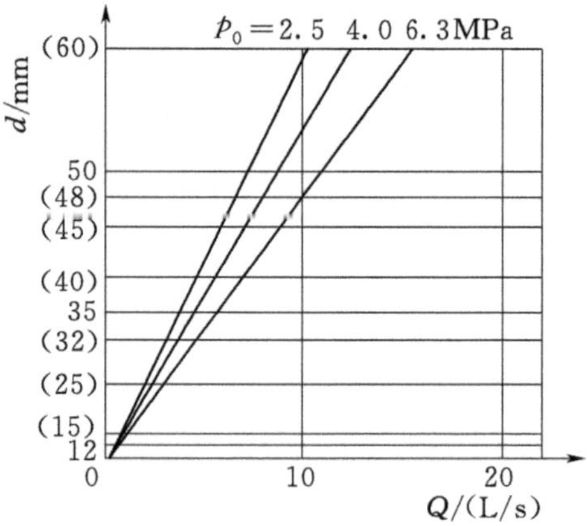

Fig. 5.11 Nomograph for governor capacity selection of medium and small turbines

Bibliography

1. Zhongyuan, Jin. 1992. *Hydraulic machinery (in Chinese)*, 2nd ed. Beijing: Water Resources and Electric Power Press.
2. Qizhao, Liu, and H. Ming. 2010. *Hydropower station (in Chinese)*, 4th ed. Beijing: China Water and Power Press.
3. Ruyun, Luo. 1990. *Design manual for power equipment of hydropower station (in Chinese)*. Beijing: China Water Resources and Hydropower Press.
4. Mechanical and electrical design manual of hydropower station. 1983. *Hydraulic machinery (in Chinese)*. Beijing: Water Resources and Electric Power Press.
5. Yuan, Zheng. 2011. *Hydraulic turbine (in Chinese)*. Beijing: China Water and Power Press.
6. Dakai, Liu. 1996. *Hydraulic turbine (in Chinese)*, 3rd ed. Beijing: China Water and Power Press.
7. Changzhi, Zhang. 1988. *Hydraulic turbine: principle and mathematical model (in Chinese)*. Wuhan: Huazhong Institute of Technology Press.
8. Rangqin, Gong. 2006. *Basic technical conditions of hydraulic turbine (in Chinese)*. Beijing: China Standards Press.
9. Yuan, Zheng and Z. Qiang. 2003. *Power equipment of hydropower station (in Chinese)*. Beijing: China Water and Power Press.
10. Qizhang, Li, Z. Qiang, and Y. Jixing et al. 2014. *Study on hydraulic stability of Francis turbine (in Chinese)*. Beijing: China Water and Power Press.
11. Quanlong, Wang. 2009. *Basic technical conditions of Francis pump turbine (in Chinese)*. Beijing: China Standard Press.
12. Xilin, Sha. 1999. *Tubular hydropower station (in Chinese)*. Beijing: China Water and Power Press.
13. Guoxuan, Liu. 2006. *Operation and maintenance of bulb tubular turbine generator (in Chinese)*. Beijing: China Water and Power Press.
14. Bo, Yu. 2008. *Principle and operation of hydraulic turbine (in Chinese)*. Beijing: China Electric Power Press.
15. Wentao, Su. 2016. *Hydraulic stability of hydraulic turbine (in Chinese)*. Harbin: Harbin Institute of Technology Press.
16. Shuhong, Liu. 2007. *Fundamentals of hydromechanics of hydraulic machinery (in Chinese)*. Beijing: China Water Resources and Hydropower Press.
17. Jinshi, Chang. 2005. *Transition process of hydraulic machinery (in Chinese)*. Beijing: Higher Education Press.

18. Zuyi, Shen. 1988. *Turbine regulation (in Chinese)*, 2nd ed. Beijing: China Water Resources and Hydropower Press.
19. Weiyou, Cai. 2000. *Turbine governor (in Chinese).* Wuhan: Wuhan Water Resources and Electric Power Press.
20. GB/T 9652.1-1997 Technical Specification for hydraulic turbine governor and oil pressure device (in Chinese).1997. Beijing: China Standard Press.
21. Shouping, Wei. 2005. *Hydraulic turbine control engineering (in Chinese).* Wuhan: Huazhong University of science and Technology Press.
22. Yuanchu, Cheng. 2010. *Automatic regulation of hydraulic turbine (in Chinese).* Beijing: China Water Resources and Hydropower Press.
23. Cobb, Bryan R., and Kendra V. Sharp. 2013. Impulse (Turgo and Pelton) turbine performance characteristics and their impaction pico-hydro installations. *Renewable Energy* 50: 959–964.
24. Keck, Helmut, and Mirjam Sick. 2008. Thirty years of numerical flow simulation in hydraulic turbo machines. *Acta Mechanica* 201: 211–229.

Chapter 6
Water Intakes and Diversion Structures

Kai Su

Abstract This chapter focuses on water intakes and diversion structures; their functions and common design requirements are described. For low-level intakes, types of intakes with different structural forms and application conditions are introduced; main facilities and layout forms are introduced according to operational conditions, and the characteristics to be considered for layout of pumped storage power station intakes are explained. For open-flow intakes, layout forms are described according to structural conditions and the concept of settling basin is introduced. In terms of design, contents of hydraulic design and structural design of intakes are introduced. For diversion structures, types and hydraulic calculation of channels and tunnels are elaborated, and the head pond and daily run-off pond are introduced.

Keywords Intakes (outlets) · Diversion conduits · Low-level intakes · Open-flow intakes · Settling basin · Head pond · Daily run-off pond

6.1 Functions and Requirements of Water Intakes

Water intake is located at the starting position of the water conveyance system and is used for water diversion to generate hydropower. Water intake is divided into two categories: low-level intakes and free-surface flow water intakes. Low-level intake, also known as deep-water intake, is used for dam-embedded hydropower stations and pressurized flow water diversion hydropower stations; it is set below the lowest water level of power generation and the water flowing through it is pressurized flow. Free-surface flow intake is the regular intake sluice and the water flowing through it is free-surface flow.

The design of water intakes should meet the following requirements and be guaranteed by corresponding facilities:

K. Su (✉)
Wuhan University, Wuhan, China
e-mail: suker8044@163.com

Y. Cheng et al. (eds.), *Hydropower Station*,
https://doi.org/10.1007/978-981-96-2383-9_6

(1) Sufficient ability to draw in water. The necessary discharge rate should enter the intake at any working water level. Therefore, the elevation of the intake and its position in the power plant should be reasonably arranged with smooth intake channel and sufficient cross-section dimension.
(2) Water intake should be provided with facilities intercepting sediment, ice and rubbish, etc., to ensure that the water quality meets the requirements of power generation.
(3) Water intake should be in an appropriate position and has less stream velocity to minimize head loss.
(4) Water intake should be equipped with a sluice gate to cut off the water flow in case of an accident or maintenance of the water conveyance system or powerhouse. For open-flow intake, a sluice gate is sometimes used to control discharge rate.
(5) Meeting general requirements of hydraulic structure. Water intake should be of simple construction with sufficient strength, rigidity and stability, and should be easy to operate and maintain.
(6) For hydropower stations with large reservoir depth, the position of intakes could be generally lower to meet the operation requirements at the lowest water level of power generation. The intake draws the deep water of the reservoir and the water temperature is low during operation of high water level, which makes the water temperature and oxygen content in the downstream river change greatly and has a negative impact on downstream aquatic organisms. In the meantime, it is also necessary to meet the requirements of drawing water at different levels and adapting to the change in water levels.

The water outlet is located at the end of the power generation system to achieve a smooth connection between the tail water of the hydropower station and the downstream river flow; its functional requirements are less than the intake and there is no requirement for sand-prevention and decontamination. In terms of flow control, it is only necessary to install gate and hoist equipment for maintenance. Due to lower functional requirements, the structure of the outlet is relatively simple and could also be designed with reference to the intake.

6.2 Low-Level Intakes

6.2.1 Types and Application Conditions of Low-Level Intakes

The water intakes (known as deep water intakes) are usually composed of the inlet, gate and transition sections. According to the construction characteristics of water intakes, low-level intakes can be classified into 4 types: dam-embedded intake, intake of run-of-river plant, tower intake and off-stream intake. Meanwhile, there are pumped-storage station intakes with two-way water flow characteristics, and

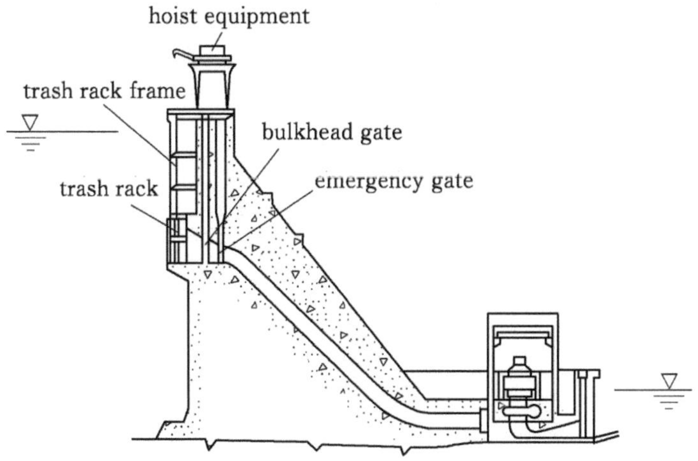

Fig. 6.1 Dam-embedded intake

some intakes for special requirements or special layouts, such as multi-level intake and siphon intake.

1. Dam-embedded intake

Water intake of a hydropower station behind the concrete dam is usually directly attached to the dam body and connected with the high-pressure penstock inside the dam, as shown in Fig. 6.1. Usually, each hydropower unit has its individual intake. The intake or its trash rack may be set on a cantilever platform overhanging the dam surface to not weaken the dam. The trash rack can be laid out in a straight line or a broken line on a plane design figure. Most of the intakes are laid out vertically and can also be set obliquely along the dam surface.

2. Intake of run-of-river plant

Intake of run-of-river plant is suitable for run-of-river plant and is a part of the dam section of the powerhouse; its typical layout is shown in Fig. 6.2. Intake of run-of-river plant has two remarkable characteristics. On the one hand, the intakes are combined with the power house to retain flood and water. On the other hand, hydropower stations are mostly medium–low water head with large discharge, and the desilting as well as sewage disposal are major troubles.

3. Tower intake

Tower intake is a tower structure which is composed of the inlet, gate, transition section and upper framework, erected in the reservoir and connecting with bankside by service bridge, and water can be entered from one side or all around, as shown in Fig. 6.3. The intake is suitable for the situation with flat slope or difficult to dig the intake along the riverbank due to bad geological condition. The inlet opening can be arranged in one or more levels according to the height. Water can be taken

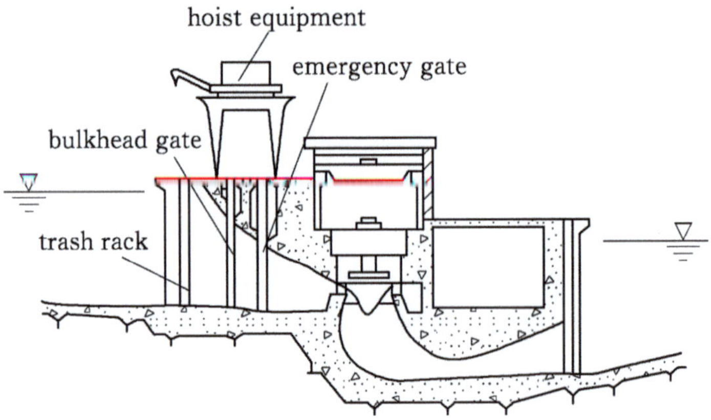

Fig. 6.2 Intake of run-of-river plant

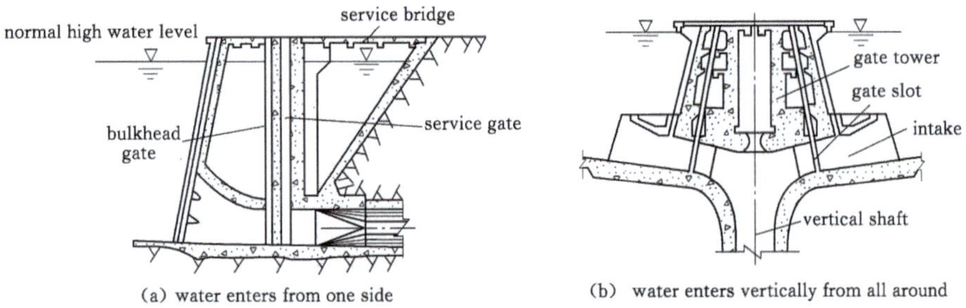

Fig. 6.3 Tower intake

from different heights according to relevant requirements in multi-level layout; it is conducive to sand prevention, sewage disposal and temporary operation at the initial stage of reservoir storage.

4. Off-stream intake

Off-stream intake is independently laid out on reservoir bank or river bank. According to the construction arrangement characteristics of water intakes, they can be classified into three types: off-stream tower intake, off-stream intake against hillside, and off-stream intake with a gate shaft. Off-stream tower intake structure is laid out close to the bank slope while the gate is arranged along the bankside, forming the structure with lower wall-building leaning against hillside and upper tower erected in the reservoir (inlet section), and this intake can also serve as a supporting structure of bankside, as shown in Fig. 6.4a. Off-stream intake against hillside is obliquely laid out on the bank slope with the gate inside the intake and the gate slot (including trash rack groove) adjoins the slope, as shown in Fig. 6.4b. The gate shaft of the off-stream intake with a gate shaft is arranged in the mountain shafts and the horn-shape section entrance is set on the bank slope. The channel between the horn-shape

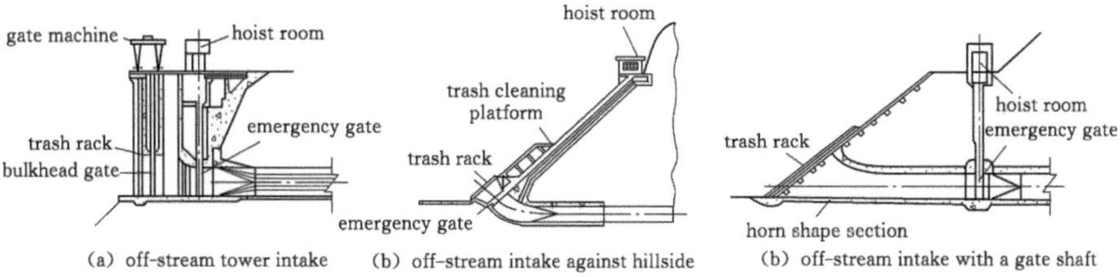

Fig. 6.4 Off-stream intake

section entrance and the gate shaft is the tunnel section, and the water flow in the tunnel is generally pressurized, as shown in Fig. 6.4c.

5. Pumped-storage station intake

Pumped-storage hydropower station has two operating conditions of power generation and pumping, and the intake has the characteristic of bidirectional overflow. For the intake of upper reservoir, it is inflow during power generation and outflow during pumping; the condition is quite opposite for the intake of lower reservoir. Pumped-storage station intake is usually divided into two types: side type and vertical shaft type. Among them, the side intake is more widely used and usually located on the bank side of reservoir, as shown in Fig. 6.5a. Its structure composition is the same as the above-mentioned off-stream intake, and the middle block and anti-vortex beam are generally set in the inlet section to meet the requirements of bidirectional flow condition and structure. Vertical shaft intake is generally located in the reservoir, as shown in Fig. 6.5b.

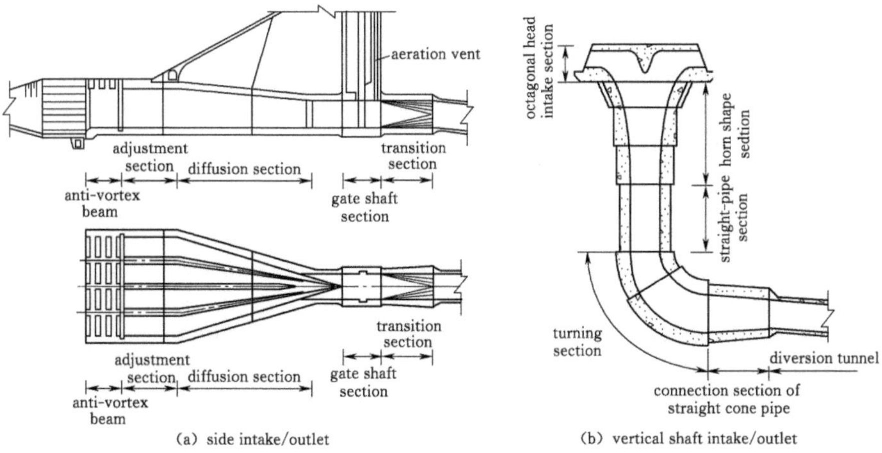

Fig. 6.5 Pumped-storage station intake/outlet

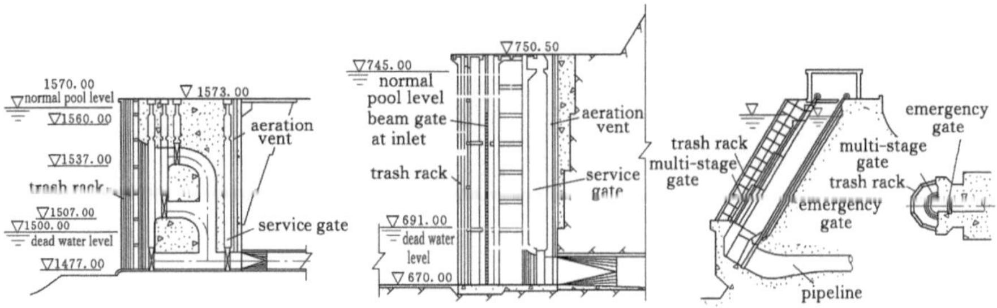

(a) multi-level intake (b) stop log gate intake (c) multi-stage gate intake

Fig. 6.6 Multi-level intake

6. Multi-level intake

To meet the requirements of drawing water in different levels, the following two types are generally adopted:

(1) Setting multi-level intake to draw water, as shown in Fig. 6.6a. This form has more restrictions on the operation of power stations and complex structure, and it is rarely adopted in actual projects.
(2) Intake is located at a low elevation using gates to control water intake (mainly including the stop log gate and the multi-stage gate). At present, the stop log gate is mostly used to control multi-level water intake, as shown in Fig. 6.6b. Its advantages include simple structure and convenient control, while it has the disadvantage of large head loss. Multi-stage gate used for multi-level water intake (as shown in Fig. 6.6c) is generally only used for small and medium-sized hydropower stations due to the complex structure. In terms of structural characteristics, multi-level intake should be a special type of dam-embedded intake, tower intake, or off-stream tower intake.

7. Siphon intake

The siphon intake is a kind of water intake that bases on the siphon principle and applies to the diversion hydropower station with little discharge. Its typical layout is shown in Fig. 6.7.

8. Outlet

The function requirement of the water outlet is less than that of the water intake, and generally it is not necessary to consider the combination with the waterflood retaining structure. And there is no requirement for sand prevention and sewage disposal. In terms of the water flow control, it only needs to be equipped with the gate and hoist equipment for inspection. Due to lower functional requirements, the structure of the outlet is relatively simple and generally adopts the following two types:

(1) Tail water platform outlet combined with the power plant, as shown in Fig. 6.8a.
(2) Off-stream tower outlet; its structure is similar to off-stream tower intake without a trash rack and only with the bulkhead gate, as shown in Fig. 6.8b.

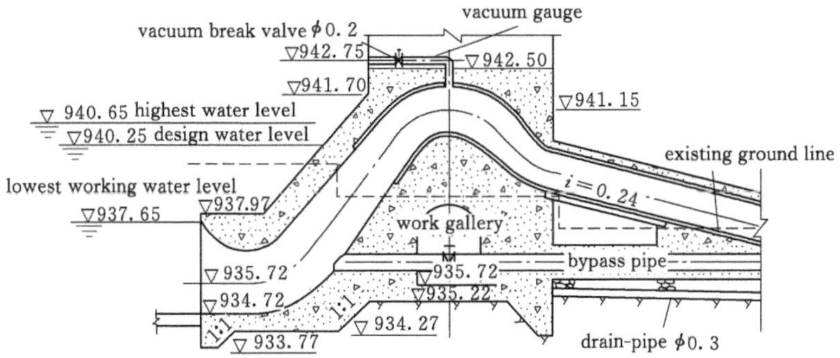

Fig. 6.7 Siphon intake

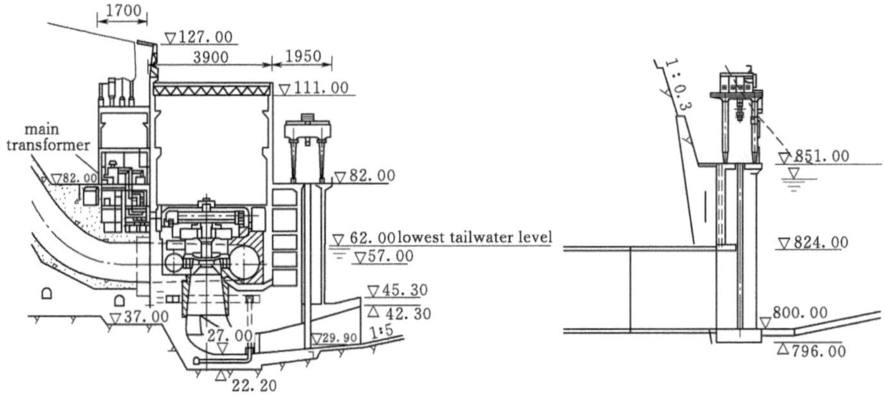

(a) tail water platform outlet (b) off-stream tower outlet

Fig. 6.8 Outlet

6.2.2 Major Equipment and Facilities of Low-Level Intakes

According to the operational conditions, the major equipment and facilities such as trash rack and trash-cleaning equipment, gate and hoist equipment, aeration vent and water-filling valve should be set at the low-level intake.

1. Trash rack

Trash rack is generally set before the water intake as the main facility to block floating objects. Trash rack can be arranged vertically or obliquely, and the intake surface can be polygonal or plane. Oblique arrangement and plane trash rack are convenient for decontamination. The layout of the trash rack and typical structure of the grid piece are shown in Fig. 6.9. Each grid piece, including the frame and a grid bar, is welded from profile steel and inserted between the block and the beam; it can be lifted for maintenance or replacement. The net distance between grid bar is determined according to the model and dimension of the turbine. For the Francis turbine and the

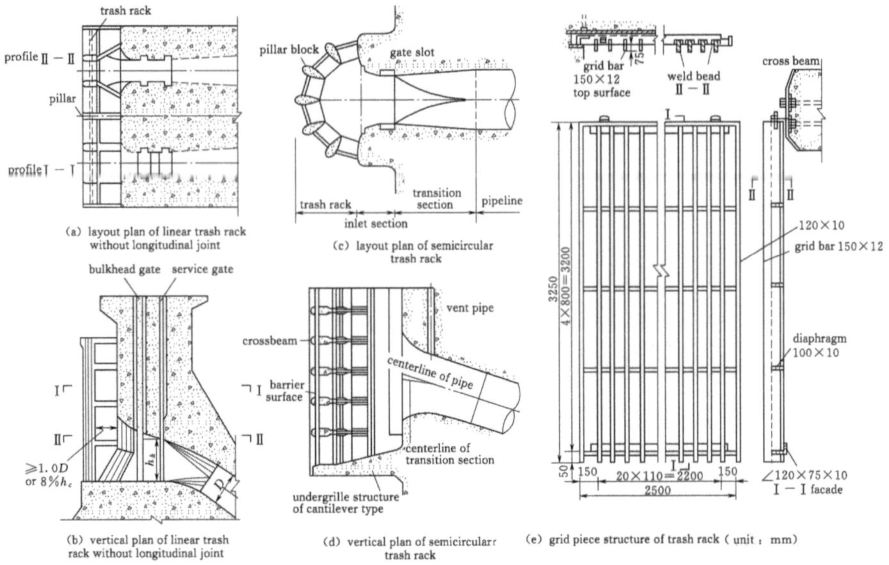

Fig. 6.9 The layout of trash rack and typical structure of grid piece

Kaplan turbine, the net distance is 1/30 and 1/20 of the runner diameter, respectively. The flow rate across the grid bar should be as low as possible and generally not more than 1.0 m/s to reduce head loss and decontamination difficulties.

When the sewage in front of the trash rack is blocked, the water level difference will be formed between the front and back of the trash rack, which not only increases the head loss but also crushes the grid piece. Therefore, the trash-cleaning equipment on the top platform should be used to complete decontamination. There are a lot of floating trashes in rivers in our country. Blockage of trash racks and difficulty in decontamination are often the key problems in the operation of hydropower stations. A streak of thick grille or the drift barrier is sometimes added in the upper stream far from the intake to block large floating objects and improve the working conditions of trash racks. In cold areas, the icing of trash rack can be prevented by heating it with a low-voltage current or introducing compressed air under the bottom. Besides, the trash rack can also be built indoors for heat preservation.

The water pressure borne by the frame and grid bar of the trash rack is determined by the difference in water head caused by the blockage of the trash rack, and the structural design is carried out accordingly. The main contents of trash rack design include the area of trash rack, the spacing of the grid bar, the size of the cross-section of the grid frame and the supporting way, etc.

2. Intake gate and hoist equipment

According to the nature of work, the intake gate can be divided into three types: service gate (opened or closed in moving water), emergency gate (opened in moving water, closed in static water) and bulkhead gate (opened and closed in static water).

The service gate or emergency gate should be arranged in the low-level intake of hydropower stations. The function of emergency gate is mainly to shut off the water passageway in case of an accident in the diversion conduit or unit to prevent the accident from spreading; it can also be used to close the intake for maintenance of the water diversion system. The gate needs to be closed quickly in these cases, so it is also called quick emergence gate. The emergency gate should be able to close rapidly in moving water (within 2–3 min), but it only needs to be opened in static water when the water pressure of the upper and lower stream of the gate is balanced. Each emergency gate is equipped with a fixed gate lifting hoist or hydraulic hoist and can be remotely controlled for emergency (quick) closing operation.

The bulkhead gate can be set in front of the emergency gate as required to overhaul the emergency gate and the gate slot, and it only needs to be opened or closed in static water. The plain batten gate or stoplog gate can be used as the bulkhead gates; one gate can be used for several intakes and operated by the traveling hoist on the platform of the intakes.

The gate of the low-level intake is rectangular, usually high and narrow, while low-level water diversion conduit is generally circular. Therefore, a transition section from rectangle to circle should be arranged at the inlet section to smooth the transition of water flow and reduce head loss. But the transition section should not be too long to facilitate the layout of the structure and improve the structural force condition.

3. Aeration vent and water-filling valve

The aeration vent should be set behind the service gate of the low-level intake, and it is used to replenish air to the pressurized water diversion system in time when the emergency gate is closed in case of causing vacuum inside the system and to vent air through the vent when fulfilling water to the pressure diversion system. The area of the aeration vent depends on the amount of air intake required to close the emergency gate. The flow rate can be taken as the maximum flow rate at the inlet with the inlet flow rate of 30–50 m/s. The top end of the aeration vent must be higher than the maximum water level of upstream to prevent water overflow. The position of the outlet should be appropriately chosen to avoid inhaling surrounding objects during air suction.

The water intake should be equipped with a water-filling valve. The function of the water-filling valve is to fulfill water into the conduit before opening the gate so that the gate can be opened under hydrostatic pressure to reduce the pressure required to open it. The water-filling valve can be set on the service gate or the by-pass pipe connecting the upper and lower sides of the gate, and the water-filling valve is arranged in the gallery.

6.2.3 Layout and Profile Dimension of Pressurized Flow Intakes

1. Layout of the low-level intake

The arrangement of a low-level intake includes the selection of an appropriate intake position and the determination of its elevation. The selection of the position of the low-level intake should be coordinated with the overall layout of the power station junction. The stream flowing through the intake should be smooth and symmetrical in case of reflux, vortex, accumulation of floating objects and sediment deposition. The water intake should be coordinated with the layout of the tunnel line when connected with the diversion tunnel, and suitable topography, geology and flow conditions should be selected.

The elevation of the intake should be below the lowest possible working level of the reservoir with a certain submerged depth, to avoid the funnel-shaped suction cyclone vortex. The submerged depth can be estimated according to the following empirical formula from J. L. Gordon:

$$S = cv\sqrt{d} \tag{6.1}$$

In the formula:

S is vertical distance from the top of the gate orifice to the lowest working water level, m; when the influence of wind and waves is taken into consideration, the lowest working water level adopted in the calculation is about half a wave height lower than the static water level;

- d is height of the gate orifice, m;
- v is flow rate of the gate section, m/s;
- c is empirical coefficient (c = 0.55–0.73); small value is taken for symmetric flowing water while large value for side flowing water.

There are many factors affecting the generation of vortex, some of which cannot be quantitatively estimated. The formula (6.1) can only be used to preliminarily estimate the submerged depth of the intake. The practice shows that it is difficult to require the intake not to generate vortex (or funnel-shaped inspiration vortex in particular) completely under various operating conditions due to the limitation of topography and the complex boundary conditions of progressive flow.

On the premise that there is no funnel-shaped inspiration vortex before the intake and no negative pressure in the diversion conduit, the elevation of the intake should be raised as much as possible to improve the structure stress condition, reduce the cost of the gate, hoist equipment and the diversion conduit, and facilitate the operation and maintenance as well.

The bottom elevation of water intake should be higher than the balanced elevation of sediment scouring and silting of reservoir. If this condition cannot be satisfied, measures combining diversion, blocking and drainage should be taken to ensure the unimpeded intake of powerhouse.

2. The profile dimension of the low-level intake

The inlet section, gate section and transition section of tower intake and off-stream intake are clearly divided. The profile dimension of the intake depends on three-section dimensions, i.e., trash rack section, gate section and tunnel section. The size of the trash rack section is usually determined by the requirement that the flow rate of the section should not exceed a certain limit value. As mentioned above, the flow rate through the trash rack should not exceed 1.0 m/s. The gate orifice is usually rectangular, and the section area of the net flow cross-section of the service gate is generally about 1.1 times of that of the tunnel section. The size of bulkhead gate orifice is usually equal to or slightly larger than the gate orifice; the width is slightly less than that of the tunnel diameter, while the height is equal to or slightly larger than the tunnel diameter. The tunnel diameter is generally determined by the comparison of kinetic energy economy, as detailed in Sect. 6.5 of this chapter.

The function of the inlet section is to connect the trash rack with the gate section. The inlet section tends to be flat base with slight shrinkage on both sides and large shrinkage on the top plate. The shrinkage curves on both sides are usually designed as circular arcs, and a quarter of elliptical curve is widely used in the design of the top plate, as shown in Fig. 6.10. There is no certain standard for the length of the inlet section, which should be as compact as possible under the condition of satisfying the engineering structure layout and smooth water flow. Generally, *a* (the long semi-axis of ellipse) can be taken 1.0–1.5 times of the height of the gate orifice while *b* (the short semi-axis of ellipse) be taken one-third of that. The length of the gate section is mainly determined according to equipment layout requirements. The transition section is a transition from rectangular to circular. Usually, the transition section adopts a rounded corner, and *r* (the radius) changes according to the rule of a straight line, as shown in Fig. 6.11. The length of the transition section is generally taken as 1.5–2.0 times of the tunnel diameter so that the shrinkage rate is not larger than 1: 8–1: 5.

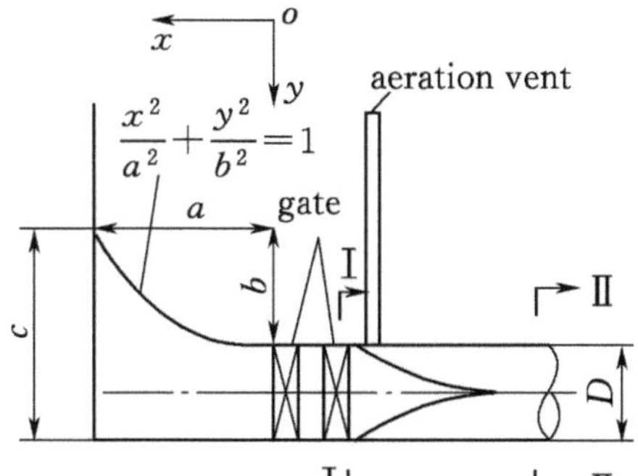

Fig. 6.10 Profile of intake

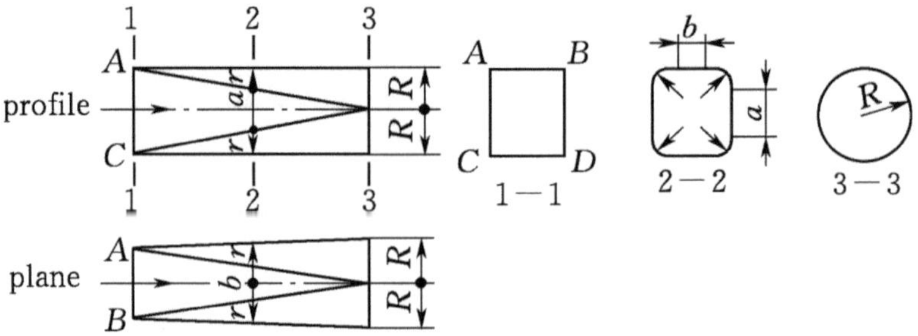

Fig. 6.11 The shape of transition section

The principle for determining the profile dimension of dam-embedded intake is the same as above, but each has its own characteristics. Dam-embedded intake is generally shaped like a rectangular bell, and the shape is usually determined by the relevant test with the principle of no negative pressure, no vortex and minimum head loss. To meet the structural requirements of the dam body, the length of the intake should be shortened. The inlet section and the gate section are often combined into one. The service gate is arranged in the middle of the bell, while the bulkhead gate is placed upstream, as shown in Fig. 6.12. The center line of the intake might be horizontal or inclined and should be determined according to the arrangement of the penstock.

Fig. 6.12 The inlet section of dam-embedded intake

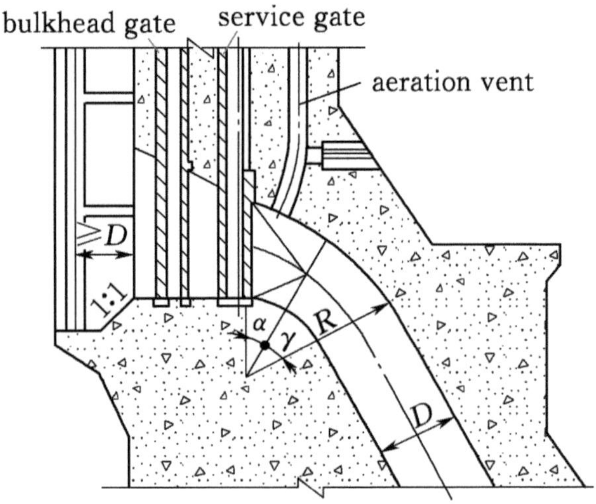

6.2.4 Layout of Pumped-Storage Station Intake

The intake of the upper reservoir of a pumped-storage station becomes the outlet when pumping, and the intake of the lower reservoir becomes the outlet when generating. The shape of the intake should be conducive to the homogeneous diffusion of the reverse flow to prevent flow separation and vortex. Therefore, the following characteristics should also be considered when the above principles are adopted in the layout design of the intake for pumped-storage station:

(1) Two-way water flow. As an intake, the water flow should be gradually smooth contraction. As an outlet, the flow of water should be spread smoothly. When the water flows in both directions, the velocity distribution of the whole section should be ensured to be even with a small head loss and no flow separation or reflux. Therefore, the design requirements for the profile dimension are stricter and the size of the intake transition section is longer. Water should be diverted with minimum head loss in the process of power generation and pumping, otherwise, the overall efficiency of the whole system will be reduced.
(2) Low submerged depth. The upper and lower reservoirs of a pumped-storage station are sometimes manually dug and filled. To minimize the engineering quantities, it is required to utilize the storage capacity as much as possible, resulting in a large working depth and a large variation of reservoir water level. When the water level of the reservoir is low, the submerged depth of the intake becomes small and the vertical-axis vortex of the inflow tends to occur. The vortex needs to be prevented and reduced by structural measures such as vortex elimination beam, grid, and plate.
(3) High flow rate. Unit capacity and flow rate in channel are often large in a pumped-storage station, resulting in a high flow rate through the waterway and asymmetrical distribution of flow rate. Water flow tends to separate at the end of the grid bar, resulting in vortex shedding, which not only increases head loss but also causes resonance if the circumfluence frequency is close to self-vibration frequency of the trash rack, causing damage to the trash rack.
(4) Erosion at the bottom and bank of the reservoir. The pumped-storage station generally has small reservoir capacity and large flow rate of intake. If water flow does not diffuse evenly, the circumfluence will be formed in the reservoir, leading to erosion at the bottom and bank of the reservoir and causing uneven distribution of intake water flow, vortex and other negative effects.

Therefore, when the layout design of the pumped-storage station intake is carried out, in order to make full use of the reservoir capacity, the orifice of the intake is often made into a larger width and a smaller height and is divided into several holes by the setting of partition blocks. Besides, multiple partition blocks are also arranged around the well loop for the off-stream intake with a gate shaft. Therefore, the width of the orifice is often large for the off-stream intake against hillside, and it is quite difficult to put the trash rack section and the diffusion section of intake into the hole, which often extends beyond the hillside instead. The trash rack tower can be arranged to open and close the trash rack. The hoist and maintenance of the trash rack can also be carried out after the water level of the reservoir drops to a certain height (such as dead water level).

In addition, the axis direction arranged at the intake should be consistent with the incoming flow direction as far as possible to better eliminate the influence of vortex at the intake. The structure on both sides of the intake should be arranged symmetrically as far as possible to improve water inflowing condition, and even anti-vortex beams, plates and other facilities should be set if necessary.

6.3 Open-Flow Intakes and Settling Basins

Open-flow intake is generally used in runoff diversion hydropower station, but also in pressurized flow diversion hydropower station of low dam reservoir. The design principle of open-flow intake is basically the same as that of the low-level intake. But because the reservoir is small, the problems of sand prevention, sewage disposal and ice prevention are prominent, so the layout and design of the intake must be considered as a whole.

6.3.1 Layout of Open-Flow Intakes

According to the layout and structural conditions, there are two types of open-flow intakes: superficial intake and intake with trash rack at bottom.

(1) The superficial intake is shown in Fig. 6.13a, b. To prevent floating objects and sediment from entering the gate while diverting surface flow of a river or reservoir, the intake is usually arranged in the concave bank and can be equipped with a floating row, breast wall, sand-sluicing gate, bottom scout gallery, etc.
(2) Intake with trash rack at bottom is shown in Fig. 6.14. It is in the sill of the over-current building, and the water diversion gallery is set in the vertical direction of the water and is covered with trash rack to divert water from the gallery. When necessary, water will flow into the diversion conduit through the sand sluicing chamber and the settling basin.

Open-flow intake should be located in the concave bank of the watercourse, where backflow is not easy to occur and floating objects are not blocked. In addition, the transverse circulation formed in the watercourse bending section guides the fresh surface water to flow into the intake, while the bottom sediment is brought to the convex bank.

In the layout of project, it should be considered to establish sand barrier in front of the intake and drainage facilities around there, such as sand sluice and sand flushing outlet, etc. in order to regularly or continuously wash away potential floating objects and silted sediment. The flow rate of sand flushing should be higher than 4 m/s. In

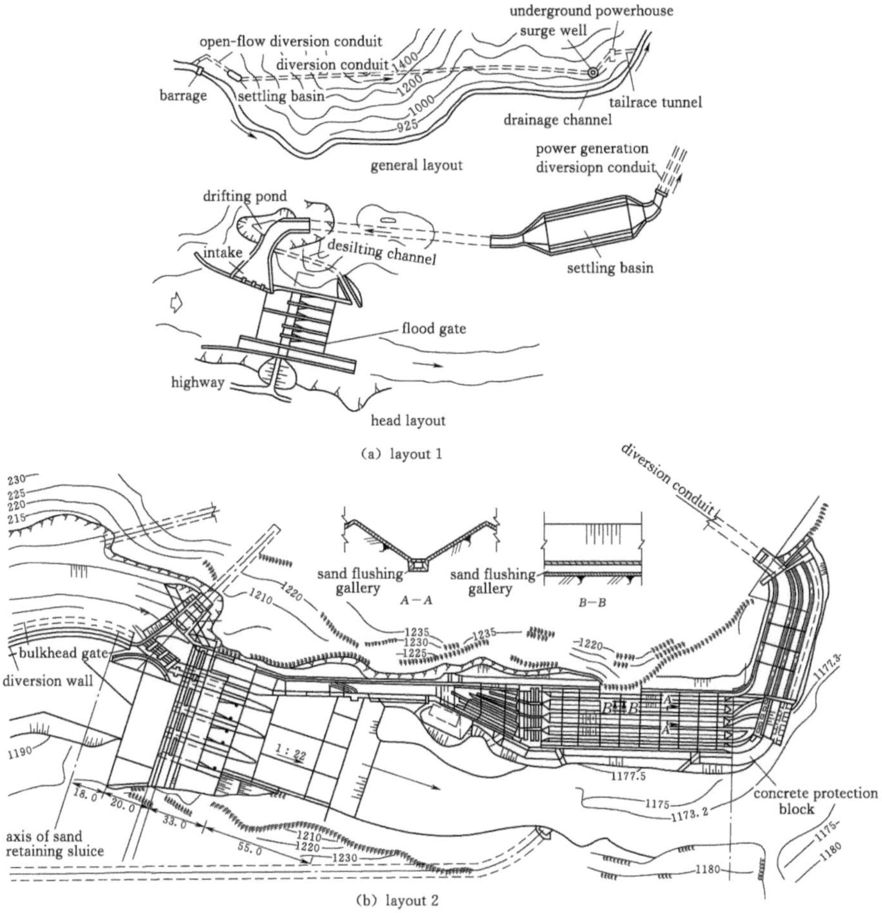

Fig. 6.13 Superficial open-flow intake

addition, settling basin can also be set behind the intake to deposit harmful sediment, and the silt in it can be removed by hydraulic or mechanical means.

6.3.2 Settling Basin

The settling basin is set up to prevent harmful particles in suspended sediment which has been introduced into the water intake from entering the diversion conduit. Harmful sediment generally refers to sediment with particle size larger than 0.25 mm, which will deposit in the diversion conduit to reduce the water-carrying capacity. Once they get inside, these harmful particles will wear and tear the wall surface of the diversion conduit to increase roughness, and even affect the strength of lining. They

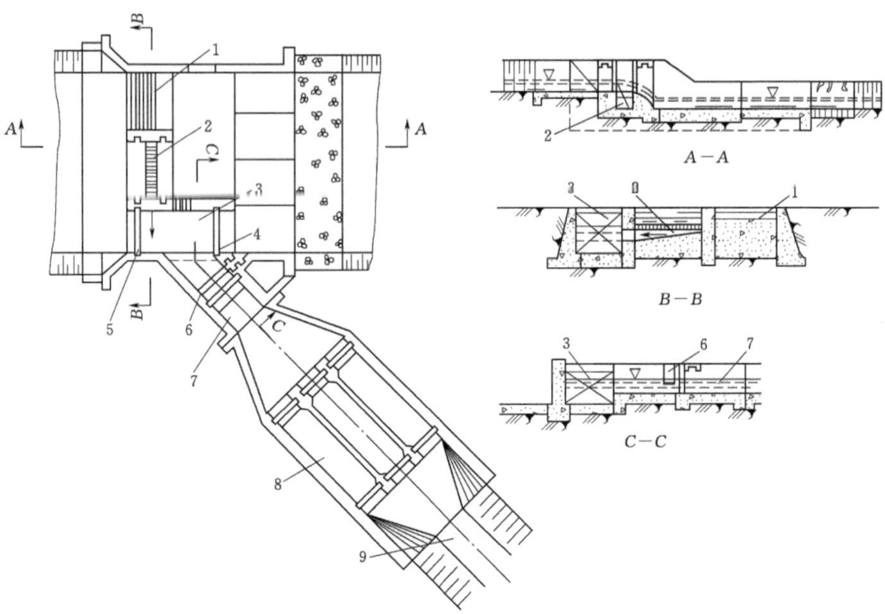

Fig. 6.14 Open-flow intake with trash rack at bottom

will also wear the turbine overflow components and aggravate cavitation, affecting the efficiency and service life of the turbine. When the water flow is in a high sediment concentration, smaller sediment particles can also cause turbine wearing out.

The settling basin is usually arranged behind the open-flow intake and before the diversion conduit. As shown in Fig. 6.13b, it is the front layout plan of a typical diversion hydropower station of medium size located on the silt river in mountain land of our country. When there is no sediment in the dry season, the sluice raises the water level and river is directly led into the tunnel. When the sediment concentration is high in the flood season, the sluice gate is opened to discharge the sandy flood. Then the inlet gate of pressure tunnel is closed and the sandy water flows into the settling basin through the left diversion conduit of the sluice gate. Clear water after sediment deposition enters another entrance of the tunnel through the diversion conduit.

The working principle of settling basin is to reduce the flow rate so as to reduce the sediment carrying capacity and make the sediment gradually deposit in the sink. Therefore, sufficient width, depth and length are required to make the flow rate smaller and uniform. According to the size of sediment particles in the flow, the average flow rate in the basin should be 0.25–0.70 m/s.

The sediment precipitated in the settling basin must be removed. The desilting method can be continuous or periodic through water flushing or mechanical equipment. As shown in Fig. 6.13b and Fig. 6.15b, the sediment is desilted continuously. The sediment gradually sinks into the channel from the bottom flushing gallery and the upper clear water flows into the diversion conduit. Figure 6.15a is an example of periodic sand flushing and the settling basin is divided into several chambers.

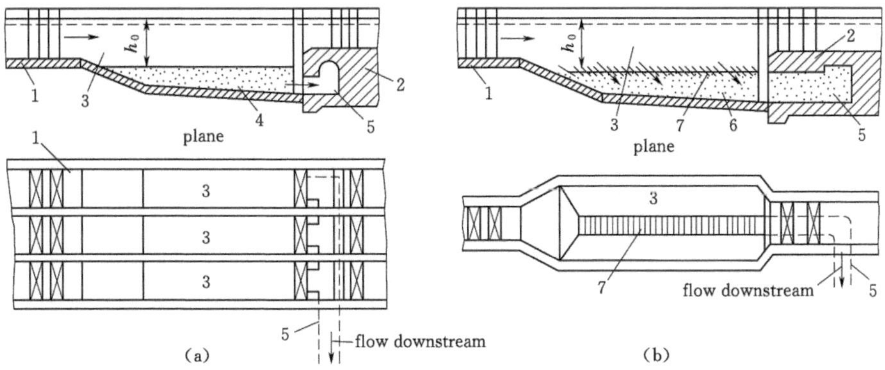

Fig. 6.15 Layout of settling basin

When the sediment in the basin is precipitated to a certain depth, the rear gate of one chamber is closed to flush sand through water flushing or mechanical equipment in several chambers by turns. As a result, power generation may not be stopped during sand flushing.

The shape, dimension and hydraulic design of the settling basin, including the method of sediment removal, should be subject to special analysis and calculation. Model tests and operation experience of the existing settling basin play an important role in the calculation method. Due to the complexity of sediment movement and operational condition of settling basin, hydraulic model tests are required in the design of settling basins for large hydropower stations.

6.4 Design of Intakes

6.4.1 Design Content

The design content includes the layout design, hydraulic design and structural design for any kind of intake. The layout design of intake has been discussed in Sect. 6.2 of this chapter. This section mainly introduces two aspects: hydraulic design and structural design.

(1) Hydraulic design. Hydraulic design should be carried out according to the type and function of intake, and three-dimensional flow field numerical computation or hydraulic model test should be used to verify the intake of large or important projects.

Hydraulic calculation of low-level intake mainly includes: ① head loss; ② discharge capacity; ③ minimum submerged depth; ④ soil foundation seepage; ⑤ cavitation number of high speed flow; ⑥ aeration vent area.

Hydraulic calculation of open-flow intake mainly includes: ① head loss; ② discharge capacity; ③ connection between upstream and downstream surface of intake; ④ soil foundation seepage; ⑤ cavitation number of high speed flow; ⑥ unstable flow.

The specific design can refer to *Design Specification for Intake of Hydraulic and Hydroelectric Engineering* (SL 285-2003), *Design Specification for Intake of Hydropower Station* (DL 5398-2007), *Hydraulic Design Manual Volume 8-Structure of Hydropower Station.*

(2) Structural design. The structural design of the intake should include the following contents: ① overall stability; ② foundation stress; ③ seepage stability and sedimentation of soil foundation; ④ overall structural design; ⑤ component design.

In the structural design, the outline of structures, the structure type and the treating measures such as seepage prevention, drainage and reinforcement of the foundation should be reasonably determined according to the geological conditions and the operational requirements for structures.

6.4.2 Loads and Load Combinations

(1) Types of loads. The loads on the intake structures can be divided into basic loads and special loads. The value of each load refers to *Design Specification for Loads of Hydraulic Structures* (DL5077-1997).

Basic loads include: ① dead weight (structure weight and permanent equipment weight); ② hydrostatic pressure at designed operating water level; ③ the designed water level difference before and after the trash rack; ④ uplift pressure at designed operating water level; ⑤ wave pressure at designed operating water level; ⑥silt pressure; ⑦ earth pressure; ⑧ ice pressure; ⑨ snow load; ⑩ wind pressure; ⑪ live load; ⑫ other loads (including grouting pressure, uneven sedimentation of foundation, etc.).

Special loads include: ① hydrostatic pressure at the verifying operating water level; ② uplift pressure at the verifying operating water level; ③ check the running water level of the wave pressure; ④ temperature load; ⑤ seismic load; ⑥other loads (including driftwood, ship impact and construction load, etc.).

(2) Loads combinations. Loads combinations can be divided into basic combination and special combination. They are listed in Table 6.1.

Table. 6.1 Table of load combinations

Load combination	Operating condition	Serial number	Types of loads											
			Dead weight	Hydrostatic pressure	Uplift pressure	Wave pressure	Silt pressure	Earth pressure	Ice pressure	Snow load	Wind pressure	Live load	Seismic load	Other loads
			1	2	3	4	5	6	7	8	9	10	11	12
Basic	Designed flood level	1	√	√	√	√	√	√	√	√	√	√		
	Normal storage level	2	√	√	√	√	√	√	√	√	√	√		
Special	Holding back water in construction period	3	√	√	√	√		√	√	√	√	√		
	Not retaining water after construction	4	√					√		√	√	√		
	Verifying flood level	5	√	√	√	√	√	√				√		
	Normal storage level + seism	6	√	√	√	√	√	√		√		√	√	
	Maintenance	7	√	√	√	√	√	√		√		√		

6.4.3 Structural Design

1. Overall stability and foundation stress

The intakes and outlets of conventional hydropower stations and pumped-storage power stations are mostly built on rock foundations at present. The stability calculation includes stability against sliding, stability against overturning and bearing stress checking of foundation.

If the intake structure is located on the sand foundation, the calculation can be more complicated, and the seepage stability and foundation sedimentation also need to be calculated. The specific calculation can be carried out by referring to *Hydraulic Design Manual Volume 8-Structure of Hydropower Station*, *Design Specification for Intake of Hydraulic and Hydroelectric Engineering* (SL 285-2003), *Design Specification for Intake of Hydropower Station* (DL 5398-2007).

2. Design for overall structure and components

The intake (outlet) of hydropower station is a typical spatial structure, and its static force calculation can be carried out by adopting means of structural mechanics. The intake (outlet) of large-scale or significant project should simultaneously conduct structural finite element analysis and calculation.

In the static calculation by adopting the means of structural mechanics, it is generally treated as plane structure, and that is to take the typical section in the vertical direction toward the water stream. And then the calculation is carried out according to the inverted frame pattern or the beam (slab) pattern of elastic foundation. For tunnel and wall intakes, the tunnel and off-stream intake with a gate pit should be designed according to *Design Specification for Hydraulic Tunnel* (SL 279-2002).

The design load of the trash rack supporting structure is mainly the water pressure difference between upstream and downstream on the grid surface, the weight of the trash-cleaning equipment, the impact force of the floating objects, the seismic inertia force, the seismic water pressure, and the structure weight, etc., of which the water pressure difference between upstream and downstream on the grid surface generally takes 2–4 m. There are many planar and three-dimensional computer programs available for the design of the supporting structure of the trash rack according to the means of structural mechanics. After obtaining the internal force of each member bar, the structure reinforcement is calculated according to the design specification of concrete structure, and it is necessary to check the instability of the member under pressure.

6.5 Diversion Conduits

6.5.1 Functions and Types of Diversion Conduits

The function of the diversion conduit is to concentrate the drop, create the head, deliver the water stream into the turbine through the penstocks, and then discharge the water downstream after the power generation. Most of the diversion conduits are set on upstream of the powerhouse, but in the underground hydropower stations constructed in the middle and rear part (as shown in Chap. 13), a considerable part of the diversion conduit is set in downstream of the powerhouse (i.e., the draft tube).

According to the working conditions and hydraulic characteristics, diversion conduit can be divided into free-surface flow diversion conduit and pressurized flow diversion conduit.

The free-surface flow diversion conduit is characterized by free water surface, and it doesn't bear high water pressure. It is suitable for directing water from river or reservoir where the water level does not change much, otherwise, it will cause large head loss at the water intake. In the structural form, the most commonly used free-surface flow diversion conduit is channel and free-surface flow tunnel as well. The channel is often arranged along the hillside contour line, which is restricted by topography and surface geological conditions, it has large length and excavation work, and it requires constant maintenance and repair during the operation period. But it is of great convenience in the ground construction. Medium and small power stations often adopt channels. The tunnel can shorten the line and has good and safe operating condition after the construction work, but it is of high requirement for the construction condition.

The characteristic of the pressurized flow diversion conduit is that the pressure flow runs in the diversion conduit and it can bear large water head and adapt to the great change in water level in the river or reservoir. Pressure tunnel is the most common structure which can take advantage of rock mass to withstand internal water pressure and prevent leakage. Under very special circumstances, the exposed penstock or exposed steel penstock in tunnel will be used in the pressurized flow diversion conduit.

6.5.2 Basic Requirements for the Design of Diversion Conduits

In addition to meeting the requirements of general hydraulic diversion conduit, special attention should be paid to the following requirements for the diversion conduit used for hydropower generations:

(1) Sufficient ability to draw in water that meets the requirements. The diversion conduit should have sufficient surface area and discharge to prevent silt and sand

from precipitating, reducing the cross-section area in the diversion conduit and preventing the surface of the diversion conduit from erosion and grass, raising the roughness and reducing the discharge. In addition, it is necessary to prevent rock-soil collapse and bank collapse from blocking the diversion conduit and minimize leakage from the diversion conduit.

(2) Less head loss. Therefore, the length of the diversion conduit, the surface roughness and the changes of the bend and cross-section should be reduced. The head loss will be small with less stream velocity, which can be realized by increasing the area of water and raising the amount of work as well. The optimal design discharge in the diversion conduit should be determined through technical and economic analysis and comparison.

(3) Water quality. At first, precautions should be taken at the intake to intercept sediment, ice and rubbish from entering the diversion conduit. As for the open channels, measures should be taken along the channel to prevent the dirt, sediment and man-made garbage on the hillside from entering the channel. Sewage that has entered the diversion conduits should be removed in time by arranging necessary facilities.

6.5.3 Channel

6.5.3.1 Types of Channels

The water diversion conduit in the hydropower station is generally made in the form shown in Fig. 6.16. The top of channel is approximately parallel to the bottom of channel. A downflow weir is set at the end of channel and it is used to control the water level at the end of channel for safety and supply water to downstream area. When the flow directed into the station is less than that through the channel, the excess water will be discharged downstream from the downflow weir. The advantage of this kind of channel is that the amount of projects is small, but the disadvantage is that the water will be discharged if there is no fixed water demand in downstream of the power station. This kind of channel is called non-automatic regulating channel.

And there is another kind of water diversion conduit. When the discharge used by units is zero, the water level in the channel is at horizontal level, so the top of channel is also horizontal, meanwhile, the cross-section of channel is becoming larger and larger, and no downflow weir is set at the end of channel, as shown in Fig. 6.17.

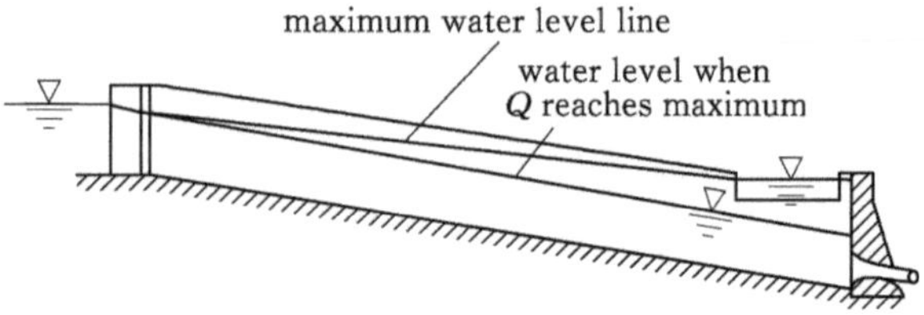

Fig. 6.16 Non-automatic regulating channel

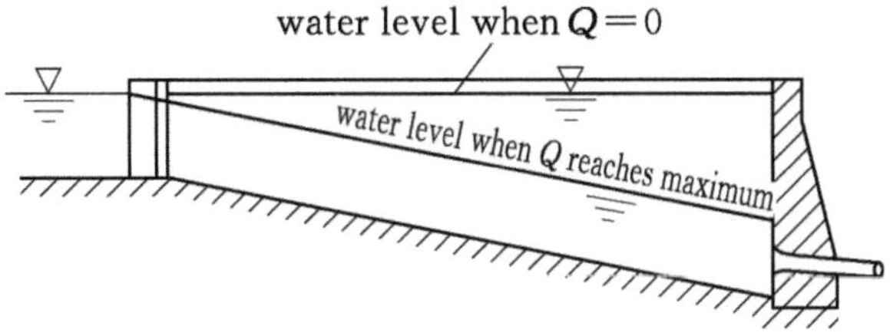

Fig. 6.17 Self-regulating channel

When the flow directed into power station changes, the channel itself plays the role of regulation and storage so that the stream flowing in and out can keep balance. This kind of channel is called self-regulating channel. Due to its large amount of projects, it can only be used when the channel has a short line, gentle slope bottom and there is no fixed water demand downstream.

6.5.3.2 Hydraulic Calculation for Channels

The main task of hydraulic design for channels is to determine the cross-section dimension, roughness, longitudinal gradient and water depth according to the designed discharge. The content of hydraulic calculation includes:

1. Calculation for steady flow

The formula to calculate the steady uniform flow in the channel can be listed as follows:

$$F = \frac{\tilde{Q}}{C\sqrt{Ri}} \quad (6.2)$$

$$C = \frac{1}{n}R^{1/6} \quad (6.3)$$

The head loss caused by the channel with length of L is:

$$\Delta H = iL \quad (6.4)$$

In the formula:

F cross-section area of water passage, m^2;
Q designed discharge, m^3/s;
C Chezy coefficient, m/s;
R hydraulic radius, m;
i longitudinal gradient;
n roughness.

The channel generally has a trapezoidal or rectangular cross-section. Roughness is determined by the type of channel lining. After determining the shape and bottom width of the cross-section (F and R is only decided by depth), the channel depth of different longitudinal gradients can be calculated according to formula (6.2). The factors mentioned above interact with each other. If the longitudinal gradient is large, the discharge will become large and the cross-section will be small causing small amount of project and large head loss as well. Artificial lining can make the surface smooth and reduce roughness, cross-section or head loss, and reduce leakage loss as well, but it increases the cost of lining. The shape of channel section, the area of water passage and the longitudinal gradient are also affected by the local topography and geological conditions. As for discharge in the channel, it cannot be so small that sediment precipitates. In the channels without artificial lining, the discharge should not be too large to prevent erosion of the channel bottom and slope. A comprehensive analysis of all factors should be carried out and a techno-economic comparison should be made to determine all parameters of the channel.

When the water level of the intake is constant, the design flow Q_{max} runs through the channel with the steady and uniform flow and the water surface is like Curve 1, as shown in Fig. 6.18, which is parallel to the bottom slope, and the head loss in the channel is ΔH_1 When flow Q directed into the hydropower station is less than Q_{max}, the intake gate can be operated to lower the front water level behind the gate and create a steady flow corresponding to the new directed flow Q with water surface like surface Curve 2 shown in Fig. 6.18, which is still parallel to the bottom slope. But the present head ΔH_2 will be bigger than ΔH_1. To increase the working head of the power station, the intake and gate of the head pond can be operated to raise the water level in the head pond to the level in proportion to Q_{max}, forming the water surface Curve 3. In this way, the head loss decreases from ΔH_2 to ΔH_1 when the directed flow is Q, which increases the power generation efficiency. Surface Curve 3 is the backwater surface line which can be calculated by the hydrodynamic method, and the corresponding flow is uniform.

The above situation refers to that the directed flow of hydropower station (that is the flow in the channel) is constant, and the current is constant as well.

2. Calculation of unsteady flow

The turbine directing flow is variable in the hydropower station, and the most typical change is the sudden disappearance and augments of the load. At this time, the flow at the end of the channel, namely the head pond, changes instantaneously, but the flow along the channel can only change gradually, resulting in the flow and water

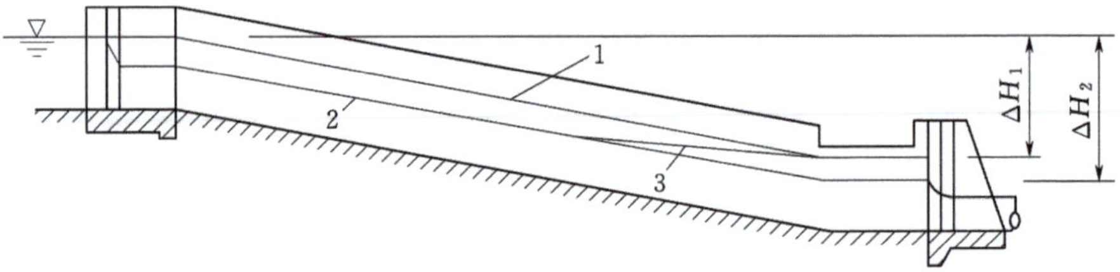

Fig. 6.18 Steady flow in exposed channel

depth of each section along the length of the channel changing with time (that is the unsteady flow in the channel).

When the power station discards the load, the turbine discharge decreases suddenly, but it is too late for the incoming flow of the channel to decrease. Then the excess water accumulates, and the water level of the channel gradually increases from the downstream to the upstream. This phenomenon gradually appears with water level increasing from the end to the head of the channel, which is called surge wave. Curve 1 in Fig. 6.19 shows the water surface line of a constant current before load changes. After the power station suddenly discards the load, the surge wave, as shown in the solid line, pushes upstream. During the process, the water stream of each section in the channel changes successively from bottom to top, and the water surface line in the channel is the surge wave line. The maximum water level along the channel should be calculated by calculating the surge wave line to determine the height of the channel and the acting head of hydropower station. To limit the water level being too high to result in a high channel top and even that water overflows at the top, downflow weir is often arranged at the end of the channel, and its top should be slightly higher than the surface of steady flow while the maximum directing flow runs through the channel so that the power station will not discard water during the normal operation. The width of downflow weir for water intake and the depth above downflow weir should be selected according to the maximum design flow of discharge channel. If the volume of head pond at the end of the channel is very large, or if there is a run-off pond connected at the end of the channel (as shown in Fig. 6.23), then the change of water level in the channel can be limited when the load is discarded.

When the power station increases the load, the discharge of the turbine will increase suddenly, but it is not enough time for the flow from the channel to increase. Then water at the end of the channel is directed, and the water level gradually decreases. This phenomenon gradually appears with water level decreasing from the end to the head of the channel, which is known as the subsidence wave as shown by the dotted line in Fig. 6.19. In this process, the discharge of each section in the channel changes, and the surface line in the channel is the ebb and flow line. The lowest water level should be determined by calculating the falling wave to decide the intake elevation of the penstocks at the end of the channel. The downflow weir

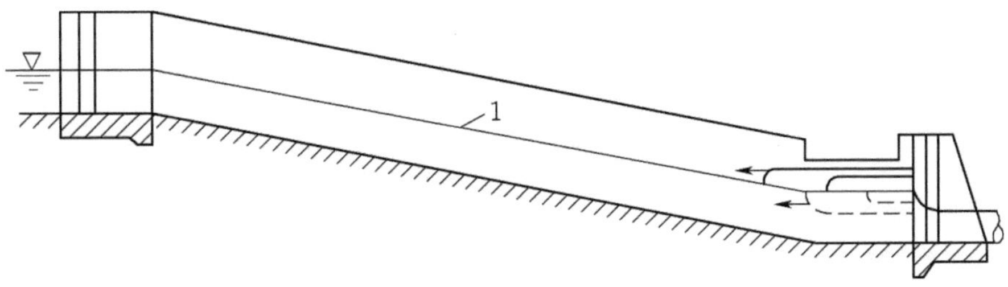

Fig. 6.19 Unsteady flow in exposed channel

does not work when the power station bears the increasing load. But the regulation storage capacity at the end of the channel can reduce the decrease in water level.

The hydraulic calculation methods of steady flow and unsteady flow are described in detail in the course of *hydraulics* and the design manual. During the calculation, the change of loads on the hydropower station should conform to the actual operating conditions of the power station and the relevant regulations. The structural design and calculation for diversion conduits of power generation are taught in the course of *hydraulics*, just like all hydraulic channels.

6.5.4 Tunnel

1. Types and characteristics of the tunnel

The tunnel used for the diversion conduit in the hydropower station can be either the upstream tunnel of the powerhouse (usually called the diversion tunnel) or the downstream tunnel (usually called the tailrace tunnel). The tunnel can be either free-surface flow or pressurized flow.

Compared with the channel, the tunnel has the following advantages.

(1) A shorter route can be adopted to avoid adverse terrain and geological conditions along the route.
(2) The pressure tunnels can adapt to the large rise or fall of water level in the reservoir and the rapid change of the discharge in the hydropower station.
(3) It is not affected by freezing, and there is no water pollution along the route.
(4) The operation is safe and reliable.

The main disadvantages of the tunnel are higher requirements on geological conditions, construction technology, and mechanization with higher unit price and longer construction period. However, with the continuous development of modern construction technology and equipment as well as the improvement of tunnel lining design theory, these shortcomings are gradually being overcome. Therefore, the tunnel has been more and more widely used.

2. Hydraulic calculation for the tunnel

The working conditions of the free-surface flow tunnel are similar to those of the channel and the content of hydraulic calculation is also the same. The purpose of steady flow calculation is to study the relationship among tunnel sections, longitudinal gradient and head loss at a certain discharge to determine these parameters and lining types. Unsteady flow also has to be simulated for the free-surface flow tunnel.

The purpose of steady flow calculation of pressure tunnel is to determine the cross-section area (or discharge), roughness and head loss according to the given design flow. The calculation formula is shown as follows:

$$\Delta H = \left(\frac{2gL}{C^2 R} + \sum \xi_i \right) \frac{v^2}{2g} \tag{6.5}$$

In the formula:

ΔH head loss, m;
L tunnel length, m;
R, C same as formula (6.2) and (6.3);
ξ_i local hydraulic loss coefficient;
v the velocity in the tunnel, m/s.

For the fixed discharge and tunnel length, the larger the tunnel area is, the smaller the discharge, surface roughness and local loss become, but the amount of project will increase. Technical and economic comparisons should be made to correctly determine these parameters.

The hydraulic calculation of a pressure tunnel is characterized by the fact that the tunnel's water passing capacity only depends on the head loss of the whole length of the tunnel, and has nothing to do with the longitudinal gradient and the buried depth of the tunnel (that is, the depth of the tunnel under the upstream level). Generally, the tunnel line of the pressure tunnel should be raised as far as possible to reduce the internal water pressure load, but the water pressure at the top of the tunnel must be no less than 2.0 m under any working condition, including the case of unsteady flow.

The unsteady flow analysis of the pressurized flow diversion conduit is closely related to the operation of units, penstocks and surge tank, etc., which will be elaborated on in Chaps. 9 and 10. Like other tunnels, the structural design of the power generation tunnel is the same as that of other tunnels, which is taught in the course of *hydraulics*.

6.6 Head Pond and Daily Run-Off Pond

6.6.1 Functions and Composition of Head Pond

A head pond is a reservoir at the end of a free-surface flow diversion conduit, provided above a power plant and serving to convey the water to penstocks and also to direct some of the flow into the lower pool around, instead of through the turbines. The function of the head pond is:

(1) During normal operation of the plant, discharge is distributed to the penstock as required and head loss is minimized.
(2) When loads change suddenly or accidents occur, the head pond can cooperate with the diversion conduits to regulate the flow rate and discharge excess water to avoid excessive water level in the free-surface flow diversion conduits. When the penstock is closed, the necessary flow can be supplied downstream. The water flow can be cut off urgently in case of an accident in the penstock to prevent the accident from spreading.
(3) Dirt, ice and sediment are prevented from entering the penstock.

According to the required functions, the head pond is composed of the following parts, as shown in Fig. 6.20:

(1) Forebay and diffusion section. Widening and deepening the end of the channel to satisfy the requirements of intake layout. Therefore, its width and depth are larger than those of the channel, and the diffusion section needs to be set between channels and the bottom slope also needs to be raised. The pond body has a certain volume and can be used to regulate discharge. Large run-off storage is required for hydropower stations with the task of peak regulation or even daily regulation. At this time, the volume of the head pond can be enlarged or the run-off pond can be arranged to connect with the head pond. The volume of the head pond and run-off pond can reduce the water level fluctuation of the free-surface flow diversion conduit when the load changes.
(2) The intake and equipment of penstocks. The intake generally adopts the water retaining wall type, and its structure, trash rack, service gate, bulkhead gate

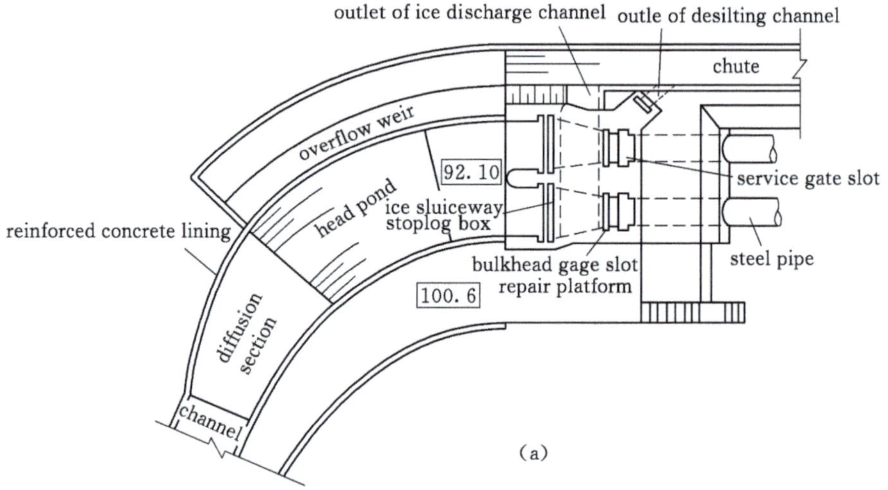

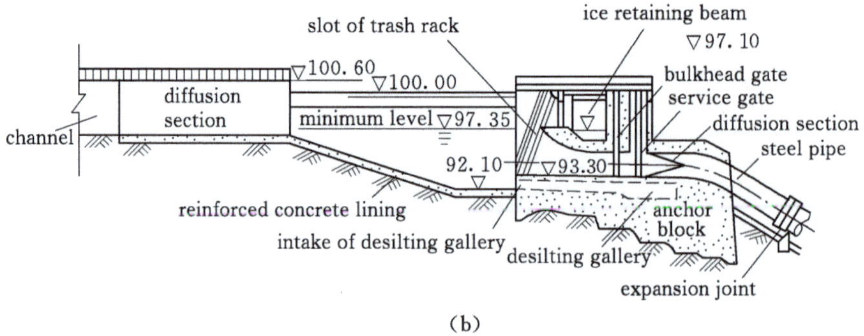

Fig. 6.20 Layout of the head pond

and aeration vent, etc. are similar to those of the pressured off-stream intake described in Sect. 6.2.

(3) Outlet structure and dirt, sediment and ice ejection equipment. Water discharge structure plays the role of discharging water, sediment, ice, etc. Overflow weir is often adopted, which is connected with chute and plunge pool, etc. In addition, the overflow can also be directly discharged into gully and watercourse. The controlling gate can be set at the top of overflow weir or not. The flow rate decreases and the sediment is precipitated after water enters the head pond from the diversion conduit. Therefore, the bottom of the penstock intake is often provided with the trash still and sand flushing pipe, and the sand flushing hole can also be used as a pond and for supplying water downstream.

6.6.2 Layout Arrangement of Head Pond

The head pond is the structure connecting the free-surface flow diversion conduit with the penstock and powerhouse. The most reasonable layout and mutual relation among these structures should be fully considered. The head pond should be as close to the powerhouse as possible to shorten the length of penstock and reduce the excavation. For this reason, the head pond is often close to steep slope with a short seepage path, so there is a possibility of instability, settlement and landslide, as shown in Fig. 6.21a, b. Therefore, good geologic condition and stability of foundation are required, and the pond is better placed on the excavated foundation to prevent leakage and damage. And sometimes, it is necessary to adopt anti-seepage and drainage facilities and carry out the design for foundation stability and structural strength.

The relative position of the head pond with the channels and penstocks can be different, as shown in Fig. 6.22. In plan a, the pipeline and the channel are in the same direction and water can be smoothly directed; in plans b and c, water-directing condition is poor, but the dirt in front of the trash rack is easy to be washed away.

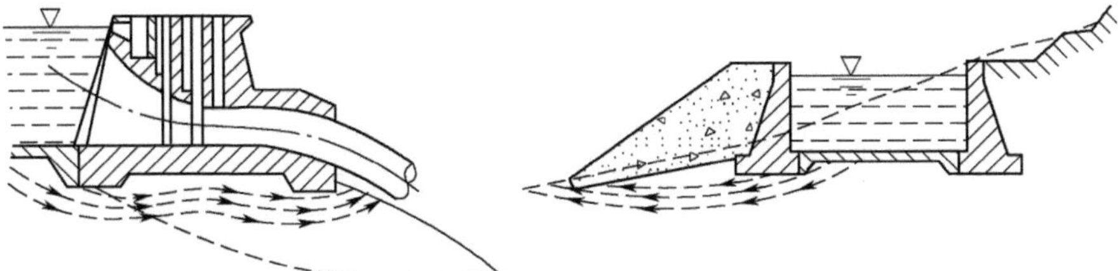

(a) Possibility of settlement (b) possibility of landslide

Fig. 6.21 The slope stability of head pond

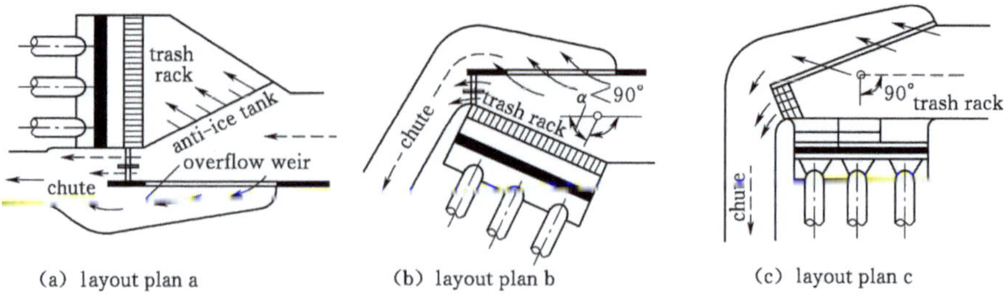

Fig. 6.22 Schematic diagram of different diversion direction in head pond

6.6.3 *Daily Run-Off Pond*

When the hydropower station conducts the daily regulation, the directed discharge for power generation will change at any time. And the diversion conduits are designed according to $Q_{\max}$ so that the channel's water passing capacity is not fully utilized for most of the day. If there is suitable topography along the downstream of the channel for constructing daily run-off pond, as shown in Fig. 6.23, the situation can be greatly improved: the channels between the daily run-off pond and the head pond are still designed according to $Q_{\max}$, but the upstream channel of the daily run-off pond can be designed according to the daily average discharge. When the discharge of hydropower station is larger than the average flow in the operation, the daily run-off pond will make up water and the water level will decrease. When the discharge of hydropower station is less than the average flow, the redundant water will be infused into the daily run-off pond to rise the water level. In this way, the upstream channel can keep around the average flow all day long. The functions of daily run-off pond will be greater when it is closer to the head pond.

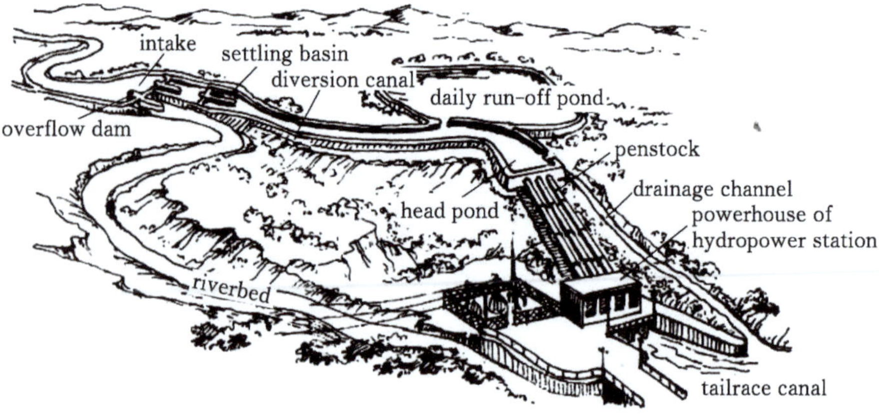

Fig. 6.23 Daily run-off pond

Questions

1. What are the types of water intakes? What are the types of low-level intakes? What are their applicable conditions?
2. What are the steps to choose and determine the position and elevation of intakes according to the requirements for design?
3. What are the functions of the service gate and bulkhead gate of intakes? What are the corresponding requirements for operation?
4. What equipment does the low-level intake include? What are the functions of the equipment?
5. What is the role and composition of head pond? What problems should be noted when laying out head pond?

References

1. Wang, Renkun and C. Zhang. 2013. *Design Manual of Hydraulic Structures: vol.8 Buildings of Hydropower Station (In Chinese)*, 2nd ed. Beijing: China Water and Power Press, 2013.
2. Ministry of Water Resources of the People's Republic of China. 2003. *SL 285-2003 Design Specification for Intake of Hydraulic and Hydroelectric Engineering (In Chinese).* Beijing: China Water and Power Press.
3. National Development and Reform Commission of the People's Republic of China. 2008. *DL 5398-2007 Design Specification for Intake of Hydropower Station (In Chinese).* Beijing: China Electric Power Press.
4. National Development and Reform Commission of the People's Republic of China. 2004. *DL 5195-2004 Specification for Design of Hydraulic Tunnel (In Chinese).* Beijing: China Electric Power Press.
5. Ma, Shanding and R. Wang. 1996. *Buildings of hydropower station (In Chinese).* Beijing: China Water and Power Press.
6. Liu, Qizhao and M. Hu. 2010. *Hydropower station (In Chinese)*, 4th ed. Beijing: China Water and Power Press.

Chapter 7
Penstock

Kai Su

Abstract This chapter focuses on penstocks of hydropower station. Firstly, functions and types of penstocks are introduced, and water supply modes and hydraulic calculation are briefly described. Then, materials and strength of steel penstocks are elaborated, facilities and structure are introduced, and design loads and load combinations of penstocks are listed. For exposed steel penstocks and other types of penstocks, including underground penstocks as well as dam body penstocks and penstocks laid on the downstream surface of dams, their respective characteristics and layout are introduced. Structural analysis as well as the stability against external pressure is carried out in details with assumptions and equations.

Keywords Penstock · Exposed steel penstock · Structural analysis · Underground penstock · Dam body penstock · Penstock laid on the downstream surface of dams

7.1 Functions and Types of Penstock

Penstock is a water pipeline from a reservoir or a flat-water building (head pond and surge tank) that supplies water directly to hydraulic turbine under pressure. It concentrates all or most of the water heads of the hydropower station. It is characterized by: ① steep gradient; ② under maximum dynamic and static water pressure; ③ close to powerhouse, threatening powerhouse if accidents happen. Therefore, its safety and economy are paid special attention to. Its materials, design methods, and techniques have special requirements different from ordinary hydraulic structures.

The main load of the penstock is the internal water pressure. The inner diameter of the penstock D and the water head H and its product value HD are the most important feature parameters indicating the scale and technique difficulty of the penstock, and the value of HD is also directly related to the installed capacity N_g provided by the penstock. If the velocity of flow in the penstock is 5–7 m/s, then $\text{HD} \approx (0.15 - 0.18)\sqrt{N_g H})$. It can be seen that when the installed capacity of the

K. Su (✉)
Wuhan University, Wuhan, China
e-mail: suker8044@163.com

Y. Cheng et al. (eds.), *Hydropower Station*,
https://doi.org/10.1007/978-981-96-2383-9_7

penstock is the same, the value of *HD* will increase with the increase of the water head of the power station. At present, the largest diameter of the penstock of the hydropower station is the diversion penstock of the underground power station on the right bank of Xiangjiaba, with a diameter of 14.4 m. The hydropower station with the highest *HD* value is the pumped-storage station, which has exceeded 5000 m^2.

Penstocks can be classified according to the layout and materials used, as shown in Table 7.1.

Buried penstocks (ditch penstocks or buried steel penstocks), earth dam culverts and wood penstocks, cast iron penstocks, etc., have been used rarely or only in small power stations.

Table 7.1 Types of penstocks

Types	Arrangement	Materials and structures
Exposed steel penstocks	Penstocks exposed to the air, Fig. 7.1	Steel penstocks, reinforced concrete penstocks, steel-lined reinforced concrete penstocks
Underground penstocks	Penstocks buried in the rock mass, Fig. 7.2	Unlined, anchor spray or concrete lining, steel-lined concrete lining
Dam body penstocks	Penstocks buried in concrete dam, Fig. 7.40	Steel-lined reinforced concrete structure
Penstocks laid on the downstream surface of concrete dams	Penstocks laid on the downstream side of concrete dams, Figs. 7.48, 7.49, 7.50, 7.51, 7.52, and 7.53	Steel penstocks, steel-lined reinforced concrete structure

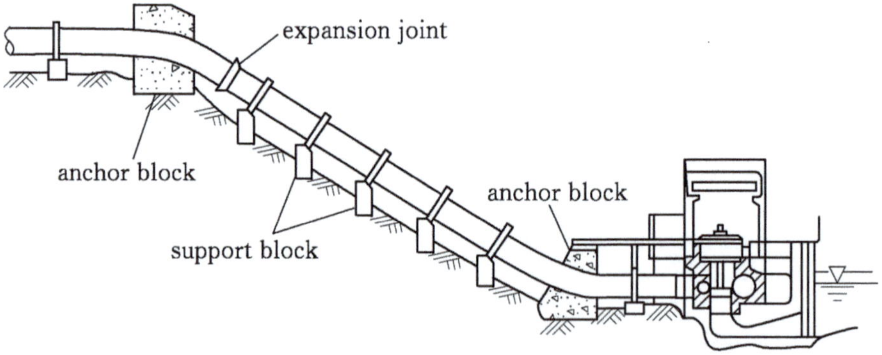

Fig. 7.1 Exposed penstocks

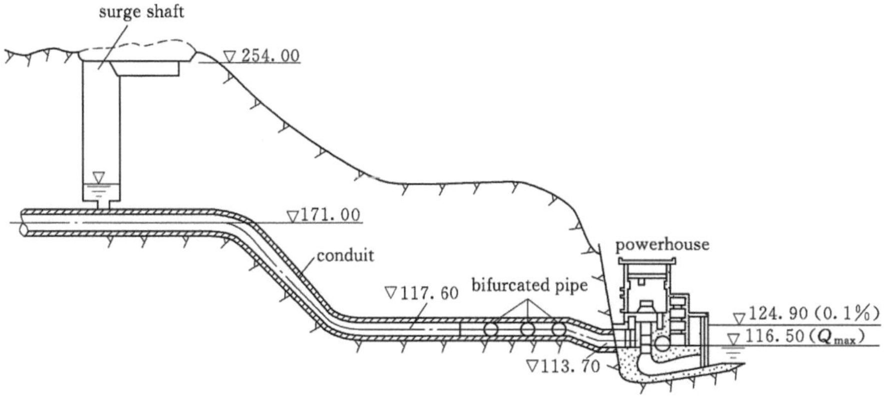

Fig. 7.2 Underground penstocks

7.2 Water Supply Modes and Hydraulic Calculation of Penstocks

7.2.1 Water Supply Modes

There are three ways for the penstock to supply water to multiple units, as shown in Table 7.2.

7.2.2 Selection of Penstock Diameter

After the water supply mode is selected, the discharge through each penstock is also determined, and then the diameter of the penstock should be selected. The materials and cost of the penstock will decrease as the diameter of the penstock decreases. However, the increased velocity in penstock will cause an increase in head loss and power loss. Therefore, the diameter of the penstock should be determined through economic comparison. There are many theoretical equations and statistical equations for calculating the economic diameter of steel penstock at home and abroad, but their essences are similar. Penstock cost and power generation loss can be calculated relatively reliably, but the principles and factors for economic comparison between the two are often difficult to determine accurately. In addition, this calculation is also difficult to consider the factors of technical conditions such as the penstock construction process, so the results calculated by the equation can often only be used as a reference. Generally, several similar diameters are selected according to the existing engineering experience and appropriate equation, the cost and electric quantity are calculated respectively, and the optimal diameter is selected based on the

Table 7.2 Water supply modes of penstocks

Water supply modes		(1) Individual water supply	(2) Centralized water supply	(3) Group water supply
Illustration		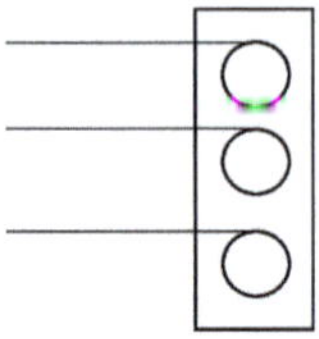	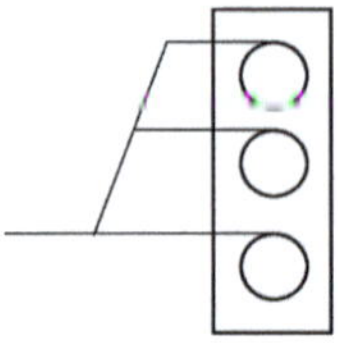	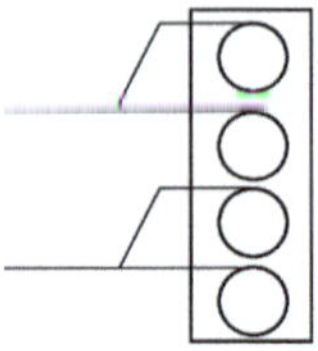
Technical and economic comparison	Operation	Flexible	General	Better
	Valve setting	Generally, the quick valve before the unit can be omitted	Quick valve is required in front of the unit	Quick valve is required in front of the unit
	Penstock structure	Small size, easy to make, no bifurcated pipes	Large size of single penstock, complicated bifurcated pipes required	Moderate size of single penstock, less bifurcated pipes
	Penstock layout	The penstock occupies a large area on the plane, and the penstock arrangement and connection with the head pond and surge tank are difficult	Easy to layout	Between individual water supply and centralized water supply
	Construction installation	Installment installation, installment investment	One-time investment	Between individual water supply and centralized water supply
	Economy	Higher cost under the same head loss; uneconomical	Lower cost under the same head loss; economical	Between individual water supply and centralized water supply
	Applicable situation	1. buried penstock, dam body penstock, 2. underground penstock or exposed penstock with large discharge or short length of single penstock	Widely used in underground penstock and exposed penstock, small number of units, small discharge of single unit, long pipeline	Widely used in underground penstock and exposed penstock, large number of units, small discharge of single unit, long pipeline

comprehensive analysis of technical problems. In the feasibility study and preliminary design stage, the following Pentaxhu equation can be used to initially determine the economic diameter of large- and medium-sized penstocks:

$$D = \sqrt[7]{\frac{5.2Q_{\max}^3}{H}} \tag{7.1}$$

In which:

D economic diameter of steel penstock, m;
$Q_{\max}$ maximum design discharge of steel penstock, m^3/s;
H design head, m.

7.2.3 *Hydraulic Calculation*

The hydraulic calculation of the penstock includes head loss and water hammer calculation to determine the working head of the hydropower station, the water pressure acting on the turbine components, and the pressure distribution of the penstock: ① the highest hydrostatic pressure line under normal working conditions; ② the highest hydrostatic pressure line under special working conditions;③ the lowest hydrostatic pressure line.

Head loss includes friction loss along the path, as well as local loss at the inlet section, trash rack, gate slot, transition section, bend section, bifurcated pipe, etc., which can be calculated according to the corresponding equation in the *hydraulics* tutorial or the design manual.

The water hammer calculation is described in detail in Chap. 9 of this book.

7.3 Materials and Strength of Steel Penstocks

The steel penstocks of the hydropower station are subject to dynamic action such as high internal water pressure and water hammer impact for a long time. It belongs to the pressure vessel type structure and has strict requirements on steel.

7.3.1 *Basic Requirements of Steel Performance*

The steel is required to have good mechanical properties and process performance, which can meet the design specifications and corresponding national standards.

Generally, the yielding strength σ_s and tensile strength σ_b, plasticity index (elongation δ and reduction of area ψ), conventional impact toughness a_k, and fracture toughness of steel are all required to be as high as possible. However, there is a contradiction between strength and plastic toughness. When the strength increases, the yielding ratio σ_s/σ_b will increase, and the plastic toughness and welding ability of the steel will decrease. For containers such as steel penstocks, good plastic toughness is extremely important. It can make the structural stress more uniform, reduce stress concentration, improve bearing capacity, prevent brittle fracture, increase safety, and facilitate cold forming and welding. Therefore, it is sometimes preferred to use steel with lower strength and higher toughness, which is more conducive to safety and even more economical.

The processing property of steel is mainly the performance of cutting, cold bending, and welding. Plastic deformation, cold work hardening, and cold brittleness during cold bending are required to be limited to the allowable range, with good weldability, simple welding process, no cracks in the weld and heat-affected zone, small residual stress, and no less than the mechanical properties of the base material.

To meet the requirements of steel performance, the suitable steel type must be selected first, and the chemical and alloy composition must comply with the relevant specifications. Different elemental compositions and contents can improve or deteriorate the properties of steel. The content of elements such as sulfur and phosphorus that are extremely harmful to the properties of steel must be strictly controlled. The same material and different heat treatment methods can get different performances. Heat treatment can also improve the properties and weldability of steel after cold working. It is stipulated that for thick plates, high-strength steels and low-temperature welding, preheating should be carried out before welding, and whole or partial heat treatment should be carried out after coiling and welding.

Hydropower station steel penstocks are large and difficult to heat treatment as a whole. Installation of the transverse seams, and even some longitudinal joints, must be carried out at the installation site. However, on-site working conditions are often unfavorable, such as narrow spaces and poor humidity and temperature conditions, which are not conducive to welding quality and heat treatment. Therefore, the choice of steel types and reasonable processing techniques is particularly important.

Quality inspection and weldability testing of steel and welding materials are required to ensure the quality of the materials used. Omitted during testing or failing to test the material as required may become one of the important causes of accidents.

The discussion about steel in the course *Hydraulic Steel Structure* applies to steel penstock, but the material requirements are more stringent for steel penstock withstanding high pressure.

7.3.2 Strength Checking and Allowable Stresses

Regarding strength checking, since steel is a metal material with relatively uniform elastic–plastic properties, practice shows that the third and fourth strength theories are more applicable. The fourth strength theory (distortion energy theory) is now adopted in China and most countries; that is, the stress at each calculation point should satisfy the following equation:

$$\sqrt{\sigma_x^2 + \sigma_r^2 + \sigma_\theta^2 - \sigma_x\sigma_r - \sigma_x\sigma_\theta - \sigma_\theta\sigma_r + 3\left(\tau_{xr}^2 + \tau_{x\theta}^2 + \tau_{r\theta}^2\right)} \le \emptyset[\sigma]\, or\, \sigma_R \quad (7.2)$$

in which:

$\sigma_\theta, \sigma_r, \sigma_x$ circumferential, radial, and axial stresses of steel penstock, N/mm^2;
$\tau_{x\theta}, \tau_{xr}, \tau_{r\theta}$ shear stress in all directions of steel penstock, N/mm^2;
$\emptyset$ welding seam coefficient, which varies from 0.9 to 0.95 related to welding method, inspection standard and structural grade;
$[\sigma]$ allowable stress, N/mm^2;
σ_R resistance limit, N/mm^2.

The value (σ_r, τ_{xr}, $\tau_{x\theta}$) of the steel penstock of the hydropower station is small and can be ignored. Therefore, the above equation can be simplified as:

$$\sqrt{\sigma_x^2 + \sigma_\theta^2 - \sigma_x\sigma_\theta + 3\tau_{x\theta}^2} \le \emptyset[\sigma]\, or\, \sigma_R. \quad (7.3)$$

In the hydraulic version of the *Design Specification for Steel Penstocks of Hydroelectric Stations* SL 281-2003, the strength is checked according to the allowable stress $[\sigma]$ and the allowable stress is often expressed as a percentage of the yielding strength of steel. For the basic load combination, $[\sigma]$ is lower for the membrane stress zone of the steel penstock and for the exposed steel penstock. For special load combinations, $[\sigma]$ is higher for the local stress zone of the steel penstock and for the buried steel penstock. For steel with a large yielding ratio, new steel and elbows, manifolds or particularly important parts can be tried, and $[\sigma]$ need to be reduced appropriately. The specific values are shown in Table 7.3.

In the electric version of the *Design Specification for Steel Penstocks of Hydroelectric Stations* NB/T 35,056-2015, the strength is checked according to the resistance limit (similar to the allowable stress $[\sigma]$). The resistance limit of steel is:

$$\sigma_R = \frac{f}{\gamma_0 \psi \gamma_d} \quad (7.4)$$

In which:

f steel design strength, N/mm^2;
γ_0 structural importance coefficient determined by the safety level of the steel penstock, the value is 1.1 for Class 1 buildings and 1.0 for Class 2 and Class 3 buildings;

Table 7.3 Allowable stresses of steel penstock

Stress area		Membrane stress zone	Local stress zone				
Load combination		Basic	Special	Basic		Special	
Stress generating force		Axial force		Axial force	Axial force and bending moment	Axial force	Axial force and bending moment
Allowable stress	Exposed penstock	0.55 σ_s	0.7 σ_s	0.67 σ_s	0.85 σ_s	0.8 σ_s	1.0 σ_s
	Underground penstock	0.67 σ_s	0.9 σ_s				
	Buried penstock	0.67 σ_s	0.8 σ_s 0.9 σ_s				

ψ design condition coefficient, the value is 1.0 under a sustained condition, 0.9 under a transient condition, and 0.8 under an accidental condition;

γ_d structural coefficient, the value can be taken from Table 7.4.

7.3.3 *Current Status of Pipeline Steel in Hydropower Stations*

The domestic steel grades that have been used widely and maturely in steel penstocks of hydropower stations in China are ordinary carbon steel Q235 (yield strength $\sigma_s = 235\ \text{N/mm}^2$) and low alloy structural steel Q345, Q345R ($\sigma_s = 345\ \text{N/mm}^2$), followed by 15 MnV and 15 MnTi ($\sigma_s = 390\ \text{N/mm}^2$). Such killed steels can be used to make pipe shells, supporting rings, branching reinforcement, etc. The production process of these steels is relatively simple, the price is not expensive, and the performance of processing and welding is good. When the plate is thinner, no preheating is needed before welding, but the yielding strength of these steels is not high enough.

With the improvement of steel penstock parameters of hydropower stations in China, the above steel products can no longer meet requirements. If steel with a lower strength grade is continued to be used, the thickness of the steel plate will increase. When the steel plate exceeds a certain thickness (for example, more than 36–38 mm), it will be difficult to process, weld, and transport, and the cost will also increase. To improve the processing conditions and reduce the cost, it is necessary to use a steel with a higher strength grade to reduce the thickness, such as high-strength steel with a tensile strength of 600 N/mm^2 ($\sigma_s = 490\ \text{N/mm}^2$) and high-strength steel with a tensile strength of 800 N/mm^2 ($\sigma_s = 690\ \text{N/mm}^2$), even high-strength steel ($\sigma_s = 880\ \text{N/mm}^2$) with a tensile strength of 1000 N/mm^2. At present, high-strength steels such as WDB620, WDL610, and Q690CF have been widely used in hydropower projects in China, and good economic benefits have been obtained.

High-strength steel has been used for steel penstocks of hydropower stations abroad. In Japan, high-strength steels of HT-60, HT-70, and HT-80 with tensile

Table 7.4 Structural coefficients of different types of penstock

Type of penstock	Type of stress		Internal force type		Structural coefficient γ_d
Exposed penstock	Overall membrane stress		Axial force		1.6
	Local stress	Partial membrane stress	Axial force		1.3
		Partial membrane stress + bending stress	Axial force + bending moment		1.1
Underground penstock	Overall membrane stress		Axial force		1.25
Buried penstock	Overall membrane stress		Axial force	Jointly loaded	1.25
				Individually checked	1.0
	Partial membrane stress + bending stress		Axial force + bending moment		1.0
Steel-lined reinforced concrete penstock	Overall membrane stress		Axial force		1.5

Notes 1. The values in the table are suitable for cases where the weld coefficient $\varphi = 0.95$. If $\varphi \neq 0.95$, γ_d should be multiplied by $0.95/\varphi$
2. For the exposed penstock in the main powerhouse, γ_d should be increased by 10–20%
For the water pressure test, γ_d should be decreased by 10%

strengths of 600–800 N/mm^2 have been widely used to manufacture steel penstocks for hydropower stations. High-strength steel contains alloying elements such as chromium, nickel, and molybdenum, which are costly and difficult to weld. But the thickness of steel penstocks can be reduced, and the material used can be reduced. The cost of transportation, processing, welding, installation, and the like can be reduced, and the total cost can be reduced. Due to the light weight of the pipe joint during installation, it is advantageous to weld the pipe joint into a longer pipe section for installation, which can shorten the installation period, so the use of high-strength steel is economical. A similar grade of high-strength steel A517 has also been used in steel penstocks of hydropower stations in the USA. High-strength steels with the tensile strength of 700 N/mm^2 such as 14 × 2 MP and И3138-2 have been used in steel penstocks of large hydropower stations in the Soviet Union, and technical and economic benefits have been achieved.

7.4 Facilities and Structure of Steel Penstocks

The exposed penstock, the buried penstock, the dam body penstock, and the underground penstock without the water intake valve in front of the turbine should be provided with a quick gate valve and necessary maintenance facilities at the head end of the penstock. For the underground penstock, if the diversion conduit from intake to the front of the penstock is long, or the internal pressure of the penstock is large and the buried depth is not large, an emergency gate valve should be set at the head end of the penstock. Gate valves are located in the pressure wall structure of the head pond or the surge tank, or the special valve chamber downstream of the surge tank. The quick gate valve at the head end of the steel penstock must be equipped with a remote (central control room) and a local operating device, which must be provided with a reliable power source. The aeration vent or aeration valve must be placed directly downstream of the quick gate valve. When arranging the pipeline, the top of the steel penstock should be at least 2.0 m below the minimum pressure line to prevent vacuum from occurring inside the penstock. Drainage facilities should be provided at the lowest point of the penstock to completely drain the water in the penstock and leak the gate valve. Penstocks should be provided with manholes and necessary facilities for inspection and maintenance. The spacing between the manholes should not exceed 200 m generally.

To reduce the head loss and facilitate production, the turning radius of the penstock should not be less than three times of the penstock diameters. To make the steel penstock have certain rigidity during construction, the minimum thickness of shell should not be less than ($D/800 + 4$) mm or 6 mm. The diameter of the same steel penstock may decrease with the increase of internal pressure, but the number of reductions should not be too much. The wall thickness of steel penstock should not change too much, generally no more than 4 mm, to facilitate welding. Longitudinal seams of steel penstock are usually welded in the factory. Welded pipe sections are transported to the site and connected by a ring seam (transverse seam). Longitudinal joints of each pipe section should be staggered to avoid sections with high stress on the vertical axis and horizontal axis of the cross section. The spacing of the circumferential welds should be spaced at least 500 mm apart to prevent the effects of the welds from interacting. After installation, the ellipticity of the steel penstock should not exceed $5D/1000$. Steel penstocks should be provided with effective antirust measures. Common antirust measures include paint protection, metallization, and electrochemical protection. Considering the corrosion, wear, and thickness error of the steel plate, the thickness of the pipe shell should be at least 2 mm larger than the calculated value.

The exposed penstock and the branching pipe should be hydrostatically tested, and the pressure value should be no less than 1.25 times of the maximum internal water pressure under normal working conditions. The purpose of the hydrostatic test includes the following points: ① Structural defects with overload internal pressure are exposed, and the overall safety of the structure is checked; ② under the condition of slow loading, the plastic deformation can occur at the defect tip, which causes the pre-stress after passivation and unloading of the defeat tip; ③ welding residual stress and peak stress of the discontinuous part are reduced by the hydrostatic test. The necessary prototype observations should be made for important steel penstocks to monitor the pipeline operation status, predict abnormal conditions, and accumulate data. For this reason, the corresponding detection equipment for measuring pressure, discharge, and stress of the pipe shell should be buried during construction.

7.5 Design Loads and Load Combinations of Penstocks

7.5.1 Design Loads

According to the provisions of *Specifications for Load Design of Hydraulic Structures* DL 5077-1997, the actions (loads) that should be included in the structural design of various penstocks are listed in Table 7.5. For *Design Specification for Steel Penstocks of Hydroelectric Stations* NB/T 35,056-2015, the design of penstock should be carried out based on the standard value of action (load) after the partial coefficient of action is considered and the design value is calculated.

7.5.2 Load Combinations

For different penstock, the corresponding calculation conditions and load combinations are shown in Table 7.6. The corresponding relationships between the calculation conditions of NB/T 35,056-2015 and SL 281-2003 are also listed in Table 7.6. The conditions in the table are adopted according to the specifications used.

7.6 Exposed Steel Penstocks

7.6.1 Layout of Exposed Steel Penstocks

1. Line selection of exposed steel penstock

   Line selection of exposed steel penstock should be considered in conjunction with the other buildings in the water diversion system of the power station, especially the head pond or surge tank, as well as the layout of the powerhouse of hydropower stations.

   (1) Pipeline lines should be as short and straight as possible to reduce cost, head loss, water hammer pressure and improve operating conditions of units. Therefore, the exposed steel penstocks are usually laid on steep ridges to avoid being placed in valleys where mountain torrents are concentrated.

Table 7.5 Action (load) classification and partial coefficient

Serial number		Action classification and name		Exposed penstock	Underground penstock	Buried penstock	Dam body penstock	Exposed bifurcated pipe	Underground bifurcated pipe	Partial coefficient of action
(1)	(1a)	Internal water pressure	Hydrostatic pressure at normal water level	√			√	√		Hydrostatic pressure $\gamma_Q = 1.0$ Water hammer pressure $\gamma_Q = 1.1$
	(1b)		Maximum operating pressure (hydrostatic pressure + water hammer pressure)	√	√	√	√	√	√	
	(1c)		Maximum pressure under special operating conditions (hydrostatic pressure + water hammer pressure)	√	√	√	√	√	√	Hydrostatic pressure $\gamma_A = 1.0$ Water hammer pressure $\gamma_A = 1.1$
	(1d)		Pressure under hydrostatic test	√				√	√	$\gamma_Q = 1.0$
(2)		Dead weight of pipe		√			√			$\gamma_G = 1.05$ ($\gamma_G = 0.95$)
(3)		Full water weight in pipe		√						$\gamma_Q = 1.0$

(continued)

Table 7.5 (continued)

Serial number	Action classification and name	Exposed penstock	Underground penstock	Buried penstock	Dam body penstock	Exposed bifurcated pipe	Underground bifurcated pipe	Partial coefficient of action
(4)	Temperature action	√			√	√		$\gamma_Q = 1.0$
(5)	The change in pipe diameter, the turning point and the internal water pressure acting on the plug, valve, and expansion joint (hydrostatic pressure + water hammer pressure)	√		√	√	√		Hydrostatic pressure $\gamma_Q = 1.0$ Water hammer pressure $\gamma_Q = 1.1$
(6)	Curved centrifugal force	√			√			$\gamma_Q = 1.1$
(7)	Force resulting from differential settlement of anchor blocks and support blocks	√						$\gamma_Q = 1.1$
(8)	Wind load	√				√		$\gamma_Q = 1.3$
(9)	Snow load	√				√		$\gamma_Q = 1.3$
(10)	Grouting pressure		√	√	√		√	$\gamma_Q = 1.3$
(11)	Earthquake load	√			√	√		$\gamma_Q = 1.0$

(continued)

Table 7.5 (continued)

Serial number		Action classification and name		Exposed penstock	Underground penstock	Buried penstock	Dam body penstock	Exposed bifurcated pipe	Underground bifurcated pipe	Partial coefficient of action
(12)		Air pressure difference caused by aeration device when the penstock is emptied		√	√	√	√	√	√	$\gamma_Q = 1.0$
(13)	(13a)	External water pressure	Groundwater pressure		√				√	$\gamma_Q = 1.0$
	(13b)		Dam seepage pressure			√	√			$\gamma_Q = 1.0$
(14)		Dam displacement				√	√			$\gamma_Q = 1.0$

Notes 1. The bracketed value of the partial coefficient of action in serial number (2) is used when the self-heavy effect is beneficial to the structure

2. The air pressure difference caused by the aeration device when the penstock is emptied in serial number (12) should not be less than 0.05 N/mm^2 and should not be greater than 0.1 N/mm^2

3. γ_G, γ_Q, and γ_A represent the partial coefficients of permanent action, variable action, and accidental action, respectively

Table 7.6 Calculation condition and load combinations of various penstock types

Types of penstock	Design conditions	Action effect combination of NB/T 35,056-2015		Calculation contents	Combination corresponding to SL 281-2003
Exposed penstock	Persistent condition	Basic combination	(1b) + (2) + (3) + (4) + (5) + (7)	Normal operating condition (1)	Basic load combination
			(1a) + (2) + (3) + (4) + (5) + (7) + (8) or (9)	Normal operating condition (2)	Basic load combination
	Transient condition		(12)	Venting condition	Basic load combination
			(1d) + (2) + (3) + (5)	Hydrostatic test condition	Special load combination
	Accidental condition	Casual combination	(1c) + (2) + (3) + (4) + (5) + (7)	Special operating condition	Special load combination
			(1a) + (2) + (3) + (4) + (5) + (7) + (11)	Earthquake condition	Special load combination
Underground penstock	Persistent condition	Basic combination	(1b)	Normal operating condition	Basic load combination
			(12) + (13a)	Venting condition	Basic load combination
	Transient condition		(10)	Construction condition	Special load combination

(continued)

Table 7.6 (continued)

Types of penstock	Design conditions	Action effect combination of NB/T 35,056-2015		Calculation contents	Combination corresponding to SL 281-2003
	Accidental condition	Casual combination	(1c)	Special operating condition	Special load combination
Dam body penstock	Persistent condition	Basic combination	(1b)	Normal operating condition	Basic load combination
			(1b)	Exposed penstock bearing calibration alone condition	Basic load combination
	Transient condition		(12) + (13b)	Venting condition	Special load combination
			(10)	Construction condition	Special load combination
	Accidental condition	Casual combination	(1c)	Special operating condition	Special load combination

(continued)

Table 7.6 (continued)

Types of penstock	Design conditions	Action effect combination of NB/T 35,056-2015		Calculation contents	Combination corresponding to SL 281-2003
Penstock laid on the downstream surface of the dam	Persistent condition	Basic combination	(1b)	Normal operating condition (1); calculate the pipe shell thickness and the number of circumferential reinforcement	Basic load combination
			(1b) + (2) + (4)	Normal operating condition (2); calculate the circumferential stress of various penstocks	Basic load combination
			(1b) + (2) + (3) + (5) + (6) + (14)	Normal operating condition (3); finite element method to calculate the upper bend section and the joint surface stress of penstock and dam	Basic load combination
	Transient condition		(12) + (13b)	Venting condition	Basic load combination
			(10)	Construction condition	Special load combination
	Accidental condition	Casual combination	(1c) + (2) + (3) + (5) + (6) + (14)	Special operating condition	Special load combination
			(1a) + (2) + (3) + (5) + (6) + (11) + (14)	Earthquake condition	Special load combination

(2) Good geological conditions should be selected to support the steel penstock on a solid foundation. In particular, areas that may cause landslides and collapses should be avoided. When dangers of flash floods, falling stones, and avalanches cannot be avoided in individual areas, practical protective measures should be taken.
(3) Ups and downs of pipeline lines should be minimized. Pipeline lines should not be crossed with traffic roads or other pipelines. When it is unavoidable, special bridges and culverts should be provided to isolate each other. Additional security measures should be taken when crossing the road.

2. Relationship between penstock and powerhouse

The relative position of the pipeline and powerhouse is mainly determined by the layout of each building in the whole plant. There are three layouts of exposed steel penstocks introduced into powerhouse:

(1) Forward entry. The axis of the steel penstock is perpendicular to the longitudinal axis of the powerhouse, as shown in Fig. 7.3a, b. The advantages are smooth flow, small head loss, parallel longitudinal axis of powerhouse with rivers, small excavation volume, and convenient transportation to the powerhouse. The disadvantage is that when the penstock is broken, the high-pressure water flow will be harmful to the powerhouse and personnel. This method is widely used in underground penstocks, dam body penstock, and ground—exposed steel penstock of middle and low head hydropower stations. Protective measures such as water retaining walls and drainage ditches should be considered when ground—exposed steel penstocks are used on the ground of high water head hydropower stations.
(2) Diagonal entry. The main pipe axis is parallel to the longitudinal axis of the power station, as shown in Fig. 7.3c, d. The advantage is to reduce the threat to the powerhouse and personnel when the steel penstock is broken; the disadvantage is that the water head loss is increased, and the longitudinal axis of the powerhouse will be arranged perpendicular to the contour because

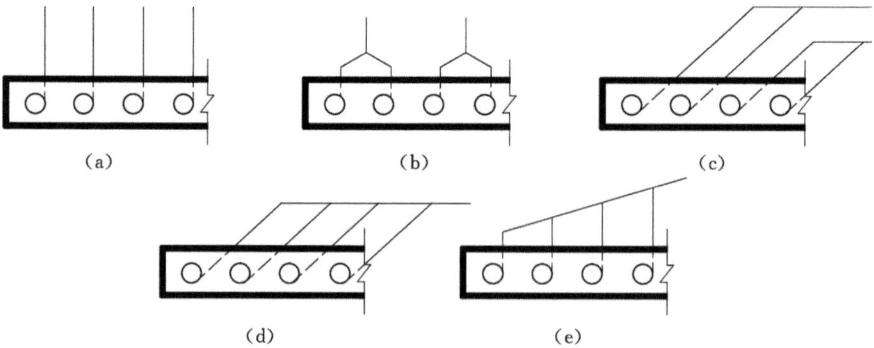

Fig. 7.3 Relative location of penstock and powerhouse

of the need for vertical contour when arranging penstocks, which will lead to the increase of excavation volume of powerhouse. This method is rarely used and is suitable for hydropower stations with high or middle water heads.

(3) Longitudinal entry. Between the forward entry and the diagonal entry, the main pipe axis intersects the longitudinal axis of the powerhouse at an angle, as shown in Fig. 7.3e. This type of layout is used when the local shape, geological conditions, water diversion system, and powerhouse layout requirements are appropriate, and it is mostly used for hydropower stations with grouped and centralized water supply.

3. Elevation of penstock

It should be ensured that there is no negative pressure in the penstock under operation, which is more important for the exposed steel penstock. It is required that the top of the penstock along the line is below 2 m of the lowest pressure slope line, and the net distance between the bottom of the steel penstock and the ground should not be less than 0.6 m.

7.6.2 Laying and Supporting Method of Exposed Steel Penstocks

Sectional laying of exposed steel penstocks is generally preferable. An anchor block is arranged at the turning point of the steel penstock. The pipe section between anchor blocks is supported by the support block, and the expansion joint is equipped between the two anchor blocks, as shown in Fig. 7.1. The anchor block fixes one end of the penstock section on the foundation to prevent displacement. The support block only supports the penstock body, and the body can be moved on the support of the support block. The function of the expansion joint is that when the temperature changes, pipes can deform freely in the pipe axis direction so that thermal stresses and forces acting on the anchor block will be reduced significantly. To reduce the water pressure in the expansion joint and facilitate the installation of steel penstock, the expansion joint is usually set at the downstream side of the upper anchor block and close to the anchor block. If the straight pipe section is too long (greater than 150 m), an anchor block may be added between them. If the longitudinal slope of the penstock is relatively slow, the anchor block may not be added, and the expansion joint is placed in the middle of the pipe section to reduce the axial force of the steel penstock caused by the friction between the penstock body and the support block. The distance between the support blocks should be determined by technical and economic comparison through stress analysis of steel penstock, considering installation conditions, support forms, foundation conditions, and other factors. When the distance between the support blocks is small, the cost will increase but the stress of steel penstock will decrease, and vice versa. The support blocks should be equidistant between two adjacent anchor blocks. There is a span of expansion joints, the pipe sections are discontinuous, and

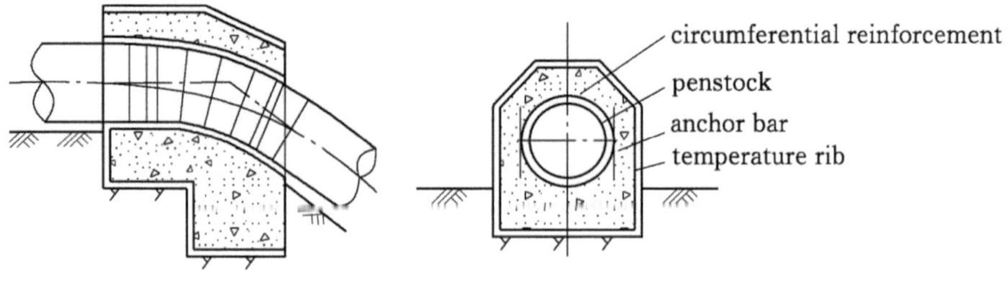

Fig. 7.4 Enclosed anchor block

the spacing of the support blocks should be shortened to reduce the bending moment. The forms of anchor block and support block are described below.

1. Anchor block

   The movements of the penstocks are restricted by the weight of anchor blocks which are usually cast by concrete. According to the fixed modes of the steel penstock, the anchor blocks are classified as follows:

   (1) Enclosed anchor block. As shown in Fig. 7.4, the whole bending section of the pipe is embedded in the concrete, which is beneficial to the fixation of the steel penstock. Temperature ribs should be arranged on the surface of the anchor block. Circumferential reinforcement and a certain number of anchor bars should be arranged around the steel penstock. For the anchor block of the high water head hydropower station, sufficient circumferential reinforcement should be arranged to limit the concrete cracks in the anchor block. This kind of anchor block has a simple structure and is steel-saving, and it is widely used in hydropower stations.
   (2) Open anchor block. As shown in Fig. 7.5, the bending section of the pipe is anchored to the concrete block by anchor rings. The stress on the pipe shell at the anchor block is uneven and the construction of the anchor ring is complicated. The advantage is that it is convenient for the maintenance of penstocks. It is rarely used in China.

2. Support block

   According to the relative displacements of supports and penstocks, support blocks can be classified as follows:

   (1) Sliding support block. When the steel penstock expands and contracts due to changes in force and temperature, it slides along the top surface of the support. The sliding support block can be divided into three types: saddle support block, saddle ring girder support block, and sliding ring girder support block.

       Saddle support block, as shown in Fig. 7.6a, the penstocks are supported on a saddle concrete support directly with a wrapping angle between 90 and 120°. To reduce the friction between the pipe shell and the support, the steel

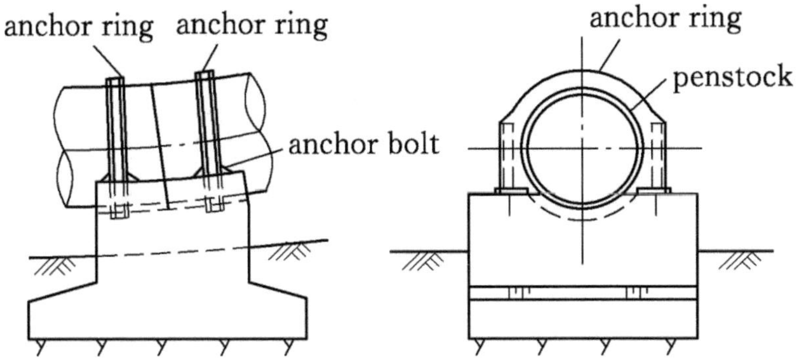

Fig. 7.5 Open anchor block

plate is laid on the support and a lubricant such as graphite is applied to the contact surface; sometimes, it is also lubricated with pressure oil. The support structure is simple, but the support section has a non-uniform force and a large friction force. Support block of this type is suitable for steel penstocks with a diameter below 1 m.

Saddle ring girder support block, as shown in Fig. 7.6b, the penstocks are supported on the concrete block through ring girders, so that the uneven loading status will be improved. Support block of this type is suitable for steel penstocks with a diameter below 2 m.

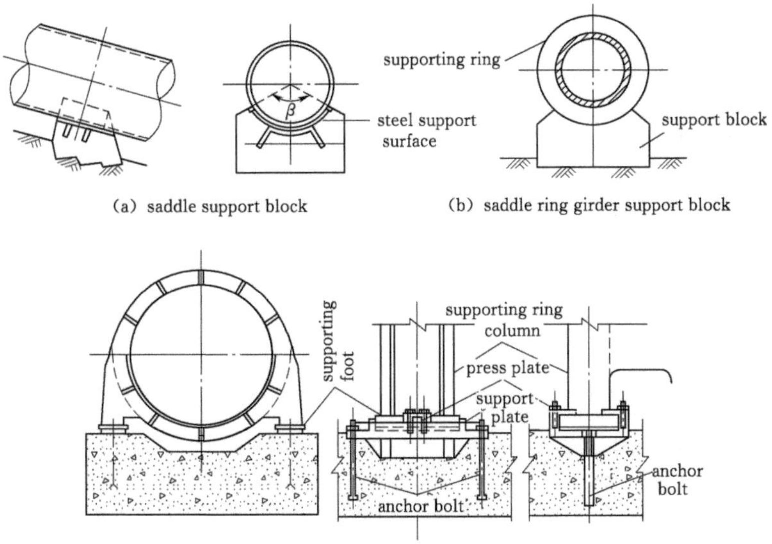

(a) saddle support block (b) saddle ring girder support block

(c) sliding ring girder support block

Fig. 7.6 Sliding support block

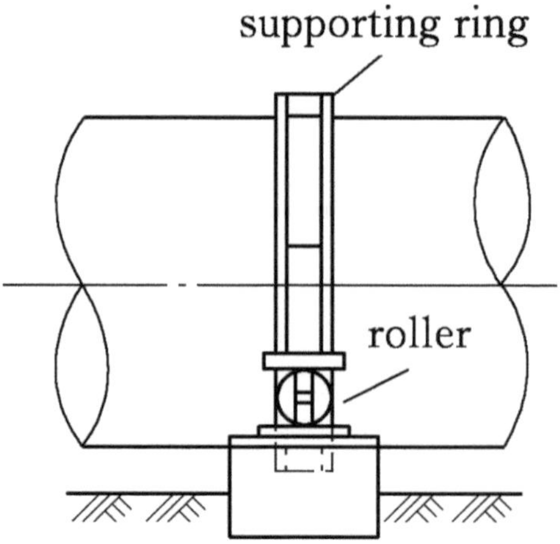

Fig. 7.7 Rolling ring girder support block

Sliding ring girder support block, as shown in Fig. 7.6c, which has less friction than the saddle support. Support block of this type is suitable for steel penstocks with a diameter of 1–3 m. In recent years, polymer composite materials have been used between the upper and lower support plates of the sliding support block, and the friction coefficient has been reduced to 0.05–0.10, which has been adapted to steel penstocks with larger diameters.

(2) Rolling ring girder support block. As shown in Fig. 7.7, a type of ring girder supports with rollers mounted on the base of legs, and the roller rolls when the steel penstock is displaced, and the friction coefficient is small. Support block of this type is commonly used for steel penstocks with small loads in the direction perpendicular to the pipe axis and a pipe diameter greater than 2 m.

(3) Rocking ring girder support block. As shown in Fig. 7.8, a type of ring girder supports with a short block that hinges to the base plate and contacts to the supporting plate of ring girder through a camber plate. When the steel penstock expands and contracts in the axial direction, the short block swings back and forth with the hinge as the center. Support block of this type has small friction and is used for steel penstocks with a pipe diameter of more than 2 m.

If the foundation is not rigid enough or the exposed steel penstock needs to be coated with concrete, the exposed steel penstock can also be laid on the continuous saddle concrete cushion to reduce the pressure acting on the foundation and the bending moment of the penstock body. In recent years, there have been some extra-large exposed steel penstocks at home and abroad. Due to the large diameter, it has been difficult to support the block by the ring girder. Instead, a saddle-shaped continuous pedestal is used. These penstocks are often outsourced with thin layers of reinforced concrete protective layers that are not subjected to internal water pressure,

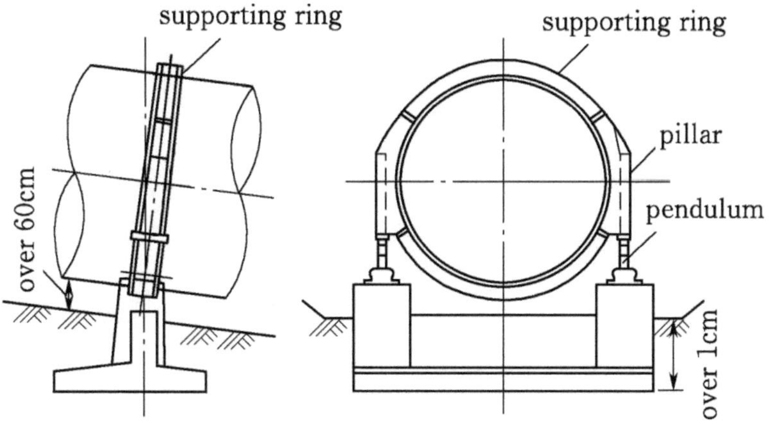

Fig. 7.8 Rocking ring girder support block

such as steel penstocks with a diameter of 12.2 m at the No. 3 Coulee Plant in the United States and steel penstocks with a diameter of 8.0 m at the Geheyan Hydropower Station in Qingjiang, China.

7.6.3 Forces and Loads Acting on Steel Penstocks and Supports

Loads acting on the exposed steel penstock are shown in Table 7.5. Calculation conditions and load combinations are shown in Table 7.6. The main forces acting on the penstock body, anchor block and support block and their calculation equations are shown in Table 7.7. Wind, snow load, earth pressure, seismic force, etc., are not included, and they can be determined by referring to the relevant provisions of *Specifications for Load Design of Hydraulic Structures* DL 5077-1997.

In the stress analysis of steel penstocks, since the actual situation is complicated, different situations may occur, and the respective forces do not occur simultaneously under any circumstances. With different conditions such as construction, maintenance or operation, temperature increases or decreases, the load can be differently combined, and the combination of forces should be determined according to the specific condition as the most unfavorable load combination.

Table 7.7 Force acting on steel penstocks and supports

Serial number	Force direction	Force name	Calculation equation	Force symbol				Forced part			Force diagram	Remarks
				Upper row		Lower part		Pipe shell	Support block	Anchor block		
				Temperature								
				Rise	Drop	Rise	Drop					
1	The axis direction of penstock	Component force of penstock gravity	$A_1 = \sum (q_5 L)$, $\sin \alpha$	+	+	+	+	√		√	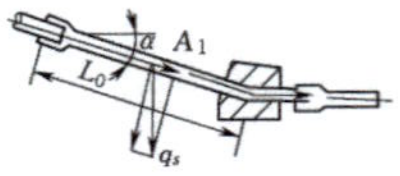	q_s—dead weight of penstock per meter
2		Force acting on the valves	$A_2 = \frac{\pi D_0^2}{4} P$	±	±	±	±	√		√	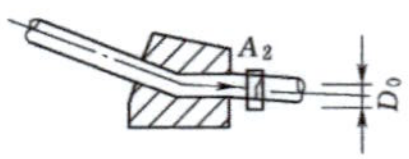	D_0—inner diameter of steel penstock P—internal water pressure; this force does not exist when valve is fully opened
3		Component force of internal water pressure of elbow pipe	$A_3 = \frac{\pi D_0^2}{4} P$	+	+	−	−	√		√	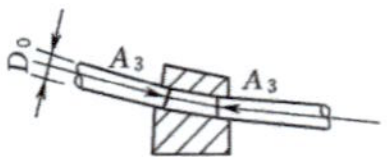	

(continued)

Table 7.7 (continued)

Serial number	Force direction	Force name	Calculation equation	Force symbol				Forced part			Force diagram	Remarks
				Upper row		Lower part		Pipe shell	Support block	Anchor block		
				Temperature								
				Rise	Drop	Rise	Drop					
4		Internal water pressure of reducing pipe	$A_4 = \frac{\pi}{4}(D_{01}^2 - D_{02}^2)P$	+	+	+	+	√		√	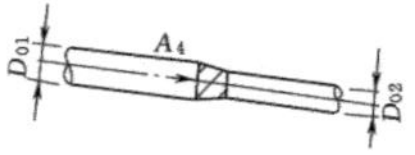	D_{01}, D_{02}—maximum and minimum inner diameter of the reducing pipe
5		Internal water pressure at the end of expansion joint	$A_5 = \frac{\pi}{4}(D_1^2 - D_2^2)P$	+	+	−	−	√		√	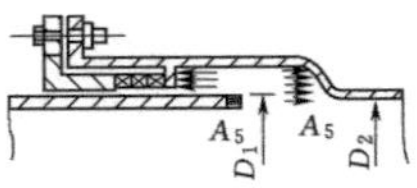	D_1, D_2—outer and inner diameter of casing in telescopic joint
6		Friction force of packing material for expansion joint during expansion variation	$A_6 = \pi D_1 b_1 \times \mu_1 P$	+	−	−	+	√		√	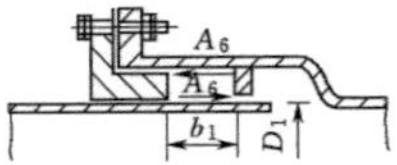	μ_1—friction coefficient between packing material of expansion joint and steel penstock; b_1—packing material along the length of the penstock axis

(continued)

Table 7.7 (continued)

Serial number	Force direction	Force name	Calculation equation	Force symbol				Forced part			Force diagram	Remarks
				Upper row		Lower part		Pipe shell	Support block	Anchor block		
				Temperature								
				Rise	Drop	Rise	Drop					
7		Friction force between the penstock and the base plate of support during temperature variation	$A_1 = \sum(qL) \times f\cos\alpha$	+	−	−	+	√	√	√	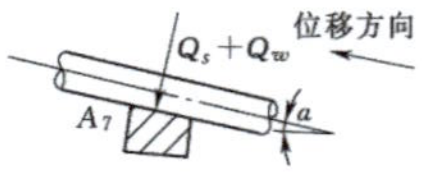	f—friction coefficient of the support to the pipe shell; L—supporting ring spacing
8		Component force of flow centrifugal force of elbow pipe	$A_8 = \frac{\pi D_0^2}{4}\frac{v_0^2}{g} \times \gamma_w$	+	+	−	−	√		√	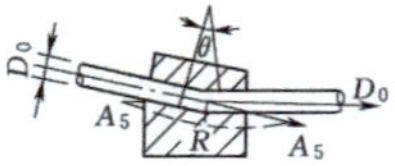	u_0—average flow rate in the penstock; R—centrifugal force; A_s—the component force of centrifugal force in the direction of penstock axis
9	Vertical penstock axis direction	Component weight of penstock	$Q_s = q_s L\cos\alpha$					√	√	√	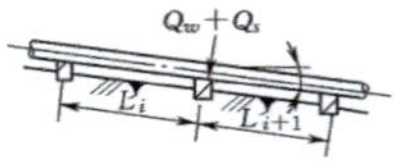	q_s—dead weight of penstock per meter
10		Component weight of water in penstock	$Q_w = q_s L\cos\alpha$					√	√	√		q_w—weight of water per meter in penstock

(continued)

Table 7.7 (continued)

Serial number	Force direction	Force name	Calculation equation	Force symbol				Forced part			Force diagram	Remarks
				Upper row		Lower part		Pipe shell	Support block	Anchor block		
				Temperature								
				Rise	Drop	Rise	Drop					
11	Radial direction	Internal water pressure	$P = H\, r_w$					√		√	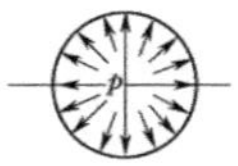	*H*—water head is counted in the center of the calculated section pipe

Notes 1. The axial force of the pipe symbol: "+" is the downward direction of the steel penstock; "–" is the upward direction of the steel penstock
2. The upper section refers to the steel penstock above the anchor block, and the lower section refers to the steel penstock below the anchor block

7.6.4 Structural Analysis of Exposed Steel Penstocks

7.6.4.1 Thickness Estimation of Pipe Shell

The wall thickness of the steel pipe shell must be set first before stress analysis of steel penstocks. The wall thickness can be estimated based on the condition that the steel penstock is subjected to the water pressure within the design. To this end, the steel penstock filled with internal water pressure per unit length can be taken out, cut along the horizontal diameter position, and the upper half is removed, and the internal water pressure is balanced with the circumferential tensile stress of the pipe shell, as shown in Fig. 7.9.

According to the force balance condition $\sum Y = 0$,

$$2T = \int_0^\pi P \sin\theta \times 1 \times \mathrm{d}s = \int_0^\pi P \sin\theta r \mathrm{d}\theta = \int_0^\pi \frac{PD}{2} \sin\theta \mathrm{d}\theta = PD.$$

Therefore,

$$T = \frac{1}{2} PD.$$

Pipe shell circumferential tensile stress is as follows:

$$\sigma_\theta = \frac{T}{t \times 1} = \frac{PD}{2t} = \frac{Pr}{t}$$

In which:

P water pressure in the center of the steel penstock, N/mm^2;

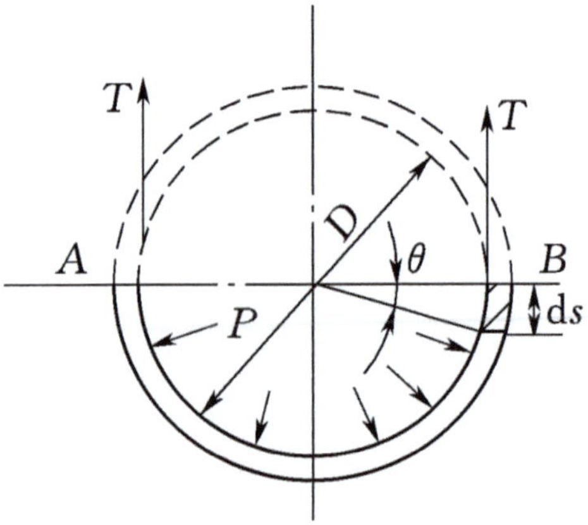

Fig. 7.9 Circumferential tensile stress of round pipe

r steel penstock radius, mm;
t pipe shell thickness, mm.

According to the condition, the steel pipe shell stress should be less than the allowable stress of the material:

$$\sigma_\theta = \frac{T}{t \times 1} < [\sigma]. \quad (7.5)$$

Equation (7.5) is called the boiler equation. Considering the strength reduction of the weld, the allowable stress should be multiplied by a weld coefficient less than 1.0. Based on the welding technique and inspection method, it can be taken from 0.9 to 0.95. Since only the internal water pressure is considered, the allowable stress is generally reduced by 5–15% when calculated according to the boiler equation. When the penstock diameter is small and the internal pressure is high, the smaller percentage is selected. When the penstock diameter is large and the internal pressure is low, the larger percentage is selected. Therefore, the wall thickness can be calculated as follows:

$$t \geq \frac{Pr}{(0.85 \text{ to } 0.95)\phi[\sigma]}. \quad (7.6)$$

The estimated thickness of the pipe shell should generally increase by 2 mm, plus the corrosive thickness, and the thickness also needs to satisfy the requirement of minimum thickness. In addition, the manufacturing process and installation, transportation, and other requirements must be considered to ensure the necessary rigidity. The minimum thickness of the pipe shell should not be less than (D/800 + 4) mm, less than 6 mm.

7.6.4.2 Stress Analysis of Steel Penstocks

The exposed penstock is laid on a series of support blocks. To make the pipe shell evenly stressed, the pipe shell at the support often adds a ring girder. Stiffener rings are sometimes required between ring girders to maintain the external pressure stability of steel penstock. Four basic calculation parts shall be selected for the calculation of pipe shell stress: mid-span section 1-1; near supporting ring section pipe shell 2-2; stiffener ring and its adjacent section pipe shell 3-3; supporting ring and its adjacent section pipe shell 4-4 as shown in Fig. 7.10. When analyzing the stress of the pipe shell, the calculated coordinates are often taken from the cylindrical coordinate system; that is, the axis direction of the pipe is the x-axis, the radial direction is the r-axis, and the tangential direction of the pipe shell is the θ-axis, as shown in Fig. 7.11.

1. Pipe shell stress of mid-span section 1-1

   Characteristics of the mid-span section are: bending moment is the maximum, and shear force is equal to zero.

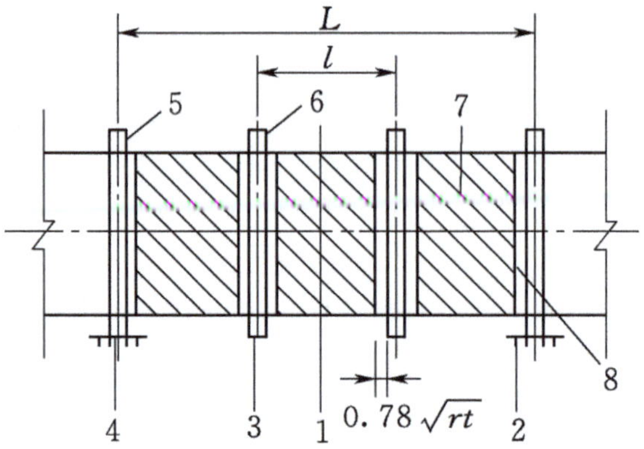

Fig. 7.10 Stress analysis calculation position of exposed penstock: 1-Mid-span section pipe shell; 2-Near supporting ring section pipe shell; 3-Stiffener ring and its adjacent section pipe shell; 4-Supporting ring and its adjacent section pipe shell; 5-Supporting ring; 6-Stiffener ring; 7-Integral membrane stress zone; 8-Local membrane stress zone

Fig. 7.11 Coordinate system of pipe shell stress analysis

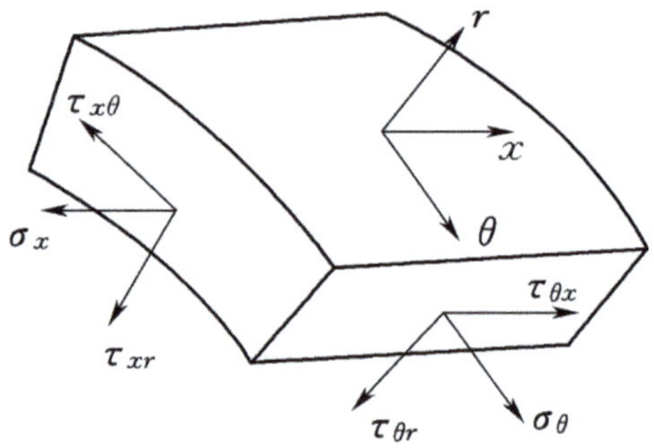

(1) Radial stress σ_r. The inner surface of the water pipe is subjected to internal water pressure and acts on the radial stress σ_r, which is equal to the internal water pressure at that point, that is, $\sigma_r = -\gamma H$. The sign "−" indicates compressive stress, and the outer surface of the pipe shell is $\sigma_r = 0$. This stress is small and can be ignored in general calculations.

(2) Circumferential stress $\sigma_{\theta 1}$. The circumferential stress of the pipe shell is mainly caused by the internal water pressure. Since the penstock is often arranged obliquely along the terrain, the axis of the pipe has an inclination angle, and the water head at the center of the penstock is H, the water head at the top of the penstock is $H - r\cos\alpha$, and the water head at the bottom of the penstock is $H + r\cos\alpha$. Set the angle of the pipe shell to any point and the top radius of the penstock to be θ, then the water head at any point on the pipe shell is $H - r\cos\alpha\cos\theta$, as shown in Fig. 7.12. We take a tiny pipe shell with a length of

$rd\theta$ along the circumference and a unit length along the axis. The circumferential tensile stress in the pipe shell per unit length is T. It can be seen in Fig. 7.13:

$$\sum y = 0$$

$$2T \sin \frac{d\theta}{2} = \Pr d\theta$$

$$P = \gamma(H - r\cos\alpha\cos\theta).$$

Because $d\theta$ is very small, $\sin\frac{d\theta}{2} \approx \frac{d\theta}{2}$, and it can be obtained as follows:

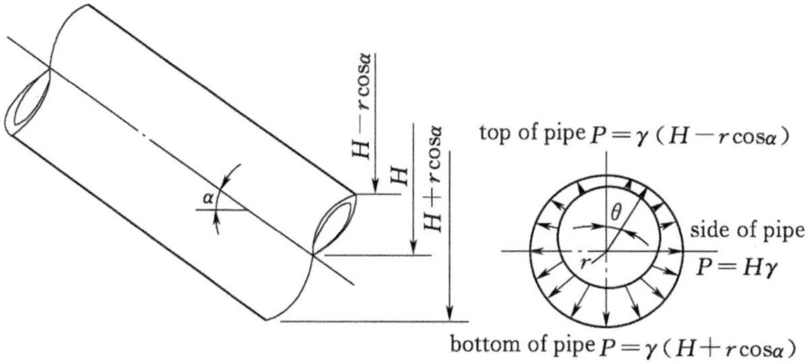

Fig. 7.12 Internal water pressure of pipe shell

Fig. 7.13 Circumferential stress of pipe shell

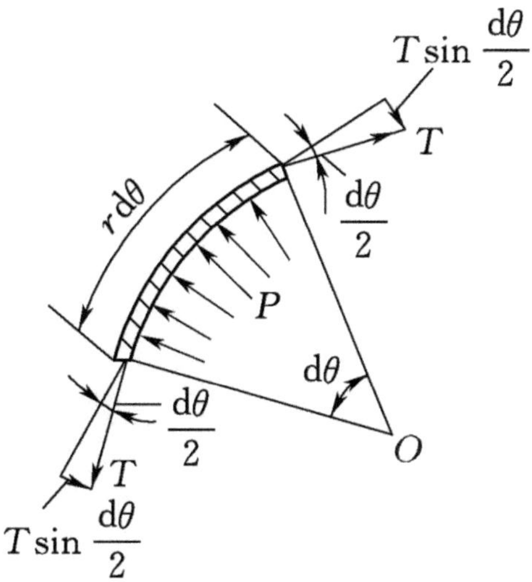

$$T = \gamma r(H - r\cos\alpha\cos\theta)$$

$$\sigma_{\theta 1} = \frac{T}{t \times 1} = \frac{\gamma r}{t}(H - r\cos\alpha\cos\theta) = \frac{Pr}{t}\left(1 - \frac{r}{H}\cos\alpha\cos\theta\right) \tag{7.7}$$

In which:

H calculation head at the center of steel penstock, mm;
α penstock axis inclination, (°);
θ the angle between the arbitrary point of the ring and the radius of the top of the penstock, (°);
γ volume weight of water, N/mm³.

Other symbols are the same as before.

(3) Axial stress σ_x. The axial stress σ_x of the mid-span section consists of two parts:

a. Axial stress from axial force σ_{x1}

$$\sigma_{x1} = \frac{\sum A}{F} = \frac{\sum A}{2\pi rt} \tag{7.8}$$

In which:

$\sum A$ the sum of all axial forces acting on the steel pipe (as shown in Table 7.5), N;
F cross-sectional area of steel pipe, mm².

b. The axial stress σ_{x2} from the normal component of pipe and water weight. The steel penstock is regarded as a continuous beam supported on the anchor blocks and a series of support blocks. The lower end of the anchor block is used as the fixed end, the upper end of the expansion joint is used as the free end, and the normal force is equivalent to the uniform load acting on the continuous beam, as shown in Fig. 7.14. The stress at each point in the direction of the mid-span section is:

$$\sigma_{x2} = -\frac{M}{W}\cos\theta = -\frac{M}{\pi r^2 t}\cos\theta \tag{7.9}$$

In which:

M bending moment calculated as multi-span continuous girder under pipe weight and water weight, N·mm, the positive and negative signs are selected according to Fig. 7.14;
W section modulus of a hollow beam, $W = \pi r^2 t$, mm³.

If the seismic force is also included, the axial stress σ_{x2} is:

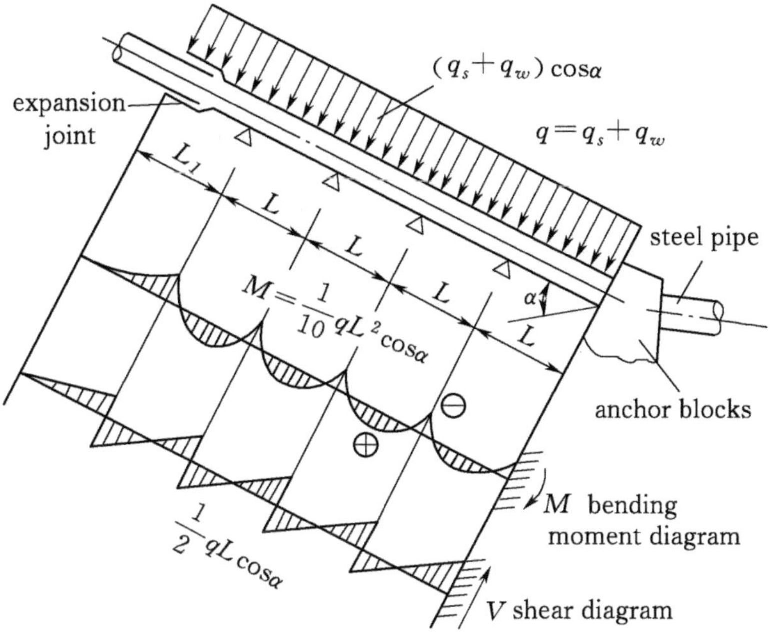

Fig. 7.14 Bending moment and shear force from normal force

$$\sigma_{x2} = \frac{1}{\pi r^2 t}(-M\cos\theta + M_e\sin\theta) \tag{7.10}$$

In which:

M_e continuous beam bending moment under seismic force, $M_e \approx \frac{n_e M}{\cos\alpha}$, N·mm;
n_e seismic coefficient, when the design intensity is 7, 8, 9°, n_e is 0.05, 0.1, 0.2, respectively.

(4) Shear stress. Because there is no shear stress in the span, so $\tau_{x\theta} = 0$

2. Near supporting ring, pipe shell section 2-2

Section 2-2 is close to the supporting ring, but not affected by the supporting ring; there is a shear force from the normal component of the pipe weight and water weight. The shear stress is on the plane of the vertical *x*-axis, and the direction is the tangential direction of the pipe shell, that is, the circumferential direction, so it is recorded as $\tau_{x\theta}$, according to the equation of the "*Material Mechanics*" tutorial:

$$\tau_{x\theta} = \frac{VS_R}{bJ}$$

Fig. 7.15 Calculation of area moment

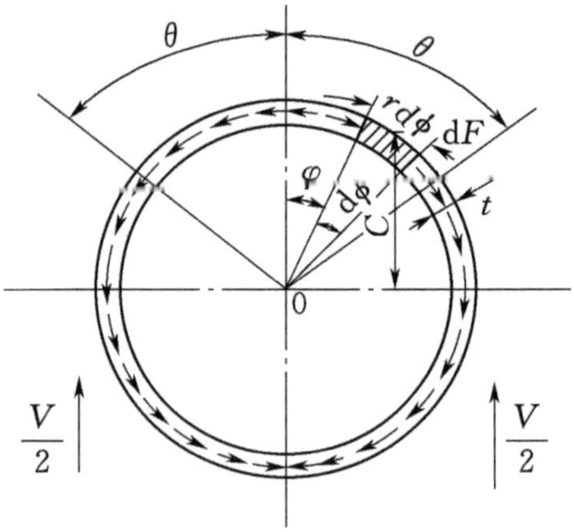

In which:

- V the continuous beam shear force under the action of the pipe weight and the water weight method, N;
- S_R area moment of annular cross-sectional area of pipe shell above computation point to the center shaft, mm^3;
- b shear section width, $b = 2t$, mm;
- J section moment of inertia, $J = \pi D^3 t/8$, mm^4.

The calculation of S_R is as follows: as shown in Fig. 7.15, a small section of the pipe shell $\mathrm{d}F = r\mathrm{d}\phi \cdot t$ is taken, and the area moment of the area to the neutral axis is:

$$\mathrm{d}S_R = \mathrm{d}F \cdot C = r\mathrm{d}\phi \cdot t \cdot r\sin(90^\circ - \phi) = r^2 t\cos\phi\mathrm{d}\phi.$$

Calculate the area moment of pipe shell area above the calculation point to the neutral axis:

$$\mathrm{S}_R = \int_{-\theta}^{\theta} \mathrm{d}S_R = 2r^2 t\sin\theta.$$

Therefore,

$$\tau_{x\theta} = \frac{V \cdot 2r^2 t \cdot \sin\theta}{2t\pi r^3 t} = \frac{V\sin\theta}{\pi rt} \tag{7.11}$$

When $\theta = 0^\circ$ (top of pipe) and $\theta = 180^\circ$ (bottom of pipe), $\tau_{x\theta} = 0$. When $\theta = 90^\circ$ (side of pipe), $\tau_{x\theta} = \frac{V}{\pi rt}$ reaches the maximum value.

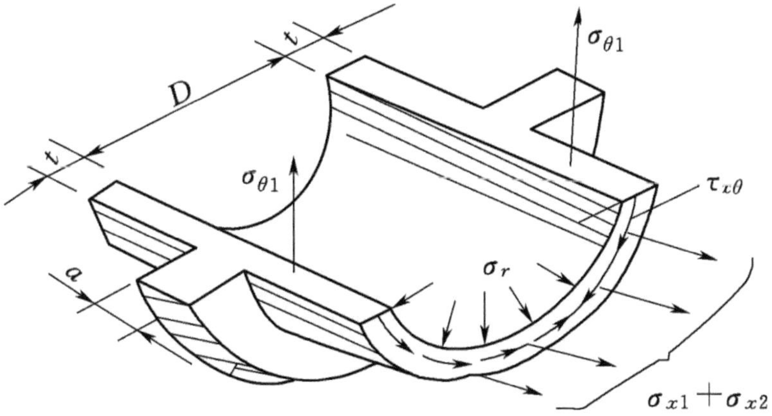

Fig. 7.16 Schematic diagram of stress distribution and direction of pipe shell

If the seismic force is also included, the shear stress is:

$$\tau_{x\theta} = \frac{1}{\pi rt} = (V\sin\theta - V_e\cos\theta) \tag{7.12}$$

In which:

V_e continuous beam shear force under the action of seismic force; $V_e \approx \frac{n_e V}{\cos\alpha}$;

other symbols (seismic coefficient n_e, continuous beam shear force V) have the same meaning as they have in previous equations.

The distribution of $\tau_{x\theta}$ is shown in Fig. 7.16, which is a comprehensive map of the above stresses.

Other normal stresses σ_r, σ_θ, and σ_x in section 2-2 are the same as section 1-1.

3. Stiffener ring and its adjacent pipe shell section 3-3

(1) Axial stress σ_{x3}. Due to the presence of the stiffener ring, the radial deformation of the pipe shell under the action of internal water pressure and pipe weight is limited, resulting in local stress, as shown in Fig. 7.17a. The influence range of the stiffener ring on the pipe shell constraint is l' on each side, which is also called the equivalent flange width. It can be obtained from the analysis of elastic theory:

$$l' = \frac{\sqrt{rt}}{\sqrt[4]{3(1-\mu^2)}} = 0.78\sqrt{\gamma t} \tag{7.13}$$

In which:

μ Poisson's ratio of steel.

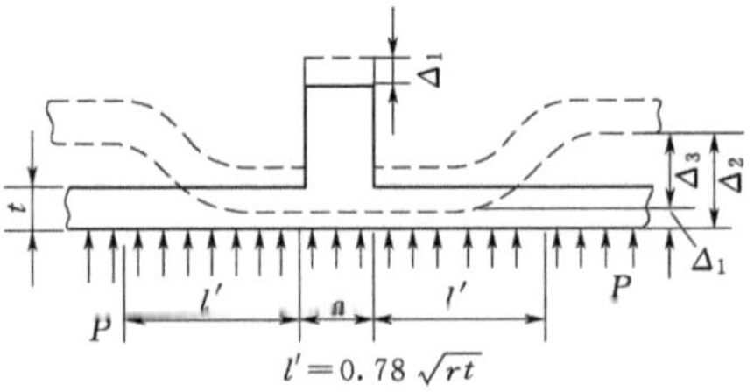

(a) local deformation of pipe shell

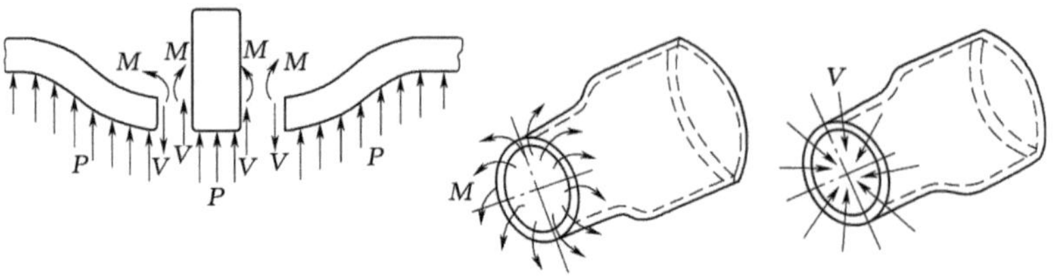

(b) radial bending moments and shear forces uniformly distributed at the incision

Fig. 7.17 Schematic diagram of the stiffener ring and its side pipe shell deformation

For the pipe shell outside the range of l', it is not affected by the stiffener ring, so there is no local stress. In the calculation, the stiffener ring effective section F is equal to its net section F' plus the shell area of $0.78\sqrt{rt}$ on both sides.

Only the effect of internal water pressure is considered in the analysis. Under the action of internal water pressure, the bending moment and shear force on the circumference of the pipe shell are equal due to the axisymmetric. It is envisaged that the stiffener ring is cut open from the pipe shell. According to the deformation compatibility conditions, there are uniform radial bending moment M' and shearing force V' at the slit, as shown in Fig. 7.17b.

Under the action of the internal water pressure P and the shear force V' from the pipe shell, the stiffener ring is radially displaced to Δ_1, and the pipe shell outside the influence range of the stiffener ring is radially displaced to Δ_2. If there is no effect of M' and V', all pipe shells will have the same displacement Δ_2. But under the action of M' and V', the displacement of the joint between the steel penstock and the stiffener ring should be the same as the displacement of the stiffener ring, equal to Δ_1. It can be seen that under the action of M' and V', the steel penstock has a displacement at section 3-3 equal to Δ_3. According to the continuous condition $\Delta_3 = \Delta_2 - \Delta_1$, the pipe shell has no angular displacement (rotation angle) under the action of M' and V'.

a. Calculate Δ_2. The displacement of the pipe shell Δ_2 outside the range of the stiffener ring is caused by uniform internal water pressure. Δ_2 is the increase of the radius. Its relative value is Δ_2/r, that is, the circumferential strain value of the pipe shell is ε. The equation $\frac{\Delta_2}{r} = \varepsilon = \frac{\sigma_\theta}{E}$ can be obtained by considering *Hooke's Law*.

Substitute the equation $\sigma_\theta = \frac{P_r}{t}$ of Eq. (7.5) into the above equation:

$$\Delta_2 = \frac{Pr^2}{\text{tE}} \tag{7.14}$$

In which:

E steel modulus of elasticity, N/mm^2.

b. Calculate Δ_1

$$\Delta_1 = \left(Pa + 2V'\right)\frac{r^2}{F'E} \tag{7.15}$$

In which:

a width of stiffener ring, mm;
F' net cross-sectional area of stiffener ring, mm^2, and the flange of pipe shell is not included.

c. Calculate Δ_3. According to the theory of elasticity, the relationship between M' and V' is as follows:

$$M' = \frac{V'}{2k}. \tag{7.16}$$

The radial displacement of the pipe shell is:

$$\Delta_3 = \frac{3V'\left(1 - u^2\right)}{k^3Et^3} \tag{7.17}$$

In which:

k the reciprocal of the equivalent flange width of the pipe shell, mm^{-1}, that is:

$$k = \frac{1}{l'} = \frac{\sqrt[4]{3\left(1 - \mu^2\right)}}{\sqrt{rt}} = \frac{1}{0.78\sqrt{rt}}.$$

According to continuous condition $\Delta_3 = \Delta_2 - \Delta_1$, substitute Eqs. (7.14), (7.15), and (7.17):

$$\frac{3V'\left(1 - \mu^2\right)}{k^3t^3E} = \frac{Pr^2}{tE} - \left(Pa + 2V'\right)\frac{r^2}{F'E}. \tag{7.18}$$

Substitute the equation $k^2 = \frac{\sqrt{3(1-\mu^2)}}{rt}$ into the above equation, and it is simplified as follows:

$$V' = \frac{P}{k}\beta. \tag{7.19}$$

Substitute Eq. (7.16) into the above equation:

$$M' = \frac{P}{2k^2\beta} \tag{7.20}$$

$$\beta = \frac{F' - at}{F' + 2t/k} = \frac{F' - at}{F' + 2tl'} = \frac{F' - at}{F} \tag{7.21}$$

In which:

F the effective cross-sectional area of the stiffener ring, mm^2, and the equivalent flange of pipe shell is included.

Finally, the axial stress σ_{x3} of the pipe shell generated by the local bending moment M' is:

$$\sigma_{x3} = \pm\frac{3Pr}{\sqrt{3(1-\mu^2)}t}\beta = \pm\frac{\sqrt{3}}{\sqrt{1-\mu^2}}\beta\frac{Pr}{t}$$

$$\mu = 0.3, \sigma_{x3} = \pm 1.816\beta\frac{Pr}{t}. \tag{7.22}$$

In the equation, the positive sign indicates that the inner edge of the pipe shell is pulled; the negative sign indicates that the outer edge of the pipe shell is pressed.

(2) Shear stress τ_{xr}. The above shear force V' produces a shear stress τ_{xr} in the side pipe shell of the stiffener ring. The direction of action is directed to the center of the penstock, and its value is $\tau_{xr} = \frac{3\beta p}{2kt}$ (middle of the pipe shell) or $\tau_{xr} = 0$ (inside and outside the pipe shell). Because the value is small, and the control point of the comprehensive stress on the pipe shell is at the inner and outer edges of the pipe shell, τ_{xr} can be ignored.
(3) Circumferential stress $\sigma_{\theta 2}$. The net cross section of the stiffener ring is subjected to the outer radial shear force 2 V' in addition to the radial uniform internal water pressure P_a, as shown in Fig. 7.17a. The total circumferential tensile stress is:

$$\sigma_{\theta 2} = \frac{r}{F'}(Pa + 2V').$$

Substitute Eq. (7.19) into the above equation:

$$\sigma_{\theta 2} = \frac{r}{F'}\left(Pa + \frac{2P}{k}\beta\right) = \frac{Pr}{F'}\left(a + 2l'\beta\right). \tag{7.23}$$

According to Eq. (7.21):

$$F' = \frac{t\left(2l'\beta + a\right)}{1 - \beta}.$$

Substitute the above equation into Eq. (7.23):

$$\sigma_{\theta 2} = \frac{Pr}{t}(1 - \beta). \tag{7.24}$$

(4) Shear stress $\sigma_{\theta x}$. The shear stress $\tau_{\theta x}$ caused by the pipe weight and water weight in the pipe shell is calculated by Eq. (7.11), and

$$\tau_{\theta x} = \tau_{x\theta}.$$

The calculation of the axial stresses σ_{x1}, σ_{x2}, and the shear stress $\tau_{x\theta}$ of section 3-3 are the same as that of section 2-2.

The stress direction and distribution of the comprehensive section 3-3 are shown in Fig. 7.18.

4. Supporting ring and its adjacent pipe shell section 4-4

The supporting ring and the stiffener ring are formally welded with a steel ring welded to the outer edge of the pipe shell, so the calculation of the pipe shell stresses

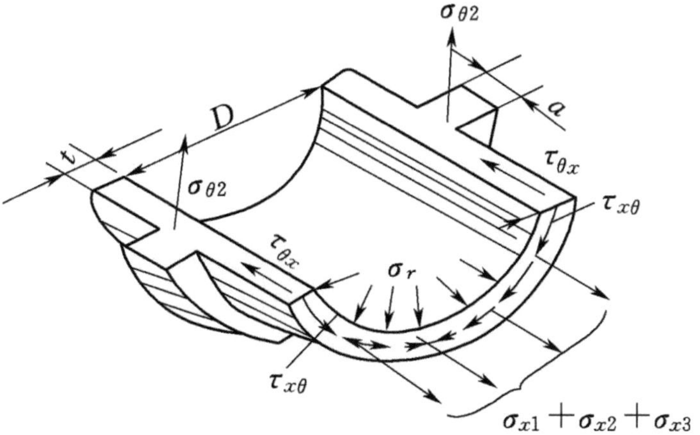

Fig. 7.18 Schematic diagram of stress distribution and direction of stiffener ring section pipe shell

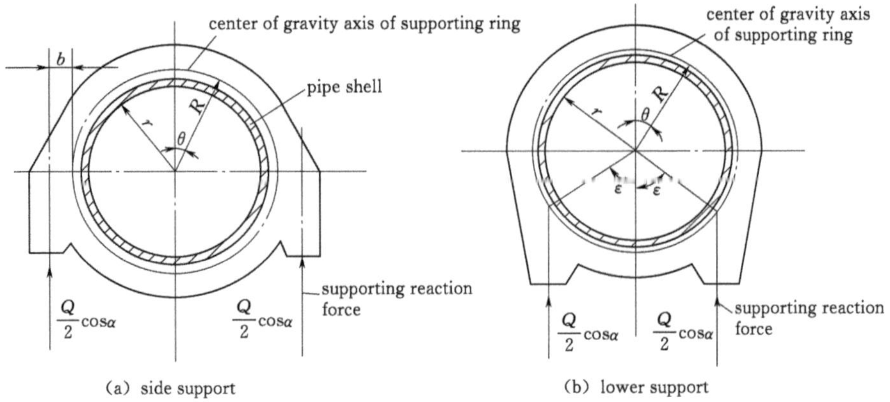

Fig. 7.19 Support modes of supporting ring

σ_{x1}, σ_{x2}, σ_{x3}, $\sigma_{\theta 2}$, $\tau_{x\theta}$, $\tau_{\theta x}$, and τ_{xr} of the section 4-4 is the same as the section 3-3. However, the supporting ring also bears the supporting reaction force caused by the transfer pipe weight and the water weight normal force to the support block, thus generating additional stress in the supporting ring. The stress state varies with the supporting method and structure.

(1) Supporting method of the supporting ring. The supporting on the exposed steel penstock of large- and medium-sized hydropower stations has two forms, side support and lower support, as shown in Fig. 7.19. The dotted line in the figure is the center of gravity axis of the supporting ring's effective section. Its distance to the center of circle is the radius *R*, and the distance to the supporting point of the support block is *b*. The supporting reaction force is $\frac{Q}{2}\cos\alpha$.
(2) Internal force calculation of supporting ring. The internal force calculation of the supporting ring is often carried out using the method in the Structural Mechanics tutorial. Because the steel penstock section is a symmetrical ring, it is a three-dimensional statically indeterminate structure. The internal force of each point on the supporting ring can be calculated by the elastic center method.

When the side support is adopted, the distance between the supporting reaction force and the center of gravity axis of the supporting ring is set to be *b*. According to the analysis, the design often takes $b = 0.04R$, so that the maximum positive bending moment on the ring is close to the maximum negative bending moment, and the utilization of steel is the most economical. Underground exposed steel penstocks are sometimes supported by lower supports, and it may be economical when $\varepsilon = 30^\circ$to90°. The meaning of symbol ε is shown in Fig. 7.19b.

a. Internal force calculation of the side support supporting ring. The load on the supporting ring is mainly: ① the shear stress from normal component of pipe weight and the water weight (expressed as the shear stress $\tau_{x\theta}$ on the two sides of the supporting ring's pipe shell); ② the reaction force $\frac{Q}{2}$ on each side of the support block; ③ the dead weight of supporting ring, but it is relatively small. The

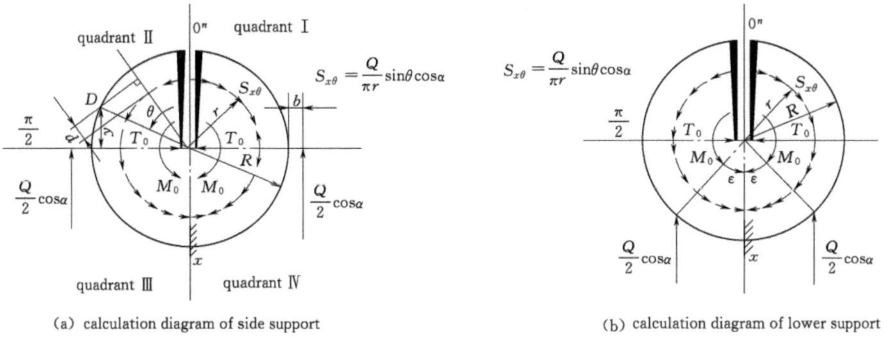

Fig. 7.20 Calculation diagram of supporting ring

steel penstocks are generally arranged in an inclined manner with a supporting reaction force of $\frac{Q}{2}\cos\alpha$. The pipe weight and water weight produce shear stress $\tau_{x\theta} = \frac{Q}{2\pi rt}\sin\theta\cos\alpha$ on both sides of the supporting ring, so the shear force acting on the supporting ring along the circumferential length of the pipe shell is:

$$S_{x\theta} = 2\tau_{x\theta} \cdot t \cdot 1 = \frac{Q}{\pi r}\sin\theta\cos\alpha. \tag{7.25}$$

To calculate the internal force on the section of the supporting ring, it is necessary to calculate the bending moment M_R, the shear force T_R, and the axial force N_R on each section of the closed ring. The calculation diagram is shown in Fig. 7.20. Based on the elastic center method, the top of the ring is cut open and the internal forces T_G and M_G are added. Since the ring is a symmetrical figure, there is no shear force at this point. The internal force is moved to the elastic center, so that the moment at the elastic center is M_0, and the thrust is T_0.

Available from the equation of the elastic center method:

$$M_0 = -\frac{\int M_s \mathrm{d}s}{\int \mathrm{d}s} \tag{7.26}$$

$$T_0 = \frac{\int M_s y \mathrm{d}s}{\int y^2 \mathrm{d}s} \tag{7.27}$$

in which:

M_s the static moment of each point on the ring, N·mm, is positive in the clockwise direction;

y the ordinate of the arc, mm;

ds the differential of the arc length, mm.

After finding M_0 and T_0 at the center of the elastic, the internal forces M_G, T_G at the top of the ring are obtained, so that the internal forces M_R, T_R, and N_R on any section of the closed ring (supporting ring) can be derived.

The derived calculation equations are listed in Table 7.8. It can be seen from the table that the internal force of the supporting ring depends on its geometry and loads Q, $S_{x\theta}$ and on position b of the fulcrum. When $b = 0.04R$, the internal force distribution of each section of the supporting ring is shown in Fig. 7.21.

In the figure, the bending moment is drawn on the side of the tension; the positive M_R indicates that the outer side of the supporting ring is pulled; the positive N_R indicates the pulling force; the direction of the positive T_R is as indicated by ↓↑ in Fig. 7.21.

b. Internal force calculation of the lower support type supporting ring. The position of the lower supporting ring fulcrum is determined by angle, as shown in Fig. 7.19b. The internal force is still calculated by the elastic center method, and the calculation diagram is shown in Fig. 7.20b. The equation for calculating the internal force in any section of the supporting ring can be found in the *Design Specification for Steel Penstocks of Hydroelectric Stations* SL 281-2003. Regardless of whether side support or lower support is adopted, it is necessary to calculate the internal force generated by the lateral seismic force when the earthquake is considered. The calculation equation is shown in the above specification.

When the bending moment M_R, the axial force N_R and the shearing force T_R generated by the supporting force of the supporting ring are calculated, the stresses generated by them are:

Normal stress:

$$\sigma_{\theta 3} = \frac{N_R}{F} \tag{7.28}$$

$$\sigma_{\theta 4} = \frac{M_R Z_R}{J_R} = \frac{M_R}{W_R}. \tag{7.29}$$

Shear stress:

$$\tau_{\theta r} = \frac{T_R S_R}{J_R a}(\text{supporting ring web}) \tag{7.30}$$

In which:

N_R axial force on the cross section of the supporting ring, N;
M_R bending moment on the cross section of the supporting ring, N·mm;
Z_R distance between the calculation point and the center of gravity axis, mm;
J_R moment of inertia of the effective section of the supporting ring to the center of gravity axis, mm^4;
W_R section moment of the effective section of the supporting ring to the center of gravity axis, mm^3;
T_R shear force on the cross section of the supporting ring, N;

Table 7.8 Internal force calculation equation for any section of side-supported supporting ring

Internal force	Quadrant	Internal force of any section	Internal force of $\theta = 0^\circ, \frac{\pi}{2}, \pi$ section	
Bending moment M_R	I or II	$M_R = \frac{QR}{2\pi}[\theta\sin\theta + (\frac{2b}{R} + \frac{3}{2})\cos\theta - \frac{\pi}{2}(1 + \frac{b}{R})]$	$\theta = 0^\circ$	$M_R = -\frac{QR}{2\pi}\left(0.07 - 0.43\frac{b}{R}\right)$
			$\theta = \frac{\pi}{2}$	$M_R = -\frac{Qb}{4}, M_E = \frac{Qb}{4}$
	IV or III	$M_R = -\frac{QR}{2\pi}[\pi - \theta\sin\theta - (\frac{2b}{R} + \frac{3}{2})\cos\theta - \frac{\pi}{2}(1 + \frac{b}{R})]$	$\theta = \pi$	$M_R = \frac{QR}{2\pi}\left(0.07 - 0.43\frac{b}{R}\right)$
Axial force N_R	I or II	$N_R = \frac{Q}{\pi}\left[\left(\frac{r-b}{R} - \frac{3}{4}\right)\cos\theta - \frac{\theta}{2}\sin\theta\right]$	$\theta = 0^\circ$	$N_R = \frac{Q}{\pi}\left(\frac{r-b}{R} - 0.75\right)$
			$\theta = \frac{\pi}{2}$	$N_R = -\frac{Q}{4}, N_E = \frac{Q}{4}$
	IV or III	$N_R = \frac{Q}{\pi}\left[\left(\frac{r-b}{R} - \frac{3}{4}\right)\cos\theta - \frac{\pi-\theta}{2}\sin\theta\right]$	$\theta = \pi$	$N_R = -\frac{Q}{\pi}\left(\frac{r-b}{R} - 0.75\right)$
Shear force T_R	I or II	$T_R = -\frac{Q}{\pi}\left[\left(\frac{r-b}{R} - \frac{5}{4}\right)\sin\theta + \frac{\theta}{2}\cos\theta\right]$	$\theta = 0^\circ$	$T_R = 0$
			$\theta = \frac{\pi}{2}$	$T_R = -\frac{Q}{\pi}\left(\frac{r-b}{R} - 1.25\right)$
	IV or III	$T_R = -\frac{Q}{\pi}\left[\left(\frac{r-b}{R} - \frac{5}{4}\right)\sin\theta - \frac{\pi-\theta}{2}\cos\theta\right]$	$\theta = \pi$	$T_R = 0$

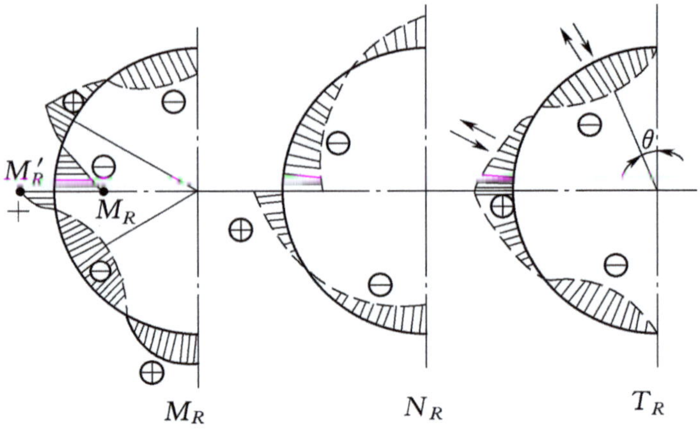

Fig. 7.21 Internal force diagram of supporting ring when $b = 0.04R$

S_R effective cross-sectional area of the supporting ring, calculate the static moment of the partial cross section of the center of gravity outside the point, mm^3;

a contact width or web thickness between contact ring and pipe shell, mm;

F effective cross-sectional area of the supporting ring, mm^2, and the equivalent flange of pipe shell is included.

The stress direction and distribution of the comprehensive section 4-4 are shown in Fig. 7.22.

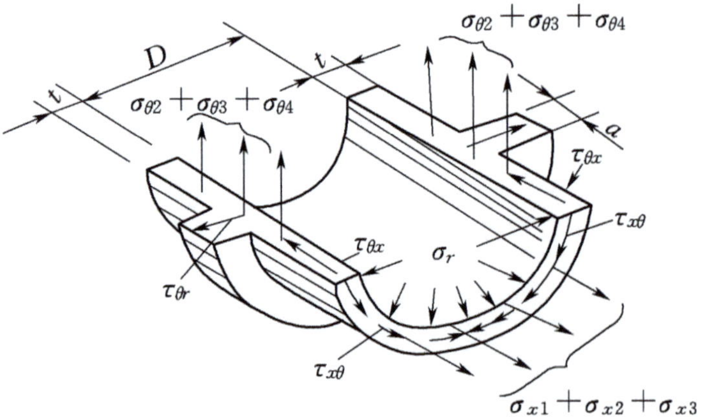

Fig. 7.22 Schematic diagram of stress distribution and direction of supporting ring section pipe shell

7.6.4.3 Strength Checking

The stress of the steel penstock belongs to the three-dimensional stress state. After various stresses are calculated, the strength checking should be carried out according to the method described in Sect. 7.3. The stress of each calculation point must meet the requirements of Eq. (7.2). If not, the pipe shell or support block spacing should be readjusted until the strength requirements are met.

7.6.5 Checking for Stability Against External Pressure of Exposed Steel Penstocks

The water diversion penstock of a hydropower station may generate a negative water hammer during the load change of units, resulting in negative pressure inside the penstock, or a vacuum may occur due to the failure of the aeration vent when the steel penstock is emptied. The steel penstock may lose stability under the action of atmospheric pressure outside the penstock, and the pipe shell is crushed. Therefore, it is necessary to check the pipe shell thickness or take measures according to the condition that the steel penstock is in a vacuum state without causing unstable deformation.

The steel penstock is approximately analyzed as a uniform and infinitely cylinder. Taking the ring of unit length into consideration, it bears a uniform external pressure P. When the value of P gradually increases to reach the critical value of P, the ring loses stability; while under the action of P_{cr}, a certain deformation state is maintained, as shown in Fig. 7.23a. The solid line is the case before the ring is displaced, and the dotted line is the case after the displacement. The radial displacement at point A is y_A, and the displacement at other points is y, as shown in Fig. 7.23b. Cutting the ring out of an arc length, and the balance condition of internal pressure on the center of the circle is considered:

$$M_A - M + \Pr(r - y_A) - \Pr(r - y) = 0.$$

So

$$M = M_A - \Pr(y_A - y). \tag{7.31}$$

Considering that the value of the displacement y is small, the external pressure P passes through the center of the circle, so there is no torque for the point O. The bending equation of the ring is

$$\frac{d^2y}{d\theta^2} + y = -\frac{Mr^2}{EJ}. \tag{7.32}$$

Substitute Eq. (7.31) into the above equation:

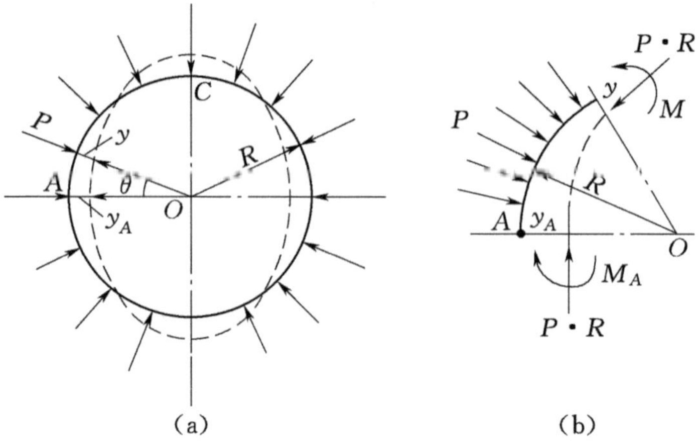

Fig. 7.23 Flexure of round pipe under external pressure

$$\frac{d^2y}{d\theta^2} + y = -\frac{r^2}{EJ}\left[M_A - \Pr(y_A - y)\right]$$

$$\frac{d^2y}{d\theta^2} + y\left(1 + \frac{Pr^3}{EJ}\right) = \frac{-M_A r^2 + Pr^3 y_A}{EJ}$$

In which:

r ring radius, mm;
E elastic modulus of steel, N/mm^2;
J moment of inertia, mm^4.

$$k^2 = 1 + \frac{Pr^3}{EJ}$$

$$\frac{d^2y}{d\theta^2} + k^2 y = \frac{-M_A r^2 + Pr^3 y_A}{EJ}.$$

The general solution of y is:

$$y = A\sin k\theta + B\cos k\theta + \frac{-M_A r^2 + Pr^3 y_A}{EJ + Pr^3}.$$

The two integral constants A and B are determined by the boundary conditions. $\theta = 0$, $\frac{dy}{d\theta} = 0$ at point A and $\theta = \frac{\pi}{2}$, $\frac{dy}{d\theta} = 0$ at point C. Thus, $A = 0$, $B\sin\frac{k\pi}{2} = 0$.

Since B cannot be zero, it must be $\frac{k\pi}{2} = 0$, that is, k should be any positive even number. The instability equilibrium condition of ring is (k is a positive even number):

$$P = \left(k^2 - 1\right)\frac{EJ}{r^3}.$$

The minimum value of the P that meets this condition, that is, the value of P at $k = 2$, is the minimum external pressure value at which the ring loses stability and flexure. It is called critical pressure, and the calculation equation is:

$$P_{cr} = \frac{3EJ}{r^3}. \tag{7.33}$$

For the thin steel penstock with long cylinder, the plane strain problem should be considered. In Eq. (7.33), E could be replaced by $\frac{E}{1-\mu^2}$, μ is Poisson's ratio of steel, and $EJ = \frac{E}{1-\mu^2} \times \frac{t^3}{12}$ is substituted into Eq. (7.33):

$$P_{cr} = \frac{3}{r^3} \times \frac{Et^3}{12\left(1 - \mu^2\right)} = \frac{2E}{\left(1 - \mu^2\right)} \times \left(\frac{t}{D}\right)^3.$$

For the sake of safety, the safety factor of K is introduced. The minimum shell thickness for unstable smooth pipes, i.e., exposed steel penstock without rigid rings, is calculated as follows:

$$P_{cr} = \frac{2E}{\left(1 - \mu^2\right)} \times \left(\frac{t}{D}\right)^3 \geq KP \tag{7.34}$$

In which:

- D steel penstock diameter, mm;
- t steel pipe shell thickness, mm;
- P design value of external pressure load of steel penstock, N/mm^2.

The design value of the external pressure load of the exposed penstock is at most one atmospheric pressure, that is, $P = 0.1$ MPa. The elastic modulus of the steel is $E = 2.0 \times 10^5$ MPa. Take $K = 2$, omit μ^2, substitute to Eq. (7.34), and the condition of no instability is obtained:

$$t \geq \frac{D}{130}. \tag{7.35}$$

If the smooth pipe cannot meet the requirement of stability against external pressure, the thickness of pipe shell could be increased by furnishing stiffener rings, and the capacity of resisting external pressure would be improved.

The rigidity of the stiffener ring should be large enough to not destabilize under the design external pressure. At this time, the smooth pipe shell between the adjacent two stiffener rings has been subjected to rigid constraints at both ends, under the external pressure, the general form of the instability deformation will not be the double waveform shown in Fig. 7.23, but multi-waveforms as shown in Fig. 7.24. The external pressure value required for multi-wave flexure is larger than the external pressure value Eq. (7.33) where double-wave flexure occurs. When the spacing of the stiffener rings is small, the pipe shell will be completely deformed with the stiffener ring, and the critical pressure of the pipe shell is the critical pressure of the stiffener ring. When the spacing between the stiffener rings is quite large, the pipe shell away from the stiffener ring can no longer be restrained by the stiffener ring. The critical pressure is the same as that of the smooth pipe without the stiffener ring. The critical pressure value of the multi-wave flexure is generated between the stiffener rings. It is calculated by *Mises Equation* in China:

$$P_\sigma = \frac{Et}{r(n^2-1)\left(1+\frac{n^2L^2}{\pi^2r^2}\right)^2} + \frac{Et^3}{12r^3(1-\mu^2)} \times \left(n^2 - 1 + \frac{2n^2-1-\mu}{1+\frac{n^2L^2}{\pi^2r^2}}\right) \tag{7.36}$$

$$n = 2.74\left(\frac{r}{L}\right)^{\frac{1}{2}}\left(\frac{r}{t}\right)^{\frac{1}{4}} \tag{7.37}$$

In which:

n number of flexure waves;
L spacing of stiffener rings, mm.

In the above equation, there is only the spacing of stiffener rings, and there is no parameter of the stiffening stiffener ring itself. This is because the condition of the above equation is that the stiffener ring is not unstable. The conditions for keeping the stiffener ring stable are described below.

When using the above equation, the value of P_{cr} should be used as the minimum value and a trial algorithm is required to determine n. In the initial evaluation, use Eq. (7.37) to calculate n, and take the integer. Then use the numbers of $n+1$, n, and $n-1$ to substitute the Eq. (7.36) to obtain P_{cr}, and the minimum value is the critical load sought.

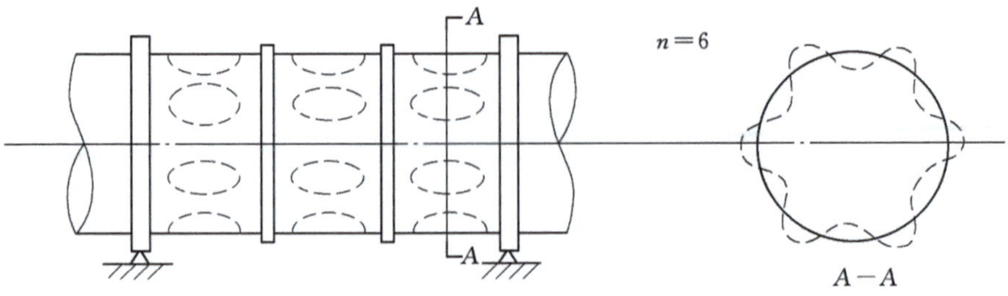

Fig. 7.24 Schematic diagram of the flexural waveform of pipe shell

It is very complicated to calculate by Eq. (7.36) and can be directly searched by the curve shown in Fig. 7.25 made according to this equation.

From the curves of Eqs. (7.36) and (7.37) and Fig. 7.25, it can be seen that when the diameter of the pipe and the thickness of the pipe shell are constant, the number of bends of the pipe shell under external pressure increases with the decrease of spacing of stiffener rings, and critical pressure value increases.

Under the external pressure of steel penstock with the stiffener ring, the stiffener ring must meet two requirements at the same time: ① the stiffener ring cannot be unstable and flexure; ② when the stiffener ring is not unstable, the compressive stress of the cross section is less than the allowable value of the material.

A section of the pipe shell adjacent to the two sides of the stiffener ring can be deformed with the stiffener ring to form an effective flange that forms part of the stiffener ring. According to the theory of elasticity, the length of the flange is $l = 0.78\sqrt{rt}$ on each side. The effective cross section of the stiffener ring is shown in Fig. 7.26. In the figure, a represents the thickness of the stiffener ring. When critical

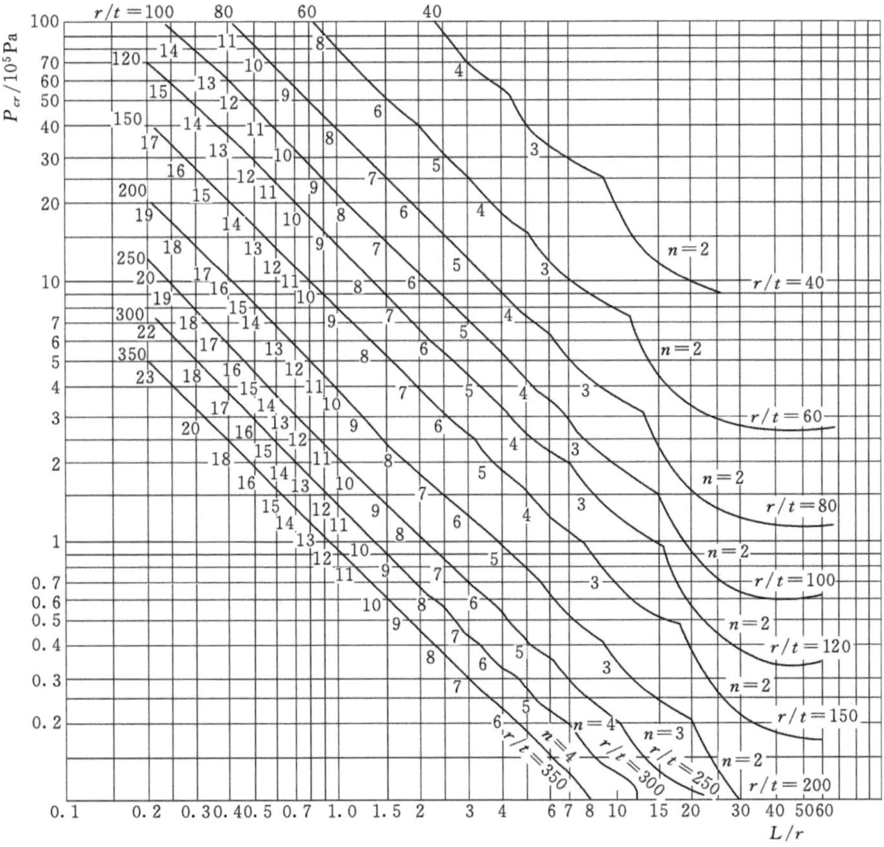

Fig. 7.25 Critical pressure curve of exposed penstock

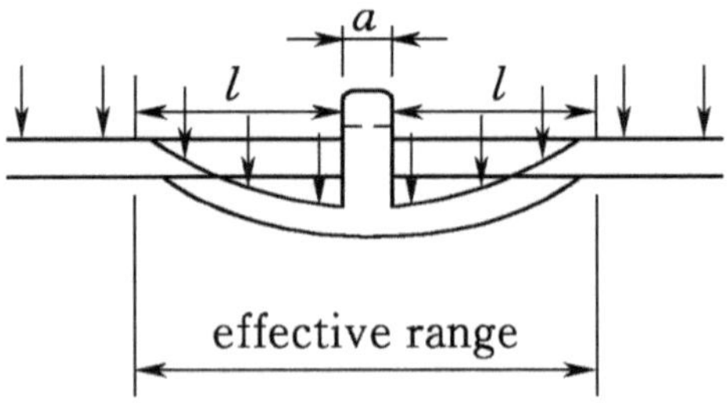

Fig. 7.26 Effective range of stiffener ring

pressure and compressive stress of stiffener ring are calculated, the effective section should be taken.

When the stiffener ring is subjected to external pressure, to maintain the critical pressure of elastic stability, Eq. (7.33) can still be used. However, each stiffener ring must withstand all external pressure on the pipe length L, so when applying Eq. (7.33), the right side of the equal sign is divided by L. The critical pressure elastic stable pressure of the stiffener ring is:

$$P_{cr1} = \frac{3EJ}{R_k^3 L}. \tag{7.38}$$

The critical pressure at which the compressive stress of the stiffener ring cross section is less than the allowable value of the material is:

$$P_{cr2} = \frac{\sigma_s F}{rL} \tag{7.39}$$

In which:

J the moment of inertia of the calculation section to its neutral axis, mm^4;
R_k the radius of the effective section center of stiffener ring, *mm*;
σ_s steel yielding strength, N/mm^2.

Other symbols are the same as before.

Therefore, the design steps for stability checking against external pressure for penstock provided with stiffener ring are as follows: Firstly, the critical pressure is calculated according to the thickness of the smooth pipe determined by the strength. If the stability requirement is not met, the stiffener ring should be set, that is, the stiffener ring spacing and the cross-sectional size should be assumed first. Then the critical pressures are calculated according to Eqs. (7.38) and (7.39), respectively, and the small value of the two is taken. If critical pressure is slightly larger than the product of the external pressure and the safety factor, the set stiffener ring parameter can meet requirements; otherwise, it should be re-assumed until the requirements are met.

7.6.6 *Structural Analysis of Anchor Blocks and Support Blocks*

7.6.6.1 Structural Analysis of Anchor Blocks

The anchor block bears the axial force, shear force, and bending moment transmitted by the exposed penstock, among which the axial force is the main external load. The anchor block must balance the external load with its self-weight to meet the requirements of stability against sliding and overturning and bearing capacity of the foundation.

The contents of the anchor block structure analysis are as follows:

1. Force analysis

The various forces acting on the exposed steel penstock and spreading to the anchor block are obtained, and the most unfavorable combination is carried out according to the water-filling or venting operation of the steel penstock when the temperature rises and drops.

2. Calculation of axial force component

Set the horizontal direction of the x-axis to be positive, the vertical direction of the y-axis to be positive, and the intersection of the water pipe axis to be the origin of the coordinate. As shown in Fig. 7.27, the component forces on the x- and y-axes of the axial resultant forces are obtained:

$$\sum X = \sum X\prime + \sum X'' = \sum A' \cos\alpha' + \sum A'' \cos\alpha''$$

$$\sum Y = \sum Y' + \sum Y'' = \sum A' \sin\alpha' + \sum A'' \sin\alpha'' \tag{7.40}$$

3. Size determination of anchor block

The size of the anchor block should be able to completely enclose the turning section of the steel penstock. The upstream surface of the anchor block is perpendicular to the pipe axis. The thickness of the concrete surrounding the penstock should not be less than 0.4 to 0.8 times the pipe diameter to make the steel penstock bear even force. For the convenience of maintenance and overhaul, the bottom of the penstock should not be less than 0.6 m from the ground. In the anchor block on the soil foundation, the bottom surface is often made horizontal. The buried depth of the anchor block should be 1 m below the freezing line and not less than 0.5 m to the rock foundation. In the seismic zone, the anchor block should be buried deep into the foundation and the foundation surface should be appropriately enlarged, and the spacing between the anchor blocks should be reduced. After the sizes are determined according to the structural requirements, the position of the center of gravity axis of the block and its weight G can be obtained.

Fig. 7.27 Force diagram of anchor block

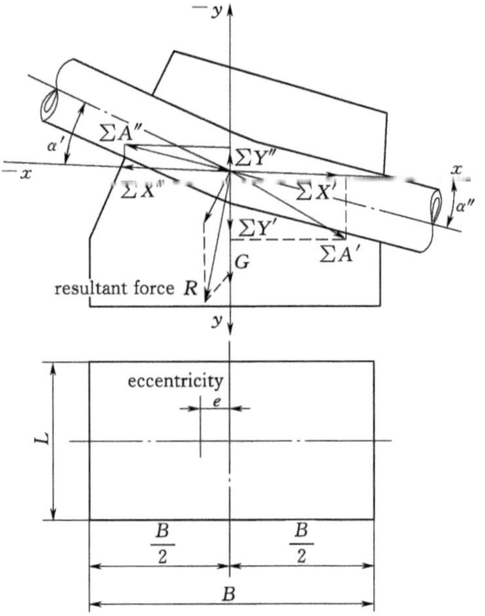

4. Calculation of the acting point of resultant force and eccentricity

The graphic method or numerical solution method is used to find the acting point position of resultant force and eccentricity e of G and $\sum A$. Acting point of resultant force should not be beyond the range of two trisection points on the bottom of the anchor block.

5. Checking of stability against sliding

Stability against sliding should meet the following requirements:

$$K_c = \frac{(\sum Y + G)f}{\sum X} \geq [K_c] \tag{7.41}$$

In which:

K_c safety factor of stability against sliding;
f coefficient of friction between block and foundation;
$[K_c]$ the allowable value of K_c and the specified values of $[K_c]$ are shown in Table 7.9.

6. Checking of bearing capacity of the foundation

Stresses on the foundation are required to be fully compressive, and the maximum value should not exceed the allowable value $[R]$ of the foundation. The foundation stress can be calculated according to the eccentric compression equation.

Table 7.9 Safety factor of stability against sliding

Structure grade	Load combination	
	Basic	Special
Grade I	1.5	1.3
Grade II	1.4	1.2
Grade III	1.3	1.1

$$\sigma = \frac{\sum Y + G}{B \cdot L}\left(1 \pm \frac{6e}{B}\right) \leq [R]. \tag{7.42}$$

The meaning of the symbol in the equation is shown in Fig. 7.27. It is assumed that compressive stress is positive, and it is required the maximum value $\sigma_{\max} \leq [R]$ the minimum value $\sigma_{\max} > 0$; that is, the tensile stress should not occur.

7.6.6.2 Structural Analysis of Support Block

The support block bears the normal component of the pipe weight and the water weight inside the pipe, and the friction between the steel penstock and the support. The structural analysis principle of the support block is similar to that of the anchor block.

The contents are as follows:

1. Force analysis

Calculate the force acting on the support block, as shown in Fig. 7.28.

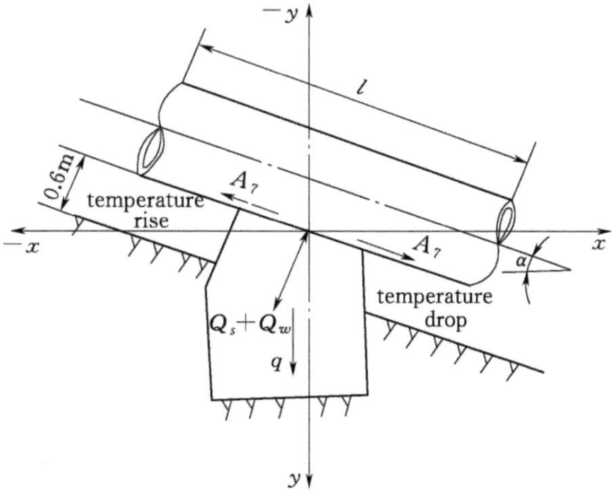

Fig. 7.28 Force diagram of support block

(1) Component of pipe and water weight of steel penstock on the support block:

$$Q_s = q_s l\cos\alpha \tag{7.43}$$

$$Q_w = q_w l\cos\alpha \tag{7.44}$$

(2) Friction between steel penstock and support block A_7.
(3) Dead weight of support block q.

2. Calculation of the horizontal and vertical force

Taking the midpoint of the top surface of the support block as the coordinate origin, the horizontal axis x is positive with the water flow, and the vertical axis y is positive. The horizontal component and vertical component of each force are obtained as follows:

$$\sum X = \pm A_7\cos\alpha - (Q_s + Q_w)\sin\alpha \tag{7.45}$$

$$\sum Y = \pm A_7\sin\alpha + (Q_s + Q_w)\cos\alpha + q \tag{7.46}$$

3. Checking of stability against sliding and overturning, and bearing capacity of the foundation

(1) Stability against sliding

$$K_c = \frac{\left[\pm A_7\sin\alpha + (Q_s + Q_w)\cos\alpha + q\right]f}{\pm A_7\cos\alpha - (Q_s + Q_w)\sin\alpha} \geq [K_c] \tag{7.47}$$

(2) Stability against overturning

$$K_0 = \frac{\sum M_1}{\sum M_2} \geq [K_0] \tag{7.48}$$

In which:

$[K_c]$ the allowable value of K_c;
$\sum M_1$ total anti-overturning moment, N·mm;
$\sum M_2$ total overturning moment, N·mm;
$[K_0]$ the allowable value of safety factor of stability against overturning, in design case $[K_0] = 1.5$, in check case $[K_0] = 1.2$.

The positive or negative sign used in the calculation is based on the most unfavorable situation for stability.

7.6.7 Expansion Joints, Manholes, and Drainage Holes of Exposed Steel Penstocks

1. Expansion joint

The expansion joint is a necessary pipe appurtenance on the segmented exposed steel penstock. Its function is to make the steel penstock expand and contract along the axial direction when the temperature rises or drops, and then the temperature stress is eliminated or reduced. The expansion joint set at the front valve of the turbine can also adapt to the angular displacement of the steel penstock caused by a small amount of uneven subsidence and at the same time provide convenience for valve disassembly and assembly.

A single axial telescopic expansion joint is shown in Fig. 7.29a. It is convenient to manufacture, install, operate, and overhaul, but it only has axial displacement. A bidirectional telescopic expansion joint with a complex structure is shown in Fig. 7.29b, and it has axial displacement or slight radial displacement. It is suitable for poor geological conditions or telescopic settlement joints with the radial displacement of steel penstock. The water-stopping fillers, such as oil-impregnated hemp ropes, rubber bands, asbestos rubber ropes, etc., are used by the expansion joint for the water stop. These fillers are placed in a water-stop sleeve after being immersed in animal oil mixed with graphite powder and pressed by a compression ring. The sliding length of the expansion joint, in addition to the expansion and contraction caused by temperature changes, should also be added to the margin of more than 50 mm.

Bellow expansion joints have been used in hydropower station water diversion penstock in recent years, as shown in Fig. 7.30. The sealed telescopic bellow expansion joints, including basic components such as bellow, wave core, and outer casing, have also successfully replaced the conventional telescopic expansion joints of many hydropower stations in China, as shown in Fig. 7.31.

2. Manhole

The function of the manholes is to enter the penstock for inspection, repair, or painting, as shown in Fig. 7.32. The manhole is arranged on the upstream side of the anchor block to fix the equipment such as cables, trolleys, and cages. The spacing of the manholes is generally not more than 200 m. The size is generally made into a circular hole larger than 45 cm or an oval hole 45 cm × 50 cm, and the hole is sealed with a detachable cover plate.

3. Drainage hole

A drainage hole should also be provided on the steel penstock. It is set at the lowest point of the steel penstock for inspection and venting to remove water and sediment in the penstock, and the structure is shown in Fig. 7.33.

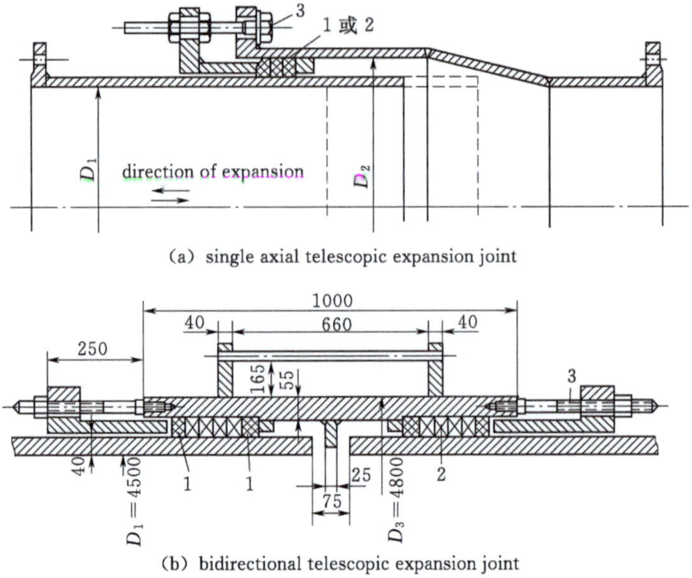

Fig. 7.29 Expansion joint (unit: mm) 1-rubber packing; 2-hemp or asbestos packing; 3-tension bolt

Fig. 7.30 Bellows expansion joints

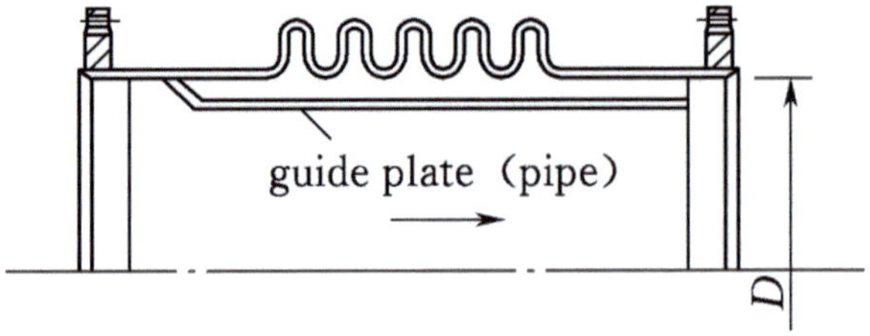

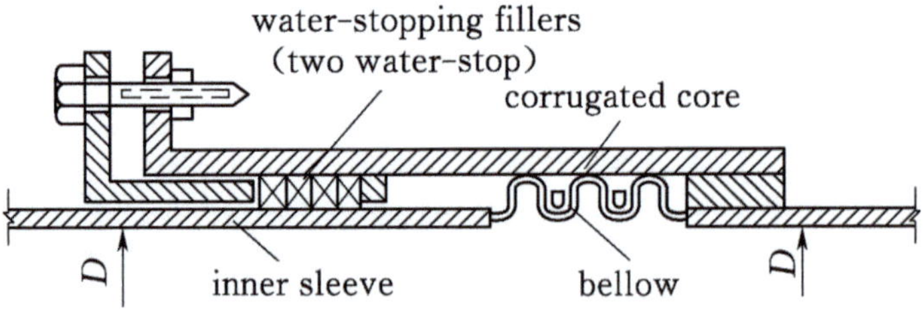

Fig. 7.31 Sealed telescopic bellow expansion joint

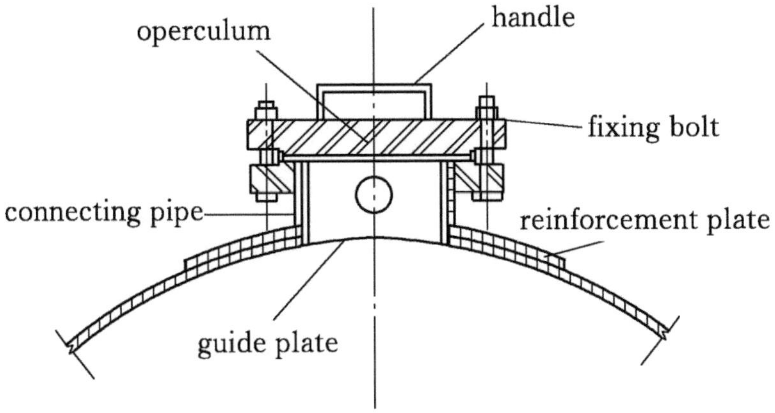

Fig. 7.32 Manhole

Fig. 7.33 Drainage hole

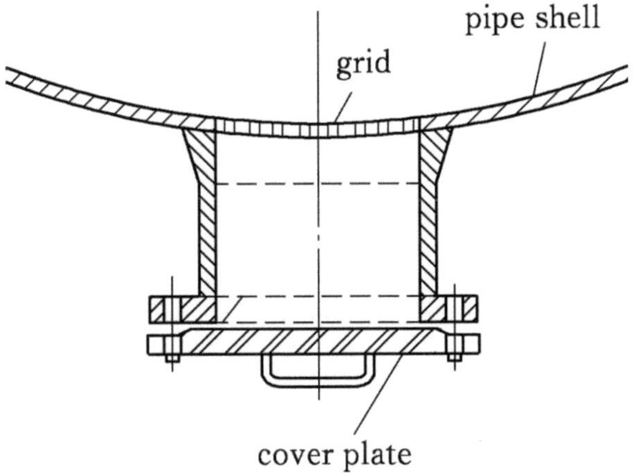

7.7 Underground Penstocks

7.7.1 Characteristics and Structure of Underground Penstocks

Underground penstock is embedded in rock directly, and the gap between the penstock and rock is filled with cement grout or concrete. The vertical arrangement can be inclined and vertical, so it is also called inclined shaft, vertical shaft, or tunnel penstock. It is made by excavating caves, installing steel linings, and then pouring concrete between the rock formations and the steel lining. Such penstocks sometimes use other lining forms instead of steel lining.

7.7.1.1 Working Characteristics and Applicable Conditions

Underground penstock is the most widely used penstock in large-sized and medium-sized hydropower stations. Among the conventional hydropower stations with installed capacity of more than 1000 MW that have been built and under construction abroad, about 40% of them are set up with underground penstocks. Most pumped-storage stations, built and under construction all over the world with capacity of more than 1000 MW, are also set up with underground penstocks. Underground penstocks also bear the highest water head and the maximum value of *HD* in the world. Among the hydropower stations built in China, the underground power station on the right bank of Xiangjiaba is the largest power station with underground penstocks (the installed capacity is 4×800 MW, the maximum designed internal pressure is 242 m, and the maximum diameter is 14.4 m). The largest underground hydropower station that has been built is Xiluodu hydropower station, and it has also set up with underground penstocks. The reason why underground penstock is so widely used is that it has some outstanding advantages compared with exposed penstock.

1. Flexible and convenient arrangement

Better topographic and geological conditions are required for exposed penstock, that is, moderate slope, sufficient strength of foundation, no landslide, debris flow, avalanche and other dangers, no strong earthquake, etc. Underground penstock is embedded in rock directly, and it is flexible to arrange the penstock alignment and easy to make the penstock short and straight because the layout can avoid the restriction of ground topography. Underground penstocks can always be arranged where the exposed penstocks are not easy to build. As for underground powerhouses, underground penstocks must be used wholly or partly. In recent decades, rock mechanics and underground engineering design and construction technology have developed rapidly. The technology for constructing pressure vertical shafts and inclined shafts is quite mature, and the construction conditions and costs begin to be superior to the ground penstocks in some countries.

2. Internal water pressure sharing by the surrounding rock to reduce the thickness of the steel lining

Solid surrounding rock can bear the internal water pressure, even if the geological conditions are poor, the bearing capacity of the surrounding rock can still be exerted after taking appropriate measures. Reducing the thickness of steel lining can not only reduce the cost, but also simplify the manufacture, welding, and installation. The superiority is even more important in ensuring the quality of the steel liner and speeding up the installation. The value of *HD* is very large for penstocks with extremely large capacity and high water head, which is difficult to realize by exposed penstock technology, but it can be solved by underground penstock.

3. Operation safety

The operation of the underground penstock is not affected by external conditions, and the maintenance is simple. The ultimate bearing capacity of the surrounding rock is generally high, and the steel has good plasticity, so the overload capacity of the penstock is very large.

Underground penstocks also have their shortcomings.

First of all, it has complex structure. There are many construction and installation procedures, high technology requirements, and poor underground construction conditions. These factors will increase the cost, thc quality of the project is not easy to guarantee and affect the construction period. Sometimes the underground penstock will become the key to the commissioning of the power station.

Secondly, particular attention should be paid to the stability of the penstock resisting external pressure when the groundwater table is high. Most of the accidents that happen at home and abroad are caused by instability under external pressure.

7.7.1.2 Layout Design

Underground penstock lines should be selected as far as possible in areas with superior topographic and geological conditions and have a good overall layout with surge tanks and powerhouses. Favorable terrain reduces the length of penstock, which saves investment and improves unit operating conditions while facilitating construction. Penstocks buried in solid, intact, and thick rock masses can make more use of the ability of the surrounding rock to bear the internal water pressure. The axis of the underground penstock should be as perpendicular as possible to the structural surface of the rock formation. To avoid the loss of stability and damage of the steel lining under the action of external water pressure, it is necessary to avoid the area with large groundwater pressure and water inflow and difficult to take effective drainage measures.

The centralized water supply mode of single pipe and multi-machine should be used in underground pipes. If the penstock is short, the quoted flow is large, the number of units is large, the construction interval by stages is long, or the engineering geological conditions are not suitable for excavating a large cross section well, the group water supply mode with two or more penstocks supplying water to several units in each direction can be adopted through technical and economic comparison. In addition to the influence of excavation blasting, the spacing between two adjacent penstocks should also be checked for rock mass strength.

The type and slope of the well (horizontal tunnel, inclined shaft, vertical shaft) should be selected according to the layout requirements, engineering geological conditions, construction machinery, and construction methods.

7.7.1.3 Requirements for Structure and Construction

The underground penstock is a combination structure composed of steel liner, backfill concrete, and surrounding rock which shares loads together, including construction procedures such as excavation, steel liner installation, concrete backfilling, and grouting.

Different methods can be used for excavation. In the process of excavation, smooth surface, and pre-crack blasting or tunneling machine should be used as far as possible to excavate, which could keep circular holes and reduce the impact of loosening. It is necessary to select the elevation and plane position of the construction hole to facilitate the transportation of the lining, transportation of steel liner, and concrete pouring and consider the possibility of being a permanent drainage hole and observation hole. The radial clearance dimension between the pipe shell and the surrounding rock depends on the construction method and structural layout (excavation, backfilling, welding, and other methods, as well as whether there are anchorage stiffener members, etc.). After the steel penstock is in place, the welding operator should leave at least 0.5 m on both sides and the top, and at least 0.6 m or more at the bottom. The stiffener ring should be kept at least 0.3 m away from the rock wall. Floor welding outside the penstock should be avoided as much as possible and reduce the height of the stiffener ring to save the amount of excavated and backfilled concrete.

The lining of steel penstock is generally welded in the processing plant by steel plate, cutting, rolling, assembling, welding, flaw detection, rust removal, antirust layer, and other processes. Penstock joints of a certain length are welded, transported into the cave, and fixed it with pre-embedded anchors. The roundness is corrected, and the cross seam is welded and installed after the seam is leveled.

The backfill concrete between the steel liner and the surrounding rock only transmits the radial internal pressure without having to withstand the circumferential tensile force, so the strength grade of concrete should not be too high, but it should not be lower than C20. More importantly, it is necessary to use suitable raw materials and grading, reasonable processes of transportation, pouring, and vibrating to ensure that the backfill concrete is dense and uniform and closely cooperates with the surrounding rock and steel liner. At the bottom of inclined penstock and flat penstock, especially near the water-stop ring and stiffener ring, the vibration must be strengthened. Loose areas or holes are not allowed. In the construction of strata with abundant groundwater, special attention should be paid to the influence of groundwater on concrete compaction by washing away cement paste. Defects of backfill concrete can cause local bending of steel liner, which is especially unfavorable, even worse than exposed penstock. All these defects are often the important causes of underground penstock accidents. Good results can be obtained by using pre-buried aggregate grouting concrete and micro-expansion cement, and the process of concrete pouring and backfill grouting can be combined into one and play a certain pre-pressing role for steel liners. The use of vertical shafts and steeper inclined shafts facilitates the quality of backfill concrete.

Grouting is divided into backfill grouting, contact grouting, and consolidation grouting. There are different opinions about the necessity of various grouting. However, the design code of steel penstock in China is stipulated: horizontal tunnel and inclined shaft should be used as top arch backfill grouting because the top arch is not easy to pour. In structural analysis, contact grouting between the steel liner and concrete should be done if the combined force between steel liner and surrounding rock is considered. It is suitable to construct in the season with the lowest temperature to reduce the gap value. Consolidation grouting can be selected according to specific conditions. During the grouting process, it should be closely monitored to prevent accidents such as instability of the steel liner. After grouting, all grouting holes must be strictly sealed to prevent the leakage of internal water during operation and cause accidents.

7.7.2 *Structural Analysis of the Underground Penstock for Internal Pressure*

7.7.2.1 Basic Working Principles

A partial cross-sectional view of the underground penstock before it is subjected to internal water pressure is shown in Fig. 7.34. Before the internal pressure is applied, there are initial gaps between steel liner and backfill concrete, between backfill concrete and rock which are assumed to be combined into an accumulative gap. There is a layer of rock fracture caused by construction blasting outside the concrete liner, and the outer part is the original rock intact area, as shown in the upper part of Fig. 7.34. Radial displacement of steel liner occurs after the internal pressure is applied to underground penstocks, and after the gap disappears, the internal pressure is continuously transmitted to the concrete liner to cause the circumferential tensile stress in the concrete, thereby generating a radial uniform crack in the liner. Backfill concrete is assumed to be radially fractured during loading so that it is only used to transmit the radial loads, rock has deformed fully already so that the concrete is not affected by the deformation of rock, which balances underground penstock under internal pressure. If the gaps are uniform and rocks are homogeneous, then the underground penstock is an axisymmetric combined cylinder structure, and the displacement and stress under uniform internal pressure can be analytically solved according to the displacement compatibility condition of the plane strain condition. The most important of these is to find the steel liner stress under the condition of the fixed wall thickness t or to find the steel liner wall thickness under the condition of the allowable stress σ_θ.

7.7.2.2 Stress Calculation and Determination of Pipe Shell Thickness

The parameters of underground penstock are shown in Fig. 7.34. To conform with the current steel penstock design specification, the unit adopts the *N-mm* system. Under the action of internal water pressure P, the stress of steel penstock can be calculated by Eqs. (7.49)–(7.52).

1. When $\Delta \geq [\sigma]\varphi r/E'$

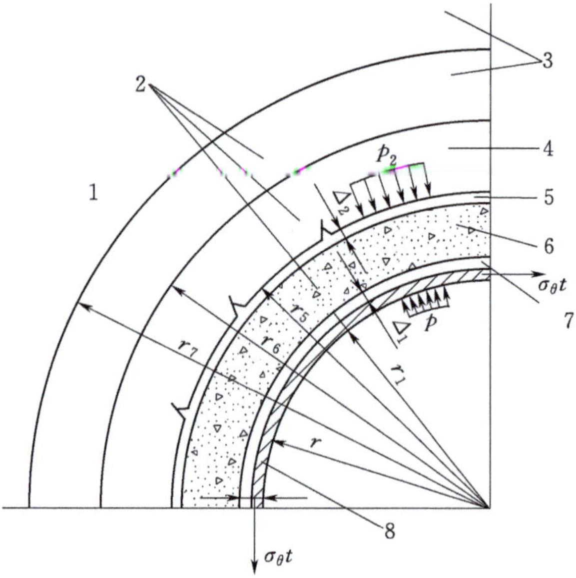

Fig. 7.34 Calculation of underground penstock 1-constant temperature zone; 2-heat-affected zone; 3-intact surrounding rock zone; 4-surrounding rock fracture zone; 5-gap between concrete liner and surrounding rock; 6-concrete liner; 7-gap between steel penstock and concrete liner; 8-pipe shell

Circumferential stress of steel penstock

$$\sigma_\theta = pr/t \tag{7.49}$$

or thickness of pipe shell

$$t = pr/[\sigma]\varphi \tag{7.50}$$

In which

$[\sigma]$ allowable stress of underground penstock, N/mm^2;
φ weld coefficient;
E' elastic modulus of steel for plane strain problem, N/mm^2, $E' = E/(1-\mu^2)$;
p uniform internal water pressure, N/mm^2;
t the calculation thickness of pipe shell, mm;
r_1 inner radius of steel penstock, mm;
σ_θ circumferential normal stress of steel penstock, N/mm^2.

2. When $\Delta < [\sigma]\varphi r/E'$

Circumferential stress of steel penstock

$$\sigma_\theta = \frac{pr + 1000K_0\Delta}{t + 1000K_0 r/E'} \tag{7.51}$$

or thickness of pipe shell

$$t = \frac{pr}{[\sigma]\varphi} + 1000K_0\left(\frac{\Delta}{[\sigma]\varphi} - \frac{r}{E'}\right) \tag{7.52}$$

In which:

K_0 rock unit resistance coefficient (lower limit), N/mm^3, used to calculate thickness of steel penstock.

7.7.2.3 Requirements for the Thickness of Surrounding Rock

(1) If the surrounding rock is considered to carry the internal pressure, two requirements should be met according to the code.

$$H_r \geq 6r_5 \tag{7.53}$$

$$H_r \geq \frac{p_2}{\gamma_r \cos\alpha} \tag{7.54}$$

$$\alpha \leq 60^{\circ}$$

In which:

H_r the minimum thickness of surrounding rock, which is perpendicular to steel penstock axis, mm, as shown in Fig. 7.35;
P_2 the maximum internal pressure shared by surrounding rock, N/mm^2;
γ_r the weight of the surrounding rock, which should be taken as a smaller value, N/mm^3;
α the angle between the axis of the underground penstock and the horizontal line, (°).

Maximum internal water pressure shared by surrounding rock:

$$p_2 = \frac{pr - \sigma_{\theta 1}t}{r_5}. \tag{7.55}$$

The minimum circumferential stress of steel penstock is:

$$\sigma_{\theta 1} = \frac{pr + 1000K_{01}\Delta_{s1}}{t + \frac{1000K_{01}r}{E'}} \tag{7.56}$$

In which:

$\sigma_{\theta 1}$ the minimum circumferential stress of steel penstock under internal water pressure, N/mm^2;
K_{01} take the maximum possible value of K_0, N/mm^3;

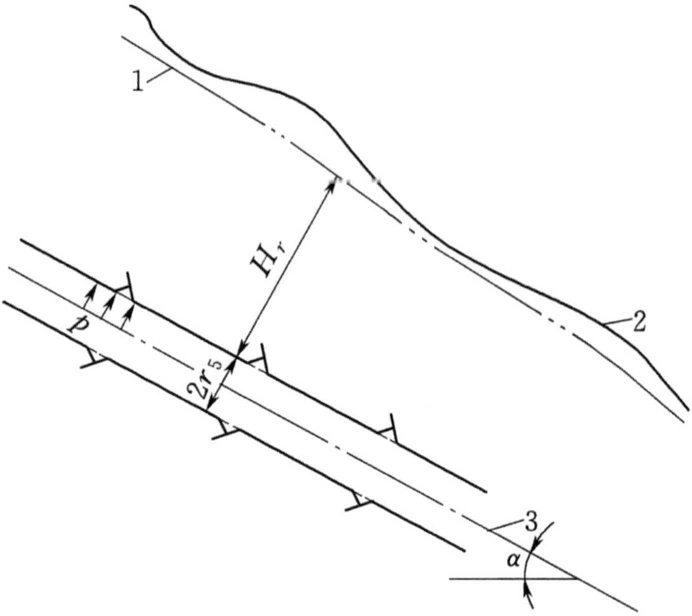

Fig. 7.35 Thickness of surrounding rock covered on underground penstock 1-calculated top surface of bedrock (excluded strata that should not be included); 2-ground line; 3-the axis of penstock

Δ_{s1} minimum cold shrinkage gap value of steel penstock, mm, construction gaps and cold shrinkage gaps are ignored.

(2) If the thickness of the surrounding rock covered on the steel penstock satisfies the requirements of Eq. (7.53) and does not satisfy the requirements of Eq. (7.54), then:

$$p_2 = \gamma_r H_r \cos\alpha \tag{7.57}$$

$$t = \frac{pr - p_2 r_5}{[\sigma]\varphi}. \tag{7.58}$$

(3) If the surrounding rock thickness covered on steel penstock does not satisfy the requirements of Eq. (7.53), then the pipe shell thickness is calculated with the allowable stress of exposed penstock, regardless of rock resistance.

For the case of $\alpha > 60°$ and steep inclined shafts or vertical shafts, the anti-uplift theory is not applicable, but the requirements of the horizontal coverage thickness of surrounding rock can be analyzed by the principle of lateral thrust balance considering the friction of surrounding rock. Generally, the internal friction coefficient is less than 1, so the horizontal coverage thickness should be greater than the minimum vertical coverage thickness.

7.7.2.4 Influencing Factors of the Steel Liner Stress

1. Influence of unit resistance coefficient of surrounding rock K_0 on steel liner stress

For example, when $r = 2000$ mm, $t = 12$ mm, $P = 2$ MPa, $\Delta = 0$ and when K_0 is equal to zero (no resistance to surrounding rock) and 4 N/mm^3. Steel liner stress is reduced from 333 to 85.5 MPa.

It can be seen that the selection of K_0 value is very important. But it is not easy to select K_0 value correctly. If the surrounding rock is regarded as an infinite homogeneous linear elastomer whose elastic modulus and Poisson's ratio are E_r and μ_r, then it can be deduced by the elastic theory:

$$K_0 = E_r/1000(1 + \mu_r).$$

That is to say, in an ideal situation, the E_r and μ_r of the rock sample can be obtained in the laboratory, and K_0 can be calculated.

However, the surrounding rock mass is composed of rock blocks, and there are always geological structures such as joints, fissures, and interlayers in the surrounding rock. When the inner hole of the surrounding rock is subjected to internal pressure, these structures are closed or opened, and deformation occurs, and the rock block is also deformed by force. The radial displacement of the rock mass orifice includes these two parts, and the former (structural deformation) is often the main. Therefore, the method of calculating K_0 based on E_r and μ_r of small rock samples is not reliable. If a large-scale test is carried out on site, such as the hydrostatic test of the test hole, the result is more realistic because the geological structure is included to some extent. However, compared with the prototype, such tests often have different locations, different sizes and loads, and the results should be carefully analyzed and used.

Increasing the resistance coefficient of surrounding rock can significantly reduce steel liner stress. Therefore, consolidation grouting is often used in engineering practice to strengthen the surrounding rock and increase the value of K_0. Under the condition of rock joint fissure development, less filling in the fissure, and good irrigation property, the effect of consolidation grouting is better; otherwise, the effect may not be good. The cost of consolidation grouting is also relatively high, and the construction period is long. The sealing of the grouting hole will affect the quality of the steel liner. Therefore, the necessity of consolidation grouting should be determined by technical and economic comparison. The improvement of rock bearing capacity after consolidation grouting must also be confirmed by experiments before it can be used in structural analysis.

2. Influence of gap value Δ on steel liner stress

For example, when $r = 2000$ mm, $t = 12$ mm, $P = 2$ MPa, $K_0 = 4$ N/mm^3, when Δ equals zero and 1 mm, respectively, the stress of steel liner increases from 85.5 to 171 MPa.

Engineering practice shows that the gap value Δ varies greatly, and there are many factors that affect the gap size. It mainly consists of the following parts:

(1) Construction gap Δ_0 is the sum of the gap between the steel liner and the backfill concrete and the gap between backfill concrete and the surrounding rock when the concrete pouring or contact grouting is completed and the temperature returns to normal (after the hydration heat dissipates). It is caused by shrinkage of concrete and grouting slurry and poor construction. Its numerical value varies with construction method and quality. Steep inclined shafts and vertical shafts are helpful to ensure the quality of backfill concrete and reduce the construction gap. After dissipation of hydration heat, serious contact grouting at low temperatures can effectively reduce the construction gap. It is also effective to reduce the warehousing temperature and cool the steel liner (such as spraying cold water) during concrete pouring. When expansive concrete is used as backfill, contact grouting cannot even be done. Construction gap is very small when pre-embedded aggregate mortar concrete is used. In general, if the concrete outside the penstock is filled with compact concrete and the contact grouting is done carefully, Δ_0 can be taken as 0.2 mm.

(2) The steel liner shrinkage gap Δ_s, also called the temperature difference gap of the steel liner, depends on the difference between the temperature when the steel liner is completed and the temperature when it is in service. The former should take the temperature in the hole during the completion of the work or the highest temperature in the hole during contact grouting, and the latter should take the water temperature. The water temperature changes year by year. The value of Δ_s changes year by year, and its value is the largest in winter.

$$\Delta_s = \alpha_s \Delta t_s r(1 + \mu).$$

If the construction season is not properly selected, it will reach a considerable value.

(3) The cold-shrinking gap Δ_R of the surrounding rock refers to the gap formed by the cooling and shrinking of the surrounding rock after the penstock is put into operation, and the water temperature is lower than the original temperature of the surrounding rock. When the temperature drops in the complete area of the surrounding rock, the radial displacement of the inner circumference is centripetal, which is beneficial to reduce the gap, the surrounding rock fracture zone, and the cracked concrete lining. After the temperature drops, the gap value is increased. Δ_R can be estimated as follows:

$$\Delta_R = \alpha_r \Delta t_R r \Delta_{R'}$$

In which

α_r expansion coefficient of surrounding rock;
Δt_R temperature difference of surrounding rock;
$\Delta_{R'}$ influence coefficient of the surrounding rock fractured zone, it can be taken as Table 7.10.

Table 7.10 Value of Δ_R

r_6/r_5	1	2	3	5	7	9	10	11
$\Delta_{R'}$	0	0.8389	1.460	2.312	2.822	3.089	3.151	3.170

The above three kinds of gaps are the main part of the underground penstock gap, and the sum of the three can be taken in the calculation, that is:

$$\Delta = \Delta_0 + \Delta_s + \Delta_R.$$

In addition, the plasticity and creep of the concrete lining ring can form a gap, but the value is generally small. The plasticity and creep of the surrounding rock also form a gap, which is an irreversible deformation. Its value is related to joints and cracks of rock mass, properties of fillings, lithology and weathering degree, influence of excavation blasting loosening, consolidation grouting, and other factors, and it is difficult to obtain reliable values. The effect of this deformation can also be considered in the calculation by appropriately reducing the resistance coefficient of the surrounding rock.

7.7.3 Stability Against External Pressure of Underground Penstocks

The steel liner of underground penstock also has the problem of instability under external pressure. This problem is even more noticeable than the internal pressure problem. In the accidents of underground penstock at home and abroad, the damage to steel liners is mostly caused by the instability of external pressure. This is because the potential safety degree of the underground penstock under the internal pressure is quite high, but the value of the external pressure may be much greater than the atmospheric pressure when the penstock is emptied.

7.7.3.1 External Pressure of Steel Liner

(1) Groundwater pressure. The value of groundwater pressure on steel liners can be selected according to the survey data. It is safe to determine the external water pressure value based on the highest groundwater level, but it often makes the design value too high. Meanwhile, it is necessary to analyze the impact of reservoir impoundment and leakage of water diversion system on the groundwater level. Groundwater level should generally not exceed the ground level.
(2) Pressure of joint grouting between steel liner and concrete. The joint grouting pressure is generally 0.2 MPa.

(3) Pressure from flowing concrete during the concrete backfilling. The value is determined by the height of the concrete in a single casting. The maximum possible value is equal to the concrete bulk density multiplied by the height of casting.

7.7.3.2 Characteristics of Stability Against External Pressure of Underground Penstock

When the external pressure value of the exposed steel penstock reaches the critical pressure value of Eq. (7.33), the steel penstock will be deformed as shown in Fig. 7.23a. Exposed penstock is located in the atmosphere without restraint, so if the external pressure slightly exceeds the critical pressure, the steel penstock will produce great flexural deformation and lose its function. The deformation of the steel liner is limited by the peripheral concrete, so the deformation performance of the steel liner of underground penstock under external pressure is quite different from that of the steel penstock.

As shown in Fig. 7.36a, there is an initial uniform gap Δ between the steel liner and the concrete lining. When the external pressure reaches the minimum critical pressure of the exposed penstock ($K = 2$), the steel liner buckles. However, when the deflection is the radial displacement to Δ, the flexural peak of the steel liner is in contact with the concrete and is resisted, and the deformation is no longer developed. If the maximum stress in the steel liner does not reach the material yield strength at this time, the steel liner can maintain a new stable state under this external pressure, without losing its performance or stability. If the external pressure continues to increase, reaching the value of P calculated by the equation $P = \left(k^2 - 1\right)EJ/r^3$ obtained in Sect. 7.6.5 corresponding to K being an integer-even number greater than 2, more waveform flexure will occur. The stress value also increases correspondingly, until it reaches the material yield value. If the external pressure continues to increase, the steel liner will be plastic flow, resulting in large deformation and the loss of performance.

In engineering, the state when the steel liner stress reaches the yield strength is called instability, and the corresponding external pressure is the critical pressure. In the critical state, there are n flexural waves, which occur simultaneously along the entire circumference, and the yield peaks are close to the concrete. The axis of the flexural wave is shown in the dotted line in Fig. 7.36a. It is different from the position of the original steel liner, because the steel liner is subjected to circumferential compression under external pressure, and the original circumferential axis is shortened.

The flexural deformation of the above steel liner under external pressure is only one possible form of steel liner flex. Another possible form is shown in Fig. 7.36b. Under the action of external pressure, several initial waves may appear on the circumference of the steel liner. But as the load increases, the steel liner will first flex at a weak part, and the possibility of flex of the pipe shell at other parts is reduced. Therefore, this is a local flexure. When the external pressure is increased until the maximum fiber stress of the steel liner after flex reaches the flexural strength of the material,

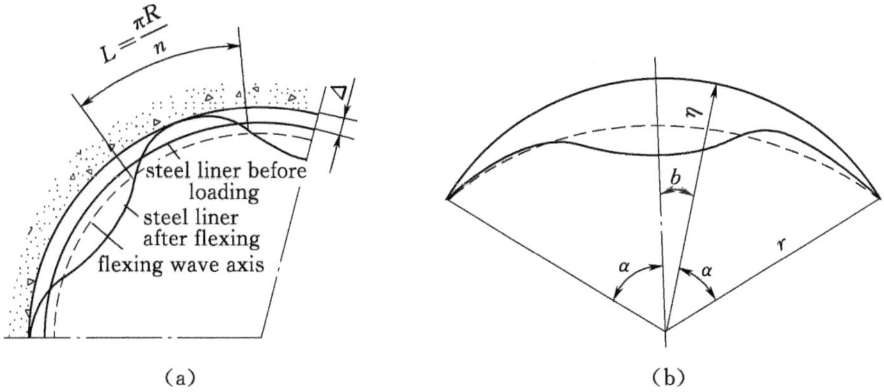

Fig. 7.36 Flexure of underground penstock steel liner under external pressure

it is considered to be an unstable state, and the corresponding external pressure is critical.

Regardless of the flexural state, the maximum fiber stress to yield strength of the steel liner is taken as the criterion for the instability of the steel liner. The flexural stress of the steel liner is related to the initial gap between the steel liner and the concrete. Therefore, the critical pressure of underground penstock steel liner must be directly related to the yield strength of material and the initial crack value, which is an important difference between the instability of underground penstock and exposed steel penstock under external pressure.

7.7.3.3 Calculation of Critical Pressure of Smooth Steel Liner

According to the first flexural form and the basic equation of elastic stability of free thin-walled circular pipe, Fuhan and Bauerot et al. obtained the equation for calculating the total radial displacement and combined stress of steel liner through a series of assumptions and deductions. When the combined stress is equal to the yield strength, the critical pressure can be solved. According to the second form of flex, Amstutz deduced the equation for the critical pressure. Some countries in the world use the first type of equation, and others use the A's equation. The critical pressure value calculated by A's equation is low. Examples of the instability failure of underground penstock in actual engineering at home and abroad show that the second type of flex is more realistic, and the first form of flex cannot be proved by many examples. Compared with the first type of equation, the calculation result of the A's equation is also close to the model test value. Therefore, A's equation is recommended as the main calculation equation in Chinese steel penstock design specifications.

The main principle of A's derivation equation is as follows. The flexural wave is composed of three half waves, two half waves outward and one half wave inward, as shown in Fig. 7.36b. According to the internal and external force balance and the curved beam deformation equation, the relationship between the external force and the displacement of the steel liner after flex is established. After flex, the steel liner and concrete begin to detach ($\Phi = \alpha$), and the steel liner displacement η, the rotation angle $d\eta/d\Phi$ and the bending moment M of the steel liner are all 0. Meanwhile, the initial gap value Δ and the circumferential deformation of the steel liner after uniform compression are considered, which is consistent with the circumferential deformation of the steel liner after local flexure. The maximum value of the bending moment occurs at the midpoint ($\Phi = 0$) of the flexural wave, and the external pressure at which the steel liner fiber stress reaches the yielding strength σ_s is the critical pressure. For a smooth pipe, the A's equation is:

$$\left(E'\frac{\Delta}{r} + \sigma_N\right)\left[1 + 12\left(\frac{r}{t}\right)^2 \times \frac{\sigma_N}{E'}\right]^{3/2} = 3.46\frac{r}{t}(\sigma_{50} - \sigma_N)\left[1 - 0.45\frac{r}{t}\frac{(\sigma_{50} - \sigma_N)}{E}\right] \tag{7.59}$$

$$P_{cr} = \frac{\sigma_N}{\frac{r}{t}\left[1 + 0.35\frac{r(\sigma_{s0} - \sigma_N)}{tE'}\right]} \tag{7.60}$$

In which $\sigma_{s0} \leq \sigma_s/\sqrt{1 - \mu + \mu^2}$

The gap Δ includes the construction gap Δ_0, the cold shrinkage gap Δ_S of the steel penstock, the shrinkage gap Δ_R of the surrounding rock, and the plastic compression gap Δ_P of the surrounding rock. In the calculation, Eq. (7.59) is used to find σ_N, and then Eq. (7.60) is used to calculate P_{cr}. A trial calculation needs to be done when calculating σ_N. For convenience, the A' equation has been made into curves (Fig. 7.37), and the value of P_{cr} can be directly obtained based on σ_s and the main parameters of the steel liner r/t, Δ/r.

It can be seen from Fig. 7.37 that the value of r/t has the greatest effect on P_{cr}. The practical range of Δ/r in engineering is (1–4) $\times$ 10^{-4}, and the change range of P_{cr} is not large. P_{cr} increases with the increase of σ_s, but it is considered that if the high-strength steel with a higher σ_s is used, the pipe shell thickness t required to withstand the internal pressure will be reduced. Compared with steel with lower strength, when steel with high strength is used as the steel liner, the critical pressure may be lowered. This is a problem that needs to be paid attention to when high-strength steel is used.

It should be pointed out that the above theoretical equation only considers the influence of r, t, Δ, and other factors on P_{cr} and assumes that the gap Δ is uniformly distributed. In fact, there are many factors for stability against the external pressure of underground penstock, and many factors are difficult to determine, such as the size and distribution of the external pressure, the size and distribution of the gap, the ellipticity and local defects of the steel liner. Therefore, the calculation results of the theoretical equation cannot accurately reflect the actual situation.

The following empirical equations can also be used in preliminary calculations:

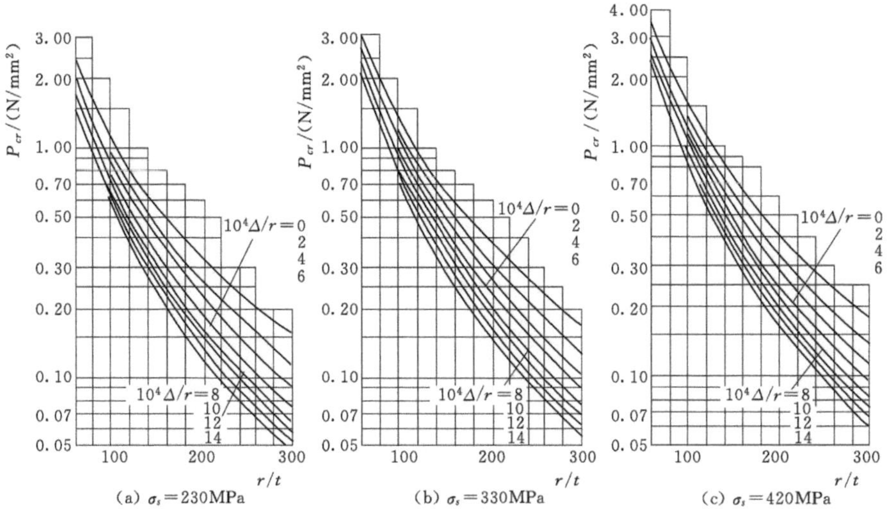

Fig. 7.37 Critical external pressure curve for underground penstock (Amstutz equation)

$$P_{\sigma} = 620\left(\frac{t}{r}\right)^{1.7}\sigma_s^{0.25}. \tag{7.61}$$

This equation is obtained from the regression analysis of 38 tests. These data are obtained by different countries and different testers at different times. This empirical equation only includes two parameters, *r/t* and σ_s, which are relatively easy to use. Since the experimental data on which the empirical equation is based objectively contains various random factors affecting the stability of the steel liner under external pressure, the equation also reflects the influence of these factors to some extent comprehensively.

7.7.3.4 Critical External Pressure Equation for Underground Penstock Provided with Stiffener Ring

Although the critical pressure of smooth pipe of the underground penstock is higher than that of the same exposed steel penstock, the external pressure of the underground penstock is often much higher, so it is often difficult for smooth pipe to meet the requirements of stability against external pressure. It is uneconomical and unreasonable to increase the shell thickness or increase the strength of the steel. The stiffener ring is widely used and is beneficial to increase the rigidity of the steel liner during transportation and construction.

Stability analysis of steel liner provided with stiffener ring under external pressure includes stability analysis against external pressure of stiffener ring and pipe shell between stiffener rings.

1. Stability calculation of stiffener ring

In theory, the stability calculation of the stiffener ring itself can also be carried out according to the equation of underground smooth pipe. Only the section characteristics of the stiffener ring need to be modified appropriately according to the effective section of the stiffener ring. Meanwhile, all the external pressure within the length of the adjacent two rings should be taken as the external pressure of the stiffener ring. However, the stiffener ring is embedded in concrete, which is constrained greatly when deforming. It is unlikely to flex inward from concrete like a smooth pipe shell. Therefore, critical pressure is calculated according to strength conditions rather than flexure in some countries, including China. The average compressive stress in the stiffener ring under external pressure should not exceed the yielding strength of the material.

$$P_{cr} = \sigma_s F / rl \tag{7.62}$$

In which

F effective section of stiffener ring, mm^2;
L spacing between stiffener rings, mm.

2. Stability of pipe shell between stiffener rings

Stability against the external pressure of pipe shell between stiffener rings of underground penstock is a space problem, and there is no perfect calculation method so far. However, within the practical range of stiffener ring spacing, due to the existence of stiffener rings, the number of flexural waves is generally large and the wave amplitude is small, and the influence of the gap between the pipe shell and concrete on the critical pressure is reduced. Therefore, in the absence of a better calculation theory, the calculation equation of exposed penstock with stiffener rings can be used (Sect. 7.6 of this chapter). That is to say, the value of the gap is considered to be large, which is more secure.

The design procedure of stability against an external pressure of steel liner provided with stiffener ring is: according to the determination of external pressure, the spacing between stiffener rings needed for the pipe shell to be stable is calculated, and then the section of stiffener rings can be designed according to the spacing.

7.7.3.5 Measures Preventing Instability of Underground Penstock Against External Pressure

Serious instability accidents of underground penstocks steel liners often occur under the action of groundwater. Taking effective measures to reduce groundwater pressure is the fundamental way to prevent the instability of steel liners. Drainage gallery combined with drainage hole is an effective measure widely adopted. Figure 7.38 is a foreign example. There are two drainage galleries on the upstream side of the penstock. The upper gallery has drainage holes with a diameter of 100 mm, a spacing of 7.5 m, and a depth of 60 m. The lower gallery has drainage holes with a spacing of

22.5 m. It is predicted that the groundwater level will be reduced by 150–160 m, and the steel liner will only withstand the external water pressure of 50–60 m. Meanwhile, drainage measures are set at the periphery of the steel liner, as shown in Fig. 7.39a. Other examples of steel liner drainage are shown in Fig. 7.39b, c.

Construction and joint grouting between steel liner and concrete should be carried out carefully to reduce the initial gap and prevent instability against external pressure. It should be noted that the grouting pressure causes the steel liner to be bulged instability during the construction period, which is also a common accident phenomenon.

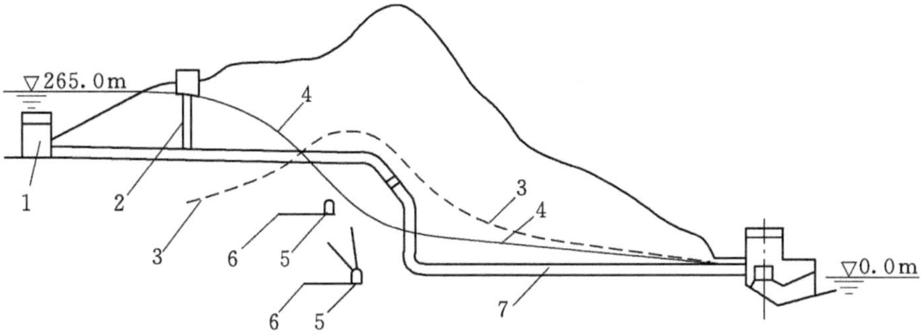

Fig. 7.38 Surrounding rock drainage of underground penstock 1-water intake; 2-gate shaft; 3-natural underground water level; 4-underground water level after reservoir repair and drainage; 5-drainage tunnel; 6-drainage hole; 7-penstock

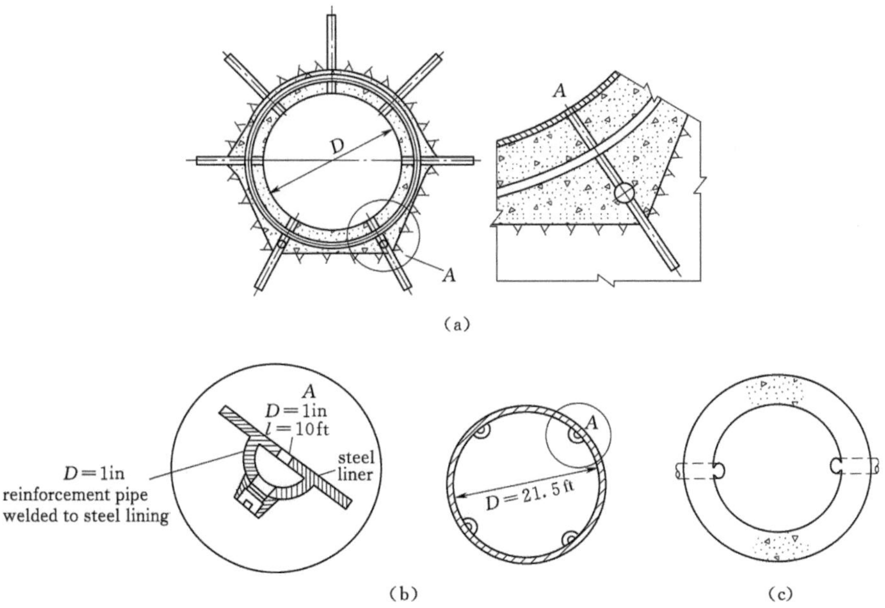

Fig. 7.39 Drainage of underground penstock steel liner

Such accidents can and should be avoided as long as the design is reasonable and avoids overfilling. The steel liner stability under the external pressure of the fluid concrete can be solved by controlling the height of the pouring layer and adopting temporary support. Generally, the permanent stiffening measures of the pipe shell should not be added to resist the external pressure load during the construction period.

7.8 Dam Body Penstocks

7.8.1 Layout Arrangement of Dam Body Penstocks

The dam body penstocks are all penstocks located in the practical section of the concrete dam. When the penstock diameter and the bearing head are large, the steel penstock structure is adopted, so it is often called the dam-embedded steel penstock. The dam body penstocks in the gravity dam are the most widely used and oldest type, as shown in Fig. 7.40a, b, c. However, the dam body penstocks can also be used in arch dams, gravity arch dams, etc., as shown in Fig. 7.40d.

The layout principles of dam body penstocks are as follows: the length of penstocks should be shortened as far as possible; adverse effects of penstocks cavity on the stress of dams should be reduced, especially the range and value of tensile stress zone in dams caused by penstocks; the interference of penstocks on dam construction should be reduced to facilitate the installation and construction of penstock itself.

1. Inclined layout

The penstocks axis is nearly parallel to the downstream dam surface and as close as possible to the downstream dam surface, as shown in Fig. 7.40a. Its advantages are that the dam-embedded intake has a high position and the water pressure is small, which is beneficial to various facilities of dam-embedded intake. The longitudinal axis of the penstock is parallel to the direction of the large principal compressive stress in the dam body, which can reduce the tensile stress in the dam around the penstocks caused by the load of dam. The position of the penstocks is high, so there is less interference in the construction of the dam. The disadvantage is that the penstock is long and there are many bends. In addition, the thickness of the concrete between the penstocks and the downstream surface of the dam is small, which is unfavorable to withstand the internal water pressure of the dam body.

2. Flat and slant layout

The penstocks are arranged in the lower part of the dam, as shown in Fig. 7.40b, d. Its advantages and disadvantages are exactly opposite to the first arrangement. For arch dams, the thickness of the dam body is not large, but when the diameter is large, it is often only possible to adopt this arrangement.

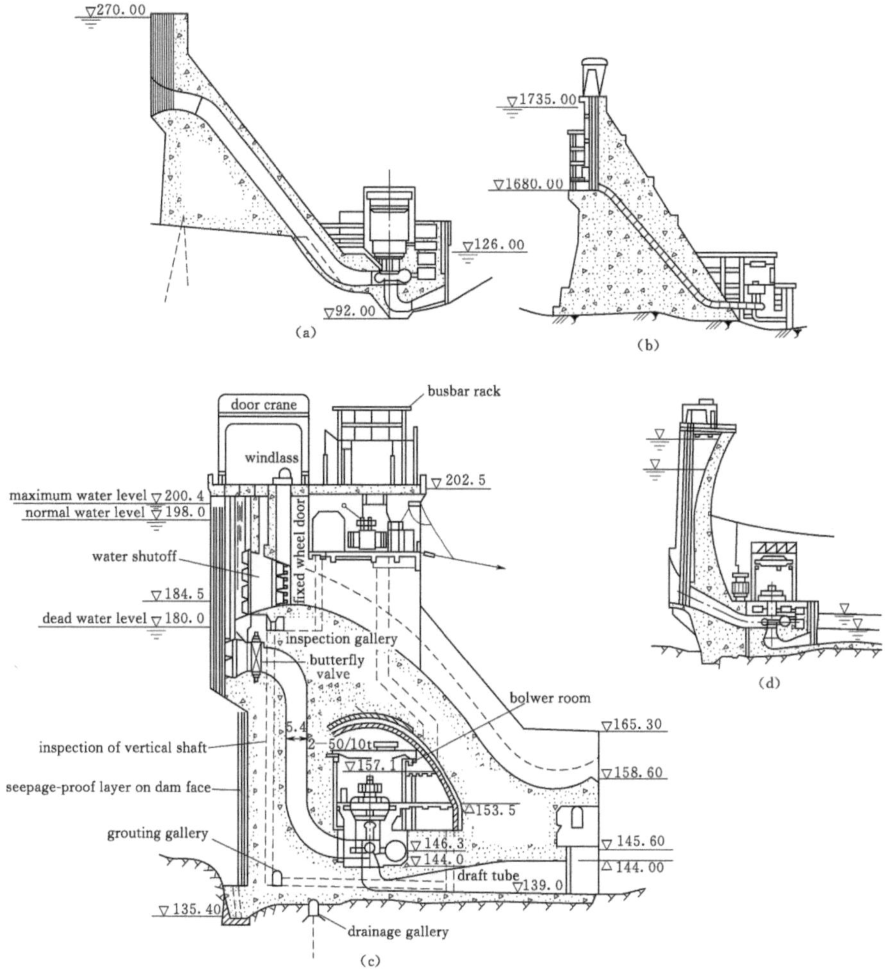

Fig. 7.40 Layout of dam body penstocks (unit: m)

3. Vertical layout

Most of the penstocks are arranged vertically, as shown in Fig. 7.40c. This arrangement is usually applied to the dam-embedded powerhouse. To avoid the interference of the steel penstock installation on the dam construction, a vertical shaft is also reserved in the dam body and a steel penstock is installed in the well later. The disadvantages are that the penstock bends greatly, the head loss is large, and the cavity caused by penstock will worsen the stress state of dam body.

In engineering practice, the first type of layout is most commonly used, because steel penstocks can be installed at a later stage and have less impact on the early pouring of the dam body. Especially for large power stations and the number of units is large, the concrete construction of the dam is often the key to the project. The progress of the dam has a very important impact on the period of power generation. Minimizing the interference of the underground penstock in the construction of the dam as far as possible is decisive in the arrangement of the underground penstocks. A steel penstock construction groove can sometimes be reserved in the dam body to avoid such interference, and the steel penstocks can be assembled once in the tank, as shown in Fig. 7.41. To meet the requirements of installing steel penstocks and backfill concrete, it is necessary to ensure that the minimum clearance on both sides and bottom is above 1 m. When the penstock diameter is large, and the reserved groove size may be large, which may cause the dam body stress to deteriorate during the construction period. The situation may be more serious if the dam body is to be loaded with water before the steel penstocks are installed. This is also a situation worthy of attention that is easy to encounter in the construction of large-scale power stations in modern times.

On the plane, the dam body penstock is preferably arranged in the center of the dam section, as shown in Fig. 7.42a. In this way, the concrete on both sides of the penstock is thicker, and the force is symmetrical. In this case, dam body and powerhouse are usually separated by longitudinal joint, and the transverse joint between powerhouse units and the transverse joint between dam sections are staggered. However, when there is no longitudinal joint between the dam and the powerhouse and they are connected as a whole, the penstock has to turn to one side on the plane because the transverse joint of the dam and the powerhouse must be in a straight line, as shown in Fig. 7.42b. At this time, the thickness of concrete wrapped on both sides of steel penstocks is different, and the left downstream may be very thin, which is unfavorable to the structural stress.

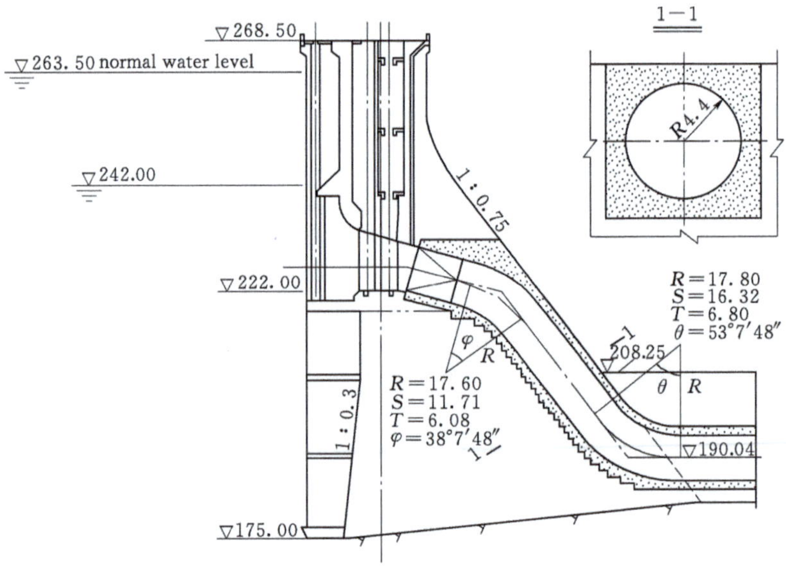

Fig. 7.41 Scheme of dam body penstock

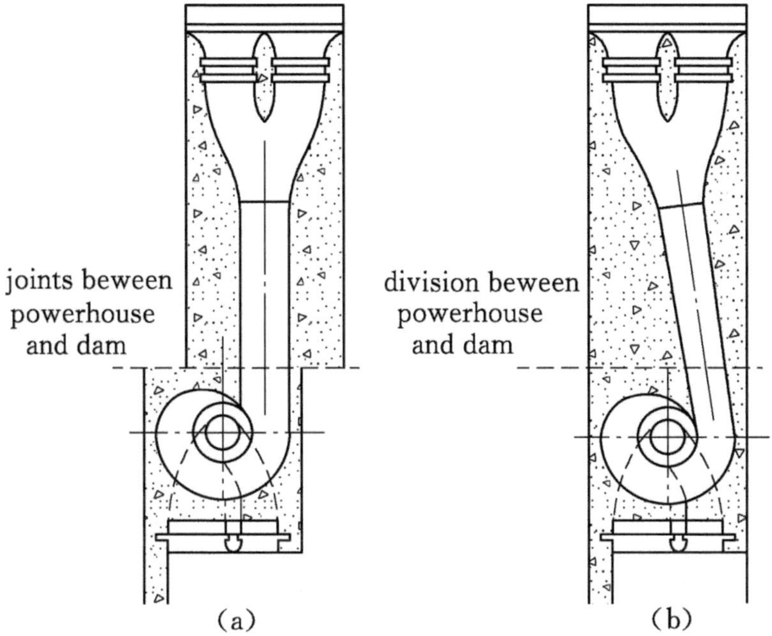

Fig. 7.42 Plane layout of the penstock in dam body penstock

The dam body penstock adopts the dam-embedded intake (Chap. 2, Sect. 7.2), and its layout and facilities must meet all the requirements of the water intake. The trash rack of the water intake is generally arranged on the cantilever of the dam body to increase the discharge area. The bulkhead gate and the service gate slot are usually arranged in the dam body, and the transition section after the gate slot is changed from a rectangular shape to a circular shape, and then it is connected to the upper horizontal section or the upper curved section of the penstock. Sometimes, the transition section can be merged with the upper curved section, and the transition section is directly connected to the inclined straight section. The water intake is located in the dam body, and the cross section of the water is large. It should be made into a narrow and high type. The transition section should be as short as possible to transition to a circular section quickly, which is beneficial to the gate structure and the dam body stress. Attention should be paid to ensuring the necessary area, the exit elevation and reasonable position of aeration vent to avoid a large inhalation airflow when inhaling, which affects the safety of equipment and operators near the outlet of the aeration vent. The area of the water-filling valve and the by-pass pipe provided in the intake space should not be too large to avoid overflowing and spraying water from the aeration vent when filling the water, which affects the normal operation of the electrical equipment between the powerhouse and the dam.

7.8.2 Structural Analysis of the Dam Body Penstocks

The transition section and the upstream part of the dam body penstocks with irregular shapes are usually not subjected to high water heads, in which reinforced concrete structures are often used; a small number of hydropower stations on silty channels are also lined with steel plates to prevent wear. Generally, starting from the transition section or below the upper bend section, the steel plate liner is used as the dam-embedded steel penstock. Usually, after the dam body concrete is poured to a certain elevation, the steel penstock is installed and positioned in the dam body, the steel bars are arranged around the steel penstock, and then the dam body concrete outside the penstock is continuously poured. In this way, the dam body penstocks and the dam body concrete become an integral steel-lined reinforced concrete structure under common stress. This part of the penstock is subject to large internal pressure, large dam body stress, and complex force, which is an important part of structure design for the dam body penstocks.

The structure of the dam body penstocks has two characteristics: ① The internal water pressure will be transmitted to the concrete of the dam outside the penstock through the steel penstock. Moreover, it can be seen from the latter analysis that for the integrally poured dam body penstocks, the steel penstock can only share a small part of the internal water pressure, and most of the internal water pressure is borne by the dam concrete. ② The existence of the penstock cavity will cause the pressure of the dam to change compared with the solid dam; that is, the tensile stress concentration is generated in the dam around the penstock. Therefore, from the structural point of view, the dam body penstocks are composed of two parts: the steel liner and the concrete outside the penstocks. Since the concrete dam is the most important building of the hydropower station behind dam, its strength and safety are very important. Therefore, when designing the structure of dam body penstock, it is not only necessary to consider the steel liner itself, but also the structural analysis and design of the dam body concrete should be put in an important position, and the analysis and design of this part is sometimes more difficult.

7.8.2.1 Loads of Dam Body Penstocks

(1) Internal water pressure.
(2) The load on the dam body or the stress of dam body. The load on the dam, such as the upstream water pressure and dead weight, will generate stress in the dam around the penstock. It can also be understood that the stress of the dam is caused by the above load in the original solid dam. Under normal circumstances, it is all compressive stress. However, due to the arrangement of the penstock cavity, the stress redistribution in the form of stress concentration of the dam body occurs in the area around the penstock. The stress redistribution occurs in a small area outside the penstock, and it produces unfavorable tensile stress in addition to the increase of the compressive stress value.

(3) Temperature load. There are mainly two types: one is the temperature force in dam concrete around dam body penstock cavity caused by temperature rise during concrete pouring; the other is the temperature force caused by the temperature difference between water in penstock or air on dam surface and concrete outside penstock when the penstock is in service.
(4) Seepage pressure of dam body. The water pressure is formed by the reservoir water seeping along the outer shell of the steel liner from the head end of the steel liner. When the penstock is emptied, the water seepage pressure will act on the outer shell of the steel liner, which may cause the instability of steel liner.
(5) Construction load. Mainly refers to grouting pressure or unsolidified concrete pressure, etc. The first three types are the loads during operation and are the most important. Different load combinations should be considered for structural analysis as shown in Tables 7.3 and 7.4 for details. The dam-embedded steel penstock is surrounded by the dam concrete, so the secondary loads such as the dead weight of steel liner and seismic force are usually not considered in the design.

7.8.2.2 Structural Design of Dam Body Penstocks

The main task of the dam body penstocks structure design is to determine the steel liner material and shell thickness, the concrete grade and reinforcement of the dam outside the penstock, and the measures preventing instability of steel liner against external pressure.

The design requirements of the dam body penstock structure are as follows:

(1) The stress of steel liner and steel bars is less than the allowable value of steel.
(2) The concrete of the dam does not crack, at least the scale and range of the crack are limited, and it does not extend to the dam surface.
(3) Steel liner is stable under external pressure.

The issue of the stability against external pressure of dam body penstocks can be determined by referring to the relevant calculation methods for underground penstocks in Sect. 7.7.

The installation and concrete pouring conditions of dam body penstock are better than those in the underground penstock. The joint between the steel liner and the dam is generally grouted, and the gap value is not large. Therefore, when the two are jointly supported, it is not difficult to achieve the requirement that the stress of steel liner and steel bar is less than its allowable value as long as there is no penetrating wide crack in the dam body concrete. However, since most of the internal water pressure is transmitted to the dam body to produce tensile stress, the stress and temperature difference of the dam body will cause tensile stress in the dam body at the same time. It is not always possible to make the tensile stress in the dam body less than the allowable tensile stress of concrete to achieve non-cracking. Increasing the thickness of steel liner and the amount of reinforcement is conducive to crack prevention, but the effect is not significant, and it is uneconomical and unreasonable.

The key to current design methods is as follows:

1. Selection of the shell thickness of steel liner

If the minimum thickness of the concrete outside the steel liner is not large, for example, less than the radius of the steel penstock, the dam body may be through crack under load. It may be considered that the steel liner alone receives the internal water pressure to determine the shell thickness for the sake of safety, and the allowable stress can be slightly larger than the exposed penstock (0.67 σ_s for the normal operating condition). If the minimum thickness of the concrete outside the steel liner is larger than the diameter of the steel liner, then the steel liner, steel bars, and concrete should be considered to jointly bear the internal water pressure, and the shell thickness of the steel liner should be selected according to the calculation of the joint-bearing principle. However, for the sake of safety, the internal water pressure under normal operating condition alone of the steel liner should be taken as a checking condition, and the allowable stress can be increased to 0.9 σ_s to determine the minimum shell thickness.

2. The concrete strength and reinforcement calculation

After selection of the shell thickness of steel liner, the amount of the circumferential steel bar should be set appropriately as shown in Fig. 7.43a. Then the steel bar is converted into a steel liner with a shell thickness of t_3. The steel liner and the peripheral dam body are simplified into an axisymmetric combined cylinder of two-layer steel liner and two-layer concrete. The minimum distance between the center of the dam body penstock and the downstream surface of the dam or the transverse joint of the dam section is taken as the outer radius of the combined cylinder, as shown in Fig. 7.43b. In this way, the analytical method can be used to calculate the stress of each component under internal water pressure. There may be three cases: the concrete is not cracked, the concrete is partially cracked, and the concrete is through cracked. Figure 7.43b is the case of partially cracked under P.

The symbols used are as follows:

- r_0 inner radius of steel liner, mm; $E_{c'} = E_c/(1-\mu_c^2)$, N/mm^2;
- r_3 centerline radius of circumferential steel bar; $\mu_{c'} = \mu_c/(1-\mu_c)$;
- r_4 outer radius of concrete cracked zone; E=elastic modulus of steel, N/mm^2;
- r_5 outer radius of concrete complete zone;
- E_c elastic modulus of concrete, N/mm^2;
- t shell thickness of steel liner, *mm*; μ_c = Poisson's ratio of concrete;
- t_3 converted thickness of circumferential steel bar, mm;
- $[\sigma_l]$ allowable stress of concrete, N/mm^2;
- Δ gap value between steel liner and concrete, mm;
- $\sigma_{\theta 1}$ circumferential tensile stress of steel liner, N/mm^2;
- P internal water pressure, N/mm^2;
- $\sigma_{\theta 2}$ maximum circumferential tensile stress of concrete, N/mm^2;
- P_l Radial contact pressure between steel liner and concrete, internal water pressure transmitted from steel liner to dam concrete, N/mm^2;

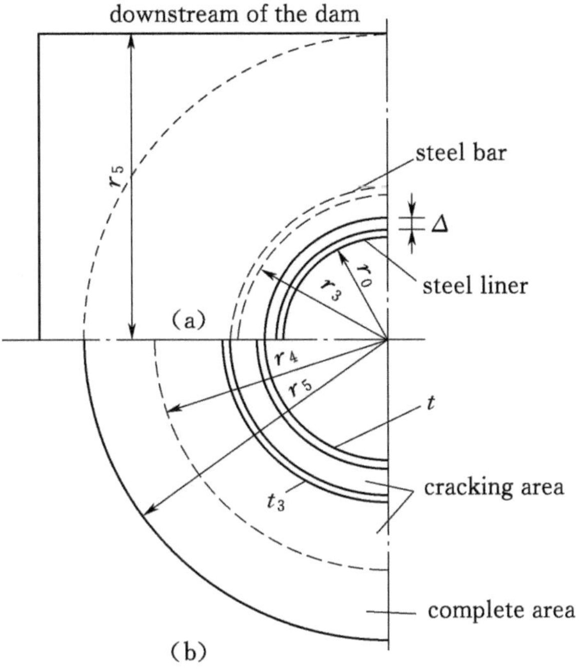

Fig. 7.43 Calculation diagram of dam body penstock

$\sigma_{\theta 3}$ tensile pressure of steel bar, N/mm^2.

(1) If the maximum tensile stress in concrete is less than the allowable tensile stress and the concrete is not cracked, then

$$P_1 = (P - \frac{E_{\prime}\Delta t}{r_0^2})/\left[1 + \frac{E_{\prime}t}{E_{c^{\prime}}r_0}\left(\frac{r_5^2 + r_0^2}{r_5^2 - r_0^2} + \mu_{c^{\prime}}\right)\right] \tag{7.63}$$

$$\sigma_{\theta 1} = (P - P_1)r_0/t \tag{7.64}$$

$$\sigma_{\theta 2} = P_1(r_5^2 + r_0^2)/(r_5^2 - r_0^2). \tag{7.65}$$

In such case, the concrete does not crack, so

$$\sigma_{\theta 2} \leq [\sigma_l].$$

Generally, $\sigma_{\theta 1}$ is not large, less than its allowable stress. The stress of steel bar is very small, so it cannot be calculated.

(2) The concrete is partially cracked, the cracked zone only transmits radial pressure, and the un-cracked zone still participates in the bearing. At this time, the maximum circumferential tensile stress $\sigma_{\theta 2}$ of the concrete is located at the inner

edge of un-cracked zone (the radius is r_4), and its value is exactly equal to the allowable stress, then:

$$\sigma_{\theta 1} \leq \sigma_{\theta 3} r_3 / r_0 + E\prime \Delta / r_0 \tag{7.66}$$

$$\sigma_{\theta 3} = \frac{E\prime r_5}{E_c\prime r_3}[\sigma_l]\left\{m\left[\ln\left(\psi \frac{r_5}{r_3}\right) + n\right]\right\} \tag{7.67}$$

$$P_1 = P - \sigma_{\theta 1} t / r_0 \tag{7.68}$$

In which $m = \psi \frac{1-\psi^2}{1+\psi^2}$; $n = \frac{1+\psi^2}{1-\psi^2} + \mu_c\prime$;

$\psi = r_4/r_5$, it is called the relative cracked depth of concrete. It needs to be obtained by trial calculation or by checking ready-made curve. See Appendix C of *Design Specification for Steel Penstocks of Hydroelectric Stations* SL 281-2003.

In this case, $\sigma_{\theta 1}$ is generally not large, small than its allowable stress. The stress of steel bar $\sigma_{\theta 3}$ is very small.

(3) The concrete is through cracked and does not participate in the bearing. The internal water pressure is shared by steel liner and steel bar, then

$$P_1 = \left(P - \frac{E\prime \Delta t}{r_0^2}\right) / \left(1 + \frac{t r_3}{t_3 r_0}\right) \tag{7.69}$$

$$\sigma_{\theta 1} = (P - P_1) r_0 / t \tag{7.70}$$

$$\sigma_{\theta 3} = P_1 r_0 / t_3. \tag{7.71}$$

In this case, the stress of steel bar $\sigma_{\theta 3}$ must be less than the steel liner stress $\sigma_{\theta 1}$.

It must be noted that the above equation is derived from the following premises:

a. Only a single internal water pressure effect.
b. Simplify the structure to an axisymmetric combination ring with uniform gap.
c. Uniform radial cracks with equal length are formed after cracking of steel liner concrete.

According to the above calculation under the action of the internal water pressure, the stress of the steel liner, the steel bar, and the internal pressure P_1 transmitted from the steel liner to the dam body concrete can be obtained under the set amount of circumferential reinforcement.

This is the design calculation of dam body penstock bearing internal water pressure. In addition to the internal water pressure, the dam body penstock is also subjected to the dam body load and temperature load, which should be considered simultaneously with the internal water pressure in the structural design. The current structural design method is: the internal pressure transmitted from the steel liner to the concrete of the dam body calculated under the above internal water pressure load is used as the uniform internal pressure acting on the inner circumference of the concrete orifice, and the concrete is completely homogenized regardless of whether the concrete is cracked or not. The inner circumferential stress in the concrete can be calculated by looking up table according to the manual or the finite element method. Similarly, the dam body stress and the circumferential stress in the concrete caused by temperature differences can be calculated by looking up table according to the manual or the finite element method. The circumferential stress generated by all loads is superimposed to obtain the total hoop stress distribution in the concrete, as shown in Fig. 7.44. Then, the reinforcement is carried out according to the calculation principle of reinforcement for reinforced concrete structure of no-bar system in the *Hydraulic Concrete Structure Design Code* SL191-2008. That is, the total tensile stress of the region where the tensile stress is greater than $0.45 f_t$ is borne by the steel bar. The area of the reinforcement within the unit length of the penstock is

$$A_S = KT/f_y \tag{7.72}$$

In which

A_s the area of the reinforcement within the unit length of the penstock, mm^2;
T the total tensile force shared by steel bar, N;
K strength safety coefficient of tensile reinforcement;
f_y design value of steel bar tensile strength, N/mm^2.

The calculated value of the temperature stress is often large, but it is difficult to get it accurately. Therefore, in actual projects, efforts are often made to do temperature control, maintenance, etc., and reduce the f_y value appropriately without calculating the temperature stress; that is, the amount of reinforcement is appropriately increased to solve the problem.

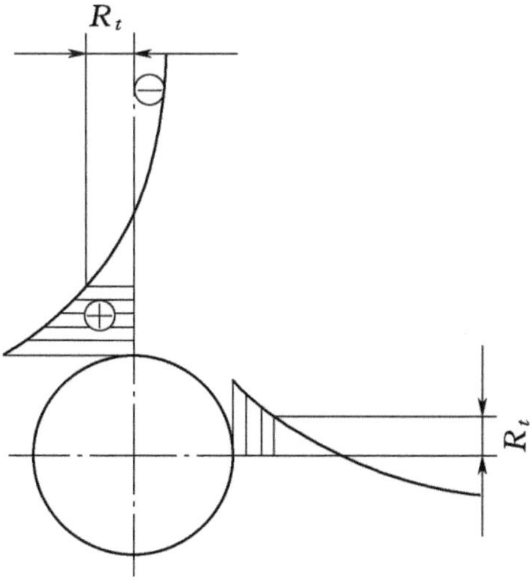

Fig. 7.44 Reinforcement method for non-bar system

If the amount of reinforcement obtained according to the above method is slightly smaller than the original amount of reinforcement, it will meet the requirements. Otherwise, the amount of reinforcement will be reset and the calculation will be repeated until the requirements are met.

Although the above design methods of the dam body penstock are applied in engineering, there are also contradictions and problems. First of all, when analyzing whether the dam body is cracked or not and the cracking shape, which is a very important problem for the stress state of the dam body penstock, only the internal water pressure is considered, and the structure is simplified into a combined ring. Therefore, it is considered that when the concrete around the steel liner is cracked without through cracked, a uniform cracking zone with equal thickness in the radial direction is formed, which is not practical. Secondly, when the reinforcement is made according to the stress figure method, the concrete stress surrounding the steel liner is calculated according to the homogeneous elastic body. Whether the concrete cracks under the action of load is not fully considered; and when the reinforcement is used, the stress of steel bar takes its design strength. When the stress of steel bar reaches the design strength, the concrete may have cracked, and no longer is the original homogeneous elastic body, and the stress state has also changed, which indicates there is a contradiction between the stress figure and the calculated stress of the steel bar. Therefore, it is difficult to give the real stress of steel liner and steel bars based on the current design method.

The existing engineering practice shows that the above common structural design method can generally ensure safety when the diameter and water head of the dam body penstock are not too large, but it is not economic and reasonable. The performance is as follows:

a. The operating stress of steel liner is mostly low.
b. The bearing capacity of the structure is not fully exerted.
c. The steel bar is too dense, which affects the construction.

Moreover, with the increasing scale of the dam-embedded power station, and the continuous improvement of dam height, diameter of the dam body penstock and water head, the current design method has difficult to solve the technical and economic problems of the project. For example, if the thickness of the steel liner is too large, it is difficult to ensure the quality and economy of the fabrication and installation process. The tensile stress value and the tensile stress zone caused by the dam body penstock are too large, and the crack resistance safety of the dam body is difficult to guarantee. Therefore, the improvement of the design method and structural measures of the dam body penstock has been paid more and more attention. Domestic research in this area has progressed in recent years. They mainly are:

a. Considering all the loads of the dam body penstock, and analyzing the stress according to the actual shape of the dam body penstock, it is no longer simplified as an axisymmetric thick-shelled cylinder.

b. Three-dimensional numerical analysis method is used to analyze the stress of the dam body penstock.
c. Considering that concrete is a nonlinear elastic material, using its real elastic–plastic properties, it analyzes the crack resistance ability of the concrete surrounding the steel liner to ensure the crack resistance safety of concrete.
d. Steel liner and steel bars are arranged according to the strength requirement of concrete cracking in such an extreme state to ensure the bearing safety of the dam body penstock.

Based on the above concepts, the design calculation principles and methods can not only overcome the defects that contradict each other in theory existing in current design method, but also can reliably judge the real crack resistance of the concrete and the bearing capacity of dam body penstock. Steel liner and steel bars can be configured more reasonably and effectively, so structural safety is increased and steel consumption is saved significantly.

7.8.2.3 Stability Against External Pressure of the Dam Body Penstock Steel Liner

The stability analysis principle and method of dam body penstock steel liner against external pressure are the same as the steel liner of underground penstock. The external pressure load of the dam body steel liner mainly includes external water pressure, fluid concrete pressure during construction and grouting pressure. The temporary load during the construction period should not be used as the design control condition and should be solved by adding temporary support and controlling the concrete pouring height. The external water pressure on the steel liner is derived from the downward seepage from the beginning of the steel liner along with the outer shell of the steel liner. Seepage water pressure can be assumed to vary linearly along the axis of the penstock, and the forward end of the steel liner is αH. The seepage water pressure where the steel liner passes through the joint between powerhouse and the dam is zero. H is the hydrostatic pressure from the upstream normal storage level to the forward end of the steel liner. α is the reduction coefficient, which can be taken from 1.0 to 0.5 according to the adopted measures such as anti-seepage, drainage, and grouting. For safety measures, the minimum external pressure of all steel liners is not less than 0.2 MPa. The upstream section of the steel liner is subjected to small internal pressure, and the pipe shell is thin, but the seepage pressure outside the steel liner is large, which is the key to the stability against external pressure. Anti-seepage measures such as water-blocking ring should be adopted at the forward end of the steel liner, and drainage measures should be provided behind the water-blocking rings, so that the osmotic pressure of the steel liner can be effectively reduced. Joint grouting can reduce the gap value and also beneficial to the stability against the external pressure of steel liner. Accidents of external pressure instability of the dam body penstock steel liner are relatively rare when it is emptied.

7.8.3 Dam Body Steel Penstocks with Cushion

The joint-bearing dam body penstocks, which is poured with the concrete of the dam body as a whole, will transmit most of the internal water pressure to the dam body, causing tensile stress in the dam body, which will affect the safety of the dam body and even crack and increase the amount of steel bars. Soft cushion is set between the steel liner and the dam body concrete to reduce the pressure taken by dam, increase the stress of steel liner and prevent the cracking of dam body. The soft cushion can absorb the radial displacement of the steel penstock under the action of internal pressure, so that only a part or a small part of the internal pressure is transmitted to the dam body. In the past, this measure was common in a small part of the penstock length at the joint between the dam body steel penstock and the powerhouse to meet the requirements of different displacements of the powerhouse and dam. With the increase of the diameter and water head of the dam body steel penstock, this type of structure has begun to be applied to the main part of the dam body steel penstock. The Toktogul Hydropower Station of the Soviet Union (Fig. 7.45), which was put into operation in the 1970s, adopted this structure in the vertical section of the dam body steel penstock in the gravity dam. The soft cushion was laid at the full circumference of the steel penstocks.

Some large-scale dam-embedded hydropower stations in Brazil also adopt this type. For example, Itaparica Hydropower Station has a unit capacity of 250 MW, a diameter of 9.5 m steel penstock in gravity dam and a maximum hydrostatic head of 60.4 m. The steel penstock is located in the center of the dam section with a length of 31 m, which is parallel to the downstream dam surface and buried very shallow. The minimum thickness of concrete on the top of the steel penstock is only more than 2 m. The maximum shell thickness of the steel penstock is 19 mm, and the upper half of the steel penstock is within the range of 204°, including the outside of the T-shaped stiffener ring, and is covered with a soft cushion. The lower half of the steel penstock including the rectangular stiffener ring and the steel support during lifting is all poured into the concrete dam body, and the grouting hole is reserved, and finally grouting. There is no soft cushion at the upper and lower ends of the whole steel penstock (where the upper part is connected with the concrete transition section, including the water-stop ring, and the lower part is connected with the spiral case transition section), so it becomes two fixed ends of the steel penstock. The cushion material is expanded polystyrene with a thickness of 20 mm and a design compression of 3 mm. The cushion material is glued to the outer surface of the steel penstock, and then concrete is poured on it.

It is the first time for Shuikou Hydropower Station in Fujian (Fig. 7.46) and Manwan Hydropower Station in Yunnan to adopt the dam body steel penstock with a soft cushion in China. Shuikou Hydropower Station is a type of power station behind the gravity dam with 7 units installed and a single unit capacity of 200 MW. The length of the diversion dam section is 20.5 m. The water diversion system consists of trash rack, intake, transition section, and steel penstock section. The end of the transition section to the volute intake is a steel penstock section with a total length of

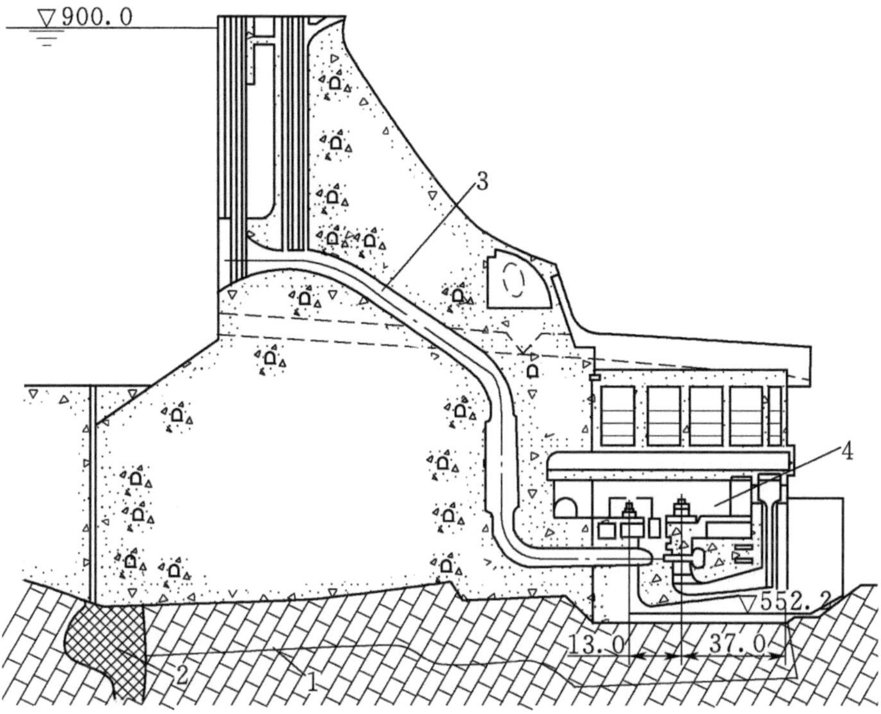

Fig. 7.45 Toktogul hydropower station 1-the boundary of consolidation grouting; 2-grouting curtain; 3-penstock; 4-powerhouse with two rows of units

59 m, a diameter of 10.5 m, a normal hydrostatic head of 61 m at the volute intake, and a concrete thickness of 2.25 m covered on the lower horizontal section and the elbow section. At a distance of about 20 m below the intake of the steel penstock to the intake of the volute, a soft cushion is laid in the range of 180° in the upper half of the steel penstock, and the thickness is 30 mm, and the cushion material is polystyrene (PS) foam plate. The shell thickness of the steel penstock is 23–24 mm. The soft cushion is pre-formed into a curved tile and is glued to the outer surface of the installed steel penstock, which is simple and rapid. Then tie the steel bars and pour the concrete outside the penstock.

The diameter of the dam body steel penstock of Manwan Hydropower Station is 7.5 m. A soft cushion layer is set from the starting point of the lower bend section to the intake of the spiral case. The normal hydrostatic pressure in the penstock is 90–104 m, and the maximum thickness of penstock shell is 28 mm. The upper part of the steel penstock is the auxiliary powerhouse, the thinnest part of the concrete outside the penstock is only about 2 m. After the soft cushion is installed, it not only reduces the reinforcement compared with the design scheme of joint-bearing capacity, but also prevents the concrete outside the penstock from cracking and ensures the safety of the dam.

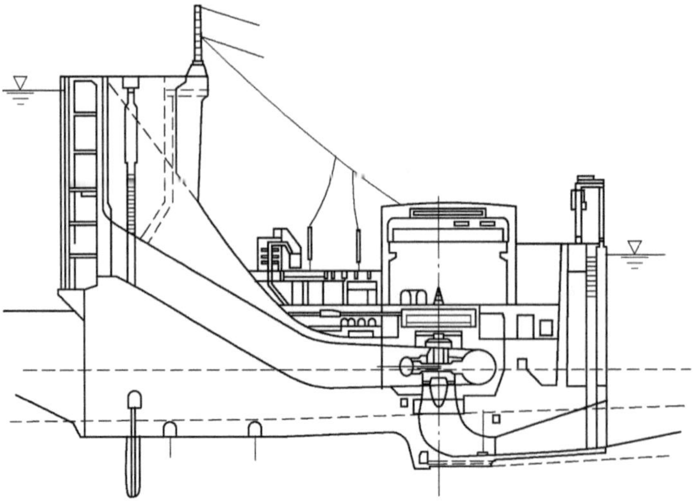

Fig. 7.46 Shuikou Hydropower Station

The stress analysis content of the dam body steel penstock with cushion is the same as the dam body steel penstock without cushion, but the effect of soft cushion must be considered when analyzing the stress of steel penstock and concrete of dam body under internal water pressure. Plane numerical method is used to calculate stress in domestic engineering. Because the deformation properties of the cushion material used at present are nonlinear elasticity under compression and the deformation modulus is not easy to be determined, the deformation modulus of the cushion should be the lowest possible value when calculating the stress of the steel penstock, and the highest value when calculating the stress of the concrete outside the penstock. The allowable stress of steel penstock is slightly higher than that of exposed steel penstock. At present, soft cushion (deformation modulus is 1to 2 MPa) and thicker cushion (more than 30 mm) is used in engineering to minimize the internal water pressure transmitted to concrete as far as possible.

In the case where the soft cushion is only laid in part of the penstock circumference, the end of the soft cushion, as shown in the middle point *A* of Fig. 7.47a, there is local stress in the circumferential direction of the steel penstock shell, which deserves attention. To reduce the local stress, the cushion thickness can be gradually reduced to zero at this point, as shown in Fig. 7.47b. Because the allowable stress in the local stress zone can be increased, the shell thickness of steel penstock should not be increased due to its existence.

The dam body steel penstock with soft cushion can fully exert the function of bearing internal pressure, greatly reduce the tension stress in dam body concrete, increase crack resistance safety, reduce the amount of reinforcement, and allow the steel penstock to be buried shallowly to reduce the interference of dam construction and steel penstock installation. These advantages are obvious. However, compared with the joint-bearing dam body steel penstock without cushion, the shell thickness

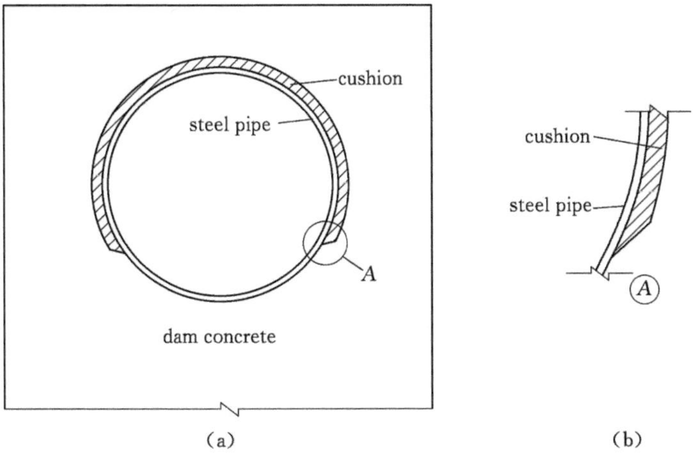

Fig. 7.47 Schematic diagram of cushion pipe section

of steel penstock will increase; there is certain local stress at the end of cushion, which forms a bidirectional local stress zone with longitudinal local stress at the pipe shell of the stiffener ring, which deserves attention, and appropriate measures should be taken to alleviate it. At present, the cushion material used is far from the linear elastic material, and there is still a lack of understanding on how to change its performance in the long-term buried environment. Therefore, it is difficult to calculate the distribution of the internal water pressure between the steel penstock and the concrete and the stress state very reliably. These issues need to be further studied and accumulated experience. It can be considered that the steel penstock with a soft cushion will have good application prospects in the dam body steel penstock of large diameter without too high water head. However, when the water head is large, the steel penstock shell thickness of this type may be too large, which will cause technical difficulties in material, processing technology and transportation and installation of the steel penstock itself.

7.9 Penstocks Laid on the Downstream Surface of Dams

Since the 1960s, the penstocks have been laid on the downstream surface of the dam in large-scale concrete dam-type hydropower stations with higher heads at home and abroad. The water intake is located on the upstream dam surface, the penstock passes through the upper part of the dam near horizontally and then is laid along the downstream surface of the dam outside the dam and is connected with the turbine spiral case by curved and horizontal sections, and the dam body and the downstream surface penstock are successively constructed, and the penstock is fixed on the dam body by keyed grooves and anchorage bars, as shown in Fig. 7.48.

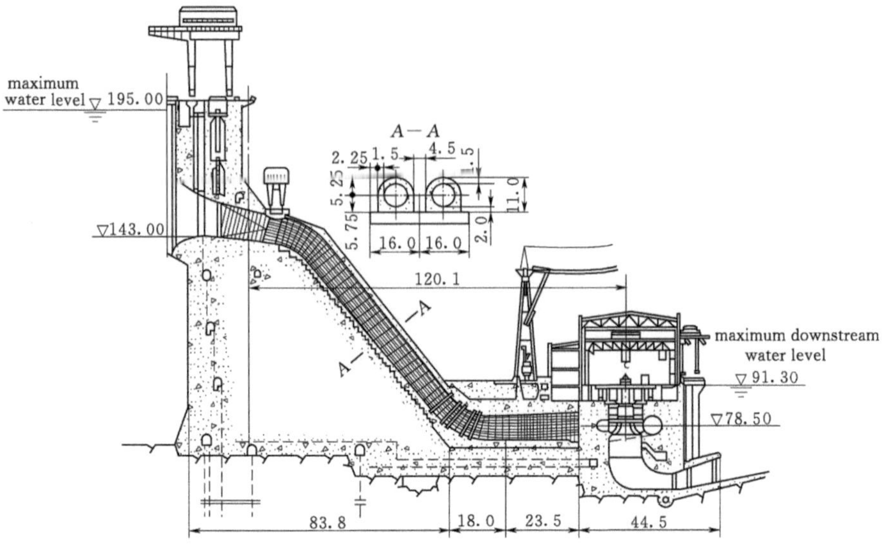

(a) Krasnoyarsk Hydropower Station

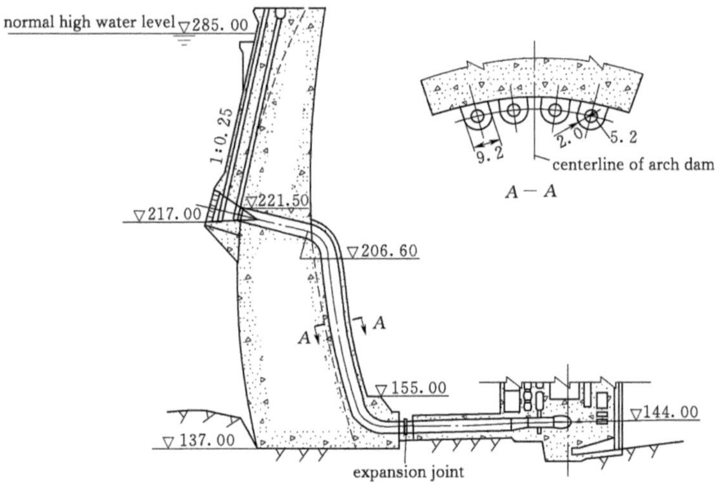

(b) Dongjiang Hydropower Station

Fig. 7.48 Layout of penstock laid on the downstream surface of the dam (unit: m)

Compared with the dam body penstock, its advantages are as follows:

(1) Easy to arrange. For thin concrete high dams, if large-diameter penstocks are laid in the dam body, it is often necessary to lower the intake and lay the penstock in the lower part of the dam body, which increases the difficulty of intake facilities. The intake can be raised as much as possible by using the downstream surface

penstock of the dam. Because the main part of the penstock is on the downstream dam surface, it is easy to maintain.

(2) Reducing the weakening of penstock cavity to the dam body is beneficial to dam safety.
(3) The construction of the dam body is not disturbed by the construction and installation of the penstock, which can improve the construction quality of the dam body, speed up the progress, and generate electricity in advance.
(4) Penstock can be constructed in stages along with the production of the units, which is conducive to rationally arranging the construction progress and reducing the backlog of investment. When the number of units is large, the benefits are more significant.

For high arch dams and large-diameter penstocks, these advantages are more obvious.

The disadvantage of this kind of penstock is that due to the dam body section above the upper bend section being lengthened, the inclined penstock section is shifted downstream, making the powerhouse move downstream, increasing the distance between the axis of the powerhouse and dam and increasing the engineering quantity. In addition, construction trestles and construction machinery are often arranged on the downstream surface of concrete dams, which will interfere with penstock construction.

Penstock laid on the downstream surface of the dam, referred to as the penstock laid on the downstream surface of the dam, has two structural types: the exposed steel penstock laid on the downstream surface of the dam and the steel-lined reinforced concrete penstock laid on the downstream surface of the dam.

7.9.1 Exposed Steel Penstocks Laid on the Downstream Surface of Dams

After the steel penstock passes through the dam body, it is connected with the upper bend penstock, which is anchored on the dam body. It is the upper fixed end of the downstream surface penstock of the dam and then connected with the exposed steel penstock through the expansion joint. Exposed steel penstock is supported by the support block on the downstream surface of the dam. Its layout and structure are similar to ordinarily exposed steel penstock. The inclined straight section of the steel penstock is connected with the lower bend penstock. The downward bend penstock is anchored on the dam body, which is the lower fixed end of the downstream surface penstock of the dam, and then enters the powerhouse to connect with the spiral case. The River Stoke Hydropower Station in Canada (Fig. 7.49), the Itaipu Hydropower Station (Fig. 7.50) built by Brazil and Paraguay, and the Big Bird Hydropower Station in Japan all adopt this kind of penstock. The Rivel Stoke Power Station is a gravity dam with a dam height of 136 m and a steel penstock diameter of 7.9 m. The upper bend section is covered with pre-stressed concrete, and the inclined straight section is an exposed penstock. The Itaipu Power Station is a double-block gravity dam with a dam height of 190 m and a steel penstock diameter of 10.5 m. The inclined straight section is an exposed penstock with a shell thickness of 30–65 mm. The Big Bird Power Station is an arch dam with an exposed steel penstock diameter of 7.5 m on the downstream side and a calculated head of 54 m. There is no use of exposed steel penstock laid on the downstream surface of dam.

The on-site installation workload of the exposed steel penstock laid on the downstream surface of the dam is small, the progress is fast and has little interference with the dam construction. However, when the diameter of the steel penstock and the water head is large, it may induce difficulties in steel material, rolling, and welding technology. If an accident of the exposed penstock laid on the downstream dam

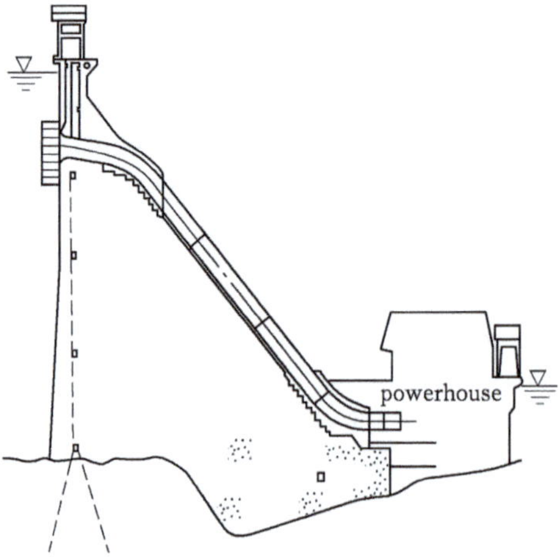

Fig. 7.49 River stoke hydropower station

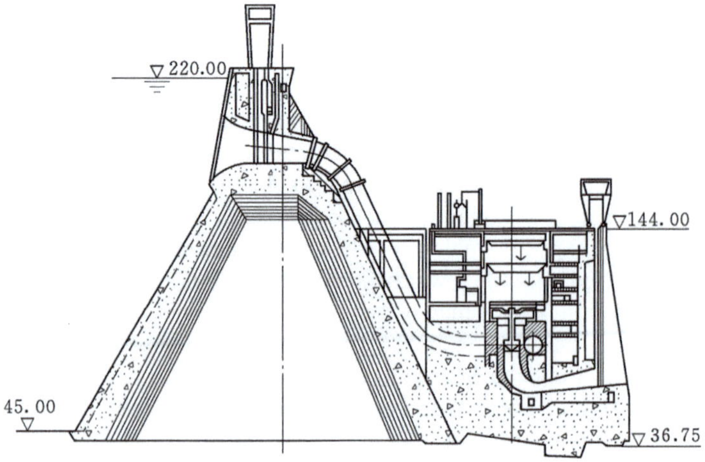

Fig. 7.50 Itaipu hydropower station

surface happens, the flow will rush into the powerhouse directly and cause serious consequences. Therefore, the steel penstock must have high safety and reliability.

7.9.2 *Steel-Lined Reinforced Concrete Penstock Laid on the Downstream Surface of Concrete Dams*

Steel-lined reinforced concrete penstock laid on the downstream surface of the dam is a combination of inner steel liner and reinforced concrete in which steel liner is encased. The penstock is fixed on the dam by keyed grooves and anchorage bars. There is no cushion set between steel liner and reinforced concrete, so the steel liner can contact with concrete compactly, which makes the two components bear the water pressure together.

The essence of this penstock structure is to replace part of the steel plates with reinforced concrete. Its advantages are:

(1) The penstock is located outside the dam body, so that concrete encasement is permitted to crack, leading the bearing capacity of steel liner and steel bar can be fully utilized.
(2) Due to the load-bearing of steel bars, the thickness of steel shell can be reduced, avoiding technological and economic problems caused by the use of high-strength steel.
(3) The joint of circumferential reinforcement can be staggered, so the technological defects will not be concentrated in one section; thus, the serious consequence of concentrate fracture caused by the defects of welding seam and material can be avoided.
(4) It is beneficial to reduce the damage from external environment to the penstock and anti-freezing in severe cold areas.

Compared with the exposed penstock, the steel-lined reinforced concrete penstock laid on the downstream surface of the dam increases the engineering amount of concrete in the pipe shell, and the penstock structure is more complicated, and there are many processes of construction and installation. If the piping arrangement is reasonable, it will not interfere with the construction of the dam or increase the additional period of construction.

This penstock was first used in the former Soviet Krasnoyarsk Hydropower Station, which was put into operation in 1967, as shown in Fig. 7.48a. The gravity dam is 124 m high and has 10 units with a single unit capacity of 500 MW. Each unit uses two penstocks with an internal diameter of 7.5 m for water diversion. The two penstocks are combined at the horizontal section in front of the powerhouse to form a single penstock connected to the volute intake. In the former Soviet Union, there were about 10 built and designed power stations using this type of penstock. Typical representative of the current level is the dam downstream penstock of the Sayano-Shushenskaya Hydropower Station (Fig. 7.51), which was commissioned in 1978. The gravity arch dam of the power station is 242 m high and has a single unit capacity of 640 MW, with a total of 10 units. Each unit uses a penstock with an internal diameter of 7.5 m for water diversion. The steel liner thickness of the penstock is 16–30 mm, and the thickness of the wrapped reinforced concrete is 1.5 m. The diameter of the circumferential reinforcement used is 60 mm and 70 mm. The Dongjiang Hydropower Station (Fig. 7.48b), the Jinshuitan Hydropower Station (Fig. 7.52), the Lijiaxia Hydropower Station (Fig. 7.53), and the Wuqiangxi Hydropower Station have been built in China all adopted such penstock. The arch dam of the Lijiaxia Hydropower Station is 165 m high, and the inner diameter of the penstock is 8 m. The gravity dam of Wuqiangxi Hydropower Station is 87.5 m high, and the inner diameter of the penstock is 11.2 m.

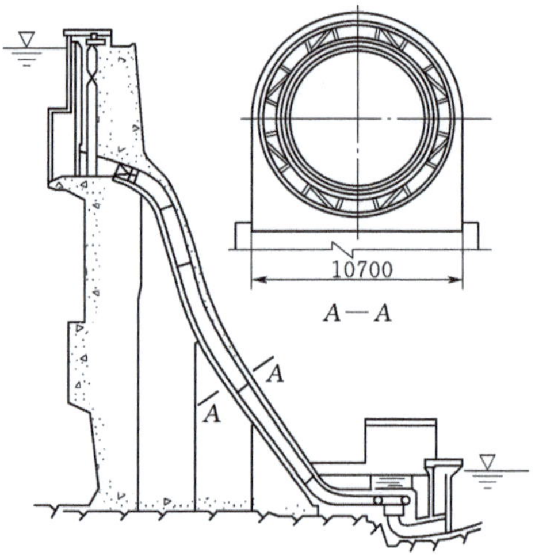

Fig. 7.51 Sayano-Shushenskaya hydropower station

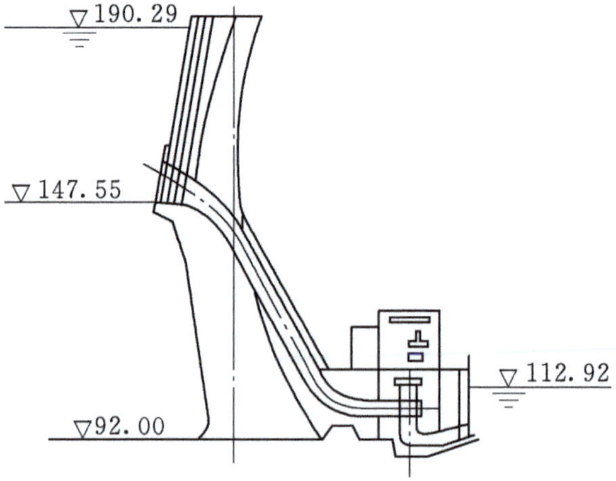

Fig. 7.52 Jinshuitan hydropower station

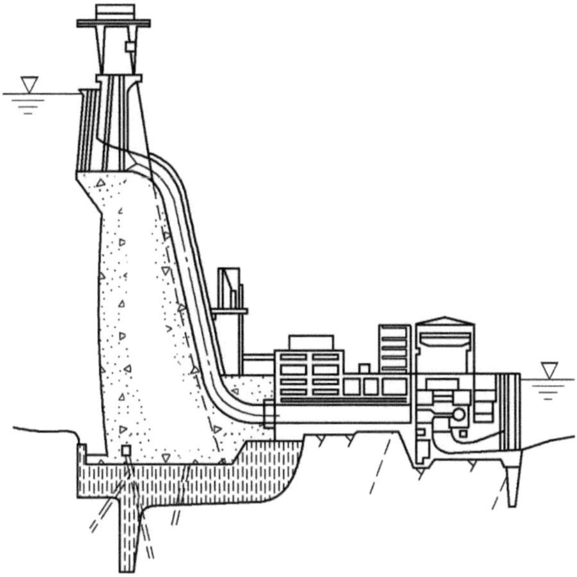

Fig. 7.53 Lijiaxia hydropower station

7.9.2.1 Requirements of Load and Structure Design

Load of steel-lined reinforced concrete penstock laid on downstream surface of dam:

(1) Internal water pressure.
(2) Axial force caused by the deformation of dam body.
(3) Temperature load.
(4) Vibration load.

Internal water pressure, including hydrostatic pressure and water hammer pressure, is the main load on the downstream surface penstock of the dam, which is borne by steel liner and reinforced concrete together. When the load is small and the concrete does not crack, the circumferential force caused by internal water pressure is mostly borne by concrete, and the stress on steel liner and steel bar is not large. When the load is large, the radial cracks occur in the shell of concrete penstock, and the circumferential tension at the crack is jointly borne by the steel liner and the steel bar.

Under the loads, axial force acting on the penstock, the shear force and normal force between the penstock and the dam body surface are caused. The axial stress of the downstream surface penstock of the dam is mainly compressive stress, which distributes unevenly along the cross section of the penstock and exists in both steel liner and penstock shell concrete. Then an adverse three-dimensional composite stress state is formed in the penstock, and it should be calculated in the design. The stress between the penstock and the dam surface should be solved with reasonable penstock and dam connection measures. In addition, the arch stress of the arch dam can cause additional stress in the penstock, but it is generally neglected in the design.

The penstock fixed on the dam locally increases the rigidity of the dam and also affects the deformation and stress of the dam. The impact is related to the dimension of the penstock. This effect should be considered in the design of the dam.

Temperature loads include the temperature difference between construction period and operation period, as well as the temperature difference between inside and outside of pipe shell during operation. This type of penstock belongs to thick wall pipe, and there is no insulating layer set on the surface of penstock, so the temperature stress cannot be ignored. The concrete of pipe shell is allowed to crack, so only the additional stress of steel liner and steel bar caused by temperature load and its effect on crack width should be considered.

Vibration of penstock induced by the water hammer is eliminated by fixing the penstock to the dam. Penstock and dam are assumed to be an integral structure when earthquake load considered. The shear stress and normal tension stress on the joint surface between penstock and dam caused by earthquake should be paid special attention to.

Under all possible combinations of the above loads, the penstock structure should satisfy the following requirements: ① sufficient strength; ② the concrete of penstock shell is permitted to crack, but the width of crack should be limited within an allowable value which usually takes 0.3 mm; ③ reliable connection between the penstock and dam body; ④ simple structure, convenient installation and construction, and economical.

7.9.2.2 Structure

The inner layer of the downstream surface penstock of the dam is steel liner, which is rolled and welded by steel plate. The penstock section of a certain length is made according to the transportation and hoisting conditions and transported to the site for circumferential seam welding assembly. Because the steel penstocks are embedded in concrete without external water pressure, if necessary measures (such as adding temporary support structure in the steel penstocks) can be taken to ensure adequate rigidity of steel penstocks during construction (when transporting and pouring out-wrapped concrete), there is no need or only a small amount of stiffening measures such as stiffener rings on the outer shell of steel penstocks.

The circumferential reinforcement should be arranged in at least two layers inside and outside. The diameter of steel bars and the gradation of concrete aggregate should be reasonably chosen so that the necessary spacing of steel bars can be ensured to facilitate concrete pouring, and the steel bars should not be too sparse to reduce the width of cracks. Welding quality of circumferential reinforcement joints must be strictly guaranteed and tested. In the former Soviet Union, special devices have been used to weld large diameter (up to 70 mm) steel bars and roll them into spiral steel ring corresponding to the length between steel liner, which not only saves steel but also reduces welding joints and improves quality and construction speed. The inner and outer steel bars are fixed by the support truss made of angle steel, which is arranged at both ends of each skeleton. To facilitate the erection of circumferential reinforcement

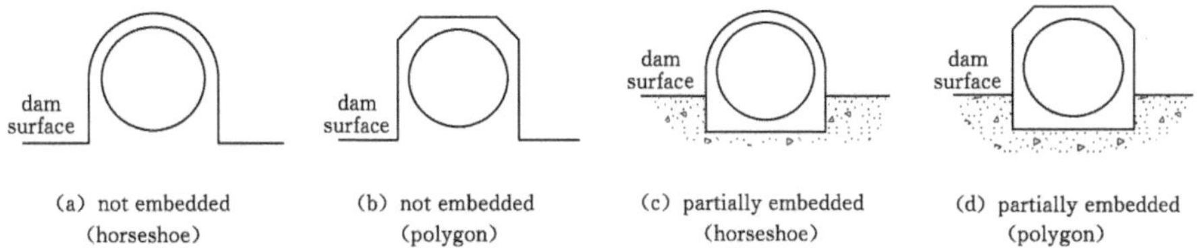

Fig. 7.54 Section type of penstock laid on the downstream surface of the concrete dam

and improve the bearing performance of downstream surface penstock under axial force, longitudinal reinforcement should be added. Corresponding to each penstock section, a steel bar frame with enough rigidity is formed, which is lifted to the site as a whole and assembled in coordination with the steel liner penstock section. A small amount of radiation is used to locate the steel support between the inner steel bar and the steel liner, and a clearance of more than 150 mm is left between the two to facilitate the pouring of concrete. The structure is shown in Fig. 7.51. It must be pointed out that the joint stress of steel liner and inner and outer steel bars depends on the concrete combined as a whole rather than connecting them with steel. Therefore, the support of fixing steel bar and steel liner should be set as few as possible, so as not to affect the quality of concrete pouring and avoid the local stress at the connection between steel liner and support.

There is shear stress between the penstock and dam surface. For an arch dam, there may be normal tension stress on joint surface between the penstock and dam. To make the penstock and dam firmly connected without being destroyed, the joint surface should be chiseled, and keyed grooves should be arranged, and sometimes the normal reinforcement should be inserted, even grouting the joint surface.

The cross-sectional outline of the downstream dam surface penstock can be horseshoe-shaped or polygonal. To shorten the distance between the powerhouse and dam, the downstream surface penstock of the dam can adopt the cross-sectional type partially embedded in the dam, as shown in Fig. 7.54. The thickness of concrete pipe shell should be as thin as possible on the premise that the steel bars can be reasonably arranged and conducive to concrete pouring.

7.9.2.3 Structure Strength Analysis

The main purpose of strength analysis of the downstream dam surface penstock structure is to determine the thickness of steel liner and the arrangement of circumferential reinforcement. The analytical method of Eq. (7.73) or Eq. (7.74) can be used.

According to the water conservancy trade standard *Design Specification for Steel Penstocks of Hydroelectric Stations* SL 281–2003, the design of steel-lined reinforced concrete penstock should follow Eq. (7.73):

$$K \cdot P \cdot r \le t \cdot \sigma_s \cdot \varphi + t_3 \cdot f_{yk} \tag{7.73}$$

In which

K general structure safety factor; not less than 2.0 under maximum pressure of normal condition; not less than 1.6 under maximum pressure of special condition; it can be reduced by less than 10% by demonstration;
f_{yk} characteristic value of tensile strength of steel bar;
σ_s yielding strength of steel plate, N/mm^2, if the steel yield ratio σ_s/σ_b is more than 0.7, then σ_s=0.7σ_b; σ_b is the tensile stress of steel;
φ welding seam coefficient, 0.9 for single-sided welding and 0.95 for double-sided welding.

According to the electric industry-standard specification: *Design Specification for Steel Penstocks of Hydroelectric Stations* NB/T 35,056-2015, the shell thickness of steel penstocks and the converted thickness of the circumferential reinforcement should satisfy Eq. (7.74).

$$P \cdot r \leq \frac{t \cdot f_s + t_3 \cdot f_y}{\gamma_0 \cdot \psi \cdot \gamma_d} \tag{7.74}$$

In which

P design value of internal water pressure, N/mm^2;
r inner radius of steel penstock, mm;
t shell thickness of penstock, mm, not less than the minimum thickness, and should meet the requirements for stability against external pressure;
t_3 reduced thickness of circumferential reinforcing steel, mm, should not be less than half of the shell thickness of steel penstock t_3=$n \cdot F_S$/1000;
f_s design value of tensile strength of steel plate, N/mm^2;
f_y design value of tensile strength of circumferential reinforcement, N/mm^2;
n number of circumferential reinforcement in the range of unit penstock length;
F_s section area of single circumferential reinforcement, mm^2;
γ_0, ψ they are the structural importance coefficient and design situation coefficient, respectively;
γ_d structure coefficient, taken as 1.5.

There are also simplified calculation methods for temperature stress and axial reinforcement configuration.

The steel liner and steel bar of the downstream surface penstock of the dam can be reliably jointly loaded with high safety. Steel liner can be as thin as possible under the condition of meeting the minimum thickness of the structure, and steel bar can be added to meet the strength requirement, without checking the condition of steel liner bearing internal pressure alone. This not only reduces the steel plate consumption and achieves economic purposes, but also ensures the process quality of thin plate easily. The amount of steel bar is high, which is beneficial to the prevention of penstock cracking, thus increasing the safety and reliability of the structure. The high consumption of reinforcement is also beneficial to reduce the crack width of concrete pipe shell.

The function of the pipe shell concrete is to integrate the steel liner and the steel bar into a whole and joint bearing, so the strength grade should not be lower than C20, nor should it be higher than C30. Concrete is allowed to crack to fully exert the effect of steel. Concrete only transmits radial force and bears axial force. Therefore, the concrete pipe shell should be as thin as possible under the condition of ensuring the reasonable arrangement of the steel bars and the quality of the concrete construction, which can save the amount of concrete work, increase the reinforcement rate, and reduce temperature stress and crack width.

The concrete crack width should be limited to ensure the durability of the structure. The crack width of the downstream surface penstock of the large dam should be limited to within the specified 0.3 mm, which may be difficult. Reducing the concrete thickness of the pipe shell, increasing the number of steel bars, reducing the diameter of the steel bars, increasing the density of the steel bars, and arranging more steel bars in the outer layer are all beneficial to reducing the crack width. Reasonable structural measures (such as pre-cracks and water stops) can also be used to reduce the adverse effects of excessive crack width on structural durability.

7.9.3 The Penstock Laid on the Upstream Surface of Concrete Dams

To overcome the difficulties in the arrangement and structural strength of the dam body steel penstock, the coordinate layout pattern of the upstream surface penstock of the concrete dam and the dam body steel penstock appeared. Below the water intake of the dam surface, the penstock is arranged along the upstream dam surface, to a certain depth, and it is connected with the horizontally arranged steel penstock section buried in the bottom of the dam body through bend penstock. This type is essential to move the water intake of dam body steel penstock from the lower part to the upper part. Its characteristics are that it not only keeps the short steel penstock in the deep dam, but also has less construction interference and avoids the technical difficulties of the pressurized flow intake. This arrangement is particularly suitable for slim arch dams, because it is difficult to arrange large-diameter penstocks in the dam in this case, except that the penstocks are horizontally passed through the dam.

A typical example of this type of penstock is the penstock of the Kabir Hydropower Station in Iran, as shown in Fig. 7.55. The Kabir Dam is a 200 m high double-curvature dam with 4 units and a single unit capacity of 250 MW. Each unit has its water intake and penstock. The penstock diameter is 6.5 m, and the thickness of pipe shell concrete is 2.1 m. There is no steel liner in the vertical section of the penstock, which is closely attached to the upstream curved surface of the arch dam and is connected with the dam body as a whole, and embedded in the dam body with a depth of 1.2 m. The downward bend section connects the vertical section with the horizontal section passing through the bottom of the dam. Steel liner is provided in the downward bend section and the horizontal penstock section.

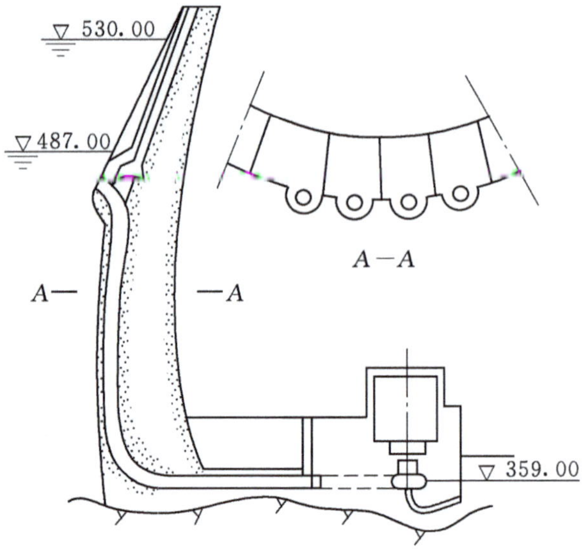

Fig. 7.55 Penstock laid on the upstream surface of the dam (unit: m)

The part located outside the dam of penstock laid on the upstream surface of the dam is in the reservoir. The two conditions of water filling and venting should be considered. When the penstock is emptied, the outside of the penstock is the reservoir water pressure, which generally causes the pipe shell to be pressed without generating large circumferential tensile stress. When the unit is in operation, and the penstock is filled with water, the internal and external pressures are balanced except for the small head velocity head and head loss. Therefore, the penstock laid on the upstream surface of the dam is under good stress conditions, and the steel liner can be omitted, which is an advantage. However, since one side of the penstock is fixed on the dam body and the arch dam is deformed by force, it will cause tangential tensile stress in a local area of the penstock. This type of tensile stress can be reduced if the depth of the penstock embedded in the dam is reduced. Circumferential reinforcement should still be provided at the inner and outer shells of the penstock. In addition, since the dam surface initially cools faster than the interior of the dam, axial tensile stress may occur in the pipe shell of the penstock, and longitudinal reinforcement should also be provided. Practice has proved that the horizontal construction joints of the vertical section without steel liner are easy to leak, and effective water-stopping measures should be taken. The penstock with steel liner located in the dam is general the dam body penstock. It is characterized by deep burial and high pressure, and its initial end is deep under the reservoir water level, and the seepage water pressure outside the steel liner is large. Therefore, it is necessary to pay attention to the stability of steel liner against external pressure when the penstock is emptied.

The advantage of penstock laid on the upstream surface of the dam is that it is easy to arrange, less interference between the penstock and the dam body, and the penstock is under good stress conditions, which can save steel. However, most of the penstock is located in the reservoir, and it is difficult to maintain; the dam body steel penstock is located in the high-stress area at the bottom of the dam, which is unfavorable to the structure, and the steel penstock needs to be buried at the initial stage of the dam body pouring. The penstock laid on the upstream surface of the dam has not been widely used in foreign countries, and it has not been applied in China.

Questions

1. What are the types of penstocks and their applicable conditions?
2. What are the main water supply modes for penstocks in hydropower stations? What are their advantages, disadvantages, and applicable conditions?
3. What is the role of anchor block and support block on the exposed steel penstocks? What are the types and characteristics of anchor block and support block?
4. Why are expansion joints installed on penstocks? Where are they located? Please explain the reasons.
5. Which control sections should be selected when designing the exposed steel penstocks? Please show the position of each control section in the way of diagram. What are the characteristics of the force?
6. Dimensions of a ground penstock are set as follows:

   Inner diameter $D = 3$ m, spacing of anchor blocks is 85 m, distance between rolling ring girder support blocks is 10 m, expansion joints are located 2 m below the upper anchor block, and spacing of pipe sections where the expansion joints are located is 5 m. Friction coefficient of rolling ring girder support blocks is 0.1, inclination angle of the axis of penstock and ground is 30°, and length of expansion joint sealing fillers $b_1 = 0.3$ m, $\mu = 0.25$. At the end of the span, internal water pressure (including water hammer pressure) is 200 m, and internal water pressure (including water hammer pressure) at expansion joints is 160 m. The steel is Q345C, and yielding strength is 325 MPa.

   Considering the temperature rise condition, the strength check is carried out on the cross section of the steel penstock at the end of the mid-span section 1-1 and the section 2-2 near supporting ring, and the mid-span section 1-1 is subjected to the external pressure stability check. The specific contents include:

   (1) Initial evaluation of the pipe shell thickness.
   (2) Calculate the pipe weight and water weight per unit length, calculate the axial force $\sum A$, shear force Q, and bending moment M of the 1-1 and 2-2 sections. The stress calculation is as follows, and it is required to be accurate to 3 significant digits after the decimal point.
   (3) Calculate the stress components at the top of the 1-1 section, the waist of the penstock, and the bottom of the penstock, and use the fourth strength theory to check.
   (4) Calculate the stress components at the top of the 2-2 section, the waist of the penstock, and the bottom of the penstock, and use the fourth strength theory to check.
   (5) Calculate the critical external pressure of section 1-1, first check the smooth surface penstock, the requirement is $p_{cr} \geq 2.0 p_{\text{outer}}$ (one atmospheric pressure, 0.1 MPa). If the requirements are not met, use the stiffener ring,

assuming the spacing of the stiffener ring $L = 1.0\text{to}2.0\text{m}$, the height of the stiffener ring section $h = 150 \sim 300\text{mm}$, and take the pipe shell thickness as the thickness of the stiffener ring. The critical external pressure between the stiffener ring and the stiffener ring is calculated, $p_{cr} \geq 2.0 p_{\text{outer}}$ required.

Note (M, V calculation method): M, V can be calculated according to multi-span continuous beam.

Above the three spans of the expansion joint, you can calculate by thinking both ends are fixed:

M value: Mid-span:$M = 0.0416QL\cos\alpha$;

Support:$M = -0.08333QL\cos\alpha$.

V value: Support: $V = 0.5Q\cos\alpha$

where Q is the sum of the weight of the single span and the weight of the water, L is the span of each span.

7. What is the main reason for the instability of underground penstocks against external pressure? What are the measures to improve the stability against external pressure?
8. What are the main structural types of dam body penstocks? What are the differences in force characteristics of these penstocks?
9. What are the advantages and disadvantages of the penstocks laid on the downstream surface of dams compared with the dam body penstocks? What are the applicable conditions?

Bibliography

1. Renkun, Wang and Z. Chunsheng. 2013. *Design manual of hydraulic structures: Vol.8 Buildings of hydropower station*, 2nd. Beijing: China Water and Power Press.
2. Ministry of Water Resources of the People's Republic of China. 2003. SL 281-2003 Design specification for steel penstocks of hydroelectric stations. Beijing: China Water and Power Press.
3. National Energy Administration. 2016. *NB/T 35056–2015 Design specification for steel penstocks of hydroelectric stations*. Beijing: China Electric Power Press.
4. Ministry of Electric Power Industry of the People's Republic of China. 1998. DL 5077-1997 Specifications for load design of hydraulic structures. Beijing: China Electric Power Press.
5. Ministry of Water Resources of the People's Republic of China. 2003. SL 279-2002 Specification for design of hydraulic tunnel. Beijing: China Water and Power Press.
6. General Administration of Quality Supervision, Inspection and Quarantine of the People's Republic of China, Standardization Administration of the People's Republic of China. 1999. GB/T 12777-1999 General specification of metallic bellows expansion joints. Beijing: The state bureau of quality and technical supervision.
7. National Development and Reform Commission of the People's Republic of China. 2009. DL/T 3057-2009 Design specification for hydraulic concrete structures. Beijing: China Electric Power Press.
8. Design manual and Atlas Compilation Group for dam body penstocks of hydropower station. 1988. Design manual and Atlas for dam body penstocks of hydropower station. Beijing: Water Resources and Electric Power Press.

9. Jiazheng, P. A. N. 1982. *Steel penstock*. Beijing: Publishing House of Electronics Industry.
10. Steel Pipe Test Group of Northwest Institute of Survey and Design, Ministry of Water Resources and Electricity. 1983. Test and discussion on combined action of steel pipe and concrete for hydropower station. *Journal of Hydraulic Engineering* 6: 61–65.
11. Hegao, Wu, S. Xiaogao and L. Zhiming. 2000. Steel-lined reinforced concrete penstock for hydropower station. Beijing: China Water and Power Press.
12. Shanding, M. A., and W. A. N. G. Ruze. 1996. *Buildings of hydropower station*. Beijing: China Water and Power Press (In Chinese).
13. Qizhao, L. I. U., and H. U. Ming. 2010. *Hydropower station*, 4th ed. Beijing: China Water and Power Press.

Chapter 8
Bifurcated Pipes

Kai Su

Abstract This chapter focuses on bifurcated pipes of hydropower station. Firstly, the functions as well as characteristics and requirements of bifurcated pipes are introduced. Meanwhile, the layout arrangement and structure modes are elaborated. Then, loads and structural design requirements of bifurcated pipes are presented. Furtherly, the design contents of them are introduced. For different types of bifurcated pipes, layout forms are introduced and stress analysis is carried out, respectively, including three-girders reinforced bifurcated pipes, crescent-rib reinforced bifurcated pipes, hem reinforced bifurcated pipes, spherical bifurcated pipes, and shell type bifurcated pipes.

Keywords Bifurcated pipes · Stress analysis · Three-girders reinforced bifurcated pipes · Crescent-rib reinforced bifurcated pipes · Hem reinforced bifurcated pipes · Spherical bifurcated pipes · Shell type bifurcated pipes

8.1 Introduction

8.1.1 Functions, Characteristics, and Requirements of Bifurcated Pipes

When using centralized and group water supply, the penstock works often involve branching the tunnel or penstock into two or more pipes for multiple units, thus bifurcated pipes are required. Bifurcated pipes are located at the upstream side of the powerhouse, usually at the bottom of the surge tank or downstream of the surge tank. The draft tube of several units often synthesizes a pressure tail water hole on downstream, and the junction is also in the form of a bifurcated pipe, but the water flow is in the opposite direction. The bifurcated pipe of pressure water diversion conduit on the downstream tends to be larger in dimension but lower in internal

K. Su (✉)
Wuhan University, Wuhan, China
e-mail: suker8044@163.com

Y. Cheng et al. (eds.), *Hydropower Station*,
https://doi.org/10.1007/978-981-96-2383-9_8

pressure. Bifurcated pipes in front of the powerhouse are mainly discussed in this chapter.

The flow distribution is proportioned at the bifurcated pipe, so the flow direction and flow pattern should change a lot, which will cause large head loss. The safety of the bifurcated pipe is very important because it bears the largest hydrostatic pressure and water hammer and is near the powerhouse. The bifurcated pipe is generally composed of thin shells and stiffening members with a large rigidity. The pipe shell is thick, the dimension of the member is large, and sometimes forging is required. The requirements of welding process are high and the cost is relatively high.

The bifurcated pipe should meet the following requirements:

(1) Smooth flow and small head loss, avoiding or reducing vortex and vibration

The main factors affecting the head loss of bifurcated pipe are as follows: The ratio of the main and branch pipe cross-sectional area, the flow distribution ratio, the semi-cone angle α_1, α_2 of the main and branch pipes, the bifurcated pipe angle β, the corner angle λ, and the turning angle of the obtuse angle area θ, as shown in Fig. 8.1. The experimental study shows that when the average flow rate across the bifurcated pipe in each section is nearly equal, or the water flow is slowly accelerated (the area before the bifurcated pipe is larger than the sum of the area after the bifurcated pipe), the vortex can be avoided and the head loss can be reduced, the bifurcated pipe should adopt cone pipe transition and the half cone angle is generally 5–10°. It is advisable to use smaller deflection angles β, smaller corner angle β, and obtuse angle area θ. However, these requirements are sometimes contradictory. For example, increasing α_2 can reduce θ, but it will increase γ, so it is necessary to fully consider the choice.

(2) Reasonable structures, good stress conditions, free from excessive stress concentration, and deformation.

(3) Convenient for fabrication, transportation, and installation.

There is always a conflict between hydraulic conditions and structural requirements for bifurcated pipe. For example, it is beneficial to the water flow with the decrease of deflection angle. But at this time, the breaking of the main bifurcated

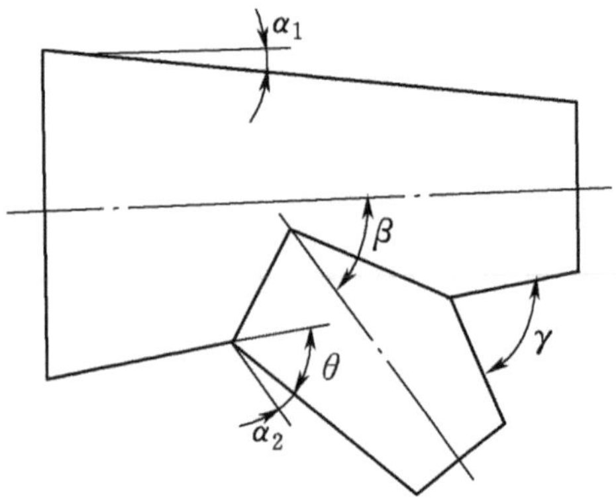

Fig. 8.1 Diagram of bifurcated pipe

pipe will be increased, which is harmful to the structure, and will increase the welding difficulty at the fork crotch. For low head station, head loss is a more important factor; for high-head station, a bit larger head loss is allowed to make the structure simpler.

8.1.2 Typical Layout Arrangement of Bifurcated Pipes

There are three typical arrangements for bifurcated pipes:

(1) Asymmetric Y-shaped arrangement, as shown in Fig. 8.2a. If the side-inlet water entry mode is adopted or the axis of the two bifurcated pipes cannot be symmetric for any reason, a smaller pipe is branched from the main pipe.
(2) Symmetric Y-shaped arrangement, as shown in Fig. 8.2b. Two same pipes are branched from the main pipe, and it is adopted when forward-inlet water entry mode is used.
(3) T-shaped arrangement, as shown in Fig. 8.2c. The main pipe is directly branched into three same pipes.

If the project is multi-unit, the following two combined layouts can be adopted: symmetric Y-shaped and asymmetric Y-shaped; symmetric Y-shaped and T-shaped arrangement.

For small-sized and medium-sized bifurcated pipes, the arrangement type is more flexible, and both symmetric Y-shaped arrangement and asymmetric Y-shaped arrangement can be used. However, for bifurcated pipes which have high heads and large values of *HD*, the technical feasibility often becomes a constraint to a certain extent. Therefore, the symmetric Y-shaped arrangement should be adopted as much as possible. If the symmetrical arrangement is difficult to be adopted for the bifurcated pipes when the overall arrangement of the water conduit diversion system is analyzed, the axial direction of the main and branch pipes can be adjusted locally by the variable cone. The main body of the bifurcated pipes can be arranged in a symmetrical form, and connected to the main bifurcated pipes through the bend pipe or the cone pipe. The increased head loss of the joint arrangement between the bifurcated pipe and the bend pipe is relatively limited, and the force condition of the bifurcated pipe is greatly improved from the aspect of structural. The thickness of the shell

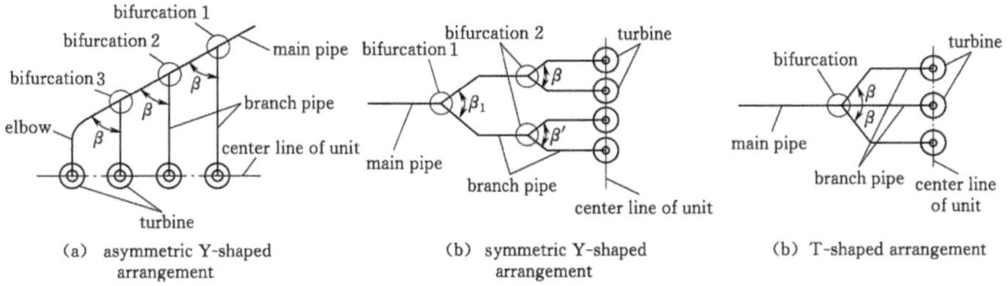

Fig. 8.2 Arrangement of bifurcated pipes

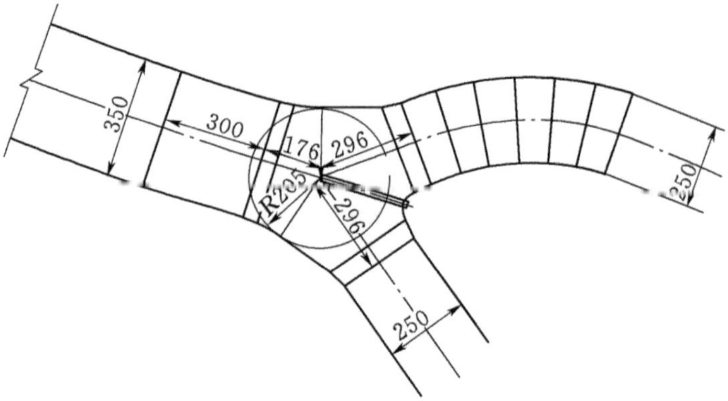

Fig. 8.3 Bifurcated pipe arrangement of Xilongchi pumped-storage power station (unit: cm)

and the rib is greatly reduced, which not only saves the quantities of engineering, but also brings convenience to manufacture and installation and is conducive to structural safety. This practice has many engineering examples at home and abroad, such as the domestic bifurcated pipe of Xilongchi pumped-storage power station (the diameter of the main pipe is 3.5m, the diameter of the two pipes is 2.5m, and the designed internal water pressure is 10.15MPa), as shown in Fig. 8.3.

Asymmetric Y-shaped arrangement is the main layout of already-built steel bifurcated pipes in China. In addition to the flexible and simple merit of the asymmetric Y-shaped arrangement, but also because of the steel bifurcated pipe built in the past is smaller in scale and more reinforced with gussets, which is more suitable for asymmetric Y-shaped arrangement.

The central line of the main pipe and branch pipe for the bifurcated pipe should be arranged in the same plane to make the structure shape simple.

The line where the main pipe and branch pipe cut each other is called intersection line. Since the main branch pipes are cut each other at the intersection line, it is often necessary to strengthen the members along the intersection lines. To facilitate the manufacture and welding of the reinforcement, the intersection lines should be plane curves. It can be geometrically proved that the necessary and sufficient condition for the intersection line to be a plane curve is that the main pipe and branch pipe has a common tangent ball, as shown in Fig. 8.4. On the plane, the common tangent ball is a common tangent circle, the center of the circle O, contrary to the main pipe and the branch pipe I $\alpha - \alpha$ contrary to branch pipe II $b - b$, straight line aa intersects straight line bb at point. AD and BD are the plane where the intersection line is, the intersection line AD and BD are two elliptical curves.

If the diameters of the main and branch pipes are greatly different, or for other reasons, it is difficult for the main and branch pipes to cut into one ball, the intersection line will be on the curved surface, and the reinforcement along the intersection line will be a curved member, which will increase the difficulty of calculation, manufacture and installation.

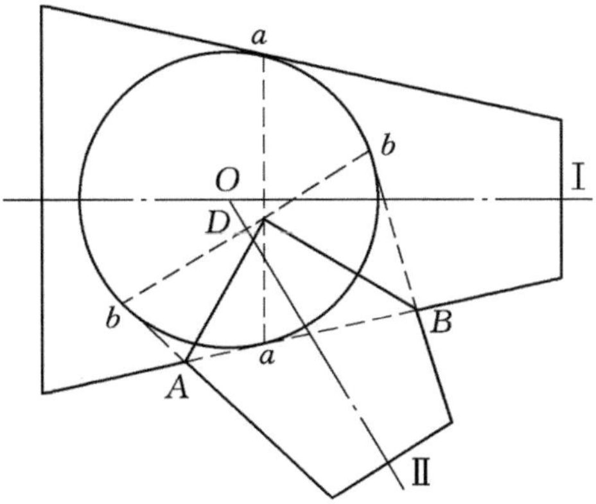

Fig. 8.4 Common cut ball of cone pipe in bifurcation

The engineering practice shows that although the central line of main pipe and branch pipe for a bifurcated pipe arranged in the same plane can make the structure simple, but it is not conducive to the drainage of the water diversion system during the overhaul. Therefore, some scholars have proposed the design idea of using the horizontal bottom steel bifurcated pipe, as shown in Fig. 8.5, and researched the corresponding hydraulic and structural characteristics. The research shows that both the hydraulic characteristics and the structural characteristics, the horizontal bottom steel bifurcated pipe can meet the relevant requirements of the specification, which needs to be promoted and applied in practical engineering.

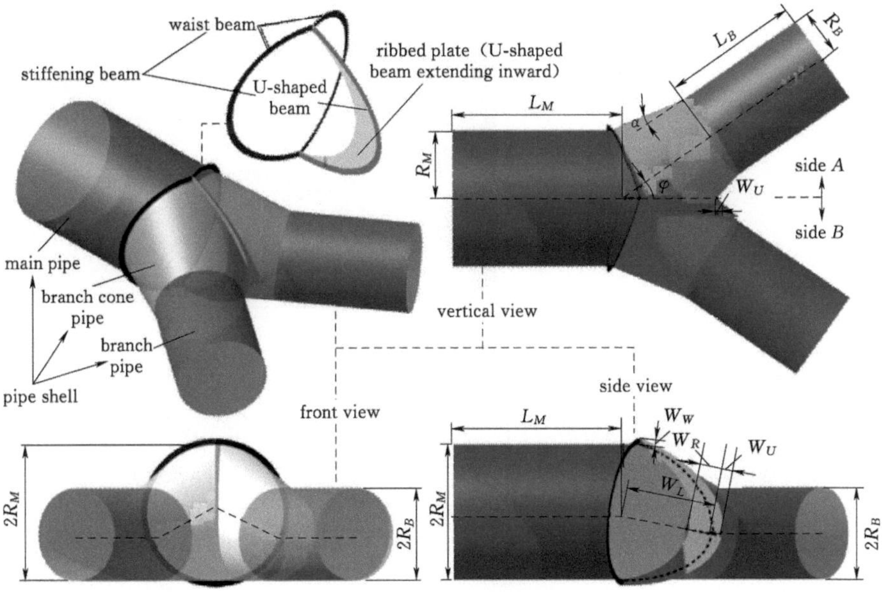

Fig. 8.5 Schematic of horizontal bottom steel bifurcated pipe

8.1.3 Structural Types of Bifurcated Pipes

There are many structural types of bifurcated pipes. It is more convenient to classify the bifurcated pipes according to their material and strengthening method, and it is also advantageous for structural analysis.

8.1.3.1 Exposed Steel Bifurcated Pipes

The bifurcated pipe is welded with steels that carry all the loads, as shown in Fig. 8.6.

According to the reinforcing method or bearing characteristics, exposed steel bifurcated pipes can be classified as the following structural types:

(1) Three-girders reinforced bifurcated pipe. As shown in Fig. 8.7, the bifurcated pipe is reinforced by the U-shaped girder and breast beams at the external side of the intersection lines to form the combination structure of shell and spatial beam system.
(2) Crescent-rib reinforced bifurcated pipe. As shown in Fig. 8.8, enlarging section of main cone and reducing section of branch cones which have a common tangent ball form a cone shell reinforced by a crescent-rib internally along the intersection line of branch cones.

Fig. 8.6 Exposed steel bifurcated pipe

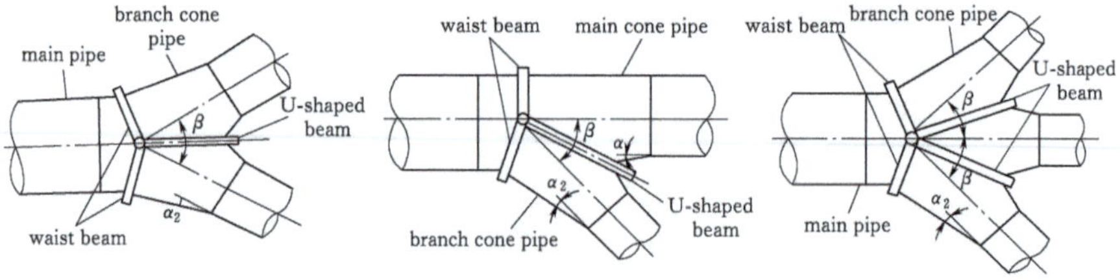

(a) symmetric Y-shaped (b) asymmetric Y-shaped (c) T-shaped

Fig. 8.7 Three-girders reinforced bifurcated pipe

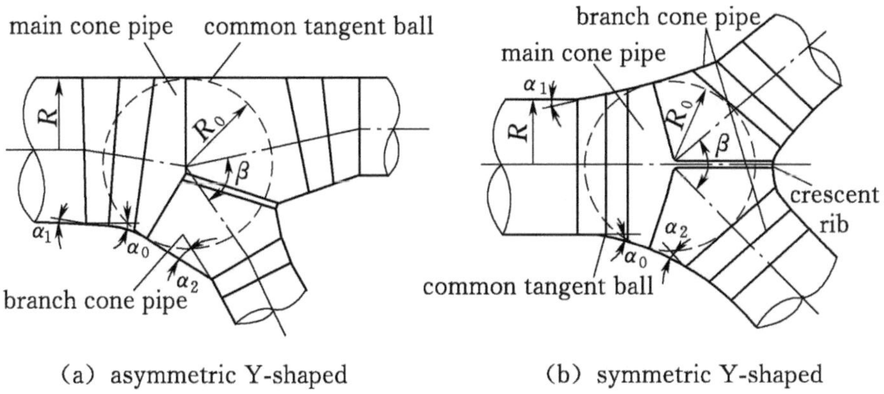

Fig. 8.8 Crescent-rib reinforced bifurcated pipe

(3) Hem reinforced bifurcated pipe. As shown in Fig. 8.9, the bifurcated pipe is reinforced with a simple curved plate welded at both sides of the intersection line of main pipe and branch pipe (there is a distinct diameter difference between the main pipe and branch pipe).
(4) Shell type bifurcated pipe. As shown in Fig. 8.10, the taper shell gradually enlarged by the main pipe and the branch pipe is continuously and smoothly coupled with the center ball shell, and no reinforcing beam is provided.
(5) Spherical bifurcated pipe. As shown in Fig. 8.11, branching is carried out through a spherical shell, and circumferential reinforced rings are required along the intersection line between the main pipe, branch pipes and the spherical shell.
(6) Partition type bifurcated pipe. As shown in Fig. 8.12, partition type bifurcated pipe is composed of diffusion section, partition section, and deformation section which are all integral shells without openings. There is no other reinforcement except partition plate.

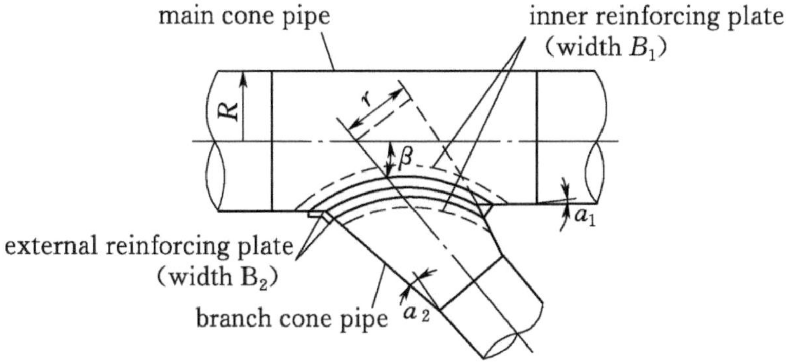

Fig. 8.9 Hem reinforced bifurcated pipe

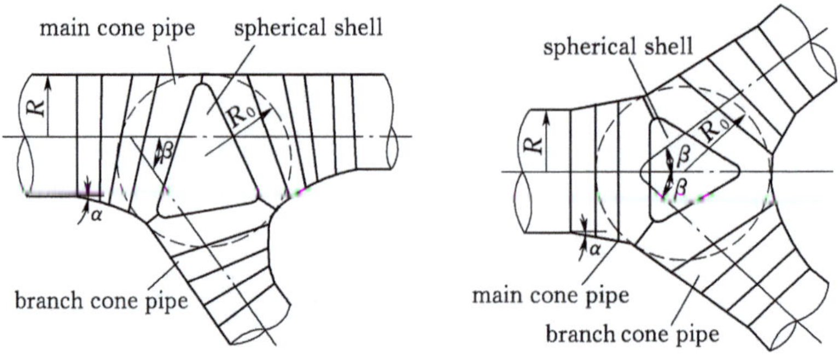

(a) asymmetric Y-shaped (b) symmetric Y-shaped

Fig. 8.10 Shell type bifurcated pipe

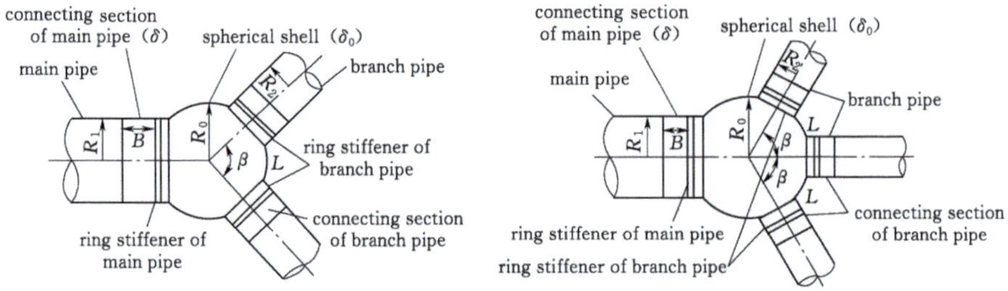

(a) symmetric Y-shaped (b) T-shaped

Fig. 8.11 Spherical bifurcated pipe

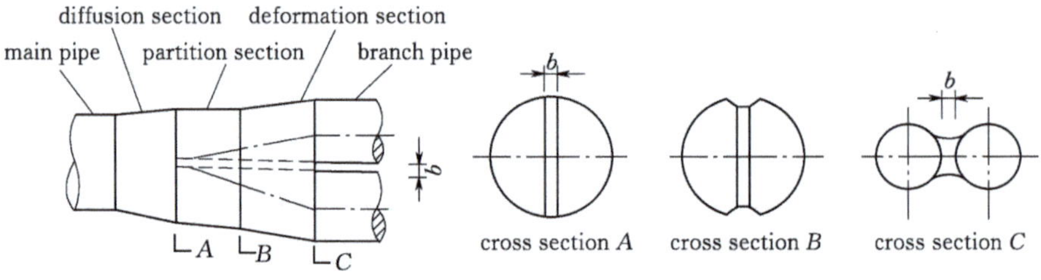

Fig. 8.12 Partition type bifurcated pipe

The dimension and internal pressure of the bifurcated pipes built in China in the 1950s are small, mostly hem reinforced bifurcated pipes; in the 1960s, due to the emergence of the high-head hydropower station, the three-girders reinforced bifurcated pipes are used frequently. With the increase in the scale of steel penstock, the three-girders reinforced bifurcated pipe with a large diameter and high internal pressure is difficult to manufacture and install, and the technical and economic indicators are not good. Three-girders reinforced bifurcated pipes are substituted by crescent-rib reinforced bifurcated pipe in recent years, especially in underground layouts. A few projects also use spherical bifurcated pipe and shell type bifurcated pipe. Partition type bifurcated pipe is a newly developed foreign-made adaptable, well-flowing, and well-stressed condition without the need for extra forgings, but there is no practice in China.

In most cases, the unbalance axial force of bifurcated pipe will be transferred to the foundation through the anchor block surrounding the bifurcated pipe. A certain number of steel bars are also arranged in the anchor block, but the internal water pressure is not considered in the design of the anchor block. In this case, it is still regarded as a penstock steel bifurcated pipe.

8.1.3.2 Underground Bifurcated Pipe

Underground bifurcated pipe is embedded in the rocks, located at the end of underground penstock. Its main characteristic is that the surrounding rock will assist bifurcated pipe in bearing the internal pressure. According to the structure, it can be divided into two types:

1. Underground steel bifurcated pipe

The steel bifurcated pipe and the surrounding rock are filled with concrete. Part of the internal water pressure is transmitted to the surrounding rock through the steel liner and concrete. The steel liner is impervious and bears part of the internal water pressure. The structure and force are the same as the underground penstock, but the structure calculation is very complicated and generally requires finite element method for analysis. The more simplified method is to estimate the percentage of internal water pressure in the surrounding rock according to the straight pipe section calculation method. The remaining internal water pressure of steel bifurcated pipe is considered, and the stress and strength of the steel bifurcated pipe are analyzed according to the method of exposed bifurcated pipe.

In principle, the steel bifurcated pipe of this kind of bifurcated pipe can adopt any type of structure of the above-mentioned steel bifurcated pipe. However, in order to reduce the diameter of the excavation hole and the size of the traffic hole and exert the joint bearing capacity of the surrounding rock better, it is preferable to use a hem reinforced bifurcated pipe or crescent-rib reinforced bifurcated pipe which has no reinforcing beam or a small reinforcing beam.

2. Underground reinforced concrete bifurcated pipe

The bifurcated pipe is a reinforced concrete structure that is integrated with the surrounding rock. When the dimension of the bifurcated pipe is small and the internal water pressure is low, depending on the surrounding rock to share the internal water pressure, the reinforced concrete cannot crack and has the function of impermeability and bearing loads. When the bifurcated pipe is large, especially when the internal pressure is very high, since the surrounding rock must be quite deformed after being loaded, the shape of the bifurcated pipe is complicated, and there will be stress concentration. Therefore, it is difficult for the reinforced concrete to escape cracking. Under the high internal pressure, it becomes a "water-permeable liner," and the internal water pressure will be transferred to the surrounding rock to bear. In this way, the surrounding rock becomes the main body that withstands internal pressure and leakage. Reinforced concrete liner can reduce the inner shell roughness and head loss, and the steel bar can reduce the crack width, which is conducive to reducing leakage. When the dimension of the bifurcated pipe is large, the underground reinforced concrete bifurcated pipe is the most advantageous structural type because if the underground steel bifurcated pipe is used, transportation and installation will be difficult to achieve. Obviously, when adopting this type of bifurcated pipe, it is necessary to require high bearing capacity and impermeability of the surrounding rock. Therefore, it is often required to have sufficient burial depth. The bifurcated pipe should have sufficient distance from the powerhouse, high-pressure grouting should be done, and the drainage of the rock mass should be well prepared. This kind of bifurcated pipe has been widely used in many high-head large-capacity pumped-storage power stations.

8.1.3.3 Steel-Lined Reinforced Concrete Bifurcated Pipes

The steel bifurcated pipe is encased in the surrounding reinforced concrete, and they bear the loads together. The inner steel bifurcated pipe, that is, the steel liner, can simultaneously prevent the seepage so that the external reinforced concrete is allowed to crack and the steel liner and the steel bar can be fully utilized. The steel liner can adopt some types of exposed steel bifurcated pipe according to the need, but since the reinforced concrete participates in the bearing, the burden of the steel liner is reduced so that the pipe shell can be thinned compared with the separately supported exposed steel bifurcated pipe. Reinforcement can also be lightened or even canceled. This structure can not only save the amount of steel plates, but also makes the material selection and process requirements simpler. The cost is reduced and the safety of structure is higher than the exposed steel bifurcated pipe.

When exposed steel bifurcated pipe is used, it is generally necessary to add a concrete anchor block outside the steel bifurcated pipe. Compared with this type of bifurcated pipe, the steel-lined reinforced concrete bifurcated pipe does not even need to increase the amount of extra concrete works, and the technical and economic benefits are more obvious.

Steel-lined reinforced concrete bifurcated pipes have been used abroad in the 1960 and 1970s. In the late 1990s, this type of bifurcated pipe has been successfully applied in China to Daqikong Hydropower Station in Guizhou (Fig. 8.13, main bifurcated pipe diameter of 2.4 m, design water pressure of 400 m, *HD* value of 960 m^2), Chaishitan Hydropower Station in Yunnan (Fig. 8.14, main bifurcated pipe diameter of 6.4 m, design water pressure of 130 m, *HD* value of 832 m^2), and many pumping stations of Shanxi Yellow River Diversion Project. This type of bifurcated pipe structure currently lacks a mature design calculation method, and is generally designed based on the method of steel-lined reinforced concrete penstock. For more important projects, the reinforced concrete nonlinear finite element method or model test method should be used for further demonstration.

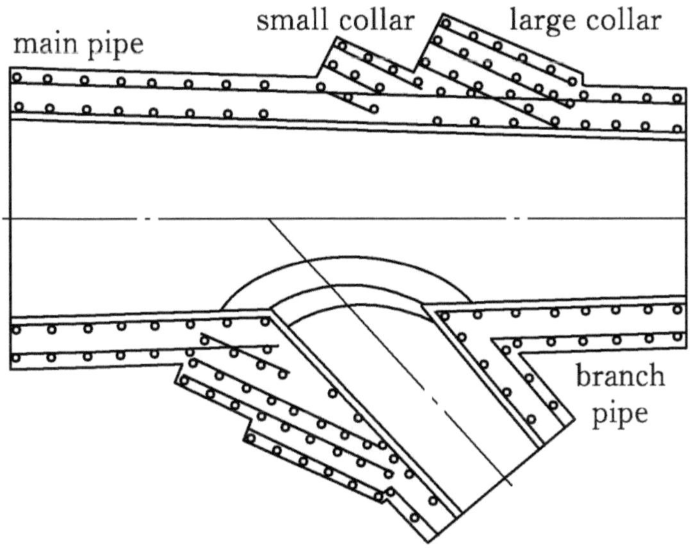

Fig. 8.13 Steel-lined reinforced concrete branch pipe diagram of Daqikong hydropower station

Fig. 8.14 Steel-lined reinforced concrete branch pipe diagram of Chaishitan hydropower station

8.1.4 *Loads and Structural Design Requirements of Bifurcated Pipes*

The load and its combination of the bifurcated pipes are the same as those of the main penstock, but the secondary loads such as dead weight, full water weight, wind load, snow load, and construction hoisting can be ignored. For underground bifurcated pipes, the pressure of the rock is carried by the concrete structure; the external pressure stability should be checked. The temperature load is generally not considered in the design of the bifurcated pipe, but for large exposed bifurcated pipe without expansion joints, the difference between the installation closing temperature and the operating temperature should be minimized to reduce the temperature stress. The load combination of the bifurcated pipe is shown in Table 8.1, and the load type is shown in Table 7.5.

Table 8.1 Calculation conditions and load combination of the bifurcated pipe

The style of bifurcated pipe	Design conditions	NB/T 35,056-2015 effect combination		Calculation contents	Corresponds to SL 281-2003 combination
Exposed bifurcated pipe	Persistent condition	Basic combination	(1b) + (4)	Normal operating conditions	Basic loads combination
	Transient condition		(12)	Venting condition	Basic loads combination
			(1d)	Hydrostatic test conditions	Special loads combination
	Accidental condition	Accidental combination	(1c) + (4)	Special operating conditions	Special loads combination
			(1a) + (4) + (11)	Earthquake conditions	Special loads combination
Underground bifurcated pipe	Persistent condition	Basic combination	(1b)	Normal operating conditions	Basic loads combination
	Transient condition		(12) + (13a)	Venting condition	Basic loads combination
			(1d)	Hydraulic pressure test conditions	Basic loads combination
			(10)	Construction condition	Basic loads combination
	Accidental condition	Accidental combination	(1c)	Special operating conditions	Basic loads combination

Table 8.2 Allowable stress of steel bifurcated pipe in specification SL281-2003

Stress area	Stress point	Load combination	
		Basic	Special
Membrane stress zone $[\sigma]_1$	The pipe shell of membrane stress zone and the reinforcement with small eccentricity	$0.5\sigma_s$	$0.7\sigma_s$
Local stress zone $[\sigma]_2$	The pipe shell of the reinforcement within $3.5\sqrt{rt}$ m bearing the bending moment and the turning point	$0.8\sigma_s$	$1.0\sigma_s$
	The reinforcement bearing bending moment	$0.67\sigma_s$	$0.8\sigma_s$

Note σ_s is the yielding strength of the steel plate; when the peak stress is calculated by the finite element method, the allowable stress value can be increased as appropriate according to this table

The allowable stress (resistance limit) of the bifurcated pipe should be slightly lower than that of the straight pipe section. This is because the structure of the bifurcated pipe is complicated and it is difficult to accurately analyze the stress. The specific values are listed in Tables 8.2 and 8.3. The pipe shell thickness of the bifurcated pipe is larger than that of the main and branch pipe, so the pipe joints should be thicker. The pipe shell thickness difference between adjacent pipe joints should not be greater than 4 mm to facilitate welding and avoid stress concentration.

A baffle should be provided inside the crescent-rib reinforced bifurcated pipe, shell type bifurcated pipe, and spherical bifurcated pipe to reduce hydraulic loss and vibration. The shape of these bifurcated pipes is all convex above and concave below. When filling water, the top air cannot be exhausted. When the water starts to run, the water flow will be mixed with air, which is unfavorable to the operation of the units. Therefore, the exhaust equipment should be installed at the top. Drainage pipe should be provided at the lowest point, or a horizontal bottom steel bifurcated pipes should be arranged as described above to evacuate the bifurcated pipes during maintenance.

8.1.5 Design Content and Method of Bifurcated Pipes

For each type of steel bifurcated pipes, the design content and steps include:

(1) Selection of bifurcated pipes type: according to the topographic and geological conditions, scale and construction technology level of the project, select the structure type of steel bifurcated pipe suitable for each specific project.
(2) Shape design: After the type of the bifurcated pipe is determined, according to the specific arrangement of the project and the diameter of the main bifurcated pipe, select the asymmetric Y-shaped arrangement, symmetric Y-shaped

Table 8.3 Steel bifurcated pipe structure coefficient γ_d of NB/T 35,056-2015 specification

Type of bifurcated pipe	Type of stress	Position	Structure coefficient γ_d ($\varphi = 0.95$)
Exposed bifurcated pipe	Overall membrane stress	Pipe shell of membrane stress zone	1.76
	Local membrane stress	Rib plate, reinforced ring	1.43
		Within $3.5\sqrt{rt}$ from the reinforcement that is subjected to bending moment and the middle of the pipe shell at the corner point, the reinforcing beam	1.43
	Local membrane stress + bending stress	Within $3.5\sqrt{rt}$ from the reinforcement that is subjected to bending moment, pipe shell surface at the corner point, reinforcing plate	1.21
Underground bifurcated pipe	Overall membrane stress	The pipe shell of membrane stress zone	1.50
	Local membrane stress	Rib plate, reinforced ring	1.35
		Within $3.5\sqrt{rt}$ from the reinforcement that is subjected to bending moment and the middle of the pipe shell at the corner point, the reinforcing beam	1.20
	Local membrane stress + bending stress	Within $3.5\sqrt{rt}$ from the reinforcement that is subjected to bending moment, pipe shell surface at the corner point, reinforcing plate	1.10

Notes 1. The table applies to the case where the welding seam coefficient $\phi = 0.95$. If $\phi \neq 0.95$, then should be multiplied by $0.95/\phi$
2. In the case of the hydraulic pressure test, the value of γ_d should be reduced by 10%

arrangement or T-shaped arrangement. Determine the shape of the steel bifurcated pipe preliminarily, including the deflection angle, cone angle and the size of reinforcing beam. Specific methods include geometric analysis methods and various computer-aided design systems, such as Auto CAD, Solidworks, CATIA, and specialized design software for the shape of steel bifurcated pipe.

(3) Structural design: For smaller steel bifurcated pipe, the analytical method provided by the relevant specifications or manuals can be used to determine the pipe shell thickness of the steel bifurcated pipe and the size of the reinforcing beams to meet the needs of engineering construction. However, for larger steel bifurcated pipe, the computer-aided design software mentioned above can be

used to determine the shape of the bifurcated pipe, further meshing and three-dimensional finite element calculation to determine the pipe shell thickness of the steel bifurcated pipe and the size of the reinforcing beam. With the construction of a large number of pumped-storage power stations, *HD* of steel bifurcated pipe is getting larger, and a new design method for joint bearing of steel bifurcated pipe and surrounding rock has been promoted. This design can only be completed based on finite element numerical methods.

(4) Expansion calculation of pipe joint: Steel bifurcated pipe is a space structure composed of a plurality of pipe joints and reinforcing beams, and each pipe joint is completed by a steel plate through scribing, cutting, coiling, welding, etc. Therefore, the expansion calculation of the pipe joint is an indispensable part of the steel bifurcated pipe design. The methods used mainly include geometric analytical calculation, mapping method and expansion method based on computer-aided design software.

8.2 Three-Girders Reinforced Bifurcated Pipes

The typical arrangement of three-girders reinforced bifurcated pipes is asymmetric Y-shaped arrangement, symmetric Y-shaped arrangement and T-shaped arrangement, as shown in Fig. 8.7.

From the perspective of bearing uniform internal water pressure, it is most advantageous for the cylindrical shell to produce a circumferential uniform film tensile stress. However, the shell is no longer a complete circle due to the mutual cutting of the shell at the bifurcated pipe, as shown in Fig. 8.15a. The cross-section of the bifurcated pipe *A-A*, as shown in Fig. 8.15b, is already a combination of two incomplete torus shells.

Although the structure of this shape is closed and the pipe can bear a certain internal water pressure, a large bending moment will generate under the action of internal pressure. The deformation will be large, and the bearing capacity will be limited. This is mainly because the cylindrical shell of each bifurcated pipe has been cut off by a circular arc in points *E* and *F*. Circumferential tensile force *T* that

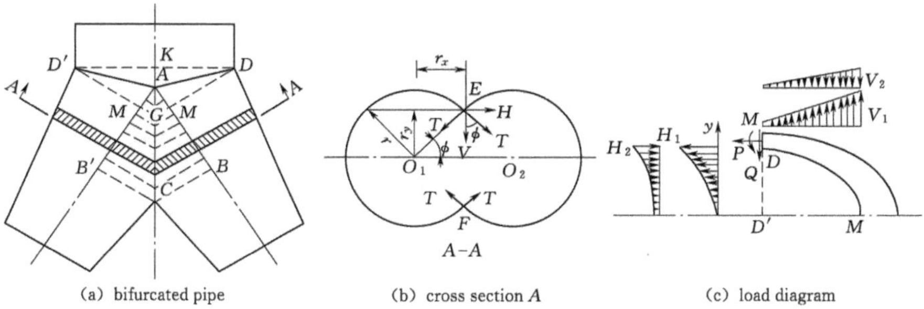

Fig. 8.15 Force analysis of bifurcated pipe

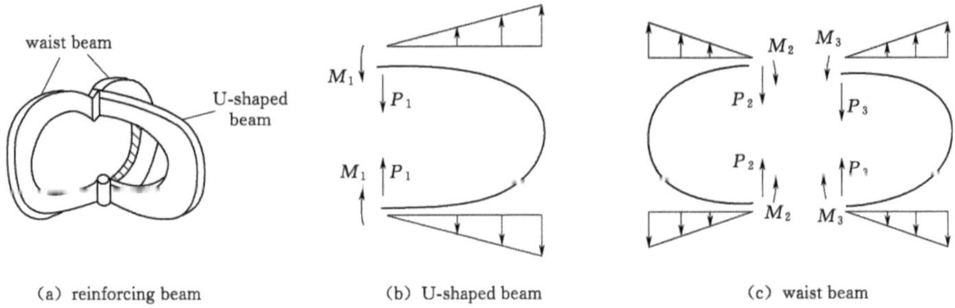

Fig. 8.16 Stress analysis of reinforcing beam

the circular arc originally applied to the cylindrical shell at two points E and F no longer exist, so that the ideal stress conditions are destroyed. This circumferential tensile force T is often referred as an unbalanced force. If the shell at points E and F is reinforced with a component, under the action of internal water pressure, the force state of the incomplete cylindrical shell will be improved by approaching the complete cylindrical shell if the reaction force applied by the reinforcement to the shell can restore the unbalanced force T as much as possible. This is the basic principle of the three-girders reinforced bifurcated pipe.

The specific method is to strengthen along the intersection line of the main pipe AD and the bifurcated pipe AD' through the waist beam, and strengthen along the intersection line of the two bifurcated pipes AC through the U-shaped beam. Then weld the joints of the U-shaped beam and the end of the waist beam together using a short column as the connection point, as shown in Fig. 8.16a. The cross-section form of the beam is more commonly used in rectangular and T-shaped. Sometimes for simplification, the two waist beams can be combined into a circular beam arranged along the edge and the beam can be extended to the point to connect with the circular waist beam.

A simplified method can be used to analyze the force system of the three-girders reinforced bifurcated pipe. Taking the symmetric Y-shaped bifurcated pipe of Fig. 8.15a as an example, the main pipe is cylindrical, and the two bifurcated pipes are the same cone pipe. Cut unit length strip perpendicular to the axis of the conical branch, and the section is flattened as shown in Fig. 8.15b.

1. Load caused by the circumferential tension of the pipe shell

The circumferential tension of the pipe shell along the axis of the cone pipe under the action of internal water pressure P.

$$T = \frac{Pr}{\cos^2 \alpha} \tag{8.1}$$

In which,

α the half cone angle of the cone pipe, degree (°);
r the radius of the cone at the section intercept, mm.

The vertical component V and horizontal component H of force along the per unit length of the intersection line are:

$$\begin{aligned} V &= T\cos\phi\cos\beta = \frac{Pr}{\cos^2\alpha}\cos\phi\cos\beta = \frac{Pr_x}{\cos^2\alpha}\cos\beta \\ H &= T\sin\phi\cos\beta = \frac{Pr}{\cos^2\alpha}\sin\phi\cos\beta = \frac{Pr_y}{\cos^2\alpha}\cos\beta \end{aligned} \tag{8.2}$$

In which,

ϕ the angle between the branch radius O_1E passing through the point E and the horizontal line, degree (°);
β the half-divided angle of bifurcated pipe, degree (°);
r_x, r_y projection of radius r in horizontal and vertical Directions, mm.

In Fig. 8.15b, the horizontal component force H caused by the unbalanced force at the point E can be decomposed into the component force along the AC direction (indicated by H_1) and the component force along the vertical AC direction. The component forces of two bifurcated pipes perpendicular to AC are on the same plane and substantially cancel each other out. The component forces along the direction of AC are superimposed. This horizontal load points upstream, and the bending moment generated in the U–shaped beam is opposite to that generated by the vertical load V, which is advantageous. H_1 will be represented in the intersection line plane along the vertical axis y, then the one side bifurcated pipe will produce H_1.

$$H_1 = \frac{a}{b}P\frac{y^2}{\sqrt{b^2 - y^2}}\frac{\sin\beta\cos\beta}{\cos^2\alpha} \tag{8.3}$$

In which,

y the vertical coordinate of the intersection line, mm, as shown in Fig. 8.15c;
a, b the long semi-axis and the short half-axis of the intersection line, mm, the value is:

$$a = R\frac{\cos\alpha\sin\beta}{\cos^2\alpha - \cos^2\beta},\ b = R\frac{\sin\beta}{\sqrt{\cos^2\alpha - \cos^2\beta}}$$

In which,

R is the diameter of the main pipe, mm.

The above V and H are the vertical load and horizontal load of a bifurcated pipe acting on the reinforcing beam on the unit length of the intersection line. For the symmetric Y-shaped bifurcated pipe, multiply by 2 to obtain the total load; for the asymmetric Y-shaped bifurcated pipe, the parameters of the two bifurcated pipes are substituted into the Eqs. (8.2) and (8.3), respectively, to find the corresponding load

V and H_1, and then superimposed separately to get the total load, its direction and distribution, as shown in Fig. 8.15c.

2. Load caused by axial force of the pipe shell

The axial force of the pipe shell has the following conditions: there are bulkheads, cone pipes, expansion joints and underground penstocks.

The vertical and horizontal components of the axial force of bifurcated pipe on the intersection line are:

$$V_2 = T\prime \sin\alpha \sin\beta[1 + \frac{(u - b')\tan\alpha\cos\beta}{r}] \tag{8.4}$$

$$H_2 = \frac{a^2}{b^2}T\prime R\frac{\sin\beta}{\cos\alpha}[\frac{\cos\alpha\cos\beta}{u} - \frac{(u - b')\sin\alpha\sin^2\beta}{ur}] \tag{8.5}$$

In which,

u	the abscissa value of the calculation point on the inner curve $\frac{u^2}{a^2} + \frac{y^2}{b^2} = 1$ of the U-shaped beam, mm, $b' = a - \frac{R}{\sin(\alpha+\beta)}$;
$T\prime$	the axial stress along the busbar direction of the unit perimeter of pipe shell of the conical tube, N/mm^2, when the nozzle is closed for hydrostatic testing, $T\prime = \frac{Pr_x}{2\cos\alpha}$; when it is underground penstock, $T\prime = \frac{v_s Pr_x}{\cos\alpha} - \Delta T\alpha_s t E_s$;
ΔT	temperature difference, °C;
v_s, E_s, α_s	the Poisson's ratio, modulus of elasticity and coefficient of linear expansion of the steel.

Other symbols are the same as before. And the load V_2, H_2 caused by a bifurcated pipe, the direction is shown in Fig. 8.15c.

The above is the load on the intersection line AC, and the load on the intersection line AD and AD' can also be obtained.

After the load of the U-shaped beam and the waist beam is determined as described above, the internal force analysis of the reinforcing beam system can be performed. The U-shaped beam and the waist beam are welded at the endpoint A, which is a spatial statically indeterminate beam structure, which can be completely solved by the structural mechanics' method. The three beams are, respectively, taken as spacers, and the respective forces are as shown in Fig. 8.16b, c. According to the consistent displacement of the joint point and the force balance condition, the internal force of the joint point can be obtained, thereby finding each internal force and stress of the beam. In the actual engineering calculation, it can be further simplified: for the exposed bifurcated pipe, the endpoint of the U-shaped beam and the waist beam is regarded as a hinge point, that is, for the underground bifurcated pipe, the termination point is regarded as a fixed connection, that is, the corner is 0, this calculation can be greatly simplified.

In the calculation, the cross-sectional dimension of the reinforcing beam needs to be set first, and the above internal force analysis is performed to check the strength.

If the requirements are not met, the cross-section should be reset and calculated. The model test, prototype observation and three-dimensional finite element calculation all show that the stress value of the reinforcing beam obtained by the simplified method above is relatively large, and the pipe shell of the adjacent reinforcing beam has a large edge effect stress in addition to the membrane stress. The test shows that the measured maximum stress of the pipe shell can reach 1.5–2.0 times of the theoretical membrane stress. Therefore, the pipe shell of the main and branch pipes of the bifurcated pipe should be appropriately thicker than the straight pipe section. For important bifurcated pipes, a three-dimensional numerical analysis or model test should be performed to verify the design.

Three-girders reinforced bifurcated pipe is a pipe type commonly used at home and abroad and can be used in large and medium-sized hydropower stations. Beams are under bending loads so the strength of steel cannot be used completely which causes large beam section. When the value of *HD* is large, the dimension of reinforcing beam will be large which induces difficulties in material, manufacture and transportation. The excavation will increase when used for underground penstock. Recently, three-girders reinforced bifurcated pipe is substituted by crescent-rib reinforced bifurcated pipe, especially in underground layouts.

8.3 Crescent-Rib Reinforced Bifurcated Pipes

The typical arrangement of crescent-rib reinforced bifurcated pipes is symmetric Y-shaped and asymmetric Y-shaped.

The crescent-rib reinforced bifurcated pipe is a kind of development of the three-girders reinforced bifurcated pipe. It replaces the U-shaped beam of the three-girders reinforced bifurcated pipe with a crescent-shaped rib embedded in the bifurcated pipe. The waist beam is eliminated so that the position and dimension of the crescent-rib can be determined, which can make the crescent-rib close to the axial tension component. Enlarging section of main cone and reducing section of branch cones which have a common tangent ball form a cone shell reinforced by a crescent-rib internally along the intersection line of branch cones, as shown in Fig. 8.8. By setting the reverse cone pipe, the downstream angle θ of the inflection point of pipe shell can be decreased (Fig. 8.1), The local stress and flow state of turning point of pipe shell can be improved, gradually expanding the flow area at the branching point, and can reduce velocity and head loss.

The basic force principle of the crescent-rib reinforced bifurcated pipe can be briefly described as follows:

Figure 8.17 shows a symmetric Y-shaped bifurcated pipe, where *AB* is an intersection curve, and a crescent-rib is placed on this intersection curve. Method of analyzing the load on the U-shaped beam of the three-girders reinforced bifurcated pipe is used to consider the load on the crescent-rib, that is, to reverse the "unbalanced ring tension" of the broken shell on the intersection line as the load of the crescent-rib. Therefore, there are loads generated by the circumferential forces of

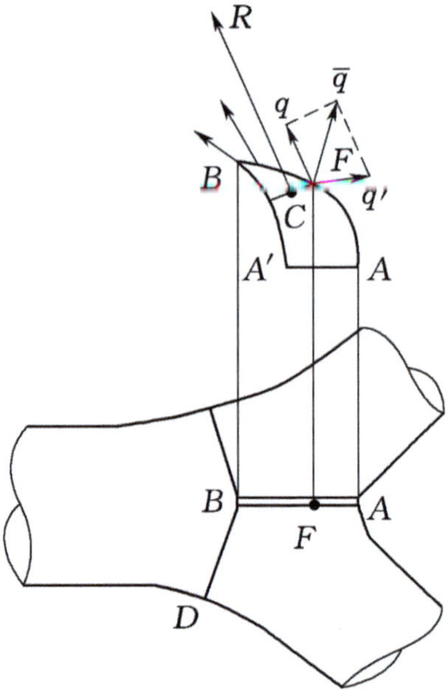

Fig. 8.17 Force analysis of crescent-rib

the two bifurcated pipes at any point on the intersection line F, and their combined values are q. At this point F, there is also an axial force from the end of bifurcated pipe, which can also be synthesized as q', q and q' will be combined with the vector to obtain the resultant load $\overline{q}$ acting on the point F, and its size and direction can be determined. When the point F is different along the intersection line, $\overline{q}$ also change. If the loads from the rib ends B to the points F are added, the total load R acting on this rib and its action line is obtained. A straight line through point F that is orthogonal to this action line is used to obtain a foot C. If the trajectory is used as the axis of the reinforcing rib, the reinforcing rib will only bear the axial tension.

The specific design method points are, first find out the trajectory of the point C, and then determine the width of the rib (that is, find the inner and outer edge line of the rib). The outer edge of the rib must be welded to the shell, so the rib width b at the point F should be at least twice as long as $\overline{FC}$, that is $b \geq 2\overline{FC}$. Once b is selected, the thickness of the ribs can be determined according to the specified strength theory, the thickness of the ribs is $t = \frac{R}{b[\sigma]_l} + C$.

However, the thickness of the ribs in the various sections is likely to be different when we calculate in this way. In order to make the ribs of equal thickness, the most important section can be selected to determine the rib thickness. Generally, the horizontal symmetry section AA' of the ribs has the largest combined force V, and the rib width $B_T = AA'$ is determined according to the section, and the rib thickness t_w is determined according to the combined force and strength requirements.

$$t_w = \frac{V}{B_T[\sigma]_l} + C \tag{8.6}$$

In which,

- $[\sigma]$ allowable stress of ribs, N/mm^2; when in the condition of basic loads, $0.5\sigma_s$ is taken; when in the condition of special loads, $0.7\sigma_s$ is taken;
- C corrosion margin, mm;
- t_w the thickness of ribs, mm; generally, take the thickness of the pipe shell 2.0–2.5 times.

Then the width of the ribs of other sections is equal to:

$$b = B_T R / V$$

However, the rib width b thus determined is not necessarily equal to $2\overline{FC}$ in the cross-section. If $b > 2\overline{FC}$, the outer edge b of the rib protrudes outside the pipe shell; If $b < 2\overline{FC}$, the width of the rib is appropriately enlarged so that the outer edge of the rib at least reaches the intersection line for welding with the shell.

The reinforcing ribs thus obtained are of equal thickness, and most of the ribs are located inside the intersection line, and because of their shape, such as crescents, they are called crescent-rib reinforced bifurcated pipe. In the actual design process, it is too cumbersome to calculate the rib size by the above method, which is generally carried out by a simplified method. The specific method is as follows:

The central section width B_T of the ribs can be initially determined from the empirical curve in Fig. 8.18a based on the branching angle. Curve I_1 is used for hydrostatic test conditions and curve II_2 is used for operating conditions. After B_T is determined, the inner edge dimension of the rib can be connected to a parabola according to the three points $BAB\prime$ in Fig. 8.18b. The equation is $y^2 = y_0^2(x_0 - x)/x_0$ that the outer edge of the rib is generally increased by 50–100 mm based on the intersection line to meet the welding needs. The thickness of the ribs can still be calculated according to Eq. (8.6).

In the above analysis, only the circumferential stress and the axial stress in the membrane stress are considered approximately, and the shear stress in the pipe shell and the local stress at the joint of the shell and the rib plate are neglected. Therefore, there must be bending stress and shear stress in the crescent-rib. For asymmetric bifurcated pipe, the force condition is more complicated. In addition, with the above method, the stress distribution and local stress of the bifurcated pipe shell cannot

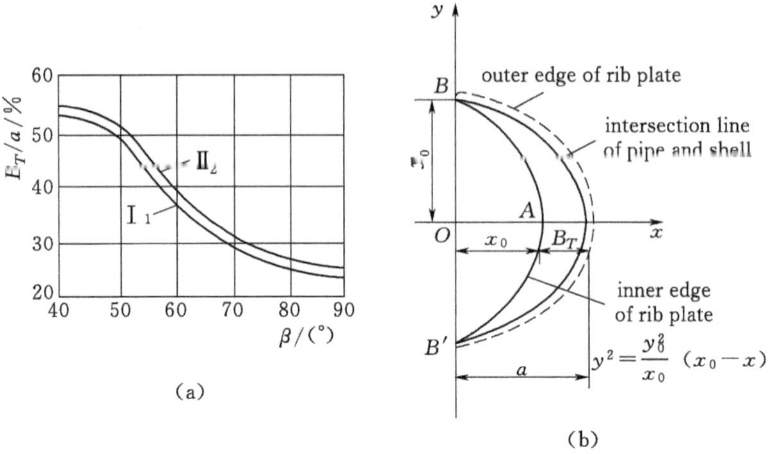

Fig. 8.18 Design reference curve of rib plate width

be calculated. Therefore, for important projects, numerical analysis or model tests should be used for further verification.

The crescent-rib reinforced bifurcated pipe has the following characteristics:

(1) Rib is under tensile force mainly, so it has a small size, and there is no breast beam. According to the analysis, the weight of the crescent-rib is usually only 1/5 to 1/4 of the three-girders reinforced bifurcated pipe reinforcing beam. The crescent-rib is a uniform thick plate, and the manufacturing process is relatively simple.
(2) Bifurcated pipe is reinforced internally so that the size of bifurcated pipe is small which can reduce the excavation, besides, the surface of the bifurcated pipe is smooth which is an advantage for rock to bear loads.
(3) Head loss is small under normal operation condition because the flow rate will decrease at the main cone and rib serves to partition the discharge.
(4) There is obvious stress concentration at the turning points of pipe shell and intersection points of shell and rib which are the weaken zones. Mitigating or reinforcing measures need to be taken, such as reducing the break-angle of the body during the design of the bifurcated pipe and increasing the pipe shell thickness or partial welt.

The application of the crescent-rib reinforced bifurcated pipe has been more than 40 years old and accumulated considerable experience. In recent years, it has basically replaced the three-girders reinforced bifurcated pipe in China.

8.4 Hem Reinforced Bifurcated Pipes, Spherical Bifurcated Pipes, and Shell Type Bifurcated Pipes

8.4.1 Hem Reinforced Bifurcated Pipes

The typical arrangement of hem reinforced bifurcated pipes is asymmetric Y-shaped.

When the dimension and the water head of steel penstock are not large and the small bifurcated pipe is required to be separated (the diameter ratio of the branch pipe and the main pipe is not more than 0.7), the special reinforcing beam system may not be provided, and the pipe shell of the main pipe and the branch pipe may be added as appropriate thick. The bifurcated pipe is reinforced with a simple curved plate welded at both sides of the intersection line of main pipe and branch pipe, which is called hem reinforced bifurcated pipe. The welted reinforcing plate can be welded to the outer pipe shell or inner pipe shell, or the reinforcing plate can be provided on the inner and outer pipe shells. The "unbalanced force" at the openings will be carried by the pipe shell and reinforcing plate, reinforcing plate will have a larger radial deformation under internal water pressure because of a small stiffness. Therefore, it can be used for the underground bifurcated pipe to facilitate the resistance of the surrounding rock and also facilitate manufacture of steel-lined reinforced concrete bifurcated pipe.

The stress state of the hem reinforced bifurcated pipe is very complicated. At present, there is no good analytical method to calculate the stress, and the design is often based on the experience accumulated by the finite element method, model test and engineering practice. The following are several ways to design a reinforcing plate by experience.

1. The area compensation method

The area compensation method is a commonly used calculation method for the hem reinforced bifurcated pipe. As shown in Fig. 8.19, when the bifurcated pipe radius r_d is not more than one-half of the main pipe radius r, the cross-sectional area A_d of the reinforcing plate along the longitudinal section of main axis line should be slightly larger than the cross-sectional area A_c of the main pipe, and the thickness of the pipe t_d is 1.0–1.3 times thicker than the thickness of the main pipe.

2. The ring method

The ring method is another calculation method for the hem reinforced bifurcated pipe. The calculation diagram is shown in Fig. 8.20. A ring is used as the stiffening structure, the inner diameter of the ring is L, and the width b is calculated along the radial direction to be the width of the reinforcing plate, and the thickness is calculated as the sum of the pipe shell thickness t and the thickness of the reinforcing plate t_d. The vertical load is the unbalanced force pr at the main pipe breach, and the transverse load $pr/2$ is the binding force generated by the main pipe shell on the deformation of the reinforcing plate.

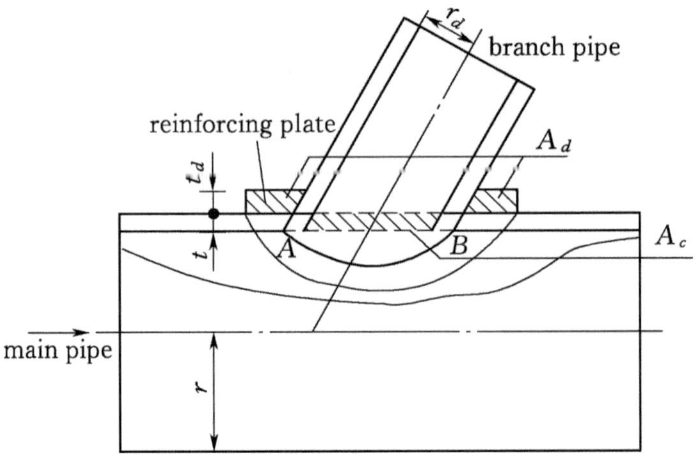

Fig. 8.19 Calculation diagram of the area compensation method

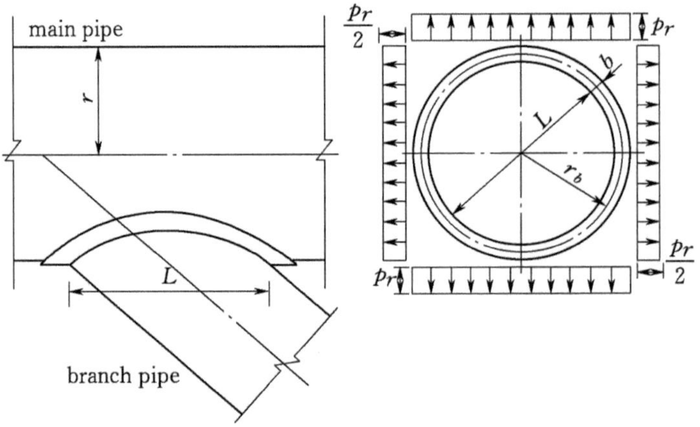

Fig. 8.20 Calculation diagram of the ring method

The maximum stress of the ring stiffening structure occurs at the waist, and the inner edge bending moment is:

$$M = \frac{prr_b^2}{8} \tag{8.7}$$

Inner edge shaft tension:

$$N = prr_b \tag{8.8}$$

Maximum stress:

$$\sigma_{\max} = \frac{M}{W} + \frac{N}{F}, \quad \sigma_{\max} \le \sigma_R \text{ or } \sigma_{\max} \le [\sigma] \tag{8.9}$$

In which,

p internal water pressure design value, N/mm^2;
r the main radius, mm;
r_b radius of the center of the ring, mm, $r_b = (L + b)/2$;
b the width of the annular plate, mm;
W annular section bending section modulus, mm^3;
F the cross-sectional area of the ring plate, mm^2.

In China, there are many applications of hem reinforced bifurcated pipe in the middle and low head underground penstocks. Practical experience is accumulated, and the largest value of *HD* has reached 990 m^2.

8.4.2 Spherical Bifurcated Pipes

Typical arrangement of spherical bifurcated pipes is symmetric Y-shaped and T-shaped.

Branching is carried out through spherical shell, and the structure is simple: a spherical shell is arranged at the branching point, and circumferential reinforced rings are required along the intersection line between the main pipe, branch pipes and the spherical shell, and the axis of each pipe should pass through the center of the ball. The main pipe is generally a cylinder, the bifurcated pipe can be a cylinder or a cone, and the intersection line of the pipe and the spherical shell is a complete circle. In order to reduce the head loss of the branching, a deflector is arranged inside the shell. The structural layout of the spherical bifurcated pipe is shown in Fig. 8.21.

The minimum radius of the spherical shell is such that the local stress of the spherical shell after the opening of the main and branch pipes does not affect each other, and it is necessary to have a welding space, which is usually taken as about 1.3–1.6 times the main radius, not less than 0.3 m, and it is taken as a small value when the following conditions are met:

(1) The shortest arc length between two adjacent holes on the spherical shell

$$L \ge 2.43\sqrt{Rt_s} \tag{8.10}$$

In which,

t_s the thickness of the spherical shell, mm.

(2) The minimum space between the spherical shell and the internal deflector is not less than 0.3 to 0.5 m.

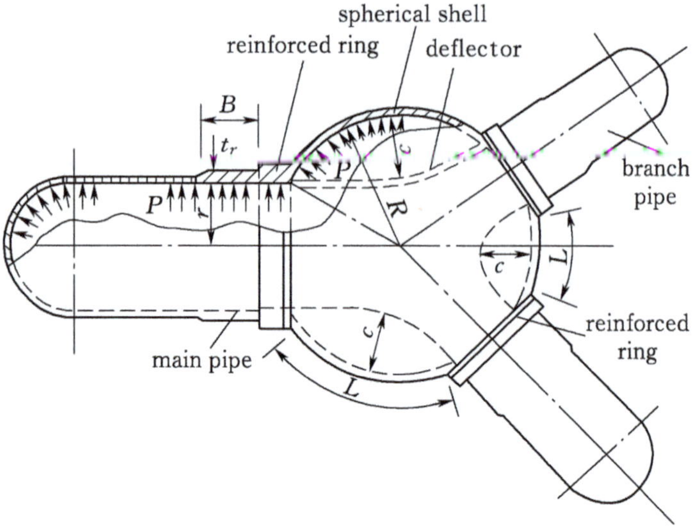

Fig. 8.21 Configuration of spherical bifurcated pipes

Under the action of uniform internal water pressure, the membrane stress of the spherical shell is uniform in all directions and only half of the circumferential stress of the cylindrical shell with the same radius. Therefore, the thickness of the spherical shell can be taken as:

$$t_s = \frac{K_1 PR}{2\varphi[\sigma]} \tag{8.11}$$

In which,

K_1 can be taken as 1.1 to 1.2.

Where the main and branch pipes are connected to the spherical shell, the complete spherical shell is broken and needs to be reinforced with a reinforced ring. The reinforced ring is a circular ring beam. The cross-section design should be such that the reinforced ring does not twist after being stressed and the radial displacement is similar to the displacement of the spherical shell before the spherical shell is broken and the displacement of the main and branch pipe. Thus, the spherical shell and the main and branch pipes are close to the ideal state of membrane stress of the spherical shell and cylindrical shell.

The pipe shell should be slightly thickened into a transition section at the connection between the main and branch pipes and reinforced ring.

The stress calculation of the reinforced ring, the spherical shell and the cylindrical shell can be analyzed by static and indeterminate structures of the mechanical properties of the rod and the shell. The reinforced ring, the spherical shell and the pipe shell are taken as three partitions, and the internal forces (axial force, shear force, bending moment) and external load (internal water pressure) where the three are connected are used as the load of the isolator. Then, according to the condition of the consistent displacement (radial displacement, angular displacement) and the balance condition of the force, the internal force at the joint is obtained, and the stress of the spherical shell, the reinforced ring, the main and branch pipe shells are obtained.

For important projects, overall structural stress analysis or model testing should be performed.

The advantages of the spherical bifurcated pipe are flexible arrangement and the axis of the bifurcated pipes can be arbitrary in plane or space, which is convenient for arrangement. The force structure of the branching part is the spherical shell, and the main and branch pipes remain as a complete cylindrical shell. The stress conditions are good, the reinforced ring is a ring beam, and the stress analysis method is relatively simple and reliable. Spherical bifurcated pipe also has certain shortcomings: the spherical shell should be molded and annealed integrally, so the construction is difficult and cost is high especially when the diameter is large.

The spherical bifurcated pipe is a mature pipe type that is widely used in foreign countries, and it is applied more when it is high head. This pipe has been successfully applied to large and medium-sized high-head hydropower stations in China, but not many, mainly due to the limitations of process and transportation conditions. As materials and processes improve, they will be more widely used.

8.4.3 Shell Type Bifurcated Pipes

The typical arrangement of shell type bifurcated pipes is symmetric Y-shaped and asymmetric Y-shaped, and can also be arranged in a T-shaped.

The shell type bifurcated pipe is developed based on a spherical bifurcated pipe. In the spherical bifurcated pipe, where the round pipe is combined with the spherical shell, the spherical shell is perforated, and a reinforced ring is required, and there is stress concentration at the joint. In order to improve the stress conditions, the larger diameter cone pipe and the spherical shell can be connected along the tangential direction. Therefore, the spherical shell only has two upper and lower small triangles, and a few sections of the gradually expanding transition section are inserted between the main pipe and the branch pipe to form a relatively smooth body, without too much of a discontinuous joint line, this is the shell type bifurcated pipe.

When designing a shell type bifurcated pipe, the following principles should be followed:

(1) The joints between the pipe joints of the shell type bifurcated pipe and between the spherical shell and its adjacent cone pipe should conform to the principle of the common tangent ball. The radius $R_i(r_s)$ of the common tangent ball can be taken as 1.15 to 1.30 times than the main radius for the symmetric Y-shaped arrangement and 1.20 to 1.35 times than the main radius for the asymmetric Y-shaped arrangement. A small ratio can be taken when the main radius is larger.

(2) The pitch number of main and branch cones is determined by the ratio of the diameter of the main and branch pipes to the diameter of the common tangent ball, the allowable turning angle of the pipe shell and the minimum spacing of the weld. The length of the pipe joint should make the local stress caused by the turning angle does not affect each other, and generally should not be less than 300 to 500 mm.
(3) The deflection angle ω should be 80–120°. The waistline turning angle C_i of the pipe shell should not be greater than 15°, and at the maximum diameter should not be greater than 12°. If the thickness of pipe shell is constant, where the pipe diameter is small, the waistline turning angle can be increased to 18°.

The shell type bifurcated pipe is subjected to internal pressure through a relatively continuous combination of shells and membrane stress and a smaller bending stress without the reinforcement being installed. Except for the two triangular spherical shells with small area, the others are conical shells, which are small in size and convenient for transportation and installation. When it is made into an underground bifurcated pipe, it is beneficial to exert the surrounding rock resistance. However, this kind of bifurcated pipe is complicated in shape, the forming process of the spherical surface is still difficult, there are many welds, and there is no reliable analytical method for structural stress intensity analysis or sufficient experience. Therefore, the design of large-scale shell type bifurcated pipe requires reference to numerical analysis method or model test. The cross section of the water at the branching point of the shell type bifurcated pipe is sharply increased, and vortex is easy to occur. The internal diversion structure should be adopted to improve the flow state.

The shell type bifurcated pipe is a promising type of bifurcated pipe that can be used for large and medium underground penstocks. This pipe has been successfully applied to Xinjiang Kashi two-stage Hydropower Station and Yunnan Chaishitan Hydropower Station in China.

Question

1. What are the structural types of bifurcated pipes? What are the characteristics and applicable conditions?

Bibliography

1. Renkun, Wang and Z. Chunsheng. 2013. *Design manual of hydraulic structures: Vol.8 buildings of hydropower station*, 2nd ed. Beijing: China Water and Power Press.
2. Zhiguo, Wang and C. Yongxing. 2007. Study on hydraulic characteristic of Escher-Wyss wyepiece of Xilongchi pumped storage power station. *Journal of Hydroelectric Engineering* 26 (1): 42–47.
3. Zhiguo, Wang, D. Yunling and G. Guibiao, et al. 2006. Experimental study on site structural model for design of surrounding rock to share internal water pressure for high pressure Y-pipe of Xilongchi pumped-storage power station. *Journal of Hydroelectric Engineering* 25 (6): 55–60.

4. Yang, Wang, W. Hegao, and D. Fangqin. 2014. Design theory and application of a new type of steel bifurcation with horizontal bottom. *Journal of Hydraulic Engineering* 45 (1): 96–102.
5. Jinglong, Luo, and W. Hegao. 2005. Program design for automatic mesh partition of crescent rib bifurcation finite element. *China Rural Water and Hydropower* 2: 86–87.
6. Shan, Fu., W. Hegao, and W. Yang. 2013. Development and application of computer aided design system for rib reinforced bifurcation based on CATIA. *Water Power* 39 (7): 73–76.
7. Ruxiang, Song, W. Hegao, and S. Kai. 2009. Development and application of crescent rib bifurcated tube joint deployment procedure. *Yangtze River* 40 (13): 34–37.
8. Shanding, M.A., and W. Ruze. 1996. *Buildings of hydropower station*. Beijing: China Water and Power Press.
9. Qizhao, L.I.U. and H.U. Ming. 2010. *Hydropower station*, 4th ed. Beijing: China Water and Power Press.

Chapter 9
Guaranteed Regulation and Water Hammer

Yongguang Cheng

Abstract In this chapter, we will discuss the following key points: Concepts and standards of calculation of guaranteed regulation; water hammer phenomenon and objectives of studying water hammer; governing equations and wave speed of water hammer; analytical solution of single pipe water hammer; water hammer calculation for complex pipes; calculation of rotational speed rise; calculation conditions for water hammer; measures to handle the contradiction between water hammer pressure and rotational speed rise. Students should be acquainted with the concept and standards of the calculation of guaranteed regulation; master the calculation of the direct and indirect water hammer pressures in single pipe; master the calculation of water hammer pressure in complex pipes; master the calculation of rotational speed rise; be familiar with the measures to reduce water hammer pressure and control rotational speed rise.

Keywords Calculation of guaranteed regulation · Water hammer · Rotational speed rise · Analytical solution · Numerical solution · Measure to solve the contradiction between water hammer and rotational speed rise

9.1 Transient Processes and Calculations of Guaranteed Regulation of Hydropower Stations

9.1.1 Phenomena and Features of Transient Processes in Hydropower Systems

During actual operations of hydropower plant, sharp load variations such as emergent load rejection, unit starting, load decrease, and load increase, happen frequently. During these events, the turbine guide vanes (or wicket gates) close or open within very short periods, in order to regulate the discharge. The pressure in penstocks,

Y. Cheng (✉)
Wuhan University, Wuhan, China
e-mail: ygcheng@whu.edu.cn

Y. Cheng et al. (eds.), *Hydropower Station*,
https://doi.org/10.1007/978-981-96-2383-9_9

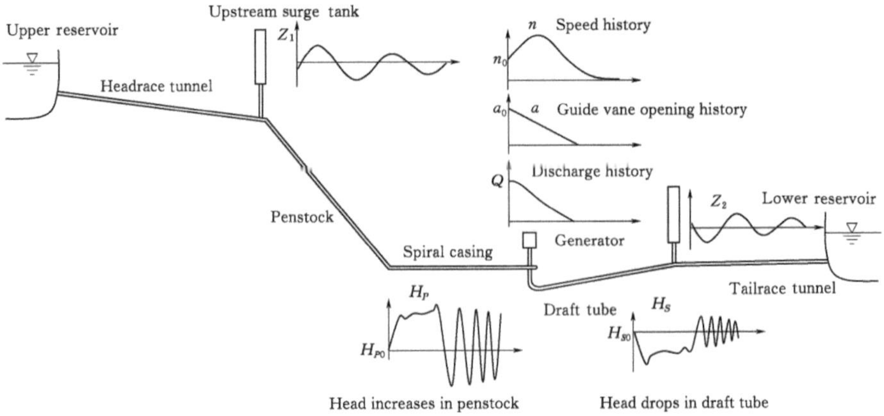

Fig. 9.1 Schematic of transient processes after load rejection of a hydropower plant

the water level in surge chambers, and the rotational speed of units thereby change. These are all called the transient processes in hydropower systems. The definition of transient processes may be stated as the unsteady processes between two steady states. Take the typical pumped-storage system shown in Fig. 9.1 as an example. After a load rejection of the unit, the guide vanes close quickly obeying the control of the governor, the discharge through the turbine decreases to zero, the rotational speed increases first and then decreases, the pressure in the spiral casing first increases and then fluctuates, the pressure in the draft tube first decreases and then fluctuates, the water level in the upstream surge tank (or surge chamber) first ascends and then oscillates, and the water level in the downstream surge tank first descends and then oscillates. That is to say, after the change of load of the unit, there will be three types of transient processes happening.

9.1.1.1 Variation of Rotational Speed of the Unit

Normally a load rejection or acceptance happens in a very short time, but the closing or opening of guide vanes needs a relative long time. Therefore, the mismatch of the output power of turbine and the load of generator will cause imbalance. From the dynamic equation for unit rotation, i.e., Eq. (4.1) in Chap. 4, we know that the temporary surplus energy after a load rejection transfers into the kinetic energy of rotating parts, therefore the rotational speed increases; on the contrary, the temporary energy deficit after load acceptance may cause decrease of the rotational speed.

9.1.1.2 Water Hammer in the Conduit System

If the discharge at the end of a pipe changes suddenly, the velocity in the pipe will change accordingly. As a result, the pressure in the pipe will change also. The change of pressure may be called as water hammer. The occurrence of water hammer conforms to Newton's second law. Water inertia is the intrinsic reason, and flowing state change is the cause. The acceleration or deceleration of the water in pipes must result in pressure rise or drop in the pipes. When turbine guide vanes close, turbine discharge decreases. And because the pipe water decelerates, the spiral casing pressure and penstock pressure increase, and the draft tube pressure decreases. When turbine guide vanes open, turbine discharge increases. Then the pipe water accelerates, the spiral casing pressure and penstock pressure decreases, and the draft tube pressure increase.

9.1.1.3 Surging Waves in Surge Chambers

The discharge changes in turbines are normally faster, and the discharge variations in headrace and tailrace tunnels are always slower. Because of the imbalance of flow rates in the pipes connecting to the surge chamber and the water inertia in the tunnels, the water levels in the surge chambers must fluctuate. After a load rejection, the water in the headrace tunnel flows into the upstream surge chamber, causing the water level rise; the water in the tailrace tunnel flows out of the downstream surge chamber, causing the water level drop. When differences between the surge chamber levels and reservoir levels reach to certain values, the water in the tunnels will be driven by the level differences to flow in reverse directions. Thus, the reciprocating water flows in the tunnels and the oscillating water levels in the surge chambers occur. The general phenomena are that for a load rejection condition, the level in the upstream chamber rises first while the level in the downstream chamber drops first. Contrastively, for a load acceptance condition, the level in the upstream chamber drops first while the level in the downstream chamber rises first.

The water hammer and surge wave phenomena mentioned above are classified as hydraulic transients, while the variations of rotational speed and guide vane opening are classified as mechanical transients.

Water hammer fluctuation and rotational speed variation are contradictory, and they are all related to the closing or opening speed and procedure of guide vanes. The shorter the closing time, the larger the water hammer pressure but the smaller the rotational speed change; on the contrary, the longer the closing time, the smaller the water hammer pressure but the larger the rotational speed change.

The task of calculating water hammer fluctuation and rotational speed variation after sudden and large load variations, and finding means to coordinate their contradiction, is called the calculation of guaranteed regulation. In this chapter, we will discuss this topic in detail.

Water hammer in pipes and surge waves in surge chambers are all due to the inertia of water, but their features are different. Water hammer pressure fluctuation is a kind of elastic wave with shorter period, larger amplitude, and faster attenuation; on the other hand, the surge chamber wave oscillation is a kind of mass wave with longer period, smaller amplitude, and slower attenuation. Because these two phenomena do not happen synchronously, and the extremal values of them have relative small interference, we normally treat them separately in preliminary calculation. Therefore, the basic concepts, principles, and calculating methods of water hammer and guaranteed regulation will be introduced in this chapter, and the types, working principles, and selection of surge chambers and the calculating methods of surge waves will be addressed in Chap. 10.

9.1.2 Basic Concepts and Standards on Guaranteed Regulation

The calculation of guaranteed regulation (CGR) is an important task during the design process of hydropower stations, because it is not only related to the strength design and strength verification of the flow passage parts of penstock, turbine, valve, etc., but also related to the stability of unit operations and the safety of hydropower plants.

The more concrete definition of CGR is as follows. Calculate water hammer pressure and rotational speed rise; select reasonable guide vane closing time and procedure; coordinate the relationship among guide vane regulation, water hammer pressure and rotational speed rise; so as to guarantee the values of pressure and speed are all within the permitted ranges specified by relevant standards.

Water hammer pressure calculation is the most important part in CGR. The purposes are to obtain the maximum pressure head and provide basic information for strength design and strength verification of penstock, spiral casing, and turbine; obtain the minimum pressure head and provide basic information for laying pipelines, avoiding negative pressure in pipes, and verifying vacuum value in draft tubes; analyze the influences of water hammer to unit operation such as unit speed change.

The standards on CGR provide the technically and economically reasonable values of water hammer and rotational speed change. They are different in different countries and times, because the establishment of standards depends on technical advance and people cognitive level development. Namely, the defined values are always changing, and we should select proper values based on specific circumstances.

The maximum water hammer pressure is normally expressed by relative value $\xi_{\max} = (H_{\max} - H_0)/H_0$, where H_0 and $H_{\max}$ are the water head in steady condition and the maximum head in spiral casing during transient condition, respectively.

The maximum rotational speed rise is normally expressed by relative value $\beta_{\max} = (n_{\max} - n_0)/n_0$, where n_0 and $n_{\max}$ are the rated speed of unit and the maximum speed during load rejection transients, respectively.

The currently in effect industrial standards concerning CGR is DL/T5186-2004, *Electrical–Mechanical Design Code of Hydropower Plant*, in which the following articles are stated.

(1) In unit load rejection condition, the permitted value of the maximum rotational speed rise ratio [β_{max}] can be chosen based on the following classifications.

a. When the unit capacity occupies a relative large proportion in the total capacity of the electrical grid, or the unit works as a frequency regulation tool, the value should be smaller than 50%;
b. When the unit capacity occupies a relative small proportion in the total capacity of the electrical grid, or the unit does not work as a frequency regulation tool, the value should be smaller than 60%;
c. For the unit with a tabular turbine, the value should be smaller than 65%;
d. For the unit with an impulse turbine, the value should be smaller than 30%
e. For the reversible pump-turbine unit, the value is normally similar to that of Francis turbine unit, but if the capacity is large and the head is high, the value can be smaller than 45%.

(2) In unit load rejection condition, the permitted value of the maximum pressure head rise ratio [ξ_{max}] in the spiral casing (before guide vanes for tabular turbine unit) can be chosen based on the following classifications.

a. If the rated head is lower than 20 m, the value should be between 100 and 70%;
b. If the rated head is between 20 to 40 m, the value should be between 70 and 50%;
c. If the rated head is between 40 and 100 m, the value should be between 50 and 30%;
d. If the rated head is between 100 and 300 m, the value should be between 30 and 25%;
e. If the rated head is higher than 300 m, the value should be smaller than 25% (30% for the reversible pump-turbine unit).

(3) In the sharp load increase or decrease condition, the minimum pressure head at the highest point along the longitudinal profiles of hydraulic system should be larger than 0.02 MPa, and there is no negative pressure and backflow vertex.

In the unit load rejection condition, the maximum vacuum at the inlet section of draft tube should be smaller than 0.08 MPa.

There are some annotations that we should pay attention to.

(1) The permitted value of maximum pressure head rise ratio in spiral casing is specified based on the comprehensive considerations of structure strength and economic efficiency. The definition of ξ_{max} has different forms in different references, and H_0 sometimes is defined as the working head of turbine.
(2) The permitted value of the maximum rotational speed rise ratio is specified to ensure unit safety and electricity quality. In the full load rejection

condition, the unit undergoes risks of severe vibration, structure damage, and generator insulation damage.

(3) For exposed steel penstocks, the smallest pressure at the highest point should be larger than 0.02 MP, because exposed steel penstocks have the instability problem due to external pressure. But for concrete-lined or bare rock penstocks, the requirement can be modified to no negative pressure and backflow vortex.

(4) The calculation of the maximum vacuum at draft tube inlet should include the influence of altitude by $H_{b\max} - \frac{\nabla}{900}$, in which $H_{b\max}$ is the permitted maximum vacuum value for zero altitude, and ∇ is the installation elevation of turbine.

9.2 Water Hammer Phenomenon and Governing Equations

9.2.1 Water Hammer Phenomenon and Intrinsic Attribution

Water hammer is the faster pressure change phenomenon in a pipe. It is caused by the acceleration or deceleration of the flowing water due to certain reasons. To correctly comprehend and explicate water hammer phenomenon and essence, the compressibility of water and elasticity of pipe wall must be considered. For convenience, we will analyze the water hammer propagation and reflection processes caused by a sudden closure of valve in the simple single pipe system. The single pipe is defined as the pipe with properties such as diameter, wall thickness and wall material constant along the length, with a large reservoir connected at the upstream end, and a valve installed at the downstream end. Figure 9.2 is the schematic, in which A denotes the valve, B denotes the reservoir, and L is the pipe length. The diameter of pipe is d_0 and the density of water is ρ_0. In the initial steady condition, the valve is fully open, the steady velocity in the pipe is V_0, and the pressure head along the pipe is H_0 if the resistance is neglected. The propagating and reflecting processes of water hammer after the sudden closing of the valve (closing time approaches zero) can be described as follows.

(1) At instant $t = 0$, the velocity at the valve section changes instantly from V_0 to 0 due to the sudden closing of the valve. The water in the pipe is still flowing toward the valve with velocity V_0 because of its inertia. The pressure head at point A rises instantly to $H_0 + \Delta H$, with $+\Delta H$ being the water hammer head and H_0 the initial steady head. After $t = 0$ and within $dt_1 > t > 0$, the water state within segment dx_1 has some changes: velocity becomes zero, pressure head becomes $H_0 + \Delta H$. The water is compressed and the density is increased to $\rho_0 + \Delta\rho$. The pipe wall is inflated with the diameter increasing to $d_0 + \Delta d$. The compressed water and inflated pipe produce a space to admit the incoming upstream water that is still flowing with velocity V_0. That is to say, during the time interval dt_1, the upstream part of pipe segment dx_1 is not influenced by water hammer pressure. After that, during intervals dt_2, dt_3, ..., the velocity

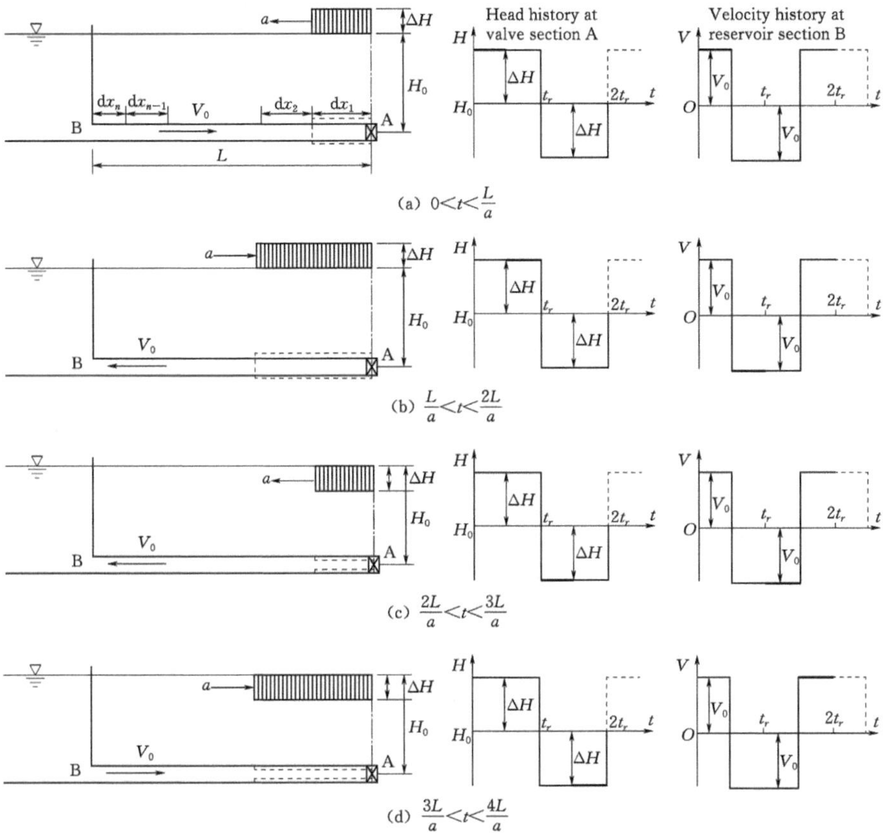

Fig. 9.2 Propagating and reflecting processes of water hammer waves in a single pipe

and pressure within dx_2, dx_3, …, will change successively like those during interval dt_1, as shown in Fig. 9.2a. It is shown that the water hammer pressure head $+\Delta H$ does not occur instantly in the whole reach of the pipe after valve closing, but propagates gradually upstream from the valve, demonstrating the feature of pressure wave. This wave is called the water hammer wave and can be defined by $a = \mathrm{d}x_i/\mathrm{d}t_i$. During $0 < t < L/a$, the water hammer wave $+\Delta H$ propagates from section A to section B, keeping the pressure head $H_0 + \Delta H$ at A and velocity V_0 at B unchanged (Fig. 9.2a).

(2) At instant $t = L/a$, the water hammer wave $+\Delta H$ reached at section B, where the velocity becomes 0 instantly from V_0. At this time, the pressure head on the right side of B is $H_0 + \Delta H$, higher than the reservoir level H_0 on the left side. Because of the head difference ΔH, the water at this reservoir boundary cannot keep balance and changes its flowing direction toward the reservoir. During $L/a < t < L/a + dt_n$, the water in the segment dx_n adjacent to B first has some changes: velocity changes from 0 to $-V_0$, pressure head from $H_0 + \Delta H$ to H_0,

density from $\rho_0 + \Delta\rho$ to ρ_0, diameter from $d_0 + \Delta d$ to d_0, and the pipe wall and water pressure recover to the initial states. The state recovery in segment dx_n can supply the water flowing toward the reservoir and make the states in segments dx_{n-1}, dx_{n-2}, ..., being $V = 0$, $H = H_0 + \Delta H$, $\rho = \rho_0 + \Delta\rho$, and $d = d_0 + \Delta d$, which are kept until the end of dt_n, as shown in Fig. 9.2b. In a like manner, in the later time intervals dt_{n-1}, dt_{n-2}, ..., similar changes will happen in segments dx_{n-1}, dx_{n-2}, ... At $t = 2L/a$, the whole pipe will be in the state with $V = -V_0$, $H = H_0$, $\rho = \rho_0$, and $d = d_0$. Just before and after $t = L/a$, the velocity at B changes instantly from V_0 to $-V_0$, and the pressure head at the pipe section just adjacent to B changes from $H_0 + \Delta H$ to H_0. That is because the water hammer wave $+\Delta H$ (incoming wave) is reflected by the reservoir boundary and a new water hammer wave $-\Delta H$ (reflected wave) is generated. These two waves cancel out one another. When water hammer wave approaches any boundary (reservoir, bifurcation, serial junction, valve, etc.), pressure and velocity will change owing to the specific restrictions by the boundary, and their values are defined by the initial water hammer wave (incoming wave) and the newly generated wave (reflected wave). For the reservoir section, head $H = H_0$ is the must kept condition. In other words, the reflected wave must be $-\Delta H$, which has the same value but opposite sign with the incoming wave $+\Delta H$. Therefore, the reflecting characteristics of reservoir to water hammer wave are normally called the "same value opposite sign" reflection. During $L/a < t < 2L/a$, the water hammer wave $-\Delta H$ propagates from section B to section A, with the pressure head $H_0 + \Delta H$ kept at A and velocity $-V_0$ kept at B (Fig. 9.2b).

(3) At instant $t = 2L/a$, the $-\Delta H$ wave reaches at section A. Because the valve is closed, the velocity at A is 0. But the velocity just on the left side of A is still $-V_0$. Because of the pulling effect, the water is inflated, and the density becomes to $\rho_0 - \Delta\rho$, diameter becomes to $d_0 - \Delta d$, and pressure head becomes to $H_0 - \Delta H$ from $H_0 + \Delta H$. This change is because the reflecting characteristics of the closed valve are "same value same sign" reflection. In other words, when the incoming wave $-\Delta H$ reaches at A, a new reflected wave $-\Delta H$ is generated due to the flow characteristics of the valve. The incoming wave and the reflected wave are the same in value but opposite in sign. At this time, there are three waves at the valve section A. They are the water hammer wave $+\Delta H$ that is generated by the valve closing, the incoming wave $-\Delta H$ that comes from the reservoir, and the reflected wave $-\Delta H$ that is generated by the reflection to the incoming wave by the closed valve. The pressure head at A is the summation of these three waves, with $+\Delta H - 2\Delta H = -\Delta H$. During $2L/a < t < 3L/a$, water hammer wave $-\Delta H$ propagates from A to B, with pressure head $H_0 - \Delta H$ kept at A and velocity $-V_0$ kept at B (Fig. 9.2c).

(4) At instant $t = 3L/a$, water hammer wave $-\Delta H$ reaches at B. Because of the "same value opposite sign" reflection by the reservoir, a reflected wave $+\Delta H$ is generated and begins to propagate toward A. The velocity at B changes from $-V_0$ to V_0, the pressure head $H_0 - \Delta H$ changes to H_0, and the pipe just adjacent to B recovers its initial diameter. During $3L/a < t < 4L/a$, water hammer wave

$+\Delta H$ propagates from B to A, with the pressure head $H_0-\Delta H$ at A and velocity V_0 at B kept (Fig. 9.2d). At instant $t = 4L/a$, water hammer wave $+\Delta H$ arrives at A, and a new reflected wave $+\Delta H$ generated by the "same value same sign" reflection. And the pressure head at the valve section A changes to $H_0 + \Delta H$ again, and the state are the same as those at $t = 0$.

We can define the phase time as $t_r = 2L/a$, which is the time during which the water hammer wave propagates from A to B and back to A again. We can also define the cycle as $T = 2t_r = 4L/a$, which is two times of the phase time.

The above descriptions and Fig. 9.2 are about a cycle of propagation and reflection of water hammer waves in the single pipe system after the sudden closure of valve.

To sum up, we can draw the following characteristics for water hammer.

(1) Water hammer pressure is actually the inertial force generated by change of water velocity. When a valve is opened or closed quickly, because the operating time is short, the change rate of velocity is large and the water hammer pressure is large.
(2) Water hammer wave travels as an elastic wave, because water is compressible and pipe wall is elastic. The value of wave speed is determined by water compressibility and pipe wall elasticity.
(3) Like other elastic waves, water hammer wave can be reflected when it reaches at boundaries where flow conditions have changes. The reflecting characteristics, namely the value and direction of the reflected wave, are determined by the boundary physical characteristics.

9.2.2 *Propagating Speed of Water Hammer Wave*

Wave speed of water hammer is an important parameter in analyzing water hammer phenomenon and calculating water hammer pressure. Its value is defined by pipe wall material, pipe wall thickness, pipe diameter, pipe cross section type, pipe supporting type, water elastic modulus, etc. Considering water compressibility and pipe wall elasticity, one can derive the equation for wave speed from the continuity equation of fluid flow. It can be written as

$$a = \frac{\sqrt{E_w g/\gamma}}{\sqrt{1 + \frac{2E_w}{Kr}}} \tag{9.1}$$

where E_w is the elastic modulus, with value 2.06×10^{3} MPa in normal temperature and pressure; γ is the volume-weight, with normal value 9.81 kN/m^3; r is the radius of round pipe; K is the elastic resistance coefficient of the pipe.

In Eq. (9.1), the numerator $\sqrt{E_w g/\gamma}$ is the wave speed in unbounded water body, and the value has increasing tendency when temperature and pressure are increasing. We normally choose this value as 1435 m/s in normal condition.

The elastic resistance coefficient K is different in different conditions, as follows.

9.2.2.1 Homogeneous Thin-Walled Round Section Steel Pipe

$$K = E_s\delta_s/r^2 \tag{9.2}$$

where E_s and δ_s are the elastic modulus and wall thickness, respectively. If the pipe is fixed in longitudinal direction (plane deformation problem), the parameter E_s should be replaced by $E_s/(1-\mu^2)$, where μ is the Poisson's ratio。If stiffening ring is added, one can use $\delta_s = \delta_0 + F/l$, in which δ_0 is the real wall thickness, and F is the cross section of stiffening ring, and l the spacing of two stiffening rings.

9.2.2.2 Unlined Tunnel with Round Section in Solid Rock

$$K = 100K_0/r \tag{9.3}$$

where K_0 is the unit resistance coefficient of rock.

9.2.2.3 Underground Steel Pipe or Reinforced Concrete Lining Round Section Pipe (Fig. 9.3)

$$K = K_s + K_c + K_f + K_r \tag{9.4}$$

$$K_c = \frac{E_c}{r_1(1-\mu_c^2)} \ln\frac{r_2}{r_1} \tag{9.5}$$

$$K_f = \frac{E_s f}{r_1 r_f} \tag{9.6}$$

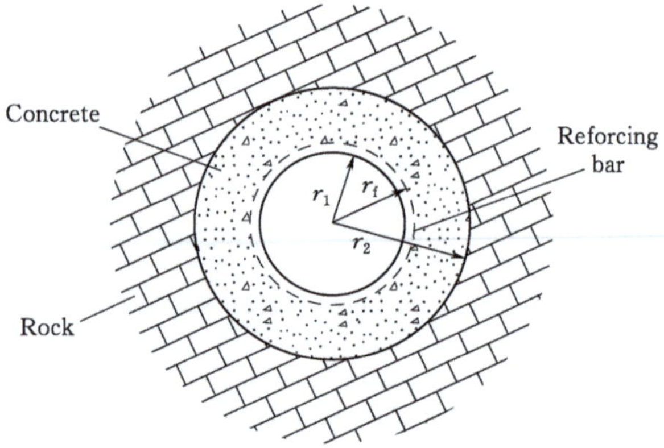

Fig. 9.3 Schematic of the section of underground pipe

In Eq. (9.6), K_s is the resistance coefficient of steel lining. K_s can be calculated based on Eq. (9.2), in which r is the inner radius, i.e., $r = r_1$, and E_s should be replaced by $E_s/(1-\mu^2)$. K_c is the resistance coefficient of the backfilled concrete. If the concrete is cracked, the radial compression can be negligible and $K_c = 0$ can be assumed. If the concrete is not cracked, Eq. (9.5) can be used. E_c is the elastic modulus of concrete. μ_c is the Poisson's ratio of concrete. r_1 and r_2 are the inner radius of steel pipe and the outer radius of lining, respectively. K_f is the resistance coefficient of circumferential reinforcing bar, and can be calculated by Eq. (9.6). f is the cross section of reinforcing bars per centimeter in pipe length. r_f is the radius of the reinforcing bar circle. K_r is the resistance coefficient of surrounding rock, and can be calculated by Eq. (9.3), in which r uses r_1.

In *Design Manual for Hydraulic Structures* (2nd Edition), some equations for other pipe types are given.

When water entrains air bubbles, wave speed will decrease sharply, because the elastic modulus is reduced by small bubbles. Theoretical calculation and experimental measurement showed that the wave speed decreases 4/5 if the air volume ratio is 0.5%. The elastic modulus of water containing air is

$$E_a = \sqrt{\frac{E_w}{1 + \frac{V_g}{V}(E_w/E_g - 1)}} \tag{9.7}$$

where E_g is the elastic modulus of air, normally valued 0.15 MPa; V_g/V is the air volume ratio.

Except for the wave speed equation for the homogeneous thin-walled pipe, the above equations for other types of pipe can only be used for estimation, because the accuracy of primary parameters, e.g., the elastic resistance coefficient K_0 of the surrounding rock, cannot be guaranteed.

However, the influence of wave speed error is not obvious for the situation in which the maximum water hammer head occurs near the ending of guide vane closing (similar to the last phase water hammer type, which will be discussed later in this chapter). Only in some special situations, e.g., the maximum water hammer head occurs at the ending of first phase time (similar to the first phase water hammer type, which will be discussed later in this chapter), the accurate calculation of wave speed become important.

When data for wave speed calculation are incomplete, one can choose 1000 m/s for homogeneous thin-walled round section steel pipe, and 1200 m/s for underground steel pipe, which are the normal values.

9.2.3 *Basic Equations for Water Hammer*

9.2.3.1 The Governing Differential Equations

Within a pressurized pipe, the water must follow the momentum and continuity equations of fluid flow. Referring to the textbook *Hydraulics*, one can derive the equations of one-dimensional unsteady pipe flow as follows.

The momentum equation

$$\frac{\partial V}{\partial t} + V\frac{\partial V}{\partial x} + g\frac{\partial H}{\partial x} + \frac{f}{2D}V|V| = 0 \tag{9.8}$$

and the continuity equation

$$\frac{\partial H}{\partial t} + V\frac{\partial H}{\partial x} + \frac{a^2}{g}\frac{\partial V}{\partial x} + V\sin\alpha = 0 \tag{9.9}$$

where V is the velocity, with the positive direction downstream; H is the pressure head; x is the distance from the upstream end of the pipe, with the positive direction downstream; t is the time; D is the pipe diameter; α is the longitudinal slope angle, with the falling slope positive; a is the wave speed; g is the gravitational acceleration; f is the Darcy-Weisbach resistance factor.

The absolute value of V in the resistance term is specified to ensure the frictional force is in the reverse direction of velocity and the energy is dissipated by frictional resistance.

In Eqs. (9.8) and (9.9), the dependent variables are V and H, while the independent variables are x and t. This set of quasi-linear hyperbolic partial differential equations describes the relations of the V and H values at any location x and any time t. Because of the non-linear terms, analytical solutions to these equations are not easy. To simplify the solutions, the convection terms $V\frac{\partial V}{\partial x}$ and $V\frac{\partial H}{\partial x}$, frictional resistance term $\frac{f}{2D}V|V|$, and slope term $V\sin\alpha$ can be omitted, because they are relatively small. Therefore, Eqs. (9.8) and (9.9) can be simplified to the typical hyperbolic linear partial differential equations

$$\frac{\partial V}{\partial t} + g\frac{\partial H}{\partial x} = 0. \tag{9.10}$$

$$\frac{\partial H}{\partial t} + \frac{a^2}{g}\frac{\partial V}{\partial x} = 0. \tag{9.11}$$

9.2.3.2 The General Solutions to the Governing Differential Equations

Equations (9.10) and (9.11) have a general solution, which can be expressed as the following form.

$$\Delta H = H - H_0 = F\left(t + \frac{x}{a}\right) + f\left(t - \frac{x}{a}\right) \tag{9.12}$$

$$\Delta V = V - V_0 = -\frac{g}{a}\left[F\left(t + \frac{x}{a}\right) - f\left(t - \frac{x}{a}\right)\right] \tag{9.13}$$

in which, H_0 and V_0 are the initial steady head and velocity; F and f are the wave functions. The particular forms and values are determined by boundary conditions。

Investigating Eqs. (9.12), (9.13) and Fig. 9.4, we can know the follow features.

(1) Wave functions F and f have the same dimension as head H, and therefore can be called as pressure waves. Wave function F propagates along $-x$ direction with speed a. Because F is a water hammer pressure wave traveling upstream, it is termed a countercurrent wave, which is the summation of many countercurrent waves. Similarly, wave function f propagates along x direction with speed a. Because f is a water hammer pressure wave traveling downstream, it is termed a cocurrent wave, which is the summation of many cocurrent waves.

(2) If the pipe properties are the same along the length, countercurrent wave $F\left(t + \frac{x}{a}\right)$ is constant when update t to $t + \Delta t$ and x to $x - a\Delta t$, because $F\left[(t + \Delta t) + \frac{(x - a\Delta t)}{a}\right] = F\left(t + \frac{x}{a}\right)$. That is to say, the countercurrent wave has constant value in the upstream propagating process. Similarly, cocurrent wave $f\left(t - \frac{x}{a}\right)$ is constant when update t to $t + \Delta t$ and x to $x + a\Delta t$, because $f\left[(t + \Delta t) - \frac{(x + a\Delta t)}{a}\right] = f\left(t - \frac{x}{a}\right)$. Namely, the cocurrent wave has constant value in the downstream propagating process. We can illustrate these by Fig. 9.4. At instant t the pressure wave $F\left(t + \frac{x}{a}\right)$ at section A can be denoted as F_t^A. After

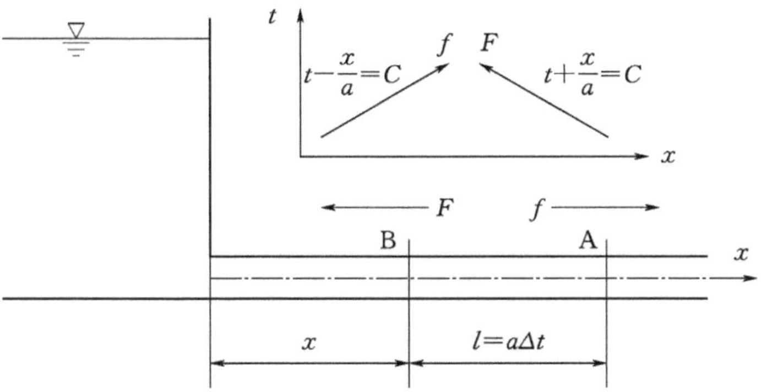

Fig. 9.4 Schematic of characteristics of wave functions

time $\frac{l}{a}$, it propagates along distance l from A to B, leading to $F^B_{t+\frac{l}{a}} = F^A_t$. In the similar manner, $f^B_t = f^A_{t+\frac{l}{a}}$. In the $x - t$ plane of Fig. 9.4, F. is the wave traveling along $t + \frac{x}{a} = C$ line, and f is the wave traveling along $t - \frac{x}{a} = C$, in which C is a constant.

(3) At any section and any instant, the water hammer pressure head ΔH equals to the summation of countercurrent wave F and cocurrent wave f. And the velocity change ΔV equals to the subtraction of cocurrent wave f from countercurrent wave F, and multiplying g/a. Note that, when the water hammer waves propagate for a long time and are reflected by boundaries, there will be many waves at any section, however, they can be summed and sorted to two waves, namely the total countercurrent wave F and the total cocurrent wave f, in which F is the summation of all countercurrent waves and f is the summation of all cocurrent waves.

(4) Finding the mathematical forms of wave functions F and f is quite difficult. In practical applications, we normally transfer Eqs. (9.12) and (9.13) to the relations of head and velocity at two adjacent sections, and then solve the equations by uniting the boundary conditions.

If we know the head H^A_t and velocity V^A_t at section A and time t, we can insert them into Eqs. (9.12) and (9.13) and cancel out f to obtain

$$H^A_t - H_0 - \frac{a}{g}\left(V^A_t - V_0\right) = 2F\left(t + \frac{x}{a}\right)$$

in which the subscript defines the time, and superscript defines the section. Similarly, we can obtain the head velocity relation at section B and time $\Delta t = \frac{l}{a}$ as follows

$$H^B_{t+\Delta t} - H_0 - \frac{a}{g}\left(V^B_{t+\Delta t} - V_0\right) = 2F\left[(t + \Delta t) + \frac{(x - l)}{a}\right]$$

Because $F\left[(t + \Delta t) + \frac{(x-l)}{a}\right] = F\left(t + \frac{x}{a}\right)$, the subtraction of the above two equations yields

$$H^B_{t+\Delta t} - H^A_t = \frac{a}{g}\left(V^B_{t+\Delta t} - V^A_t\right) \tag{9.14}$$

If we know the head H^B_t and velocity V^B_t at section B and time t, with the similar approach, we can obtain

$$H^A_{t+\Delta t} - H^B_t = -\frac{a}{g}\left(V^A_{t+\Delta t} - V^B_t\right) \tag{9.15}$$

Equations (9.14) and (9.15) are normally called the basic interlocking equations of water hammer. They describe the relationship of pressure and velocity between any two sections in a pipe. These two equations are valid for any pipe with constant material, diameter and wall thickness along length.

For application convenience, the general Eqs. (9.12) and (9.13) and the basic interlocking Eqs. (9.14) and (9.15) for water hammer can be normalized. Defining relative water hammer $\xi = \frac{\Delta H}{H_0}$, relative velocity $v = \frac{V}{V_{\max}}$, relative countercurrent wave function $\Phi = \frac{F}{H_0}$, relative cocurrent wave function $\varphi = \frac{f}{H_0}$, and factor $\rho = \frac{aV_{\max}}{2gH_0}$, and inserting them into Eqs. (9.12)–(9.15), we finally get

$$\xi = \Phi + \varphi \tag{9.16}$$

$$v - v_0 = -\frac{1}{2\rho}(\Phi - \varphi) \tag{9.17}$$

$$\xi_{t+\Delta t}^{B} - \xi_{t}^{A} = 2\rho\left(v_{t+\Delta t}^{B} - v_{t}^{A}\right) \tag{9.18}$$

$$\xi_{t+\Delta t}^{A} - \xi_{t}^{B} = -2\rho\left(v_{t+\Delta t}^{A} - v_{t}^{B}\right) \tag{9.19}$$

where $\rho = \frac{aV_{\max}}{2gH_0}$ is called the pipe characteristic factor, which reflects the ratio of kinetic energy to potential energy in a pipe and is an important influencing factor of water hammer type and history characteristics.

9.2.4 *Reflecting Characteristics of Water Hammer Wave*

At the changing sections of pipe properties, such as water intake, pipe bifurcation, serial junction, and valve, water hammer wave is reflected, so as to keep the balance of head and continuity of flow rate. Reflection is an important aspect of water hammer wave characteristics. Generally speaking, when the incoming wave reaches at a section where pipe properties change, one part comes back in the form of reflected wave, and the other part continues to propagate forward in the form of transmitted wave. The ratio of reflected wave to incoming wave is called the reflection coefficient, denoted by r; the ratio of transmitted wave to incoming wave is called the transmission coefficient, denoted by S. The relation of the reflection coefficient and the transmission coefficient is

$$s - r = 1 \tag{9.20}$$

The reflection coefficients for different boundaries can be derived based on the general solution equations of water hammer and their specific boundary conditions.

9.2.4.1 Reservoir Section

Because a reservoir (or head pond and surge chamber) has a very large volume and water surface, it can be assumed that the head of reservoir is constant when water hammer approaches. Therefore, $H = \text{const.}$, resulting in $\xi = 0$ With incoming wave Φ and reflected wave φ, we can obtain the reflection coefficient $r = \frac{\varphi}{\Phi} = -1$ by manipulating Eq. (9.16). Thus, the reflecting feature of reservoir is "same value opposite sign."

9.2.4.2 Closed Valve or Dead End

A closed valve is the same as a dead end in flow characteristics. Because velocity at a dead end section is 0, $v - v_0 = 0$ is kept. With incoming wave φ and reflected wave Φ we can obtain the reflection coefficient $r = \frac{\Phi}{\varphi} = 1$ by manipulating Eq. (9.17). Thus the reflecting feature of dead end or closed valve is "same value same sign."

9.2.4.3 Opened Valve

The discharge equation of valve can be found in previous course *Hydraulics*, in which it is in the form of orifice flow as follows

$$Q_{\max} = \phi_0 \omega_{\max} \sqrt{2gH_0}$$

where $\omega_{\max}$ is the maximum opening area, and ϕ_0 is the discharge coefficient for full opening.

If the opening is ω, and the head in pipe is H, then the discharge is

$$Q = \phi\omega\sqrt{2gH} = \phi\omega\sqrt{2g(H_0 + \Delta H)}$$

where ω is the opening area and ϕ is the corresponding discharge coefficient.

Dividing the above two equations, we can get the normalized equation for valve flow

$$v = \frac{AV}{AV_{\max}} = \frac{Q}{Q_{\max}} = \frac{\phi\omega\sqrt{2g(H_0 + \Delta H)}}{\phi_0\omega_{\max}\sqrt{2gH_0}} = \tau\sqrt{1+\xi} \tag{9.21}$$

in which A is the pipe cross sectional area, and $\tau = \phi\omega/\phi_0\omega_{\max}$ is the hydraulic relative opening. Assuming discharge coefficient is constant for different openings, i.e., $\phi = \phi_0$, we have the geometrical relative opening $\tau = \omega/\omega_{\max}$.

When $\xi \leq 0.5$, taking the first and second terms of Taylor expansion of $\sqrt{1+\xi}$ in Eq. (9.21) we can obtain appropriate equation for valve

$$v = \tau(1 + \xi/2) \tag{9.22}$$

Assuming the relative valve opening τ is constant and the relative water hammer $\xi_0 = 0$ then we have $V_0 = \tau$. When the incoming wave φ arrives at the valve, the reflected wave Φ generates. Submitting Eqs. (9.22) into (9.17) yields

$$v - v_0 = \left[\tau\left(1 + \frac{\xi}{2}\right) - \tau\right] = \frac{\tau\xi}{2} = -\frac{1}{2\rho}(\Phi - \varphi)$$

Combining with Eq. (9.16) and eliminating ξ, we obtain the reflection coefficient for an opened valve

$$r = \frac{\Phi}{\varphi} = \frac{1 - \rho\tau}{1 + \rho\tau} \tag{9.23}$$

With the pipe characteristic factor ρ and the valve opening τ at any time known, we can calculate the reflection coefficient of the corresponding time by using Eq. (9.23). For example, if the valve is closed, we see $\tau = 0$ and find $r = 1$, which is the "same value same sign" reflection and identical to that for a dead end.

Equations (9.21)–(9.23) are defined for valves, and also suitable for impulse turbines. But for reaction turbines, they are just an approximation. The following analysis will use these equations.

9.2.4.4 Serial Junction

For the serial junction shown in Fig. 9.5, if the incoming wave Φ_1 is from pipe 1, then the reflected wave is φ_1 and the transmitted wave is Φ_2. Combing Eqs. (9.16) and (9.17) with the momentum and continuity equations for serial junction, we can obtain

$$r_1 = \frac{\varphi_1}{\Phi_1} = \frac{\rho_2 - \rho_1}{\rho_1 + \rho_2} \tag{9.24}$$

where $\rho_1 = \frac{a_1 V_1}{2gH_0}$ and $\rho_2 = \frac{a_2 V_2}{2gH_0}$ are the pipe characteristic factors for pipe 1 and pipe 2, respectively.

If r_1 is positive, the reflected wave has the same sign with the incoming wave, resulting in the increase of water hammer pressure in pipe 1. On the contrary, if r_1 is negative, the reflected wave has the opposite sign with the incoming wave, resulting

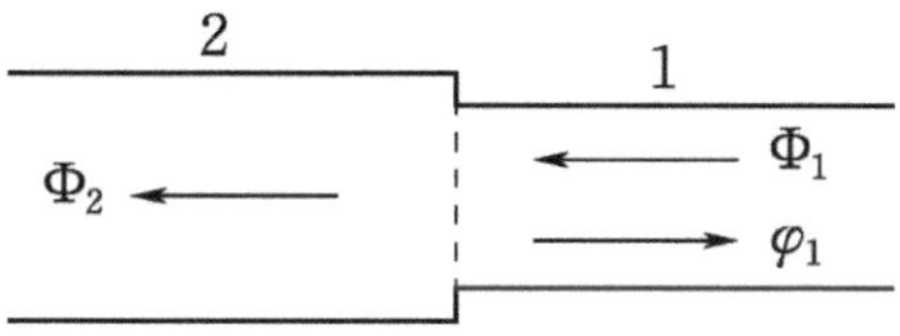

Fig. 9.5 Schematic of wave reflection at a serial junction

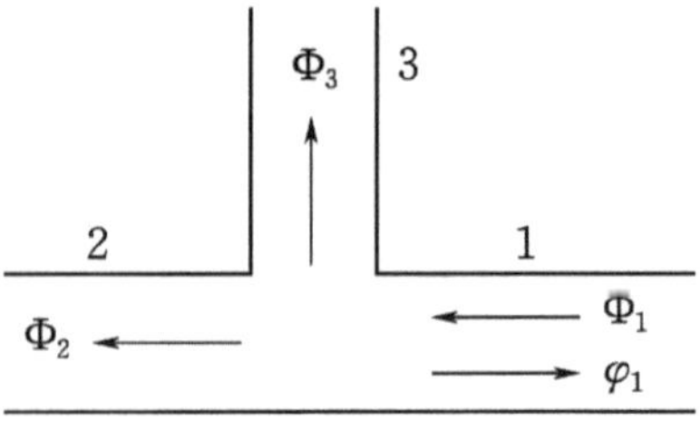

Fig. 9.6 Schematic for wave reflection at a T bifurcation

in the decrease of water hammer pressure in pipe 1. If the section area of pipe 2 approaches 0, we have $\rho_2 \to \infty$ and $r_1 = 1$, which is identical to that of the dead end. If the section area of pipe 2 approaches infinity, we have $V_2 = 0$, $\rho_2 = 0$ and $r_1 = -1$, which is identical to that of the reservoir.

9.2.4.5 Bifurcation

For the bifurcation shown in Fig. 9.6, if the incoming wave Φ_1 is from pipe 1, then the reflected wave is φ_1 and the transmitted waves are Φ_2 and Φ_3. Combing Eqs. (9.16) and (9.17) with the momentum and continuity equations for bifurcation, we can obtain

$$r_1 = \frac{\varphi_1}{\Phi_1} = \frac{\rho_2\rho_3 - \rho_1\rho_2 - \rho_3\rho_1}{\rho_1\rho_2 + \rho_2\rho_3 + \rho_3\rho_1} \tag{9.25}$$

where $\rho_i = \frac{a_i Q}{2gH_0A_i}$, in which Q is the discharge of the main pipe (using other discharge does not influence the value of the reflection factor), A is the cross sectional area, and $i = 1, 2, 3$ are the subscriptions labeling the pipes.

9.3 Analytical Solutions to Water Hammer and Rotational Speed Rise

9.3.1 Direct Water Hammer and Indirect Water Hammer

9.3.1.1 Direct Water Hammer

The phase time $t_r = 2L/a$ is defined as the time within which the water hammer wave propagates forth and back in a pipe. If the closing/opening time of the valve $T_s \le t_r$, the head at the valve is only influenced directly by the water hammer caused by the closing action, because the valve has already been closed before the reflected water hammer by the reservoir reaches at the valve. We customarily call this a direct water hammer. Because the cocurrent wave does not arrive, we have $f\left(t - \frac{x}{a}\right) = 0$ at the valve. Applying the water hammer general Eqs. (9.12) and (9.13) to cancel out $F\left(t+\frac{x}{a}\right)$, we obtain the equation for direct water hammer

$$\Delta H = H - H_0 = -\frac{a}{g}(V - V_0) \tag{9.26}$$

This equation can be normalized to $\xi_d = 2\rho(v_0 - v).\xi_d = 2\rho(v_0 - v)$.

Investigating Eq. (9.26), we know that closing the valve results in positive water hammer head ΔH, because $(V - V_0)$ is negative (velocity decreases); On the contrary, opening the valve results in negative water hammer head ΔH, because $(V - V_0)$ is positive (velocity increases). The value of direct water hammer head is only related to the velocity change $(V - V_0)$ and pipe characteristics (expressed by wave speed a), and has nothing to do with the changing speed and law of valve opening and the length of pipe.

The pressure rise from direct water hammer is enormous. For example, given the initial velocity $V_0 = 5\,\text{m/s}$, finial velocity $V = 0\,\text{m}$, and wave speed $a = 1000$ m/s, we may obtain $\Delta H \approx 500\,\text{m}$. Therefore, in designing hydropower stations we should always try best to avoid the occurrence of direct water hammer.

9.3.1.2 Indirect Water Hammer

If the closing/opening time of valve $T_s > t_{r,}$ the head at the valve is influenced not only by the water hammer caused by closing action, but also by the reflected water hammer from the reservoir, because the reflected wave reaches at the valve before it is closed. We customarily call this an indirect water hammer. The water hammer head at the valve is the summation of the upstream propagating countercurrent wave F and the downstream propagating cocurrent wave f. The pressure rising from an indirect water hammer is smaller than that from a direct water hammer.

When an indirect water hammer happens, the propagating and reflecting processes of water hammer waves are very complex. The head change at the valve is related to many aspects, in which the time and law of valve closing/opening are the main factors.

9.3.2 Interlocking Equations for Water Hammer in Single Pipe

The maximum water hammer head is the main concern in engineering applications. It always occurs at the pipe section just adjacent to the valve (may be roughly called the valve section), because the water hammer is generated by the valve and the reflected wave from reservoir arrives at the valve later. In the following, we will derive the equations for water hammer head at the valve in the single pipe system, through combing the basic interlocking Eqs. (9.18) and (9.19) and the boundary conditions of valve and reservoir.

Consider the single pipe system in Fig. 9.2, in which B is the reservoir section, A is the valve section, L is the pipe length, and phase time $t_r = L/a$. For convenience, we use the subscript to denote the occurrence time and superscript to denote the occurrence location. For example, time $t = 0.5t_r$ is corresponding to subscript 0.5, and time $t = t_r$ is corresponding to subscript 1.

Based on the previous derivation is Sect. 2.4, the boundary conditions for the reservoir and valve are $\xi^B = 0$ and $v^A = \tau\sqrt{1+\xi^A}$, respectively.

With the relative valve openings at the end of every phase given, the water hammer pressure heads at section A for different phases can be derived successively.

9.3.2.1 Water Hammer Head at the End of the First Phase

(1) Observing Eq. (9.18), we can easily write out the equations for the water hammer wave from A to B

$$\xi_{0.5}^B - \xi_0^A = 2\rho\left(v_{0.5}^B - v_0^A\right)$$

At $t = 0$ instant, water hammer head $\xi_0^A = 0$, because the valve opening at A has no change. Also noting $\xi_{0.5}^B = 0$ at B, we have

$$2\rho\left(v_0^A - v_{0.5}^B\right) = 0$$

which yields $v_0^A = v_{0.5}^B$

Because the boundary condition at A holds $v_0^A = \tau_0\sqrt{1+\xi_0^A} = \tau_0$, we obtain

$$v_{0.5}^B = v_0^A = \tau_0$$

(2) Observing Eq. (9.19), we can easily write out the equation for water hammer wave from B to A

$$\xi_1^A - \xi_{0.5}^B = -2\rho\left(v_1^A - v_{0.5}^B\right)$$

Inserting $\xi_{0.5}^B = 0$, $v_{0.5}^B = \tau_0$, and $v_1^A = \tau_1\sqrt{1+\xi_1^A}$ into this equation, we obtain

$$\xi_1^A = -2\rho\left(\tau_1\sqrt{1+\xi_1^A} - \tau_0\right)$$

which can rewritten to

$$\tau_1\sqrt{1+\xi_1^A} = \tau_0 - \frac{\xi_1^A}{2\rho} \tag{9.27}$$

The root ξ_1^A of Eq. (9.27) has two versions, and the one has physical meanings is

$$\xi_1^A = 2\rho\left[\left(\tau_0 + \rho\tau_1^2\right) - \sqrt{\left(\tau_0 + \rho\tau_1^2\right)^2 - \left(\tau_0^2 - \tau_1^2\right)}\right] \tag{9.28}$$

This is the equation for the water hammer head at the end time of the first phase.

9.3.2.2 Water Hammer Head at the End of the Second Phase

(1) Write out the equations for the water hammer wave from A to B

$$\xi_{1.5}^B - \xi_1^A = 2\rho\left(v_{1.5}^B - v_1^A\right)$$

Because $\xi_{1.5}^B = 0$, we obtain

$$\xi_1^A/2\rho = \left(\tau_1\sqrt{1+\xi_1^A} - v_{1.5}^B\right)$$

which yields $v_{1.5}^B = \tau_1\sqrt{1+\xi_1^A} - \frac{\xi_1^A}{2\rho}$

(2) Write out the equations for the water hammer wave from B to A

$$\xi_2^A - \xi_{1.5}^B = -2\rho\left(v_2^A - v_{1.5}^B\right)$$

Inserting ξ_1^A, $v_{1.5}^B = \tau_1\sqrt{1+\xi_1^A} - \frac{\xi_1^A}{2\rho}$, and $v_2^A = \tau_2\sqrt{1+\xi_2^A}$ into this equation, we obtain

$$\xi_2^A = -2\rho\left[\tau_2\sqrt{1+\xi_2^A} - \left(\tau_1\sqrt{1+\xi_1^A} - \frac{\xi_1^A}{2\rho}\right)\right]$$

Considering Eq. (9.27), we finally obtain

$$\tau_2\sqrt{1+\xi_2^A} = \tau_0 - \frac{\xi_2^A}{2\rho} - \frac{\xi_1^A}{\rho} \tag{9.29}$$

9.3.2.3 Water Hammer Head at the End of the *n*th Phase

Following the similar principles and procedures, we can obtain the equation for the water hammer head at the end of the nth phase as follows

$$\tau_n\sqrt{1+\xi_n^A} = \tau_0 - \frac{\xi_n^A}{2\rho} - \frac{1}{\rho}\sum_1^{n-1}\xi_i^A \tag{9.30}$$

Investigating into Eq. (9.30), we may find that the water hammer head at the end of the nth phase is defined not only by the relative valve opening τ_n, but also by the water hammer heads at the ends of all previous phases.

Equations (9.27)–(9.30) are conventionally called the interlocking equations for single pipe water hammer. When the openings of valve action law ($\tau_n, i = 0, 1, \ldots, n$) are given, we can solve water hammer heads ξ_1^A, ξ_2^A, …, and ξ_n^A successively, and obtain the water hammer head history at the valve section finally. It should be noted that the above derivations are based on several assumptions. First, obtaining the interlocking Eqs. (9.18) and (9.19) from (9.12) and (9.13) is based on the no frictional resistance assumption, therefore they are only valid when frictional resistance can be neglected. Second, the equations are suitable for impulse turbines and their usage for reaction turbines should be modified, because the flow characteristics of orifice are used for the valve in the derivation. However, these equations are valid for any closing/opening laws (also called regulating procedures), and can be used to analyze the influence of non-linear law to water hammer head.

9.3.3 *Water Hammer Caused by Linear Closure of Valve*

The main task of calculating water hammer is to figure out the maximum water hammer head. If a linear regulating procedure is used, we can calculate the maximum water hammer head by more convenient equations, without needing to solve the interlocking equations step by step.

The actual closing law of turbine wicket gate (guide vanes) is similar to the solid line in Fig. 9.7. The total time from full opening ($\tau_0 = 1$) to fully closed ($\tau = 0$) is T_z. Because of the inertia of regulating system, the curve has a nearly horizontal segment at the beginning, in which the opening change is slow and the corresponding water hammer is small. This hysteresis effect is negligible in water hammer calculation. Because of the buffer effect of regulating system, the curve has a slowdown tail as the opening approaches 0. This only influences the water hammer history just before the opening is closed. Therefore, to simplify calculation, we normally extent the straight line part to get a linear closing curve, which is corresponding to the effective closing time T_s. The ratio T_s/T_Z is normally between 0.6 and 0.95, and we can use $T_s = 0.7T_z$ if no exact information provided.

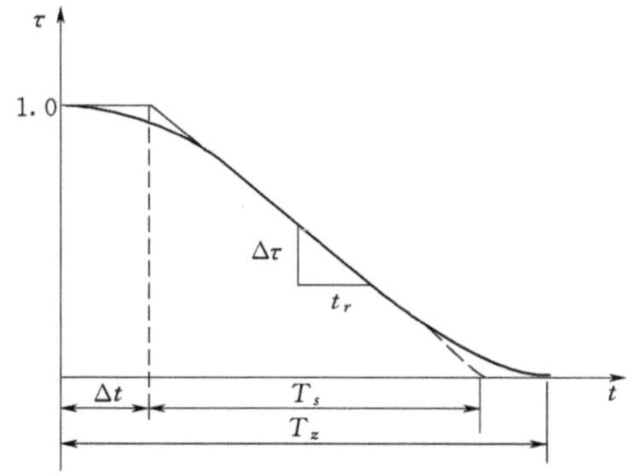

Fig. 9.7 Schematic of closing curve of opening

9.3.3.1 First Phase Water Hammer and Last Phase Water Hammer

When the valve closes according to a linear low, the water hammer pressure history curve at the valve can be classified to two types, as shown Fig. 9.8a, b. These can be proved by using the reflection and superposition characteristics of water hammer waves. The main reason of producing these two types is their different reflection characteristics.

(1) The first phase water hammer

The maximum water hammer head $\xi_{\max}$ occurs at the end of the first phase, with $\xi_{\max} = \xi_1^A$, as shown in Fig. 9.8a. From Eq. (9.23), we know that if $\rho\tau_0 < 1$ (τ_0 is the initial opening), refection factor r is positive and the reflection characteristics at the valve belong to the same sign reflection. During the closing process of valve, there are three water hammer waves at the valve at any instant: the pressure increasing wave by continuing closing, the reflected wave by the valve, and the incoming wave from the reservoir. The former two are countercurrent waves, and the last is the cocurrent pressure decreasing wave that is reflected by the reservoir. During the first phase, the cocurrent pressure decreasing wave does not arrive at the valve. Therefore, the water hammer head at the valve is defined only by the pressure increasing wave, which is ξ_1^A at the end of the first phase. At the end of the second phase, the cocurrent pressure decreasing wave has arrived at the valve. If the reflection factor of valve is positive, the reflected wave by the valve is also a pressure decreasing wave. Thus, the summation of the two pressure decreasing waves will exceed the pressure increasing wave by the continuing closing in the second phase, and results in $\xi_2^A < \xi_1^A$ at end of the second phase. At the end of the third phase, the pressure increasing wave that is generated by reservoir reflection to the pressure decreasing wave from the valve in the second phase has arrived at the valve, and a new pressure increasing wave is generated by valve reflection. These two pressure increasing waves make the water hammer at the valve increased again, and result in $\xi_3^A > \xi_2^A$. Like these manners, a serial of water hammer heads at the ends of the following phases will fluctuate around a limit ξ_m^A, and finally approximate to ξ_m^A after many phases.

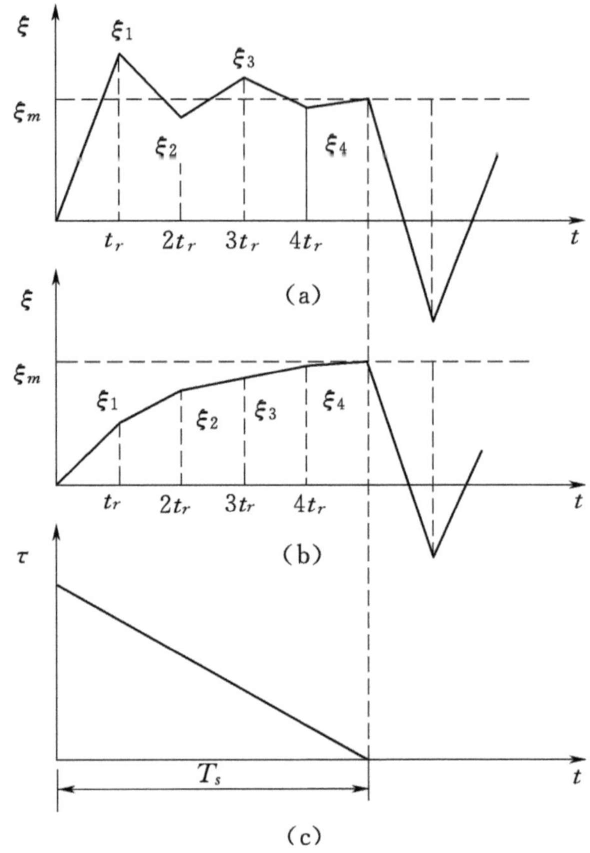

Fig. 9.8 Water hammer types under linear closing law

$\rho\tau_0 < 1$ is the approximate discriminant for the first phase water hammer. For load rejection condition, if $\tau_0 = 1$, $\rho = aV_{\max}/2gH_0 < 1$, $V_{\max} = 5\,\text{m/s}$, and $a = 1000\,\text{m/s}$, then $H_0 > 250\,\text{m}$. That is to say, the first phase water hammer only happens in high head hydropower stations. In other words, the first phase water hammer is one of the features of high head hydropower stations.

(2) The last phase water hammer

The maximum water hammer head $\xi_{\max}$ occurs at the end of valve closing, with $\xi_{\max} \approx \xi_m^A > \xi_1^A$, as shown in Fig. 9.8b. It is also called the limit water hammer, because $\xi_{\max}$ is approximate to the limit ξ_m^A. Based on (9.23), we know that when $\rho\tau_0 > 1$, the reflection factor r is negative, and the reflected wave changes sign at the valve. During the second phase, the pressure decreasing wave from the reservoir is reflected by the opposite sign reflection, and a new pressure increasing wave is generated. The superposition of the pressure increasing waves generated by both reflection and continuing closing makes the water hammer head at the valve larger, namely $\xi_2^A > \xi_1^A$ at the end of the second phase. In the same manner, as long as

$\rho\tau > 1$ is satisfied for the later phases, the water hammer head at the valve will continue to rise and finally approach to limit ξ_m^A。

$\rho\tau_0 > 1$ is the approximate discriminant for the last phase water hammer. The last phase water hammer is one of the features of low head hydropower stations.

9.3.3.2 Analytical Equations for Water Hammer

For linear closing/opening laws, the opening change in one phase time t_r, is $\Delta\tau = \pm t_r / T_s$, in which the minus sign "–" indicates opening decrease when valve closes and the plus sign "+" indicates opening increase when valve opens. Defining $-\rho\Delta\tau = \sigma$, we have

$$\sigma = \pm\frac{LV_{\max}}{gH_0T_s}$$

Here, σ is called the water hammer characteristic factor, which reflects the relative change rate of water flow momentum in pipe and has a direct effect on water hammer pressure head. When the valve closes, σ is positive, and when the valve opens, σ is negative.

The equation for the first phase water hammer can be found in Eqs. (9.27) or (9.28). Considering that the water hammer heads in hydropower stations are normally smaller than 50% of the steady heads and the term $\left(1 + \xi_1^A/2\right)$ can be used to replace $\sqrt{1 + \xi_1^A}$ in such situation, we can simplify Eq. (9.27) to

$$\xi_1^A = \frac{2\sigma}{1 + \rho\tau_0 - \sigma} \tag{9.31}$$

Equation (9.31) is convenient for applications. For the condition of valve closing, σ and ξ_1^A are positive, while for the condition of valve opening, σ and ξ_1^A are negative.

We can obtain the approximate equation for the last phase water hammer by the following derivation.

From Eq. (9.30), we can write out the equations for the nth and $(n + 1)$th water hammer phases

$$\tau_n\sqrt{1 + \xi_n^A} = \tau_0 - \frac{\xi_n^A}{2\rho} - \frac{1}{\rho}\sum_1^{n-1}\xi_i^A$$

$$\tau_{n+1}\sqrt{1 + \xi_{n+1}^A} = \tau_0 - \frac{\xi_{n+1}^A}{2\rho} - \frac{1}{\rho}\sum_1^{n}\xi_i^A$$

Subtracting these two equations and knowing $\sum_{1}^{n} \xi_i^A = \sum_{1}^{n-1} \xi_i^A + \xi_n^A$, we get

$$\tau_{n+1}\sqrt{1+\xi_{n+1}^A} - \tau_n\sqrt{1+\xi_n^A} = -\frac{1}{2\rho}\left(\xi_{n+1}^A - \xi_n^A\right) - \frac{1}{\rho}\xi_n^A \tag{9.32}$$

If the phase number n is large enough, we can let $\xi_n^A = \xi_{n+1}^A = \xi_m^A$. Thus Eq. (9.32) becomes

$$\sigma\sqrt{1+\xi_m^A} = \xi_m^A \tag{9.33}$$

Further, we obtain

$$\xi_m^A = \frac{\sigma}{2}\left(\sigma + \sqrt{\sigma^2+4}\right) \tag{9.34}$$

From Eq. (9.34), we know that the last phase water hammer head ξ_m^A is a function of σ, and has nothing to do with the initial opening τ_0 and wave speed a.

Inserting $\sqrt{1+\xi_m^A} = 1 + \xi_m^A/2$ into Eq. (9.33), we can also obtain a simpler equation

$$\xi_m^A = \frac{2\sigma}{2-\sigma} \tag{9.35}$$

Formulas (9.31), (9.34), and (9.35) not only can be used for valve closing condition, but also can be used for valve opening condition. In valve closing conditions, $\sigma = +\frac{LV_{\max}}{gH_0T_s}$ is positive and water hammer head is also positive; in valve opening condition, $\sigma = -\frac{LV_{\max}}{gH_0T_s}$ is negative, and water hammer head is also negative.

Table 9.1 is the summary of calculations for many types of water hammer. These equations are valid not only for linear closing and opening of valve, but also for any closing/opening laws, however the selection of σ should be cautious.

9.3.3.3 Type Discrimination of Water Hammer

The discrimination of water hammer type is the first step to calculate water hammer. When the type is known, we can select the right equation to obtain the water hammer head.

The discrimination of the first and last water hammer types by $\rho\tau_0 < 1$ and $\rho\tau_0 > 1$ is approximate. Actually, water hammer type is related not only to $\rho\tau_0$, but also to σ. In the following, we will derive the detailed discriminating equations.

Table 9.1 Summary of equations for water hammer calculation

		Opening Initial	Opening Final	Occurrence instant of the maximum value	Calculation equations	Approximate equations
Valve closing	Direct	τ_0	τ_c	End of closing	$\tau_c\sqrt{1+\xi} = \tau_0 - \frac{1}{2\rho}\xi$	$\xi = \frac{2\rho(\tau_0-\tau_c)}{1+\rho\tau_c}$
		τ_0	0	End of closing	$\xi = 2\rho\tau_0$	$\xi = 2\rho\tau_0$
		1	0	End of closing	$\xi = 2\rho$	$\xi = 2\rho$
	Indirect	τ_0	0	Limit water hammer	$\xi_m = \frac{\sigma}{2}\left(\sqrt{\sigma^2+4}+\sigma\right)$	$\xi_m = \frac{2\sigma}{2-\sigma}$
		τ_0		End of the first phase	$\tau_1\sqrt{1+\xi_1} = \tau_0 - \frac{1}{2\rho}\xi_1$	$\xi_1 = \frac{2\sigma}{1+\rho\tau_0-\sigma}$
		1		End of the first phase	$\tau_1\sqrt{1+\xi_1} = 1 - \frac{1}{2\rho}\xi_1$	$\xi_1 = \frac{2\sigma}{1+\rho-\sigma}$
		τ_0		End of the n-th phase	$\tau_n\sqrt{1+\xi_n} = \tau_0 - \frac{1}{\rho}\sum_1^{n-1}\xi_i - \frac{1}{2\rho}\xi_n$	$\xi_n = \frac{2\left(n\sigma-\sum_1^{n-1}\xi_i\right)}{1+\rho\tau_0-n\sigma}$
Valve opening	Direct	τ_0	τ_c	End of closing	$\tau_c\sqrt{1-y} = \tau_0 + \frac{1}{2\rho}y$	$y = \frac{2\rho(\tau_c-\tau_0)}{1+\rho\tau_c}$
		τ_0	1	End of closing	$\sqrt{1-y} = \tau_0 + \frac{1}{2\rho}y$	$y = \frac{2\rho(1-\tau_0)}{1+\rho}$
		0	1	End of closing	$\sqrt{1-y} = \frac{1}{2\rho}y$	$y = \frac{2\rho}{1+\rho}$
	Indirect	τ_0	1	End of the first phase	$y_m = \frac{-\sigma}{2}\left(\sqrt{\sigma^2+4}-\sigma\right)$	$y_m = \frac{-2\sigma}{2-\sigma}$
		τ_0	1	End of the first phase	$\tau_1\sqrt{1-y_1} = \tau_0 + \frac{1}{2\rho}y_1$	$y_1 = \frac{-2\sigma}{1+\rho\tau_0-\sigma}$
		0	1	End of the first phase	$\tau_1\sqrt{1-y_1} = \frac{1}{2\rho}y_1$	$y_1 = \frac{-2\sigma}{1-\sigma}$
		τ_0	1	End of the n-th phase	$\tau_n\sqrt{1-y_n} = \tau_0 + \frac{1}{\rho}\sum_1^{n-1}y_i + \frac{1}{2\rho}y_n$	$y_n = \frac{-2\left(n\sigma+\sum_1^{n-1}y_i\right)}{1+\rho\tau_0-n\sigma}$

Note y_i is the relative pressure drop defined by $y_i = (H_0 - H_i)/H_0$ which is always positive. σ is positive for closing and negative for opening, following it definition

The boundary condition between the first and last phase water hammer types is $\xi_1^A = \xi_m^A$. Replacing ξ_1^A in Eq. (9.27) by ξ_m^A yields

$$\tau_1\sqrt{1+\xi_m^A} = \tau_0 - \frac{\xi_m^A}{2\rho} \tag{9.36}$$

Inserting Eq. (9.33) into Eq. (9.36), we have

$$\frac{\xi_m^A}{\sigma}\tau_1 = \tau_0 - \frac{\xi_m^A}{2\rho} \tag{9.37}$$

Further inserting $\tau_1 = \tau_0 + \Delta\tau$ and $\sigma = -\Delta\tau\rho$ into Eq. (9.37), we get

$$\xi_m^A = \frac{2\sigma}{2 - \sigma/\rho\tau_0}$$

Then, inserting this ξ_m^A into Eq. (9.33), we have

$$\sigma = \frac{4\rho\tau_0(1 - \rho\tau_0)}{1 - 2\rho\tau_0} \tag{9.38}$$

Obviously, when Eq. (9.38) is satisfied, $\xi_1^A = \xi_m^A$ is correct. In other words, Eq. (9.38) is the boundary condition of the first phase water hammer and the last phase water hammer.

Figure 9.9 is the water hammer type discriminant diagram, in which $\rho\tau_0$ is the horizontal coordinate and σ is the vertical coordinate. The curve $\xi_1^A = \xi_m^A$ is defined by Eq. (9.38), which separates the first phase and the last phase water hammer (limit water hammer) types. The curve $\rho\tau_0 = \sigma$ is a straight line with angle 45° against the horizontal coordinate, which divides the direct and indirect water hammer regions. The horizontal coordinate separates the valve closing and opening regions, with $\sigma > 0$ in the upper region for closing and $\sigma < 0$ in the lower region for opening. Thus, the plane of $\rho\tau_0$ against σ is divided to five regions: the positive limit water hammer region I, within which $\xi_m^A > \xi_1^A$; the positive first phase water hammer region II, within which $\xi_1^A > \xi_m^A$; the direct water hammer region III, within which ξ_d^A is only influenced by wave speed and velocity change; the negative limit water hammer region IV, within which $y_m^A > y_1^A$; the negative first phase water hammer region V, within which $y_1^A > y_m^A$. Note that y here denotes the absolute form of relative water hammer decreases.

From Fig. 9.9, we know the following:

If $\rho\tau_0 < 1.0$, $\xi_1^A > \xi_m^A$ normally satisfies, and the first phase water hammer occurs;

If $\rho\tau_0 > 1.5$, $\xi_m^A > \xi_1^A$ always satisfies, and the last phase water hammer occurs.

If $1 < \rho\tau_0 < 1.5$, either the first phase water hammer or last phase water hammer can occur depending on different σ values, and in a few cases the direct water hammer can occur.

When the valve opens, if the initial opening is zero, the negative first phase water hammer must occur. This is similar to the load acceptance condition. If the initial opening of wicket gate is zero, then the first phase negative water hammer occurs. However, there is a speed-no-load opening τ_x, in which the output power is zero, and load acceptance normally starts from an opening around τ_x Therefore, in such situation, Fig. 9.9 should be used to find which negative water hammer will occur,

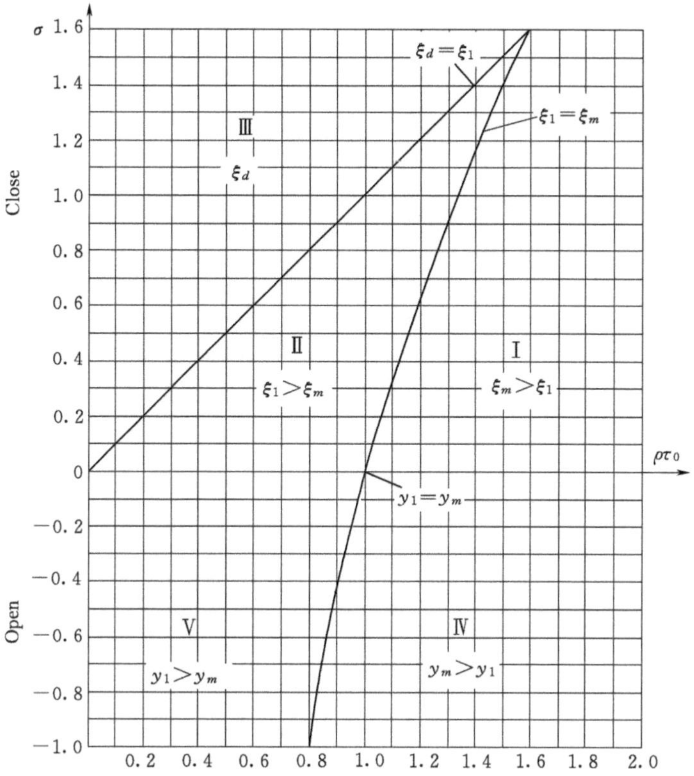

Fig. 9.9 Chart for type discrimination of water hammer

the first phase negative water hammer or the last phase negative water hammer. The value of τ_x is related to turbine types, with $\tau_x = 0.08$ to 0.12 for Francis turbines, $\tau_x = 0.07$ to 0.10 for movable-bladed Kaplan turbines, and $\tau_x = 0.20$ to 0.25 for fixed-bladed Kaplan turbines.

After discriminating the type of water hammer according to Fig. 9.9, we can select the corresponding equation in Table 9.1 to calculate the pressure head value of water hammer.

9.3.4 Influence of Initial Opening and Closing Procedure on Water Hammer

9.3.4.1 Influence of Initial Opening on Water Hammer

Hydropower units may operate under different load conditions. When the unit runs at full load, the initial opening $\tau_0 = 1$; when the unit runs at part load, $\tau_0 < 1$.

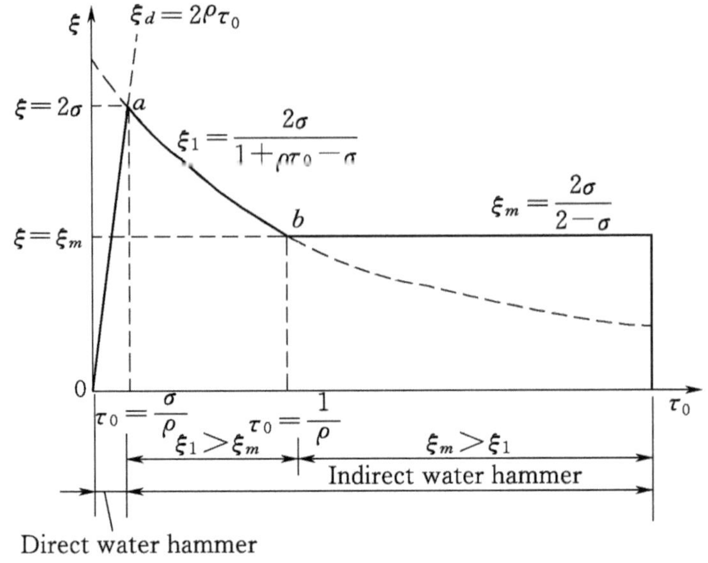

Fig. 9.10 Water hammer values according to different initial openings

Therefore, during load rejection transients, the initial opening τ_0 of the unit may have a variety of values.

According to Eqs. (9.34) and (9.35), ξ_m is only related to σ, and has nothing to do with τ_0. Therefore, the ξ_m curve in the ξ-τ_0 plane of Fig. 9.10 is a horizontal line.

According to Eq. (9.31), ξ_1 increases if τ_0 decreases. Therefore, the ξ_1 curve in the ξ–τ_0 plane is a dropping curve.

If the valve is closed before the first phase time, and direct water hammer happens with $\tau_1 = 0$. According to Eq. (9.27), $\xi_d = \xi_1 = 2\rho\tau_0$. The ξ_d curve in the ξ–τ_0 plane is a straight line passing through the origin of coordinates with a slope of 2ρ.

In Fig. 9.10, point a is the intersection point of ξ_1 curve with ξ_d curve. Applying Eq. (9.31) and $\xi_d = 2\rho\tau_0$, and giving $\xi_1 = \xi_d$, we can obtain the initial opening corresponding to point a as

$$\tau_0^a = \sigma/\rho$$

This opening is the dividing point between the direct water hammer and indirect water hammer, and therefore called the critical opening.

Point b in Fig. 9.10 is the intersection point of ξ_1 curve and ξ_m curve. Applying Eqs. (9.31) and (9.35), and giving $\xi_1 = \xi_m$, we can obtain the initial opening corresponding to point b as

$$\tau_0^b = 1/\rho$$

Analyzing Fig. 9.10, we can draw the following conclusions.

(1) If the initial opening $\tau_0 > 1/\rho$, namely $\rho\tau_0 > 1$, the maximum water hammer occurs at the ending time of valve closure, with $\xi_m > \xi_1$. And ξ_m has nothing to do with τ_0.
(2) If $\sigma/\rho < \tau_0 < 1/\rho$, the maximum water hammer occurs at the first phase time, with $\xi_1 > \xi_m$. The smaller τ_0, the larger ξ_1.
(3) If $\tau_0 \leq \sigma/\rho$, the direct water hammer occurs, but the value is not the maximum.
(4) If the opening is the critical opening $\tau_0 = \sigma/\rho$, the maximum direct water hammer occurs, with $\xi_{\max} = \xi_d = 2\rho\tau_0 = 2\sigma$.

The thick solid curves in Fig. 9.10 represent the maximum water hammer pressure for various initial openings. It can be seen that the maximum water hammer pressure does not occur in full load rejection condition, but in smaller load rejection condition. For low head power stations, though ρ is large, the first phase water hammer may occur because of a small initial opening τ_0. Here the following points should be noted.

(1) The turbine running in the speed-no-load condition has no output power. Therefore, the unit shutting down from speed-no-load can use very slow guide vane closing law. By the slow closing the water hammer pressure can be greatly reduced, because direct water hammer $\xi_d = 2\sigma$ cannot occur if $\tau_x > \sigma/\rho$. The actual maximum water hammer is related to τ_x.
(2) The above discussions are based on the assumptions that the opening changing law is linear and the closing time is proportional to the initial opening. Actually, the change of opening is determined by the characteristics of the speed governing system, and there is a delay at the end of closing, as shown in Fig. 9.7. In small load rejection condition, the actual closing time is larger than $\tau_0 T_s$, the closing time proportional to the idealized straight line closing. Therefore, the load rejections with small initial openings are normally not the critical condition for the maximum water hammer. However, the issue on small initial opening inducing large water hammer should be considered in design.

9.3.4.2 Influence of Opening Change Law on Water Hammer

The above concepts and calculation equations for the first phase and last phase water hammer types are deduced under certain assumptions: simple single pipe system, no frictional loss, using orifice discharge characteristics for the valve, and most importantly linear opening change law.

Opening change law actually has a significant effect on pressure change history of water hammer. Figure 9.11 shows three different closing laws with the same closing time, and their corresponding water hammer history curves. The closing curve II is a first-faster-late-slower law. The water hammer curve II rises rapidly and reaches at the maximum in the beginning stage because of the faster closing speed, and then falls gradually because of the slower closing speed in the late stage. The closing curve III is a first-slower-late-faster law, opposite to the closing curve II. The water hammer curve III rises slowly in the first stage due to slower closing speed, climbs

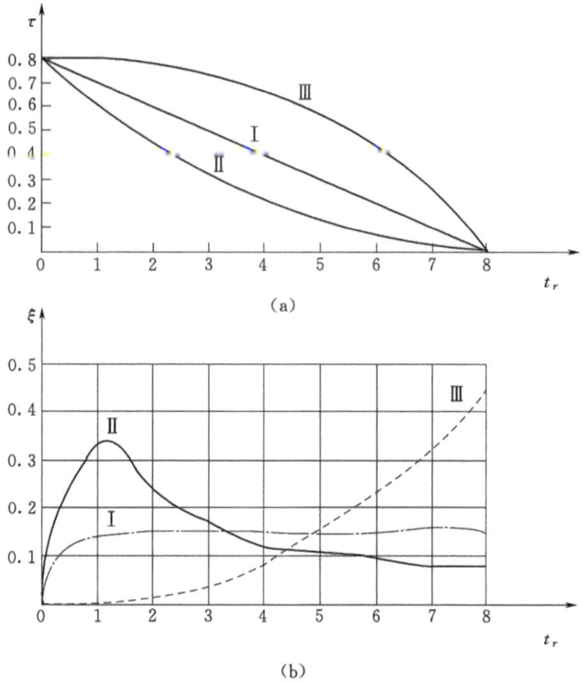

Fig. 9.11 Influence of closing law on water hammer

up fast in the late stage due to faster closing speed, and reaches at the maximum near the end of closing. The closing curve I is a uniform speed law, and its corresponding water hammer curve I is mean and smooth. The closing law I is relatively reasonable, because the maximum water hammer is small, with $\xi_{max} = 0.16$. The closing law III is the worst one with $\xi_{max} = 0.47$, which is 3 times of that of law I. The closing law II is also worse, with $\xi_{max} = 0.35$. This comparison shows that water hammer pressure can be reduced by reasonably adjusting opening and closing laws of wicket gate.

The wicket gate closing law affects the extreme and history of water hammer pressure. But the selection of closing law is limited by the characteristics of governing system. Within the adjustable range of governor, a reasonable closing law requires that the water hammer pressure is as small as possible within a certain closing time. Reasonable closing law is an economical and effective measure to effectively reduce water hammer pressure without increasing extra investment. The water hammer similar to the first phase water hammer often occurs in the high head hydropower stations, for which the first-slower-late-faster closing law can be selected to reduce the water hammer value at the first phase time so as to reduce the maximum value. The water hammer similar to the last phase water hammer often occurs in the low head hydropower stations, for which the first-faster-late-slower closing law can be selected to reduce the water hammer value near the last phase time so as to reduce the maximum value.

9.3.4.3 Influence of Final Opening on Water Hammer

When the opening change is ended, water hammer phenomenon will normally retain for a certain period of time. It is sometime necessary to investigate this phenomenon when setting speed governor parameters and designing penstock. For example, positive water hammer can be reflected by a just closed valve to negative water hammer, which may be larger in absolute value than that caused by opening valve.

The water hammer phenomenon after opening change is ended is determined by the reflecting characteristics of valve at its final opening. The reflection characteristics can be discriminated by Eq. (9.23). Substituting the final opening τ_c into Eq. (9.23), we obtain Eq. (9.39) as follows.

$$r = \frac{1 - \rho\tau_c}{1 + \rho\tau_c} \tag{9.39}$$

(1) If the valve is closed completely at the n-th phase time, then $\tau_c = 0$ and $r = 1$, which indicates a same value same sign reflection. The pressure increasing wave generated at the end of valve closing propagates upstream to the reservoir and is reflected to the pressure decreasing wave. The pressure decreasing wave travels downstream to the valve and reflected to a new pressure decreasing wave, which again travels upstream. The superposition of the two pressure decreasing waves and the pressure increasing wave can lead to the situation that the absolute value of water hammer pressure at the $(n + 1)$th phase time is the same as that of the water hammer pressure at the nth phase time, but their signs are opposite. When flow resistance is neglected, the water hammer pressure after the valve is closed shall be a periodic non-damped sawtooth-shaped oscillation, as shown in Fig. 9.12a.
(2) If the final opening of valve is not zero with $\tau_c > 0$ and $\rho\tau_c < 1$, then $0 < r < 1$, which indicates a reducing value same sign reflection. Based on analysis similar to the first case, we can infer that the water hammer pressure after the value operation is finished is a periodic damped sawtooth-shaped oscillation, as shown in Fig. 9.12b.
(3) If the final opening of valve is not zero with $\tau_c > 0$ but $\rho\tau_c = 1$, then $r = 0$, which indicates a non-reflection condition. When the reservoir reflected wave reaches at the valve the water hammer wave at that section disappears, as shown in Fig. 9.12c.
(4) If $\tau_c > 0$ and $\rho\tau_c > 1$, then $-1 < r < 0$, indicating a reducing value opposite sign reflection. The negative cocurrent wave from the reservoir is reflected by the valve, leading to the generation of a positive countercurrent wave with reduced value. Thus the water hammer at the valve cannot be negative after the closing process is ended. Due to this reducing value opposite sign reflection, the water hammer pressure decreases gradually as the phase number increases, as shown in Fig. 9.12d.

Fig. 9.12 Water hammer history after opening change is ended

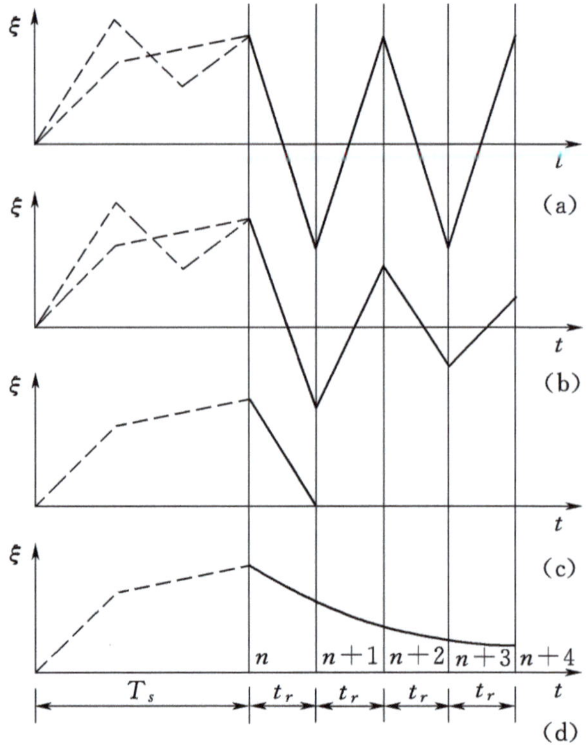

Similar conclusions can be obtained for the valve opening condition (corresponding to load-increase condition for hydropower plant). However, because τ_c is relative larger, the happening possibility of the latter two cases is larger.

9.3.5 Water Hammer Pressure Distribution Along Pipe Length

The above discussion is about the water hammer history at the valve end of a pressure pipe. When designing the pressure pipe, not only the pressure history at the valve end is to be calculated, but also the pressure distribution along the pipe length is to be calculated. The design of pipe strength needs to know the maximum water hammer pressure value at each section along the pipe; the design of pipe layout needs to know the minimum water hammer pressure at each section along the pipe to check the possibility of vacuum in the pipe.

When the valve is opened and closed following a straight line law, the water hammer pressure history at any section along the pipe is similar to that at the valve

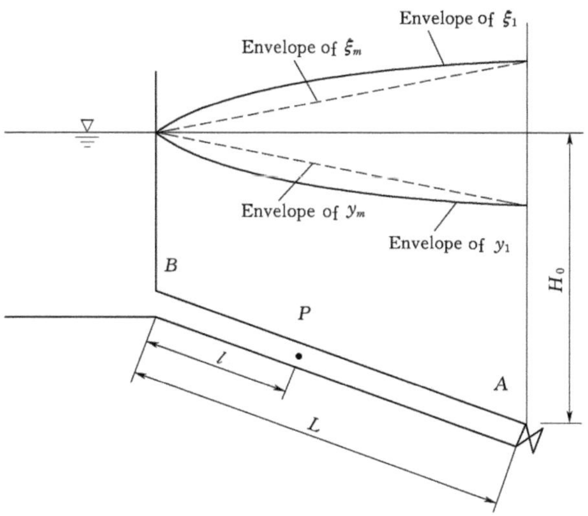

Fig. 9.13 Distribution of water hammer pressure along the pipe

end, having the first phase type and the last phase type. The pressure distribution laws for these two types are different, as shown in Fig. 9.13.

9.3.5.1 Distribution Law for the Last Phase Water Hammer Type

It is proved that when the last phase water hammer (limit water hammer) appears at the valve end of the pipe, the distribution of water hammer pressure along the pipe is a straight line, no matter the water hammer is positive or negative, as shown in by the dotted lines in Fig. 9.13. If the maximum water hammer pressure at the valve end is ξ_m^A or y_m^A, then the maximum water hammer pressure at any point P in the pipe can be calculated with proportional equation

$$\xi_m^P = \frac{l}{L}\xi_m^A \; or \; y_m^P = \frac{l}{L}y_m^A \tag{9.40}$$

9.3.5.2 Distribution Law for the First Phase Water Hammer Type

Studies show that the first phase water hammer is not distributed in straight line law. The distribution curve for positive water hammer is an upward convex, and the distribution curve for negative water hammer is a concave, as shown by solid curves in Fig. 9.13. It is not safe to assume that the distribution is in a straight line law.

The maximum water hammer pressure rise at any point P occurs at the moment when the maximum water hammer rise at point A is transmitted to the point P, namely, its occurring moment lags behind $(L - l)/a$ of that of point A. The maximum water hammer pressure at P can be calculated by using

$$\xi_1^P = \frac{2\sigma}{1+\rho\tau_0-\sigma} - \frac{2\sigma_{AP}}{1+\rho\tau_0-\sigma_{AP}} \tag{9.41}$$

where $\sigma = \frac{LV_{\max}}{gH_0T_s}$ and $\sigma_{AP} = \frac{(L-l)V_{\max}}{gH_0T_s}$.

It can be seen that the first term on the right side of Eq. (9.41) is the water hammer pressure at the first phase time for point A when the pipe length is L, the second term is the water hammer pressure at the first phase time for point A with the pipe length $L-l$ (equivalent to the reservoir moving to point P), and the maximum water hammer pressure at point P is the difference between the above two.

As for the negative first phase water hammer, the maximum pressure drop at P can be calculated by using

$$y_1^P = \left|\frac{2\sigma}{1+\rho\tau_0-\sigma} - \frac{2\sigma_{AP}}{1+\rho\tau_0-\sigma_{AP}}\right| \tag{9.42}$$

where $\sigma = -\frac{LV_{\max}}{gH_0T_s}$ and $\sigma_{AP} = -\frac{(L-l)V_{\max}}{gH_0T_s}$, with y_1^P is the absolute value of the relative pressure decrease.

9.3.6 *Analytical Solution of Water Hammer in Complex Pipe System*

In the above sections, we have been discussing the water hammer problem in a single pipe. The single pipe refers to the pipe whose diameter, wall thickness and material do not change along the pipe length, and the characteristics of the whole pipe (including wave speed) remain unchanged. In practical engineering, single pipe is rare, and complex pipe systems are often encountered. There are three types of complex pipe systems as follows.

(1) Serial pipes: the thickness of pipe wall increases gradually with the increase of water head, the diameter decreases with the increase of water head, or the pipe structural type changes along the length.
(2) Bifurcated pipes: one main pipe is connected to two or more branch pipes. They are often encountered in grouping water supply and combined water supply.

    Whether in serial pipes or bifurcated pipes, the water hammer wave will be reflected and transmitted at the locations where pipe characteristics change, making the water hammer phenomenon and calculation more complex.
(3) Pipe system equipped with a reaction turbine. In this case, the influence of spiral casing and draft tube, as well as the difference between turbine flow characteristics and orifice flow characteristics should be considered. The discharge of turbine is not only related to the working head and wicket gate opening of turbine, but also to the speed and type of turbine.

9.3.6.1 Calculation of Water Hammer Pressure in Serial Pipes

The pipe characteristic factor ρ and water hammer characteristic factor σ of the pipes in series are different, because the velocity V and wave speed a of each pipe are different. In practice, serial pipes are often converted into an equivalent single pipe for simple calculation. This method is called the "equivalent pipe method."

As shown in Fig. 9.14, the length, velocity and wave speed of each pipe of the serial pipes are labeled. Using the "equivalent single pipe" approach, we need to define the velocity V_e and wave speed a_e for the equivalent single pipe based on the following requirements: the pipe length is the same as the original, the phase time is the same as the original, and the water kinetic energy in the pipe is the same as the original.

According to the same phase time condition, we can get

$$a_e = L/\sum_{1}^{n} \frac{L_i}{a_i} \tag{9.43}$$

According to the same water kinetic energy condition, we can get

$$V_e = \sum_{1}^{n} L_i V_i / L \tag{9.44}$$

Then the characteristic factors for the equivalent single pipe can be calculated by using

$$\rho_e = \frac{a_e V_e}{2gH_0},\ \sigma_e = \pm \frac{LV_e}{gH_0 T_s} \tag{9.45}$$

Knowing ρ_e and σ_e, we can use the previous analytical equations for water hammer in a single pipe to calculate the water hammer values in a serial pipe system.

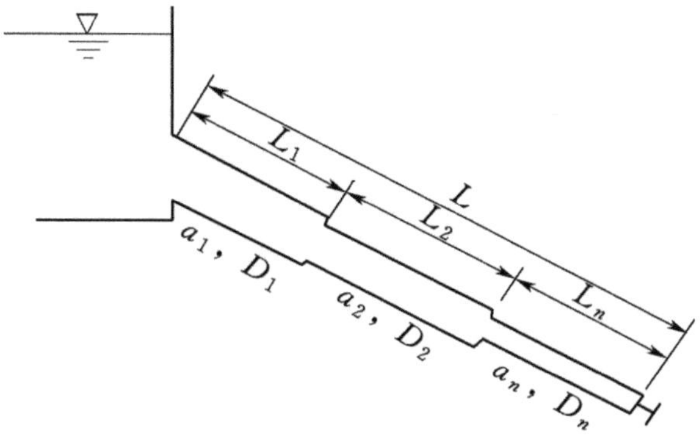

Fig. 9.14 Schematic of serial pipes

This approximate approach ignores the local reflection of water hammer wave at the inner boundaries between pipes, and assumes that the water hammer pressure value is controlled only by the transformation of the water kinetic energy to pressure energy in the pipes. It is proved that the error of calculating a last phase water hammer under linear opening/closing law of valve is small, while the error of calculating a first phase water hammer or under non-linear opening/closing law is large.

9.3.6.2 Calculation of Water Hammer Pressure in Bifurcated Pipes

The water hammer calculation for bifurcated pipes is more complex than that for serial pipes, because the structure of bifurcated pipes is more complex. Normally several branches are connected with the main pipe symmetrically or asymmetrically, and the lengths of the branch pipes are often different, as shown in Fig. 9.15. During the wicket gate opening/closing process, the water hammer waves caused at a pipe end travel upstream along the branch pipe to reach the bifurcation point, at which the reflected wave are folded back to the original branch pipe, and the transmitted waves continue to propagate into the main and other branch pipes. These innumerable water hammer waves constantly interfere with each other, resulting in complicated conditions. This phenomenon can be solved by using the water hammer interlocking equations, together with the boundary condition equations for bifurcation, valve and reservoir. However, this kind of calculation is rather complicated. For rough analytical estimation, there are the following two approaches.

(1) Method of joining branch pipes. Combine all units to a large unit, and locate this unit at the end of the longest branch pipe. Choose the summation of the discharges of all units as the discharge of the large unit, and the summation of the cross sectional areas of all branch pipes as the section area of the longest combined pipe. By this way, the bifurcated pipes become serial pipes, which can be treated by "equivalent single pipe" approach for water hammer calculation.
(2) Method of cutting branch pipes. Select a branch pipe and cut off other branch pipe, resulting in serial pipes with the original cross sectional areas and

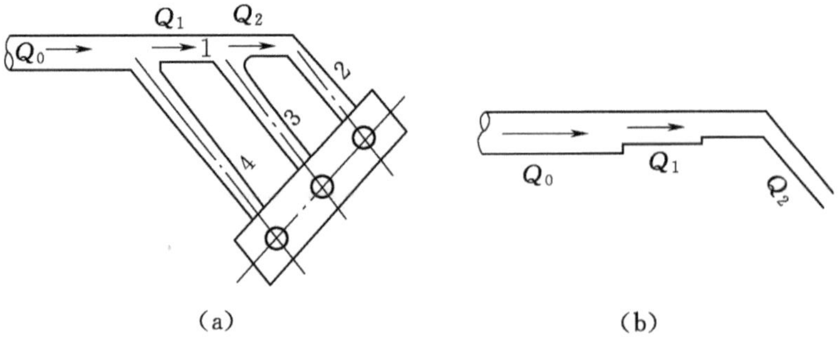

Fig. 9.15 Schematic of the cutting branch method for bifurcated pipes

discharges in different pipe segments (Fig. 9.15b). By this way, we can calculate the water hammer values in every branch pipes.

The reflection of water hammer wave at bifurcation is much more complicated than that at the serial pipes, therefore, the simplified methods mentioned above are rough. When the main pipe is longer and the branch pipes are shorter (for example, the lengths of the branch pipes are less than 10% of the main pipe), the error of calculation result is not large, otherwise the error will be large.

9.3.6.3 Calculation of Water Hammer Pressure in Pipe System with Reaction Turbine

Up to now, we have discussed the water hammer problem caused by valve opening and closing, which is similar to the water hammer phenomenon caused by impulse type of hydraulic turbines. The above equations are still usable when concerning reaction type of turbines. However, the error is large sometimes. The main differences of reaction turbine and impulse type turbine in terms of water hammer calculation are the following.

(1) A reaction turbine has a spiral casing and a draft tube, while an impulse turbine has water distribution pipes (Fig. 9.16).
(2) The discharge law of reaction turbine is different from that of an orifice. Reaction turbine's discharge is not only related to wicket gate opening and water head, but also related to turbine type and rotational speed. These relations cannot be accurately expressed by analytical equations.
(3) When the discharge of reaction turbine is changed, the water hammer phenomenon will occur in both the spiral casing and the draft tube.

The spiral casing is equivalent to the extension part of the penstock, and its water hammer phenomenon is similar to that in the penstock. The draft tube is behind the wicket gate; therefore, the water hammer phenomenon is opposite to that in the penstock: negative water hammer occurs in the draft tube when the wicket gate is closing, and positive water hammer occurs in the draft tube when the wicket gate is opening.

Therefore, the calculation of water hammer pressure concerning a reaction turbine is more complex. In preliminary analysis, the following approximate analytical methods can be used.

First, view the spiral casing as an extension of the penstock, and move the wicket gate to the outlet of the draft tube. In this way, the draft tube also becomes an extension part of the penstock, and the influence of the draft tube water hammer on the penstock water hammer is approximately included. Second, using the boundary condition for the impulse turbine, namely $v = \tau\sqrt{1+\xi}$, to calculate the water hammer value ξ at the pipe end (before the imaginarily moved wicket gate). Here, ξ incudes the water hammer components of the penstock, spiral casing, and draft tube. ξ will be positive for wicket gate closing and negative for wicket gate opening.

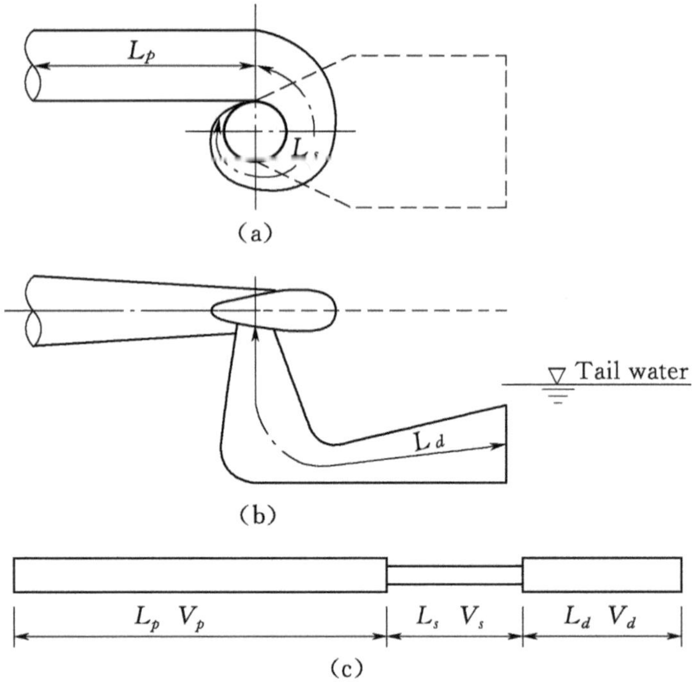

Fig. 9.16 Schematic of a reaction turbine and its simplification

Thus, the penstock, spiral casing, and draft tube are combined to serial pipes (Fig. 9.16c), for which an equivalent single pipe can be applied in water hammer calculation. The equivalent length L, wave velocity a_e, and velocity V_e of the equivalent single pipe can be calculated by

$$L = L_p + L_s + L_d \tag{9.46}$$

$$a_e = L / \left(\frac{L_p}{a_p} + \frac{L_s}{a_s} + \frac{L_d}{a_d} \right) \tag{9.47}$$

$$V_e = \left(L_p V_p + L_s V_s + L_d V_d \right) / L \tag{9.48}$$

where the subscripts p, s and d are corresponding to penstock, spiral casing and draft tube, respectively. The calculation of L_s and L_d may refer to Fig. 9.16a, b. The velocity in the spiral casing

$$V_s = \phi_{\max} Q_0 / 360^\circ F_s$$

in which $\phi_{\max}$, Q_0, and F_s are the scroll angle, the maximum discharge, and the inlet section area of the spiral casing, respectively. The velocity V_d can be determined by averaging the velocity values at the inlet and outlet. Therefore, the pipe characteristics constant ρ_e and water hammer characteristics constant σ_e of the equivalent pipe can be calculated. Further, the maximum water hammer rise ratio ξ can be obtained based on the equation for the first phase or last phase water hammer.

The distribution of water hammer along the system can be estimated based on the water momentum values in the penstock, spiral casing, and draft tube.

The relative water hammer at the penstock end uses

$$\xi_p = \frac{L_p V_p}{(L_p + L_s + L_d)V_e}\xi \tag{9.49}$$

The relative water hammer at the spiral casing end uses

$$\xi_s = \frac{L_p V_p + L_s V_s}{(L_p + L_s + L_d)V_e}\xi \tag{9.50}$$

The relative water hammer at the draft tube inlet uses

$$y_d = \frac{L_d V_d}{(L_p + L_s + L_d)V_e}\xi \tag{9.51}$$

With the value y_d obtained, the vacuum value at the draft tube inlet can be evaluated by

$$H_d = H_s + y_d H_0 + \frac{V_d^2}{2g} < 8\text{m} \tag{9.52}$$

where H_s is the suction height of turbine, m; $y_b H_0$ is the head drop in the draft tube, m; $\frac{V_d^2}{2g}$ is the velocity head when y_b occurs, m.

The value of H_d should be limited to a value smaller than 8 m, to prevent water column separation.

For medium and high head hydropower stations, the penstocks are usually longer, and the influences of the spiral casing and draft tube on water hammer are smaller. In this case, the spiral casing and draft tube effects can be neglected. However, for small head hydropower stations, these two parts should be considered, and the effect of the draft tube are normally more significant.

9.3.7 Calculation of Rotational Speed Rise

The rotational speed rise of turbine-generator unit after a load rejection can be expressed as a relative value, namely the speed rise ratio $\beta_{\max}$, which is defined as

$$\beta_{max} = \frac{n_{max} - n_0}{n_0} \tag{9.53}$$

where n_0 is the rated rotational speed, n_{max} is the highest speed.

The calculation of speed rise ratio in load rejection condition is normally conducted by using the following two equations.

9.3.7.1 The Leningrad Metal Factory Equation of the Former Soviet Union for Speed Rise Ratio

After the full load is rejected, the wicket gate closes from the full opening to the speed-no-load opening, lasting time T_{s1}, during which the output power decreases from N_0 to 0. In this process, the surplus energy of unit may be illustrated as the shaded area in Fig. 9.17. The influences of the action delay of wicket gate and the head increase due to water hammer make the *N*-*t* curve convex. The surplus energy will be converted into the unit's rotational speed rise. If the moment of inertia of the unit is I, the initial angular speed before the load rejection is ω_0, and the maximum angular speed after the load rejection is ω_{max}, then the theorem of angular momentum can be expressed as

$$\int_0^{T_{s1}} N\mathrm{d}t = \frac{1}{2}I\left(\omega_{max^2} - \omega_0^2\right) \tag{9.54}$$

where $\omega_0 = \frac{\pi n_0}{30}$; $\omega_{max} = \frac{\pi(n_0+\Delta n)}{30} = \frac{\pi n_0}{30}(1+\beta)$;$I = 1000 \times \frac{G}{g} \times \left(\frac{D}{2}\right)^2 = 1000\frac{GD^2}{4g}$; n_0 is the rated rotational speed, $r/\min$; I is the moment of inertia, kgms; GD^2 is the flywheel moment, tm^2.

The left hand term of Eq. (9.54) defines the shaded area in Fig. 9.17, and it can be expressed as $\frac{1}{2}N_0T_{s1}f$, in which N_0 is the rated output power, f is correction factor accounting for the influence of water hammer. Inserting this expression into Eq. (9.54), and simplifying the equation, we obtain

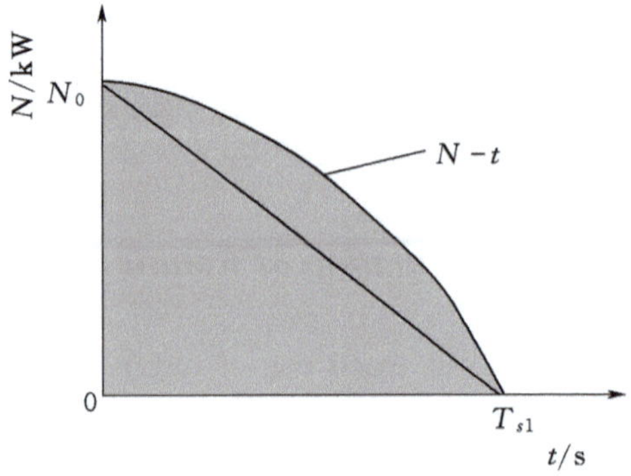

Fig. 9.17 Output power curve during load rejection transients

$$\beta_{\max} = \sqrt{1 + \frac{365 N_0 T_{s1} f}{n_0^2 GD^2}} - 1 \tag{9.55}$$

For Francis and impulse turbines, $T_{s1} = (0.8 - 0.9)T_s$; for Kaplan turbines $T_{s1} = (0.6 - 0.7)T_S$. The correction factor $f = -1.196\sigma^2 + 1.967\sigma + 0.967$, which shows a downward concave curve. This equation is relatively accurate for the conditions with the last phase water hammer.

9.3.7.2 The Changban Equation of China for Speed Rise Ratio

When the delay time of wicket gate action is considered, we can obtain the Changban equation as follows:

$$\beta_{\max} = \sqrt{1 + \frac{365 N_0}{n_0^2 GD^2}(2T_c + T_n f)} - 1 \tag{9.56}$$

where, T_c is the regulation delay time, (s); $T_c = T_A + 0.5\delta T_a$, in which T_A is the delay time of wicket gate action (normally 0.1 s for electric governor and 0.2 s for mechanical governor), δ is the residual unevenness of governor (normally 0.02 to 0.06), and T_a is the acceleration time constant of the unit, with $T_a = \frac{n_0^2 GD^2}{365 N_0}$; T_n is the speed rise time, defined as the time from the load rejection instant to the peak speed instant, during which the unit is accelerating. From statistics, $T_n = (0.9 - 0.00063 n_s)T_s$, in which n_s is the specific speed, with $n_s = n_0\sqrt{N_0}/H^{5/4}$.

The value of T_c is normally smaller than 0.2 s, and the maximum value should be within 0.5 s.

The influence of T_n to $\beta_{\max}$ is related to specific speed n_s; smaller n_s has larger influence, however, which is normally within 3.5%.

The correction factor f for water hammer influence is a new one, with $f = 1.003 + 0.414\,\sigma + 0.589\,\sigma^2$, which shows an upward concave curve.

The Changban equation is relatively accurate for the conditions with the first phase water hammer.

9.4 Numerical Solution to Water Hammer and Rotational Speed Rise

The basic equations for water hammer, as shown in Eqs. (9.8) and (9.9), cannot be solved analytically, because they are quasi-linear hyperbolic partial differential equations. In addition, the boundary conditions of practical water conveyance systems, such as bifurcations, surge chambers, and turbines, are complex and difficult to be calculated by analytical methods. Therefore, it is necessary to calculate the hydraulic and mechanical transient processes in complex systems numerically by computer. The features of numerical methods are as follows:

(1) Fewer simplifications and assumptions are needed in the calculation. For example, pipe friction can be taken into account, and the opening and closing of turbine wicket gate can be non-linear laws.
(2) Complex boundary conditions can be easily treated. For example, turbine characteristic curves can be considered, and the calculation of water hammer pressure, surge chamber water level oscillation and unit speed change can be synchronously conducted.
(3) Faster calculation with high accuracy.

9.4.1 *Method of Characteristics for Water Hammer Equations*

There are many numerical methods for solving the quasi-linear hyperbolic partial differential equations, in which the method of characteristics (MOC) is widely used because of its clear physical concepts, strict procedures, and high accuracy.

The basic idea of the method of characteristics is transforming the partial differential Eqs. (9.8) and (9.9) into two ordinary differential equations that satisfy some specific conditions and then solving them by finite difference methods.

9.4.1.1 The Characteristic Equations and Characteristic Line Equations

The dependent variables in Eqs. (9.8) and (9.9) are H and V, and the independent variables are x and t. If we can find specific relations between x, and t, we will be able to transform the partial differential equations to equivalent ordinary differential equations. Under the specific conditions, the solutions to the ordinary differential equations can be the solutions to the partial differential equations.

Linearly combining Eqs. (9.8) and (9.9), namely multiplying Eq. (9.9) by a constant λ and adding Eq. (9.8), we can obtain

$$\lambda\left[\frac{\partial H}{\partial t} + \left(V + \frac{g}{\lambda}\right)\frac{\partial H}{\partial x} + V\sin\alpha\right] + \frac{\partial V}{\partial t} + \frac{\partial V}{\partial x}\left(V + \frac{a^2\lambda}{g}\right) + \frac{f}{2D}V|V| = 0 \tag{9.57}$$

Knowing $H = H(x, t)$ and $V = V(x, t)$ are the solutions to Eq. (9.57) and they satisfy

$$\frac{dH}{dt} = \frac{\partial H}{\partial t} + \frac{\partial H}{\partial x}\frac{dx}{dt} \tag{9.58}$$

$$\frac{dV}{dt} = \frac{\partial V}{\partial t} + \frac{\partial V}{\partial x}\frac{dx}{dt} \tag{9.59}$$

we can obtain the following relations by comparing Eq. (9.57) with Eqs. (9.58) and (9.59).

$$\frac{dx}{dt} = V + \frac{g}{\lambda}, \frac{dx}{dt} = V + \frac{a^2\lambda}{g} \tag{9.60}$$

Given

$$\lambda = \pm\frac{g}{a} \tag{9.61}$$

Equation (9.60) becomes

$$\frac{dx}{dt} = V \pm a \tag{9.62}$$

That is to say, if $\lambda = g/a$ is satisfied, Eq. (9.57) can be written as

$$C^+\begin{cases} \frac{dx}{dt} = V + a \\ \frac{g}{a}\left(\frac{dH}{dt} + V\sin\alpha\right) + \frac{dV}{dt} + \frac{f}{2D}V|V| = 0 \end{cases} \tag{9.63}$$

If $\lambda = -g/a$ is satisfied, Eq. (9.57) can be written as

$$C^-\begin{cases} \frac{dx}{dt} = V - a \\ -\frac{g}{a}\left(\frac{dH}{dt} + V\sin\alpha\right) + \frac{dV}{dt} + \frac{f}{2D}V|V| = 0 \end{cases} \tag{9.64}$$

These two sets of equations are the ordinary differential equations, with t the independent variable and V and H the dependent variables. That is to say, through the above treatment, the problem of partial differential equations is transferred to the problem of ordinary differential equations. Equations (9.61) and (9.63), called the characteristic line equations, are the specific conditions for the transformation. Equations (9.62) and (9.64), called the characteristic equations, are equivalent to Eqs. (9.8) and (9.9). The essentials of these two sets of equations are as follows. The solutions $H(x, t)$ and $V(x, t)$ to Eq. (9.62) are also the solutions to Eq. (9.8), but correct only when they fit the line defined by Eq. (9.61). Similarly, the solutions $H(x, t)$ and $V(x, t)$ to Eq. (9.64) are also the solutions to Eq. (9.9), but correct only when they fit the line defined by Eq. (9.63).

Geometrically speaking, Eqs. (9.61) and (9.63) define two clusters of lines on the $x - t$ plane, seeing Fig. 9.18. All variables H and V along line $\frac{dx}{dt} = V + a$ satisfy Eq. (9.62); All variables H and V along line $\frac{dx}{dt} = V - a$ satisfy Eq. (9.64). The slopes of the two clusters of straight lines represent the directions of characteristic lines. Characteristic line $\frac{dx}{dt} = V + a$ is called the forward wave characteristic

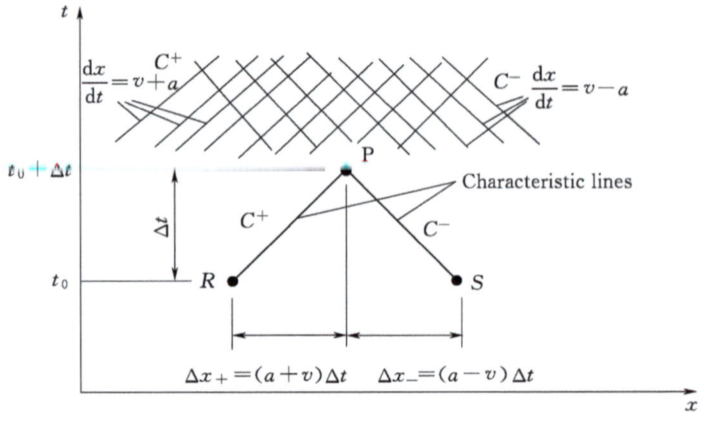

Fig. 9.18 Characteristic lines on the $x - t$ plane

line, symbolled by C^+, because it represents that the direction of water hammer propagation is consistent with the positive direction of coordinate x. Characteristic line $\frac{dx}{dt} = V - a$ is called the backward wave characteristic line, symbolled by C^-, because it represents that the direction of water hammer propagation is opposite to the positive direction of coordinate x.

Equations (9.61)–(9.64) are the basic equations for solving water hammer problem in pipes by the method of characteristics.

9.4.1.2 The Differential Equations and Computational Grid

Because velocity V is far smaller than wave speed a, the characteristic line equations can be simplified as $\frac{dx}{dt} = \pm a$. Also, the inclination term $V \sin \alpha$ in characteristic Eqs. (9.62) and (9.64) is normally small, and can be omitted. In simulation, the pipe should first be discretized as sections, with the segment length Δx. Based on difference theories, to keep stability of the simulation scheme, the Courant condition should be satisfied

$$\Delta t \leq \frac{\Delta x}{a} \tag{9.65}$$

This stability condition means the timestep Δt is constrained. To simplify simulation, one normally chose Δt as a constant and let segment length $\Delta x = \Delta t \cdot a$. This simulation scheme is called constant timestep grid scheme, as shown in Fig. 9.19. The coordinates of any point on $x - t$ plane correspond to the location of the pipe and the time of simulation. For example, the point P is corresponding to the flow state at section i of the pipe when time is $t + \Delta t.t + \Delta t$.

By using Eqs. (9.62) and (9.64) and the spatial segment Δx and timestep Δt, we can build the relations of head and discharge between adjacent sections.

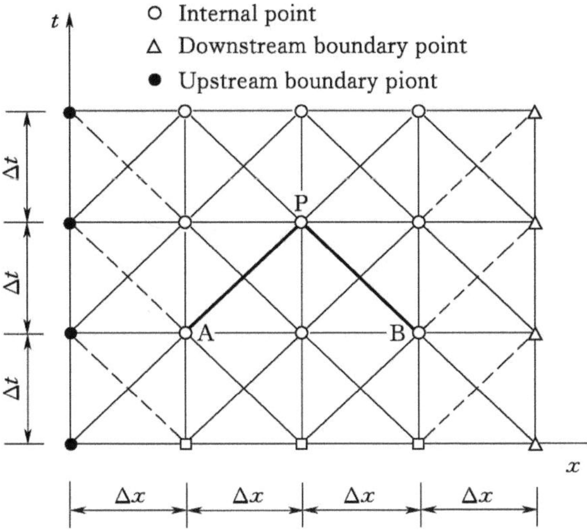

Fig. 9.19 Constant timestep grid

Integrating Eq. (9.62) from point A to point P along C^+ line yields

$$\frac{g}{a}(H_P - H_A) + (V_P - V_A) + \frac{f}{2D} V_P |V_A| \Delta t = 0 \tag{9.66}$$

Similarly, integrating Eq. (9.64) from point B to point P along C^- line yields

$$-\frac{g}{a}(H_P - H_B) + (V_P - V_B) + \frac{f}{2D} V_P |V_B| \Delta t = 0 \tag{9.67}$$

Equations (9.66) and (9.67) are the difference equations of H and V between points A and P, and points B and P, respectively. The trapezoidal equation is applied in integrating the resistance term.

Knowing $V = Q/A$, in which A the cross sectional area of pipe, we can simplify and regularize Eqs. (9.66) and (9.67) as

$$C^+ : Q_P = Q_{CP} - C_{QP} H_P \tag{9.68}$$

$$C^- : \quad Q_P = Q_{CM} + C_{QM} H_P \tag{9.69}$$

where the coefficients are defined by $Q_{CP} = \frac{Q_A}{1+S|Q_A|} + C_{QP} H_A$ $Q_{CM} = \frac{Q_B}{1+S|Q_B|} - C_{QM} H_B$, $C_{QP} = \frac{C}{1+S|Q_A|}$, $C_{QM} = \frac{C}{1+S|Q_B|}$, $C = \frac{gA}{a}$, and $S = \frac{f \Delta t}{2DA}$.

During the computation of every timestep, the head and discharge at instant t are always known, and the values at instant $t + \Delta t$ can be calculated promptly. The values at instant t can be defined by the initial condition or calculated by the previous timestep. For example, knowing the head and discharge values of points A and B at t

instant, namely H_A, H_B, Q_A, and Q_B, one can first calculate coefficients Q_{CP}, Q_{CM}, C_{QP}, C_{QM}, S, and C, and then calculate the head H_P and discharge Q_P of point P at $t + \Delta t$ instant by using Eqs. (9.68) and (9.69). With similar procedures, one can calculate the head and discharge values of every grid point at $t + \Delta t$ instant.

With the values of all grid points at instant $t + \Delta t$ known, one can follow the same procedures to obtain the values of all grid points at instant $t + 2\Delta t$. The whole simulation is based on these procedures and equations, calculating step by step from the initial state, until reaching at the given terminal time.

It should be noted that, because the direction of coordinate x is the same with the flow direction, the intersection angle of the $\frac{dx}{dt} = a$ line and the x coordinate is an acute angle; thus point A is called the upstream point of P, because it is located on the left of P; when using Eq. (9.68) to calculate Q_{CP} and C_{QP}, the physical quantities (discharge, head, wave speed, cross sectional area, etc.) should chose those at the upstream section. Similarly, when using Eq. (9.69) to calculate Q_{CM} and C_{QM}, the quantities at the downstream section should be selected.

When the head and discharge values at the internal points are known, the remaining task is to calculate the values at the boundaries. In case of P is located at a boundary, only one between Eqs. (9.68) and (9.69) can be used. Additional boundary condition equations should be provided to make the equations closed.

9.4.2 Boundary Conditions

Mathematically speaking, the calculation of a pressurized unsteady pipe flow is actually equal to solving a set of partial differential equations under the definite solution conditions, which include initial conditions and boundary conditions. The initial conditions for unsteady flows are normally defined by the steady state, which are relatively easy to solve. On the other hand, the boundary conditions are complex and have different forms for different types.

Boundaries can be divided into inner boundary type and outer boundary type, according to the position of boundary in the whole system. The boundaries that are located in the internal part of system are called the inner boundaries, such as serial junction and bifurcation, while those located at the ends of system are called the outer boundaries, such as reservoir and downstream pond.

On the other hand, boundaries can also be divided into power boundary type and non-power boundary type, from the point of view of whether fluid energy (excluding loss) is input or output. For example, the turbine, pump and so on are the power boundaries; the surge chamber, gate, valve and so on are the non-power boundaries.

Fluid parameters have different changing laws for different boundaries. In treating boundary conditions, the mathematical models or boundary equations should be established based on the physical properties of fluid flow. The actual boundaries in engineering are always complex, therefore, appropriate simplifications that are not affect the real physical states are normally conducted.

9.4.2.1 Non-power Boundaries

(1) Reservoir

One of the characteristics of reservoir boundary (Fig. 9.20) is its large free surface. During transient process, the water level or head can be treated as constant. The boundary condition is

$$H_P = H_0 \tag{9.70}$$

where H_0 is the initial head and H_P is the head at the present time.

For upstream reservoir, Eqs. (9.70) and (9.69) can be utilized to solve discharge Q_P; for downstream reservoir, Eqs. (9.70) and (9.68) can be utilized.

(2) Dead end

Dead end is normally located at the end of a pipe (Fig. 9.21). When wicket gate or main valve is closed, or penstock outlet is plugged during building period, dead end boundary can be encountered. The boundary condition can be

$$Q_P = 0 \tag{9.71}$$

By using Eq. (9.69) or (9.70), we can calculate the head at the upstream end (or downstream end) as follows

$$\left.\begin{array}{l} H_P = Q_{CP}/C_{QP} \text{ Upstream} \\ H_P = Q_{CM}/C_{QM} \text{ Downstream} \end{array}\right\} \tag{9.72}$$

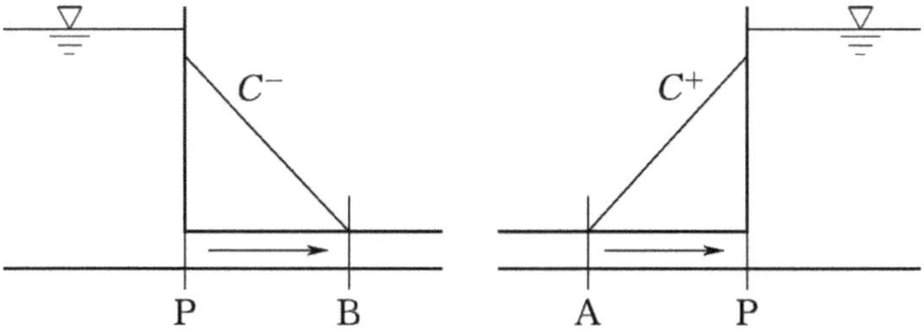

Fig. 9.20 Schematic of reservoir boundary condition

Fig. 9.21 Schematic of dead end boundary

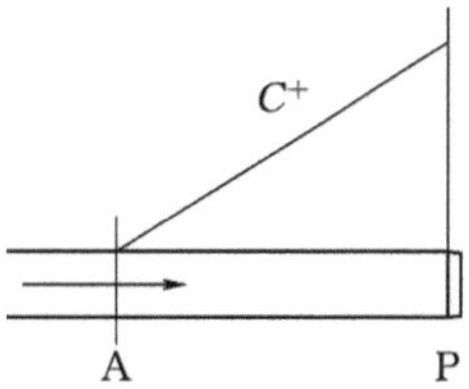

where the subscripts "Upstream" and "Downstream" indicate the upstream and downstream sections of the dead end section, which are normally different.

(3) Bifurcation and serial pipes

For a bifurcation with n pipes connected, there are $2n$ unknown variables. Because only n characteristic equations are provided, we need to find n boundary equations to solve problem. The continuity condition at the bifurcation is

$$\sum_{i=1}^{n} Q_i = 0 \tag{9.73}$$

With local head loss and velocity head neglected, the momentum condition can be

$$H_{P1} = H_{P2} = \cdots = H_P \tag{9.74}$$

Equations (9.73) and (9.74) provide the additional n equations; therefore, the variables at the bifurcation can be solved.

For the specific bifurcation shown in Fig. 9.22, Eq. (9.73) becomes

$$Q_{P1} - Q_{P2} - Q_{P3} - Q_{P4} = 0 \tag{9.75}$$

And Eq. (9.74) becomes

$$H_{P1} = H_{P2} = H_{P3} = H_{P4} = H_P \tag{9.76}$$

The characteristic Eqs. (9.68) and (9.69) are

$$\left.\begin{array}{l} Q_{P1} = Q_{CP1} - C_{QP1}H_P \\ Q_{P2} = Q_{CM2} + C_{QM2}H_P \\ Q_{P3} = Q_{CM3} + C_{QM3}H_P \\ Q_{P4} = Q_{CM4} + C_{QM4}H_P \end{array}\right\} \tag{9.77}$$

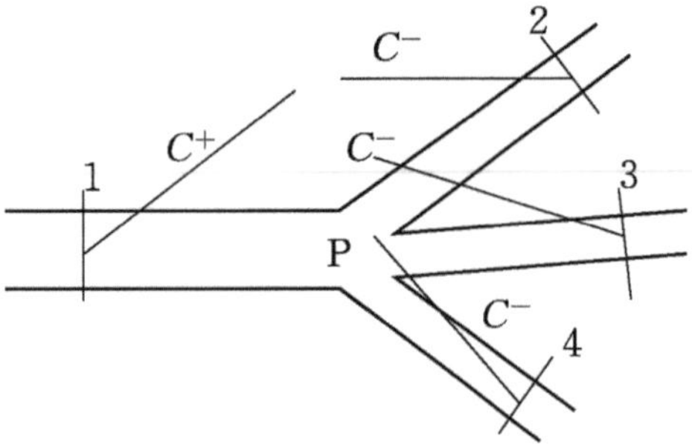

Fig. 9.22 Schematic of bifurcation

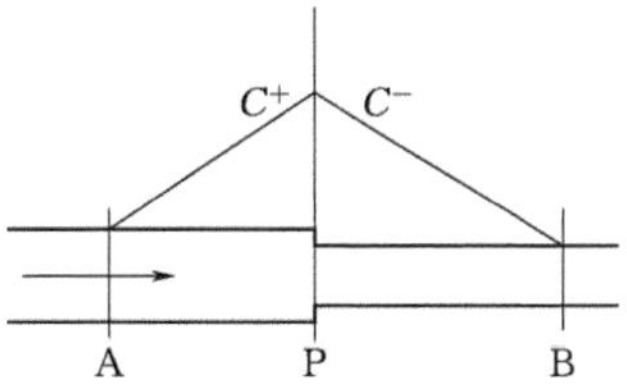

Fig. 9.23 Schematic of serial pipes

Submitting Eqs. (9.77) into (9.75) and consolidating the result, we can obtain the head at the bifurcation

$$H_P = \frac{Q_{CP1} - \sum_{i=2}^{4} Q_{CMi}}{C_{QP1} + \sum_{i=2}^{4} C_{QMi}} \tag{9.78}$$

With the head known, one can further obtain the discharges of all connected pipes by Eq. (9.77).

Serial pipe junction can be viewed as a specific condition of bifurcation junction (Fig. 9.23). If $n = 2$, the bifurcation junction becomes the serial pipe junction. Therefore, the head Eq. (9.78) becomes

$$H_p = \frac{Q_{CP1} - Q_{CM2}}{C_{QP1} + C_{QM2}} \tag{9.79}$$

Then, the discharge can be solved by using Eq. (9.68) or Eq. (9.69).

(4) Surge chamber

The hydraulic system with a surge chamber is shown in Fig. 9.24. It can be virtually treated as the bifurcation with free surface in one of the pipes. In view of the height of surge chamber is shorter comparing to the hydraulic pipeline, the water elasticity in the surge chamber can be neglected in calculations. If the connecting pipe between the surge chamber and the hydraulic pipeline is short (Fig. 9.24), the influence of water inertia in the connecting pipe can also be neglected. Therefore, from energy equation, we can obtain

$$Z_P = H_P - \alpha Q_{TP}|Q_{TP}| \tag{9.80}$$

where Z_P is the water level at the present time; H_P is the piezometric head at point P in surge chamber the bottom; Q_{TP} is the discharge flowing into or out of surge chamber (positive for flowing into); α is the coefficient of impedance.

For the continuity, we can write

$$Q_{TP} = Q_{1P} - Q_{2P} \tag{9.81}$$

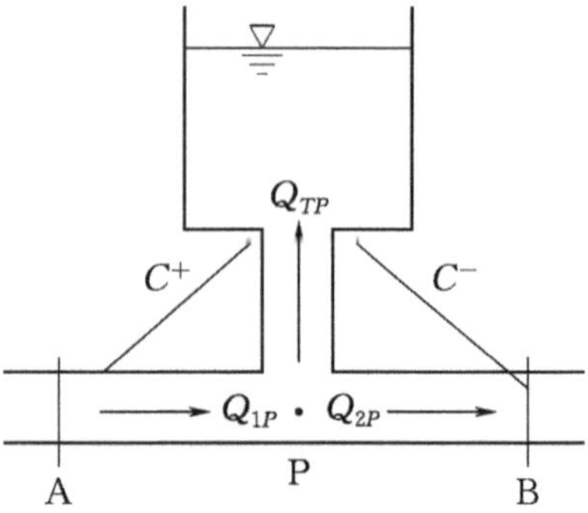

Fig. 9.24 Schematic of surge chamber

The characteristic equations of the upstream and downstream pipes are

$$Q_{1P} = Q_{CP} - C_{QP1}H_P \tag{9.82}$$

$$Q_{2P} = Q_{CM} + C_{QM2}H_P \tag{9.83}$$

When water enters the surge chamber, the water level rise velocity can be expressed as

$$\mathrm{d}Z_P/\mathrm{d}t = Q_{TP}/F = \frac{1}{F}(Q_{1P} - Q_{2P}) \tag{9.84}$$

in which F is the sectional area of surge chamber.

For the surge chamber layout shown in Fig. 9.4, there are five unknowns: Z_P, H_P, Q_{TP}, Q_{1P}, and Q_{2P}. Through combining and solving the five Eqs. (9.80)–(9.84), we can easily obtain the surge chamber water level Z_p.

Equation (9.84) is a first-order ordinary differential equation, which can be solved by the prediction-correction method in satisfactory accuracy. If the timestep Δt is small or the water level oscillation is slow, the first-order difference scheme, namely the rectangular equation can be used in approximation, by which Eq. (9.84) can be rewritten as

$$Z_p = Z + \frac{1}{F}Q_{TP}\Delta t \tag{9.85}$$

9.4.2.2 Power Boundary-Turbines

Hydraulic turbines belong to power boundary type. The following will discuss impulse turbine boundary and reaction turbine boundary separately.

(1) Impulse turbine

Impulse turbine is installed at the end of penstock; therefore, it belongs to external boundaries. The jet nozzle is an orifice with a needle valve. The discharge Q_P is defined by the type of nozzle, the discharge area ω of nozzle, and water head H_P, having nothing to do with unit speed. Therefore, the boundary condition of an impulse turbine is the same as that of a valve at pipe end. The expression is

$$Q_P = \phi\omega\sqrt{2gH_P} \tag{9.86}$$

in which ϕ is the flow coefficient of the nozzle, depending on the type of nozzle.

Jointing Eqs. (9.68) and (9.86), we can solve the head by

$$H_P = \left[Q_{CP}C_{QP} - \phi\omega\sqrt{(g\phi\omega)^2 + 2gQ_{CP}C_{QP}} + g(\phi\omega)^2\right]/C_{QP}^2 \tag{9.87}$$

(2) Reaction turbine

For reaction turbine, the discharge is not only related to the wicket gate opening α and head H, but also influenced by the unit speed n. Therefore, the equation defining the rotation of unit must be included in building the boundary conditions, making the solution more complicated. In addition, reaction turbine belongs to the inner boundary type, because it is normally located in between a long penstock and a long tailrace conduit, as shown in Fig. 9.25. Therefore, for a reaction turbine boundary, four unknowns, namely H_P, H_S, Q_P and Q_S, are need to be solved. The upstream penstock and downstream tailrace conduit can provide two characteristic equations, and the remaining two equations should be built based on the discharge and torque characteristics of reaction turbine.

The continuity condition is

$$Q_P = Q_S \tag{9.88}$$

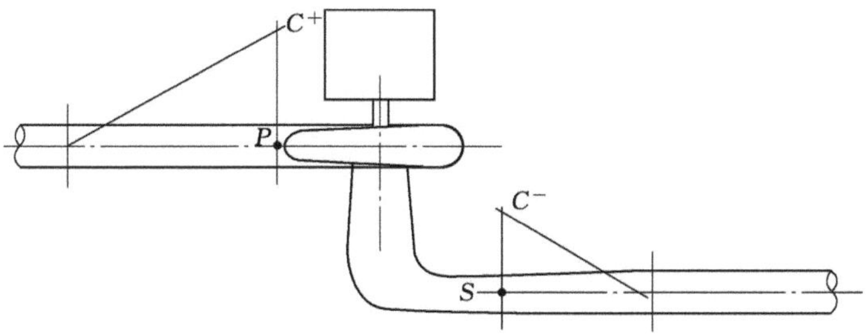

Fig. 9.25 Schematic of reaction turbine

in which the subscripts P and S are used to label the physical variables of upstream pipe and downstream pipe, respectively.

According to the similarity law of hydraulic turbine, the discharge Q_P of turbine can be expressed as

$$Q_P = Q_1' D_1^2 \sqrt{H_P - H_S} \tag{9.89}$$

where Q_1' is the unit-discharge, and D_1 is the runner diameter.

The characteristic equations of the upstream and downstream pipes are

$$Q_P = Q_{CP} - C_{QP} H_P \tag{9.90}$$

$$Q_S = Q_{CM} - C_{QM} H_P \tag{9.91}$$

The four Eqs. (9.88)–(9.91) are closed in principle for solving the four unknown variables. However, the problem is that the unit-discharge Q_1' should be first provided. Q_1' depends on the characteristics of turbine, and normally be expressed as the function of relative opening τ and unit speed n_1' as follows

$$Q_1' = f_1(\tau, n_1') \tag{9.92}$$

which is provided by the comprehensive characteristics curves from model test measurements.

The relative opening of wicket gate changes with a given law, namely $\tau = f(t)$ depends on the closing laws of wicket gate. On the other hand, the unit speed n_1' is defined by the similarity law of hydraulic turbine

$$n_1' = nD_1 / \sqrt{H_P - H_S} \tag{9.93}$$

where the rotational n changes accompanied by water hammer fluctuates. The equation governing the rotational speed is

$$\frac{GD^2}{4g} \cdot \frac{2\pi}{60} \cdot \frac{\mathrm{d}n}{\mathrm{d}t} = M_t - M_g \tag{9.94}$$

where M_g is the retarding torque of generator, being zero for load rejection transients. The turbine driving torque is defined by

$$M_t = M_1' D_1^3 (H_P - H_S) \tag{9.95}$$

where the unit-torque is also provided by the comprehensive characteristics curves and expressed by

$$M_1' = f_2(\tau, n_1') \tag{9.96}$$

From the above discussion, the number of unknowns becomes nine: Q_P, Q_S, H_P, H_S, Q_1', n_1', n, M, and M_1', and the number of available equations also becomes nine: Eqs. (9.88)–(9.96). These the boundary condition equations of a reaction turbine can be solved by iteration.

9.4.3 *Simulation Flow Chart for Guaranteed Regulation*

Figure 9.26 shows the general flow chart for the calculation procedures of hydraulic transient processes of hydropower stations. Normally, it is divided into four steps: input of the original parameters, calculation of steady flow, calculation of unsteady flow, and output of the results. Within the four procedures, the calculation of unsteady flow is an iteration procedure.

9.4.4 *Simulated Results and Analysis*

9.4.4.1 Collation and Display of Results

The results of the calculation of guaranteed regulation include two parts: steady state results and unsteady state results (hydraulic transients). The display of results is very important for analyzing basic rules.

For steady state, the results needed to be displayed are the following. Reservoir water level, tailrace water level; discharge, head, output power, wicket gate opening and rotational speed of each unit; discharge and flowing direction in each pipe; energy grade line and piezometic head line (hydraulic grade line) along all pipes; surge chamber and gate shaft water levels. Figure 9.27 is a typical chart for displaying. Based on the displayed chart, one can check whether the results are correct and meet the requirements of the simulated conditions. For example, are the output power and rotational speed correct? Are the energy and piezometric head distributions reasonable? Is the flow continuity at a bifurcation satisfied? If one of these questions has a NO answer, then there would be some input errors or mistakes. In this case, one should carefully verify the related input data.

For unsteady state flow, the results needed to be displayed are the following three aspects. The guaranteed regulation parameters of unit, the maximum and minimum pressure distributions along pipelines, the maximum and minimum water levels in surge chamber, and the pressure difference across the bottom plate of surge chamber.

Normally, the time histories of the water hammer pressure intensity (or piezometric head or pressure increase ratio) at the nose of spiral casing, the water hammer pressure intensity (or piezometric head) at the inlet of draft tube, the rotational speed or speed rise ratio of unit (for load-increase condition, the output power), and the opening of wicket gate can be drawn in one chart for analyzing the physical processes,

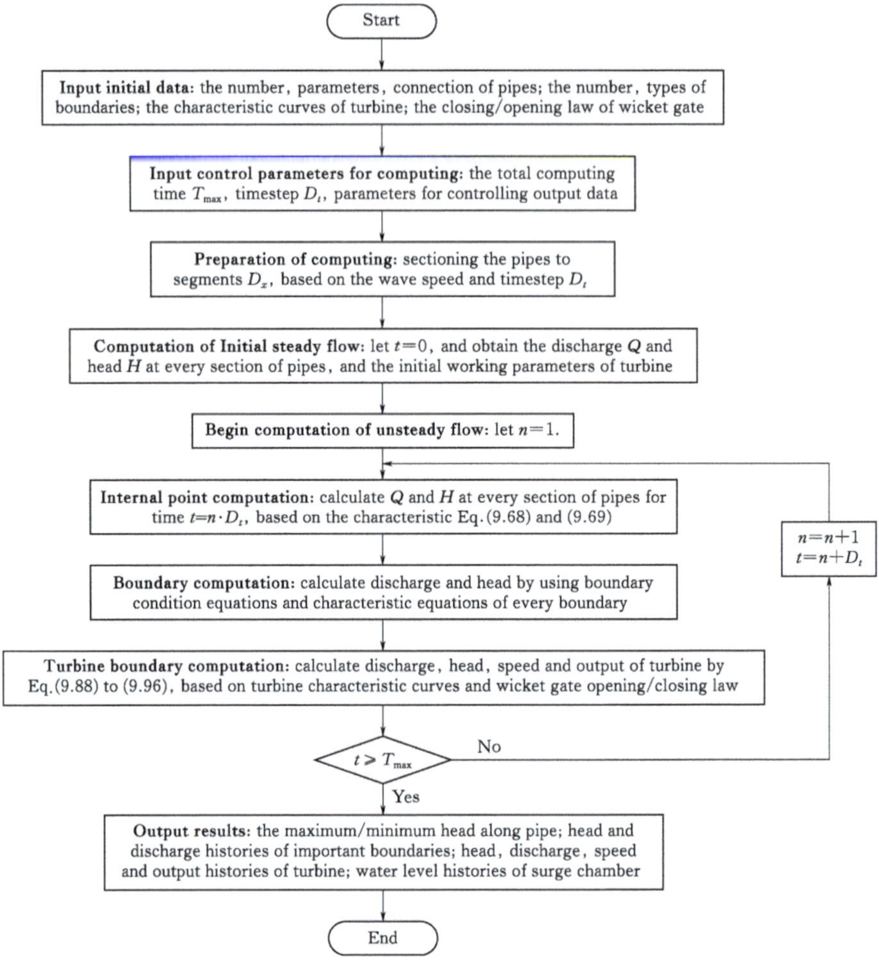

Fig. 9.26 Simulation procedures

as shown in Fig. 9.28. The extremes of the parameters can also be denoted in the figure.

The envelope lines of the maximum and minimum pressures (or piezometric heads) together with the pipeline elevation line are normally drawn along the headrace pipes and tailrace pipes, as shown in Fig. 9.29.

The histories of surge chamber water level and pressure difference across the bottom plate are normally drawn in a figure like Fig. 9.30. If necessary, the inflow and outflow rates can also be displayed in this figure.

In addition, the histories of pressure and discharge at any concerned point in the hydraulic system can also be displayed in proper styles.

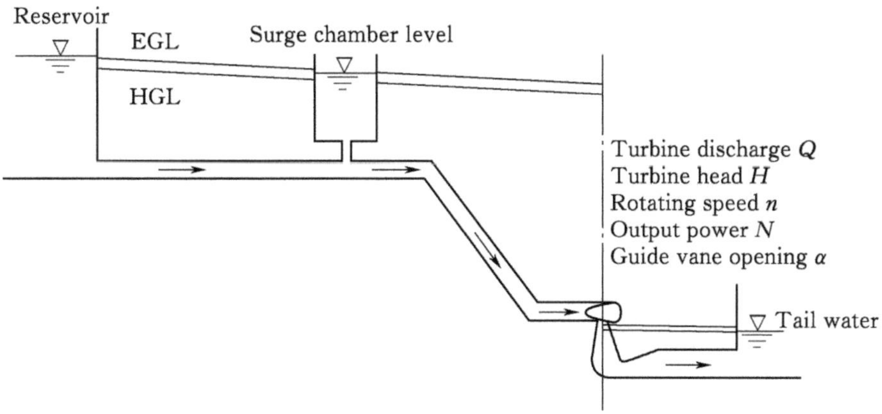

Fig. 9.27 Schematic of displaying the steady state results

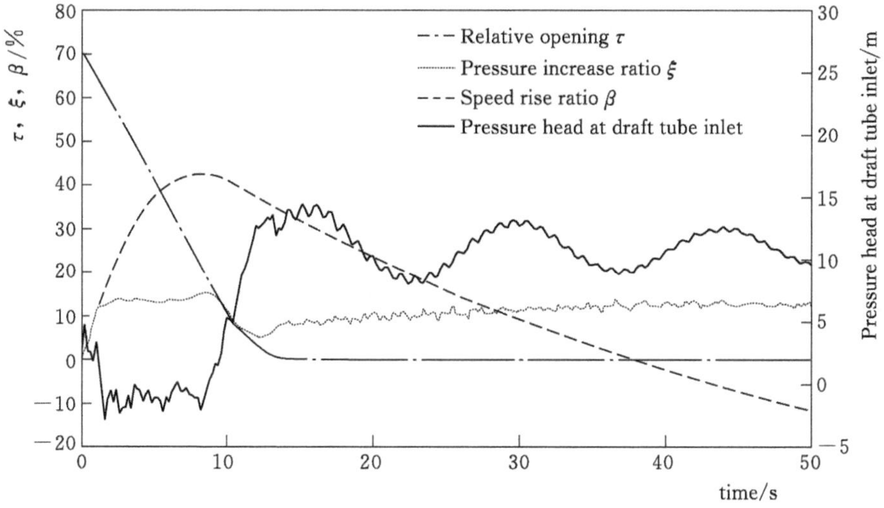

Fig. 9.28 Histories of the guaranteed regulation parameters of a practical hydropower station

The extremes of the above variables and their happening times can be listed in tables for easy lookup and analysis, like Table 9.2.

9.4.4.2 Analysis of Results

Under different operating conditions and for different hydropower stations, the calculation results of guaranteed regulation must be different. However, some general features reflecting the inherent physical laws exist. The deep understanding of the

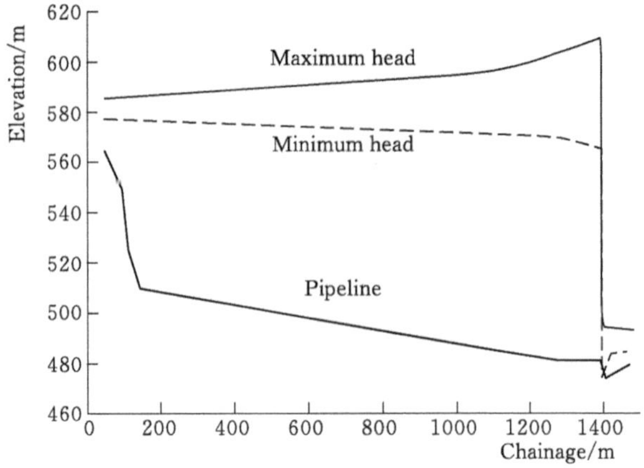

Fig. 9.29 Envelope lines of the maximum and minimum piezometric heads along the pipeline

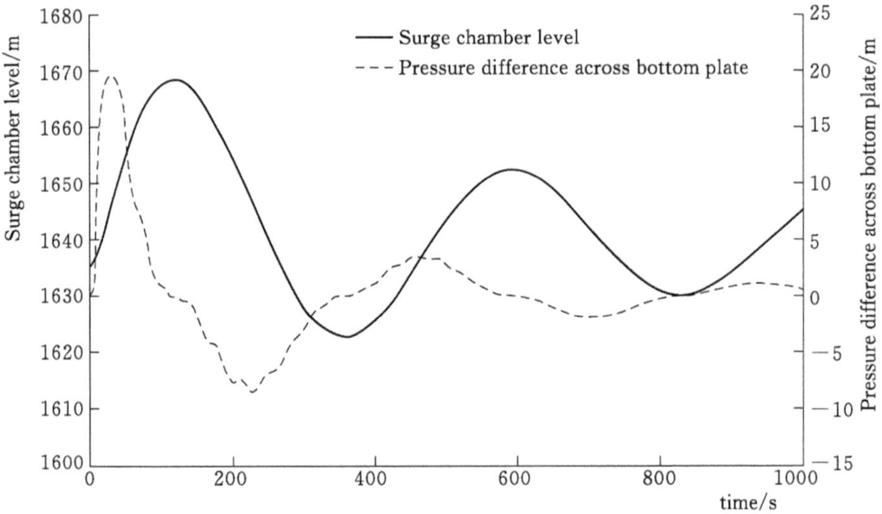

Fig. 9.30 Surge tank water level and pressure difference across the bottom plate

commonness is not only conducive to the analysis and judgment of the calculation results, but also conducive to the adoption of effective engineering measures to improve the guaranteed regulation parameters and further ensure the safe operation of hydropower stations.

When the pipeline system is determined, the opening and closing laws of wicket gate plays the most important role in regulating the guaranteed parameters and the distribution of water hammer pressure along the pipeline. The basic rule is that the shorter the closing time of wicket gate, the larger the water hammer pressure and the smaller the speed rise ratio. On the contrary, the longer the closing time of wicket

Table 9.2 Extremes of key variables in the calculation of guaranteed regulation

Variables	Values	Happing time (s)	Corresponding work condition
Maximum water hammer pressure at the spiral casing outlet/m			
Minimum water hammer at the draft tube inlet/m			
Maximum speed rise ratio/%			
Maximum surge chamber level/m			
Minimum surge chamber level/m			
Minimum pressure difference across surge chamber bottom plate/m			

gate, the smaller the water hammer pressure and the larger the speed rise ratio. There is also a similar rule when choosing piecewise closing laws of wicket gate.

When the opening and closing laws of wicket gate are determined, the envelopes of the maximum and minimum pressures along the headrace pipes, together with the maximum and minimum water hammer pressures in the spiral casing depend mainly on the flow inertia in the headrace pipes, namely the pipe characteristic coefficient $\rho_u = \frac{\sum_{i=1}^{n} a_i V_i}{2gH_0}$ and water hammer characteristic coefficient $\sigma_u = \frac{\sum_{i=1}^{n} L_i V_i}{gH_0 T_s}$. They are also related to the discharge characteristic curves of turbine, but almost independent of the inertia ρ_d and σ_d in the tailrace pipes. Similarly, the envelopes of the maximum and minimum pressures along the tailrace pipes, together with the maximum and minimum water hammer pressures in the draft tube depend mainly on the flow inertia in the tailrace pipes, and almost independent of the flow inertia in the headrace pipes.

Furthermore, three hints may be concluded: (1) water hammer pressure mainly depends on the water flow inertia in the pipeline, therefore, numerical solutions should be basically consistent with analytical solutions; (2) In the absence of real turbine characteristic curves, the similar turbine characteristic curves can be used in numerical calculation of guaranteed regulation, which will not cause obvious deviation in water hammer pressure and will not affect the scheme of bidding design and primary design of hydropower stations; (3) To control the maximum pressure at the end of spiral casing or the maximum vacuum at the inlet of draft tube, optimizing the layout and sizes of the headrace pipes or the tailrace pipes is the basic means, because other engineering measures have not obvious effects.

If reasonable laws of opening and closing of wicket gate are chosen, the characteristics of water hammer pressure of a high head hydropower station should be similar to those of the first phase water hammer. That is, the maximum water hammer pressure at the spiral casing end occurs during the closing process, and the occurrence time is in the early period; the envelope lines of the maximum and minimum pressures along the pipeline are close to the quadratic curves. For middle and low head hydropower stations, the characteristics of water hammer should be similar to those of the last phase water hammer, in which the maximum water hammer pressure at the spiral casing end occurs at the ending time of wicket gate closure and the envelope lines of the maximum and minimum pressures along the pipeline are linear curves.

When wicket gate is closed by linear laws, whether analytical or numerical calculation is conducted, the resulting water hammer pressure will oscillate back and forth, and the attenuation will be very slow. However, for practical hydropower stations, there is no such obvious oscillating phenomenon of water hammer pressure in prototype tests, because the real closing laws are not linear but have a short buffer before the wicket gate is closed (Fig. 9.7) and there is a certain flow leakage when the opening is zero. Therefore, to eliminate this specific phenomenon after wicket gate is closed, we can simulate the load rejection transients by giving a short buffer in the closing law or leaving a very small unclosed opening.

The maximum value of speed rise occurs during the closing of wicket gate. That is, the speed rising time T_n is smaller than the closing time of wicket gate. When the closing law of wicket gate is given, the maximum speed rise ratio mainly depends on the torque characteristics of turbine, and influenced by water hammer pressure. Normally, the higher the efficiency of turbine, the larger the maximum speed rise ratio, and the larger the pipeline water inertia, the greater the maximum speed rise ratio. Therefore, the deviations between analytical calculations and numerical calculations are usually obvious, and the maximum speed rise ratio of a load rejection transient process obtained by an analytical calculation is often smaller than that by a numerical simulation.

Whether the calculation results are reasonable depends not only on the correctness of input parameters, but also on the accuracy of computing schemes adopted by simulation program. For example, in some special cases (such as unreasonable treatment of turbine characteristic curves), the obtained root of the high-order equations has no physical meanings. The resulting spurious numerical disturbance will affect the subsequent transient processes and lead to unreasonable results.

The unreasonable computing results are normally as follows: the history curves of pressure, discharge, rotational speed, pressure difference and other variables fluctuate with strange jumps or undamped oscillations showing timestep scale zigzaging, which have magnitudes close to or far beyond the extreme values and deviate from the occurrence times and laws of the extreme values. In this situation, the trajectory of working point in the characteristics chart of turbine should be checked. If the trajectory line is beating or folding, rather than smoothly sliding, the calculation result is usually wrong and should not be adopted.

9.4.5 Simulation Example of Water Hammer in Single Pipe

9.4.5.1 Software Interface and Usage Instructions

The teaching demonstration software described below was programmed under the Delphi 4.0 of object-oriented programming environment and can be run directly in windows environment. It can simulate and display the water hammer propagation and reflection phenomena in the “reservoir-pipe-valve” system, namely a single pipe system. The simulation functions can meet the preliminary design requirements of hydropower station, and the demonstration functions is effective for deepening students’ understanding of the physical essence of water hammer. Corresponding to this textbook, the software can be used to analyze and demonstrate the following knowledge points: (1) propagation and reflection of water hammer wave; (2) the direct and indirect water hammer types; (3) the first phase and last phase water hammer types; (4) the influence of initial valve opening to water hammer history and extreme values; (5) the influence of valve closing and opening laws to water hammer history and extreme values; (6) the water hammer history after valve operation.

The interface for input is shown in Fig. 9.31. The operation steps are as follows. First, we input the primary parameters of the “reservoir-pipe-valve” system, and press button【数据确定】to accept, and further press button【进行计算】 to calculate. Second, click the heading tab “波动过程” to show the display and output interface (Fig. 9.32), press button 【开始演示】 to display the animation of water hammer propagation and reflection process, after which the extreme values and the envelope lines are printed on the figures. Third, press button【打印屏幕】 to copy the final result interface as Bitmap format like Fig. 9.32, or press button 【打印曲线】 to draw the pressure history at valve and the envelope lines of the maximum and minimum pressures along the pipe in formal style, like Figs. 9.33, 9.34 and 9.35.

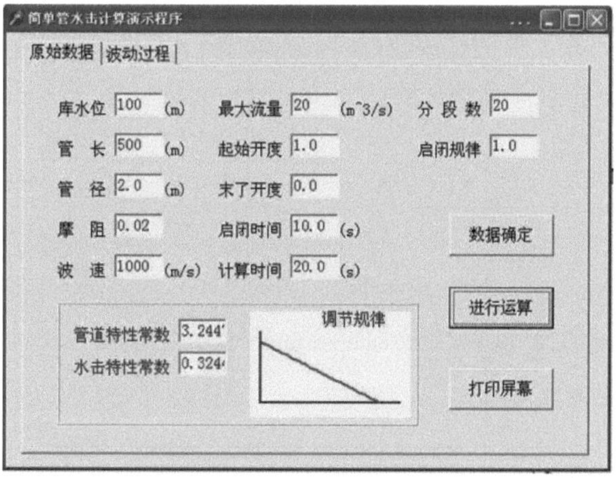

Fig. 9.31 Interface for parameter input

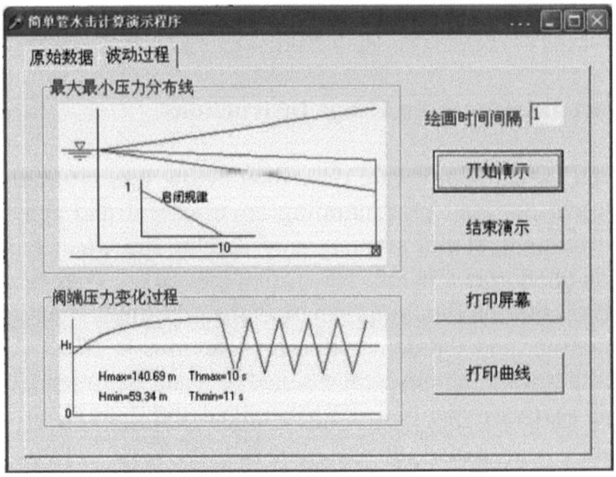

Fig. 9.32 Interface for displaying results

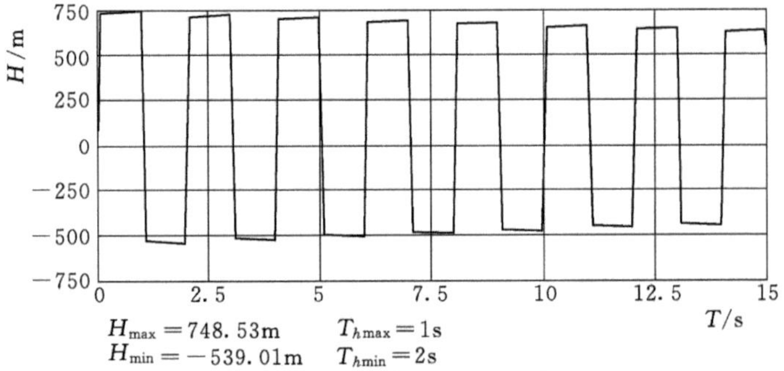

(a) Pressure history at the valve

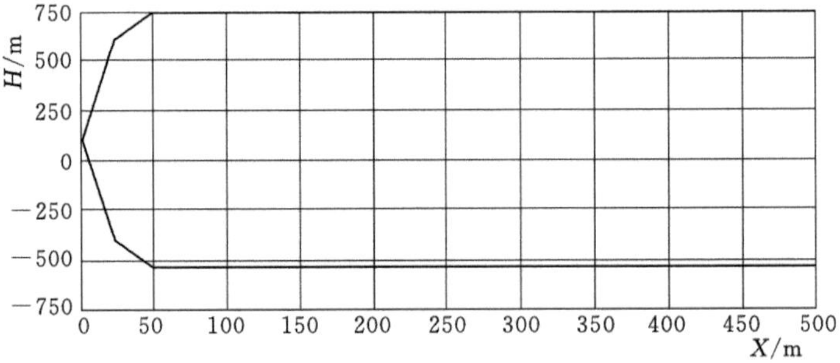

(b) Envelope lines of the maximum and minimum pressures along the pipe

Fig. 9.33 Direct water hammer results of the single pipe water hammer problem

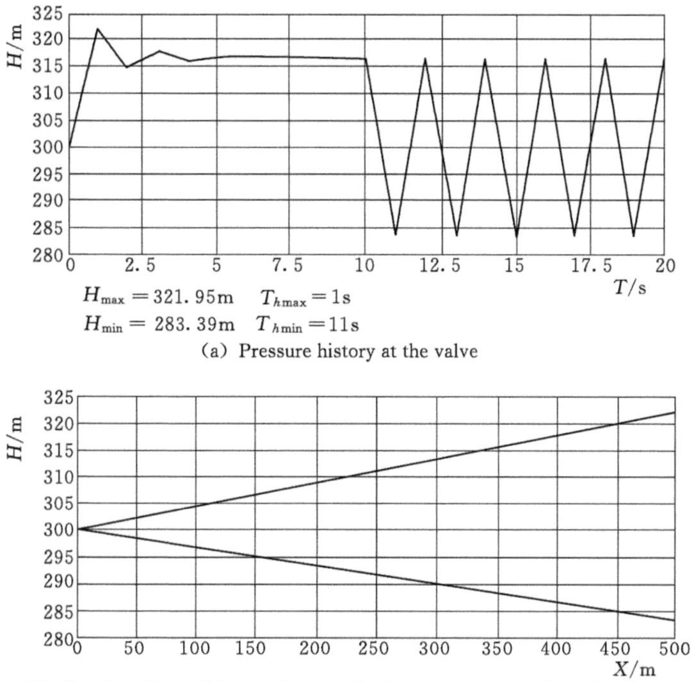

(a) Pressure history at the valve

(b) Envelope lines of the maximum and minimum pressures along the pipe

Fig. 9.34 First phase water hammer results of the single pipe water hammer problem

9.4.5.2 Parameter Definitions and Original Codes

(1) Parameter definitions

Hr	Reservoir water level with the valve elevation as datum, m;
H0	Piezometric head at the valve with the valve elevation as datum, m, corresponding to Q0;
Hi	Initial head when the transient simulation begins, m;
Q0	Maximum discharge in the pipe, m^3/s, corresponding to the maximum hydraulic opening of valve, Tau = 1;
Qi	Initial discharge when the transient simulation begins, corresponding to H_i, m^3/s;
A	Wave speed in the pipe, m/s;
F	Darcy-Weisbach frictional factor, corresponding to f in Eq. (9.8);
Xl	Length of the single pipe, m;
D	Diameter of the single pipe, m;
Tmax	Physical problem time for simulation, s;
N	Segment number of spatial discretization;
Dt	Time step,s, with Dt = Xl/N/A;
Tc	Time of valve operation, s;

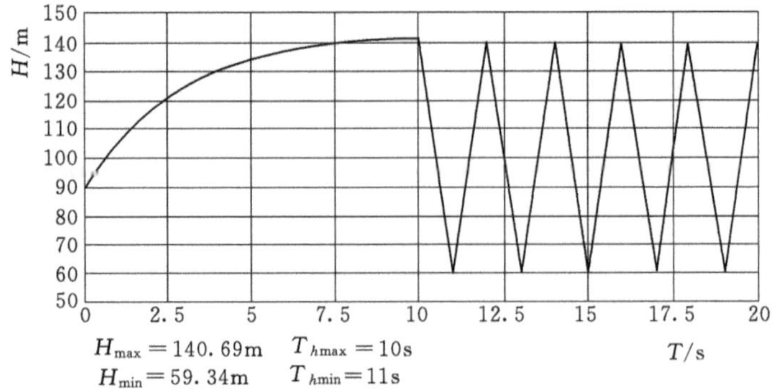

(a) Pressure history at the valve

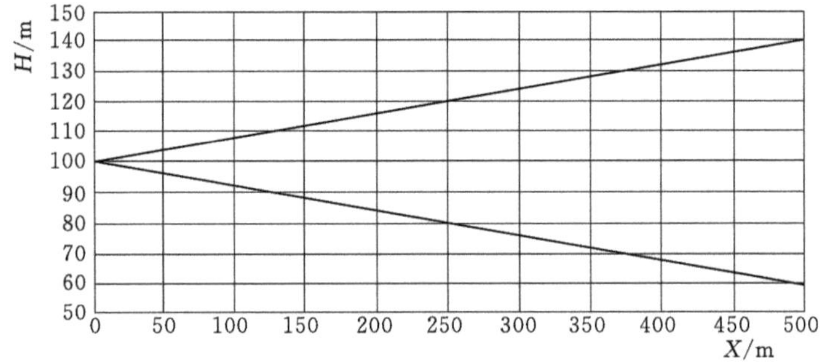

(b) Envelope lines of the maximum and minimum pressures along the pipe

Fig. 9.35 Last phase water hammer results of the single pipe water hammer problem

Ti	Initial hydraulic opening;
Tf	Final hydraulic opening;
Em	Control factor for opening law, with Em = 1 for linear law, Em < 1 for first fast late slow law, and Em > 1for first slow late fast law;
Tau	Relative hydraulic opening of valve at time T, displayed as τ in equation $\tau = \phi\omega/\phi_0\omega_{\max}$. The value can be determined by $T_{au} = T_i + (T_f - T_i)\left(\frac{T}{T_c}\right)^{Em}$, in which Ti, Tf, and Em are defined above;
Hmax	Maximum head at the valve, m;
Hmin	Minimum head at the valve, m;
Thmax	The time when the maximum head occurs, s;
Thmin	The time when the minimum head occurs, s.

(2) Original codes in Delphi 4.0 format

```
Var    /*Define Variables*/
  i,k,N, Kmax: integer;
   Hr,Xl,D,A,G,F,Q0,Tmax,Ti,Tf,Tc,Em,Ru,H0,Hi,Qi,Dt,Cvp,Tau,T,Cv,Thmax,Thmin,Qcp,Qcm,Cqp,Cqm,C,S,R:
   double;
  H,Q: array[0..40,0..10000] of double;
  Hmax,Hmin: array[0..40] of double;
begin     /*Beginning of the Simulation*/
  G:=9.81;
  Dt:=Xl/(A*N);
  S:=F*Dt/(2.0*D*0.7854*D*D);
  C:=(G*0.7854*D*D)/A;
  Kmax:=Round(Tmax/Dt)+1;
/*Steady State Simulation at T=0, for Obtaining the Discharge and Head at Each Node along the Pipe and
the Initial Opening of the Valve. */
  R:=F*Xl/(2.0*G*D*D*D*D*D*0.7854*0.7854*N);
  H0:=Hr-R*N*Q0*Q0;
  Hi:=Hr/(1.0+R*N*Q0*Q0*Ti*Ti/H0);
  Qi:=Sqrt(Hr*Q0*Q0*Ti*Ti/(R*N*Q0*Q0*Ti*Ti+H0));
  for i:=0 to N do
    begin
      H[i,0]:=Hr-i*R*Qi*Qi;
      Q[i,0]:=Qi;
      Hmax[i]:=H[i,0];
      Hmin[i]:=H[i,0];
        end;
      Tau:=Ti;
    /*Beginning of Transient Simulation*/
      for k:=1 to Kmax do
        begin
          T:=k*Dt;
   /*Calculations for Inner Nodes. Using Characteristics Equations (9-68) and (9-69) to Calculate the Discharge
and Head at Each Inner Nodes, For Time T:=k*Dt. */
          for i:=1 to N-1 do
            begin
              Cqp:=C/(1+S*ABS(Q[i-1,k-1]));
              Cqm:=C/(1+S*ABS(Q[i+1,k-1]));
              Qcp:=Q[i-1,k-1]/(1+S*ABS(Q[i-1,k-1]))+Cqp*H[i-1,k-1];
              Qcm:=Q[i+1,k-1]/(1+S*ABS(Q[i+1,k-1]))-Cqm*H[i+1,k-1];
              H[i,k]:=(Qcp-Qcm)/(Cqp+Cqm);
              Q[i,k]:=Qcm+Cqm*H[i,k];
            end;
```

```
/*Calculations for the Reservoir Boundary Node. Obtain the Discharge and Head by Solving Equations (9-70)
and (9-69) */
            Cqm:=C/(1+S*ABS(Q[1,k-1]));
            Qcm:=Q[1,k-1]/(1+S*ABS(Q[1,k-1]))-Cqm*H[1,k-1];
            H[0,k]:=Hr;
            Q[0,k]:=Qcm+Cqm*H[0,k];
/* Calculations for the Valve Boundary Node. Obtain the Discharge and Head by Solving Equations (9-86)
and (9-68) */
             if (T<Tc) then
               Tau:=Ti-(Ti-Tf)*Exp(Em*ln(T/Tc))
             else
               Tau:=Tf;
             Cv:=Tau*Tau*Q0*Q0/H0;
             Cqp:=C/(1+S*ABS(Q[N-1,k-1]));
             Qcp:=Q[N-1,k-1]/(1+S*ABS(Q[N-1,k-1]))+Cqp*H[N-1,k-1];
             H[N,k]:=(Qcp*Cqp+Cv/2.0-Sqrt(Cv*Cv/4.0+Qcp*Cqp*Cv))/(Cqp*Cqp);
             Q[N,k]:=Qcp-Cqp*H[N,k];
        /*Find the Maximum and Minimum Heads and Their Happing Times */
             if (H[N,k]>Hmax[N])then THmax:=T;
             if (H[N,k]<Hmin[N])then THmin:=T;
        /*Determine the Envelope Lines of the Maximum and Minimum Heads along the Pipe */
             for i:=0 to N do
               begin
                 if (H[i,k]>Hmax[i])then Hmax[i]:=H[i,k];
                 if (H[i,k]<Hmin[i])then Hmin[i]:=H[i,k];
                 end;
               end;
        end;
        /*Simulation Finish, and Output Results */
```

9.4.5.3 Simulation Examples

(1) **Direct water hammer**

a. Input Data: $Hr = 100\,\text{m}$; $X1 = 500\,\text{m}$; $D = 2\,\text{m}$; $F = 0.02$; $A = 1000\,\text{m}$; $Q0 = 20\,\text{m}^3/\text{s}$; $Ti = 1.0$; $Tf = 0.0$; $Tc = 0.1$; $T\max = 20\,\text{s}$; $N = 20$; $Em = 1.0$.
b. The printed pressure history at the valve and the maximum and minimum pressure envelopes along the pipe are shown in Fig. 9.33.

(2) **First phase water hammer**

a. Input data: $Hr = 300\,\text{m}$; $X1 = 500\,\text{m}$; $D = 2\text{m}$; $F = 0.00002$; $A = 1000\,\text{m}$; $Q0 = 10\,\text{m}^3/\text{s}$; $Ti = 1.0$; $Tf = 0.0$; $Tc = 10$; $T\max = 20\,\text{s}$; $N = 20$; $Em = 1.0$.

b. The printed pressure history at the valve and the maximum and minimum pressure envelopes along the pipe are shown in Fig. 9.34.

(3) **Last phase water hammer**

a. Input data: $Hr = 100\,\text{m}$; $X1 = 500\,\text{m}$; $D = 2\,\text{m}$; $F = 0.02$; $A = 1000\,\text{m}$; $Q0 = 20\,\text{m}^3/\text{s}$; $Ti = 1.0$; $Tf = 0.0$; $Tc = 10$; $T\max = 20\,\text{s}$; $N = 20$; $Em = 1.0$.
b. The printed pressure history at the valve and the maximum and minimum pressure envelopes along the pipe are shown in Fig. 9.35.

9.5 Calculation Conditions and Improvement Measures for Guaranteed Regulation

9.5.1 Tasks of Calculation of Guaranteed Regulation

As stated in the introduction section of this chapter, the tasks of calculating the water hammer pressure and rotational speed rise, optimizing the operating laws of wicket gate, and coordinating their contradictions to keep both water hammer pressure and rotational speed rise in the allowable ranges, are called the calculation of guaranteed regulation (CGR).

The detailed tasks of CGR can be summarized as follows.

(1) Calculate the maximum pressures in penstock and spiral casing, and provide the basic loads for strength design and strength check of penstock, spiral casing and turbine.
(2) Calculate the minimum pressures in penstock and draft tube, and provide the basic data for playout and negative pressure protection of penstock and vacuum check of draft tube.
(3) Calculate the maximum rotational speed rise ratio, and check whether it is within reasonable or allowable limit.
(4) Appropriately select the closing and opening times and laws of wicket gate to make water hammer pressure and speed rise ratio within allowable ranges, and to control water hammer pressure as small as possible to reduce project investment.
(5) In case of the water hammer pressure and speed rise ratio cannot meet the requirements at the same time, find economic and effective measures to solve the contradiction.

9.5.2 Critical Conditions for Calculation of Guaranteed Regulation

In practical operations, turbine units always work in different conditions, namely the water heads, discharges, output powers, and guide vane openings are different. Obviously, not all load rejection conditions experience the most critical values of water hammer pressure and rotational speed rise, even if their guide vane closing laws are the same. Therefore, to minimize the calculation workload while do not miss the most critical values, we need to figure out the most critical conditions.

First, we should know that the maximum and minimum pressures in the hydraulic system are the summation of the static pressure and water hammer pressure. The static pressure is determined by the reservoir water level or tailrace water level, and the water hammer pressure is related to the water head, discharge, regulation time and laws.

Second, it should be noted that the water hammer pressure is also related to the layout of the hydraulic system and the operation schedule of different working conditions or different units.

Therefore, finding the most critical conditions is a complicate task that needs trial and error process.

For the one-pipe-one-unit hydraulic system shown in Fig. 9.36a, the critical conditions are normally as follows.

(1) **The maximum rotational speed rise**. It normally happens in the full load rejection from the rated operating point with rated water head. The reasons are as follows. At the rated operating point with the rated head and rated power, the discharge and wicket gate opening are the largest; after load rejection, the closing time of wicket gate is the longest, and the temporary surplus output power of the turbine is the largest, therefore the speed rise should be the largest.

(2) **The maximum transient pressure head in the spiral casing and penstocks**. During hydraulic transients, the transient pressure head at any location is composed of two parts: the steady head in the previous steady operating condition and the water hammer head produced by the transient process. Therefore, the maximum transient pressure heads in the spiral casing and penstock normally happen in the load rejection condition under the rated power with the rated head

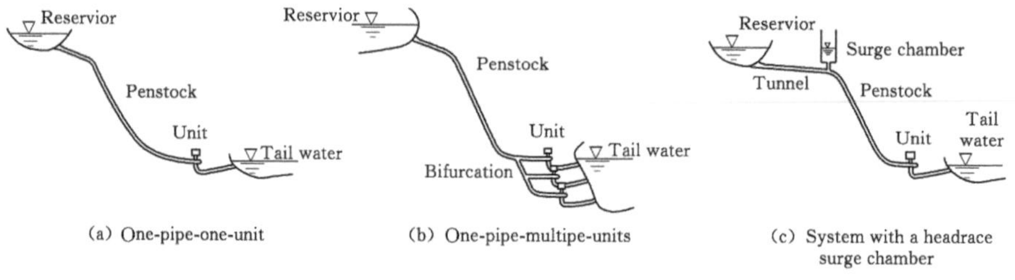

Fig. 9.36 Schematic of typical hydraulic systems

or largest head, or under the largest power, and in the condition of reservoir normal level or higher levels. The reasons are as follows. In the condition of reservoir normal level or higher levels, the steady pressure head in spiral casing and penstock are normally the largest; under the rated power with the rated head or largest head, or under the largest power, the water hammer pressure head is normally the largest; the summation of the static and dynamic pressures might be the largest.

Note that: with the same head and reservoir level, partial load rejection (for instance 70–90% of the rated power) may result larger water hammer pressure head than full load rejection does. The reason is that the changing rate of discharge in the partial load rejection condition may be larger than that of the full load rejection condition. However, considering this possibility in the preliminary design phase is not necessary, because the differences are normally small.

(3) **The minimum transient pressure head in the draft tube and its extension conduit**. It normally happens in the full load rejection condition under the rated head or largest head, and in the lowest tail water level condition, which may be obtained based on the relation curve of tail water level and discharge. The reasons are as follows. The steady pressures in the draft tube and its extension conduit are the lowest in the lowest tail water level condition; the negative water hammer pressure head under the rated head or largest head is the relative larger or largest; the summation of these two parts always results in the lowest transient pressure head.

Similarly, cautions should be paid to the partial load rejection conditions (for instance 70–90% rated power) because the negative water hammer pressure head may be larger.

(4) **The minimum transient pressure head in the spiral casing and penstocks**. It normally happens in the load acceptance transient process under the rated head or a lower head, in the condition of reservoir dead level or low levels. The reasons are as follows. In the condition of reservoir dead level or low levels, the steady pressure heads in the spiral casing and penstock are normally relatively small; the negative water hammer pressure head is always relatively larger during the load acceptance transient process under the rated head or a lower head; the summation of these two parts always results in the lowest transient pressure head

(5) **The maximum transient pressure head in the draft tube and its extension duct**. It normally happens in the load acceptance transient process under the rated head or a lower head, in the condition of the design tail water level or highest operational tail water level. Because the transient pressure heads in the draft tube and its extension conduit are much smaller than those in the spiral casing and penstock, the structure strength is easy to meet design standards, and we normally do not analyze this maximum pressure head in CGR. But in the special case that the tailrace conduit is long and the unit weight is light, the maximum transient pressure head in the draft tube should be considered to prevent the unit-lifting accident.

For the one-pipe-multiple-unit hydraulic system (Fig. 9.36b) and the system with surge chambers (Fig. 9.36c), the conditions in CGR are more complicated.

The complications are because of the superposition of multiple waves and the discharge shifts between the units sharing one pipe. For these systems, not only the combinations between the reservoir water level and tail water level, and the simultaneous load rejections of the units in the common hydraulic system should be considered, but also the successive load rejections or load acceptances, or load rejection after load acceptance, and the superposition of water hammer and surge wave should be considered. If other conditions are the same, the critical results of CGR in the systems sharing common pipe (Fig. 9.36b) or having surge chamber (Fig. 9.36c) are larger than those in the simple system (Fig. 9.36a).

It should be noted that, because every hydropower station has unique conditions and requirements for operation, the selection of calculation conditions for the critical parameters should be cautious and based on specific circumstances.

9.5.3 Measures to Solve the Contradiction in Guaranteed Regulation

Reducing water hammer pressure is of great significance to reduce the costs of water conveyance system and turbine-generator units and to improve the operating conditions of the units. In normal situations, water hammer pressure and speed rise ratio must meet the guaranteed values specified in the standards. The main measures to reduce water hammer pressure are as follows.

9.5.3.1 Choosing Appropriate Regulation Laws

Figure 9.11 shows the effect of different valve opening laws on water hammer history and extreme value. It can be seen that using appropriate regulation law can effectively reduce the extreme value of water hammer pressure. Although linear laws are simple (Fig. 9.37a), piecewise closing laws (Fig. 9.37b) is often used in engineering to effectively reduce water hammer pressure. In middle and low head hydropower stations, the last phase water hammer generally appears, the first-faster-late-slower closing laws should be reasonably adopted; in high head hydropower stations, the first phase water hammer generally appears, the first-slower-late-faster closing laws should be reasonably adopted. Selecting appropriate regulation laws to effectively reduce water hammer pressure is the priority measure, because it is simple and easy in implementation without increasing too much cost.

In practice, the piecewise linear closing laws are normally used. The basic principles for optimizing the laws are as follows.

(1) On the premise that the maximum speed rise ratio is slightly smaller than or equal to the allowable value specified by standards, and the closing time of

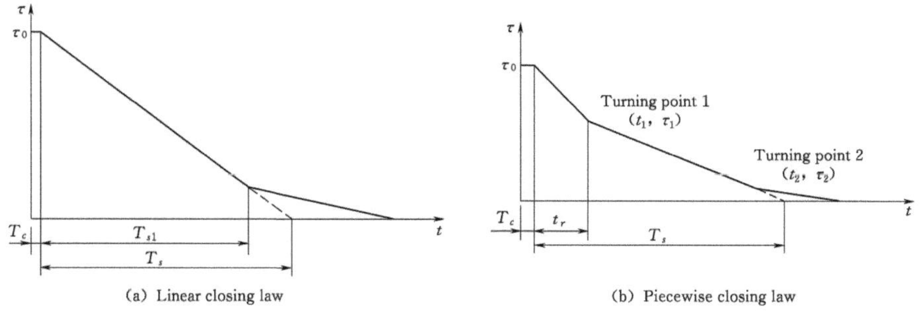

Fig. 9.37 Schematic of linear and piecewise linear closing law

wicket gate should be prolonged as far as possible to reduce water hammer pressure.

(2) Select appropriate turning points to stagger the occurrence times of the maximum water hammer pressure and the maximum speed rise ratio, so that both parameters can meet the requirements of standards. As shown in Fig. 9.37b, normally, the first turning point is located at time $t = T_c + t_r$, in which T_c and t_r are the delay time of closing law and the phase time of penstock, respectively. The second turning point is selected to meet the speed-no-load opening of wicket gate.

9.5.3.2 Shortening Penstock Length

Shortening penstock length can let the reflected wave from the reservoir travel back to the penstock end earlier, thus water hammer value can be reduced. Investigating the water hammer coefficient $\sigma = \frac{LV_{\max}}{gH_0T_s}$, we know that reducing L can get smaller σ; base on the approximate equations $\xi_1^A = \frac{2\sigma}{1+\rho\tau_0-\sigma}$ and $\xi_m^A = \frac{2\sigma}{2-\sigma}$, smaller σ will give smaller ξ_1 or ξ_m. Therefore, as long as the terrain, geological conditions are permitted, adopting shorter penstock layout is the priority.

9.5.3.3 Reducing the Velocity in Penstock

Reducing penstock velocity means reducing the momentum of water in penstock, and smaller momentum will result in smaller water hammer. However, reducing velocity means increasing penstock diameter because the discharge of a hydropower station is given, and the diameter increase may lead to a cost increase. Under certain conditions, if a proper increase of penstock diameter can cancel the need for a surge chamber, then this measure in reasonable.

9.5.3.4 Reducing the Changing Rate of Penstock Velocity

To reduce the changing rate of penstock velocity, the most direct method is to increase wicket gate closing time T_s, however, this further leads to the increase of rotational speed rise ratio β. Therefore, the increase of T_s is always limited. In order to resolve this contradiction, the following measures can be taken.

(1) Setting a pressure relief valve. This valve is a bypass drain device installed in the spiral casing of a reaction turbine, as shown in Fig. 9.38. After load is rejected, the governor automatically closes the wicket gate with given law and time to control the maximum speed rise ratio β_{max} within allowable limit. In the same time, the governor automatically opens the pressure relief valve to release a part of the flow to the tailrace pipe. Releasing water can reduce the changing rate of penstock velocity, and further reduce water hammer in the penstock and spiral casing. After the wicket gate is closed, the pressure relief valve is closed slowly at a speed controlling the subsequent water hammer pressure within the allowable value.

Compared with setting a surge chamber, adopting a pressure relief valve is cheaper. However, the pressure relief valve does not work the during load increase and normal operation conditions, and cannot improve the operation stability of the unit. In addition, the size and cost of powerhouse may need increase. Furthermore, when load change is small (lower than 15% of the rated power), the pressure relief valve does not act, and the wicket gate should close alone with slow speed, which deteriorates the stability of the unit. For high head hydropower stations with few units and small discharges, if setting pressure relief valves can economically cancel setting surge chamber, the scheme is reasonable.

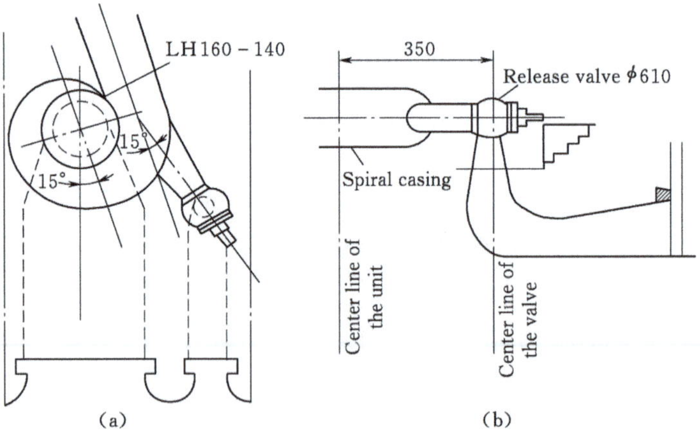

Fig. 9.38 Schematic of a pressure relief valve (unit in cm)

(2) Setting water resistor. Water resistor is the device that uses water resistance to consume energy. It is connected to generator busbar, and operated by governor based on the law linked to the closing law of wicket gate. After load is rejected, the governor automatically connects the generator to the water resistor, which works as a load by consuming power in water resistance. This makes the slow closing of wicket gate possible, which can control both water hammer and speed rise ratio in allowable values. Water resistor is cheap in price, convenient in manufacture, simple in installation, but poor in reliability. In addition, water resistor cannot work during load-increase conditions and short-circuit accidents. Therefore, this means only suitable for small hydropower stations.

(3) Setting deflector for Pelton turbine. Deflector is a device installed behind the needle valves of a Pelton turbine for changing jet directions after load rejection. The governor can operate the reflectors within 1 to 2 s to prevent strong jets striking on the runner, and further control speed rise ratio. With deflectors for controlling speed rise ratio, the needle valves can close slowly for controlling water hammer. Deflector is simple in structure, cheap in price, and no need in enlarging powerhouse. However, reflector does not work in load-increase conditions.

9.5.3.5 Increasing the Rotational Inertia of Unit in Reasonable Range

The larger the flywheel moment GD^2, the smaller the maximum speed rise ratio β_{max}. With GD^2 increased, the closing time of wicket gate can be prolonged to reduce water hammer pressure. But with the development of science and technology, the rotational inertia of generator decreases gradually, therefore, increasing GD^2 normally means needing larger cost of generator.

9.5.3.6 Setting a Surge Chamber in Penstock

For the hydropower stations with long water conveyance systems, the above measures may not be effective and economical to make the CGR parameters meeting the specification requirements. In this situation, setting surge chambers in the water systems may be the feasible choice. Surge chamber can reflect water hammer waves by its large water surface, which can shorten penstock length and further reduce water hammer pressure. The contents about surge chamber will be discussed in Chap. 10.

Questions

1. Please analyze the propagation, reflection, and superposition processes in the single pipe (reservoir-single pipe-valve system) after a sudden closure of valve, and draw the histories of pressure head at the valve and velocity at the reservoir with a cycle $T = 4L/a$.
2. For the hydraulic system with upstream and downstream surge chambers, after a load is rejected, what kinds of water hammer will occur in the spiral casing and

in the draft tube, respectively? And what kinds of surge waves will occur in the upstream surge chamber and downstream, respectively? How will the rotational speed and discharge change? Please draw a schematic to show the history curves. How about the situation in load acceptance condition?

3. Why water hammer and rotational speed rise are contradictory in calculation of guaranteed regulation? Do you know the main measures to solve the contradiction?
4. In the analytical solution of water hammer, pipe characteristic factor ρ and water hammer characteristic factor σ are important. What about their physical meanings and effects in water hammer calculation?
5. The analytical solution of water hammer equations can be expressed as wave functions. How about the characteristics of wave functions? Please derive the water hammer reflection factor of a T-shaped junction or a partially opened value, by using the general solution equations in the form of wave functions.
6. Please derive the interlocking equations for indirect water hammer. Are there any requirements on the closing and opening procedures of valve?
7. How the first phase, last phase, and direct water hammer types are defined? Are their boundaries between them distinct and absolute? How about the influence of pipe characteristic factor ρ, water hammer characteristic factor σ, and valve opening τ?
8. How about the influence of valve opening/closing time on the history and maximum value of water hammer pressure? What about the features of valve closing procedures for different hydropower stations with different heads.

Bibliography

1. China Electricity Council. 2004. *Electrical-mechanical design code of hydropower plant DL/T 5186–2004*, 6. Beijing: China Electric Power Press.
2. Renkun, Wang and Z. Chunsheng. 2013. Design manual of hydraulic structures, 2nd ed., Vol. 8 Buildings of Hydropower Station. Beijing: China Water and Power Press, Beijing.
3. Mingdeng, Zhao, and Y. Zhonghua. 2021. *Hydraulics*, 1. Beijing: China Water and Power Press.
4. Jiandong, Yang, X. Ming, W. Hegao and C. Yongguang. 2017. Hydropower station, 3rd ed. Beijing: China Water and Power Press.
5. Jushan, Tang. 2007. *Typical design drawings of middle and small water conservancy and hydropower projects, water diversion buildings of hydropower stations, diversion tunnels and surge chambers.* Beijing: China Water and Power Press.
6. Кривченко, Г.И. 1981. Transient process in power equipment of hydropower station Translated by Chang Zhaotang. Beijing: Water Resources and Electric Power Press.
7. Rongqiao, Wu and C. Jianzhi. 1997. *Hydraulic transients of hydropower station.* Beijing: China Water and Power Press.
8. Jaeger C. 1987. *Fluid transients in hydroelectric engineering practice*, Translated by Wang Shuren et al. Dalian: Dalian Institute of Technology Press.
9. Wylie E.B. and V.L. Streeter. 1993. *Fluid transients in systems*, 3rd ed. Prentice Hall Englewood Cliffs.
10. Chaudhry M.H. 2014. *Applied hydraulic transients*, 3rd ed. New York: Springer.

Chapter 10
Surge Chamber

Yongguang Cheng

Abstract In this chapter, we will introduce the functions, types, working principles, setting conditions and layouts of surge chambers. Then we will discuss the basic equations of surge chamber waves, and their analytical and numerical solution methods. Further, we will analyze the stability problem of surge chamber wave oscillations. Students are encouraged to schematically draw the water level oscillating histories in headrace and tailrace surge chambers of hydropower stations and pumped storage power stations, under typical transient conditions, such as load increase, load rejection, pump trip, and pump starting. And when the main parameters of a specific hydropower station are given, students should be able to decide whether a surge chamber is needed, select a proper surge chamber type, calculate the highest and lowest surge waves, analyze the stability of surge wave oscillation, and finally design the dimensions of the surge chamber.

Keyword Surge chamber · Functions of surge chamber · Setting condition of surge chamber · Calculation of surge chamber wave · Stability of surge chamber wave

10.1 Functions, Requirements, and Setting Conditions of Surge Chamber

10.1.1 Functions of Surge Chamber

As stated in Chap. 9, surge chamber is a typical water hammer pressure suppression building. To reduce the water hammer in headrace pipes, a headrace surge chamber can be set at the junction of headrace tunnel and penstock (Fig. 10.1); to control the draft tube vacuum, a tailrace surge chamber can be set at the junction of draft tube extension conduit and tailrace tunnel (Fig. 10.2). Taking advantage of the larger

Y. Cheng (✉)
Wuhan University, Wuhan, China
e-mail: ygcheng@whu.edu.cn

Y. Cheng et al. (eds.), *Hydropower Station*,
https://doi.org/10.1007/978-981-96-2383-9_10

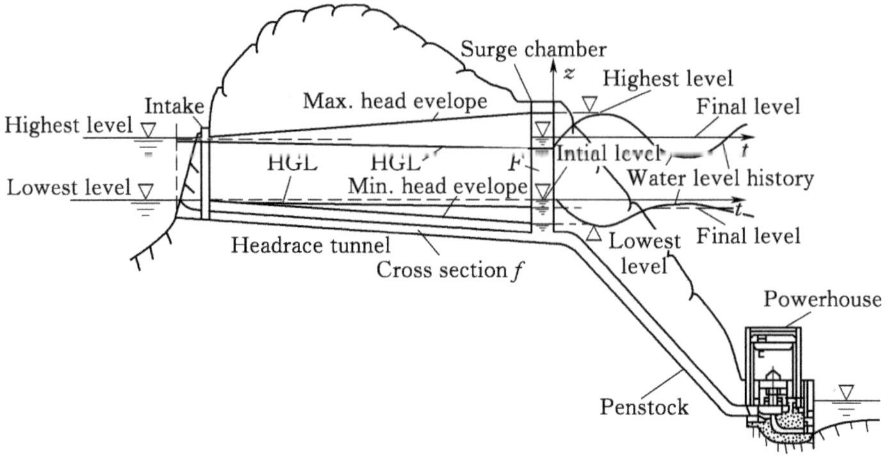

Fig. 10.1 Schematic of the hydraulic system with a headrace surge chamber

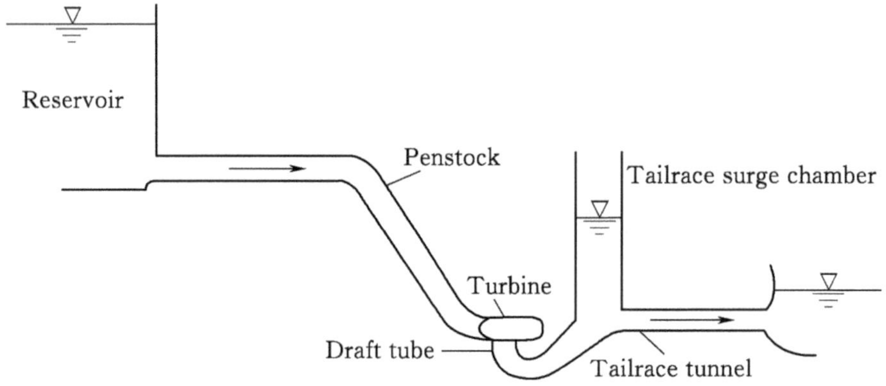

Fig. 10.2 Schematic of the hydraulic system with a tailrace surge chamber

cross section (compared to the connected pipes and tunnels) and free surface, a surge chamber can reflect water hammer waves (similar to a reservoir when the free surface is very large), which can not only reduce water hammer in the penstock and draft tube, but also prevent water hammer penetration into the headrace tunnel and tailrace tunnel. In addition, thanks to the decreases of water inertia in the penstock and draft tube extension conduit, the operation condition of the turbine-generator unit can be improved. Therefore, the main functions of surge chamber can be expressed as follows.

(1) Improve the calculation of guaranteed regulation (CGR) parameters by reflecting water hammer and shortening the lengths of penstock and draft tube extension conduit.

(2) Improve the operation conditions of unit and provide qualified electricity for the grid, in peak load fitting and frequency regulating.

In one word, surge chambers play an important role in keeping the stability, safety, and regulation quality of hydropower stations that have long hydraulic systems.

10.1.2 Requirements for a Surge Chamber

To better play its roles, the design of a surge chamber should meet the basic requirements as follows:

(1) Setting the surge chamber as closer as possible to the powerhouse, to shorten the length of penstock or draft tube extension conduit. The shorter the penstock/draft tube extension conduit, the smaller the water hammer head at the spiral case/draft tube inlet.
(2) A surge chamber should have good reflection characteristics and can fully reflect the water hammer waves coming from the penstock/draft tube extension conduit, to avoid as far as possible the water hammer penetration into the headrace/tailrace tunnel.
(3) The working state of surge chamber must be stable, and the water level oscillations after load changes must damp soon. Otherwise, the stable operations of hydropower stations will be endangered and even the power grids will be influenced.
(4) In normal working conditions of units, the head loss under the surge chamber should be small, which is beneficial to the economic operation of hydropower stations.
(5) Reasonable type, safe structure, convenient construction, and low cost are also important for surge chamber design.

10.1.3 Setting Conditions for a Surge Chamber

Whether a surge chamber should be set is influenced by technological and economic conditions. Here, we only discuss technical issues. Based on the operation goal of safe, stable, and good regulating quality, the criteria are classified as three types.

(1) The criteria based on the CGR.
(2) The criteria based on the stability of hydropower station.
(3) The criteria based on the regulation quality of hydropower station.

The last two types are related to the theory of governor regulation of hydropower station, and will be introduced in postgraduate courses. Here only the criteria based on CGR are discussed. The setting criteria for headrace surge chamber and tailrace surge chamber are different.

10.1.3.1 Setting Criterion for Headrace Surge Chamber

In preliminary design phase, the criterion of setting a surge chamber is the water inertia time T_w in the pressurized headrace pipe system:

$$T_w = \frac{\Sigma L_i V_i}{g H_p} \tag{10.1}$$

where L_i are the lengths of the pressurized headrace pipes, including spiral casing. Subscript i is the pipe label. If there is a bifurcation, considering the longest branch pipe, m; V_i is the average velocity values in headrace pipes, corresponding to L_i, m/s; g is the gravitational acceleration, 9.81 m/s^2; H_p is the rated head, m.

In the process of calculating T_w, the discharge and head should correspond to each other. Normally the largest discharge and the corresponding rated head are selected. If the lowest head is used, then the discharge should be the corresponding one.

If $T_w > [T_w]$, then a headrace surge chamber should be set. Based on collection and analysis of the information about design and operation experience in China and available standards all over the world, the Chinese standard *Specifications for Design of Surge Chamber of Hydropower Stations* NB/T35021-2014 proposes that the criterion $[T_w]$ should be within 2 to 4 s. For high head hydropower stations, smaller $[T_w]$ should be selected; for low head hydropower stations, larger $[T_w]$ should be selected. The selection of $[T_w]$ is also related to the importance of the hydropower station in the power grid. If the hydropower station operates alone or its capacity is larger than 50% of the grid's capacity, selecting small $[T_w]$ is appropriate; if the hydropower station's capacity is smaller than 10–20% of the grid's capacity, selecting relatively larger $[T_w]$ is appropriate.

10.1.3.2 Setting Criterion for Tailrace Surge Chamber

The tailrace surge chamber is used to shorten the length of pressurized tailrace system, reduce the vacuum degree in the draft tube during load rejection transient process, and avoid water column separation, which is assumed very dangerous. Because the flow patterns in the draft tube are very complex, the occurring condition of water column separation is not easy to determine accurately. Generally, the critical length of pressurized tailrace pipe system is used to define the necessity of setting a tailrace surge chamber in preliminary design phase. The corresponding equation is

$$L_w > \frac{5T_s}{V}\left(8 - \frac{\nabla}{900} - \frac{V_d^2}{2g} - H_s\right) \tag{10.2}$$

where L_w is the length of pressurized tailrace pipe system, m; T_s is the closing time of wicket gate during load rejection transient process, s; V is the velocity in the pressurized tailrace pipe system in normal operation condition, m/s; ∇ is the setting elevation of the turbine, m; V_d is the velocity at the draft tube inlet, m/s; H_s is the suction height of the turbine, m.

It should be noted that many engineering examples show that the criterion (10.2) is too conservative, because the actual lengths of the pressurized tailrace systems of some hydropower stations far exceed the critical length calculated by Eq. (10.2), but there is no liquid column separation happened in the draft tube during load rejection transient processes. The main reason is that the formula does not consider the time relationship between the maximum water hammer vacuum and the maximum flow velocity. Therefore, the criterion (10.2) is only for rough evaluation, and the final check should be made by accurate water hammer calculation. In the detailed analysis, the maximum vacuum at draft tube inlet must be kept smaller than 8 m.

10.2 Working Principles and Basic Equations of Surge Chamber

10.2.1 Working Principles

To discuss the working principles of surge chamber, we should first discuss the water level oscillations in surge chamber. In the discussion of water level oscillations, we normally assume that before the oscillation phenomenon in the "tunnel-surge chamber system" happens, the water hammer phenomenon in the penstock has vanished. This is because water hammer is usually a short process compared to surge chamber water level oscillation, and the former takes tens of seconds to vanish but the latter often takes minutes, even tens of minutes to attenuate.

Taking the load rejection and load increase transient processes as examples to discuss the surge chamber water level oscillation processes, as shown in Fig. 10.1.

The initial condition of a load rejection is a normal operation condition, in which the surge chamber water level is lower than the reservoir level, with the level difference equal to the summation of the headrace tunnel head losses and the velocity head under the surge chamber (Fig. 10.1). When load is rejected, the wicket gate is closed quickly and the discharge in the penstock becomes zero consequently. Due to the head difference at both ends of the headrace tunnel, the water in the tunnel flows into the surge chamber, and the water level rises continuously to reach at and exceed the reservoir water level. After the surge chamber water level is higher than the reservoir water level, driven by the water inertia in the tunnel, the level still rises until reach at the highest level, at which the velocity in the tunnel becomes zero. After that, driven by the opposite head difference, the water flows reversely from the surge chamber to the reservoir, and the water level drops continuously to reach at the reservoir water level. Again driven by the inertia, the water level drops until reach at the lowest level, at which the velocity in the tunnel becomes zero. The lowest surge chamber water level is generally much lower than the reservoir water level and the initial water level. After that, driven by the head differences and water inertia, the above water level rising and dropping processes repeat, and because of head losses the oscillation attenuates and finally stabilizes at the static water level, which is consistent with the reservoir water level. The water level history curve for the load rejection condition is demonstrated in the upper part of Fig. 10.1.

Similarly, the water level oscillation of surge chamber after a load increase can be shown in the lower part of Fig. 10.1 and described as follows. In load increase transient process, the turbine discharge increases as the wicket gate opens. Because of the water inertia in the headrace tunnel, the sudden increased discharge in the penstock cannot be supplied instantly by the water in the tunnel. Therefore, the water in the surge chamber flows out to compensate the insufficiency, leading to the decrease of the surge chamber water level. This forms a head difference between the reservoir water level and the surge chamber level. Then the new head difference drives the water in the tunnel to accelerate and flow to the surge chamber. When the discharge in the tunnel is equal to that in the penstock, the surge chamber water level reaches at the lowest point. After that, the increasing discharge in the tunnel continues and the extra water flows into the surge chamber, leading to the rise of surge chamber water level. The water level reaches at the highest point when the tunnel discharge and penstock discharge are balanced. Then the opposite head difference between the reservoir and the surge chamber forces the tunnel discharge to decrease and the surge chamber water level to drop. Again, owing to head losses in the tunnel, this water level oscillation gradually attenuates and finally stabilizes at the static water level, which is lower than the reservoir level. This final water level difference is equal to summation of the headrace tunnel head losses and the velocity head under the surge chamber when the hydropower is in the final steady working condition.

Analyzing the history curves of the surge chamber water level Z and the headrace tunnel velocity V in Fig. 10.1, we can conclude as follows.

(1) For load rejection transient process, the instants when the tunnel velocity V=0 are corresponding to the extremes of the surge chamber water level Z, namely $Z = Z_{\max}$ or $Z = Z_{\min}$; For load increase transient process, the instants when the tunnel discharge Vf (f is the cross-sectional area of the tunnel) equals to the turbine discharge Q (the same as the penstock discharge), are corresponding to the extremes of the surge chamber water level Z. These rules are important for understanding the physical nature, analyzing the oscillation characteristics, and designing appropriate surge chambers.

(2) The water level oscillation is actual the energy exchange and dissipation process in the "tunnel-surge chamber system." Water level difference between the reservoir and the surge chamber (or head difference at both ends of the tunnel) stands for the potential energy, and the velocity in the tunnel stands for the kinetic energy. The oscillation is the exchange of potential energy and kinetic energy, which is dissipated by the head losses in the system. If the head losses are absent, the oscillation would continue without attenuation.

(3) The attenuation of the oscillation of surge chamber water level is conditional. When the wicket gate is completely closed or the governor is not in action, the attenuation is definite. But if the governor is working to regulate the turbine

discharge following the oscillating head for keeping constant output power, the oscillation of surge chamber water level may be excited and undamped. This is the instability of surge chamber water level and will be discussed later.

(4) The oscillation of surge chamber water level is a type of mass oscillations, and it is featured by massive water flowing into and out of the surge chamber. The difference between mass wave and water hammer wave is obvious, and water hammer wave is elastic wave, which is related to the compressibility of water body and pipe wall. The continuity equation and momentum equation of incompressible fluid can be used to describe the water level oscillation of surge chamber.

(5) According to the requirement of grid dispatching, the long period oscillation of surge chamber wave is unfavorable, because surge wave superposition may happen due to restarting the load rejected unit. There must be a superposition between the oscillation caused by the initial load rejection and the oscillation caused by the subsequent restarting and load increase. This superposition may lead to larger amplitude and slower attenuation of the surge chamber wave, which are related to the superposition instant and superposition mode. The surge wave superposition problem should be investigated in the design and operation of the hydropower stations with surge chambers.

The main task of this chapter is to teach students how to determine the type and size of surge chamber by conducting hydraulic calculations. Therefore, the following two aspects should be discussed.

(1) Calculate the water level oscillation process and amplitude, then, determine the top elevation and bottom elevation of the surge chamber.

(2) Study the stability of surge chamber wave, determine the critical stable cross-sectional area, and further select the appropriate cross-sectional area of the surge chamber.

10.2.2 Basic Equations for Surge Chamber

Figure 10.1 is the schematic of a hydraulic system with a headrace surge chamber. In the steady generating condition, the discharge of turbine Q_0 is constant, and the flows in the headrace tunnel and penstock are steady. The discharge in the tunnel and penstock equals to Q_0, with the velocity in the tunnel $V_0 = \frac{Q_0}{f}$, in which f is the cross-sectional area of the tunnel. There is no outflow in the surge chamber and the water level Z_0 is static. The head distribution along the system can be described by the Bernoulli equation.

When turbine discharge Q changed, the water level in the surge chamber and the water in the headrace tunnel will oscillate, as described in the previous section. With the tunnel velocity V and surge chamber water level Z defined as the dependent variables of time t, the following momentum equation for the tunnel water can be obtained, if the compressibility of water, elasticity of tunnel wall and water inertia in the surge chamber are neglected.

$$\frac{L}{g}\frac{\mathrm{d}V}{\mathrm{d}t} = -Z - h_w - h_r \tag{10.3}$$

where L is the length of the headrace tunnel, m; Z is the surge chamber water level with the reservoir water level as the datum, upward direction is positive, m; h_w is the total head loss in the tunnel, m; h_r is the local head loss when water flows out or into the surge chamber, m.

From the continuity principle, a part of the tunnel water flows into the surge chamber and the remaining part flows into the penstock and then turbine. Therefore, the continuity equation reads

$$fV = F\frac{\mathrm{d}z}{\mathrm{d}t} + Q \tag{10.4}$$

where F is the cross-sectional area of the surge chamber, can be described as a function of elevation, m^2; Q is the discharge in the penstock and turbine, can be described as a function of time, m^3/s.

The oscillation of surge chamber water level results in the change of turbine head, which leads to the change of output power. If we assume that constant output power can be kept by the governor, then the constant output power equation can be written as

$$\frac{Q}{Q_0} = \frac{(H_0 - h_{w0} - h_{mo})\eta_0}{(H_0 + Z + h_r - h_m)\eta} \tag{10.5}$$

where H_0 is the difference of the reservoir and tailrace levels, namely the gross head, m; h_{m0} is the head loss in the headrace tunnel when flowing discharge Q_0, m; h_m is the head loss in the penstock when flowing discharge Q, m.

Equations (10.3) and (10.4) are the momentum equation and continuity equation for surge chamber water oscillation. In the above derivation, the water in the headrace tunnel is viewed as the control body and the directions of V and Z are carefully defined. If the surge chamber is set in the tailrace system or the directions are redefined, the equations can be derived by the similar procedures but the forms of Eqs. (10.3) and (10.4) may be changed a little. These two equations can be used in calculating the water level oscillating process when the initial and boundary conditions are given.

Equation (10.5) is the equation describing the relation of the output powers in different working points for a hydropower system with a headrace surge chamber. It is used to analyze the stability of surge wave in small disturbance conditions.

10.3 Typical Layout and Basic types of Surge Chamber

10.3.1 Typical Layout of Surge Chamber

According to the relative locations of the surge chamber and the powerhouse, the typical layouts of surge chamber can be summarized as follows.

10.3.1.1 Headrace (Or Upstream) Surge Chamber

The upstream surge chamber, located on the upstream side of the powerhouse, dividing the headrace pressure hydraulic system into the headrace tunnel and penstock, as shown in Fig. 10.1. This layout is widely used in the conduit type hydropower stations and the riverbank underground hydropower stations with posterior powerhouse. The purpose of setting an upstream surge chamber is to control the maximum water hammer pressures in turbine spiral casing and penstock to meet the requirements of design and relevant specifications.

10.3.1.2 Tailrace (Or Downstream) Surge Chamber

The downstream surge chamber, located on the downstream side of the powerhouse, dividing the tailrace pressure hydraulic system into the draft tube extension conduit and tailrace tunnel, as shown in Fig. 10.2. This layout is normally used in the riverbank underground hydropower stations with anterior powerhouse. The purpose of setting a downstream surge chamber is to control the minimum water hammer pressure in the draft tube inlet to prevent from water column separation and to ensure the safe operations of units and the power station.

10.3.1.3 Headrace and Tailrace Two Surge Chambers

Some underground hydropower stations, especially pumped storage stations, have long headrace pressure system and tailrace system on both sides of powerhouse. To reduce the water hammer pressure and improve the operating conditions of units, an upstream surge chamber and a downstream surge chamber are all set in the system, forming a system of headrace and tailrace two surge chambers, as shown in Fig. 10.3.

In load rejection transient processes, the water level oscillations in the headrace and tailrace two surge chambers do not affect each other, and the highest water levels and the lowest water levels of the two surge chambers can be obtained, respectively. However, in load increase transient processes, the water levels of the headrace and tailrace two surge chambers oscillate because of the change of turbine discharge. And the oscillation of water level in any surge chamber causes a new change of turbine discharge, which affects the water level oscillation in another surge chamber. This

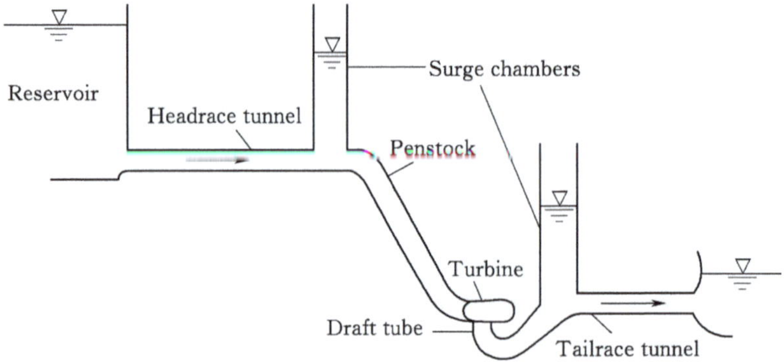

Fig. 10.3 Hydraulic system with headrace and tailrace two surge chambers

hydraulic phenomenon is complex, especially when the hydraulic characteristics of the headrace tunnel are close to those of the tailrace tunnel, in which resonance may occur. When designing such surge chambers, special attention should be paid to the analysis of whole history and damping features of surge wave oscillations.

10.3.1.4 Headrace Double Surge Chambers

In case the headrace pressure system is very long and the topographic and geological conditions are limited, two surge chambers have to be set on the upstream side of powerhouse, as shown in Fig. 10.4. The surge chamber near the powerhouse is called the main surge chamber because it plays a leading role in reflecting water hammer. The other surge chamber located upstream the main surge chamber is called the secondary or auxiliary surge chamber because it can reduce the oscillation amplitude in the main surge chamber by bearing a part of water inertia in the tunnel. The closer of the auxiliary surge chamber to the main surge chamber, the larger the effect of the auxiliary surge chamber. The oscillation damping of the hydraulic system is borne together by the two surge chambers. The sum of the cross-sectional areas of the two surge chambers is larger than the needed area if a single surge chamber is set. This layout can be used when a construction shaft is already available, or new units are added to the original hydropower station.

10.3.2 Basic Types of Surge Chamber

Since the first adoption of surge chamber in 1895 in Switzerland, the sizes of surge chambers are becoming larger and larger and the types of surge chambers are also varying, with the development of hydropower technology. The following are the basic types of surge chambers from the point of view of hydraulic performance.

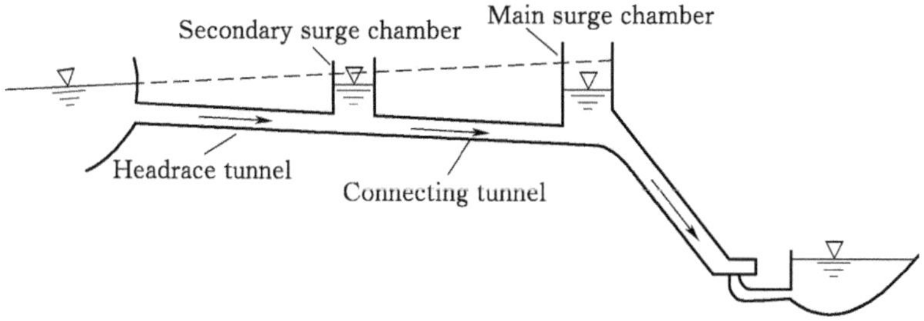

Fig. 10.4 Hydraulic system with headrace double surge chambers

10.3.2.1 Simple Surge Chamber

As shown in Fig. 10.5a, the merits of simple surge chamber are that it has simple shape with the same section from the top to bottom, and its function of reflecting water hammer wave is good. Its drawbacks include that the head loss at the junction of the tunnel and the surge chamber is large in normal operation conditions, the oscillation amplitude is large, the oscillation attenuation is slow, and the chamber volume is large. This type usually is adopted in low head or small discharge hydropower stations. To reduce the head loss in the bottom of the surge chamber, a short connecting pipe with cross-sectional area not smaller than that of the tunnel can be installed between the tunnel and surge chamber, as shown in Fig. 10.5b.

10.3.2.2 Throttled Surge Chamber

If the surge chamber is connected to the tunnel by a small orifice or a thin connecting pipe with cross-sectional area smaller than that of the tunnel, the surge chamber is defined as the throttled surge chamber, as shown in Fig. 10.5c, d. The thin connecting pipe or the small orifice can increase the throttling head losses when water flows into and out the surge chamber. These losses dissipate a part of energy and make the oscillation amplitude smaller and the oscillation attenuation faster than those of the simple surge chamber. The volume of the throttle surge chamber is smaller than that of the simple surge chamber in the same condition. The head loss under the throttled surge chamber is small in normal operation condition. However, because of the large throttling effect, the water hammer reflection of this surge chamber is not complete, and there is penetrating water hammer wave in the tunnel. Therefore, the selection of orifice or connecting pipe cross-sectional area is important in designing a throttled surge chamber.

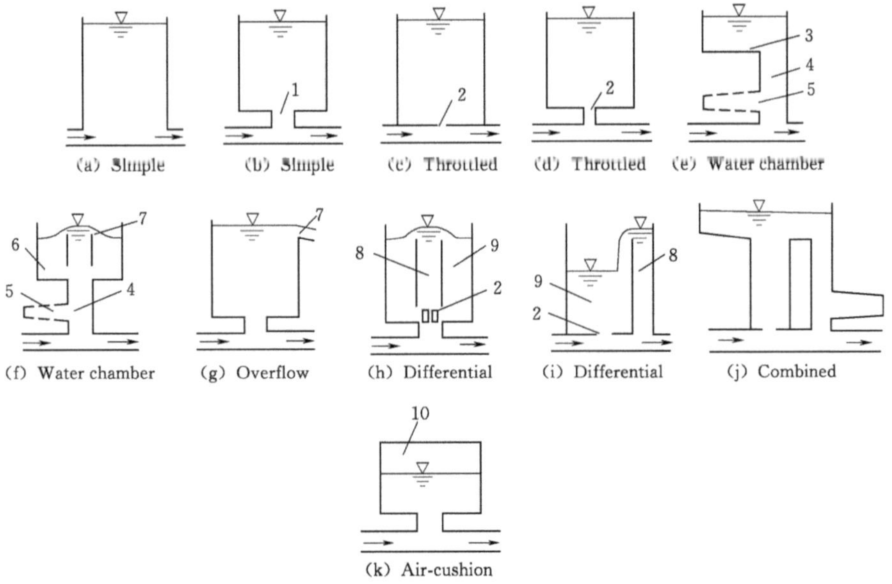

Fig. 10.5 Basic types of surge chambers 1-connection pipe; 2-throttle orifice;3-upper chamber; 4-shaft; 5-lower chamber; 6-water chamber; 7-overflow weir;8-rising pipe; 9-large well;10-compressed air

10.3.2.3 Water Chamber Surge Chamber

The water chamber surge chamber is normally composed of a vertical shaft and an upper water chamber or a lower water chamber or both the upper and lower water chambers, as shown in Fig. 10.5e, f. The water chamber has larger cross-sectional area, which can effectively suppress water level oscillation amplitude. If the upper and lower water chambers are all set, the surge chamber is called the double water chamber surge chamber. After load rejection happens, the water level in the shaft rises rapidly, and once it enters the upper water chamber the rising speed slows down immediately due to larger cross-sectional area. The upper water chamber limits the highest surge water level. In load increase conditions, the water level in the shaft drops rapidly, and once it enters the lower water chamber the dropping speed slows down immediately due to larger cross-sectional area. The lower water chamber limits the lowest surge water level. Because the center of gravity of the water pouring into the upper chamber is higher in load rejection transients, and the center of gravity of the water flowing out of the lower chamber is lower in load increase transients, the same energy can be stored in a relatively smaller space. Therefore, the total space of this surge chamber is relatively small. This type of surge chamber is suitable for hydropower stations with high water head and large reservoir working depth, and should be built underground.

10.3.2.4 Overflow Surge Chamber

There is an overflow weir on the top of the overflow surge chamber, as shown in Fig. 10.5g. In load rejection transients, the water level rises rapidly, once it exceeds the weir crest, overflow begins, which suppresses the continuing increase of water level and keeps the highest water level just a little higher than the crest elevation. The overflow water can be discharged to the downstream river (Fig. 10.5g) or stored in the upper chamber (Fig. 10.5f). When the water level of the shaft drops, the stored water flows back to the shaft through the bottom hole (Fig. 10.5f).

10.3.2.5 Differential Surge Chamber

The differential surge chamber consists of a small shaft, a large well and some impedance holes. The small shaft is called the rising pipe, which has an overflow top. The large well (or outer well) and the rising pipe can be arranged concentrically to form easily supported structure and connected by the impedance holes in the bottom for water body exchange, as shown in Fig. 10.5h. In normal working conditions of the hydropower station, the water levels in the rising pipe and large well are the same. During transient processes, the water levels in the rising pipe and the large well oscillate differentially. This type of surge chamber takes into account the advantages of the throttled type and the overflow type. The required volume is small, the effect of reflecting water hammer wave is good, and the attenuation of water level oscillations is faster, but the structure is complex. To simplify and strengthen the structure, the rising pie can be arranged on one side of the large well or outside the large well, as shown in Fig. 10.5i. The differential surge chamber is normally adopted when the required stability cross-sectional area is large and the variation of reservoir water level is small.

10.3.2.6 Combined Surge Chamber

In some situations, a combined surge chamber that combines the characteristics of two or more typical types of surge chambers can be used based on specific conditions and requirements of the hydropower station, as shown in Fig. 10.5j. The structure form and water level oscillation features are more complex. The combined surge chamber is mostly used in pumped storage power stations, which need fast water level oscillation attenuation.

10.3.2.7 Air-Cushion Surge Chamber

Air-cushion surge chamber is a closed chamber that uses the elasticity of air to suppress water level oscillation amplitude. As shown in Fig. 10.5k, the lower part of the chamber is filled with water, and the above part is filled with compressed air. Because the closed air pressure is generally higher than one atmospheric pressure, the water level in the surge chamber is lower than that of a normal surge chamber, which reduces the height of the surge chamber. This surge chamber can be located underground and close to the powerhouse to shorten the penstock. The drawbacks of this surge chamber include needing larger stable cross section, and needing compressed air machine to replenish the air chamber regularly, which increases the building and operating costs. The air-cushion surge chamber is an alternative type when ground geological and topographic conditions are not suitable for conventional surge chamber types or the ventilation shaft is too long or the cost is too high. This type is mostly adopted in the hydropower stations with high water head, good geological condition, and deep buried headrace system. Figure 10.6 compares the layout differences of the air-cushion and normal open surge chambers.

10.3.3 Location Selection of Surge Chamber

The factors to be considered in the selection of surge chamber location are discussed as follows, taking the location optimization of tailrace surge chamber as an example.

When the length of the tailrace system exceeds the critical length defined in Eq. (10.2), setting a tailrace surge chamber becomes necessary. The tailrace surge chamber should make the lowest pressure in the draft tube higher than the vaporizing pressure of water to avoid water column separation during load rejection transient processes. The previous understanding is that the closer the tailrace surge chamber to the powerhouse, the more favorable the pressure value in the draft tube, but this

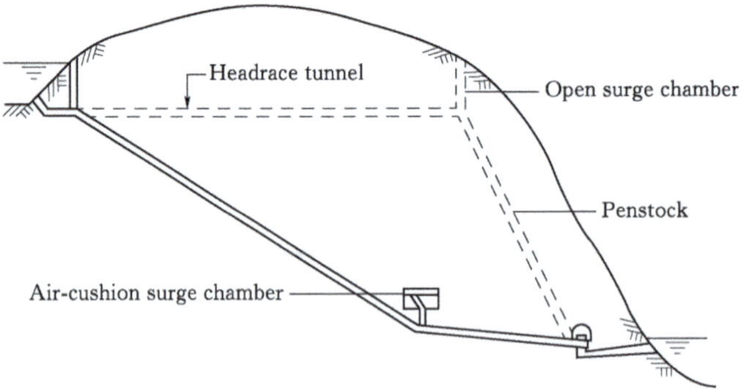

Fig. 10.6 Comparison of layouts of open surge chamber and air-cushion surge chamber

is not conducive to the stability of the surrounding rock of the powerhouse cavern. Therefore, the layouts of underground hydropower stations are often restricted by this contradiction.

Theoretically, for the tailrace system with a surge chamber, the minimum pressure of the draft tube depends not only on the water hammer pressure in the draft tube extension conduit, but also on the mass oscillation in the tailrace surge chamber and the tailrace tunnel. It is the superposition of the two fluctuations. On the premise of the same length of the tailrace system, the shorter the draft tube extension conduit, the smaller the water hammer pressure, but the longer the tailrace tunnel, the larger the water level oscillation in the surge chamber. On the contrary, the longer the draft tube extension conduit, the larger the water hammer pressure, but the shorter the tailrace tunnel, the smaller the water level oscillation in the surge chamber. Therefore, in optimizing the location, we can move the tailrace surge chamber back and forth in a certain range to find the favorable location where the draft tube minimum pressure does not change obviously. By this way, the distance between the powerhouse and the tailrace surge chamber can be increased.

Similarly, for the headrace surge chamber, the favorable location for controlling the maximum pressure in the spiral casing is not that most closer to the powerhouse. The most favorable location of the headrace surge chamber is within a certain range, in which the adjustment of location does not obviously affect the maximum pressure in the spiral casing.

10.4 Calculation of Surge Chamber Waves

The basic equations for calculating the water level oscillation of surge chamber are the momentum equation and continuity equation expressed by Eqs. (10.3) and (10.4), respectively. The equations are nonlinear ordinary differential equations, and the accurate analytical solutions can be obtained only in some special cases, such as the case with constant section surge chamber and during load rejection transient processes. For most cases, various assumptions have to be introduced to obtain approximate solutions. There are approximate solutions by graphical method or asymptotic method, numerical solutions by the Runge-Kutta method, and joint numerical simulation of water hammer and surge chamber wave. In this section, we present the analytical solutions, the Runge-Kutta method, and the joint numerical simulation method.

10.4.1 Analytical Solutions to Surge Chamber Waves

10.4.1.1 Throttled Surge Chamber with Constant Cross Section

1. **The highest water level**

If sudden closing of wicket gate is assumed for load rejection transient processes, the discharge in the penstock Q=0. Introducing this condition into Eq. (10.4), one obtains

$$fV = F\frac{\mathrm{d}Z}{\mathrm{d}t} \tag{10.6}$$

Define $h_w = \alpha V^2 = h_{w0}\left(\frac{V}{V_0}\right)^2$ and $h_r = \beta\left(F\frac{dZ}{dt}\right)^2 = \beta(fV)^2 = h_{r0}\left(\frac{V}{V_0}\right)^2 = h_{r0}\left(\frac{Q_T}{Q_0}\right)^2$, in which α is the head loss coefficient of the tunnel, with local and frictional losses included; β is the throttle coefficient; h_{ro} is the throttle head loss when discharge is Q_0. Substituting these definitions into Eq. (10.3), one obtains

$$\frac{L}{g}\frac{\mathrm{d}V}{\mathrm{d}t} = -Z - (h_w + h_r)\left(\frac{V}{V_0}\right)^2 \tag{10.7}$$

Let $y = V/V_0$, then $V = V_0 y$ and $dV = V_0 dy$. Substituting these into Eq. (10.7) yields

$$\frac{L}{g}\frac{V_0 \mathrm{d}y}{\mathrm{d}t} = -Z - h_{w0}y^2 - h_{r0}y^2$$

If the two sides are divided by h_{w0} and let $\eta = h_{r0}/h_{w0}$, the above expression can be rewritten as

$$\frac{L}{gh_{w0}}\frac{V_0 \mathrm{d}y}{\mathrm{d}t} = -\frac{Z}{h_{w0}} - (1+\eta)y^2$$

$$\frac{L}{gh_{w0}}\frac{V_0 \mathrm{d}y}{\mathrm{d}Z}\frac{\mathrm{d}Z}{\mathrm{d}t} = \frac{L}{gh_{w0}}\frac{V_0 \mathrm{d}y}{\mathrm{d}Z}\frac{fV}{F} = \frac{LfV_0^2}{gFh_{w0}}y\frac{\mathrm{d}y}{\mathrm{d}Z} = -\frac{Z}{h_{w0}} - (1+\eta)y^2$$

Let $\lambda = \frac{LfV_0^2}{2gFh_{w0}}$, $X = -\frac{Z}{\lambda}$ and $X_0 = -\frac{Z_0}{\lambda} = \frac{h_{w0}}{\lambda}$, one gets

$$\frac{\mathrm{d}(y^2)}{\mathrm{d}X} = \frac{X}{X_0} - (1+\eta)y^2 \tag{10.8}$$

Equation (10.8) is a first-order linear differential equation, and its integral can be obtained as follows:

$$y^2 = \frac{(1+\eta)X+1}{(1+\eta)^2 X_0} + Ce^{(1+\eta)X} \tag{10.9}$$

From the initial conditions $t = 0, V = V_0, y = 1, Z = h_{w0}, X = X_0$ to determine the integral coefficient C as

$$C = \frac{\eta(1+\eta)X_0 - 1}{(1+\eta)^2 X_0} e^{-(1+\eta)X_0}$$

Then

$$y^2 = \frac{(1+\eta)X+1}{(1+\eta)^2 X_0} + \frac{\eta(1+\eta)X_0 - 1}{(1+\eta)^2 X_0} e^{-(1+\eta)(X_0 - X)} \tag{10.10}$$

When the water level reaches the highest, $Z_{\max} = -\lambda X_{\max}$ and $V = 0$, namely $y = 0$. Substituting this into Eq. (10.10) yields

$$1 + (1+\eta)X_{\max} = [1 - \eta(1+\eta)X_0]e^{-(1+\eta)(X_0 - X_{\max})}$$

Taking the logarithm of the above equation, we get

$$\ln[1 + (1+\eta)X_{\max}] - (1+\eta)X_{\max} = \ln[1 - (1+\eta)\eta X_0] - (1+\eta)X_0 \tag{10.11}$$

Equation (10.11) is for the throttled surge chamber.

For simple surge chamber, $\eta = 0$ because of no throttling effect. Then Eq. (10.11) becomes

$$X_0 = -\ln(1 + X_{\max}) + X_{\max} \tag{10.12}$$

With the initial value X_0 know, one can solve the maximum value $X_{\max}$ from the above implicit Eq. (10.11) or (10.12) for the throttled surge chamber or simple surge chamber, respectively.

2. **The lowest water level**

For the headrace surge chamber, the lowest water level can be the second extreme on the history curve of water level in the load rejection condition or the first extreme on the history curve in the load increase condition. These two should be calculated and compared.

(1) The second extreme of water level in load rejection condition.

Equation (10.10) can also be used here. However, X here should use the inverse sign, and the initial condition $y = 0$ and $X = X_{\max}$ should be used when determining the integral coefficient C. After similar derivation, one can obtain the following equation for the throttled surge chamber:

$$\ln[1-(1+\eta)X_2]+(1+\eta)X_2=\ln[1-(1+\eta)X_{\max}]+(1+\eta)X_{\max} \tag{10.13}$$

For simple surge chamber, inserting $\eta = 0$ into (10.13) yields

$$X_2+\ln(1-X_2)=X_{\max}+\ln(1-X_{\max}) \tag{10.14}$$

That is to say, we should first calculate the first extreme $X_{\max}$ by using Eq. (10.11) or (10.12) then can calculate the second extreme X_2 by using Eq. (10.13) or (10.14).

(2) The first extreme of water level in load increase condition.

For simple surge chamber, if the discharge of unit increases from mQ_0 to Q_0, with load coefficient $m < 1$, the first extreme can be calculated by the Vogt approximate equation :

$$\frac{|Z_{\min}|}{h_{w0}}=1+(\sqrt{\varepsilon}-0.275\sqrt{m}+0.05/\varepsilon-0.9)(1-m)\left(1-m/\varepsilon^{0.62}\right) \tag{10.15}$$

where $\varepsilon=\frac{LfV_0^2}{gFh_{w0}^2}=2\lambda/h_{w0}$.

10.4.1.2 Asymptotic Solutions to Surge Chamber Wave

According to the asymptotic method for nonlinear vibration, the basic equations of water level oscillation in surge chamber can be rewritten as follows:

$$\frac{\mathrm{d}^2Z}{\mathrm{d}t^2}+\omega^2Z=-\omega^2h_{w0}\left(\frac{Q}{fV_0}+\frac{F}{fV_0}\frac{\mathrm{d}Z}{\mathrm{d}t}\right)\left|\frac{Q}{fV_0}+\frac{F}{fV_0}\frac{\mathrm{d}Z}{\mathrm{d}t}\right|-\eta\omega^2h_{w0}\left(\frac{F}{fV_0}\right)^2\frac{\mathrm{d}Z}{\mathrm{d}t}\left|\frac{\mathrm{d}Z}{\mathrm{d}t}\right| \tag{10.16}$$

where Z is the surge chamber level with the reservoir level as the datum, and its positive direction is upward; F is the cross-sectional area; L, f, and V are the length, cross-sectional area and velocity of the tunnel, respectively; Q is the discharge of penstock; $\omega=\sqrt{\frac{gf}{LF}}$; h_{w0} and V_0 are the head loss and velocity in the tunnel when discharge is Q_0; $\eta=h_{r0}/h_{w0}$; h_{r0} is the throttle head loss when discharge Q_0 goes through the throttle orifice or connection pipe.

Let $\beta = \eta h_{w0}/V_0^2$, namely $\beta = \eta\alpha$, in which α and β are the head loss coefficients for the tunnel and the throttle orifice, respectively. With the positive direction of discharge in the tunnel defined as flowing to the penstock or surge chamber, the first-order approximate solution of (10.16) can be expressed as

$$Z = A\cos\varphi \tag{10.17}$$

$$\frac{\mathrm{d}A}{\mathrm{d}t} = -\frac{\varepsilon}{2\pi\omega}\int_0^{2\pi} f(-A\sin\varphi)\sin\varphi \mathrm{d}\varphi \tag{10.18}$$

$$\frac{\mathrm{d}\varphi}{\mathrm{d}t} = \omega - \frac{\varepsilon}{2\pi A\omega}\int_0^{2\pi} f(-A\sin\varphi)\cos\varphi \mathrm{d}\varphi \tag{10.19}$$

The following will provide the analytical expressions of the first extreme and the second extreme of the air-cushion surge chamber under load rejection and load increase conditions.

1. **The first and second water level extreme for air-cushion surge chamber after load rejection** (with the finally stabilized water level as the datum)

$$Z_i = (-1)^i a_i + \frac{1}{6}\frac{m(m+1)p_s}{\gamma l_s^2\sigma_s}a_i^2 \quad (i = 1,\ 2) \tag{10.20}$$

where $a_0 = \sqrt{\left(\frac{h_{w0}}{\sigma_0}\right)^2 + \left(\frac{Q_0}{F\omega_s}\right)^2}$, $\theta_0 = \arccos\left(\frac{h_{w0}}{\sigma_0 a_0}\right)$, $\varepsilon = \frac{Lfv_0^2}{2gFh_{w0}}$,$a_i = \frac{a_0}{1+\frac{2a_0(\eta+1)}{3\pi\varepsilon}(i\pi-\theta_0)}$, $\eta = \frac{Q_0^2}{2g\varphi^2S^2h_{w0}}$, $T_s = 2\pi\sqrt{\frac{LF}{gf}}/\sqrt{\sigma_s}$, $\omega_s = \frac{2\pi}{T_s}$, $\sigma_s = 1 + \frac{mp_s}{\gamma l_s}$, and $\sigma_0 = 1 + \frac{mp_0}{\gamma l_0}$.

$$l_s = l_0 - h_{w0}/\sigma_0 \tag{10.21}$$

$$p_s = \frac{p_0 l_0^m}{(l_0 - h_{w0}/\sigma_0)^m} \tag{10.22}$$

where γ is the volume weight of water; m is the polytropic constant of ideal gas, which is 1.0 in isothermal state and 1.4 in isentropic state; p_0 and l_0 are the absolute pressure in the chamber and the converted height of air chamber (air volume divided by the actual cross-sectional area) under the full load operation condition of the units, respectively; p_s and l_s are the absolute pressure and air chamber height when the units are stopped,p_0, l_0, p_s, and l_s can be calculated by (10.21) and (10.22), according to the given operation conditions;φ is the discharge coefficient of the throttle orifice, which should be obtained by measurement and can be chosen within 0.60–0.80 in primary calculation.

2. **The lowest water level for air-cushion surge chamber in load increase transient process** (with the unit stopped water level as the datum)

When the unit discharge increases from mQ_0 to Q_0, the water level in the air chamber drops. The lowest water level $Z_{\min a}$ can be calculated by

$$z_{\min a} = a + \left(\frac{2}{3}\frac{gFh_{w0}}{Lfv_0^2} + \frac{1}{6}\frac{m(m+1)p_0}{\gamma l_0^2 \sigma_0}\right)a^2 + \frac{h_{w0}}{\sigma_0} \tag{10.23}$$

where $a = \frac{a_1}{1-\mu_2 a_1}$, $a_1 = \frac{a_0}{1+\mu_2 a_0} e^{\mu_1(\pi-\theta_0)}$, $a_0 = \sqrt{\left(\frac{(1-m'^2)h_{w0}}{\sigma_0}\right)^2 + \left(\frac{(1-m')Q_0}{F\omega_0}\right)^2}$, $\theta_0 = \arccos\left(\frac{(1-m'^2)h_{w0}}{\sigma_0 a_0}\right)$, $\mu_1 = \frac{-gh_{w0}}{\omega_0 L v_0}$, $\mu_2 = \frac{4}{3\pi}\eta\frac{F}{Q_0}\omega_0$, $\omega_0 = \frac{2\pi}{T_0}$, and $T_0 = 2\pi\sqrt{\frac{LF}{gf}}/\sqrt{\sigma_0}$.

The definitions of symbols are similar to the previous. However, attention should be paid to the water level datum. Here, the datum is the water level when the unit is stopped. If the operation water level H_{T0} is known, the unit stopped water level should be $H_{T0} + \frac{h_{w0}}{\sigma_0}$.

10.4.2 *Numerical Solutions to Surge Chamber Waves*

10.4.2.1 Runge-Kutta Method

The Runge-Kutta method is suitable for solving first-order ordinary differential equations, and can be easily adopted in calculating surge chamber waves. The truncation error of this method is proportional to the fifth power of timestep size. With proper timestep given, the calculation accuracy can be guaranteed. To explain the procedures, we take the following two-dimensional first-order ordinary differential equations as an example:

$$\begin{aligned} \frac{\partial u}{\partial t} &= f(t, u, v) \quad u(t_0) = u_0 \\ \frac{\partial v}{\partial t} &= g(t, u, v) \quad v(t_0) = v_0 \end{aligned} \tag{10.24}$$

The classical fourth-order Runge-Kutta method is used to solve the problem as follows

$$
\begin{aligned}
\begin{bmatrix} u_{n+1} \\ v_{n+1} \end{bmatrix} &= \begin{bmatrix} u_n \\ v_n \end{bmatrix} + \frac{h}{6}\begin{bmatrix} k_{11} + 2k_{12} + 2k_{13} + k_{14} \\ k_{21} + 2k_{22} + 2k_{23} + k_{24} \end{bmatrix} \\
k_1 &= \begin{bmatrix} k_{11} \\ k_{21} \end{bmatrix} = \begin{bmatrix} f(t_n, u_n, v_n) \\ g(t_n, u_n, v_n) \end{bmatrix} \\
k_2 &= \begin{bmatrix} k_{12} \\ k_{22} \end{bmatrix} = \begin{bmatrix} f\left(t_n + \frac{h}{2}, u_n + \frac{hk_{11}}{2}, v_n + \frac{hk_{21}}{2}\right) \\ g\left(t_n + \frac{h}{2}, u_n + \frac{hk_{11}}{2}, v_n + \frac{hk_{21}}{2}\right) \end{bmatrix} \\
k_3 &= \begin{bmatrix} k_{13} \\ k_{23} \end{bmatrix} = \begin{bmatrix} f\left(t_n + \frac{h}{2}, u_n + \frac{hk_{12}}{2}, v_n + \frac{hk_{22}}{2}\right) \\ g\left(t_n + \frac{h}{2}, u_n + \frac{hk_{12}}{2}, v_n + \frac{hk_{22}}{2}\right) \end{bmatrix} \\
k_4 &= \begin{bmatrix} k_{14} \\ k_{24} \end{bmatrix} = \begin{bmatrix} f(t_n + h, u_n + hk_{13}, v_n + hk_{23}) \\ g(t_n + h, u_n + hk_{13}, v_n + hk_{23}) \end{bmatrix}
\end{aligned} \tag{10.25}
$$

where h is the timestep size, with current time $t_n = t_0 + nh$, in which subscript n stands for the previous timestep and n+1 for the current timestep.

The basic equations for water oscillation in the tunnel-surge chamber system can be rewritten as

$$\frac{\mathrm{d}Z}{\mathrm{d}t} = \frac{Q - Q_m}{F} = f(t, Z, Q), \quad Z(t_0) = Z_0 = H_u - h_{w0} \tag{10.26}$$

$$\frac{\mathrm{d}Q}{\mathrm{d}t} = \frac{gA}{L}(H_u - Z - \alpha Q|Q| - \beta Q_T|Q_T|) = g(t, Z, Q), \quad Q(t_0) = Q_0 \tag{10.27}$$

where Q is the discharge in the tunnel, m^3/s; Q_m is the discharge in the penstock, m^3/s; Q_T is the discharge flowing into the surge chamber, m^3/s; Z is the surge chamber water level, m; H_u is the reservoir water level, m; A is cross-sectional area of the tunnel, m^2; L is the length of the tunnel, m; F is the cross-sectional area of the surge chamber, m^2; g is the gravitational acceleration, m/s^2; α is the head loss coefficient of the tunnel; β is the head loss coefficient of the throttle orifice.

Equations (10.26) and (10.27) can be solved by the Runge-Kutta method demonstrated by Eqs. (10.24) and (10.25). The following is the code.

1. Codes in C programming language

```
#include<stdio.h>
#include<math.h>
double f(double,double,double);
double g(double,double,double);
void main(void)
{
    double t0,z0,v0,k11,k21,k12,k22,k13,k23,k14,k24,t1,z1,v1,h,zmax=0,tmax=0;
    int n,i,m=0;
     double z[10000],t[10000];
     FILE *fp;
     fp=fopen("Results for throttle surge chamber after load rejection.txt","w");
     printf("Input data t0 z0 v0 h n=\n");   //Initial time, initial surge chamber level, initial tunnel velocity, timestep
size and calculation steps.
     scanf("%lf%lf%lf%lf%d",&t0,&z0,&v0,&h,&n);
     fprintf(fp,"%15s%15s%15s\n","            n","t            ","z            ","v            ");
     fprintf(fp,"            ");
     for(i=0;i<50;i++)


                                   fprintf(fp,"%c",'-');
                              fprintf(fp,"\n");
                              for(i=0;i<=n;i++)
                              {
                                   t1=t0+h;
                                   k11=f(t0,z0,v0);
                                   k21=g(t0,z0,v0);
                                   k12=f(t0+h/2,z0+h*k11/2,v0+h*k21/2);
                                   k22=g(t0+h/2,z0+h*k11/2,v0+h*k21/2);
                                   k13=f(t0+h/2,z0+h*k12/2,v0+h*k22/2);
                                   k23=g(t0+h/2,z0+h*k12/2,v0+h*k22/2);
                                   k14=f(t0+h,z0+h*k13,v0+h*k23);
                                   k24=g(t0+h,z0+h*k13,v0+h*k23);
                                   z1=z0+h*(k11+2*k12+2*k13+k14)/6;
                                   v1=v0+h*(k21+2*k22+2*k23+k24)/6;
                                   fprintf(fp,"%15d%15.6lf%15.6lf%15.6lf\n",i,t0,z0,v0);
                                 t[i]=t1;
                                   z[i]=z1;
                                   t0=t1;
                                   z0=z1;
                                   v0=v1;
                              }
```

```
        zmax=z[0];
        for(i=0;i<n-1;i++)
        if(z[i]>zmax)
        {
            zmax=z[i];
            m=i;
        }
        tmax=t[m];
        fprintf(fp,"zmax=%lf,tmax=%lf\n",zmax,tmax);
        fclose(fp);
}
double f(double t,double z,double v)    //The continuity equation of surge chamber wave
{
        double z1;
        double F=113.0973;              //The cross sectional area of surge chamber
        double f1=15.904;               //The cross sectional area of the tunnel
        double Q=0;                     //The discharge in penstock
        z1=(f1*v-Q)/F;                  //The continuity equation
        return z1;
}
double g(double t,double z,double v)    //The momentum equation of surge chamber wave
{
        double v1;
        double Q=0;                     //The discharge in penstock
        double L=7905;                  //The length of the tunnel
        double f1=15.904;               //The cross sectional area of the tunnel
        double k1=0.00516;              //The head loss coefficient of the tunnel
        double k2=0.0049809;            //The head loss coefficient of the throttle orifice
        v1=-(z+k1*f1*f1*v*fabs(v)+k2*(f1*v-Q)*fabs(f1*v-Q))*9.81/L;  //The momentum equation
        return v1;
}
```

2. **Engineering example and results**

A practical hydropower station with a headrace throttle surge chamber is chosen as the example. The given parameters in the code are as follows: the tunnel length $L = 7905$ m; the cross-sectional area of the tunnel $f_1 = 15.904$ m^2; the head loss coefficient of the tunnel k_1=0.00516; the cross-sectional area of the surge chamber $F = 113.0973$ m^2; the head loss coefficient of the throttle orifice $k_2 = 0.0049809$. The input data include the initial time of calculation t0=0, the initial surge chamber water level $Z_0 = 594.153$ m, the initial tunnel velocity $V_0 = 3.09$ m/s, calculation timestep size $h = 0.02$ s, calculation steps $n = 12000$. The surge chamber wave caused by a full load rejection is calculated and the resulting water level oscillating history is shown in Fig. 10.7.

10.4.2.2 Joint Calculation of Surge Chamber Wave and Guaranteed Regulation Parameters

The joint calculation of surge chamber wave and guaranteed regulation parameters is mature in this computer popular time. Compared to the separate calculations of surge chamber wave and water hammer, the joint calculation has many advantages.

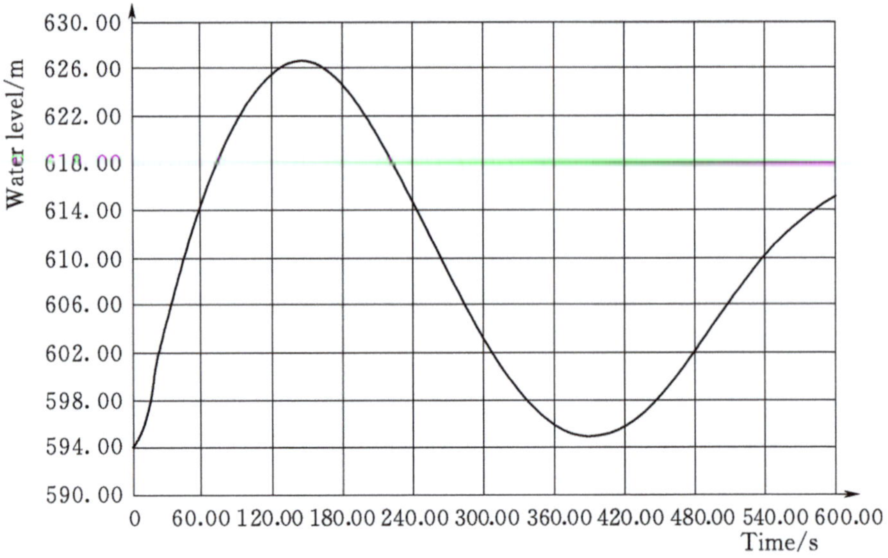

Fig. 10.7 Surge chamber water level history after load rejection calculated by Runge-Kutta method

(1) The influence of surge chamber wave on the guaranteed regulation parameters can be accurately taken into account.
(2) The combined transient processes caused by multiple operations of hydropower plant can be easily simulated to consider the superimposition of surge chamber waves and its influence to guaranteed regulation parameters.
(3) It is easy to calculate the complex hydraulic systems with many different components, such as the "one-tunnel-three-unit" tailrace system in Fig. 10.8, in which three draft tube extension conduits join by the bifurcation under the surge chamber and lead to the one tailrace tunnel.

The joint calculation of surge chamber wave and guaranteed regulation parameters include several basic treatments: ① providing the boundary condition equations for the surge chamber; ② linking the boundary condition equations with the characteristic equations of the pipe sections just connected to the surge chamber for forming a posed equation set; ③ adding the surge chamber equation set into the calculation procedures of hydraulic transient simulations; ④calculating the transient process of the hydraulic system with surge chambers. Here, we only introduce the numerical models of several surge chamber types.

1. **Simple surge chamber**

As shown in Fig. 10.9, there are 7 unknowns for the simple surge chamber: the piezometric head H_{P1} and discharge Q_{P1} at the last section of Pipe 1; the piezometric head H_{P2} and discharge Q_{P2} at the first section of Pipe 2; the piezometric head H_{TP} at the surge chamber bottom; the discharge Q_{TP} flowing into the surge chamber; the surge chamber water level Z.

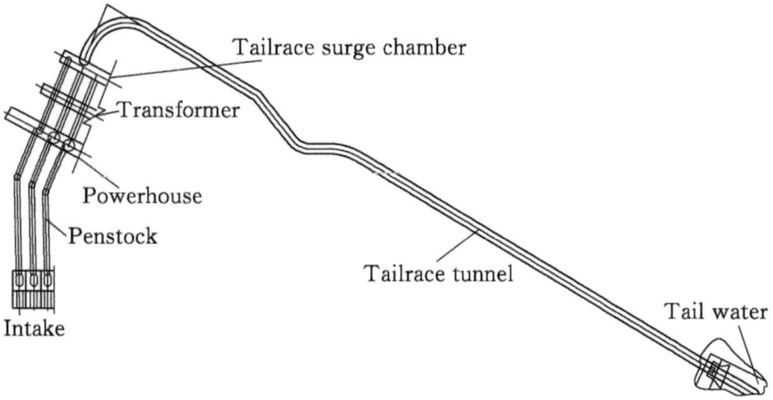

Fig. 10.8 "One-tunnel-three-unit" hydraulic system of a practical hydropower station

Fig. 10.9 Schematic for simple surge chamber boundary calculation

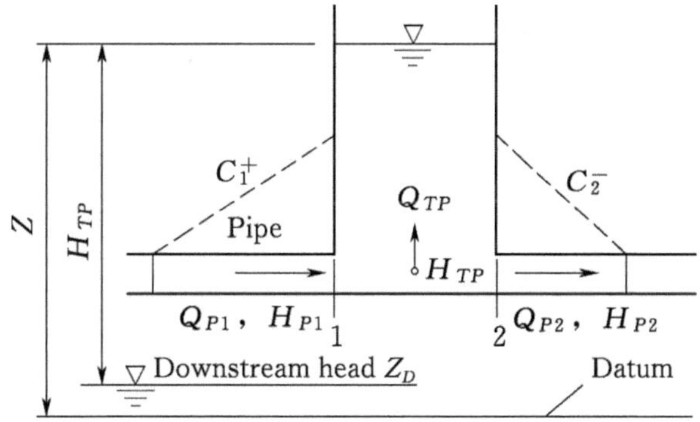

There are 7 equations available.

The characteristic equation at the last section of Pipe 1:

$$C_1^+ : \quad Q_{P1} = Q_{CP} - C_{QP} \cdot H_{P1} \tag{10.28}$$

The characteristic equation at the first section of Pipe 2:

$$C_2^- : \quad Q_{P2} = Q_{CM} + C_{QM} \cdot H_{P2} \tag{10.29}$$

The continuity equation in the bottom of the surge chamber:

$$Q_{P1} = Q_{P2} + Q_{TP} \tag{10.30}$$

If the head loss at the surge chamber bottom and velocity head are neglected, then

$$H_{P1} = H_{TP} \tag{10.31}$$

$$H_{\mathrm{P2}} = H_{\mathrm{TP}} \tag{10.32}$$

If the water inertia and the frictional loss in the surge chamber are neglected, we have

$$Z = H_{TP} + Z_{\mathrm{D}} \tag{10.33}$$

in which the tailrace water level Z_{D} is the datum for piezometic head H_{TP} and other piezometric heads.

The equation for surge chamber water level can be

$$Z^n = Z^{n-1} + \frac{Q_{TP}^n + Q_{TP}^{n-1}}{2} \frac{\Delta t}{F} \tag{10.34}$$

where Δt is the timestep size; F is the cross-sectional area, which may be the function of elevation; the superscript $n - 1$ stands for the value of the previous timestep.

Equations (10.28), (10.29)–(10.33), (10.34) are the 7 equations for the 7 unknowns, therefore, the solutions can be obtained.

2. **Differential surge chamber (throttle surge chamber is similar)**

Figure 10.10 is the schematic for differential surge chamber calculation, in which Z_{sD} is the top elevation of the rising shaft, Z_{DB} is the bottom elevation of the outer well, Z_{RP} is the water level in the rising pipe, and Z_{MP} is the water level in the outer well. The calculation of surge wave can be classified to the following four situations.

(1) When the water level is under the bottom plate of the outer well, the condition is the same as the throttle surge chamber.

In this situation (Fig. 10.9), there are 5 unknowns: discharge Q_{P1} at the last section of Pipe 1, discharge Q_{P2} at the first section of Pipe 2, piezometric head H_{TP} at the bottom of surge chamber, discharge Q_{TP} flowing into the surge chamber, and surge chamber water level Z.

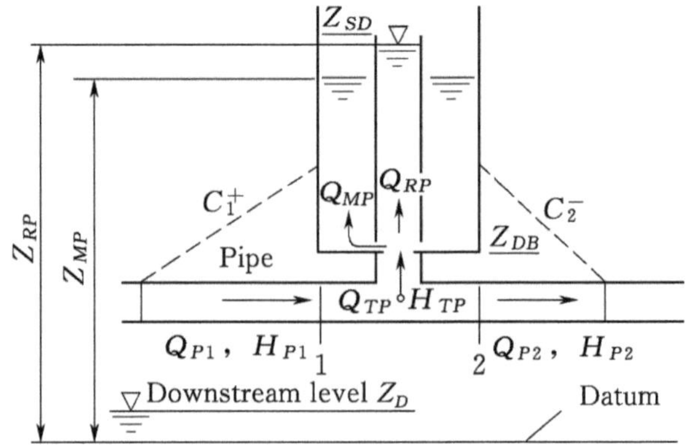

Fig. 10.10 Schematic for differential surge chamber boundary calculation

The available equations are similar to those for the simple surge chamber. To save space, we only list the one different. Equation (10.33) should be rewritten as

$$Z = H_{TP} + Z_D - \beta Q_{TP}|Q_{TP}| \tag{10.35}$$

to include the throttling effect, in which β is the throttle head loss coefficient.

(2) When there is no overflow at the top of the rising shaft, namely $Z_{SD} < Z_{RP}$ and $Z_{MP} > Z_{DB}$

As shown in Fig. 10.9, there are 8 unknowns for this situation: the discharge Q_{P1} at the last section of Pipe 1, the discharge Q_{P2} at the first section of Pipe 2, the piezometric head H_{TP} at the bottom of surge chamber, the discharge Q_{TP} flowing into the surge chamber, the water level Z_{MP} in the outer well, the discharge Q_{MP} flowing into the outer well, the water level Z_{RP} in the rising shaft, and the discharge Q_{RP} flowing into the rising shaft.

There are following 8 equations in Eqs. (10.36)–(10.43) for the 8 unknowns.

The characteristic equation at the last section of Pipe 1:

$$C_1^+ : Q_{P1} = Q_{CP} - C_{QP} \cdot H_{TP} \tag{10.36}$$

The characteristic equation at the first section of Pipe 2:

$$C_2^- : Q_{P2} = Q_{CM} + C_{QM} \cdot H_{TP} \tag{10.37}$$

The continuity equations:

$$Q_{P1} = Q_{P2} + Q_{TP} \tag{10.38}$$

$$Q_{TP} = Q_{RP} + Q_{MP} \tag{10.39}$$

The flow equation through the orifice between the outer well and the rising shaft:

$$\begin{aligned} Q_{MP} &= m_1 \times A_C\sqrt{2g(Z_{RP} - Z_{MP})} \quad \text{when} \quad Z_{RP} \geq Z_{MP} \\ Q_{MP} &= -m_2 \times A_C\sqrt{2g(Z_{MP} - Z_{RP})} \quad \text{when} \quad Z_{RP} < Z_{MP} \end{aligned} \tag{10.40}$$

where m_1 and m_2 are the discharge coefficients for flowing into and out, respectively, and A_C is the orifice area.

The water level equations:

$$Z_{RP} = H_{TP} + Z_D - \beta Q_{TP}|Q_{TP}| \tag{10.41}$$

$$Z_{RP}^{n} = Z_{RP}^{n-1} + \frac{Q_{RP}^{n} + Q_{RP}^{n-1}}{2}\frac{\Delta t}{A_S} \tag{10.42}$$

$$Z_{MP}^{n} = Z_{MP}^{n-1} + \frac{Q_{MP}^{n} + Q_{MP}^{n-1}}{2}\frac{\Delta t}{A_J} \tag{10.43}$$

where A_S is the cross-sectional area of the rising pipe, A_J is the cross-sectional area of the outer well, and superscript $n-1$ denotes the previous timestep.

(3) When there is overflow at the top of the rising pipe, namely $Z_{RP} > Z_{DB}$ and $Z_{MP} > Z_{DB}$.

In this situation, there are 9 unknowns: the discharge Q_{P1} at the last section of Pipe 1, the discharge Q_{P2} at the first section of Pipe 2, the piezometric head H_{TP} at the bottom of surge chamber, the discharge Q_{TP} flowing into the surge chamber, the water level Z_{MP} in the outer well, the discharge Q_{MP} flowing into the outer well, the water level Z_{RP} in the rising pipe, the discharge Q_{RP} flowing into the rising shaft, and the discharge Q_{CP} of the overflow.

Compared to situation (2), the overflow discharge Q_{CP} is the additional unknown. To close the equations, an overflow equation is added:

$$\begin{aligned} Q_{CP} &= \mu_1 \times L_C\sqrt{2g}(Z_{RP} - Z_{SD})^{\frac{3}{2}} \quad Z_{RP} > Z_{SD} \text{ and } Z_{MP} \le Z_{SD} \\ Q_{CP} &= -\mu_2 \times L_C\sqrt{2g}(Z_{MP} - Z_{SD})^{\frac{3}{2}} \quad Z_{RP} \le Z_{SD} \text{ and } Z_{MP} > Z_{SD} \end{aligned} \tag{10.44}$$

where μ_1 and μ_2 are the overflow discharge coefficients for flowing into and out the rising pipe, respectively, and L_C is the length of the overflow weir.

Equations (10.42) and (10.43) should be rewritten as

$$Z_{RP}^{n} = Z_{RP}^{n-1} + \frac{Q_{RP}^{n} + Q_{RP}^{n-1} - Q_{CP}^{n} - Q_{CP}^{n-1}}{2}\frac{\Delta t}{A_S} \tag{10.45}$$

$$Z_{MP}^{n} = Z_{MP}^{n-1} + \frac{Q_{MP}^{n} + Q_{MP}^{n-1} + Q_{CP}^{n} + Q_{CP}^{n-1}}{2}\frac{\Delta t}{A_J} \tag{10.46}$$

(4) When the water level is higher than the rising shaft.

In this situation, the equations and the solutions are the same with situation (1).

3. **Air-cushion surge chamber**

The schematic of air-cushion surge chamber is shown in Fig. 10.11. There are 6 unknowns, which are the discharge Q_{P1} at the last section of Pipe 1, the discharge Q_{P2} at the first section of Pipe 2, the piezometric head H_{TP} at the bottom of surge chamber, the discharge Q_{TP} flowing into the surge chamber, the water level Z in the air chamber, and the pressure head H_q in air.

Compared to the throttle surge chamber, only the air pressure head H_q is added. To close the equations, a gas state equation should be supplemented and Eq. (10.35) should be rewritten.

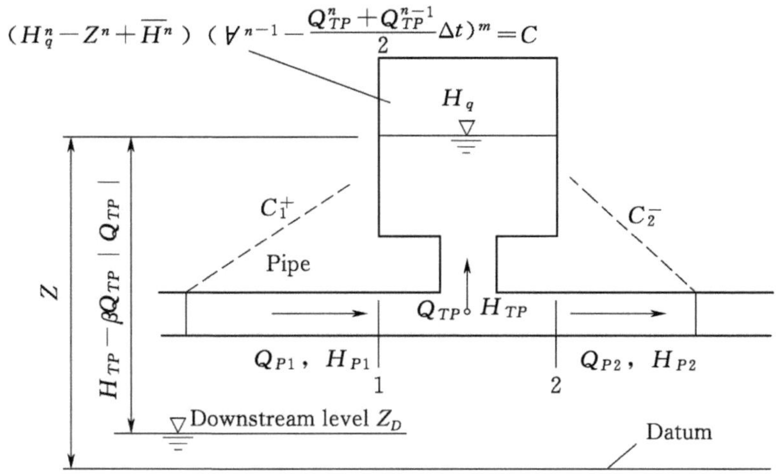

Fig. 10.11 Schematic of the air-cushion surge chamber

The gas state equation can be

$$\left(H_q^n - Z^n + \overline{H}^n\right)\left(\forall^{n-1} - \frac{Q_{TP}^n + Q_{TP}^{n-1}}{2}\Delta t\right)^m = C \qquad (10.47)$$

where $\forall$ is the volume of the air chamber, m is the polytropic constant of idea gas with value 1.0 to 1.4, $\overline{H}$ is the atmospheric pressure expressed in head form, and superscript $n-1$ denotes the previous timestep. Equation (10.35) becomes

$$H_q = H_{TP} + Z_D - \beta Q_{TP}|Q_{TP}| \qquad (10.48)$$

10.4.3 Critical Working Conditions for Calculation of Surge Chamber Wave

10.4.3.1 The Highest Water Level in the Headrace Surge Chamber

The design condition is normally as follows. When the reservoir level is at the normal level (or the design flood level for the powerhouse), all the units (n units) that are connected to the common surge chamber reject loads from the full load operation conditions, wicket gates close quickly.

The checking conditions are as follows.

(1) When the reservoir level is at the highest generating level, all the units (n units) that are connected to the common surge chamber reject loads from the full load operation conditions, wicket gates close quickly.
(2) When the reservoir level is at the normal level (or the design flood level for the powerhouse), the (n–1) units that are connected to the common surge chamber increase loads to the full load conditions and the final one unit increases load to the full load; the surge chamber water level oscillates because of the load increase; when the discharge flowing into the surge chamber reaches the largest, all the n units reject loads and all wicket gates close quickly.

10.4.3.2 The Lowest Water Level in the Headrace Surge Chamber

The design conditions are normally as follows.

(1) When the reservoir level is at the lowest generating level, the (n–1) units that are connected to the common surge chamber are operating in the full load conditions and the final one unit increases load to the full load.
(2) When the reservoir level is at the lowest generating level, all the n units that are connected to the common surge chamber increase loads from 2/3 loads to the full loads.
(3) When the reservoir level is at the lowest generating level, all the n units reject loads and all wicket gates close quickly; the second extreme of water level history curve may be the lowest water level of the surge chamber.

The checking condition is normally as follows. When the reservoir level is at the lowest generating level, all the n units reject load and all wicket gates close quickly; the surge chamber water level oscillates because of the load rejection; when the discharge flowing out the surge chamber reaches the largest, one of the n units increases load from speed-no-load to full load.

10.4.3.3 The Highest Water Level in the Tailrace Surge Chamber

The design conditions are normally as follows.

(1) When the tailrace water level is at the design tailrace level for the powerhouse, the (n–1) units that are connected to the common surge chamber are operating in the full load conditions and the final one unit increases load to the full load.
(2) When the tailrace water level is at the design tailrace level for the powerhouse, all the n units that are connected to the common surge chamber increase loads from 2/3 loads to the full loads.
(3) When the tailrace level is at the design tailrace level for the powerhouse, all the n units reject loads and all wicket gates close quickly; the second extreme of water level history curve may be the highest water level of the surge chamber.

The checking condition is normally as follows.

(1) When the tailrace water level is at the checking tailrace level for the powerhouse, the (n–1) units that are connected to the common surge chamber are operating in the full load conditions and the final one unit increases load to the full load.
(2) When the tailrace water level is at the checking tailrace level for the powerhouse, all the n units that are connected to the common surge chamber increase loads from 2/3 loads to the full loads.
(3) When the tailrace water level is at the checking tailrace level for the powerhouse, all the n units reject loads and all wicket gates close quickly; the second extreme of water level history curve may be the highest water level of the surge chamber.
(4) When the tailrace water level is at the design tailrace level for the powerhouse, all the n units that are connected to the common surge chamber reject the full loads and all wicket gates close quickly; the surge chamber water level oscillates because of the load rejections; when the discharge flowing into the surge chamber reaches the largest, one of the n units increases load from speed-no-load to full load.

10.4.3.4 The Lowest Water Level in the Tailrace Surge Chamber

The design conditions are normally as follows.

(1) When the tailrace water level is at the level corresponding to the n units in operation, all the n units that are connected to the common surge chamber reject full loads and all wicket gates close quickly.
(2) When the tailrace water level is at the lowest tailrace level, the operating units reject loads and wicket gates close quickly.

The checking condition is normally as follows. When the tailrace water level is at the lowest tailrace level, the (n–1) units that are connected to the common surge chamber are generating and the final one unit increases load; these cause the oscillation of surge chamber water level; when the discharge flowing out the surge chamber reaches the largest, all the n units reject loads and all wicket gates close quickly.

10.4.3.5 Other Considerations

If the lowest water level in the headrace surge chamber and the highest water level in the tailrace surge chamber in the checking conditions cannot meet the requirements of standards, adjusting the time interval of successive unit start-ups, or controlling the load increase increment, or restricting the restart-up time after load rejection can be adopted to solve the problem.

If the unit number and power outlet routes are more, and the transmission bus bars are sectioned, the analysis and evaluation on the possibility of simultaneous load rejections of all units should be conducted. If the situation of simultaneous load rejections is impossible, only the load rejections of some units should be considered in the calculation of the highest surge water level.

If the hydropower station stores water by stages, analyses should be conducted for different stages.

For surge chambers in large hydropower stations or complex surge chamber types, sometimes model tests should be conducted to investigate the detailed hydraulic characteristics.

10.4.4 Example of Joint Calculation of Surge Chamber Wave and Guaranteed Regulation Parameters

10.4.4.1 Parameters of the Practical Hydropower Station

Jinping II Hydropower Station, located on the main stream of Yalong River in Sichuan Province, China, installs 8 Francis turbine units, with capacity of 600 MW for each. The hydraulic system adopts "4 tunnels to 8 units" layout, which is mainly composed of the intakes, accident gates, headrace tunnels, headrace surge chambers, penstocks, etc. Figure 10.12 shows one of the four hydraulic systems. The average length of the systems is 16.67 km. The 1# and 3# headrace tunnels are constructed to four-center circular section of diameter 10.8 m in the middle reach, and four-center horseshoe section of diameter 11.8 m in other reaches. The 2# and 4# headrace tunnels are constructed to four-center horseshoe section of diameter 11.8 m in all reaches. At the end of the headrace tunnels are four differential surge chambers, with cross-sectional area (including the rising pipe) 415.64 m^2 and total height 140 m. The 4 headrace tunnels are branched to 8 penstocks by Y-shaped bifurcations under the surge chambers. The diameter of penstocks changes from 6.5 to 6.05 m along the length. The tailrace hydraulic systems are "one-unit-one tunnel" layout, which is composed of tailrace tunnels, accident gates, and outlets. The total lengths of tailrace tunnels are between 212.4 to 261.3 m. The tailrace tunnels have archway shaped section of width 9.50 m and height 12.80 m.

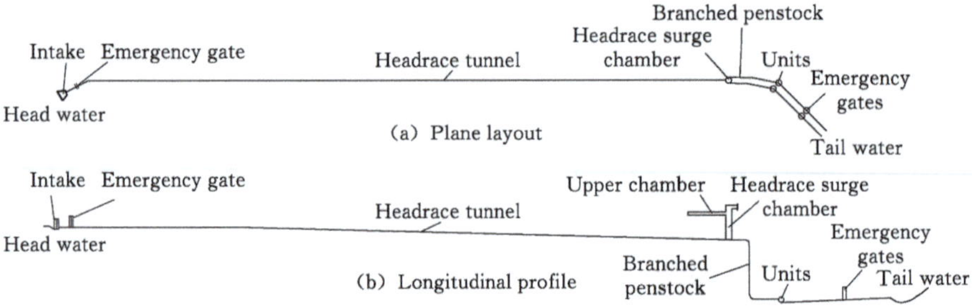

Fig. 10.12 Hydraulic system layout of one of the four systems of Jinping II Hydropower Station

10.4.4.2 Conditions for the Joint Calculation of Surge Chamber Wave and Guaranteed Regulation Parameters

The reservoir level is the normal water level 1646.00 m; the tailrace level is 1333.74 m, which is corresponding to 8 units in operation; the two units sharing a common hydraulic system reject loads simultaneously from the rated operation, and the wicket gates close quickly.

10.4.4.3 Results

The results for 1# and 4# hydraulic systems are listed in Table 10.1. Figure 10.13 is the histories of guaranteed regulation parameters for 1# turbine-generator unit, Fig. 10.14 is the histories of surge chamber parameters for 1# hydraulic system, and Fig. 10.15 is the envelope curves of pressure head along 1# hydraulic system.

Table 10.1 Extreme values for 1# and 4# hydraulic systems

	Largest pressure at spiral case/m		Lowest pressure at draft tube inlet/m		Maximum speed rise ratio/%		Water level difference between outer well and rising shaft/m	
Hydraulic system	1#	4#	1#	4#	1#	4#	1#	4#
Extreme value	365.96	364.98	6.31	5.04	47.07	9.00	-31.53 14.36	-29.42 14.48
Happening time/s	27.26	192.24	5.46	5.04	2.88	45.87	20.40 376.80	19.20 347.40
	Highest water level/m				Lowest water level/m			
	Outer well		Rising shaft		Outer well		Rising shaft	
Hydraulic system	1#	4#	1#	4#	1#	4#	1#	4#
Extreme value	1681.86	1681.00	1681.86	1681.00	1607.39	1608.36	1607.38	1608.35
Happening time/s	201.96	191.54	201.96	191.54	498.56	481.20	499.08	479.50
	Upward pressure difference cross bottom plate/m				Downward pressure difference cross bottom plate/m			
	Outer well		Rising shaft		Outer well		Rising shaft	
Hydraulic system	1#	4#	1#	4#	1#	4#	1#	4#
Extreme value	14.75	14.68	2.94	2.61	46.81	33.77	2.87	9.86
Happening time /s	381.82	346.86	339.04	306.26	17.62	17.30	2.89	77.40

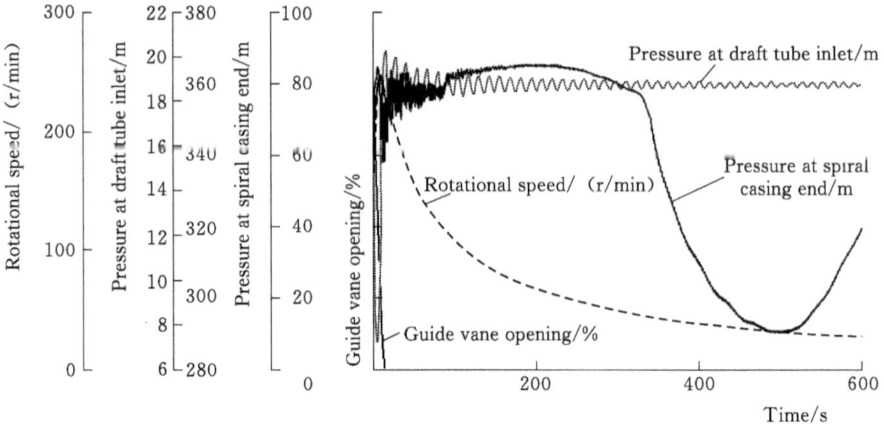

Fig. 10.13 Histories of guaranteed regulation parameters for 1# turbine-generator unit

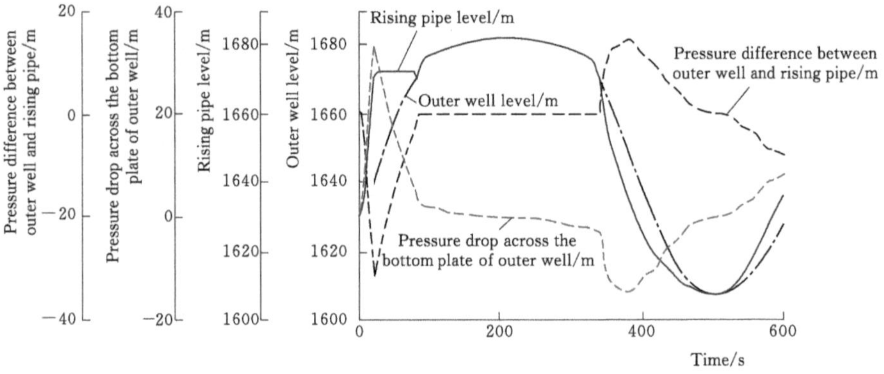

Fig. 10.14 Histories of surge chamber parameters for 1# hydraulic system

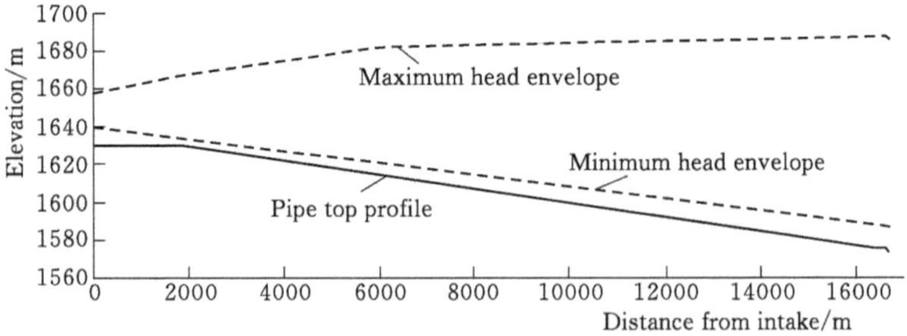

Fig. 10.15 Envelope curves of pressure head along 1# hydraulic system

10.4.4.4 Conclusions

From Table 10.1, Figs. 10.13 and 10.14, one can conclude as follows.

(1) The maximum dynamic pressure at spiral casing may be controlled either by the water hammer pressure generated by the closure of the wicket gate or by the maximum surge chamber wave. For example, the maximum spiral casing pressure of the 1# hydraulic system happens at an early time, and it is caused by water hammer fluctuation. On the other hand, the maximum spiral casing pressure of the 4# hydraulic system happens at a later time which is closer to the highest surge wave time, therefore, it is caused by the highest surge chamber wave.
(2) The minimum pressure at the draft tube inlet is controlled by the closing law of wicket gate, because the tailrace system is short in this example.
(3) The happening time of the maximum speed rise ratio is corresponding to the time at which the working point slides to the runaway curve. This ratio is controlled by the closing time and law of wicket gate and the inertia GD^2 of turbine-generator unit.
(4) The water level changes in the outer well and rising pipe can be described as follows. During the water level increase period, the rising rate in the rising pipe is faster because of its small cross-sectional area. When the overflow from the rising pipe begins, the water level difference between the outer well and the rising pipe reaches the largest. Then the water level in the rising pipe is stabilized, while the water level in the outer well rises continuously. After the outer well level reaches the top of the rising pipe, the two levels meet together and rise with no difference. In the water level decrease period, the level drop in the rising pipe is also faster.

10.5 Stability of Surge Chamber Water Oscillation

10.5.1 Surge Wave Stability of Headrace Surge Chamber

When the load of hydropower station changes, for example, from a partial load to the full load, it will inevitably cause the changes of turbine discharge and working head. The change of turbine discharge will lead to the oscillation of water level in surge chamber. And the change of surge chamber water level will cause the change of turbine working head. In order to keep constant output power that is corresponding to the full load, the governor automatically regulates the wicket gate opening to change the discharge for fitting the output power. The change of discharge conversely causes the change of surge chamber water level. This mutual excitation may lead to the magnification of surge chamber water oscillation, which is absolutely not allowed in actual hydropower stations. In order to maintain the stable operations of hydropower station, turbine-generator unit and power grid, it is necessary to require that the water level oscillation in surge chamber is attenuated and the attenuation should be fast. The equal amplitude oscillation of surge chamber water level is also not allowed, because it is only a critical stable state.

The task of this section is to theoretically derive the critical conditions for stability of surge chamber water level oscillation, thus provide guides for hydraulic design of surge chamber.

10.5.1.1 Condition of Stable Surge Chamber Water Oscillation

The unstable phenomenon of surge chamber water oscillation was first found in Haimbach Hydropower Station, Germany. Through hydropower expert Thoma's intensive study, a famous equation for calculating the critical stable section of surge chamber water oscillation, the Thoma equation was put forward in 1910.

The hypotheses of deriving the equation are as follows: (1) the oscillation amplitude of surge wave is infinitely small, thus the problem is linear; (2) the water inertia in the penstock can be neglected; (3) the governor regulation is ideal to keep constant output power; (4) the unit works alone and its efficiency remains unchanged. These four assumptions are collectively referred as the Thoma's hypotheses.

The derivation is based on the previous introduced equations for surge chamber wave, namely the momentum equation (10.3), continuity equation (10.4) and constant output power equation (10.5). Following perturbation method, by letting $V = V_0 + \Delta V$, $Q = Q_0 + \Delta Q$ and $Z = Z_0 + \Delta Z$ and inserting them into these equations, we can obtain

$$f\Delta V = F\frac{\mathrm{d}\Delta Z}{\mathrm{d}t} + \Delta Q \tag{10.49}$$

$$\frac{L}{g}\frac{\mathrm{d}\Delta V}{\mathrm{d}t} = -\Delta Z - 2\alpha' V_0 \Delta V \tag{10.50}$$

$$\Delta Q H_0' = -\left(\Delta Z + \frac{2k_1 V_0 \Delta V}{2g}\right) Q_0 \tag{10.51}$$

where $\alpha' = \alpha + \frac{k_1}{2g}$ and $H_0' = H_0 - h_{w0} - 3h_{m0}$.

Letting $\Delta Z = x$, using differential operators to replace ΔQ, ΔV and their derivatives, and combining Eqs. (10.49)–(10.51), we have

$$\frac{LF}{gf}H_0'\frac{\mathrm{d}^2x}{\mathrm{d}t^2} + \left(\frac{2\alpha' V_0 F}{f}H_0' - \frac{LQ_0}{gf}\right)\frac{\mathrm{d}x}{\mathrm{d}t} + \left(H_0' - 2h_{w0}\right)x = 0 \tag{10.52}$$

Rewrite Eq. (10.52), we finally obtain

$$\frac{\mathrm{d}^2x}{\mathrm{d}t^2} + 2n\frac{\mathrm{d}x}{\mathrm{d}t} + P^2x = 0 \tag{10.53}$$

where $n = \frac{V_0}{2}\left(\frac{2\alpha' g}{L} - \frac{f}{FH_0'}\right)$ and $P^2 = \frac{gf}{LF}\left(1 - 2\frac{h_{w0}}{H_0'}\right)$.

The ordinary differential equation (10.53) represents a damped free vibration whose damping term may be positive or negative. If the damping is zero ($n = 0$), the vibration will never decay and become a continuous periodic oscillation. If the head loss is not taken into account, the water level oscillation in surge chamber is an undamped vibration, and the period is

$$T = 2\pi\sqrt{\frac{LF}{gf}} \tag{10.54}$$

In practical situations, damping always exists, and the influence of damping on the oscillation period T is very small, therefore, Eq. (10.54) is often used in analysis.

Assuming the solution of Eq. (10.54) is $x = \mathrm{e}^{\lambda t}$, one can obtain the characteristic equation:

$$\lambda^2 + 2n\lambda + P^2 = 0 \tag{10.55}$$

which has roots $\lambda_1 = -n + \sqrt{n^2 - P^2}$ and $\lambda_2 = -n - \sqrt{n^2 - P^2}$.

The solutions to Eq. (10.53) can be classified to the following three situations.

(1) $n^2 < P^2$. λ has two complex roots: $\lambda_1 = -n + \mathrm{i}\sqrt{P^2 - n^2}$ and $\lambda_2 = -n - \mathrm{i}\sqrt{P^2 - n^2}$. Inserting the roots into $x = e^{\lambda t}$, one obtains the two special solutions to Eq. (10.53):

$$x_1 = \frac{C_1}{2}\left(\mathrm{e}^{\lambda_1 t} + \mathrm{e}^{\lambda_1 t}\right) = C_1\mathrm{e}^{-nt}\cos\sqrt{P^2 - n^2}t,$$
$$x_2 = \frac{C_2}{2i}\left(\mathrm{e}^{\lambda_1 t} - \mathrm{e}^{\lambda_2 t}\right) = C_2\mathrm{e}^{-nt}\sin\sqrt{P^2 - n^2}t.$$

Therefore the general solution is

$$x = \mathrm{e}^{-nt}\left(C_1\cos\sqrt{P^2 - n^2}t + C_2\sin\sqrt{P^2 - n^2}t\right) = x_0\mathrm{e}^{-nt}\cos\left(\sqrt{P^2 - n^2}t - \theta\right) \tag{10.56}$$

This equation defines the periodical oscillation of surge chamber wave.

When $n > 0$, the factor e^{-nt} decreases with time t, and the oscillation is damping; when $n < 0$, the factor e^{-nt} increases with time t, and the oscillation is also damping; when $n = 0$, the resistance of system is zero, and (10.56) is a cosine curve without damping.

From the above discussion, we know that the necessary condition for damping oscillation represented by Eq. (10.56) is $n > 0$. This condition is obviously satisfied, because Eq. (10.56) is obtained under the conditions $n^2 < P^2$.

(2) $n^2 = P^2$. The general solution to Eq. (10.55) is $x = e^{-nt}(C_1 t + C_2)$, which presents an aperiodic oscillation with damping condition $n > 0$.

(3) $n^2 > P^2$. When the damping is large, the roots of Eq. (10.55) are two real ones. Inserting the roots into $x = e^{\lambda t}$ yields the general solution of Eq. (10.53): $x = C_1 e^{\lambda_1 t} + C_2 e^{\lambda_2 t}$. This solution has no periodic factor, indicating the oscillation is aperiodic. The damping condition is $\lambda_1 < 0$ and $\lambda_2 < 0$, namely $n > 0$ and $P^2 > 0$.

According to the above discussion, we know that the necessary and sufficient conditions for the oscillation in the "tunnel-surge chamber" system to be attenuated in any case are $n > 0$ and $P^2 > 0$.

Condition $n > 0$ yields

$$F > \frac{Lf}{2\left(\alpha + \frac{k_1}{2g}\right)g(H_0 - h_{w0} - 3h_{m0})} \tag{10.57}$$

This equation shows that one of the damping conditions is that the cross section area of surge chamber must be greater than a certain value, which is

$$F_{th} = \frac{Lf}{2\left(\alpha + \frac{k_1}{2g}\right)g(H_0 - h_{w0} - 3h_{m0})} \tag{10.58}$$

where F_{th} is the stable critical cross section area of surge chamber, normally called Thoma area.

The condition $P^2 > 0$ requires

$$H_0 > 3(h_{w0} + h_{m0}) \tag{10.59}$$

For any hydropower station, the sum of tunnel head loss and penstock head loss should be far smaller than 1/3 of the hydrostatic head, otherwise, the construction of the hydropower station is uneconomic and infeasible.

10.5.1.2 Discussion of Stability Condition of Surge Chamber Water Oscillation

1. **The parameter selection for Thoma equation**

From the numerator of Eq. (10.58), we know that the longer the tunnel length L or the larger the tunnel cross section area f, the larger the critical surge chamber area F_{th}. From the denominator of Eq. (10.58), we know that the smaller the head H_0 or the smaller head loss coefficient α, the larger the critical surge chamber area F_{th}. Therefore, when calculating F_{th}, the minimum head appearing in the normal operation of the hydropower station, and the lowest water level in the reservoir that is generally the dead water level should be selected. However, the special minimum water level should also be considered if the hydropower station has the task of preliminary stage generation and war time generation. The minimum roughness should be chosen when calculating α. The definition of $\alpha = h_w/V^2$ should be paid special attention, in which h_w includes all head losses from the tunnel inlet to the surge chamber bottom, namely the summation of frictional losses and local losses. For the penstock, on the other hand, the average roughness can be selected. The discharge value in the calculation should correspond to the minimum head of the hydropower station.

In addition, it can be seen from Eq. (10.58) that the stable surge chamber areas required by high head hydropower stations are small, and the amplitudes of the water level oscillations play a controlling role in the design of surge chamber. Therefore, for the condition with high head and large reservoir level variation, the double chamber surge chamber can be used. But the cross area of the vertical shaft that connects the two chambers must meet the requirement of Thoma stability. On the other hand, for the low head or middle head hydropower stations, the stability of water level oscillation plays a controlling role, therefore, simple, throttled or differential surge chamber can be used. The stable area of water level oscillation in a differential surge chamber is guaranteed by the area summation of outer well and rising pipe.

2. **Influence of velocity head $\frac{k_1}{2g}$ under the surge chamber**

For a simple surge chamber, the velocity head at the bottom of surge chamber is approximately zero, thus $\alpha' = \alpha$. It should be noted that α includes the head loss coefficient of sudden expansion at the bottom of surge chamber $\frac{1}{2g}$.

For throttled surge chambers with orifice or connecting pipe, the velocity head at the bottom of the surge chamber is not zero, normally $k_1 = 0.7 \sim 1.0$ to 1.0. For this condition, α does not includes the head loss coefficient $\frac{1}{2g}$. Therefore, under the same conditions, the stable areas of simple surge chamber and throttled surge chamber are similar. However, the head loss of throttled surge chamber $\frac{V^2}{2g}$, is smaller than that of simple surge chamber, which improves the effective head of hydropower station.

3. **Influence of water inertia in the penstock**

Thoma equation does not consider the effect of water inertia in the penstock. Actually, the water inertia in the penstock is not conducive to the stability of surge tank water oscillation, and the buffer effect of the governor is conducive to the stability. Therefore, these two effects cancel out each other.

4. **Influence of turbine efficiency**

Constant turbine efficiency is assumed in the derivation of Thomas equation. In fact, turbine efficiency varies with head and discharge. Inserting $\eta = \eta_0 + \frac{\partial \eta}{\partial Q}\Delta Q_0 +$

$\frac{\partial \eta}{\partial H}\Delta H_0$ into the constant output power equation and repeating the derivation processes, one can obtain

$$F_{th} = \frac{Lf(1+\Delta)}{2\alpha' g[H_0 - 2h_{w0}(1+\Delta)]} \tag{10.60}$$

$$\Delta = \frac{H_0}{\eta_0}\left(\frac{\partial \eta}{\partial H}\right)_0 \Big/ \frac{Q_0}{\eta_0}\left(\frac{\partial \eta}{\partial Q}\right)_0 \tag{10.61}$$

in which subscript 0 stands for the steady working point (H_0, Q_0); $\frac{\partial \eta}{\partial H}$ and $\frac{\partial \eta}{\partial Q}$ stand for the changing rates of efficiency with head and discharge, respectively. These changing rates can be calculated based on the characteristic curves, as shown in Fig. 10.16.

The most adverse effect of efficiency on stable section happens when η increases with the increase of H and decreases with the increase of Q.

The changing trends of efficiency are different for different working points. Therefore, when considering the influence of turbine efficiency, it is necessary to select the minimum values of F_{th} by comparing different heads and different discharges.

5. **Influence of the power grid**

For hydropower station operating alone, the constant output power must be guaranteed by the units in the only power station, when surge chamber water level changes. For hydropower station operating within the power grid, the constant output power can be guaranteed by all units in the all power stations connected to the grid. Therefore, the changing range of discharge in the grid connected hydropower station is reduced, which is beneficial to the attenuation of surge chamber water level oscillation. This indicates that F_{th} can be reduced for hydropower stations participating in the grid. However, the influence of power grid is very complex, the influence of the grid can be considered only under the premise of full argumentation.

6. **Stability condition for large oscillation**

When the oscillation amplitude of surge chamber water is large, the differential equations for the motion are no longer linear. Therefore, Thoma equation cannot be directly applied to large oscillation problem. At present, we have no rigorous

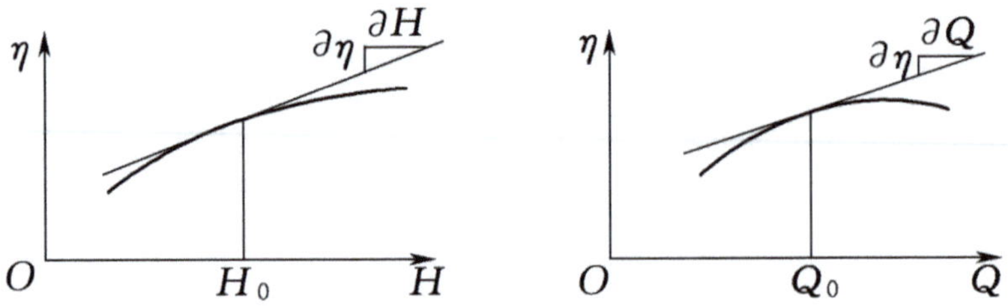

Fig. 10.16 Curves of turbine efficiency against head and discharge

analytical solutions to the nonlinear oscillation stability problem. To analyze the problem, only graphical method and numerical approximation can be used. However, research results show that if the stability of small oscillation cannot be guaranteed, large oscillation must not decay. In order to ensure the attenuation of large oscillation, the cross-sectional area of surge chamber must be larger than the critical stable section and have a certain safety margin.

10.5.2 Surge Wave Stability of Tailrace Surge Chamber

In deriving the stable condition for tailrace surge chamber wave, Thoma hypothesis is still the basis. However, for the tailrace surge chambers with connecting pipe or throttle orifice, the velocity head at the bottom of the surge chamber is negative, namely $\alpha' = \alpha - \frac{k_1}{2g}$. Therefore, the critical stable area is

$$F_{th} = \frac{Lf}{2g\alpha'(H_0 - h_{w0} - 3h_{m0})} \tag{10.62}$$

in which, F_{th} is the Thoma critical stable area, m^2; L is the length of pressure tailrace tunnel, m;f is the cross-sectional area of tailrace tunnel, m^2; H_0 is the smallest gross head, m; α is the head loss coefficient from the surge chamber to the lower reservoir or tailrace channel, with $\alpha = h_{w0}/V^2$ including local losses and frictional loss; V is the section-averaged velocity in tunnel, m/s; h_{w0} is the head loss value, m; h_{m0} is the head loss value in the upstream of the tailrace surge chamber, including the losses in the penstock and the draft tube extension conduit, m.

From Eq. (10.62), we know that for hydropower stations with medium or low head, the stable cross-sectional area for tailrace surge chamber with connecting pipe is much larger than that for simple surge chamber, because of the minus sign before the velocity head in the head loss coefficient. Thus, it becomes the main obstacle of adopting other tailrace surge chamber types beyond simple surge chamber. Before the constructions of Dachaoshan, Longtan and Xiaowan hydropower stations in China, the tailrace throttle surge chamber was rarely used in the world.

Using Gardel's head relations for bifurcation, reference [5] analyzed the influence of velocity head and momentum exchange term on the critical area of tailrace surge chamber. It was proved theoretically that the velocity head is not conducive to the stability and the momentum exchange term is conducive to the stability. Considering the two effects, the authors proposed

$$F_{th} = \frac{Lf}{g\left[2\left(\alpha - \frac{k_1}{2g}\right) + \frac{\overline{\sigma}}{2g}\right](H_0 - h_{w0} - 3h_{m0})} \tag{10.63}$$

where $\overline{\sigma} = 0.5\left[0.4\left(1+\frac{1}{\varphi}\right)\text{ctg}\frac{\theta}{2} + 0.36 + 3.84 - \varphi\right]$, which is defined by the area ratio φ between the connecting pipe and tailrace tunnel, and by the intersection angle θ between the connecting pipe and the tailrace tunnel axis. In the specific conditions of φ=1 and $\theta = 90^{\circ}$, $\overline{\sigma} = 2.0$.

Inserting $k_1 = 1$ and $\overline{\sigma} = 2.0$ into Eq. (10.63), one can obtain

$$F_{th} = \frac{Lf}{2g\alpha(H_0 - h_{w0} - 3h_{m0})} \tag{10.64}$$

Equation (10.64) is the formula for calculating the stable cross-sectional area of simple type tailrace surge chamber. It shows that for tailrace surge chamber with connecting pipe, the effects of velocity head and momentum exchange term at the connecting pipe can cancel out each other, and the resulting formula is close to that for the tailrace simple surge chamber.

For simple application, the new standard *Design Code for Surge Chamber of Hydropower Stations* NB/T35021-2014 explicitly claims that the calculation of critical cross-sectional area of tailrace surge chamber should use Eq. (10.64) without needing to distinguish surge chamber structural type.

It is worth noting that if the size and type of the connecting pipe are reasonably selected, the cross-sectional area needed for stability of tailrace surge chamber can be further reduced.

10.5.3 Surge Wave Stability of Air-Cushion Surge Chamber

For a high head hydropower station with long headrace tunnel, if the terrain is limited or the environmental protection requirements are high but the geological conditions are favorable, an air-cushion type surge chamber may be considered. The air-cushion surge chamber is normally avoided in tailrace system, because of small gas pressure in the chamber.

Based on the Thoma hypothesis and the gas state equation (10.47), the equation for calculating the critical stable area of air-cushion surge chamber can be derived as

$$F_{sV} = F_{th}\left(1 + \frac{nP_0}{l_0}\right) = \frac{Lf}{2g\alpha'(H_0 - h_{w0} - 3h_{m0})}\left(1 + \frac{np_0}{\gamma_0}\right) \tag{10.65}$$

where F_{SV} is the critical stable area for air-cushion surge chamber, m^2; F_{th} is the critical stable area for normal surge chamber types, m^2; n is the polytropic constant of idea gas, n=1.0 to 1.4; p_0 is the absolute pressure in the chamber, N/m^2; γ is the unit weight of water, kg/m^3; l_0 is the converted height from dividing the air volume in the chamber by area F_{SV}, m.

From Eq. (10.65), we know the following.

(1) Strictly speaking, the stable condition of surge wave in air-cushion surge chamber depends on the surface area, absolute pressure of gas, and gas volume of the surge chamber, rather than just one of these factors.
(2) The critical stable area of air-cushion surge chamber is larger than that of the conventional surge chamber, by the multiplier $\left(1+\frac{np_0}{y_0}\right)$. That is, the critical section area is inversely proportional to the gas volume (l_0); when the gas volume is equal to infinity, it will be equivalent to a conventional surge chamber, and the required stable sectional area will be equal to the Thoma area; the stable sectional area is proportional to the absolute pressure of gas (p_0) in chamber, and the larger the absolute pressure, the larger the stable sectional area.
(3) Rewriting Eq. (10.65) yields $F_{SV}\left(\frac{1}{F_{th}}-\frac{nP_0}{\gamma \forall_0}\right)=1$, in which $\forall_0 = l_0 F_{SV}$. This expression shows that the gas volume $\forall_0$ should not be too small, otherwise the water oscillation in the air-cushion surge chamber would be unstable.
(4) Because $\frac{P_0}{\gamma} = Z_u - Z_0 - h_{w0} + \overline{H}$, the critical stable volume of gas can be derived by inserting this relation into (10.65)

$$\forall_{SV} = \frac{\left[l_0 + n\left(Z_u - Z_0 - h_{w0} + \overline{H}\right)\right]Lf}{2g\alpha'(H_0 - h_{w0} - 3h_{wm0})} = l_0 F_{th} + \frac{np_0}{\gamma} F_{th} \tag{10.66}$$

10.5.4 Surge Wave Stability for System with Both Headrace and Tailrace Surge Chambers

For the system with both headrace and tailrace surge chambers, the water level oscillations of two surge chambers affect each other during transient conditions. The cross-sectional area of surge chamber obtained by Evangelisti equation in the early years is obviously too large. Later, Jaeger introduced dimensionless parameters to study the stable area of surge wave in headrace and tailrace double chamber systems directly, but the assumed conditions are quite different from the actual conditions. In reference [6], the stability conditions and the variation laws of stability region of water level oscillations in the headrace and tailrace double surge chambers were analyzed in depth. The analytical equations for the interference point and resonance point in the stable domain are derived theoretically. It was pointed out that the sectional areas of the headrace and tailrace surge chambers cannot be Thoma section at the same time.

10.5.4.1 Basic Equations and Stability Criterion

For the system with both headrace and tailrace surge chambers, as shown in Fig. 10.17, the basic equations can be listed as follows:

The momentum equation for headrace surge chamber:

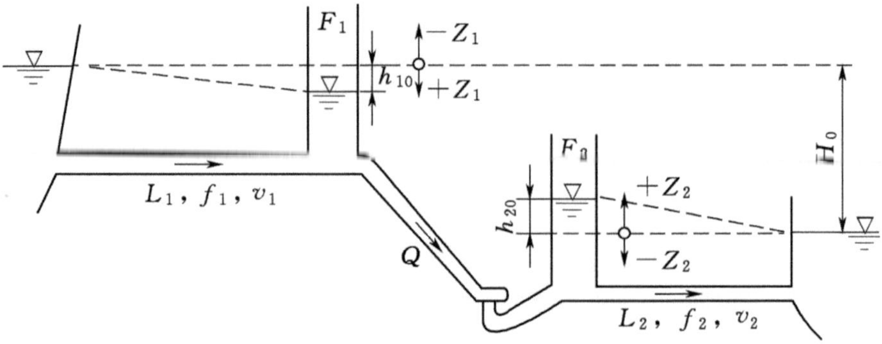

Fig. 10.17 Schematic of the hydraulic system with both headrace and tailrace surge chambers

$$\frac{L_1}{g}\frac{\mathrm{d}V_1}{\mathrm{d}t} = Z_1 - h_1 - \frac{k_1 V_1^2}{2g} \tag{10.67}$$

The continuity equation for the headrace surge chamber:

$$Q_1 = f_1 V_1 + F_1 \frac{\mathrm{d}Z_1}{\mathrm{d}t} \tag{10.68}$$

The momentum equation for the tailrace surge chamber:

$$\frac{L_2}{g}\frac{\mathrm{d}V_2}{\mathrm{d}t} = Z_2 - h_2 + \frac{k_2 V_2^2}{2g} \tag{10.69}$$

The continuity equation for the tailrace surge chamber:

$$Q_1 = f_2 V_2 + F_2 \frac{\mathrm{d}Z_2}{\mathrm{d}t} \tag{10.70}$$

The constant output equation

$$Q_{10}(H_0 - h_{10} - h_{m0} - h_{20}) = Q_1(H_0 - Z_1 - h_m - Z_2) \tag{10.71}$$

The symbols in the above equations are defined in Fig. 10.17.

For further derivation, the following parameters are introduced.

(1) The stable sectional area and corresponding Thoma cycle for each surge chamber when it operates alone.

$$F_{thi} = \frac{L_i f_i}{2g\alpha_i' H_1'}, i = 1, 2 \quad T_{thi} = 2\pi\sqrt{\frac{L_i F_{thi}}{g f_i}}, i = 1, 2 \tag{10.72}$$

in which, $H_1' = H_0 - h_{10} - 3h_{m0} - h_{20}, \alpha_1' = \alpha_1 + k_1/2g, \alpha_2' = \alpha_2 - k_2/2g$, $H_1' = H_0 - h_{10} - 3h_{m0} - h_{20}, \alpha_1' = \alpha_1 + k_1/2g, \alpha_2' = \alpha_2 - k_2/2g$, and $H_1' = H_0 - h_{10} - 3h_{m0} - h_{20}, \alpha_1' = \alpha_1 + k_1/2g, \alpha_2' = \alpha_2 - k_2/2g$.

(2) The amplification factor of sectional area of surge chamber and the corresponding actual cycle

$$n_i = F_i/F_{thi}, i = 1, 2 \quad T_i = 2\pi\sqrt{\frac{L_i F_i}{g f_i}} = T_{thi}\sqrt{n_i}, i = 1, 2 \tag{10.73}$$

(3) The relative head loss coefficient

$$r_i = \frac{2\alpha_i' V_{i0}^2}{H_1'}, i = 1, 2 \tag{10.74}$$

From the above equations and definitions, the fourth-order differential equation for water level oscillation stability of the headrace and tailrace surge chamber system can be derived as follows:

$$A_0 x_1^{(4)} + A_1 x_1^{(3)} + A_2 x_1^{(2)} + A_3 x_1^{(1)} + A_4 x_1^{(0)} = 0 \tag{10.75}$$

where $A_0 = 1, A_1 = a_1 + a_2, A_2 = b_1 + b_2 + a_1 a_2 - c_1 c_2, A_3 = a_1 b_2 + b_1 a_2 - c_1 d_2 - c_2 d_1, A_4 = b_1 b_2 - d_1 d_2, a_i = \frac{2\pi}{T_{thi}}\sqrt{r_1\left(1 - \frac{1}{n_i}\right)}, b_i = \frac{4\pi^2}{n_i T_{thi}^2}(1 - r_i), c_i = \frac{2\pi}{n_i T_{thi}}\sqrt{r_i}$, $d_i = \frac{4\pi^2 r_i}{n_i T_{thi}^2}$, and i=1 and 2.

According to the stability theory of ordinary differential equation, when the coefficients of Eq. (10.75) satisfy the Routh-Herwitz stable condition, namely

$$\begin{array}{l} A_1 > 0, A_2 > 0, A_3 > 0, A_4 > 0 \\ \Delta_3 = A_1(A_2 A_3 - A_1 A_4) - A_3^2 > 0 \end{array}, \tag{10.76}$$

the oscillation governed by Eq. (10.75) will be stable.

10.5.4.2 Interference Point, Resonance Point, and Stability Domain

According to detailed analysis of the variation laws of stability domain for the system with both headrace and tailrace surge chambers (the processes are omitted here), it can be seen that two asymptotes and three characteristic points can determine the boundary of stability domain, as shown in Fig. 10.18, in which the magnification factors of the two surge chambers are the two coordinates. The two asymptotes are approaching $n_i = 1, i = 1, 2$; the three characteristic points are the interference points (appear in a pair at the intersections of $A_1 = 0$ and $A_2 = 0$) and resonance point.

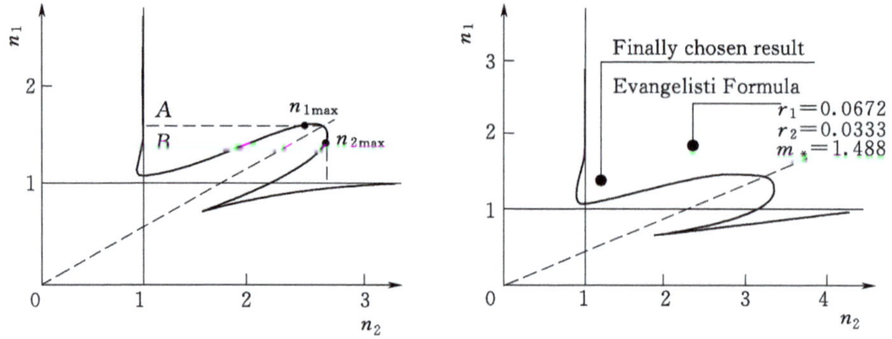

Fig. 10.18 Stability domain and parameters for a specific hydropower station

The interference points can be calculated by

$$n_2 = \frac{\left(m_* + \frac{1}{m_*} + 2K\right) \pm \sqrt{\left(m_* + \frac{1}{m_*} + 2K\right)^2 - 4(1 - r_1)\left(K + \frac{1}{m_*}\right)(K + m_*)}}{2(1 - r_1)\left(K + \frac{1}{m_*}\right)} \tag{10.77}$$

$$n_1 = \frac{\left(m_* + \frac{1}{m_*} + 2K\right) \pm \sqrt{\left(m_* + \frac{1}{m_*} + 2K\right)^2 - 4(1 - r_1)\left(K + \frac{1}{m_*}\right)(K + m_*)}}{2(1 - r_1)\left(K + \frac{1}{m_*}\right)} + \frac{1 + \frac{K}{m_*}}{1 - r_2} \tag{10.78}$$

where $m_* = T_{th1}/T_{th2}$ is the ratio of Thoma cycles; $K^2 = r_2/r_1$ is the ratio of head loss coefficients.

When the cycles of headrace and tailrace surge chambers are equal, the water level oscillations will resonate. Therefore, the point of equal cycles on the stability boundary curve will be the resonance point. The equation of resonance point is as follows.

In general, the area magnification factors of the two surge chambers approximately satisfies the following relation:

$$(n_1 - 1)(n_2 - 1) = 1 \tag{10.79}$$

In order to calculate the resonance point accurately, one can set $n_1 = 1/m_*^2 + 1$ as the initial value and conduct the following iteration:

$$n_{1,j+1}^* = n_{1,i}^* - \frac{E_0 n_{1,i}^3 + E_1 n_{1,i}^2 + E_2 n_{1,i} + E_3}{3E_0 n_{1,i}^2 + 2E_1 n_{1,i} + E_2} \tag{10.80}$$

where $E_0 = m_*\left(\frac{X_1}{X_2} - \sqrt{r_1 r_2}\right)\frac{X_1}{X_2}$, $E_1 = -(2 - r_1 - r_2)\frac{X_1}{X_2} - \left(1 + \frac{1}{m_*^2}\right)m_*\left(\frac{X_1}{X_2} - \sqrt{r_1 r_2}\right)\frac{X_1}{X_2}$,

$E_2 = \frac{1-r_1-r_2}{m_*} + \left(1 + \frac{1}{m_*^2}\right)\left(2\frac{X_1}{X_2} - \sqrt{r_1 r_2}\right)$, $E_3 = -\left(1 + \frac{1}{m_*^2}\right)\frac{1}{m_*}$, $X_1 = \sqrt{r_1} + m_*\sqrt{r_2}$ and $X_2 = \sqrt{r_2} + m_*\sqrt{r_1}$.

In normal conditions, one or two iterations are sufficient.

10.5.4.3 Design Criteria and Calculation Method for Stable Section

According to the above equations and methods, the approximate contour of the stable boundary can be drawn. When the point is located on the top right of the boundary curve, the system is stable. Considering the requirements of stability and margin, it is suggested that the design criteria for stable sections of water level oscillations in the headrace and tailrace chambers are as follows.

When $m_* \geq 1$, choosing $n_1 = (1 \sim 1.05)n_{1\,\max}$ to $1.05)n_{1\,\max}$ and $n_2 = 1.05 \sim 1.1$ to 1.1 is enough; when $m_* < 1$, choosing $n_2 = (1 \sim 1.05)n_{2\,\max}$ to $1.05)n_{2\,\max}$ and $n_1 = 1.05 \sim 1.1$ to 1.1 is proper. Here, $n_{1\,\max}$ and $n_{2\,\max}$ are defined as

$$n_{1\,\max} = \frac{n_{1r}}{1 - r_2},\ n_{2\,\max} = \frac{n_{2r}}{1 - r_1} \tag{10.81}$$

where n_{1r} and n_{2r} are the values of the resonance point.

The above selection methods can result in reasonable sectional areas for the headrace and tailrace surge chambers, because both cost issues and safety margins are considered.

Questions

1. Why can the surge chamber shorten the penstock and reflect the water hammer wave? Please deduce the reflection coefficient and penetration coefficient of water hammer in throttle surge chamber.
2. Please draw the highest surge water level fluctuation history curve for upstream surge chamber and downstream surge chamber. Attention should be paid to use the correct conditions: the transient process, the reservoir (tail water) water level, the initial water level of the surge chamber, the final stable water level, and the maximum pressure envelope of the tunnel.
3. Please draw the lowest surge water level fluctuation history curve for upstream surge chamber and downstream surge chamber. Attention should be paid to use the correct conditions: the transient process, the reservoir (tail water) water level, the initial water level of the surge chamber, the final stable water level, and the maximum pressure envelope of the tunnel.
4. What are the differences between the setting conditions for the upstream surge chamber and downstream surge chamber?

5. What kind of physical conservation phenomena does the fluctuating process of water level in surge chamber reflect? Can you obtain the analytical solution of water level fluctuation? What are the differences between large fluctuation and small fluctuation?
6. Defining the directions of water level and discharge is very important when deriving the basic equations of surge chamber water level fluctuation. Please derive the basic equation of water level fluctuation in upstream surge chamber, under the condition that the surge chamber water level is equal to the reservoir water level, and downward fluctuation is positive and upward fluctuation is negative; the flow direction in tunnel is positive if toward the surge chamber, and negative if leaving the surge chamber; the discharge of unit is positive if toward the unit.
7. Please derive the basic equation of water level fluctuation in downstream surge chamber, under the condition that the surge chamber water level is equal to the reservoir water level, and downward fluctuation is positive and upward fluctuation is negative; the flow direction in tunnel is positive if leaving the surge chamber, and negative if toward the surge chamber; the discharge of unit is positive if leaving the unit.
8. Using mathematical software Mathcad, Matlab, etc., we can directly solve the basic equations of surge chamber, and easily obtain the extreme water levels and fluctuating history curves. Please use one of the mathematical software to calculate the water level fluctuation process in the upstream throttle surge chamber.
9. What calculation conditions should be selected when determining the highest and lowest surge water levels? Please discuss these separately for the upstream surge chamber, downstream surge chamber, one water way one unit, one water way two units and so on.
10. What are the characteristics and applicable conditions of different types of surge chambers? How to coordinate the critical stable sectional area and surge wave amplitude when selecting surge chamber type?
11. What are the idealized assumptions in deriving Thoma's stable sectional area of surge chamber? What factors have influence on the critical sectional area of surge chamber, and how to consider these factors when calculating the critical sectional area?
12. What factors should be considered in determining the sectional area of surge chamber? Why?

Bibliography

1. National Energy Administration. 2015. *Specification for design of surge chamber in hydropower station: NB/T35021-2014 (In Chinese)*, 6. Beijing: China Electric Power Press.
2. Jiandong, Yang, X. Ming and W. Hegao et al. 2017. *Hydropower station*, 3rd ed. (In Chinese). Beijing: China Water and Power Press.

3. Jiandong, Yang, W. Zhengchun and Z. Jiajia. 2008. Discussion on the setting conditions of upstream surge chamber (In Chinese). *Journal of Hydroelectric Engineering* 27 (5): 114–117.
4. Zhao, Guilian, and J. Yang. 2004. Research on the condition to set a tailrace surge tank. *Journal of Hydrodynamics* 16 (4): 486–491.
5. Haiyan, Bao, Y. Jiandong and F. Liang et al. 2011. Discussion on the setting condition of surge Chamber based on regulating quality of Hydropower Station (In Chinese). *Journal of Hydroelectric Engineering* 30(3): 180–1860.
6. Jiandong, Yang, X. Lai, and C. Jianzhi. 1995. The effect of velocity head and momentum distribution on critical stable sectional area of the downstream surge tank (In Chinese). *Journal of Hydraulic Engineering (SHUILI XUEBAO)* 7: 59–66.
7. Jiandong, Yang, C. Jianzhi, and X. Lai. 1993. Stability of water level fluctuation in hydraulic system with both upstream and downstream surge chambers (In Chinese). *Journal of Hydraulic Engineering (SHUILI XUEBAO)* 7: 55–56.
8. Qizhao, Liu. 1995. *Surge chamber of hydropower station (In Chinese).* Beijing: China Water and Power Press.
9. Yuanlin, Hao and Y. Ting et al. 2017. *Design of air-cushion surge chamber for hydropower station* (In Chinese). Beijing: China Water and Power Press.
10. Jushan, Tang. 2007. *Typical design drawings of middle and small water conservancy and hydropower projects, water diversion buildings of hydropower stations, diversion tunnels and surge chambers (In Chinese).* Beijing: China Water and Power Press.
11. Кривченко, ГИ. 1981. *Transient process in Power equipment of hydropower station (In Chinese).* Translated by Chang Zhaotang. Beijing: Water Resources and Electric Power Press.
12. Rongqiao, Wu., and C. Jianzhi. 1997. *Hydraulic transients of hydropower station (In Chinese).* China Beijing: Water Resources and Electric Power Press.
13. Jaeger, C. 1987. *Fluid transients in hydroelectric engineering practice (In Chinese).* Translated by Wang Shuren et al. Dalian: Dalian Institute of Technology Press.
14. Wylie E.B. and V.L. Streeter. 1993. *Fluid transients in systems*, 3rd ed. New York: Prentice Hall Englewood Cliffs.
15. Chaudhry M.H. 2014. *Applied hydraulic transients*, 3rd ed. New York: Springer.

Chapter 11
Layout Design of Powerhouse

Juntao Chen

Abstract In this chapter, the important place of hydropower generation–the powerhouse of hydropower station is introduced. First of all, the four basic types of powerhouse, and the composition of buildings and equipment, are introduced in detail. Secondly, the main hydraulic machinery and electrical equipment in the powerhouse is summarized, and the layout of the main structural layer of the main powerhouse is presented, including the generator layer, hydraulic turbine layer, and spiral case layer. On this basis, the spatial structure design of the main powerhouse, including the calculation method of the main elevation and the determination of the plane size, is discussed. In addition, the auxiliary equipment of the powerhouse is introduced, including hydraulic machinery auxiliary equipment, electrical auxiliary equipment and operation and maintenance equipment, and the compositions of each equipment are explained. Then the function and classification of the auxiliary powerhouse are introduced, as well as the corresponding layout principle of the powerhouse. Finally, the influence factors of the layout of the whole powerhouse project and the layout principles of the main buildings are summarized.

Keyword Main powerhouse · Auxiliary powerhouse · Fundamental types of powerhouse · Buildings of powerhouse · Electromechanical equipment · Mechanical equipment · Auxiliary equipment · Size design of powerhouse · Influencing factors and layout principles for powerhouse arrangement

J. Chen (✉)
Wuhan University, Wuhan, China
e-mail: chenjt@whu.edu.cn

Y. Cheng et al. (eds.), *Hydropower Station*,
https://doi.org/10.1007/978-981-96-2383-9_11

11.1 Fundamental Types and Components of Powerhouses

11.1.1 Functions

The hydroelectric station powerhouse is a complex of hydraulic buildings, electromechanical equipment, and electrical equipment layout. From the perspective of energy utilization, it is a place where hydropower can be converted into electricity. Therefore, the main function of the powerhouse is to channel the concentrated water from the water-retaining or diversion channels into the hydraulic turbine through the penstock, effectively convert the hydropower into electricity by the electric generator, and deliver the electricity to the user reliably and economically as required.

In hydroelectric power generation, the potential and kinetic energy of water flow are first converted into the turbine's mechanical energy and then converted into electric energy by the generator. Hydropower plant powerhouse's mission is to arrange main equipment and auxiliary devices (which ensure the safe operation of the equipment) properly and economically through a series of engineering measures and provide effective operation conditions and place for the installation and maintenance of the equipment and devices. At the same time, it creates a good working environment for operators, achieving a safe and efficient transformation of hydropower into electrical energy.

11.1.2 Fundamental Types of Powerhouse

Due to the different development mode and layout of hydro projects, the types of powerhouses are also different. According to the structural characteristics of hydropower stations, hydropower stations can be divided into the ground powerhouse and the underground powerhouse.

11.1.2.1 Ground Powerhouse

Ground powerhouse means whose structure and buildings are all arranged on the ground. According to the location of the plant in the hydro project, the ground powerhouse can be divided into three types: powerhouse at the dam-toe, water-retaining powerhouse, and riverside powerhouse.

1. Powerhouse at dam-toe

This type of powerhouse is laid out at the rear of a non-overflow dam and connected with the dam body. In general, the hydropower station has a high water head and a large storage capacity. The dam blocks water, and the powerhouse is located behind the dam without playing a role in blocking water. The water for power generation is introduced into the powerhouse directly through the dam body. The powerhouse

Fig. 11.1 Powerhouse at dam-toe

at dam-toe is generally located behind the retaining dam section (Fig. 11.1). The powerhouse is separated from the dam by permanent joints, and the powerhouse is not subject to the hydraulic thrust of the dam body. The powerhouse at dam-toe is the most common form of layout. Depending on the location of the dam section where the powerhouse is located, there are several other forms as follows:

(1) Overflow type powerhouse. The powerhouse is arranged at the rear of the overflow dam monolith with the flood discharged through the top of the powerhouse. In the case of flood discharge, the vibration is relatively great under the hydrodynamic pressure of water flow. Generally, the dam is not very high. This type is used when the riverbed is relatively narrow, the discharge is large, the number of units is large and the geological conditions are poor, thus there is no place for the layout of the powerhouse dam monolith. The overflow section at the top of the dam is combined with the plant to form an integral building, which is called the overflow type powerhouse (Fig. 11.2).
(2) Ski-jump type powerhouse. The powerhouse is arranged at the dam-toe of the overflow dam monolith. When a high dam is built in the valley, high water head and large flow result in the flood jumping over the top of the powerhouse into the downstream river. As shown in Fig. 11.3, before the beginning and ending of the flood discharge, a small flow nappe will impinge on the roof of the powerhouse, but the time is short and the load is small, which will not threaten the safety of the powerhouse.

Fig. 11.2 Overflow type powerhouse

Fig. 11.3 Ski-jump type powerhouse

(3) Powerhouse within dam. The powerhouse is arranged inside the cavity of the dam and connected with the dam body. The size and shape of the cavity exert a great influence on the stress of the dam body. This type is used when the riverbed is relatively narrow and the discharge is large, thus there is no place for the layout of the powerhouse dam monolith, as shown in Fig. 11.4. This kind of powerhouse is placed in the dam body, which causes great construction disturbance, poor stress condition, and poor operation condition.

Fig. 11.4 Powerhouse within dam

Fig. 11.5 Water-retaining powerhouse

2. Water-retaining powerhouse

When the water head is low and the unit capacity is large, the powerhouse is located in a river and acts as a water-retaining structure. The reservoir volume of the water-retaining powerhouse (Fig. 11.5) is small, the reservoir depth is small, and the pollutants drifts seriously during the flood period. The tidal power plant is also a water-retaining powerhouse, as shown in Fig. 11.6. It is built in the bay and uses tubular turbine units of two-way power generation to generate electricity using the seawater's rise and fall. The units commonly used in water-retaining powerhouses are vertical axial flow turbines and horizontal tubular turbines. Different types of units adopted, the structural features of the powerhouse are also different. According to the arrangement of the unit and the discharge structure, the water-retaining type has the following forms:

(1) Water-retaining powerhouse for axial flow turbines with vertical shaft. The main powerhouse and the water intake are connected into an integral structure (Fig. 11.7) and the powerhouse is located in the centre of the riverbed. It is used in the condition of low water head and high flow. The powerhouse plays a role of water-retaining. Generally, the capacity of reservoir is not great and the reservoir is not deep. The turbines used are mostly axial turbines.

It is divided into three parts along the direction of water flow:

a. Intake section: determined by layout requirements for intakes, trash racks, and the upper lip curve of intakes and spiral case.

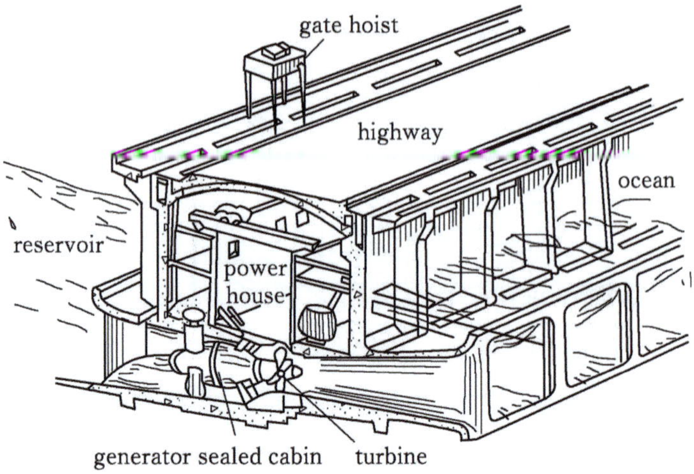

Fig. 11.6 Tidal powerhouse

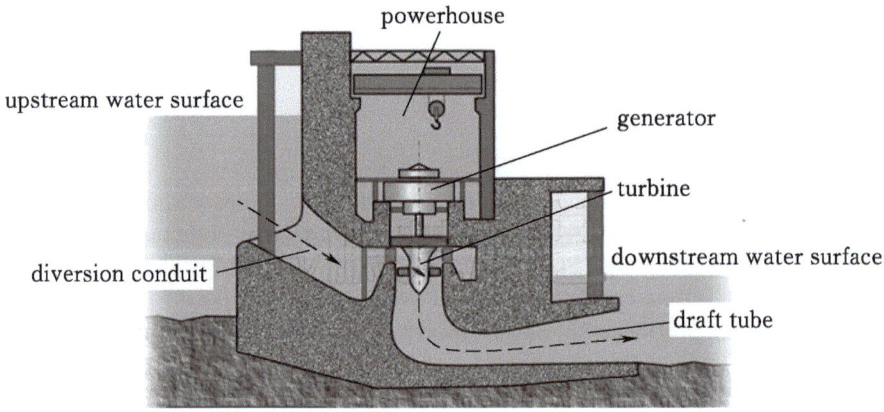

Fig. 11.7 Cross-section of water-retaining powerhouse with vertical axial flow turbine

b. Main powerhouse section: determined by powerhouse equipment layout and stability requirements.
c. Draft tube section: as the water head is relatively low and the discharge is relatively high and as required by the diffusion of the cross-section of the draft tube, the draft tube is relatively long, therefore the tailrace platform is normally equipped with main transformers (Fig. 11.8).

There are some problems for water-retaining powerhouse:

a. For the water-retaining powerhouse, stability is the key and the seepage pressure of the powerhouse foundations should be reduced to the greatest extent to avoid differential settlements.

Fig. 11.8 Layout of water-retaining powerhouse with vertical axial flow turbine

b. Because of low water head and high flow, it's difficult for trash interception and removal. In cold areas, trash racks should be prevented from freezing.
c. The reservoir is small and the sediment is likely to be deposited in front of the powerhouse.

(2) Water-retaining powerhouse with bulb turbine (Fig. 11.9). Water flows into and out of the powerhouse in cross-flow. Bulb turbine units are equipped in the flow passage of the powerhouse. Inside the bulb, there are generators which are fixed by radial concrete supporting structures in the flow passage. The concrete supporting structures can be designed as cavity to connect the plant and the bulb turbine.

Bulb turbine units have high efficiency, and they are only suitable for plants with low water head (lower than 20 m).

As great vibration occurs to the bulb, it is difficult to seal the main shaft when the unit is running.

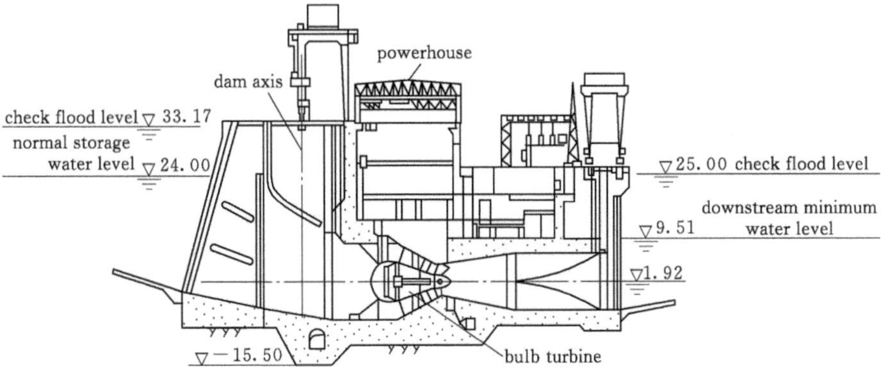

Fig. 11.9 Water-retaining powerhouse with bulb turbine

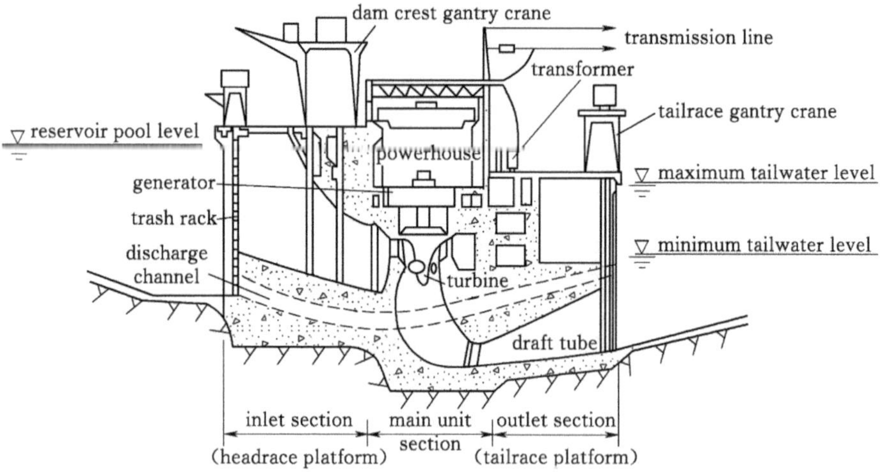

Fig. 11.10 Discharge tunnel is arranged between the spiral case and the draft tube

(3) Water-retaining powerhouse with discharge tunnel. The discharge tunnel passes through the powerhouse. The length of the discharge sluice can be reduced and the powerhouse is arranged compactly. Sediment and floater flushing can be achieved through water discharge to reduce dirt interference. In the case of flood discharge, water head can be increased with water discharge to increase the energy output. As the structure is complicated, over-discharge may cause resonance of the turbine and it will be inconvenient for operation.

Depending on the difference of location, there are three layout types for discharge tunnel:

a. The tunnel is located between the spiral case and the draft tube (Fig. 11.10), and water is discharged through the bottom outlet.
b. The tunnel is located above the top board of the spiral case (the upper part of the generator floor.as shown in Fig. 11.11), and water is discharged by weir flow.
c. The tunnel is located between two draft tubes, and sluices and units are arranged alternately, which is called the pier-type powerhouse.

3. Riverside powerhouse

The riverside powerhouse is separated from the dam, located by riverside, and its water head is created by conduits which bring the water for power generation. This kind of powerhouse mainly uses the diversion channel to concentrate all the water head (diversion type hydropower station) or part of the water head. Since the powerhouse is far from the dam, sluice gates and other buildings, the diversion channel is relatively long. Therefore, the riverside powerhouse is also known as the diversion type powerhouse, as shown in Fig. 11.12a. The diversion type hydropower station consists of three major parts: the headworks, the diversion buildings and the powerhouse project, as shown in Fig. 11.12b.

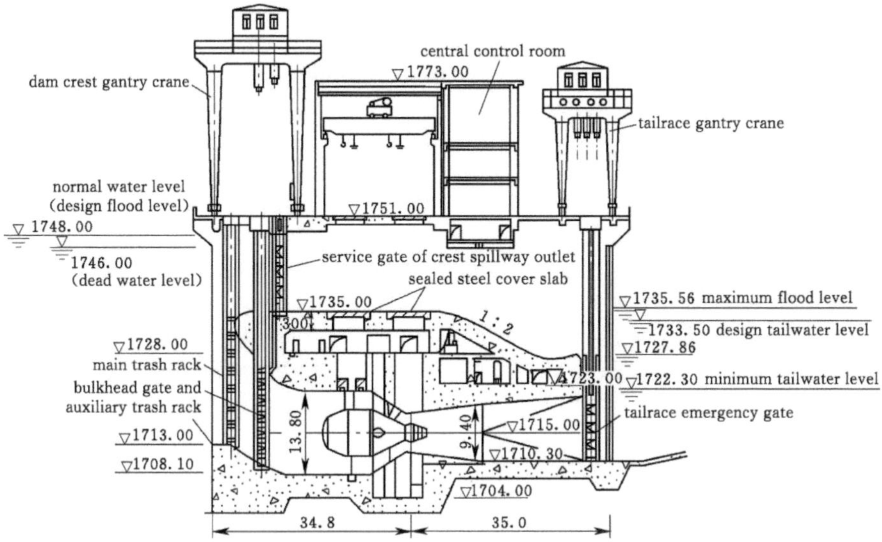

Fig. 11.11 Discharge tunnel is arranged at the top of the generator floor

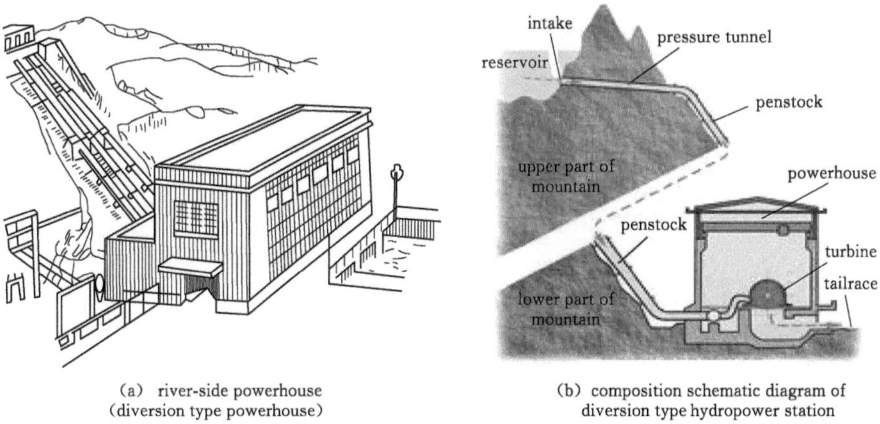

Fig. 11.12 Riverside powerhouse

Headwork structures: there are water-retaining and diversion structures to raise the water level and direct the water to the diversion channel, buildings for the removal of dirt, debris and sediment. Sometimes there're also anti-icing facilities and ice discharge structures, such as dams, river gates, water inlets of diversion channel, trash racks, desilting basin, flushing, and ice discharge facilities.

Water intake structures: in the free-surface flow diversion channel, there're usually arranged with lateral overflow weir, sand-guide sill, and protective engineering measures for preventing rock collapse and debris flows. Fore bay and daily regulation pond are usually built at the end of the diversion channel. In order to reduce the impact of water hammer and improve the operating conditions of the unit, surge chamber is usually set between the end of the pressurized flow diversion channel and the penstock.

Powerhouse project: the powerhouse project includes the end of penstock and a complete set of buildings thereafter, mainly including the main powerhouse, the auxiliary powerhouse, the booster switching station, the tailrace (the free-surface flow channel or the tunnel). The specific layout method is selected according to the topographic and geological conditions.

The layout position of powerhouse at dam-toe and water-retaining powerhouse is relatively unitary. The riverside powerhouse layout position is more flexible.

11.1.2.2 Underground Powerhouse

When the structure of the powerhouse is all arranged underground, it is called underground powerhouse. This kind of powerhouse is completely separated from the dam building, with less environmental interference and no interaction between the powerhouse and dam construction, which is advantageous to early electricity generation. But its operating conditions are poor, which requires high geological conditions.

Underground powerhouse layout are classified as follows.

According to the location of powerhouse relative to the diversion channel, there are headrace powerhouse, mid-section powerhouse and tail powerhouse.

According to the buried way of powerhouse, there are fully underground powerhouse, cavern powerhouse and the semi-underground powerhouse.

1. Fully underground powerhouse

The fully underground powerhouse chamber structure is all located in the underground, as shown in Fig. 11.13. The powerhouse and the outside world are connected with access tunnels, outlet tunnels, etc. Located in the deep part of the rock mass, the stability of the chamber is relatively good, but the transportation is not very convenient and the ventilation and lighting conditions are poor.

2. Cavern powerhouse

The main chamber of the underground powerhouse is directly connected to the ground at one end. The underground powerhouse is exactly a cave-house, as shown in Fig. 11.14. The external traffic of the powerhouse is directly connected with the external highway. The rock mass of the entrance of the powerhouse is shallow, so the stability of the surrounding rock around the entrance is relatively poor. But the ventilation and lighting conditions are better, and the transportation and the operation management are more convenient as well.

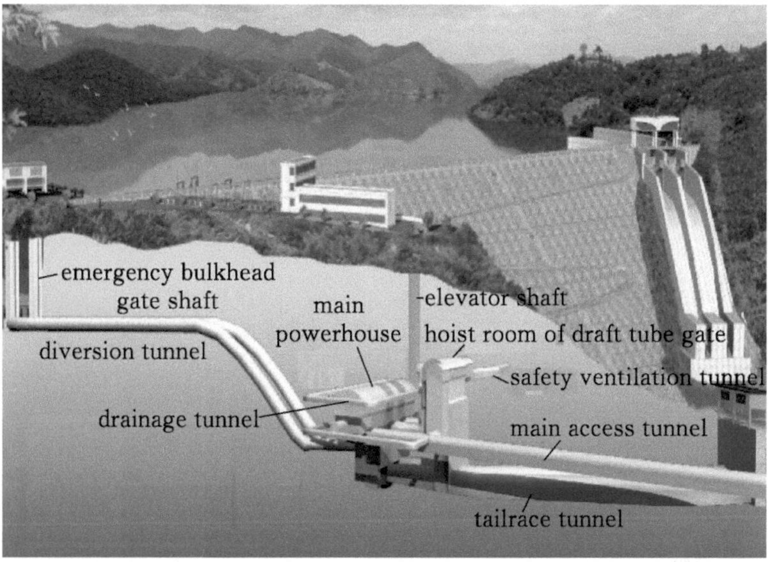

Fig. 11.13 Fully underground powerhouse

Fig. 11.14 Cavern powerhouse

3. Semi-underground powerhouse

Semi-underground powerhouse (Fig. 11.15), also called shaft powerhouse, is a special form of underground powerhouse. When the terrain excavation requires deeper or the terrain condition doesn't meet the requirement of the fully underground powerhouse, the powerhouse can be arranged in the underground shaft. The semi-underground plant consists of an underground shaft which opens on the top and an above-ground structures. The hydraulic turbine-generator units and their ancillary

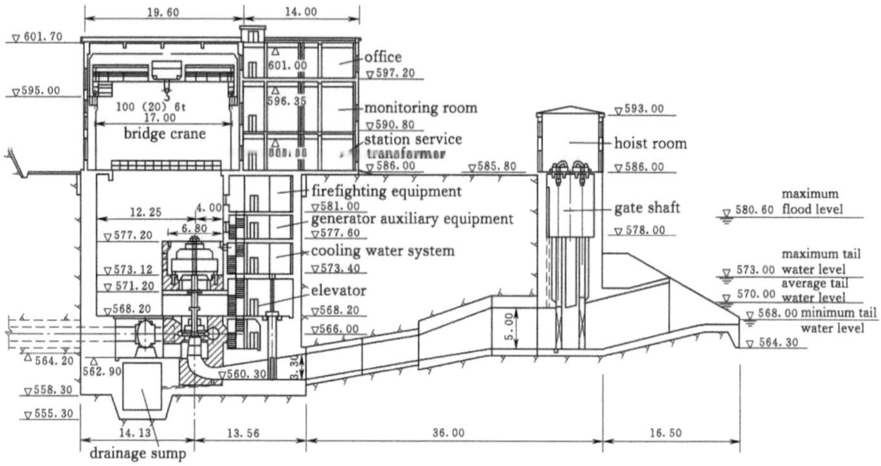

Fig. 11.15 Semi-underground powerhouse

equipment are arranged in the shaft, and assembly site and auxiliary powerhouse in charge of operation and management are arranged on the ground.

11.1.3 Components of Powerhouse

11.1.3.1 Construction Composition

1. Main powerhouse

Main powerhouse is the main component of the powerhouse. It consists of generator hall where the units and other auxiliary equipment are arranged, and assembly site for equipment installation and overhaul.

From the facade, the main building is divided into the superstructure and the substructure by the generator storey. The structure above the generator storey is similar to that of the industrial building, with plate, beam, and column as the main structure. The structure below the generator storey is similar to that of the hydraulic structure, with the block concrete entity structure as the main structure. From the plane, the main plant is divided into the main unit section and installation field. The main unit section is mainly the place where hydraulic turbine-generator unit and auxiliary equipment are arranged, and the assembly site is the place where the units are installed and overhauled.

2. Auxiliary powerhouse

The auxiliary powerhouse is the layout room of control equipment, electrical equipment and auxiliary equipment which are required to ensure the normal operation of the main unit of the hydropower station. At the same time, it is the room for power station operation, control, monitoring, communications, tests, management, and the operation personnel work.

3. Main transformer field

It is an area for arranging the main transformer. The main transformer's function is to raise the voltage from the generator to the specified voltage level and then leads it to the switch station.

4. Switch station

It is a place of high-voltage transmission and distribution with arranging high-voltage distribution equipment, high-voltage switches and busbar protection equipment.

11.1.3.2 Equipment Composition

1. Waterway system

From entering the plant to the outflow in turn including: penstock→ inlet valve→ spiral case→ draft tube→ draft tube gate→ tailwater canal.

2. Electromechanical system

Including hydraulic turbine-generator unit, excitation system, governing system, oil, gas, and water system, hoisting device and various test and maintenance equipment.

3. Electrical equipment system

Including primary equipment and secondary equipment according to operation and management requirements.

Primary equipment: generator → switch room → main transformer →switch station.

Secondary equipment: including control, protective, testing and monitoring equipment centred on the central control room, such as machine panels, tele automatics system, communication system, dispatching system, and direct current (DC) system.

The relationships of equipment systems are shown in Fig. 11.16.

11.1.4 Layout Requirements of Powerhouse

11.1.4.1 Main Buildings

1. Main powerhouse

It mainly meets the requirements of hydraulic turbine-generator unit arrangement, overhaul and operation, and also meets the requirements of auxiliary equipment (governor, machine panels, inlet valve, hoisting device, etc.) which are used for the operation of the main unit.

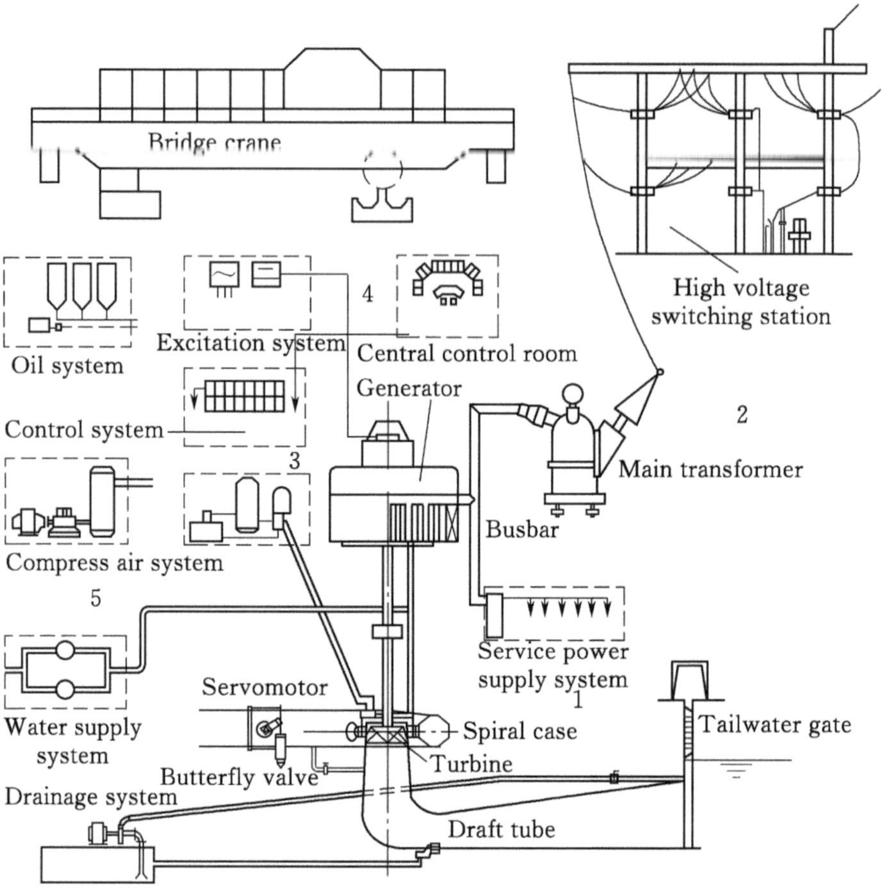

Fig. 11.16 Components of powerhouse

2. Auxiliary powerhouse

The auxiliary powerhouse is usually arranged around the main powerhouse to meet the arrangement of control equipment, electrical equipment, and auxiliary equipment to ensure the normal operation of the main unit. The windows in the auxiliary powerhouse should avoid the western sun as far as possible, ensure ventilation and daylighting and make the outgoing wire, mechanical and electrical equipment's operation and maintenance reasonable.

3. Main transformer

As the busbar which link generator to main transformer is expensive, it is more economical to shorten the outgoing wire, the layout should be as close as possible to the main powerhouse. The main transformer of the ground powerhouse is usually arranged in three ways: downstream of the main powerhouse (located on the tailrace

Fig. 11.17 Main transformer layout in the tailrace platform

(a) outdoor switch station (b) indoor switch station

Fig. 11.18 Example of switch station layout of a hydropower station

platform); upstream of the main powerhouse (located between the powerhouse and dam); and at the end of the main powerhouse. Main transformer's layout aims for safety, reliability and easy maintenance. The layout of the main transformer on the tailrace platform is shown in Fig. 11.17.

4. Switch station

There are two layout ways: indoor and outdoor. Indoor switch station's equipment is small with high cost. However, the layout is compact, which can be considered in combination with the layout of the auxiliary powerhouse. Outdoor switch station equipment is larger with larger area requirement. But the layout is flexible. In accordance with local conditions in the powerhouse, it can be arranged near the open space, in order to facilitate wiring and outgoing wire layout. An example of switch station layout of hydropower station is shown in Fig. 11.18.

11.1.4.2 Requirements of Structural Design

Hydropower powerhouse has different characteristics compared with general industrial plant, so the structural design requirements of the plant are different.

1. The hydraulic structure characteristics of the powerhouse

The substructure of the powerhouse is all mass concrete, accounting for more than 90% of the total. As the substructure is subjected to water pressure and earth pressure, it needs water stop and drainage, and its mechanical characteristics are similar to the hydraulic structure. Therefore, the design of the substructure should be in accordance with hydraulic standards.

2. Architectural characteristics of the powerhouse

The superstructure of powerhouse is plate, beam, column and frame structure, which is similar to industrial plant, and the structure must be stable and reliable. Therefore, the superstructure should be designed in accordance with the building standards.

3. The operation characteristics of the powerhouse

The electromechanical, electrical and hydraulic equipment are concentrated in the powerhouse. In order to operate, monitor and manage the facilities of the powerhouse more conveniently, it is required that the buildings are beautiful and practical, and the site has good working and sanitary conditions. Therefore, the powerhouse should be designed according to the standard of living housing.

11.2 Major Equipment and Structural Arrangement

The equipment and structure size directly affect the layout and investment of the main powerhouse. The reasonable layout of the equipment in the powerhouse makes them interrelated and coordinated to form an organic whole. And these are very important to ensure the completion of production tasks and to achieve comfort, economic, aesthetic purposes for the powerhouse.

11.2.1 Electromechanical Equipment

11.2.1.1 Hydraulic Turbine

It has been introduced in Chaps. 1–4. As we have learned, the sizes of spiral case and draft tube are the primary factors that affect the powerhouse size.

11.2.1.2 Governing System

Its composition includes manipulation cabinet, electrical cabinet (electronic governor), servomechanism, rod and lever mechanism, and oil pressure equipment.

It has been introduced in hydraulic machinery.

11.2.1.3 Generator

Main components of generator include rotor, stator, upper and lower bracket, thrust bearing, guide bearing, excitation system, etc.

Generators are divided into suspended type generators and umbrella typc generators.

1. Suspended type generator (Fig. 11.19)

The thrust bearing of a suspended type generator is located on the upper bracket and the rotating part of the whole generator is suspended. Its advantages include little abrasion of the thrust bearing, convenient assembly and stable operation. And its disadvantages include large size of the upper bracket with considerable consumption of steel. Its speed is generally above 150 r/min.

2. Umbrella type generator (Fig. 11.20)

The thrust bearing of an umbrella type hydro-generator is located on the lower bracket and it is like an umbrella handle that supports the rotating part of the unit.

It has light upper bracket and can reduce the unit height (and the powerhouse height) and save steel. The thrust bearing is allowed not to be removed when the generator is being repaired, which reduces maintenance time.

Its disadvantages include large diameter of thrust bearing, easy abrasion and complexity of design and manufacturing. Its speed is mostly below 150 r/min.

The umbrella type generators can be subdivided into umbrella type generators and semi-umbrella type generators.

According to arrangement form, generator includes: ① the stator exposed; ② the stator buried; ③ the upper bracket buried; ④ other layout types.

The excitation system is the direct current power supply that provides the rotor with magnetic field. If the excitation is interrupted, the generator instantly rejects the

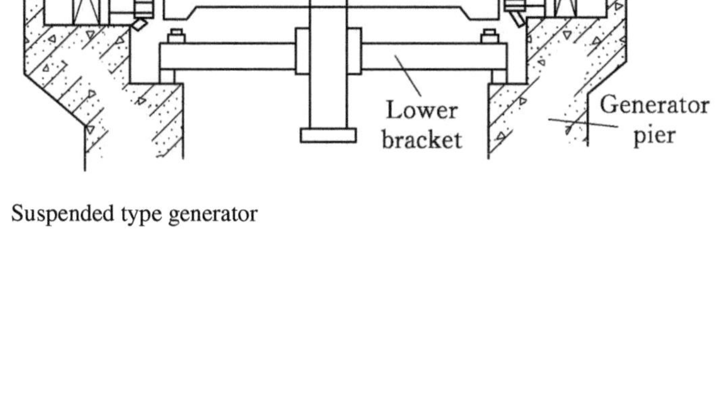

Fig. 11.19 Suspended type generator

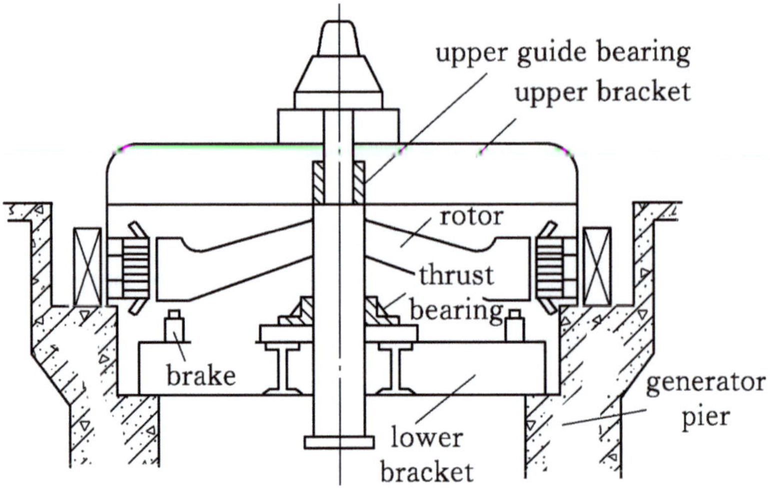

Fig. 11.20 Umbrella type generator

load completely. Therefore, the excitation system must be reliable. There are two types of excitation:

(1) Exciter (Fig. 11.21): it is a DC generator and is generally coaxial with the hydraulic generator. It is located on the top of a generator and provides the generator rotor with direct current to form magnetic field.
(2) Thyristor excitation: some of the output current of the generator passes the silicon-controlled rectifier. After voltage reduction, the current is sent back to the generator as exciting current.

Generally, three to five exciter panels are configured beside each unit in order to control and adjust the exciting current of the hydro-generator.

Fig. 11.21 Excitation transformer

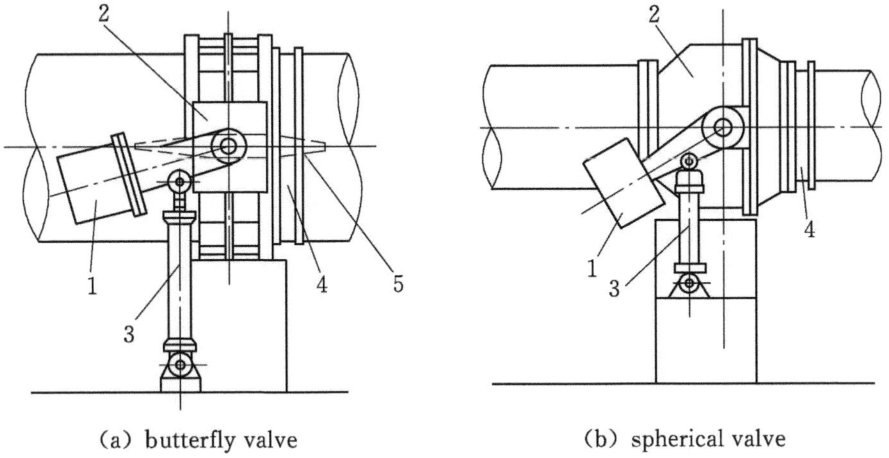

Fig. 11.22 Inlet valve 1-weight hammer; 2-valve body; 3-Servomotor; 4-expansion join; 5-valve

11.2.2 Mechanical Equipment

11.2.2.1 Inlet Valve

The functions of inlet valve include:

(1) Stopping the unit for maintenance or checks.
(2) Stopping the unit for a long period of time (reducing water leakage).
(3) Cutting off the flowing water when the unit rejects load and the turbine governor fails at the same time.

Its main types include butterfly valve and spherical valve. (Fig. 11.22). It has some associated equipment: expansion joint, gas charging valve, bypass valve, and drainage valve.

11.2.2.2 Hoisting Equipment

Generally, bridge crane is used. In few cases, gantry hoist is used. Their functions include hoisting equipment used during the installation and maintenance (Fig. 11.23).

Its main operating parameters include:

(1) Lifting capacity: rated value is usually the rotor weight of generator with shaft (or main transformers, or turbine runner with shaft).
(2) Span: the distance between the track central lines of bridge cranes.
(3) Hoisting height: meeting the requirements of unit installation and maintenance, usually controlled by rotors, runners, etc.
(4) Working range of bridge cranes: namely the hook limit line.

Fig. 11.23 Double-trolley bridge crane

The selection of types and number will affect the upper structure size of the powerhouse, investment and progress of installation and maintenance. It can be determined according to the unit number and the maximum weight of lifting objects.

11.2.3 Main structures of main powerhouse

Main powerhouse can be divided into superstructures and substructures according to the vertical distribution of main powerhouse structure. Above the floor slab of generator storey are the superstructures. Below the floor slab of generator storey are the substructures, which include the generator pier, spiral case, and draft tube (all are mass concrete).

11.2.3.1 Superstructure

The superstructure is the slab-beam-column system, which includes roof structure (roof, roof truss or roof girder), crane beam, and structural column.

11.2.3.2 Generator Pier

The generator pier must meet the requirements of load-bearing, load transfer, torsional resistance and seismic resistance.

Fig. 11.24 Columnar pier

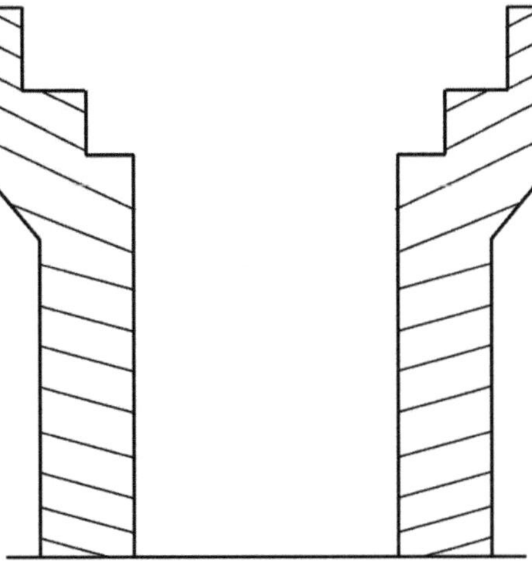

There are 5 types.

1. Columnar pier (Fig. 11.24)

It is widely used on medium-sized units in China. Its interior is a circular turbine well and the exterior is round or octagonal cylinder, with the wall thickness being 1.5 m or more.

The capacity of compression strength and torsional resistance is good, with good rigidity. It is generally less-reinforced concrete and steel-saving.

On the other hand, the internal space of the turbine well is small, therefore the installation, repair and maintenance of the turbine are not so convenient.

2. Block pier (Fig. 11.25)

The parts below the generator are all concrete block, except the reserved space for turbine well and the necessary channel. This sort of pier base is often used on large units.

It has strong strength and stiffness, but it also has large concrete consumption so that the concrete is not fully utilized.

3. Frame pier (Fig. 11.26)

For medium-sized unit, the generator is supported in the ring beam, and the ring beam is supported on 4 to 6 columns; the columns will transmit the load to the lower part of the block structure, so it is also called column type pier.

It has small concrete consumption. The top cover of turbine is relatively spacious. The layout, installation, maintenance, and overhaul of the equipment are more convenient.

The performance of torsion resistance is worse than that of the columnar pier, and the rigidity is also smaller.

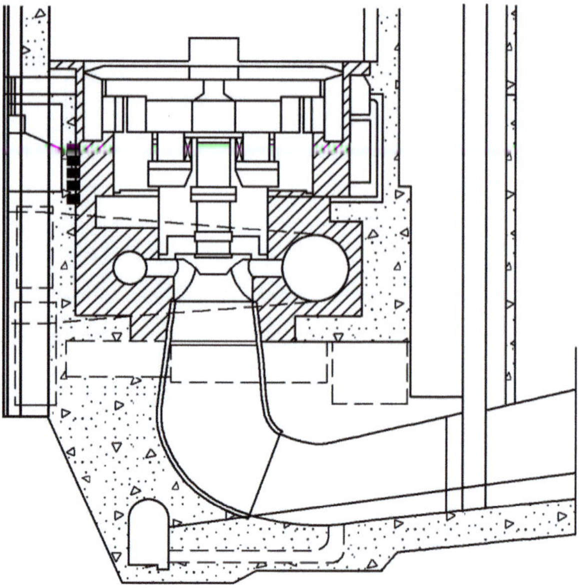

Fig. 11.25 Block pier

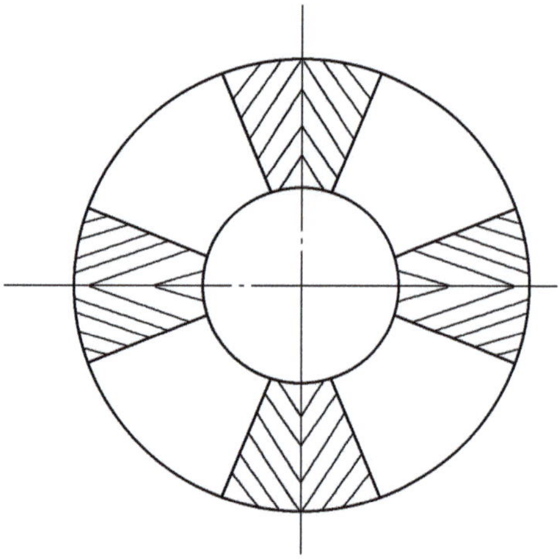

Fig. 11.26 Frame pier

4. Parallel wall pier (Fig. 11.27)

The pier base consists of two blocks of parallel load-bearing reinforced concrete walls and the two beams between them; the generator is directly supported on the parallel walls and beams. It is suitable for large units.

It has spacious top cover and is convenient to operate. The turbine runner can be hung out between two parallel walls without removing the generator.

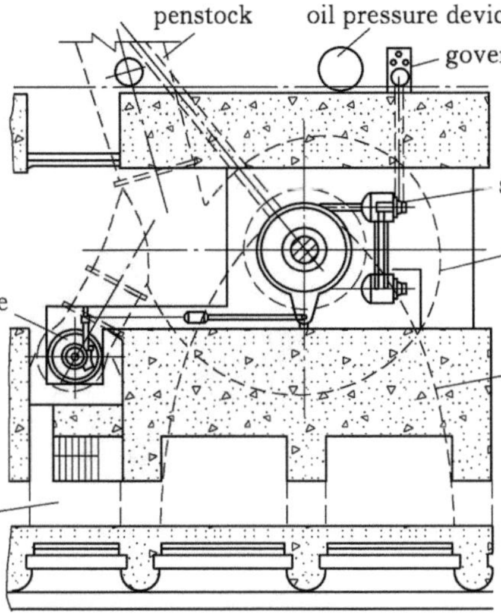

Fig. 11.27 Parallel wall pier

5. Steel pier

It uses the steel structure to support the generator and transmits the load to the top cover, the stay ring or the spiral case.

The generator and the turbine directly form a complete and compact set. It is easy and quick to install and reduces the complex reinforced concrete works. But it has large steel consumption.

11.2.3.3 Spiral Case and Draft Tube

As introduced in previous chapter, the sizes of spiral case and draft tube are the primary factors that affect the powerhouse size.

When $H \leq 40$ m with large flow, the concrete spiral case (Fig. 11.28) is commonly used.

When $H \geq 40$ m with low flow, the metal spiral case (Fig. 11.29) is commonly used.

Draft tube is the discharge component that directs the flow of the turbine downstream. There are conical draft tube and elbow draft tube, as shown in Figs. 11.30 and 11.31. Conical type is mostly used in small power station. In order to reduce foundation excavation, elbow draft tube is usually used in large and medium-sized hydropower plants. It is composed of straight cone section, elbow section, and divergent section. In order to reduce the span of divergent section, 1–2 dividing piers should be installed. Draft tube shall be provided with manholes of a size of 600 mm × 600 mm or ϕ650 mm for access by maintenance personnel. The size and shape of draft tube are usually determined by model test or nominal size.

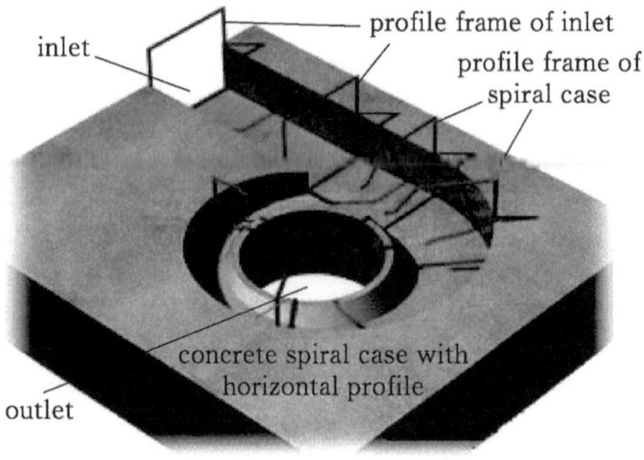

Fig. 11.28 Reinforced concrete spiral case structure

Fig. 11.29 Spiral case of Xiaowan hydropower station

11.2.3.4 Drainage Structure

Generally, drainage gallery and drainage sump are set at lowest part of foundation block, so as to pump service drainage or leak drainage to draft tube. The bottom elevation of the sump shall be lower than the lowest elevation of the powerhouse.

11.2.4 Layout Design of Main Powerhouse

The task of the layout design of main powerhouse is to determine the outline size of the main powerhouse economically and reasonably, and to arrange all kinds of equipment according to the equipment layout requirements of the main powerhouse, so as to ensure the installation, operation, maintenance and management of all kinds

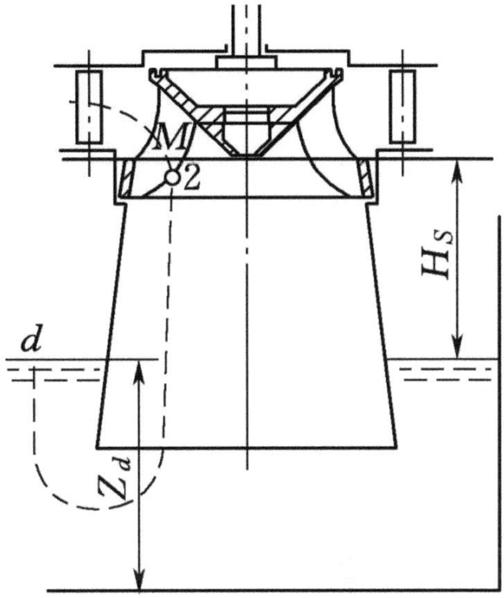

Fig. 11.30 Conical draft tube

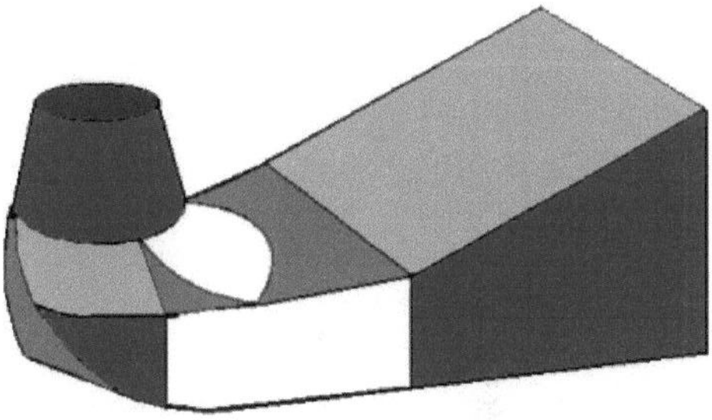

Fig. 11.31 Elbow draft tube

of equipment. In the layout of the main powerhouse, the superstructure, substructure should be considered together with length and width of the main powerhouse. The layout design of the main powerhouse mainly considers the layout of generator storey, turbine storey and spiral case storey.

11.2.4.1 Generator Storey

In this storey, we should consider the following arrangement (Fig. 11.32).

(1) Generator: it includes the stator exposed, the stator buried, the upper bracket buried, and other layout types.
(2) Governor system: it includes operation cabinets, electrical cabinets, oil pressure units, servomotor, and so on, with one set for each generator. Every generator is arranged correspondingly, the upper corresponds to the lower; the electrical

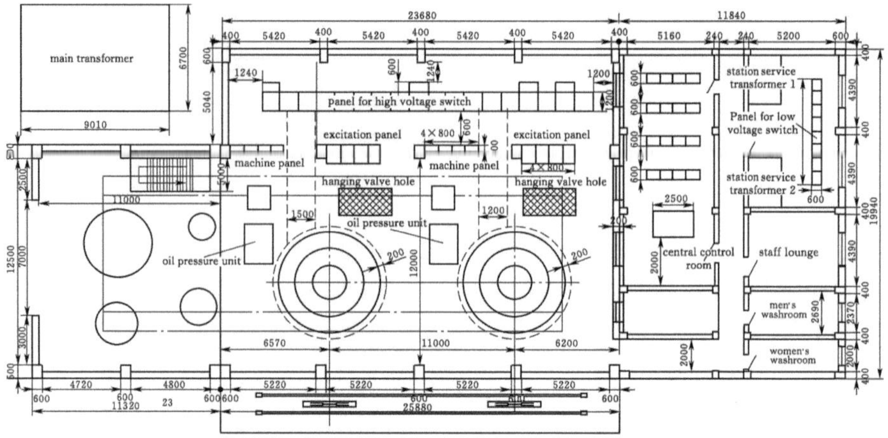

Fig. 11.32 Layout of the generator storey

cabinet is next to the machine panels; the operation cabinet is next to the oil pressure units.

(3) Machine panel: each unit is arranged correspondingly with 3 to 5 blocks (braking panel, relay protection panel, temperature detecting panel, power panel, etc.).
(4) Excitation panel: the excitation panel can generally be put together with the machine panels.
(5) Stairs: one stair is set for every one or two units, with a clear width of 1 to 1.2m and the slope 34–38° for each stair.
(6) Hanging valve hole: it corresponds to the inlet valve position and its size is slightly larger than the inlet valve, with a plus of 0.25 m on each side.
(7) Transportation corridor: the main corridor of 2–2.5 m is located on the upstream side or the downstream side; the clear distance among devices is about 1.5 m for inspection.
(8) Hoisting device: crane operating range and the hoisting height limit of the crane hook.

11.2.4.2 Turbine Storey

Besides generator pier, the device layout of turbine storey both on the upstream and the downstream sides consists of oil-gas-water system piping and the generator main outlet, neutral point, transformers of the primary electrical system, etc.

There are some other devices on the turbine storey (Fig. 11.33): rod and lever mechanism, governor servomotor, vertical deep well pump, inlet valve operating mechanism, etc.

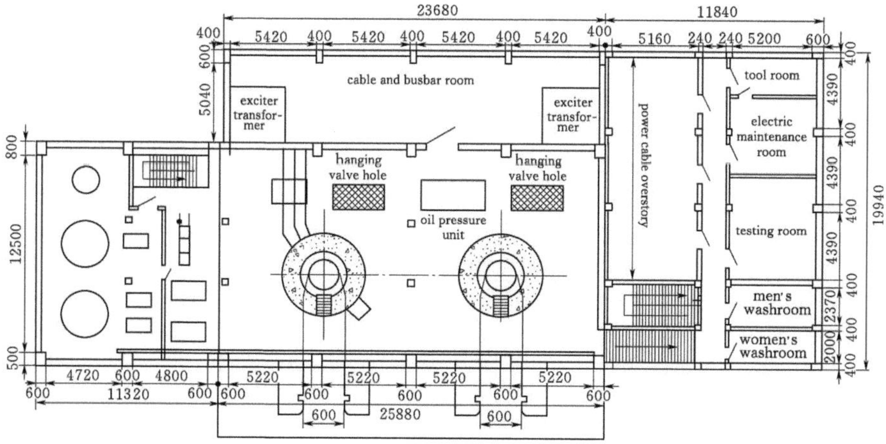

Fig. 11.33 Layout of the turbine storey

11.2.4.3 Spiral Case Storey

In the grouped water supply and centralized water supply hydropower station, butterfly valve or spherical valve is generally deployed in front of the spiral case inlet (Fig. 11.34).

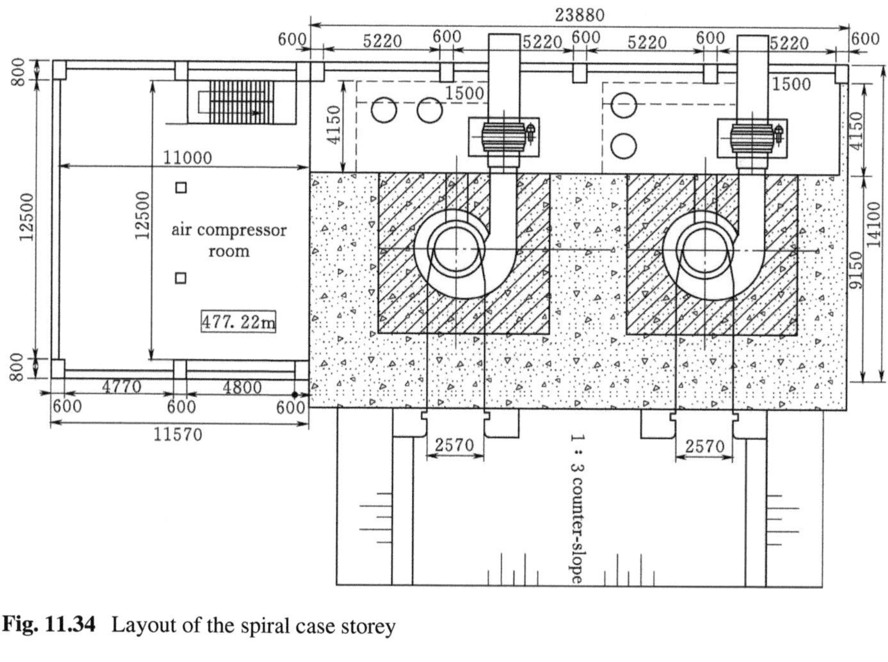

Fig. 11.34 Layout of the spiral case storey

11.3 Profile Design of Main Powerhouse

The main task of the profile design of main powerhouse is to determine the elevation of the powerhouse reasonably according to the data of the hydraulic turbine—generator unit, so as to meet the requirements of the layout of the main and auxiliary equipment of the main powerhouse and the ventilation and lighting needs of the powerhouse structure. Arrange power generation and distribution reasonably, to provide a good working environment for operators. The elevation of the main powerhouse can be divided into two parts: underwater and overwater. Each elevation of the underwater part is determined by the size of turbine, spiral case, draft tube, and so on; the overwater part is determined by the size of generator, hoisting height and mode. For main powerhouse profile design, the installation elevation of the turbine should be determined first, and then the other elevation should be determined on this basis (Fig. 11.35).

11.3.1 Turbine Installation Elevation

For vertical mixed flow turbine:

$$\nabla_T = \nabla_\omega + H_s + \frac{b_0}{2}$$

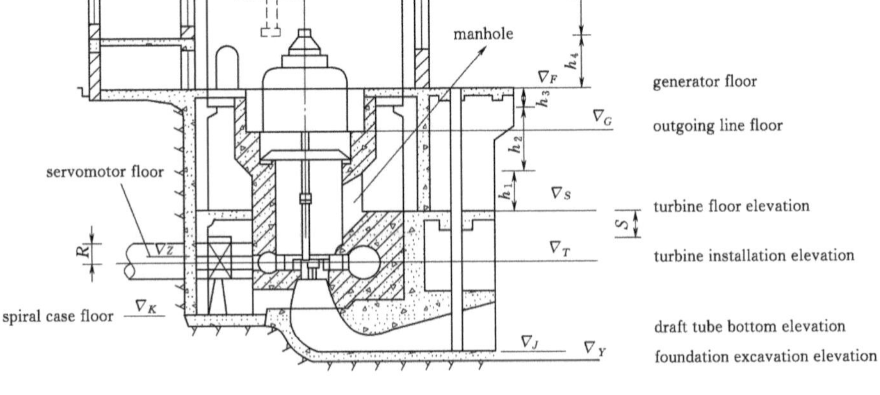

Fig. 11.35 Elevation of powerhouse section

For vertical axial flow turbine:

$$\nabla_T = \nabla_\omega + H_s + xD_1$$

In the formula:

∇_ω The minimum tailwater level of the downstream;
b_0 The height of the guide vanes;
x The coefficient, which is about 0.41.

11.3.2 Turbine Floor Elevation

$$\nabla_S = \nabla_T + R + S$$

In the formula:

R The largest radius of the spiral case;
S The thickness of the concrete at the top of the spiral case, as to the metal spiral case, it is generally more than 1m.

11.3.3 Generator Device Elevation

$$\nabla_G = \nabla_S + h_1 + \Delta h$$

In the formula:

h_1 The height of the pier manhole, which is general 2m;
Δh The thickness of the concrete at the top of the manhole, which is 1m on average.

11.3.4 Generator Floor Elevation

$$\nabla_F = \nabla_S + h_1 + h_2 + h_3$$

In the formula:

h_1 The height of the pier manhole, general 2 m;
h_2 The operation space;
h_3 The height of lintel, general 1–2 m.

There are several main points of the generator floor elevation design.

(1) When applying existing units, the dimensions between the generator and turbine is given.
(2) The buried mode of the generator should be considered.

(3) The clear height between the turbine storey and the generator storey is no less than 3.5–4 m.
(4) The generator floor is the best to be the same height with the assembly site, and be higher than the downstream design flood level. If they cannot be taken into account at the same time, other flood control measures should be considered.

11.3.5 Crane Rail Top Elevation

$$\nabla_D = \nabla_F + h_4 + h_5 + h_6 + h_7$$

In the formula:

h_4 Vertical safety distance of the hook;
h_5 The maximal height of the lifting object;
h_6 The height of sling (or spreader);
h_7 The distance from the highest position of main hook to the rail top.

The crane rail top elevation depends on the following factors:

(1) The biggest lifting object: generator rotor (Fig. 11.36), turbine runner, the main transformer (Fig. 11.37).
(2) Hoisting way: rigid or flexible lifting, horizontal and vertical safe distance of lifting objects (For flexible sling, the vertical safe distance is 0.6–1.0 m, and that of steel sling is 0.25–0.5 m; the horizontal safety distance is 0.3 m).
(3) Crane type: double trolley or single trolley.

There are two aspects should be paid attention to:

(1) The elevation of the crane rail is designed by drawing instead of calculating.
(2) The route when lifting objects should be straight.

Fig. 11.36 Rotors' hoisting

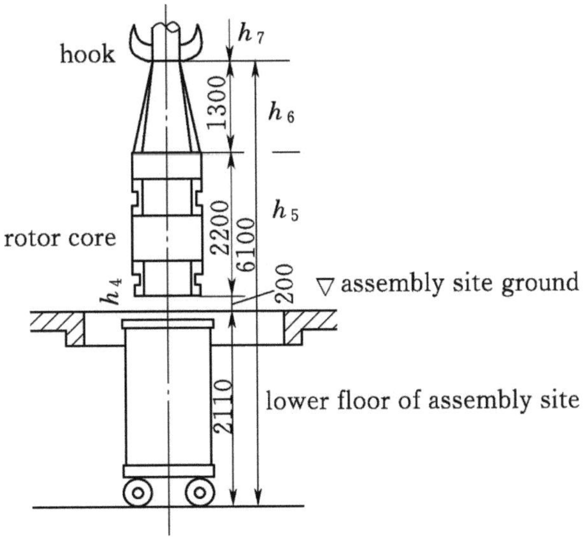

Fig. 11.37 Lifting diagram of main transformer

When hoisting main transformer, there are two types of transformers: bell-type and barrel-type. For transformer maintenance, the bell-type transformer hangs the shell while the barrel-type hanging the iron core.

Besides, the probability of transformer repair is low. In order not to increase the powerhouse height, a transformer maintenance pit is often set up in the assembly site.

11.3.6 Roof Elevation

$$\nabla_w = \nabla_D + h_8 + h_9 + h_{10}$$

In the formula:

h_8 The distance from the top of the crane rail to the top of the trolley.
h_9 The distance from the top of the trolley to the bottom of the roof girder (generally 20–50 cm);
h_{10} The height of roof girder (truss), roof thickness and waterproof layer.

11.3.7 Powerhouse Foundation Excavation Elevation

$$\nabla_J = \nabla_T - x - h - \delta$$

In the formula:

x The height from the turbine installation elevation to inlet of draft tube;

Fig. 11.38 Meaning of h and δ

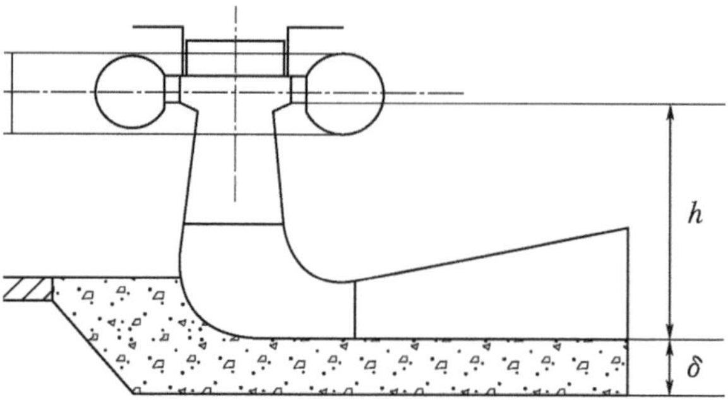

Fig. 11.39 Integral type bottom board

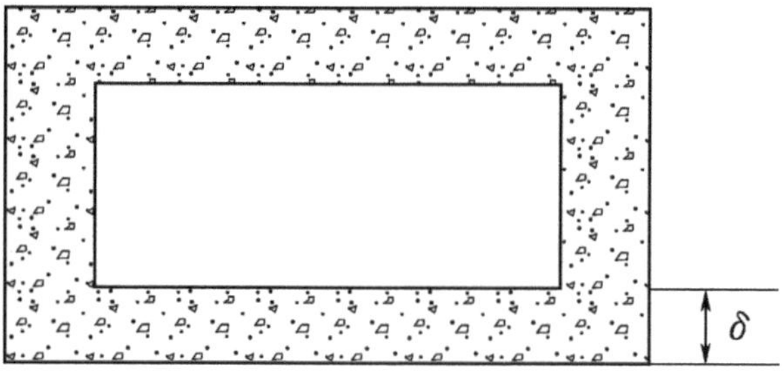

h The height of the draft tube;
δ The thickness of the draft tube.

(Fig. 11.38)

The bottom board of the draft tube is poured as a whole with the side pier or central pier, which is called integral type (Fig. 11.39). If they are poured separated, which is called separate type. The thickness of the bottom board in the integral type is general 1–2 m, while that generally is 0.5–0.8 m in the separate type.

11.3.8 Tailrace Platform

The gantry crane is usually arranged on the tailrace platform.

In low-head powerhouse, draft tubes tend to be longer. On the tailrace platform, the main transformer field can be arranged. Below the space of the tailrace platform, a variety of auxiliary powerhouse can be arranged.

Figures 11.40 and 11.41, respectively, shows the typical horizontal profile and longitudinal profile of powerhouse structure. These two diagrams show the relative relationship between the elevation of the main powerhouse and the auxiliary powerhouse and the arrangement of the equipment.

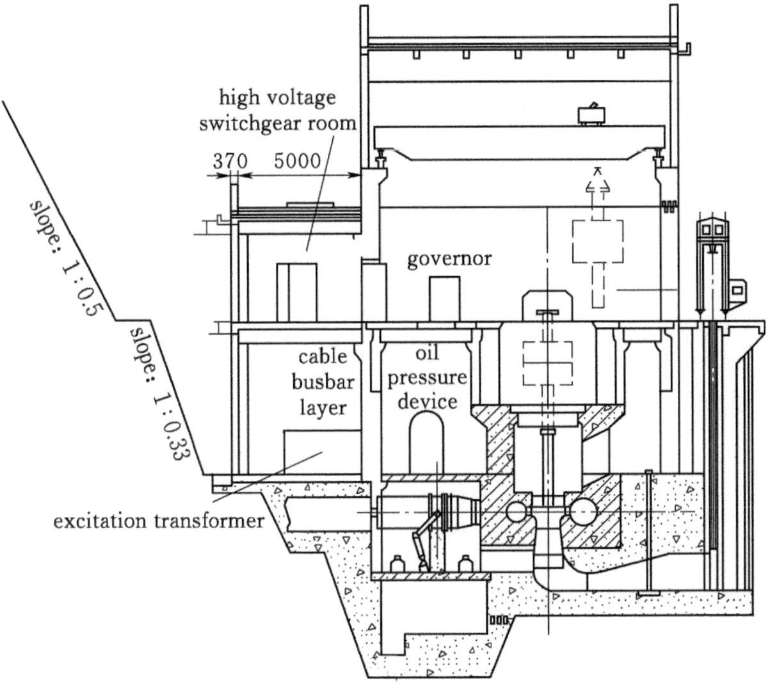

Fig. 11.40 Cross-section of powerhouse structure

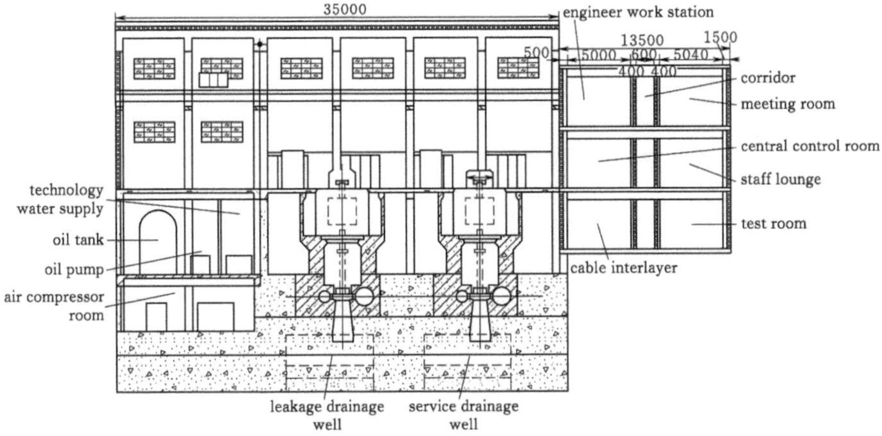

Fig. 11.41 Longitudinal section of powerhouse structure

11.4 Plane Size of Main Powerhouse

11.4.1 Length and Width of the Main Powerhouse

11.4.1.1 Unit Block Length

The distance between centre lines of two hydropower units is

$$L_1 = L_{+x} + L_{-x}$$

At the spiral case storey, the length of spiral case envelope curve is determined by its size and thickness of Phase II concrete.

$$L_{+x} = R_1 + \delta_1$$
$$L_{-x} = R_2 + \delta_2$$

δ_1 and δ_2 are the thickness of concrete around the spiral case (Fig. 11.42), which are preliminary designed to be 1.2–1.5 m.

At generator storey:

$$L_{+x} = \frac{\varnothing_3}{2} + \frac{b}{2} + \delta_3$$
$$L_{-x} = \frac{\varnothing_3}{2} + \frac{b}{2} + \delta_3$$

In the formula:

$\varnothing_3$ The inner diameter of the ventilation barrel;
δ_3 The wall thickness of the ventilation barrel, which is 0.3–0.4 m on average;
b The clear distance between the two generators' ventilation barrel outer wall, which is 1.5–2.0 m on average. But if stairs is set between two units, the distance should be an average of 3–4 m.

Based on the above calculation of L_{+x} and L_{-x}, the sum of the maximums is considered as the unit block length. The unit block length for low-head power plants is generally determined by the spiral case, the draft tube size and the thickness of the around Phase II concrete. For high head power plants, the length is generally determined by the generator size and the layout of devices on the generator storey.

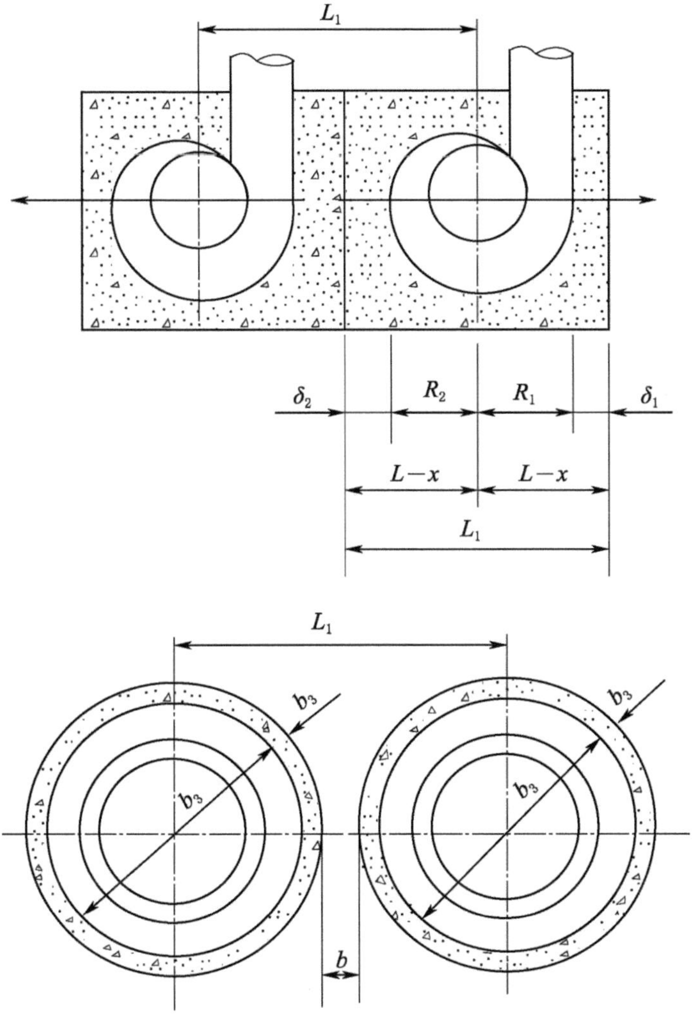

Fig. 11.42 Unit length

11.4.1.2 Length of Unit Block Terminal

$$L_2 = L_1 + \Delta L$$

ΔL mainly depends on the location of the assembly site, bridge crane width and its working range. It should be ensured that the unit centreline or the inlet valve centreline is within the hook limit line with a margin of 0.2–0.3 m.

11.4.1.3 Width of Main Powerhouse

The width of the main powerhouse is determined by the unit size and the layout of devices on the generator storey (Fig. 11.43). Generally, the width of main powerhouse can be determined by

$$B = B_S + B_X$$

(1) Upstream side width of the powerhouse B_S:

$$B_S = \frac{\emptyset_3}{2} + \delta_3 + A$$

A is the clear distance between the outer side of ventilation barrel and the inner side of upstream wall of the powerhouse, and it is determined by the layout of devices.

The devices arranged on the upstream side are as follows:

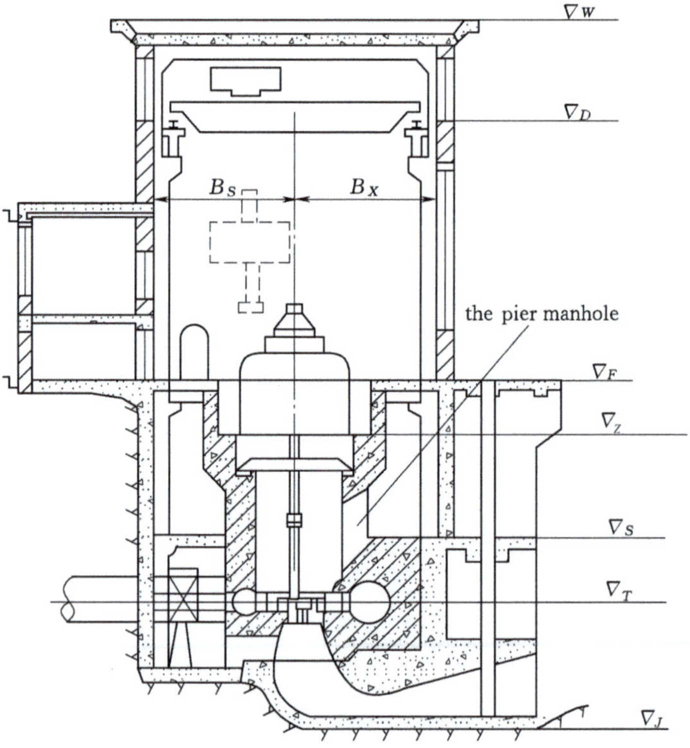

Fig. 11.43 Width of powerhouse

a. Governor (including mechanical cabinets, electrical cabinets, oil pressure devices, and servomotor). The pipe lines connected between the governors should be as close as possible.
b. Machine panels. Considering the requirements of the inspection in front of the machine panels, the margin is generally 1.5 m. Considering the requirements of inlet wire from the rear of the machine panels, the margin is generally 80cm.
c. Hole for hoisting butterfly valve. The distance between the butterfly valve centreline and the unit centreline should be more than twice as much as the diameter of the butterfly valve.

In addition, the size of ventilation barrel should be considered.

(2) Downstream side width of the powerhouse B_X.

The downstream side should mainly consider the requirements of the generator size and the main corridor of the generator storey. Generally, the main corridor width is 2–2.5 m. The requirements of the spiral case size and the thickness of the surrounding concrete should also be met (Main corridor width is the distance from the outer of the ventilation barrel to the edge of the structural column).

11.4.2 Assembly Site

Plane size of the assembly site (Fig. 11.44) is with the same width of the generator hall and its length depends on the required area of a unit's expanding overhaul.

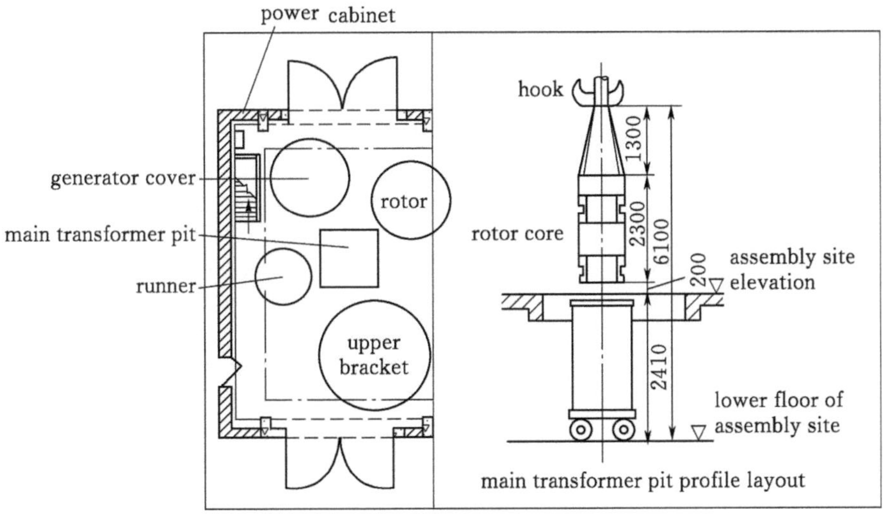

Fig. 11.44 Schematic of the assembly site

As to suspended type generator, we generally put down with four pieces–the generator rotor, upper bracket, turbine runner, and generator cover.

As to umbrella type generator, we generally put down with five pieces–add on lower bearing bracket.

The overhaul distance around the rotor is 2 m, and that of other parts is 1m. There is a rotor support platform below the rotor.

The determination of assembly site floor elevation:

(1) When the generator floor elevation is higher than the downstream design flood level, the assembly site floor elevation is the same with the generator floor ground.
(2) When ∇_F is less than the downstream design flood level.

    a. When the assembly site elevation is still the same with the generator floor ground, it's necessary to take some flood control measures to protect walls, doors, and external transportation of the assembly site.
    b. When the assembly site elevation is above the design flood level, the rail top elevation is required to be based on the assembly site.
    c. The assembly site could be divided into two parts, with one part above the generator floor elevation and the other part of the same height as the generator floor elevation, so that the assembly site area will be slightly larger.

11.4.3 Measures to Reduce Powerhouse Size

Reducing the powerhouse size can save materials and speed up the construction progress. The measures are as follows.

(1) Adopting a new type of unit. For example, the thrust bearing of the new umbrella type unit is supported on the top cover of the turbine, and the lower bracket only bears the braking reaction force. Therefore, the bracket may be small, which can shorten the main shaft, reducing the powerhouse's height.
(2) Adopting a reasonable arrangement. For example, the height of the generator storey can be reduced by adopting the isolated island arrangement, and the generator storey can be cancelled for small units.
(3) Using crane with double trolley. The double trolley can reduce the tonnage of the crane and reduce the powerhouse height. But this scheme will increase the cost of the side unit and the crane, so it is necessary to make scheme comparison.
(4) When the assembly site elevation is controlled by the main transformer shaft core, the bell-type main transformer can be selected, or the main transformer can be overhauled on site, which can reduce the powerhouse height.

11.5 Auxiliary Equipment of Powerhouse

The auxiliary equipment of powerhouse are necessary to ensure the normal operation, installation and power transmission and distribution of the units. Its type can be divided into hydraulic machinery auxiliary equipment and electrical auxiliary equipment.

11.5.1 Hydraulic Machinery Auxiliary Equipment

Hydraulic machinery auxiliary equipment are necessary to ensure the operation, maintenance, and control of hydraulic turbine, mainly including speed regulating system and oil, water, and gas system.

11.5.1.1 Governing System

Governing system is the turbine output regulation system. Its function is to ensure the stable operation of units and the quality of power supply.

Composition: manipulation cabinet, servomotor, and oil pressure equipment.

(1) Manipulation cabinet is the core part of the governing system. Generally, it is a rectangular (0.8 m × 1.9 m), located on the upstream side of the main powerhouse, as shown in Fig. 11.45.
(2) The oil pressure equipment provides the driving force for the governing system, which is composed of pressure oil drum, oil storage tank and oil pump. In the oil drum, 2/3 is high-pressure gas and 1/3 is oil, and the oil pressure is 2.5–16 MPa. The tank collects the oil return and oil leakage of the governor. The pump feeds

Fig. 11.45 Operation cabinet and oil pressure device layout of hydropower station

Fig. 11.46 Application cylinder layout of hydropower station

oil to the pressure oil drum. The oil pressure equipment is generally arranged at the generator floor, close to the manipulation cabinet, as shown in Fig. 11.45, or on the upstream side of the turbine floor.

(4) The servomotor is the force transfer structure of the governing system, which is composed of oil pressure piston. By pushing the regulating ring, the turbine guide vane opening can be controlled to adjust the turbine output. Servomotor is generally located on the upstream side of the turbine floor, as shown in Fig. 11.46.

11.5.1.2 Oil System

1. Types and functions

    (1) Turbine oil system: lubricating oil for the bearings of the units, pressure oil for operating the governors and inlet valves.
    (2) Insulating oil system: insulating oil for the electrical equipment (transformers, circuit breakers, outside of the plant).

The uses of turbine oil and insulating oil are different due to their different properties. They cannot be mixed. In order to facilitate the management, two independent oil systems are generally set up separately, so there are turbine oil depot and insulating oil depot.

2. Layout

During the transportation and storage of oil, physical, and chemical changes occur for various reasons, resulting in oxidation and deterioration of the oil, which cannot be used normally. After oil deterioration, acid value increases, flash point decreases, colour deepens, viscosity increases, and impurities increase, which affects the lubrication and heat dissipation of oil, results in the corrosion of metal and fibre, and fail the operating system. So it is necessary to treat the deteriorated oil. According to the degree of deterioration and pollution of oil, it can be divided into dirty oil and

Fig. 11.47 Oil treatment equipment layout

waste oil. Dirty oil is slightly deteriorated or contaminated by water and mechanical impurities. It can still be used after simple purification treatment in the oil treatment chamber. Waste oil is deeply degraded and deteriorated oil, which must be regenerated by chemical or physicochemical methods to restore the original properties of the oil. Due to the high cost of waste oil regeneration, large equipment investment and large land occupation, and there is not much waste oil in hydropower station, so there is generally no waste oil regeneration equipment. The oil system of a hydropower station generally consists of the following components.

(1) Oil storage room: the turbine oil storage room should be arranged in the lower storey of the assembly site, or turbine floor, or auxiliary powerhouse under turbine floor.

    The insulating oil storage room should be arranged close to the main transformer and switch stations, and the place should have convenient site access.

(2) Oil treatment room: the oil pumps and filters are placed here, and they should be arranged close to the oil storage room (Fig. 11.47).
(3) Oil laboratory: pay attention to the fire-proof, explosion-proof, moisture-proof and ventilated conditions.
(4) Turbine oil supply tank: replenish the equipment with new oil automatically.
(5) Turbine oil waste oil tank: deploy one waste oil tank for each unit to collect waste oil.

11.5.1.3 Water Supply System

1. Types and functions

   (1) Technical water supply: cooling water used for generators, main transformers and air compressors, and lubricating water used for guide bearing (The water quality, temperature and pressure should meet certain requirements).
   (2) Living water supply: living water supply including living and firefighting water. Firefighting water flow rate should be 15L/s, and water beam should be able to spray to the highest position where the building may burn. Living water depends on the number of plant operators, and water quality and temperature should be able to meet the requirements of life. If necessary, purification equipment should be set up to ensure the water quality.

2. Water supply types

   According to the water head of hydropower station, water supply is divided into direct water supply (Fig. 11.48), pump water supply and composited water supply.

   (1) Direct water supply. Water supply by gravity flow from upstream ($20\ m \leq H \leq 80\ m$).
   (2) Pump water supply. Water feed by pump from downstream ($H < 20$ m or $H > 80$ m).
   (3) Composited water supply. Deploy the water supply by gravity flow and pump at the same time when the water head has variation changes.

Fig. 11.48 Direct water supply

3. Equipment

   Pipelines and control components, water pumps, water filter (water strainer), pressure reducing valves, etc.

4. Layout

   (1) Water supply by gravity flow. Water is mainly taken from the upstream reservoir or the penstock, and this type is often arranged on upstream side of the turbine storey with lower elevation.
   (2) Water feed by pump. Water is mainly taken from the tail water, and this type is often arranged on downstream side of turbine storey with lower elevation.
   (3) Composited water feed. This type can be deployed in the lower storey of the assembly site, or auxiliary powerhouse.

11.5.1.4 Water Drainage System

The water drainage system includes leak drainage and service drainage. Leak drainage system is mainly used for the leakage water from all parts of the powerhouse that cannot be discharged by gravity to the downstream, including the leakage of technical water, living water and various components, expansion joints and settlement joints. Any leak water that can flow downstream flows downstream, the rest leaking water is introduced into the drainage sump, and then pumped to the downstream. Service drainage mainly refers to the water that cannot be discharged downstream from the spiral case, draft tube and penstock during unit maintenance. Service drainage methods are mainly as follows:

(1) Drainage sump. The draft tube of each unit is connected with the drainage sump by pipeline and controlled by valves, as shown in Fig. 11.49. The water in the draft tube can flow into the drainage sump by itself through the pipeline and then be pumped downstream. The volume of drainage sump is related to the capacity of water pump.
(2) Drainage gallery. Along the longitudinal axis of the powerhouse, a drainage gallery is set at the lowest part of the lower block structure of the powerhouse. During unit maintenance, the water collected from each draft tube is directly discharged into the drainage gallery, and then pumped downstream. Because the volume of drainage gallery is large, and the drainage is fast, it can shorten the drainage time. This kind of method is often used in the water-retaining powerhouse or the powerhouse with many units.
(3) Subsection drainage. Drainage sump and water pump room are set between two sets of units to form a service drainage system.
(4) Movable pump. No drainage sump is set. For the unit in need of maintenance, the movable water pump shall be temporarily set for drainage.

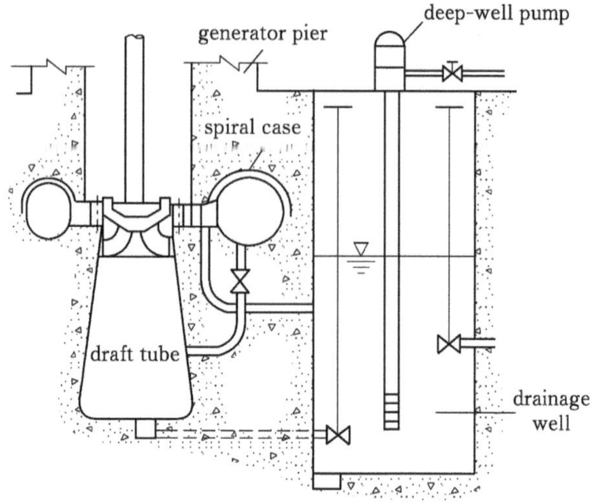

Fig. 11.49 Layout and structure of service drainage sump

Leak water and service drainage sumps are generally set up separately. For small and medium-sized power stations, leak water, and service drainage sump can be combined into one, and whether they can be combined or not should be demonstrated according to specific conditions. But retaining valves should be set up between the two system pipelines and drainage sumps to prevent the water pours back over the powerhouse during service drainage work.

11.5.1.5 Compressed Air System

1. Types and functions

   (1) High-pressure compressed air system (2.5 MPa or 4 MPa). It is used for the pressure tanks of turbine governor's oil pressure system, or the arc extinction and operation of the distribution equipment (such as air-blast circuit breakers, isolating switches and low oil circuit breakers, white thin pipes)

   (2) Low pressure compressed air system (0.8 MPa). It is used for generator braking, phase modulation of hydraulic turbine, and water stop of the butterfly valve by air inflation to circular air band.(White thick pipes).

2. Equipment

   The equipment contains: air compressors (Fig. 11.50), compressed air tanks (Fig. 11.51), pipelines, valves and control components, etc.

Fig. 11.50 Air compressor

Fig. 11.51 High- and low-pressure compressed air tank

3. Layout

   The air system should be arranged in the assembly site, the turbine storey, or the lower storey of auxiliary powerhouse under the turbine storey.

   Pay attention to the fire-proof and explosion-proof conditions. Keep away from the central control room and the relay protection room.

11.5.2 Auxiliary Equipment for Electrical Equipment

The function of auxiliary equipment for electrical equipment is to ensure that the power generated by the generator can be distributed to all users. The equipment mainly includes high-voltage electrical equipment, low voltage electrical equipment, relay protection, and DC system.

11.5.2.1 High-Voltage Electrical Equipment

1. Main transformer

Its main function is to increase the voltage from the generator to the specified voltage level for power transmission.

The arrangement of the main transformer should be as close to the powerhouse as possible to shorten the outlet wire. So as to reduce the power loss and failure rate, as shown in Figs. 11.52 and 11.53. At the same time the requirement of fire and explosion prevention, lightning and water- fog protection, ventilation and cooling, and others should be considered to ensure safety, and reliability.

2. High-voltage switch station

High-voltage distribution equipment should be arranged according to load demand. The equipment include various high-voltage switches and busbars. The layout

Fig. 11.52 Main transformer arranged on tailrace platform

Fig. 11.53 Main transformer arranged between powerhouse and dam

Fig. 11.54 Outdoor high-voltage switch station

Fig. 11.55 Indoor high-voltage switch station

includes outdoor high-voltage distribution and indoor high-voltage distribution, as shown in Figs 11.54 and 11.55.

11.5.2.2 Low-Voltage Electrical Equipment

Low voltage–the generator voltage before the main transformer. Low-voltage electrical equipment includes: generator wiring, busbar, AT (auxiliary transformer), breaker, and disconnector.

Fig. 11.56 Relay protection system

11.5.2.3 Relay Protection System

The functions of the relay equipment are as follows.

When malfunction occurs in the system, the relay will disconnect automatically to protect other equipment from being damaged. In addition, it can also display disordered working signals to the operator on duty, as shown in Fig. 11.56.

Higher requirements for equipment should be ensured:

(1) Quick action: to reduce the damage caused by short-circuit current.
(2) Good reliability: to prevent relay misoperation and rejection, resulting in equipment damage.
(3) Good selectivity: the relay should only troubleshoot faults to ensure the normal operation of non-fault sections and the correct cutting off of current.
(4) High sensitivity: high ability to reflect the fault and reduce the impact of the fault.

11.5.2.4 DC System

Functions of the DC system: Acting as the power supply of the emergency illumination and partial power-consuming devices when it comes to generator failure without external power supply, as shown in Fig. 11.57.

Composition: battery room, acid room, chamber, charging set room, ventilation room, etc.

The layout requirements: preventing acid gas from spilling, and anti-explosion. The chamber should be arranged outside of the acid room. The entire system cannot be set above the central control room and switch room. Regular ventilation must be carried out per hour.

Fig. 11.57 DC system

11.5.3 Maintenance Equipment for Machine Operation

11.5.3.1 Equipment for Central Control Room

The central control room is the centre of operation, management, control and monitoring of the entire powerhouse, as shown in Fig. 11.58.

It should have good ventilation, lighting, heating, anti-vibration, and noise controlling conditions. The indoor clear height is 4–4.5 m.

Fig. 11.58 Central control room

Fig. 11.59 Communication system

11.5.3.2 Communication System

The communication system of the hydropower station is mainly to ensure the timely contact between the hydropower station and the power grid and the dispatching command centre, which directs the operation of the hydropower station according to the requirements of power grid, as shown in Fig. 11.59. It includes computer network communication, carrier communication, microwave communication, telecommunication communication, and so on, and sets up a special room to arrange various communication equipment.

11.5.3.3 Maintenance Equipment

The maintenance equipment of hydropower station is mainly used for mechanical equipment maintenance and electrical equipment test, including mechanical, electrical testing, processing, maintenance, and other equipment. Mechanical equipment maintenance generally includes air compressor, water pump, etc.; Electrical tests include the test of relay protection and automatic devices, measuring metres, precision instruments, etc. The objects of high-voltage test are electrical equipment above 3kV, and the objects of low-voltage test are electrical equipment below 500V and electrical secondary circuit equipment.

11.6 Layout of Auxiliary Powerhouse

The auxiliary powerhouse of hydropower station is the place where various auxiliary equipment are arranged to ensure the operation, control, test, and management of mechanical and electrical equipment. And it is the working room for production, technology and operation personnel.

11.6.1 Influence Factors of Auxiliary Powerhouse

The scale of auxiliary powerhouse is related to the installed capacity and number of units of hydropower station. The main factors affecting its area are installed capacity, status, automation degree of power station and plant location, etc.

(1) The larger the hydropower station capacity is, the more complex the technology is, the more auxiliary equipment is, and the larger the area of the auxiliary powerhouse is. On the contrary, the area of auxiliary powerhouse can be reduced.
(2) The larger the proportion of hydropower station in the power system, the greater the role it plays (frequency modulation, etc.), the higher the requirements of technical protection measures, and the larger the layout area of the auxiliary powerhouse.
(3) The higher the automation degree of hydropower station is, the more sophisticated the equipment is, and the smaller the area of the auxiliary powerhouse is.
(4) The farther the powerhouse is separated from the living area, the more complete the facilities are, and the area will be increased accordingly.

11.6.2 Location of Auxiliary Powerhouse

There are generally three kinds of layout of auxiliary powerhouse: upstream side auxiliary powerhouse, downstream side auxiliary powerhouse, and end auxiliary powerhouse.

1. Upstream side auxiliary powerhouse

Advantages: compact arrangement, short cable, convenience for monitoring units, and good ventilation and lighting at the downstream side of the main powerhouse.

Disadvantages: the electrical equipment line and the water intake system equipment interfere with each other, and the water diversion pipeline may be lengthened.

Application scenes: it is suitable for the riverside powerhouse and powerhouse at dam-toe.

2. Downstream side auxiliary powerhouse

Advantages: the electrical equipment can be centrally arranged downstream, the electrical equipment line and the water intake system equipment do not interfere with each other, and it's convenient to monitor units.

Disadvantages: the ventilation and lighting of the main power house are affected. The vibration of draft tube during operation has a great impact on the electrical equipment of the auxiliary powerhouse, which is easy to cause misoperation. The draft tube is long and will increase the size of the lower structure of the powerhouse.

Application scenes: it is suitable for the water-retaining powerhouse.

3. End auxiliary powerhouse

Advantages: because the auxiliary width is the same with main powerhouse, the total powerhouse width is small. With convenient ventilation and lighting, the operation environment is better and the operation interference of the central control room is less.

Disadvantages: the busbar and cable lines are longer. When the number of units is large, it is not very convenient for monitoring, maintenance, overhaul, and test, and investment will increase as well.

Application scenes: it is suitable for the riverside powerhouse.

11.6.3 Types and layout principles

1. According to production properties

   (1) auxiliary powerhouse for direct production.
   (2) auxiliary powerhouse for auxiliary production.
   (3) auxiliary powerhouse for indirect production.

2. According to function properties

   (1) auxiliary powerhouse for control and operation.
   (2) auxiliary powerhouse for auxiliary equipment.
   (3) auxiliary powerhouse for production maintenance.
   (4) auxiliary powerhouse for production management.

For the auxiliary powerhouse layout, the direct production room is first considered from the perspective of production to ensure the operation requirements of the unit. Then the auxiliary production room is arranged, giving more priority to the rooms with more personnel in and out. Finally, the indirect production room is considered to ensure the necessary living conditions. From the perspective of equipment layout, the operation control equipment layout is first arranged, then the electrical equipment layout is considered, and then the turbine equipment layout is arranged, and finally the test, maintenance, management and other rooms are arranged.

11.6.3.1 Auxiliary Powerhouse for Control and Operation

1. Central control room

It is the neural centre of operation, management, domination, and surveillance of the entire powerhouse.

Layout principle: it should be as high as the generator floor to the greatest extent. It should have good ventilation, lighting, heating, anti-vibration, and noise controlling conditions. The indoor clear height is 4–4.5 m. And the area is corresponding with the unit number and automation level. A shifting room can be built close to the central control room.

2. Cable room

It is located directly on the lower floor of the central control room, and it is also known as cable interlayer whose clear height is about 2.5 m and the area is less than or equal to the central control room.

3. Relay protection room

When malfunction occurs in the system, the relay will disconnect automatically to protect other equipment from being damaged. In addition, it can also display disordered working signals to the operator on duty. The relay protection room is generally arranged close to the central control room.

4. Switch room (Generator voltage distribution equipment room)

The switch room should be as close to the generator as possible to shorten the generator main outlet. Pay attention to fire-proof, ventilation, and security conditions. It is allowable to set only exit when the length of switch room is not more than 7 m. Distance between two exits should be less than 30 m. The gates of exits should be open outward.

5. Cyclic detection room

Proceeding scheduled scans to the running conditions of the unit and transformer, etc., printing out the results.

Installing digital detectors in hydropower station can help find malfunctions. This not only improves the reliability of operation but also alleviates the operating crew from inspecting and metre reading.

11.6.3.2 Auxiliary Powerhouse for Auxiliary Equipment

The location and size of the auxiliary equipment rooms are mainly determined according to the operation requirements of the unit:

1. Busbar gallery

The clear distance between the busbar and the bottom of the floor should be not less than 0.8 m, and the busbar should be not less than 0.6 m from the bottom of the beam. The distance between the outer side of the three-phase busbar and the wall should generally be not less than 0.8 m. The width of maintenance gallery is 1.0 m for single-bus and 1.2 m for double-bus.

2. Cable gallery and cable channel

The number of the cables between the main powerhouse and the central control room or switch room is big, so the cable gallery is used to hold these cables. Small hydropower stations generally use cable channels.

3. Rooms for service power of power station

Rooms for service power of power station include:

(1) Auxiliary transformer (AT): in order to shorten the busbar connected to the AT, it should be arranged close to the switchgear room as near as possible.
(2) High-voltage power supply switch cabinet: it is generally arranged in the turbine storey, and it should keep a distance from the central control room.
(3) 220/400 V low-voltage distribution equipment: the power supply used for the powerhouse illumination and power equipment.

4. DC system

Functions of the DC system: acting as the power supply of the emergency illumination and partial power-consuming devices when it comes to generator failure and without external power supply; besides, acting as the power supply of the partial instruments in the powerhouse.

Composition: storage battery room, acid room, chamber, charging room, ventilator room, etc.

The layout requirements: preventing acid gas leakage, and anti-explosion. The chamber should be arranged outside of the acid room. The entire system is not allowed to be set above the central control room and switch room. Regular ventilation must be carried out per hour.

5. Air compressor room

It should be arranged in the lower storey of the assembly site, the turbine storey, or auxiliary powerhouse under the turbine storey.

6. Oil treatment room

The turbine oil storage room should be arranged in the lower storey of the assembly site, or turbine storey, or auxiliary powerhouse under turbine storey.

The insulating oil storage room should be arranged close to the main transformer and switch stations, and the place should have convenient site access.

The oil pumps and filters are placed in oil treatment room, and they should be arranged close to the oil storage room.

7. Pump room

(1) Water supply by gravity flow

Pump rooms are often arranged on upstream side of the turbine storey with lower elevation.

(2) Water feed by pump

Pump rooms are often arranged on downstream side of turbine storey with lower elevation.

(3) Composited water feed

Pump rooms can be deployed in the lower storey of the assembly site, or auxiliary powerhouse which are located at one end of the powerhouse.

8. Drainage gallery

The bottom elevation of the drainage sump should be lower than the lowest elevation of the main powerhouse. The lower part should be arranged in the concrete, and the upper part is the drainage pumping room.

The elevation of the drainage gallery is generally the same as the bottom elevation of the draft tube.

11.6.3.3 Auxiliary Powerhouse for Maintenance

Some rooms arranged for the maintenance and test of auxiliary equipment can be flexibly determined according to the specific conditions of the auxiliary powerhouse:

(1) Electrical test room. The electrical equipment for high-voltage tests is bulky and cumbersome to handle, so it is often arranged at the location where the equipment is convenient to carry and centralized; other electrical tests can be arranged according to the specific conditions of the auxiliary powerhouse, which are generally arranged above the generator floor with an area of about 20–50m^2. The laboratory should be equipped with ventilation, heating, dustproof, moisture-proof, and other measures, and the test bench should have good daylighting and local lighting.

(2) Electrical instrument repair room. It is generally arranged above the generator floor, with good ventilation, illumination, and other conditions. The area is about 20–40 m^2.

(3) Mechanical repair room. It is mostly arranged outside the factory, with an area of 40–60 m^2.

(4) Tools and warehouse room. The suitable location should not be too low, with an area of about 10–20 m^2.

(5) Operation and duty room. It is close to the central control room and has the shortest way to the main powerhouse, covering an area of about 20–25 m^2.

11.6.3.4 Auxiliary Powerhouse for Production Management

It mainly includes administrative office, meeting room, reception room, guard room, toilet, etc. The toilet and bathroom are usually set close to the installation site; the administrative office, meeting room and night rest room are arranged in a quiet place; the reception room and guard room are arranged at the entrance and exit of the plant. The auxiliary powerhouse for production management is generally determined according to the situation. For power stations with high degree of automation management, the administrative offices are mostly arranged outside the powerhouse, and the area is determined by the land occupation and economic conditions. The general rooms are 15–20 m^2.

11.7 Layout Arrangement of Powerhouse

The hydropower powerhouse consists of the main powerhouse, the auxiliary powerhouse, the main transformer field, switch station, the site access, headrace structure, and tailrace structure.

The layout of the powerhouse is directly related to the development mode of the power station, and should be based on topographical, geological, hydrological conditions to properly resolve the relationship with other structures.

The layout of the powerhouse is arranged with the main powerhouse as the centre and reasonable arrangements for the relative position of the corresponding structures.

11.7.1 Basic Information

(1) Terrain–topographic maps.
(2) Engineering geology and hydrogeology: rock stratum, lithology, structural plane, the earthquake intensity, groundwater, permeability, etc.
(3) Hydrology: various characteristic flows.
(4) Meteorology: precipitation, temperature, wind scale, wind direction, etc.
(5) Waterpower utilization planning: water level, water head, kinetic energy parameters.
(6) Electromechanical equipment.
(7) Powerhouse design criteria: depending on the project scale, the importance, and the possible harm caused by settlement.
(8) Construction materials: source, transport ways.

11.7.2 Influencing Factors

11.7.2.1 Adjacent Building

(1) Spillway: it is necessary to avoid the influence of flood discharge on the tail water, and the influence of flood discharge mist on the equipment of main transformer and switch station.
(2) Shipping channel: powerhouse layout should give consideration to the requirement for the flow of shipping channel; avoid the flow of intake and outlet affecting the waterway.
(3) Others: considering the requirements for anti-pollution and sand prevention, the intake should keep away from the import of log pass structure, and be close to the import of sediment release tunnel.

11.7.2.2 Operation Factors

(1) Water flow line. Inlet should be anti-pollution and sand prevention and the diversion conduit should be as short and straight as possible to reduce the head loss. Outlet should be silting prevention and anti-scouring to keep the smooth flow.
(2) Power line. Power line layout process: generator → switch room → main transformer →switch station. The power line should be short, straight, and smooth. And it should be convenient for the main transformer maintenance and inlet-outlet wire of switch station with less earthworks.
(3) Traffic line. It should be convenient for entrance transportation and meet the requirements of the slope gradient, the turn radius, and the transportation of heavy-outsized pieces.

11.7.2.3 Flood Control and Drainage

(1) Adopt reliable measures to ensure that the main and auxiliary powerhouse, the main transformer and the switch station are not flooded under the downstream design flood level.
(2) Adopt reliable drainage measures according to the regional rainstorm intensity and other possible water catchments.

11.7.3 *Layout Principles for Major Structures*

11.7.3.1 The Location of Main Powerhouse

The main powerhouse is the core of the powerhouse and it plays a decisive role on the plant layout. In addition to considering the coordination of each part, the following conditions need to be considered.

(1) The main powerhouse should be built on the good batholith. To reduce the excavation, it is best that the fresh bedrock surface elevation is close to the powerhouse bottom elevation.
(2) The main powerhouse location should be considered with headrace and tailrace to ensure the smoothness of inlet and outlet water.
(3) The main powerhouse location should be convenient for site access and has enough construction area.

11.7.3.2 Auxiliary Powerhouse

The role of the auxiliary powerhouse and its main types has been explained before. Here is the interpretation of the auxiliary powerhouse area and its location relative to the main powerhouse. According to the topographic, geological conditions and the main powerhouse location, the auxiliary powerhouse can be arranged at the upstream side, downstream side or one end of the main powerhouse.

The preliminary design of the auxiliary powerhouse can refer to similar hydropower stations.

11.7.3.3 Transformer Field

(1) The main transformer should be as close to the generator or generator voltage distribution equipment as possible to shorten the generator voltage bus and reduce energy loss.
(2) The main transformer should be convenient for transport, installation and maintenance. Transformer field is better to be arranged at the same elevation with the assembly site and site access, and be laid with track.
(3) The main transformer should be convenient for maintenance, inspection, and troubleshooting.
(4) The civil construction structures should be economic and reasonable with solid and stable foundation; they have the function of drainage and flood control and are in compliance with fire-proof, explosion-proof, ventilation, and heat dissipation requirements.

For the powerhouse at dam-toe, the main transformer can be arranged on the platform between the powerhouse and the dam. It can also be arranged on the tailrace platform at the downstream side of the main powerhouse. For the water-retaining powerhouse, the main transformer is generally arranged on the draft tube platform at the downstream side of the main powerhouse. For the riverside ground powerhouse, the main transformer layout is more flexible, which can be arranged at the upstream side, downstream side and end of the main powerhouse, mainly according to the landform around the main powerhouse.

11.7.3.4 High-Voltage Switch Station

High-voltage switch stations are generally open air and arranged on the riverbank or the slopes near the transformer field.

(1) High-voltage inlet and outlet wires and low-voltage control cables are required to be arranged conveniently and short.
(2) The site should be with a stable foundation and slope to avoid gully mouth and other adverse terrains.
(3) When arranged in the river valley or mountain pass, attention should be particularly paid to wind speed and frozen.
(4) The site layout should be neat, clear, and compact.
(5) The civil construction structures should be reasonable and consistent with the requirements of the fire prevention and security.

For the powerhouse at dam-toe and water-retaining powerhouse, the switch stations are generally arranged on open ground on both sides of the dam. When the slopes of the two sides are steep and open ground cannot be found, the switch stations can also be arranged on the top of the main powerhouse. The switch station of the water-retaining powerhouse is generally arranged on the open ground not far from the main powerhouse.

11.7.3.5 Headrace, Tailrace, and Transport Routes

Headrace is generally a perpendicular diversion and so is the tailrace. The tailrace canal is often an adverse slope with a slope of about 1:4 to 1:5.

Entrance traffic is usually road or rail, and the route should generally be higher than the maximum flood level. The longitudinal slope should be better than 8%; the turning radius should not be less than 20 m; turnarounds should be set up near the powerhouse.

11.7.3.6 Administrative and Living Area

To avoid interference, it should be arranged not too close to the powerhouse. However, considering the safety of powerhouse operation and management, it should not be too far to facilitate contact. According to the actual situation and local conditions, the living area can be arranged with full consideration of various factors.

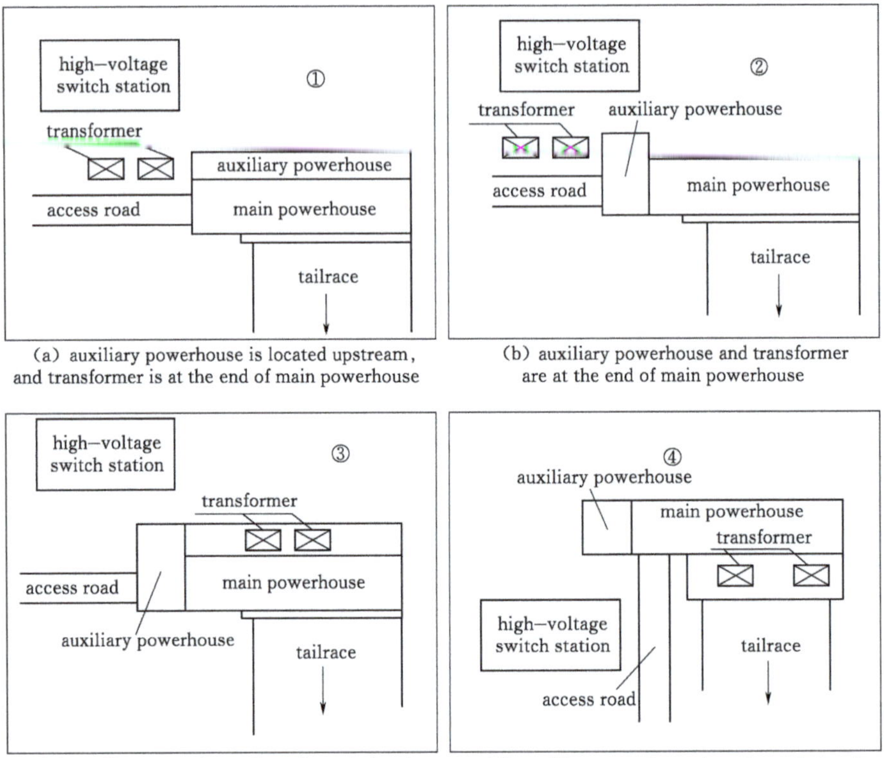

(a) auxiliary powerhouse is located upstream, and transformer is at the end of main powerhouse

(b) auxiliary powerhouse and transformer are at the end of main powerhouse

(c) auxiliary powerhouse is at the end of main powerhouse, and transformer is located upstream

(d) auxiliary powerhouse is at the end of main powerhouse, and transformer is located downstream

Fig. 11.60 Several typical layouts of powerhouse buildings

Figure 11.60 shows several typical layout plans of the main powerhouse, auxiliary powerhouse, main transformer, switch station, headrace, tailrace, and access road of the powerhouse. Figure 11.61 is a layout example of a hydropower station buildings.

(a) Auxiliary powerhouse is located upstream, and transformer is at the end of main powerhouse.
(b) Auxiliary powerhouse and transformer are at the end of main powerhouse.
(c) Auxiliary powerhouse is at the end of main powerhouse, and transformer is located upstream.
(d) Auxiliary powerhouse is at the end of main powerhouse, and transformer is located upstream.

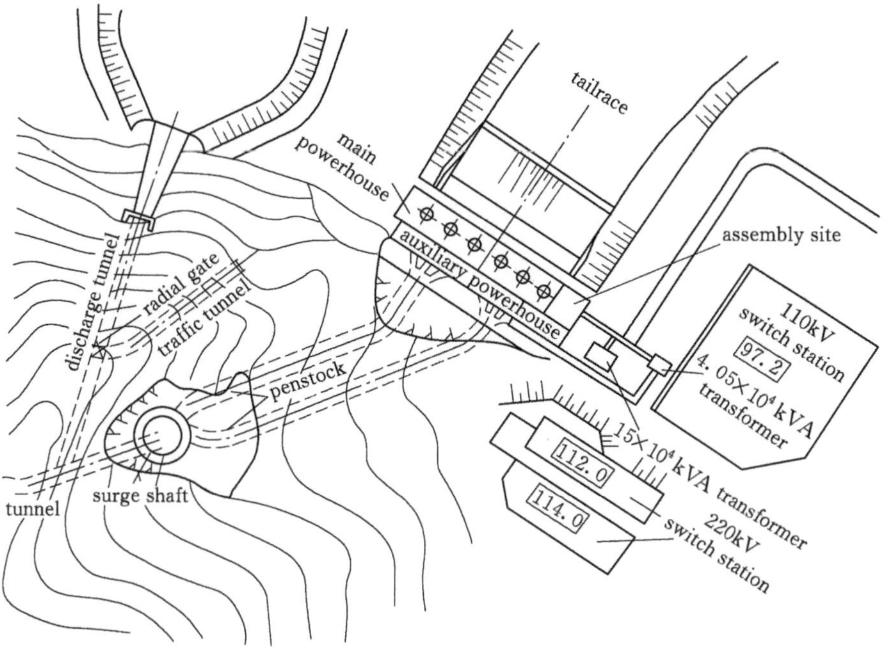

Fig. 11.61 Layout example of a hydropower station

Questions

1. Why should the hydropower station powerhouse be classified? What are the characteristics of each basic type of powerhouse?
2. Why should the powerhouse of hydropower station be classified? What are the characteristics of each basic type of powerhouse?
3. What are the functions of the powerhouse? What are the main buildings of the powerhouse?
4. What are the main electromechanical and mechanical equipment of the main powerhouse? Please list the functions and layout requirements of the equipment.
5. What are the main forms of generator supporting structure of hydropower station? Can you present the characteristics of each form?
6. What are the main controlled elevations when determining the elevation size of the main powerhouse? Which elevation is usually decided first? What factors should be taken into account when determining this elevation?
7. How to determine the elevations of the generator floor and the top of the crane rail? Do you know the main factors that affect these elevations?
8. How to determine the unit spacing and powerhouse width of the main powerhouse? What factors should be considered when determining the plane size of the main powerhouse?

9. How to determine the location, elevation and area of the assembly site? What is the impact of incoming traffic on the layout of the assembly site?
10. What is the "three supplies and one drainage" in the powerhouse? Please list their functions and layout requirements.
11. What are the electromechanical auxiliary equipment for the powerhouse and the operation and maintenance equipment for machine operation? Please list their functions and layout requirements.
12. What are the requirements and methods for lighting, ventilation, transportation, and fire prevention of powerhouse?
13. For the powerhouse project, please consider the main factors affecting its layout and summarize its main layout principles. And what are the requirements for the layout of each building?

Bibliography

1. Design code for Hydropower house. 2014. *SL 266: 2014 (In Chinese)*. Beijing: China Water and Power Press, Beijing.
2. Design code for Hydropower house. 2013. *NB/T 35011:2013 (In Chinese)*. Beijing: China Electric Power Press.
3. Ma, Zhiming, Q. Liu and M. Hu. 2010. *Hydropower station (In Chinese)*, 4th ed. Beijing: China Water and Power Press.
4. Hou, Caishui. 2011. *Hydropower station (In Chinese)*, 2nd ed. Beijing: China Water and Power Press.
5. Yu, Jianhua and S. Zhang. 2014. *Hydropower plant design and construction* (In Chinese). Zhengzhou: The Yellow River Water Conservancy Press.
6. Deng, Xiaoling and T. Jun et al. 2011. *Hydropower station buildings and their construction (In Chinese)*. Beijing: China Electric Power Press.
7. Ma, Shanding and R. Wang, 1996. *Hydropower station buildings (In Chinese)*. Beijing: China Water and Power Press.
8. Liu, Qizhao and M. Hu. 2010. *Hydropower station (In Chinese)*, 4th ed. Beijing: China Water and Power Press.

Chapter 12
Structural Design of Powerhouse

Juntao Chen

Abstract In this chapter, the structural design of the powerhouse at the hydropower station is discussed on the base of the previous chapter. Firstly, the structural composition of the main powerhouse and the types of loads on the structure are introduced, and the load transmission system of the powerhouse structure is summarized. In order to alleviate the influence of temperature and settlement on the safety of the powerhouse, it is necessary to set up structural joints and construction joints. Therefore, the construction methods of concrete pouring carried out in stages and blocks are introduced, and the important sealing structures are discussed.

Keywords Main powerhouse · Structure composition · Load · Transmission of load · Concrete pouring construction · Pouring in stages and blocks · Permanent joint · Sealing structure

12.1 The Structure Composition and Load Transmission of the Main Powerhouse

12.1.1 The Structure Composition of the Main Powerhouse

12.1.1.1 Upper Structure

The upper structure of hydropower station powerhouse usually adopts a reinforced concrete structure. The upper frame is Π-shaped structure in cross-section, and is connected by connecting beam and crane runway girder in longitudinal direction to form spatial framework, as shown in Fig. 12.1. The upper structure is generally a bar structure with definite stress, which is usually solved by structural mechanics method.

J. Chen (✉)
Wuhan University, Wuhan, China
e-mail: chenjt@whu.edu.cn

Y. Cheng et al. (eds.), *Hydropower Station*,
https://doi.org/10.1007/978-981-96-2383-9_12

Fig. 12.1 Upper structure of powerhouse

1. Powerhouse frame

It is the skeleton of the upper structure of the powerhouse, which has two kinds of structures: integral structure and assembled structure. For the integral structure, columns and beams are poured together to form a rigid frame with high stiffness and good integrity, but the construction is difficult. For the assembled structure, beams are precast, and beams and columns are assembled to form a bent frame with small rigidity and poor integrity, but convenient for construction.

2. Roof girder

It bears all the loads on the roof of the powerhouse, including wind, rain, snow, roof slab, roof truss, and beam weight, etc., and transmits the load to the frame of powerhouse. There are two types of girders: integral structure and assembled structure. Most of the integral girders are rectangular cross-section and easy to cast. Most of the assembled girders are T-shaped, with light structure and easy hoisting construction.

3. Crane beam

It bears crane load, including weight of lifting parts, weight of crane and longitudinal and transverse horizontal load produced by crane starting or braking, and transmits load to the powerhouse frame. Generally, it is reinforced concrete structure in the form of multi span continuous beam and simply supported beam. The multi-span continuous beam has good stress condition and small size, but it needs to be cast in place, so it is inconvenient to construct; the simply supported beam can be precast, which is convenient for construction and has short construction period, but the structural stress is large and needs large section size.

4. Column

It bears the load from the roof truss, roof girder, crane beam, external wall, and other upper parts and the weight of the column itself, and transfers the load to the mass concrete structure at the lower part of the powerhouse. Columns are generally rectangular cross-section and arranged at equal intervals. The foundation of columns

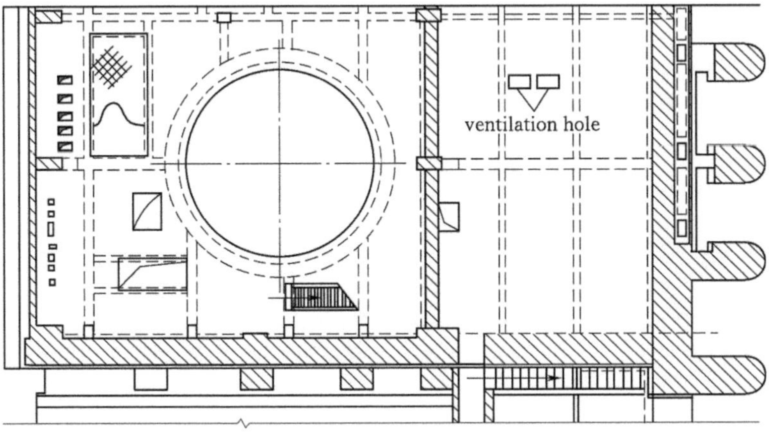

Fig. 12.2 Slab-beam structure layout of hydropower station powerhouse

must be placed on the first stage concrete. When the lower foundation stiffness is more than 12–15 times of the column stiffness, it can be considered as fixed connection; when the foundation is jointed, it is necessary to use double column with columns on both sides of the joint.

5. Floor slab

It is generally the bearing slab, slab-beam structures, which bears the equipment and live load on the floor. Because there are many hydro electromechanical equipment in a hydropower station, the layout requirements are different with many holes, large load and impact load, so the shape of floor slab is irregular. For floor openings (such as governor, oil pressure device, butterfly valve lifting hole, lifting hole, stairs), the secondary beam shall be arranged under the floor; the thickness of the floor shall be greater than 15 cm, and it is better to be designed as one way slab; when the span of slab and beam is large, column shall be added at the lower part of beam, and secondary beam shall be added at the lower part of slab. The structural layout of slab-beam is shown in Fig. 12.2.

12.1.1.2 Substructure

1. Generator pier

It is the supporting structure of generator. It mainly bears the floor load of the generator story and the self-weight and rotation load, and transfers them to the concrete around the stay ring and spiral case. The pier mainly bears compression and torsion loads, which can be simplified into plane structure for calculation (Fig. 12.3).

2. Spiral case and stay ring of hydraulic turbine

It is the guide mechanism of hydraulic turbine. It mainly bears the self-weight, water pressure and upper load, and transfers the load from the pier to the draft tube through the stay ring, and transfers the equipment weight and live load of the turbine story to the draft tube through the spiral case roof. The concrete around the spiral case is generally reinforced concrete (Fig. 12.3).

3. Draft tube

It is the discharge mechanism of hydraulic turbine. There is steel lining in the cone section, and the periphery is reinforced concrete. It bears the load from stay ring and spiral case roof, which is transmitted to the foundation through draft tube frame. With small water pressure, structural reinforcement is generally used (Fig. 12.3).

4. Foundation block concrete

It is the mass concrete block in contact with rock foundation. It mainly transfers the load from the upper structure to the foundation and bears the upper structure load (including: bent column, spiral case peripheral block structure, and draft tube load), self-weight, foundation reaction force, and temperature stress. It is generally adopts structural reinforcement (Fig. 12.3).

12.1.2 Transmission of Load

1. Upper structure load

    (1) Structural self-weight: slab, beam, column, wall, roof, etc.
    (2) Equipment weight: crane, lifting object, and auxiliary equipment.
    (3) Live load: wind, snow, visitors, temporary live load, etc.

2. Substructure load

    (1) Self-weight of concrete block: including block weight and upper transferred load.
    (2) Unit static load: weight of turbine, generator, butterfly valves, and auxiliary equipment.
    (3) Unit dynamic load: rotating load and runaway rotating load during unit operation.
    (4) Water pressure: the load transmitted to concrete by water pressure of spiral case and draft tube.
    (5) Temperature load: load of mass concrete caused by temperature change.
    (6) Foundation reaction: foundation reaction and uplift pressure.

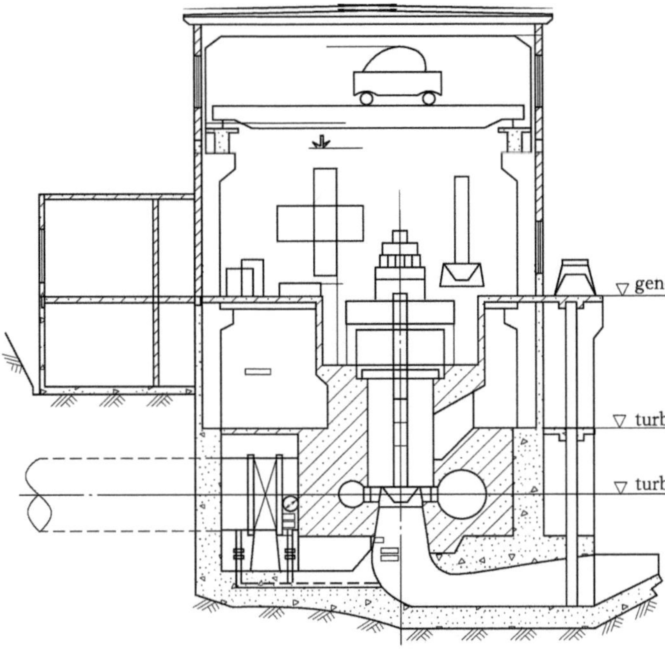

(a)

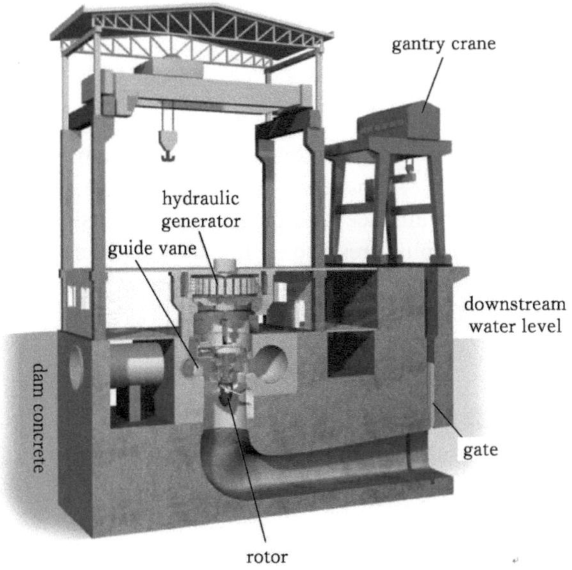

(b)

Fig. 12.3 Substructure layout of powerhouse

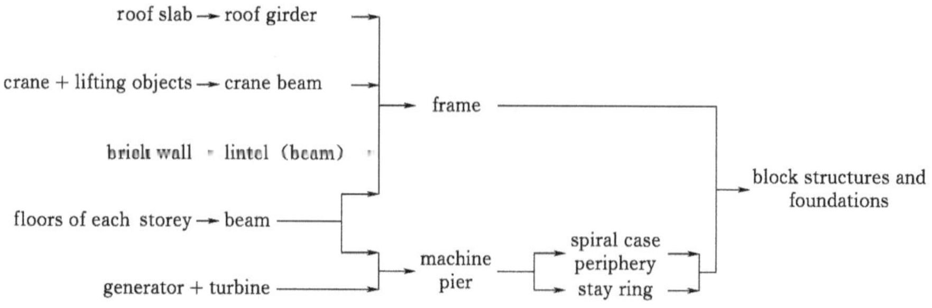

Fig. 12.4 Force transmission structure approach of powerhouse

3. Dead load and live load

   The force transmission system of powerhouse structure can be divided into two categories.

   (1) The transmission system of the hydropower unit. Such as generator pier, spiral case, and draft tube.
   (2) The transmission system of roof, beam, and column. Such as roof slab, roof truss or roof girder, crane beam, bent column, and floor slab are similar to those of general industrial plants.
   The force transmission way of the powerhouse is to transfer all kinds of dead loads and live loads acting on the plant to the foundation through various load-bearing components. The force transmission path is shown in Fig. 12.4.

4. Other loads

   Downstream water pressure → downstream wall; Upstream water pressure of water retaining powerhouse → upstream wall; Seepage pressure of River-side powerhouse → upstream wall; Seepage pressure of expansion joint → spiral case concrete; Foundation uplift pressure → foundation concrete; Wind load → enclosure structure → bent column → foundation; Snow load → roof structure → bent column → foundation.

12.2 Joints and Stage Pouring of Concrete

12.2.1 Powerhouse Joints and Sealing Structure

The functions of powerhouse joints are as follows. Expansion joint and settlement joint (permanent joint): to prevent the formation of cracks caused by temperature change and the differential settlement of foundation. The expansion joints cannot be led to the foundation, while the settlement joints must be lead to the foundation to play the role of expansion joints at the same time. Construction joint (temporary joint): to prevent the formation of cracks caused by concrete shrinkage which will affect the structural integrity.

12.2.1.1 Setting of Permanent Joints

Due to the influence of natural environmental factors, the powerhouse of hydropower station has to go through hot summer, cold winter, and difference between day and night temperatures. The concrete structure of powerhouse will expand and contract due to temperature variation. If the powerhouse length is too large, the concrete of walls and floors will crack, so it is necessary to set structural joints, called expansion joints or temperature joints. Because the plant foundation contacts with the foundation, when the geological conditions of the foundation are different, or the load gap of the powerhouse section is disparate, the uneven settlement of the foundation may be caused. In order to prevent the cracks of the walls and components caused by mutual restraint during the settlement, it is necessary to set up structural joints, which are called settlement joints. Expansion joint and settlement joint can sometimes be combined into one, which is called expansion settlement joint. The setting of permanent joints shall generally meet the following requirements.

(1) The permanent joint cannot be set in the middle of the unit section to prevent the different settlement of the unit section. Generally, one structural joint per unit section is used for large-scale powerhouse on rock foundation, and one structural joint per two unit sections is used for medium-sized powerhouse.
(2) The temperature joint can only be set above the water, and it does not need to be connected to the foundation; while the settlement joint must be connected to the foundation, which can also act as an expansion joint.
(3) The permanent joint must be provided with water-stop to prevent water from seeping into the powerhouse.
(4) Due to the great difference of the load, settlement joints must be set between the main unit room and the assembly site and at the boundary between the main and auxiliary powerhouses.
(5) Joint spacing and joint width. The joint spacing of powerhouse structure should be determined according to climate, temperature conditions, structural forms, geological conditions of foundation, and temperature control measures. The spacing between transverse structural joints also depends on the unit section length. Because the substructure is underwater all year round, the spacing of structural joints is mainly related to the foundation conditions. The maximum expansion joint spacing of the upper structure is 30–70 m. The maximum settlement joint spacing of substructure is 20–40 m on rock foundation and 45–50 m on soft foundation to reduce the influence of unit deflection and crane rail surface staggering. Generally, the structural joints on the rock foundation run through the upper structure and substructure; on the soft foundation, if the spacing between the structural joints is large, an expansion joint connecting only the upper structure of the powerhouse can be set between the two joints.

The width of permanent joint should be determined according to the foundation conditions, potential temperature deformation, joint spacing, plant height, and horizontal displacement caused by expected uneven settlement. The joint on the rock foundation is generally 1–2 cm. The joint on the soft foundation can be wider, but it cannot exceed 6 cm.

12.2.1.2 Water-Stop of Permanent Joint

When the permanent joints of the lower structure of the powerhouse are infiltrating by upstream and downstream water flows, water-stops shall be provided on the water-facing surface of the permanent joints. A closed system should be formed between the vertical and horizontal water-stops. Two water-stops shall be installed in important parts, with an asphalt well in the middle. Asphalt wells, are not necessary for the secondary parts, as shown in Fig. 12.5.

The water-stop materials include metal, rubber water-stop, and plastic water-stop, as shown in Fig. 12.6. Metal water-stop can be divided into copper sheet, galvanized iron sheet, and asphalt coated steel sheet. The copper sheet is a kind of non-ferrous metal, which is expensive and should be used as little as possible. It is only used in important parts of large hydropower stations. The galvanized iron sheet is used in general parts, and asphalt coated steel sheet is used for unimportant parts. However, rubber and plastic water-stops still have the disadvantage of aging for a long time.

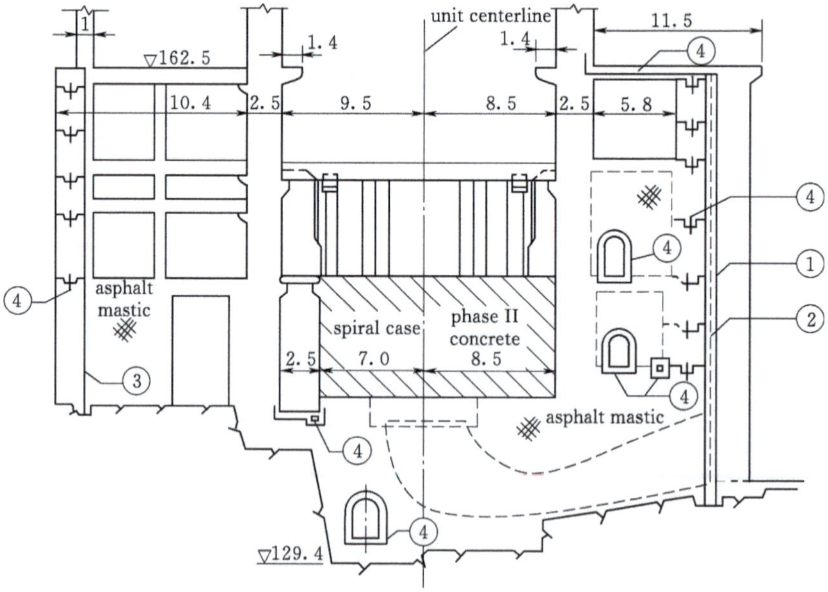

Fig. 12.5 Arrangement of water-stops structure (unit: m) ①-copper sheet A; ②-plastic water-stop A; ③-copper sheet B; ④-plastic water-stop B

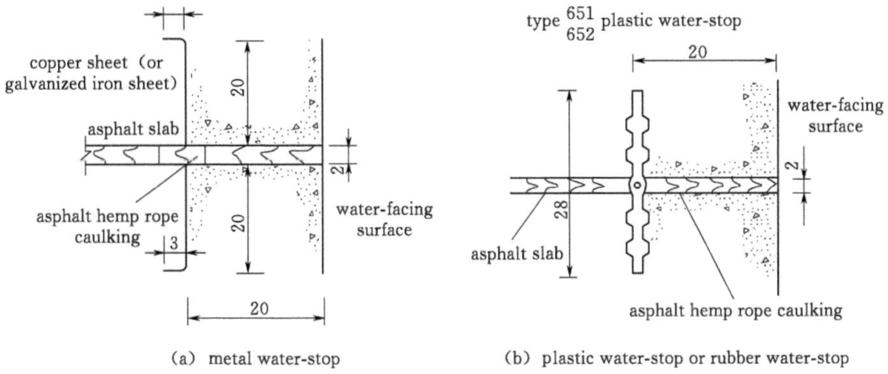

Fig. 12.6 Water-stop structure of permanent joint (unit: cm)

In order to prevent the joints from being filled with sand or debris during the construction or operation and to prevent the wind and rain from attacking the powerhouse, the permanent joints of the roof, floor, and inner and outer walls of the powerhouse should be filled with asphalt fir, multi-layer asphalt felt, light polymer material (closed cell foam board), ointment, mastic (asphalt plus talcum powder, slag powder to make asphalt glue), and asphalt fiber and other elastic waterproof materials.

12.2.1.3 Construction Joint and Water-Stop Structure

The concrete pouring joint set up according to the construction conditions is called construction joint or temporary joint. Since the construction joint is a temporary joint left by block pouring in the construction process, it damages the structural integrity and weakens the structural strength, measures must be taken to deal with the temporary construction joint. It can increase the shear resistance of the structure mainly by using key grooves and inserting reinforcing bars.

In the lower structure of the powerhouse, all construction cold joints, especially vertical cold joints, should be set with water-stops. For underwater construction joints, asphalt felt water-stops shall be pre-pasted on the surface of the old concrete before the new concrete is poured, and plastic water-stops or copper strip water-stops shall be set on the water-facing surface. Figure 12.7 shows the concrete structure joints and water-stops arrangement of a hydropower station.

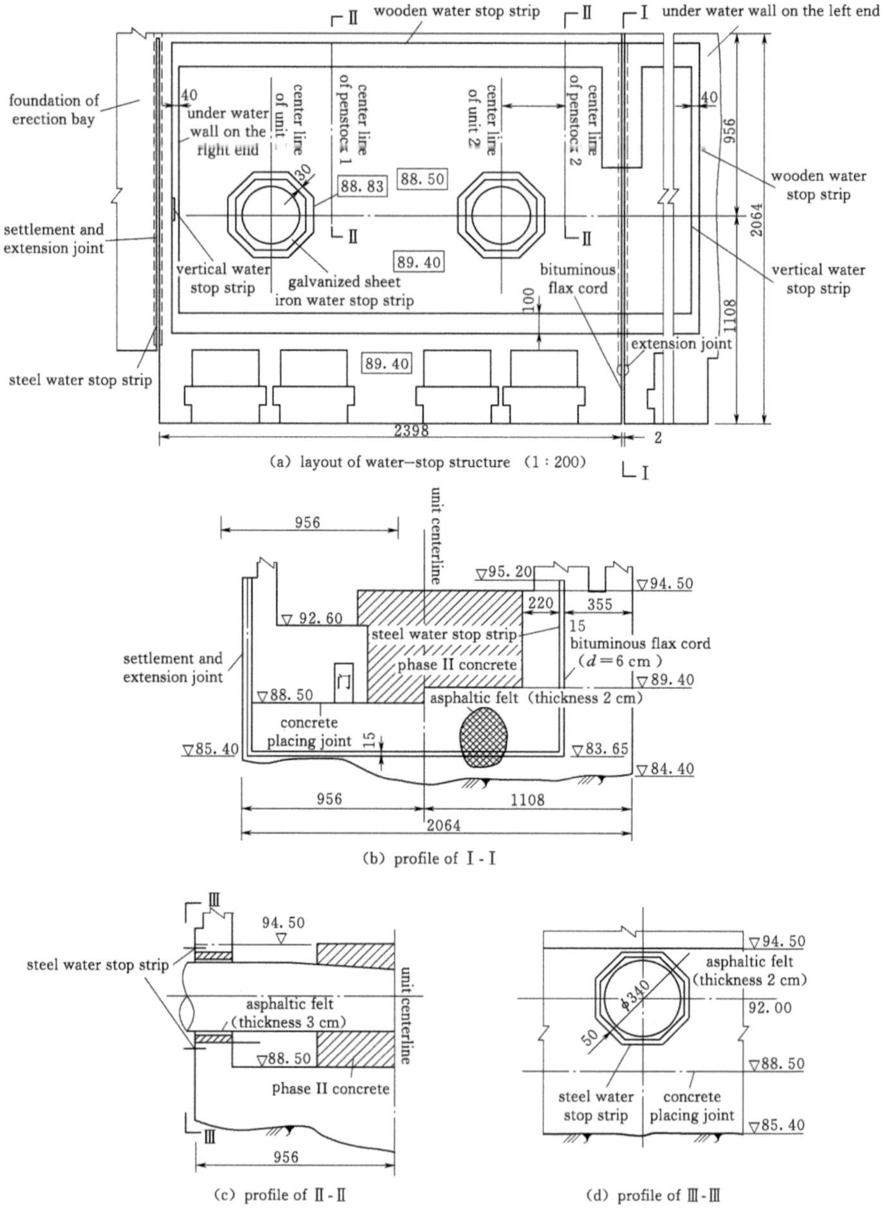

Fig. 12.7 Concrete structure joints and water-stops arrangement of a hydropower station

12.2.2 The Partition and Stage Pouring of the Concrete

12.2.2.1 Pouring in Stages

In order to meet the requirements of unit equipment installation, the powerhouse concrete is generally divided into two phases. The principle of staging is as follows:

(1) Meet the construction requirements of electromechanical equipment installation and embedded parts.
(2) For the powerhouse where the units are installed and operated in stages, it should meet the requirements of stability, strength, and anti-seepage during initial operation.

According to the equipment installation requirements, the concrete at the parts where the equipment and embedded parts need to be installed can be divided into phase II concrete, and the concrete in other parts can be divided into phase I concrete. That is, before the installation of the equipment, the floor, draft tube, upstream and downstream walls, bent column, crane beam, and some floor slab-beam concrete are poured first, which is called phase I concrete. The concrete that needs to be reserved for the unit installation and embedded parts, such as the straight section of the draft tube, spiral case, stator bracket, the outer shell of the ventilation barrel and the connected floor slab, the unit embedded parts, is called the phase II concrete. But for the low head reinforced concrete spiral case, it is generally the first stage concrete. In Fig. 12.8, the Roman numeral "I" is used to indicate the first stage concrete, and "II" refers to the second stage concrete, and the sequence of pouring is indicated by the lower subscript ordinal number.

12.2.2.2 Pouring in Blocks

In order to facilitate the construction and ensure the project quality, the concrete of each stage should be poured in layers and blocks according to the structural form, size, pouring capacity, temperature control, and other conditions during the concrete pouring to ensure the pouring quality.

1. Reasons for pouring into blocks

Because the concrete itself has the characteristics of dry shrinkage and temperature variation, it is easy to cause cracks in the concrete structure. Therefore, in order to prevent concrete cracking, the pouring block should not be too large. In addition, due to the limitation of construction pouring capacity, the concrete blocks poured at one time should not be too large. In order to meet the requirements of construction pouring capacity and prevent concrete cracking and damage, concrete pouring must be divided into blocks.

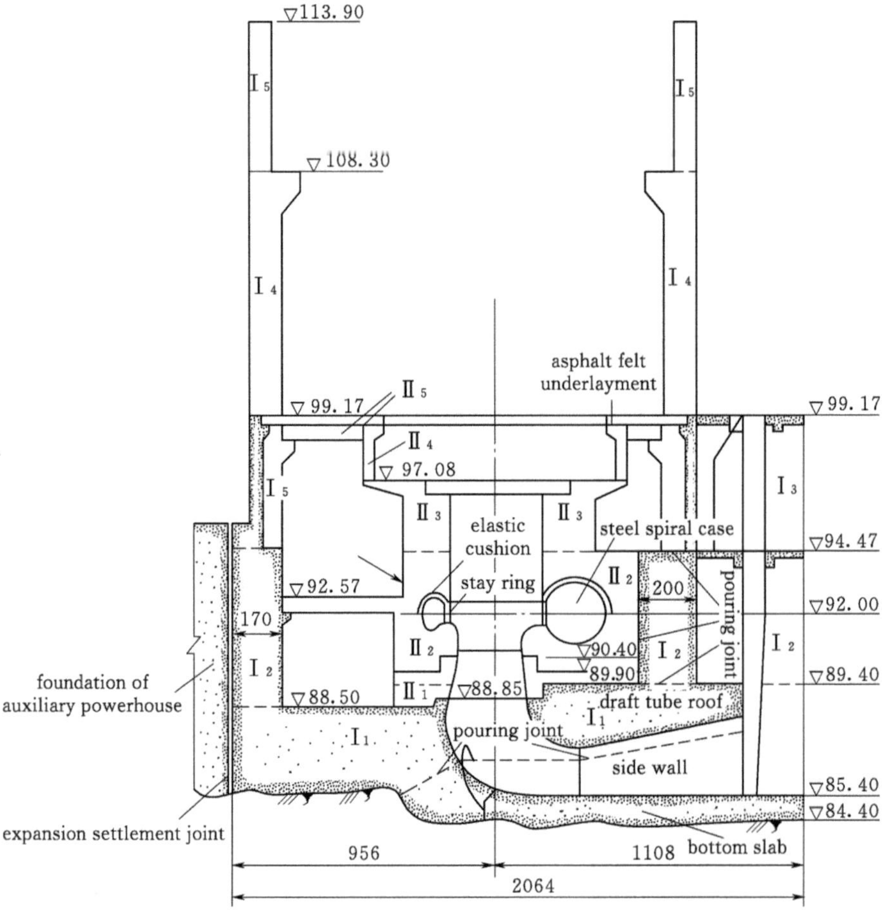

Fig. 12.8 Stage diagram of hydropower powerhouse

2. Factors of pouring into blocks

Too many pouring blocks lead to weak structure and difficult temporary joints treatment; too few pouring blocks make construction difficult and temperature control measures difficult to guarantee. Therefore, the pouring block should be divided based on the structural characteristics, construction progress, pouring capacity, temperature control measures, and other factors. It is difficult to make unified provisions for concrete pouring of powerhouse. Only some principles can be put forward: ① The division should ensure the convenience of the main equipment installation and the embedding; ② The pouring joints should be placed as far as possible in the parts with the smallest internal forces; ③ The size of blocks should be compatible with the concrete mixing production volume, vibrating work intensity, and pouring method. The block geometric size of on the same layer should be basically the same. The geometric shape should avoid thin slices or sharp corners to ensure that no cold joints occur in the concrete; ④ Under the premise of quality, the pouring blocks should be divided as large as possible; ⑤ Try to make the work process repeatable to simplify the construction and reuse the template. Use sequence placement to avoid disturbing neighboring concrete that has not yet reached sufficient strength.

3. Requirements for pouring into blocks

Generally, the block size is determined according to the pouring capacity and structural characteristics, and block heights can be different for different parts. Considering the temperature and bedrock constraints, the block height of foundation block is generally 1–2 m; the block height of other parts, such as spiral case and generator pier, is generally 3–4 m; the block height of wall and draft tube pier can be 4–6 m. Generally, the side length of the pouring block is not more than 15–20 m, and the area is not more than 300 m^2. The soft foundation can be enlarged appropriately. The draft tube pier and side wall should be small in thickness, and the side length should not exceed 15 m.

4. Arrangement and treatment of construction joints.

Pouring into blocks leads to temporary construction joints, which weakens the overall concrete structure. Therefore, construction joints should be arranged and treated reasonably. The construction joints should be arranged in places with low tensile stress to avoid sharp corners, thin-wall or thin layer shape of the block. The upper and lower construction joints should be staggered, and the staggered joint spacing should be 1/3–1/2 of the block height, but not less than 30 m.

When construction joints are arranged in places with high tensile stress or sudden changes in body shape, insertion ribs, and keyway should be set; in places prone to cracks or uneven settlement, wide grooves should be reserved; for vertical construction joints, transverse keyway should be provided on the joint surface. The area of the keyway should be about 1/3 of the joint surface, and the insertion ribs should be 3–6 cm2 per square, as shown in Fig. 12.9.

Fig. 12.9 Vertical construction joint surface with insertion ribs and keyway

Questions

1. What are the structural characteristics of the main powerhouse? What are the loads on the ground powerhouse structure and how are they transmitted to the foundation? Please draw a schematic diagram to show the load transmission system of the powerhouse structure?
2. Why should the hydropower powerhouse set up structural joints and construction joints? What are the requirements for setting the structural joints and construction joints?
3. Why should the concrete pouring of the powerhouse structure be carried out in stages and blocks? What are the requirements for staging and blocking?

Bibliography

1. Design code for Hydropower house SL 266: 2014 (In Chinese). Beijing: China Water and Power Press.
2. Design code for Hydropower house NB/T 35011: 2013 (In Chinese). Beijing: China Electric Power Press.
3. Yu, Jianhua and S. Zhang. 2014. *Hydropower plant design and construction (In Chinese).* Zhengzhou: The Yellow River Water Conservancy Press.
4. Ma, Shanding and R. Wang. 1996. *Hydropower station buildings (In Chinese).* Beijing: China Water and Power Press.

Chapter 13
Underground Powerhouse

Juntao Chen

Abstract In this chapter, the underground powerhouse is introduced. Hydropower stations are often located in hard areas such as the alpine and gorge regions, and the underground powerhouse has become an important type of powerhouses. Therefore, in this chapter the development of the underground powerhouse is introduced firstly, including the application scenarios, as well as the characteristics of layout and construction. Next, the layout form of the underground powerhouse is analyzed, including its composition, layout principles, layout types and methods. In addition, because of the complex characteristics of underground powerhouse cavern excavation, this chapter focuses on the layout of the main cavern structure of underground powerhouse, including main and auxiliary powerhouse, main transformer tunnel and switch station, assembly site, surge chamber, draft tube and tailrace tunnel, auxiliary cavern and so on. On this basis, the layout design of the underground powerhouse is introduced from the aspects of the powerhouse layout principle and the determination of structural size. Finally, the calculation methods of structure design for crane girder on rock wall in underground powerhouse structure and the surrounding rock stability and support analysis method of underground powerhouse cavern are introduced in detail.

Keywords Underground powerhouse · Development characteristics of underground powerhouse · Layout of underground powerhouse · Layout design; carven layout · Crane girder on rock wall · structural design of crane girder on rock wall · Stability of surrounding rock · Supporting design · The finite element method

---

J. Chen (✉)
Wuhan University, Wuhan, China
e-mail: chenjt@whu.edu.cn

Y. Cheng et al. (eds.), *Hydropower Station*,
https://doi.org/10.1007/978-981-96-2383-9_13

13.1 Development Characteristics of Underground Powerhouse

13.1.1 Reasons for Development of Underground Powerhouse

The main building of the powerhouse is arranged in the rock body under the ground, which is called the underground powerhouse. China is rich in hydropower and water resources. The capacity of hydropower that can be economically developed is 402 million kW, which ranks first in the world. In the development of hydropower resources, the ratio and scale of underground powerhouse becomes larger and larger, the construction conditions of underground powerhouse become more and more complicated as well. The following factors have led to the rapid development of underground powerhouse:

(1) In our country, hydropower resources are mostly concentrated in the western region, so as to promote the development of large underground powerhouse. Hydropower resources in the western region account for about 71% of all economically exploitable sources. The south-western and north-western rivers are located in the alpine and gorge regions. Most of the hydropower stations are large reservoirs with high dams. Due to the narrow riverbed, these large hydropower stations have to be developed in the form of underground powerhouse.
(2) In order to meet the demand of peak-shaving and valley-filling in the electric power system, pumped storage underground powerhouses with high head and large capacity are developing rapidly. In east and north China, because of the abundant resources of thermal power and nuclear power, a large number of pumped storage power stations with high head and large capacity have been planned and designed to ensure the stable operation of the electric power system. Because of upper and lower reservoir layout and the requirements of the pump working head, most of powerhouses of pumped storage power station are laid out in underground powerhouse.
(3) Advanced drilling and blasting technique, mechanized construction level, and means of excavating and supporting are more and more advanced. The construction and excavation speed of underground powerhouse is getting faster and faster, the cost is getting lower and lower, and the engineering project is getting more and more secure which makes the development of large underground powerhouse has developed rapidly.

13.1.2 Application Scenarios for Underground Powerhouse

(1) It is difficult for the layout of large and ultra-large power station in alpine and gorge regions (Fig. 13.1);
(2) The pumped storage power station with high water head (Fig. 13.2).

Fig. 13.1 Alpine and gorge regions

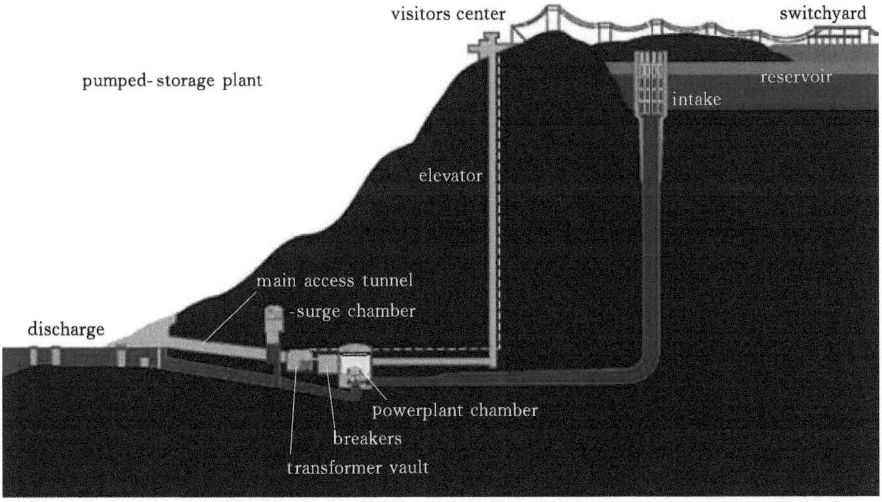

Fig. 13.2 Pumped storage power station with high water head (internet picture with unknown source)

13.1.3 Features of Underground Powerhouse

Since the first underground power station was built from 1904 to 1907, plenty of underground hydropower stations have been constructed in Europe, Canada, Japan, Soviet Union, India, etc.

In China, underground hydropower stations also start to develop rapidly. The great progress in design and construction technology of underground powerhouse, the big decrease in the cost of engineering construction and equipment as well as the acceleration of construction speed, all show the advantages of underground powerhouse.

With the use of efficient underground construction machinery sets and the continuous development of rock mechanics and shotcrete-anchorage technology, underground powerhouse has an increasing high percentage in large and ultra-large power stations constructed and under construction.

13.1.3.1 Advantages of Underground Powerhouse

(1) Underground powerhouse benefits to layout of whole project, and won't be threatened by the possible danger of landslides in valley regions.
(2) Underground powerhouse can shorten the penstock and shaft, or cancel the surge chamber, so the investment can be reduced.
(3) Underground powerhouse is free from the interference of riverbed and the climate outside, so its construction can speed up due to perennial construction.
(4) Underground powerhouse can improve the regulation capacity of hydraulic turbines. The operation of underground powerhouse is free from the influence of climate and landslide, as well as the interference of water spray and tailrace fluctuations at flood discharge.
(5) Underground powerhouse is safe when used as civil defense facility. It is antiseismic and won't damage the landscape.

13.1.3.2 Main Disadvantages of Underground Powerhouse

Compared with the ground powerhouse, the underground powerhouse has following disadvantages.

(1) Large quantities of tunnel excavation, slow construction progress and high project investment.
(2) The high requirements for geological conditions and the high engineering risks caused by the changes in geological conditions.
(3) The ventilation, lighting, transportation and other conditions of underground powerhouse are poor.
(4) The leakage and drainage problems of underground powerhouse are prominent, and they often require special engineering measures.

13.1.4 Features of Underground Hydropower Stations Under Construction

(1) Underground powerhouses are used in large hydroelectric project with high water head, large reservoirs with high dam, canyon reservoirs with high water head, and diversion hydropower stations in high mountain area. Meanwhile, both total installed capacity and unit installed capacity are increasing.
(2) The scale of large underground caverns is becoming larger. In particular, the number of pumped storage underground power stations and their size are increasing.
(3) The drill-blast technique, the mechanized construction and excavation technique and the support technique are becoming more and more advanced. Meanwhile, the process of construction and excavation speed up.
(4) The complexity of geological conditions and technical difficulties for underground powerhouses are increasing.

13.1.5 Complexity of Underground Powerhouse Cavern Group

The underground powerhouse is a complex spatial underground structure excavated from the complex rock mass structure. Due to the complexity of the rock mass structure itself, the stability problem of the surrounding rock of the underground powerhouse cavern based on the rock mass is extremely complicated. In summary, the underground powerhouse caverns have the following complex characteristics.

(1) The underground powerhouse caverns have a complex geological occurrence environment. They are based on rock mass and are supported by surrounding rock. In the long-term evolution process, the rock mass has been affected by natural actions, which include the complex geostress distribution caused by the self-weight of the rock body and the geotectonic movement, subterranean water seepage and erosion, rock mass construction and characteristic alteration, geological movement and fission, and fault cutting. These complex occurrence environments make the underground powerhouse caverns very complicated.
(2) Spatial structure of underground powerhouse caverns is complex. Underground powerhouse consists of main powerhouse, main-transformed cavern, bus–bar tunnel, headrace tunnel, tailrace tunnel, surge chamber and other cavern groups, which makes the spatial structure of underground powerhouse caverns crisscross.
(3) The rock mass characteristics of underground powerhouse caverns are complicated. The rock masses of the underground powerhouse are different, and the characteristics of the rock masses are also quite different.
(4) The disturbance mechanism of underground powerhouse caverns excavation is complicated. Most of the underground powerhouse caverns were constructed

by using drilling and blasting technique, which is strongly affected by the engineering disturbance during the construction period.

13.2 Layout of Underground Powerhouse

13.2.1 Components of Underground Powerhouse

(1) Headrace structures (water intake, headrace tunnel, surge chamber, penstock, tailrace surge chamber and tailrace tunnel).
(2) Main and auxiliary powerhouse.
(3) Step-up substation (main transformer cavern).
(4) Switch station.
(5) A series of auxiliary tunnels including busbar tunnel, headrace tunnel, tailrace tunnel, outgoing line tunnel, access tunnel, construction tunnel, ventilation tunnel, etc.

13.2.2 Layout Principle of Underground Powerhouse

(1) The layout of the underground powerhouse should be coordinated with the general layout to meet the requirements of equipment layout, production operation, energy conservation and environmental protection.
(2) According to the terrain, geological conditions, construction requirements, environmental impact factors, etc., the location of the main powerhouse and access tunnel should be determined to ensure that the external traffic of the powerhouse is convenient and in order.
(3) The main caverns of underground powerhouse should be arranged at the location with good geological environment and avoid geological fault and unfavorable geological structure zone as far as possible. When the geological discontinuity cannot be avoided, the intersection angle between the carven axis and the geological discontinuity should be maintained as far as possible.
(4) The layout of underground cavern group should follow the principle of combining temporary and permanent structures and one carven for multiple purposes. The number of auxiliary caverns and the cutting of rock mass should be reduced as far as possible, so as to facilitate the stability of surrounding rock mass.
(5) The span of underground powerhouse cavern should be reduced as much as possible. The overlying rock mass of each cavern should be of sufficient thickness, and the rock column between each cavern shall be of enough safe distance to ensure the safety of surrounding rock stability of the cavern.
(6) The layout of water inlet and outlet lines of the underground powerhouse should be simple and smooth, so as to ensure the stable operation of the turbine and the stability of the headrace tunnel and tailrace tunnel.

13.2.3 *Information Required for Layout of Underground Powerhouse*

In order to arrange the underground powerhouse reasonably, we must collect all kinds of data needed in the layout design, and further provide and supplement them according to the content and design depth requirements of different design stages.

1. Hydrological data of powerhouse area

The inlet and outlet elevations of the underground powerhouse for external traffic, ventilation and outgoing lines should be higher than the check flood level to prevent the flood from flowing back into the powerhouse. Moreover, the outlet elevation of the tailrace tunnel is affected by the backwater of downstream cascade hydropower stations and the sedimentation at the tail of the reservoir. The hydrological data of the powerhouse directly affects its layout.

2. Topographic data of powerhouse area

The underground powerhouse should be located in a section with a relatively stable mountain. The layout of the underground powerhouse caverns should not only consider whether the entrance will be affected by the dam flood or flash flood, but also consider whether the exit and traffic road will be affected by the rock collapse on the hillside. The connection with downstream cascade power stations should also be considered to make full use of water head and the off-site construction land, switch station land and other requirements should be considered as well. Therefore, the topographic data of the powerhouse area has great influence on the layout of auxiliary caverns, traffic, construction, etc.

3. Geological data of the powerhouse area

Geological data such as geological structure, earthquake intensity, rock characteristics and initial in situ stress field are the key to determine the layout of underground powerhouse. The underground powerhouse should try to select areas with good geological conditions, low seismic intensity and good rock integrity to ensure the stability of surrounding rock of underground powerhouse caverns.

4. Electromechanical equipment of the powerhouse

The form and size of electromechanical equipment are directly related to the outline size, layout and structural form of underground powerhouse.

5. Information on local environmental and social needs

Comprehensive consideration should be given to the requirements of mucking and waste disposal area during the construction period of underground powerhouse, local traffic environment and equipment transportation conditions; Give full consideration to the requirements of the destruction and influence of some necessary ground buildings on the original vegetation and natural landscape and how to restore them after completion; At the same time, we should take the development of local society and tourism needs into account.

13.2.4 *Development Modes of Underground Hydropower Station*

According to the practical conditions such as geology, terrain, construction, traffic, operation, etc., underground powerhouse layout mainly include head-type layout, tail-type layout and middle-type layout.

13.2.4.1 Layout of Head-Type Powerhouse

Location: a place near intake structure.

Features: headrace tunnel is short, penstock is composed of shafts and tailrace tunnel is long.

Advantages:

(1) Compact layout is easy for management, construction and operation. Head-type layout plan is relatively economical.
(2) The regulation of hydraulic turbine is improved and no surge chamber is required for penstock.

Disadvantages:

(1) Tailrace tunnel may be very long. Tailrace surge chamber is generally required along with access tunnel, ventilation tunnel, outlet tunnel, etc.
(2) Powerhouse is located near the reservoir. It's buried deep and is difficult for line outgoing and seepage prevention.

13.2.4.2 Layout of Tail-Type Powerhouse

Location: near tailrace outlet.

Features: there are a long pressure headrace tunnel and a shorter tailrace tunnel. Surge chamber is generally required at the end of penstock.

Advantages:

(1) Short tailrace and auxiliary tunnels are easy for construction, operation and management.
(2) Low tail water is easy for seepage prevention.

Disadvantages:

(1) Headrace conduit is long and surge chamber is required.
(2) It's difficult to arrange tailrace tunnel when tail water fluctuates greatly.

13.2.4.3 Layout of Middle-Type Powerhouse

Location: a place in the middle of headrace tunnel.

Features: Both headrace tunnel and tailrace tunnel are long.

Application: When there are no appropriate terrains and geological conditions, the head-type and tail-type layout can be adopted.

Advantages: Headrace and tailrace surge chambers may be canceled, so it may be economical.

Disadvantages: Headrace and tailrace surge chambers may be required, so it will be difficult for turbine regulation.

13.2.5 Layout Design Steps of Underground Powerhouse

13.2.5.1 Determine Layout of Underground Powerhouse

The layout of underground powerhouse is the premise of determining the overall layout of underground powerhouse project. The layout method should be comprehensively determined through technical and economic comparison according to factors such as layout of power station project, hydrology, meteorology, topography, geological conditions, operation conditions, construction conditions and eco-environmental requirements, etc.

13.2.5.2 Select Location and Longitudinal Axis Orientation of Main Powerhouse Cavern

The layout position of the main powerhouse cavern is the key to the layout of the underground powerhouse project. Only when the location and longitudinal axis orientation of the main powerhouse cavern are determined can the layout of other auxiliary caverns be carried out based on the main cavern.

13.2.5.3 Determine Relative Position of Underground Main Caverns Powerhouse

The main caverns of the underground powerhouse mainly include the main powerhouse cavern, main transformer cavern, surge chamber, etc. The layout of the underground powerhouse project is mainly to arrange the relative positions of main caverns such as the main transformer room and surge chamber around the main powerhouse cavern. Generally, it is necessary to determine the layout pattern according to technical economical comparison between various possible layout schemes. Table 13.1 shows the common layout patterns and characteristics of the main powerhouse, main transformer tunnel and downstream surge chamber.

Table 13.1 common layout patterns and characteristics of the main powerhouse, main transformer cavern and downstream surge chamber.

Pattern	Location of main transformer cavern	Diagram	Advantages	Disadvantages
1	Between the main powerhouse and the downstream surge chamber	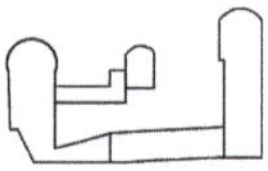	Compact layout, and convenient operation and maintenance; Reduce the harm degree of accidents	Not conducive to the stability of cavern rock
2	Located on the upstream side of the main powerhouse	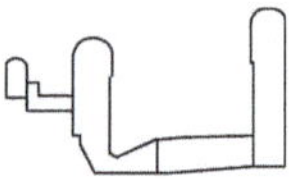	Compact layout, and convenient operation and maintenance; Reduce the harm degree of accidents; Shorten the distance between main powerhouse and surge chamber	Unfavorable to the layout of penstock, seepage control, and drainage; The surrounding rock stability of powerhouse is relatively poor
3	Located in the main powerhouse	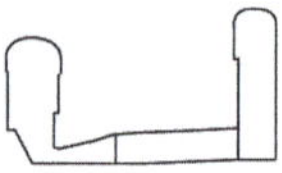	Shortest busbar; Convenient operation and management; Shorten the distance between main powerhouse and surge chamber	Fire and explosion are more dangerous; The main powerhouse building has a long length
4	The main transformer tunnel and the downstream surge chamber of the main powerhouse are triangular distribution	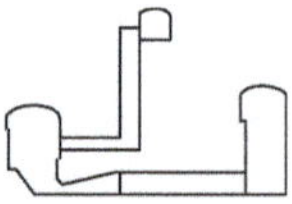	Shorten the distance between main powerhouse and surge chamber; Reduce the harm degree of accidents	Inconvenient operation and maintenance; Accessory channels are increased

13.2.5.4 Determine Outline Size, Elevation of Each Main Cavern and Spacing Between Caverns

The key to the layout of underground powerhouse is to determine the outline size, elevation and spacing between the main caverns. The size of each main cavern should be reasonably reduced and optimized as far as possible, the elevation should meet the requirements of unit operation, and the spacing should ensure the stability of surrounding rock of caverns.

13.2.5.5 Determine Layout of Water Diversion and Tailrace Routes, Incoming Direction of Penstock, and Unit Gate Valves

According to the layout pattern of each main cavern, the inlet and outlet positions of water diversion system are selected first, and then the layout of water diversion and tailrace routes is determined. Then, according to the positions of water intake and main powerhouse, the incoming direction of penstock is determined, and the layouts of unit gate valves are determined by comprehensive layout of water diversion pipeline system.

13.2.5.6 Determine the Layout and Elevation Position of Each Auxiliary Cavern

According to the general layout pattern of the powerhouse project, the routes of the auxiliary caverns such as transportation tunnel, outgoing line tunnel and ventilation tunnel are selected. According to the requirements of topography, geology, construction, and operation, the elevations of inlet and outlet of each auxiliary caverns are determined.

13.3 Cavern Layout of Underground Powerhouse

13.3.1 Layout of Main Powerhouse

13.3.1.1 Selection of Main Powerhouse Location

1. Uniform lithology, high strength, simple strata, integrated rock mass and moderate buried depth.
2. In deep canyon terrain, powerhouse location should keep away from relief joint zone and the stress concentration area.

13.3.1.2 Determination of Longitudinal Axis Direction

To choose location with good geological conditions is the key to surrounding rock mass stability of underground powerhouse.

The longitudinal axis direction should be determined based on the bedding plane and main structural plane. It should be perpendicular to the bedding plane and the main structural plane or at a great angle.

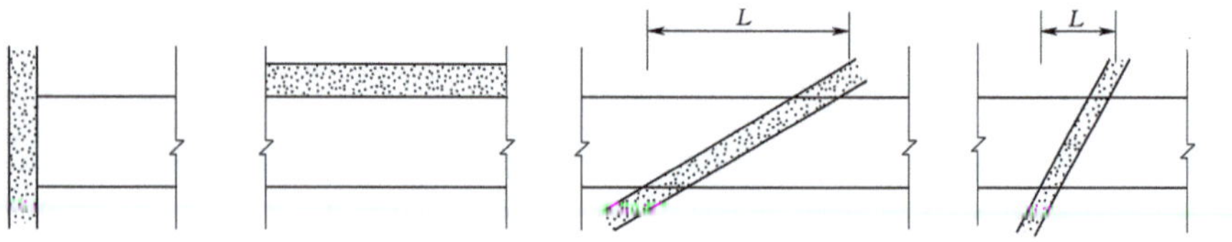

Fig. 13.3 Relationship between the longitudinal axis and the main structure of rock mass

The longitudinal axis direction should consider the principal stress direction of initial stress of rock mass. The angle between the direction of maximum principal stress and the direction of longitudinal axis is expected to be small.
The direction of longitudinal axis should not be consistent with the direction of large fracture system.

As the initial stress of rock mass is highly directional, it plays a control role in the deformation and failure of surrounding rock of underground engineering. The longitudinal axis direction of underground powerhouse should be consistent with or at a small angle with the direction of maximum principal stress of rock mass, as shown in Fig. 13.3.

13.3.1.3 Buried Depth of Powerhouse

Because of the different geological conditions, engineering layout and construction requirements, the buried depth of powerhouses is also different when selecting the location. To ensure the stability of rock mass on arch crown of caverns, a certain buried depth is required. It depends on the influence of rock stress redistribution on cavern stability after the excavation of caverns and is generally more than two times of the excavation span.

13.3.2 Layout of Main Transformer Cavern and Switch Station

13.3.2.1 Layout Methods of Main Transformer Cavern

1. Parallel with main powerhouse cavern at downstream side—a common method

It makes the arrangement of the main caverns compact, which is beneficial for the main transformer to directly enter the main powerhouse to be repaired. The main transformer carven can be free of cranes. It is beneficial to the layout, operation and maintenance of the indoor switchgear room. It is convenient for the main powerhouse and the main transformer cavern to be arranged separately, which can reduce the harm degree of accidents. When there is a downstream surge chamber, this layout increases the length of the extension section of the draft tube and increases the difficulty of surrounding rock stability and unit regulation guarantee.

2. Parallel with main powerhouse cavern at upstream side—Pengshui project

This arrangement is beneficial to the downstream surge chamber arrangement, which can shorten the distance between the main powerhouse and the surge chamber, and is beneficial to the regulation guarantee of the hydraulic turbine unit. The layout of the main caverns is compact, and the operation and maintenance are convenient, which can reduce the harm degree of accidents when the main transformer goes wrong. But it is not conducive to the anti-seepage and drainage layout of the powerhouse building. In order for the main transformer to directly enter the main powerhouse for maintenance, the elevation at the bottom of the main transformer tunnel is low, which has a great influence on the layout of high-pressure pipelines entering the powerhouse.

3. End of main powerhouse cavern—Dongpin project, Langya Mount project

The transformer is arranged in the same extended chamber at the end of the main powerhouse, and is separated from the unit by a fire-proof and explosion-proof wall. This arrangement can reduce one main transformer cavern, which is beneficial to the stability of surrounding rock of the cavern; The bus is short and the power loss is small, which is beneficial to operation and management; The length of draft tube is short, which is beneficial to the regulation and guarantee of the unit. However, the main powerhouse cavern is long, which has a great influence on the unit when there is a problem with the main transformer.

4. The triangle layout of main transformer cavern, main powerhouse and downstream surge chamber

This arrangement can reduce the distance between the main powerhouse and surge chamber, shorten the length of draft tube, and reduce the mutual interference among the three caverns. However, the bus bar is long, so the main transformer cannot enter the main powerhouse for maintenance, and lifting equipment needs to be added in the main transformer chamber, which is inconvenient for operation and maintenance.

13.3.2.2 Cavern Distance of Main Transform Cavern

The cavern spacing between the main transformer cavern and the powerhouse cavern is a key to the layout of the main transformer cavern. The determination of rock mass thickness between underground caverns involves many factors, such as geological conditions, cavern scale, in-situ stress and so on, so it is difficult to get a unified regulation. It can usually be determined according to the following principles.

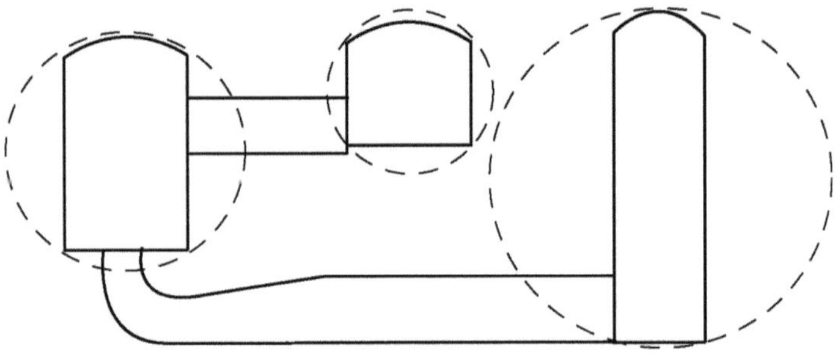

Fig. 13.4 Stress effect ranges in the arrangement of crowded caverns

(1) According to the engineering experience analogy.
(2) Demonstration by numerical methods.

In underground powerhouses, the arrangements of a large number of caverns are sometimes limited by certain conditions. Some crowded phenomena often appears both in terms of plane or section, such as shown in Fig. 13.4. The thickness of the rock mass between two adjacent caverns is typically determined by the stability calculation and experiment of surrounding rock.

13.3.3 Layout of Auxiliary Powerhouse

13.3.3.1 Layout Principle of Auxiliary Powerhouse

All auxiliary powerhouses are placed underground, which will greatly increase the excavation volume of underground cavern engineering, prolong the construction period and increase the cost. Layout should follow the principle of combining centralization and decentralization, combining ground and underground, and convenient management and operation:

(1) All auxiliary equipment close to the main unit can be centrally arranged in the underground cave close to the main unit room. For those that must be located underground, the space they occupy should be reduced as much as possible.
(2) Any equipment that can be placed far away from the main unit can be distributed or placed on the ground by utilizing existing caverns.
(3) The layout of the central control room should comprehensively consider the factors such as convenient operation, maintenance, monitoring, quick troubleshooting, convenient internal traffic, economy and rationality, etc.
(4) Attention should be paid to the safety and disaster prevention design of the main cavern. An emergency refuge room should be set up at an appropriate position in the auxiliary building for some people who have no time to evacuate in case of disaster accident.

13.3.3.2 Layout Mode of Auxiliary Powerhouse

Most of the current hydropower stations adopt remote control and remote measuring management, and computer monitoring. Therefore, the layout and area of the auxiliary powerhouse are simplified as much as possible. Due to the characteristics of the layout of the underground powerhouse, the main layout ways of the auxiliary powerhouse are as follows.

(1) The auxiliary powerhouses are arranged at both ends of the main powerhouse caverns, as shown in Fig. 13.5. The hydraulic machine auxiliary equipment is usually arranged at one end of the assembly site, a three-supply and one-drainage system is arranged at the lower level of the assembly site. At the other end of the main powerhouse, there are electrical machine auxiliary equipment, including the central control system, communication system, automatic protection system, calculation and detection system, and DC system.
(2) The auxiliary powerhouse is arranged at one end of the main powerhouse cavern, as shown in Fig. 13.6. For this arrangement, the assembly site and auxiliary powerhouse are centrally arranged at one end of the main powerhouse. Usually, the assembly site is arranged close to the unit section, the hydraulic machine auxiliary equipment is arranged on the lower floor, and then the electrical machine auxiliary equipment is arranged close to the end of the assembly site.
(3) When the assembly site is arranged in the middle of the main powerhouse cavern, the hydraulic and electrical machine auxiliary equipment can be arranged separately at both ends of the main powerhouse cavern, or the hydraulic machine auxiliary equipment can be arranged under the assembly site and the electrical machine auxiliary equipment can be arranged at one end of the powerhouse, as shown in Fig. 13.7.

13.3.4 Layout of Assembly Site of Underground Powerhouse

The assembly site layout of underground powerhouse is basically the same as that of ground powerhouse. The assembly site of underground powerhouse shall be determined according to the topographic and geological conditions of the powerhouse area and the specific location of the incoming traffic lines. There are mainly the following two kinds of arrangement:

13.3.4.1 Arranged at the End of the Main Powerhouse Cavern

This is the most common arrangement, as shown in Figs. 13.5 and 13.6. When there are few units, the assembly site can generally be arranged at both ends of the underground powerhouse to reduce the interference of unit operation, installation and maintenance.

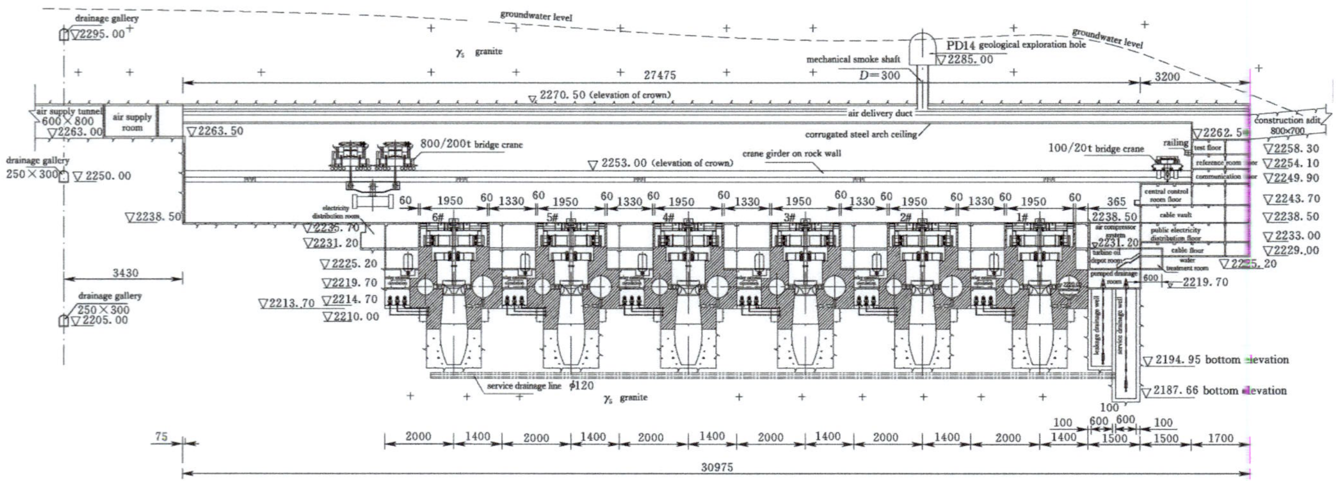

Fig. 13.5 The auxiliary powerhouse is arranged at both ends of the cavern of the main powerhouse (Elevation unit: m; Dimension unit: mm)

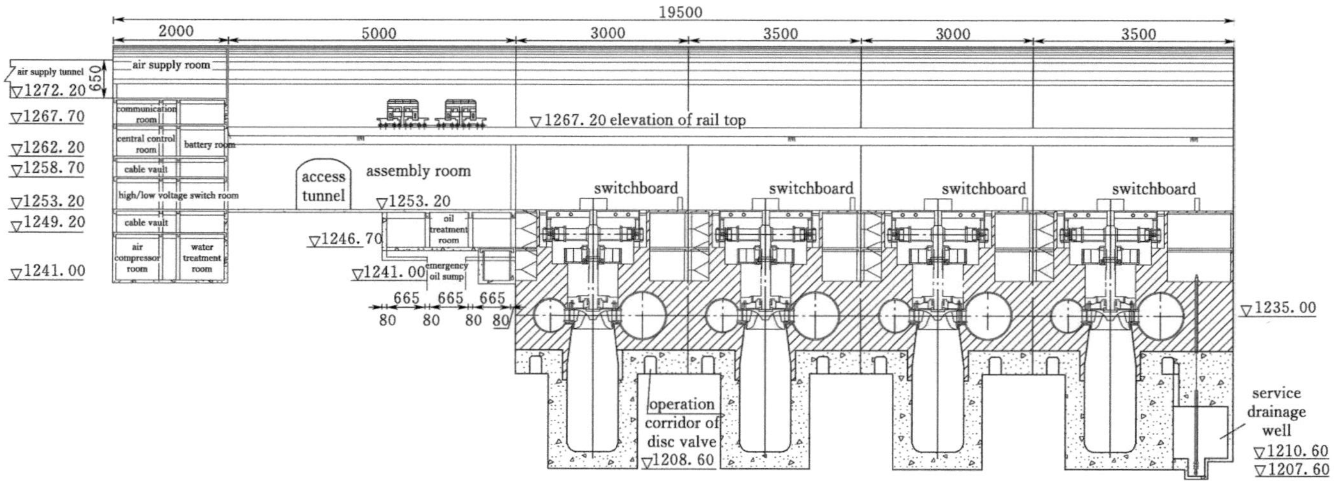

Fig. 13.6 The auxiliary powerhouse is arranged at one end of the main powerhouse cavern. (Elevation unit: m; Dimension unit: mm)

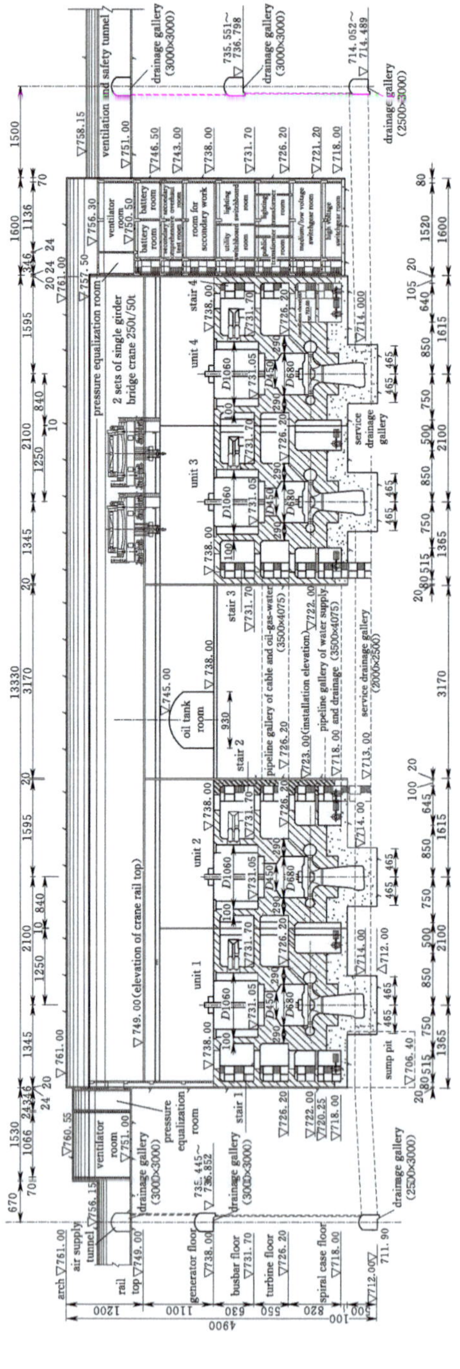

Fig. 13.7 The auxiliary powerhouse is arranged in the middle and one end of the main powerhouse cavern (Elevation unit: m; Dimension unit: mm)

13.3.4.2 Arranged Between Unit Sections

When there are many units, and the geological conditions of surrounding rock are poor, the assembly site can be arranged in the middle of the main powerhouse in consideration of limiting the deformation of the cavern side wall, as shown in Fig. 13.7. This arrangement is convenient to stage commissioning of multi-unit power stations, and the area of the assembly site can be fully utilized because it is not limited by the limit lines of the left and right hooks of the crane. The assembly site is arranged between the units, and the rock mass under the assembly site can be preserved, thus reduces the free surface length of the rock wall on both sides of the main powerhouse, which is beneficial to the stability of surrounding rock.

The determination principle of assembly site area is the same as that of the ground powerhouse, which can be determined according to the area required for the extended overhaul of one unit.

13.3.5 Layout of Surge Chamber

13.3.5.1 Location of Surge Chamber

The location of the surge chamber should be comprehensively compared and determined according to the layout of water conveyance pipeline system of hydropower station, topography and geological conditions along the water conveyance pipeline, unit operation parameters, layout of main buildings of underground powerhouse and surrounding rock stability of surge chamber cavern.

The distance between the surge chamber and powerhouse cavern should take into account of the operation conditions of hydropower station and the surrounding rock stability of cavern. The location of surge chamber should avoid unfavorable geological conditions and facilitate the layout of construction branch tunnel and transportation.

13.3.5.2 Form of Surge Chamber

The upstream surge chamber of long diversion power station generally uses cylindrical surge chamber. The shape of the downstream surge chamber of the large underground power station depends on the number of units, hydraulic stability requirements and geological conditions, and can be in the form of corridor, cylinder and semi-cylinder. The corridor shape is favorable for setting the draft tube access door, but the oversize surge chamber free surface is unfavorable for the stability of surrounding rock. Cylindrical shape is beneficial to the stability of surrounding rock, but it is not conducive to the layout of draft tube access door. Therefore, it is usually necessary to add a tailwater gate chamber, which increases difficulty of building layout. Semi-cylindrical surge chamber is a new structural form, which takes into account the advantages of both corridor type and cylindrical surge chamber, and has a better adaptability to geological conditions.

When the length of the tailrace tunnel is relatively short, the following measures can be taken to cancel the downstream surge chamber:

(1) Reduce the installation elevation of hydraulic turbine, which can reduce the vacuum degree caused by negative water hammer, and meet the safe operation of the unit.
(2) When the downstream tail water level fluctuates slightly, the free-flow tunnel with free-surface flow or free-surface flow for part of tunnel sections can be adopted.
(3) When the downstream tail water level fluctuates greatly, the sloping ceiling tailrace tunnel can be used to avoid setting up a large-scale surge chamber.
(4) Enlarge the size of tailrace tunnel and reduce the flow velocity in it, so as to achieve the goal of no downstream surge chamber.

13.3.6 Layout of Draft Tube and Draft Tunnel

13.3.6.1 Structural Layout of Draft Tube

The determination principles of draft tube structure and size in underground powerhouse are basically the same as those in ground powerhouse, but considering the stability requirements of surrounding rock of underground powerhouse cavern, narrow high draft tube is usually used in underground powerhouse. On the premise of keeping the outlet section of draft tube the same, reduce the width of draft tube and increase its height, which can increase the thickness of rock column between draft tubes and improve the stability of rock column.

13.3.6.2 Axis Layout of Tailrace Tunnel

The selection of the tailrace tunnel route is one of the important links in tailrace tunnel design. The following factors need to be fully considered.

(1) Geological conditions along the tunnel are as complete as possible.
(2) The tunnel axis should preferably be kept straight.
(3) The tunnel should have sufficient buried depth.

13.3.6.3 Structure Form of Tailrace Tunnel

The structural forms of tailrace tunnel can be roughly divided into three types.

(1) Free-flow tunnel: there is no need to set surge chamber, and it is suitable for the situation of downstream water level with small amplitude.

(2) Pressure tunnel: pressure exists in various working conditions. When the tunnel is long, the downstream surge chamber is generally set.
(3) Tailrace tunnel with mixed free-surface-pressurized flow: there is no pressure at low water level, and there is pressure at high water level. At mid water level, part of the tunnel section is under pressure and part of it is unpressurized.

13.3.6.4 Sloping Ceiling Tailrace Tunnel

The sloping ceiling tailrace tunnel is a kind of tailrace tunnel with mixed free-surface-pressurized flow. Pengshui Hydropower Station in China adopts sloping ceiling tailrace tunnel, as shown in Fig. 13.8.

13.3.7 *Layout of Underground Auxiliary Caverns*

(1) Underground hydropower station has many auxiliary caverns, such as access tunnel, outgoing line tunnel, safe pedestrian tunnel, drainage tunnel, ventilation tunnel, mucking tunnel, geological exploration tunnel, and main valve chamber. The typical layout of drainage gallery is shown in Fig. 13.9.
(2) 2. Under the premise of satisfying construction and operation, one auxiliary tunnel should be designed to play as much of its function as possible to reduce investment and speed up the schedule. This is closely related to the layout of whole powerhouse project.

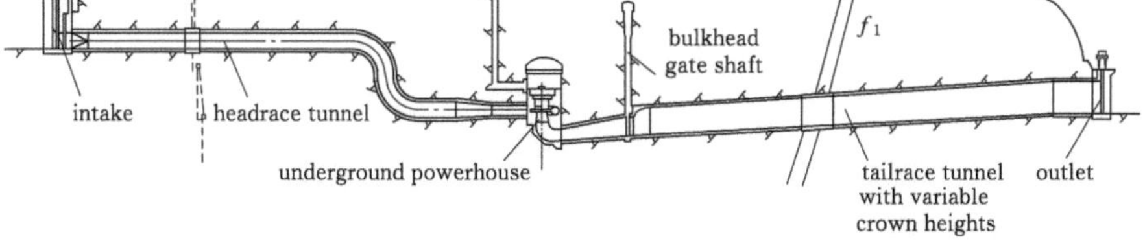

Fig. 13.8 Schematic diagram of longitudinal section of sloping ceiling tailrace tunnel in Pengshui hydropower station

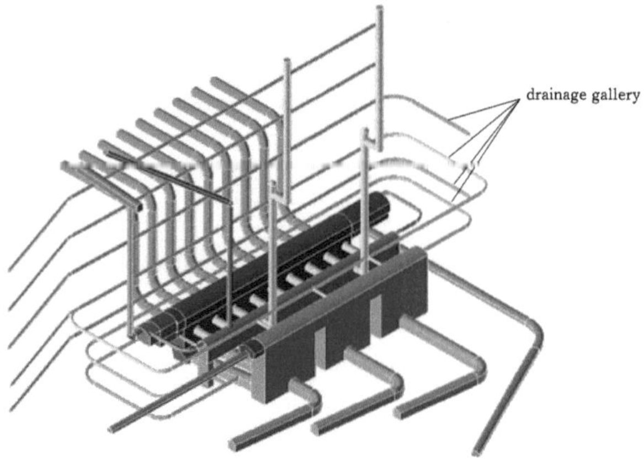

Fig. 13.9 Schematic diagram of drainage gallery layout of underground powerhouse

13.4 Layout Design of Underground Powerhouse

13.4.1 Layout Principle of Underground Powerhouse

(1) On the premise of ensuring the operation, installation, overhaul and maintenance of electromechanical equipment, minimize the plane dimension of the powerhouse.
(2) The layout of main electromechanical equipment in the powerhouse should be compact, neat, concise and practical, and the height of the powerhouse should be reduced as much as possible on the elevation.
(3) Make full use of the space of the underground powerhouse, rationally arrange the equipment of the auxiliary powerhouse, and try to use one cavern for multiple purposes. Improve the space utilization rate of underground powerhouse and ensure the stability of surrounding rock of cavern.

13.4.2 Size Determination of Underground Powerhouse

The principle for determining the size of underground powerhouse is basically the same as that of ground powerhouse. The control dimensions of the main unit room in the main powerhouse should be determined with the principle of striving for the minimum by considering some factors, such as the number of units, the flow passage components of hydraulic turbines, the control dimensions of generators and air ducts, the lifting modes and effective working range of cranes, the positions of water inlet valves and speed regulators, the structural requirements of the powerhouse, operation and maintenance, the traffic in the powerhouse, etc. When the size of the unit section is controlled by the turbine spiral case, the minimum space required for spiral case installation shall be met. If concrete is poured with water filling and pressurization, it's necessary to consider the space required for installing and disassembling the bulkhead and water filling and pressurization device. When the size of the unit section is controlled by the size of the generator and its air duct, the unit spacing should not only meet the equipment layout requirements, but also keep the necessary width of the passage. In addition, the shape of draft tube and the thickness of rock column between draft tubes should be considered, and the structural dimensions of the upper and lower parts of the main powerhouse should be coordinated with each other, so as to make comprehensive use of space to arrange auxiliary equipment as far as possible.

Inlet valve and expansion joint are generally arranged in the main powerhouse, which can be installed and overhauled by crane in the main powerhouse, with convenient operation and management and compact layout. However, the width and length of the powerhouse will be increased when the inlet valve is arranged in the main powerhouse cavern. When the inlet valve size is large, the width of the main powerhouse cavern will be too large, so it can be arranged in a separate cavern outside the main powerhouse.

13.4.3 Elevation Layout Design of Underground Powerhouse

The determination principle of underground powerhouse elevation is basically the same as that of the ground powerhouse. Since the underground powerhouse is generally narrow, as shown in Fig. 13.10, some passages are usually arranged on the elevation, so attention should be paid to the traffic arrangement of each passage on the elevation. The traffic in the powerhouse should meet the requirements of convenient management, easy maintenance and quick troubleshooting. One staircase should be set for every one to two units between main floors, and there should be no less than two stairs in the whole powerhouse. The generator floor and turbine floor should be designed with straight horizontal passageways through the whole powerhouse. The main passage size, stair width and slope, and emergency exit setting must comply with the requirements of electromechanical and fire protection design code. Each floor in the main powerhouse shall meet the requirements of unit and auxiliary equipment layout, installation and maintenance, structure size, and construction space.

13.4.4 Plane Layout Design of Underground Powerhouse

The layout principle of each floor of underground powerhouse is basically the same as that of the ground powerhouse. Generally, the underground powerhouse is arranged compactly in a narrow and long row. Considering the lifting and traffic requirements, there is usually no other auxiliary equipment arranged on the generator floor except for the unit (Fig. 13.11). If there is an outgoing line layer, the governing equipment of the unit can be arranged here (Fig. 13.12), and if there is no outgoing line layer, it can

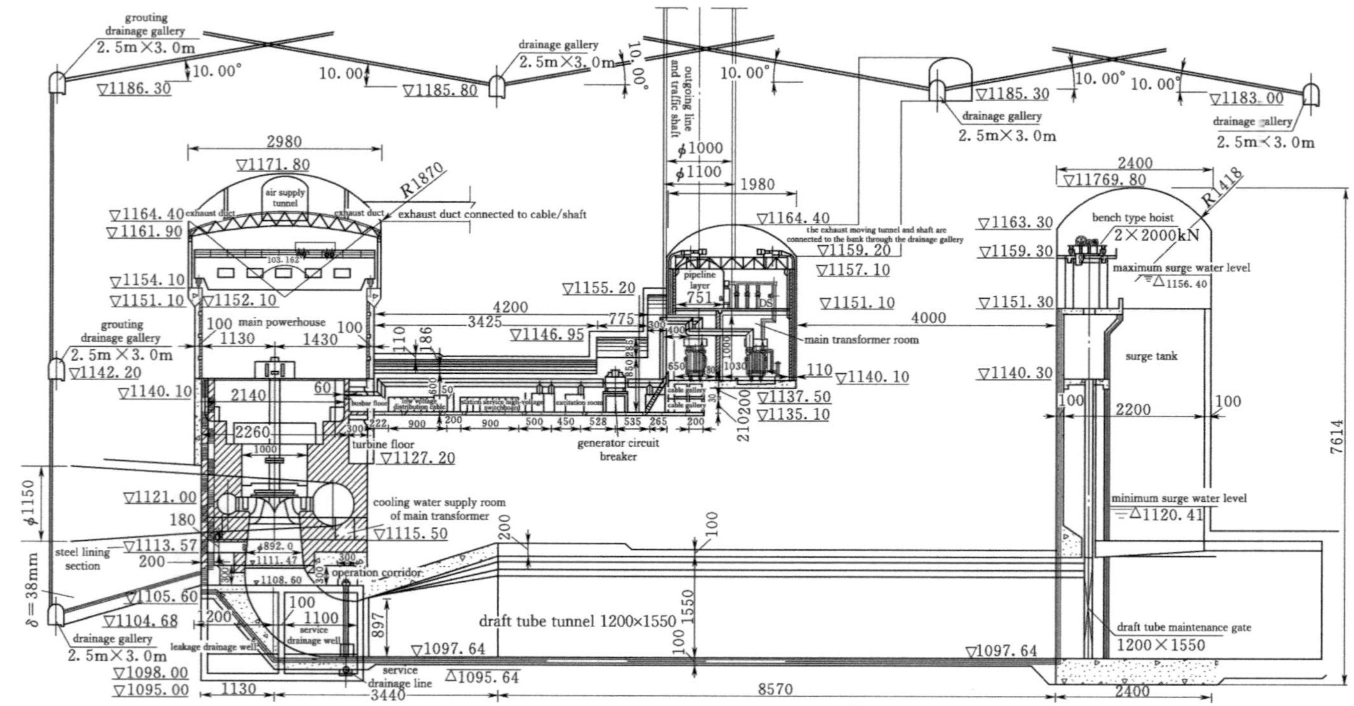

Fig. 13.10 Cross-sectional layout of underground powerhouse (Elevation unit: m; Dimension unit: mm)

be arranged in the turbine layer. The layout of butterfly valve, turbine manhole and some hydraulic machine auxiliary equipment are mainly considered in the turbine floor (Fig. 13.13). Due to the narrow width of the underground powerhouse, the spiral case floor mainly considers layout of some necessary passages and pump equipment (Fig. 13.14). Therefore, it affects the layout of auxiliary equipment and communication channels in underwater parts, and it is necessary to make full use of the space of auxiliary caverns for the layout of auxiliary equipment.

13.4.5 Layout Design of Auxiliary Powerhouse

Due to the requirements of stability and ventilation of underground powerhouse caverns, the following factors should be considered in the layout of auxiliary powerhouse.

(1) Auxiliary powerhouses are mostly located at one end of the main powerhouse, so as to avoid increasing the span of the main powerhouse and affecting the stress distribution of rock mass and the cavern stability. The hydraulic machine auxiliary equipment is mostly arranged at the other end of the powerhouse building, and is placed on the lower floor of the assembly site.
(2) The central control room is the center of operation, control and monitoring of the whole power station. In order to eliminate faults quickly, reduce the occurrence of faults, improve the reliability of operation, and shorten the length of cables, the central control room is generally arranged in the auxiliary powerhouse at the same height as the generator floor. If the central control room and the generator floor are arranged at different elevations, spacious stairs and convenient special traffic passages should be set between them.
(3) Because of the high humidity of underground powerhouse, when arranging oil depot and oil treatment room, the oil quality will be affected, and the oil smell emitted is easy to cause pollution, which also increases the complexity of fire prevention in the powerhouse. Therefore, moisture-proof and ventilation measures should be fully considered where the oil depot and oil treatment room are arranged.
(4) When the unit capacity is large, in order to avoid increasing the plane dimension of the powerhouse, the space of bus-bar cavern and main transformer cavern can be fully utilized to arrange low-voltage electrical appliances and auxiliary transformers.
(5) Full attention should be paid to the layout of battery room in underground powerhouse.

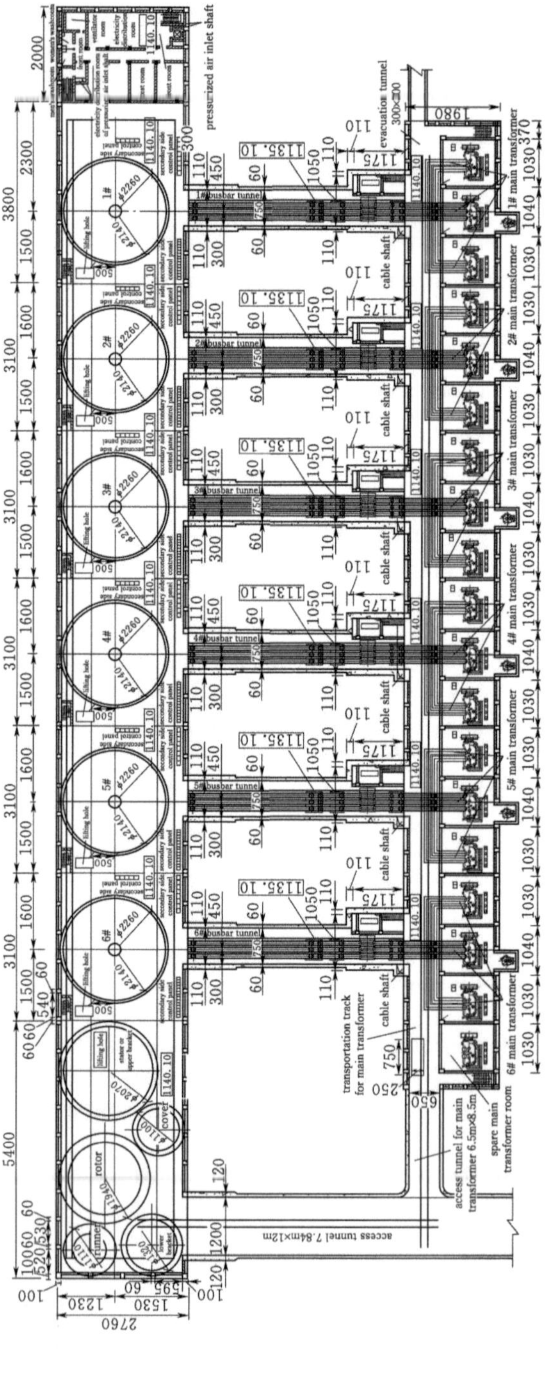

Fig. 13.11 Layout of generator floor of underground powerhouse

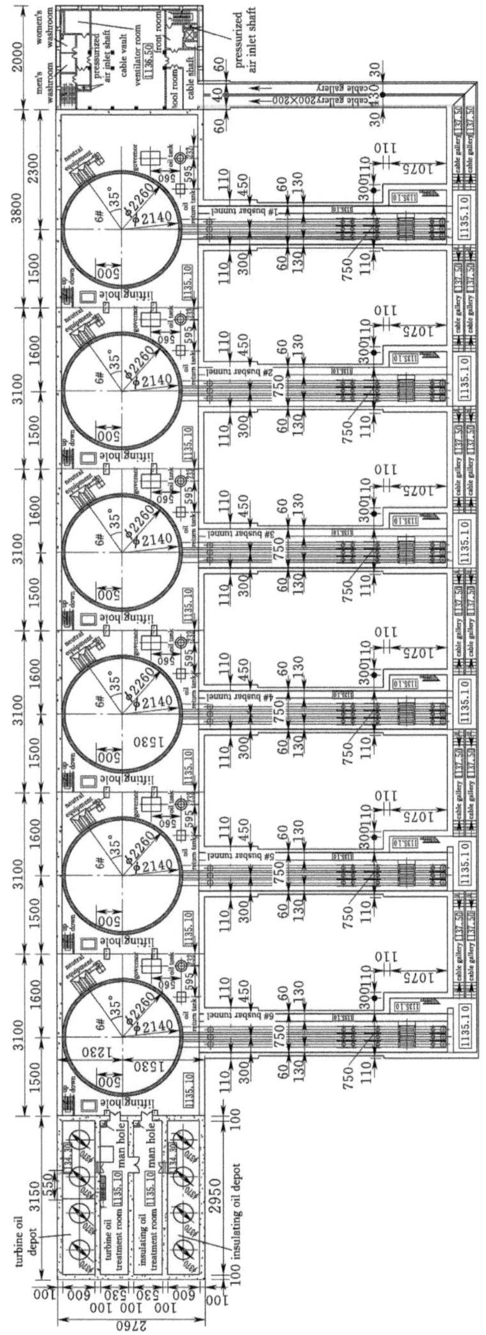

Fig. 13.12 layout of outgoing line floor of underground powerhouse

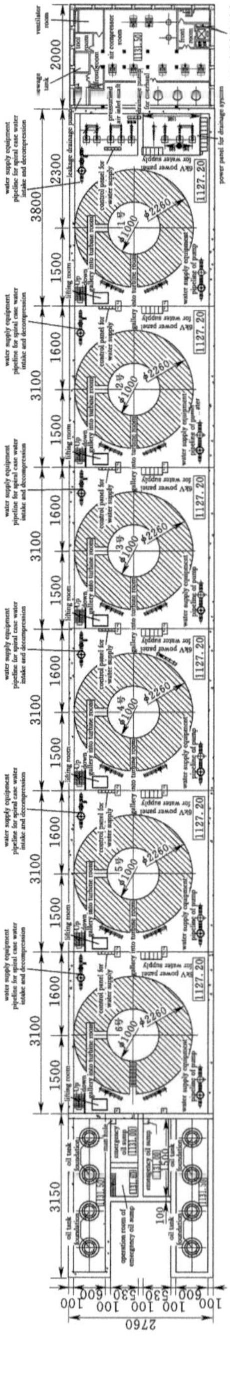

Fig. 13.13 Layout of hydraulic turbine floor of underground powerhouse

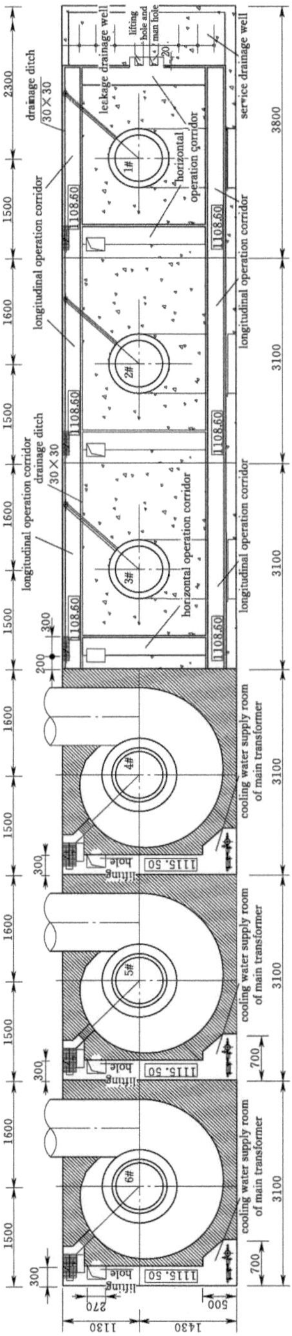

Fig. 13.14 Layout of spiral case floor of underground powerhouse

13.4.6 *Ceiling Design of Underground Powerhouse*

13.4.6.1 Function of Ceiling in Underground Powerhouse

(1) Ceiling can prevent falling rocks at the top.
(2) Ceiling can provide main ventilation channel.
(3) Ceiling is beneficial to arrangement and maintenance of drainage, moisture-proof, fire prevention, and lighting.

13.4.6.2 Types of Ceiling in Underground Powerhouse

According to the stress characteristics of the ceiling structure, its structural forms can be divided into two types: suspended structure and self-supporting arch structure.

13.4.6.3 Layout of Ceiling in Underground Powerhouse

The ceiling of underground powerhouse can be made of corrugated steel plate and galvanized profiled formwork as required, which can be used alone or combined with reinforced concrete.

13.5 Structural Design of Crane Girder on Rock Wall

13.5.1 *Form of Crane Girder in Underground Powerhouse*

According to the different stress modes of crane girder structure, the crane girder of underground powerhouse can be divided into the following three forms.

1. Rock bolt crane girder

Anchor the crane girder on the rock wall with anchor rod or anchor cable, as shown in Fig. 13.15c. For this form, the load of crane girder is mainly transmitted to the stable surrounding rock by anchor rod or anchor cable, and the stress condition is better. Rock bolt crane girder is widely used in underground powerhouse, which is applicable to a wide range of surrounding rock conditions.

2. Rock bench crane girder

The crane girder is directly located on the rock bench, and the load is transmitted to the surrounding rock through the rock bench, as shown in Fig. 13.15d. For this form, the rock bench is under large forces, requiring good surrounding rock conditions. Therefore, the construction excavation should not disturb the rock bench too much to ensure high excavation quality. This form is rarely applied in practical engineering.

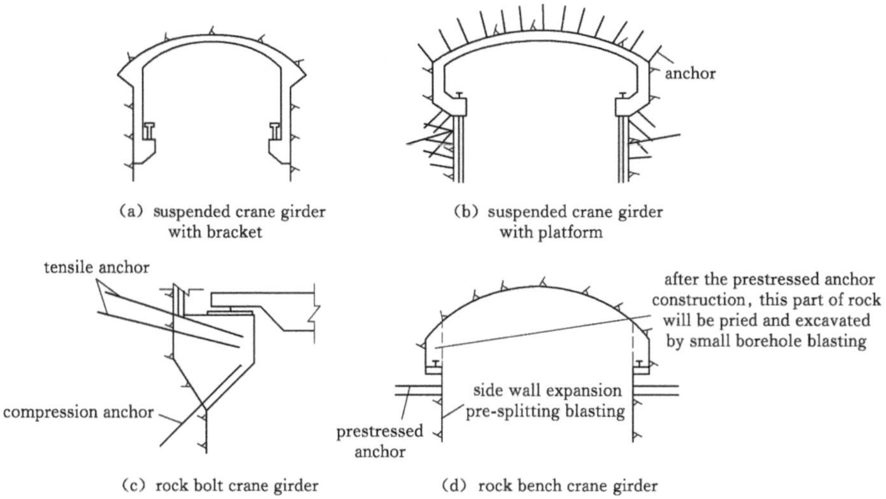

Fig. 13.15 Structure form of crane girder in underground powerhouse **a** Suspended crane girder with bracket; **b** Suspended crane girder with platform; **c** rock bolt crane girder; **d** Rock bench crane girder 分图名放各个小图下

3. Suspended crane girder

Suspended on the arch support of reinforced concrete arch crown, the load of crane girder is transmitted to surrounding rock through reinforced concrete arch support, as shown in Fig. 13.15a, b . This form is adopted when the surrounding rock condition is poor and the arch crown of the main powerhouse cavern must be supported by reinforced concrete arch.

13.5.2 Section Size Design of Crane Girder on Rock Wall

The section size of rock-wall crane girder (Fig. 13.16) can be determined by engineering analogy method according to the section size of rock-wall crane girder of existing projects. According to the operation needs and geological conditions, the following factors should also be comprehensively considered.

13.5.2.1 Top Surface Width of Crane Girder on Rock Wall—B

The top surface width shall meet the requirements of layout and operation, and be drawn up according to the following formula:

$$B \geq C_1 + C_2 \tag{13.1}$$

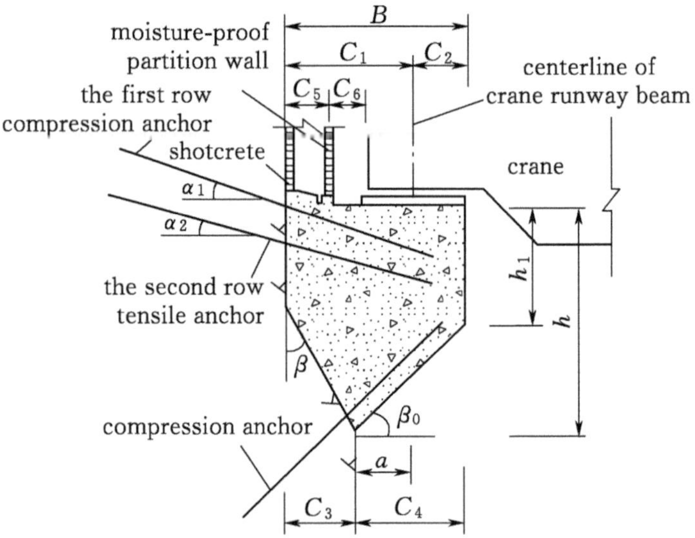

Fig. 13.16 Size determination of rock wall crane girder

In the formula:

C_1 horizontal distance from runway center line to upper rock wall edge, including shotcrete thickness of rock wall above rock wall crane girder, net width of gap in moisture-proof partition wall, thickness of moisture-proof partition wall, and minimum horizontal distance between bridge crane end and moisture-proof partition wall, mm;

C_2 the minimum horizontal distance from the runway center line to the outer edge of rock wall crane girder, mm.

13.5.2.2 Inclination Angle of Crane Girder on Rock wall—β

Comprehensively considering the influence of rock layers, main geological structure, and joints and fissures, inclination angle should be determined by the section size of crane girder on rock wall, the arrangement of anchor rods and the stress state. Generally, it is between 20 and 40°.

The dip angle of the bottom surface of rock-wall crane girder should not be too large, which can be determined based on the design experience of corbel, the design example of rock-wall crane girder and relevant research results of existing project. Generally, it is between 20 and 50°.

13.5.2.3 Cross-Section Height of Crane Girder on Rock wall—h

Cross-section height of crane girder h is related to many factors, such as maximum wheel pressure, rock wall angle, bond strength between concrete and rock wall, minimum spacing of anchor rod arrangement, stability against sliding of rock wall slope, etc. Generally, the following requirements shall be met:

$$h > 3.33(C_4C_2) \tag{13.2}$$

In the formula:

C_4 Length of cantilever, mm.

13.5.2.4 Anchor Inclination Angle of Crane Girder on Rock Wall—α

The smaller the anchor inclination angle α is, the smaller the anchor force is, but it is unfavorable to the stability against sliding. The larger the inclination angle α is, the greater the stress on the bolt is, but the thinner the rock layer on the bolt is, which is not conducive to safe and durable use.

13.5.3 Anchor Spacing Layout Design of Crane Girder on Rock Wall

According to GB 50,007-2011 *Code for design of building foundation*: the spacing of rock bolts should not be less than 6 times of the diameter of bolt holes. According to *Code for construction of crane girder on rock wall of hydropower and water* resources *projects*: bolt aperture should not be less than the bolt diameter plus 50 mm. In order to prevent the bolt construction from grout leaking within the allowable construction error, the anchor spacing of crane beam on rock wall should be more than 700 mm.

The insertion depth L of rock wall crane girder anchor rod shall be selected according to the following recommended formula:

$$L = 0.15H + 2 \tag{13.3}$$

In the formula:

H Height of powerhouse side wall, m.

13.5.4 *Structural Design of Crane Girder on Rock Wall*

13.5.4.1 Load of Crane Girder on Rock Wall

Loads borne by crane girder include vertical wheel pressure of bridge crane, lateral horizontal load, self-weight of crane girder, gravity of track and accessories, and gravity of moisture-proof partition wall on girder. The stressed anchor of rock wall crane girder also bears the release stress of surrounding rock.

13.5.4.2 Structural Design Status of Crane Girder on Rock Wall

Permanent load of crane girder includes self-weight of crane girder, gravity of track and accessories, gravity of moisture-proof partition wall on girder, and action of cavern excavation. Variable loads include vertical wheel pressure and horizontal load of the crane.

(1) Fundamental combination of permanent condition: under the design standard section, the action effects combination of permanent load and variable load when bridge crane lift at rated load.
(2) Fundamental combination of transient conditions: under the design standard section, the action effects combination of permanent load and variable load during crane dynamic load test.
(3) Accidental combination: under the non-standard excavation section where both overbreak and rock wall angle changes are within the permitted range, the action effects combination of permanent load and variable load during crane dynamic load test.

All three design conditions should be designed according to the ultimate limit states, but the rock crane girder may not be designed according to the serviceability limit states.

13.5.4.3 Structural Design Method of Crane Girder on Rock Wall

Structural design methods of rock wall crane girder mainly include rigid limit equilibrium method and finite element method.

(1) Rigid limit equilibrium method: The rock wall crane girder is regarded as a rigid body, the rock wall is regarded as an elastic foundation conforming to Winkler's hypothesis, and its anchors under tension and compression are simplified as springs conforming to Winkler's hypothesis. Under the action of load, the crane girder produces translation and rigid body rotation. According to the equilibrium equation of force and the continuous deformation condition of crane girder and foundation, the rock stress, and axial and shear forces of tension and compression bolt can be calculated.

Advantages: the stress relationship is clear and the calculation formula is simple.

Disadvantages: cavern deformation caused by stress release of surrounding rock after cavern excavation is not considered; the bonding force and friction force at the interface between the concrete of rock bolt crane girder and rock mass are not fully considered.

(2) Finite element numerical analysis method: the design of crane girder on rock wall in large underground powerhouses, complex geological conditions, high geostress areas and high seismic areas should be calculated by finite element method.

The contents of finite element calculation include the effect and influence on rock crane girder anchor rods, girders and surrounding rocks caused by excavation in the middle and lower part of underground powerhouse during construction period and the crane load during operation period, and the finite element evaluation of safety and stability of rock wall crane girder.

Stability evaluation of rock wall crane girder should be carried out from displacement, stress, safety factor of anchor bolt and safety factor of anti-sliding stability of interface:

Safety factor of anchor rod:

$$\begin{aligned} K_b &= \frac{f_{yk}}{\sigma_b} \\ \sigma_b &= \sqrt{3\tau^2 + \sigma^2} \end{aligned} \tag{13.4}$$

In the formula:

f_{yk} standard value of tensile strength of anchor rod, kN/m^2;
σ_b anchor stress, kN/m^2;
σ maximum axial tensile stress of anchor rod, kN/m^2;
τ anchor shear stress, kN/m^2.

The safety factor of anti-sliding stability of the interface between crane girder and rock wall is defined as the ratio of anti-sliding force to sliding force.

$$K_j = \frac{\sum_{i=1}^{n} (f_i\sigma_i + c_i)A_i}{\sum_{i=1}^{n} \tau_i A_i} \tag{13.5}$$

In the formula:

σ_i, τ_i normal stress and shear stress of the ith element on the interface, kN/m^2;
f_i, c_i friction coefficient and cohesion of the ith element on the interface, kN/m^2;
A_i the area of the ith unit along the slip surface, m^2;
n number of elements on the interface.

13.5.5 Structural Requirements of Crane Girder on Rock Wall

(1) Rock wall crane girder must be built on the basis of stable side wall surrounding rock. When rock mass on the side wall of crane girder is affected by joints with steep inclination or faults, and possibly unstable rock mass appears, its stability should be analyzed by block stability calculation, and prestressed anchor bolts or anchor cables should be adopted at the upper and lower parts of crane girder and long anchor bolts should be adopted in the middle part of crane girder to ensure the stability of side wall.

When a fault is exposed at the crane girder on the rock wall, the fault shall be excavated generally, and the excavated part shall be backfilled with reinforced concrete and poured integrally with the girder body. If the fault damages the rock abutment, short columns anchored to the wall can be set under the crane girder on both sides of the fault.

(2) In order to reduce the weakening of the surrounding rock foundation of the crane girder on rock wall by the intersecting tunnel, the height of the section of bus tunnel near the powerhouse should be lowered, the base slap of the bus tunnel should be lowered to the cable layer, and a certain thickness of supporting rock mass should be kept between the top of the intersecting tunnel and the bottom of the girder as far as possible.

Before excavation of intersecting tunnel under crane girder, it's necessary to set lock bolt around tunnel orifice from powerhouse side, carry out controlled blasting strictly, and bolting and making timely shotcrete support to reduce disturbance to surrounding rock foundation of crane girder. Arch sealing lining and backfill grouting should be carried out in a certain range of the tunnel intersection near the powerhouse, and the stability monitoring of surrounding rock at the intersection should be strengthened.

(3) When the rock wall crane girder has a certain longitudinal stiffness, it can ensure the effective transmission of wheel pressure along the length direction of the crane girder. When the crane girder on rock wall is long, the geological conditions are quite different, or the heights of tunnel side wall are quite different, the crane girder on rock wall should be designed with expansion joints. Usually, additional concrete and shear reinforcement should be used under the crane girder on both sides of the expansion joint, and the transverse tensile reinforcement within two meters on both sides of the concrete end of the crane girder should also be strengthened. Expansion joints shall be filled with foam plastic board or asphalt board, and rubber water stop shall be set up.

(4) Construction joints should be set for crane girder on rock wall to adapt to the deformation of surrounding rock. In order to reduce the shrinkage deformation and shrinkage cracks of concrete, the crane girders should be poured in sections according to the requirements of pouring capacity and construction period. The spacing between construction joints should be determined by considering factors such as temperature and pouring capacity during pouring.

(5) Generally, there is water seepage in the upper rock wall of the girder body, so it should be considered to set a drainage ditch on the side of the girder near the rock wall, and bury a downward drainage pipe in the girder body.

The axial tension of the stressed anchor rod of the crane girder on the rock wall should be transmitted to the stable rock mass in the deep surrounding rock as much as possible.

13.6 Stability and Support Analysis of Surrounding Rock of Underground Powerhouse Cavern

13.6.1 Basic Data and Requirements for Stability Analysis of Surrounding Rock

(1) Exploration and test data of underground powerhouse: mechanical parameters of surrounding rock, ground stress, groundwater, earthquake and other geological data.
(2) Design data of underground powerhouse layout: underground powerhouse project layout and underground powerhouse structure layout. The layout of underground powerhouse project should be comprehensively considered according to the topography, geology, construction and general layout of hydropower station in the powerhouse area.
(3) Design data of excavation and support for underground powerhouse construction: excavation methods and sequence, anchoring support schemes and support parameters.

13.6.2 Stability Analysis Methods of Surrounding Rock

At present, the stability analysis methods of surrounding rock of underground powerhouse generally include qualitative analysis and quantitative analysis. Qualitative analysis mainly includes geological analysis and engineering analogy, while quantitative analysis mainly includes numerical analysis, model test, and on-site testing and feedback analysis.

(1) Engineering analogy method based on the surrounding rock classification is still the main method of underground engineering design at present, which has been widely used in design at home and abroad.
(2) Numerical analysis is the most widely used in quantitative analysis. It mainly includes some methods used for continuous element, such as finite element method (FEM), boundary element method (BEM), fast Lagrange analysis method, and element free method (EFM). In addition, it includes some method used for non-continuum, such as key block theory, discrete element method

(DEM), discontinue deformation analysis, interface stress element method (ISEM), etc.

(3) In the numerical analysis method, the finite element method is generally used to analyze and calculate the overall stability, excavation scheme and support scheme, and seepage stability of underground powerhouse caverns. The finite difference method, boundary element method and other numerical analysis methods of continuum are supplemented.

Among all kinds of analysis methods, finite element method and fast Lagrange analysis for continuum are the most mature, widely used and stable. Discrete element method and discontinuous deformation analysis method in the calculation method of non-continuum also have a wide application.

13.6.3 Calculation Requirements for Stability Analysis of Surrounding Rock

According to the terrain, geological conditions, building layout and the purpose of the calculation of surrounding rock stability, the finite element method calculation model of surrounding rock stability can be established and the calculation range can be determined.

The calculation model should fully reflect the influence of the structure and topography of underground powerhouse caverns, the influence of artificial factors such as the shape and structure of caverns, excavation sequence, construction methods, supporting measures, supporting time, etc. It should reflect the working state of underground powerhouse at each stage of excavation, support and operation, and should fully reflect the influence of geological factors such as rock mass structure classification, rock mass fissure characteristics and spatial combination, physical and mechanical properties of rocks, initial in-situ stress field of surrounding rocks and groundwater conditions. The calculation boundary is based on the principle that the excavation of underground powerhouse caverns will not be affected, and the stability analysis of surrounding rock of underground powerhouse caverns should be calculated by three-dimensional finite element method in principle.

13.6.4 Evaluation Method of Surrounding Rock Stability

13.6.4.1 Evaluation Content of Surrounding Rock Stability

The stability analysis of surrounding rock should start from four aspects: rock mass structure, rock mass stress, groundwater, and engineering factors, and then evaluate the overall or local stability characteristics of various surrounding rocks quantitatively or qualitatively. There are two types of damage, one is integral damage, the other is local damage. Integral damage is caused by the geostress of rock mass exceeding the allowable strength of surrounding rock in a large range, and this kind of damage mostly occurs in rock strata with high stress of rock mass and low strength of surrounding rock. Local damage refers to the damage occurring in a local area, which mostly occurs in the rock mass of hard surrounding rock cut by multiple sets of structural planes.

13.6.4.2 Evaluation Method of Surrounding Rock Stability

1. Geological analysis and evaluation method

It is mainly used to find out the main factors related to the stability of surrounding rock, such as rock characteristics, geological structure, rock stress, groundwater action, etc., and determine the type of surrounding rock through comprehensive analysis, and judge the stability of surrounding rock from the geological occurrence environment of underground powerhouse. This analysis method is objective and practical, but it needs rich experience.

2. Engineering analogy evaluation method

It mainly compares the practical experience of similar projects built at home and abroad, and comprehensively analyzes the similarities and differences of various factors from the classification of surrounding rock, geological conditions, rock mass characteristics, dynamic observation data, etc., and considers the differences of engineering grades and categories. Through comprehensive analysis and comparison with existing projects with similar conditions, we can judge the stability of surrounding rock, so as to determine the structural form and excavation mode of surrounding rock of the proposed project, and design the supporting form and supporting parameters of the cavern.

3. Numerical analysis and evaluation method

Although there is still no unified and clear conclusion up on the method of evaluating the stability of surrounding rock of underground powerhouse caverns by numerical analysis, people's understanding tends to be consistent.

13.6.5 General Principles of Support Design

The support design of underground powerhouse should follow the principle of "New Austrian Tunneling Method", make rational use of the self-supporting capacity of surrounding rock, ensure the design headroom section, seal the cracks in rock mass, and prevent water leakage and weathering of rock wall. Classification of surrounding rock is the basis of supporting design of underground powerhouse, and supporting parameters can be determined according to the classification, with reference to relevant regulations and engineering analogy.

The support method should be selected according to the principle of "flexible support as the main support, rigid support as the supplement, system support as the main support, and local reinforcement as the supplement". The design of supporting parameters should implement the principle of dynamic design and follow the requirements of different design stages.

13.6.6 Classification and Applicable Conditions of Support

According to the layout characteristics, the surrounding rock support of underground powerhouse can be divided into systematic support and local support. According to its action characteristics, it can be divided into flexible support, rigid support and combined support. According to the construction characteristics, it can be divided into advanced support and random support.

System support is a general support design which arranges the rock mass of caverns according to a certain format, so as to improve the overall bearing capacity of rock mass structure of powerhouse caverns and ensure the overall stability of caverns. Local support is a strengthening support measure for local unstable blocks in underground powerhouse. The length of anchor rod of local support should go deep into stable rock mass to ensure the stability of local blocks. The local support design should be carried out for the parts where the stability of surrounding rock cannot be guaranteed only by system support, such as local poor rock mass, weak structure of cavern intersection and fault crossing.

Flexible support is a supporting structure which is closely integrated with surrounding rock and can adapt to the deformation of surrounding rock. It is usually composed of one or more of shotcrete, anchor bolt and anchor cable. Flexible support is mainly suitable for the surrounding rock above grade III as permanent support. Flexible support of grade IV to V surrounding rock can only be used as primary support, and rigid support should be used as permanent support in later period. Rigid support has a large enough section size, and its structural rigidity can limit the deformation of surrounding rock. Its forms mainly include reinforced concrete lining, reinforced concrete anchor pier and reinforced concrete replacement. Rigid support is mainly used to bear large loose pressure of surrounding rock, and is suitable for grade IV to V surrounding rock or shallow-buried cavern. Combined support is composed of flexible support and rigid support, which is also called the combination of primary support and secondary support. It is mainly used for weak surrounding rocks with poor self-stabilization ability and strata with obvious plastic rheology. When the surrounding rock is poor and rigid support is needed as permanent support, flexibility support should be adopted in time after excavation to ensure the stability of surrounding rock during construction. Secondary support is permanent rigid support applied after complete or partial excavation, and its function is to ensure the long-term stability of underground powerhouse caverns.

Questions

1. What are the advantages of underground powerhouse? Compared with the ground powerhouse, what are the complexities of the underground powerhouse?
2. What are the principles and contents of the layout of the underground powerhouse complex? How to arrange the underground powerhouse?
3. What are the development methods of underground powerhouse? What are the characteristics and layout requirements of these development methods?
4. What are the characteristics of the crane beam in the powerhouse? What should be included in the structure calculation of the crane beam?
5. Please condense the main factors considered in the layout of the main powerhouse cavern and the main transformer cavern of the underground powerhouse, and list the layout requirements.
6. What are the principles and methods for the layout of the auxiliary powerhouse and assembly site of the underground powerhouse?
7. What are the structural forms of the surge chamber and tailrace tunnel of the underground powerhouse? What points should be paid attention to when arranging them?
8. What are the principles of layout and design of underground powerhouse? What factors should be taken into account when determining the size?
9. There are some main structural forms of crane beam in underground powerhouse, do you know their characteristics? How to determine the cross-section size of the rock wall crane beam?
10. What are the requirements for anti-seepage ventilation and lighting of underground powerhouse?
11. Please discuss the analysis methods and stability evaluation methods of surrounding rock stability and support of underground powerhouse caverns.
12. What are the general principles of cavern support design of underground powerhouse? And please list the basic types of support and think about their applicable conditions?

Bibliography

1. Design code for Hydropower house SL 266: 2014 (In Chinese). 2014. Beijing: China Water and Power Press.
2. Design code for Hydropower house NB/T 35011–2013 (In Chinese). 2013. Beijing: China Electric Power Press.
3. Liu, Zhenya. 2013. *General Design of pumped storage power station project of state grid corporation of China, underground workshop volume (In Chinese).* Beijing: China Electric Power Press.
4. Yang, Shuren and W. Zhou. 1993. *Design of underground hydropower plant (In Chinese).* Beijing: Water Resources and Electric Power Press.
5. Li, Xiesheng. 1993. *Underground hydropower station construction (In Chinese).* Beijing: Water Conservancy and Electric Power Press.

6. Dang, Lincai. 2012. *Construction and Key technologies of underground hydropower projects in China (In Chinese)*. Beijing: China Water and Power Press.
7. Ma, Shanding and R. Wang 1996. *Hydropower station buildings (In Chinese)*. Beijing: China Water and Power Press.
8. Liu, Qizhao and M. Hu. 2010. *Hydropower station (In Chinese)*, 4th ed. Beijing: China Water and Power Press.

MIX
Papier aus verantwortungsvollen Quellen
Paper from responsible sources
FSC® C105338
FSC
www.fsc.org

If you have any concerns about our products,
you can contact us on
ProductSafety@springernature.com

In case Publisher is established outside the EU,
the EU authorized representative is:
Springer Nature Customer Service Center GmbH
Europaplatz 3, 69115 Heidelberg, Germany

Printed by Libri Plureos GmbH
in Hamburg, Germany